Proceedings of the Conference in Honour of

Murray Gell-Mann's 80th Birthday

Quantum Mechanics, Elementary Particles, Quantum Cosmology and Complexity

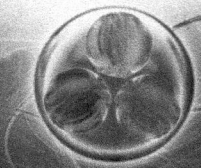

Proceedings of the Conference in Honour of

Murray Gell-Mann's 80ᵗʰ Birthday

Quantum Mechanics, Elementary Particles, Quantum Cosmology and Complexity

Nanyang Technological University, Singapore, 24–26 February 2010

Editors

H Fritzsch *(University of Munich, Germany)*
K K Phua *(Nanyang Technological University, Singapore)*

Co-editors

B E Baaquie *(National University of Singapore, Singapore)*
A H Chan *(National University of Singapore, Singapore)*
N-P Chang *(City University of New York, USA)*
S A Cheong *(Nanyang Technological University, Singapore)*
L C Kwek *(National University of Singapore, Singapore)*
C H Oh *(National University of Singapore, Singapore)*

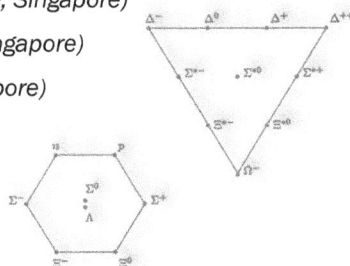

World Scientific

NEW JERSEY · LONDON · SINGAPORE · BEIJING · SHANGHAI · HONG KONG · TAIPEI · CHENNAI

Published by

World Scientific Publishing Co. Pte. Ltd.

5 Toh Tuck Link, Singapore 596224

USA office: 27 Warren Street, Suite 401-402, Hackensack, NJ 07601

UK office: 57 Shelton Street, Covent Garden, London WC2H 9HE

British Library Cataloguing-in-Publication Data
A catalogue record for this book is available from the British Library.

ISBN-13 978-981-4335-60-7
ISBN-10 981-4335-60-6
ISBN-13 978-981-4338-62-2 (pbk)
ISBN-10 981-4338-62-1 (pbk)

PREFACE

The Conference in Honour of Murray Gell-Mann's 80th Birthday was no ordinary conference. Held from 24 to 26 February 2010 at the idyllic and beautiful premises of the Nanyang Executive Centre, NTU, the conference was graced by four Nobel Laureates and several renowned scientists across the globe who celebrated Professor Murray Gell-Mann's 80th birthday and his contributions to fundamental sciences.

Jointly organized by the Chair, Professor Harald Fritzsch and Co-Chair, Professor Phua Kok Khoo, the conference attracted 249 participants from 33 countries, including the United States, Australia, Britain, China, France, Germany India, Israel, Indonesia, Malaysia and Singapore. The four Nobel Laureates present at the conference were: Professor Yang Chen Ning (Physics, 1957), Professor Kenneth Geddes Wilson (Physics, 1982) Professor Gerard 't Hooft (Physics, 1999) and Professor Murray Gell-Mann (Physics, 1969). Professor Harald Fritzsch and Professor George Zweig both shared a light-hearted account of their encounters with Professor Gell-Mann, with their lectures titled "Murray Gell-Mann — A Scientific Biography" and "Memories of Murray and the Quark Model" respectively.

Another highlight of the conference took place on the second day when all the participants were treated to a sumptuous banquet dinner at the Regent Hotel. The Guest of Honour, Mr Lim Chuan Poh (Chairman, Agency for Science, Technology and Research), paid tribute to Prof Gell-Mann for his efforts in deepening our understanding of the laws of Nature. Prof Kenneth Young (CUHK, Hong Kong), who was also Prof Gell-Mann's student, spoke affectionately about his former PhD mentor.

In celebration of Professor Gell-Mann's birthday, students from Hwa Chong Institution put up a fi ne display of Wushu performance, an elegant fan dance and a captivating fusion of Eastern-Western music. The exotic music ensemble by the Institution's Chinese Orchestra, highly commended by Professor Chang Ngee Pong (City University of New York), culminated in a "Happy Birthday" song. Everyone sang, clapped and cheered for Professor Gell-Mann who was visibly moved; he told us that he was very touched by the gesture of the audience at the Banquet.

The last day of the conference ended with parallel sessions on Particle Physics, Quantum Mechanics and Complexity. In addition, more than 80 teachers attended a Physics Education workshop which was held in conjunction with the conference.

Most of the teachers who attended the workshop thoroughly enjoyed the captivating talks and hands-on activities at the workshop.

The parallel sessions on Quantum Mechanics and Complexity began with two invited talks by Professor Kerson Huang from Massachusetts Institute of Technology (MIT) and Dr Gunnar Pruessner from the Department of Mathematics, Imperial College London. Professor Huang presented the idea of Conditioned Self-Avoiding Walk (CSAW) as a model of protein folding. Dr Pruessner gave an overview on the current development of Self-Organized Criticality (SOC). The special talks were followed by talks from researchers from Australia, Japan, the Netherlands, Indonesia, France, Russia, Brunei, Turkey and Singapore. The topics presented were wide-ranging, spanning from quantum entanglement to the complexity theory of business, social and biophysical systems.

CONTENTS

---

*Abstract only.

**Contributed Talks: Particle Physics, Cosmology and
General Relativity**

Contributed Talks: Quantum Mechanics and Complexity

MGM and his wife Margaret.

MURRAY GELL-MANN — A SCIENTIFIC BIOGRAPHY

HARALD FRITZSCH

Physik-Department, Universität München,
Theresienstrasse 37A, 80333 Munich, Germany

Keywords: Strangeness; SU(3); quarks; QCD.

Murray Gell-Mann is one of the most outstanding scientists of the last century. At this conference on the occasion of his 80th birthday, the important scientific achievements of Gell-Mann will be discussed. I shall describe in particular his contributions to particle physics.

Murray's father Arthur Gell-Mann grew up in Czernowitz, a city once belongs to the Austrian empire. Today, Czernowitz belongs to the Ukraine. Arthur Gell-Mann studied in Vienna and came to the United States in 1911. He lived and worked in New York, where he became the owner of a language school and married Pauline Reichstein.

Murray Gell-Mann was born in New York on September 15, 1929. He grew up in the area west of the central park together with his brother Ben, who was two years older. Murray was a gifted child and learned to read and write at the early age of three. When he was ten years old, he read *Finnegans Wake* by James Joyce, a difficult book, which should play a specific role later in his life.

When Gell-Mann was 15 years old, he received a scholarship from Yale University, which allowed him to study physics at Yale. Afterwards, he went to the Massachusetts Institute of Technology and worked on his Ph.D. His advisor was Victor Weisskopf. He completed his Ph.D. in 1951. Gell-Mann went as a post-doctoral fellow to the Institute for Advanced Study in Princeton. The director of the Institute at that time was Robert Oppenheimer. Frequently, Gell-Mann talked with Albert Einstein. In Princeton, he met the English woman Margaret Dow, who worked at Princeton University. In 1955, they got married.

In 1952, Gell-Mann went to the University of Chicago and joined the research group of Enrico Fermi. Gell-Mann was in particular interested in the new particles, discovered in the cosmic rays (hyperons, K-mesons). Nobody understood why these particles were created easily in collisions of protons with nuclei, but decayed rather slowly. Fermi thought that this strange property should follow from an angular momentum barrier and advised Gell-Mann to work on this problem.

In order to understand the peculiar properties of the new hadrons, Gell-Mann introduced a new quantum number, which he called strangeness. The nucleons were assigned strangeness zero. The newly discovered lambda-hyperon was assigned the strangeness −1, likewise the sigma-hyperons. The Xi-hyperons were given strangeness −2. The negatively charged K-meson had strangeness −1. Gell-Mann assumed that the strangeness quantum number was conserved by the strong and electromagnetic interactions, but violated by the weak interaction. Thus the decays of the strange particles into normal particles without strangeness could only proceed via the weak interaction. A similar idea was developed independently by Kazuhiko Nishijima in Osaka (Japan).

The idea of strangeness explained in a simple way why the new particles were produced copiously in hadronic collisions, but decayed very slowly. In a collision, a new particle with strangeness −1 could be produced by the strong interaction together with a particle with strangeness +1. For example, a negatively charged sigma-hyperon could be produced together with a positively charged K-meson. However, a positively charged sigma-hyperon could not be produced together with a negatively charged K-meson, since both particles have strangeness −1. Likewise, two neutrons could not turn into two lambda-hyperons.

Today, we describe the strange particles with the quark model. The strangeness of a particle is minus the number of strange quarks in the particle. Thus, the lambda-hyperon has the quark structure (uds) — it has strangeness −1. It would have been better to use the opposite sign for the strangeness, but this became clear only after the introduction of the quark model in 1964.

In 1954, Gell-Mann and Francis Low worked on the renormalization program of quantum electrodynamics (QED). They introduced a new method, which was later called the "renormalization group" method. Gell-Mann and Low calculated the energy dependence of the renormalized coupling constant. In QED, the effective coupling constant increases with the energy. This was observed with the LEP accelerator at CERN. It was found that the fine structure constant at 200 GeV is about 1/127, while at low energies it is close to 1/137. The observed increase agreed perfectly with the theoretical prediction.

Kenneth Wilson, a Ph.D. student of Gell-Mann, applied the method of Gell-Mann and Low very successfully to phase transitions in condensed matter physics. He was awarded the 1982 Nobel Prize in physics.

In 1955, Gell-Mann obtained an offer of associate professorship from the California Institute of Technology, Pasadena, which was initiated by Richard Feynman. In the same year, he moved to Pasadena. One year later, he was promoted to full professor — Gell-Mann became the youngest full professor in the Caltech history.

In 1957, Gell-Mann started to work with Richard Feynman on a new theory of weak interactions. They published one year later their paper "Theory of the Fermi Interaction." Feynman and Gell-Mann described the weak interaction by a universal 4 — Fermi interaction, given by the product of two $V − A$ (vector minus axial vector, currents. The lepton current is a product of a charged lepton field and an antineutrino field. The electrons emitted in a beta-decay are left-handed, the emitted antineutrinos are right-handed. Some of the experiments were in disagreement with the new $V - A$ theory. Feynman and Gell-Mann suggested in their paper that these experiments were wrong, and it turned out that they were correct.

Feynman and Gell-Mann apply their formalism to the muon decay and calculate the lifetime of the muon. It agrees very well with the observed lifetime. Feynman and Gell-Mann discuss an interaction, which is universal and which involves two-component neutrinos. It preserves invariance under CP and T, but both the C and P symmetry are maximally violated. A nonleptonic parity-violating weak interaction of hadrons is predicted, given by the product of two hadronic weak currents, multiplied with the Fermi constant.

In 1960, Gell-Mann invented a new symmetry to describe the new baryons and mesons, found in the cosmic rays and in various accelerator experiments. He used the unitary group SU(3). At the same time, such a symmetry was also considered by Yuval Neeman, who worked at the Israeli embassy in London.

In the experiments, six new baryons were found: the three sigma-hyperons, the lambda-hyperon and the two chi-hyperons. These baryons and the nucleons were placed in an octet of the group SU(3), likewise the three pi-mesons, the four K-mesons and the eta-meson. The spin-3/2 baryon resonances were placed in a ten representation. Gell-Mann and Neeman could not explain why there were no particles, which could be placed in a triplet or in a sextet representation of SU(3).

Gell-Mann described the symmetry breaking by a SU(3)-octet. He and also S. Okubo found a mass formula, which is called the Gell-Mann–Okubo mass formula. It describes the mass differences among the baryons very well. For the decuplet, Gell-Mann found a simple rule for the mass differences: the equal spacing rule. The mass difference between the sigma resonances and the delta resonances should be the same as the mass difference between the chi resonances and the sigma resonances.

Nine particles were known in 1960, the four delta resonances (strangeness zero), the three sigma resonances (strangeness minus one) and the two chi resonances (strangeness minus two). Gell-Mann predicted the existence and the mass of a negatively charged tenth particle with strangeness minus three, which he called the omega minus particle. This particle is unique in the decuplet, since due to its strangeness minus three, it could only decay by the weak interaction. Thus, it would have a relatively long lifetime. It was discovered in 1964 by Nicholas Samios and his group in Brookhaven — it had the mass that Gell-Mann had predicted. Thus, the SU(3)-symmetry was very successful. In 1969, Gell-Mann received the Nobel prize for the new symmetry, based on the group SU(3).

In the SU(3) model, one can introduce an octet of vector currents and an octet of axial vector currents. In 1962, Gell-Mann proposed the algebra of currents, which led to sum rules for cross-sections, like the Adler sum rule. Current algebra was the main topic of research in the following years. Gell-Mann wrote several successful papers with his colleague Roger Dashen on various topics of current algebra.

In 1964, Gell-Mann discussed the triplets of SU(3), which were considered as constituents of the hadrons. Gell-Mann called them quarks, using an artificial word, used by James Joyce in his book Finnegans Wake. At the same time George Zweig, who worked for a year at CERN, introduced the same triplets, which he called "aces." Gell-Mann published his results in the European journal *Physics Letters*, as he was afraid that the referees of *Physical Review Letters* would reject his article. His letter had the title: "A Schematic Model of Baryons and Mesons."

The three quarks, named u, d and s, were the constituents of the hadrons. The baryons were considered as bound states of three quarks, for example the proton and the neutron had the structure (uud) and (ddu). The lambda-hyperon consisted of u, d and s. The omega minus was a bound state of three strange quarks: (sss). The quarks had peculiar properties, in particular the electric charges $2/3$ and $-1/3$.

The quark model was not taken seriously by many physicists, due to severe problems. For example, the omega minus, a bound state of three strange quarks, placed symmetrically in an s-wave, violated the Pauli principle, since the wave function was not antisymmetric.

But in 1968, the quarks were found indirectly in the SLAC experiments. In the deep inelastic electron–proton experiments, the electron were deflected by point-like constituents, which were identified with the quarks. In 1971, Gell-Mann and Harald Fritzsch described the results of the experiments at SLAC with the light-cone algebra of currents.

William Bardeen, Harald Fritzsch and Murray Gell-Mann introduced in 1971 a new quantum number for the quarks, which was called the color quantum number. The quarks appeared in three colors: red, green and blue. Instead of three quarks, there were now nine quarks. The three colors were described by the color group SU(3). The hadrons were considered as color singlets, as "white" states. The simplest color singlets were the bound states of a quark and an antiquark (meson) or three quarks (baryon). The baryon wave functions were antisymmetric in the color index. The electromagnetic decay rate of the neutral pion agreed very well with the prediction of the color theory.

In 1972, Fritzsch and Gell-Mann introduced a gauge theory for the strong interactions. The color quantum number was considered to be a gauge quantum number, like the electric charge in QED. The color symmetry was considered to be an exact symmetry. The gauge bosons were massless gluons, which transformed as an octet of the color group. Later, they called this theory Quantum Chromodynamics (QCD). It was discussed at the 1972 Rochester conference at Fermilab. In this theory, the hadrons are considered as color singlets. The quarks have nonintegral charges and are confined. In 1973, Harald Fritzsch, Murray Gell-Mann and Heinrich Leutwyler discussed the advantages of this theory in a letter, published in *Physics Letters*.

In QCD, the gluons interact with the quarks, but also with themselves. This self-interaction leads to the interesting property of asymptotic freedom. The gauge coupling constant of QCD decreases, if the energy is increased. The first calculations in this direction were made by Iosif Khriplovich in Novosibirsk in 1969, and later by G. 't Hooft in Utrecht (unpublished). He did not realize the importance for strong interaction physics. In 1973, the calculations of David Gross and Frank Wilczek as well as of David Politzer clearly demonstrated the property of asymptotic freedom. They also suggested that at low energies the coupling constant might increase without limit, thus all colored objects, e.g. the quarks, are confined. A rigorous proof of the confinement property is still missing.

The quark–gluon coupling constant was measured in many experiments. At the energy of about 91 GeV (the mass of the Z-boson) the strong interaction analogue of the fine structure constant is about 0.12.

In QCD, the scaling property of the cross-sections, observed in deep inelastic scattering, is not an exact property, but it is violated by small logarithmic terms. The scaling violations were observed and in good agreement with the theoretical predictions. The scaling violations are described by a scale parameter "lambda," which is an energy parameter. The experiments are in agreement with the theoretical predictions, if this parameter is about 250 MeV.

In the absence of the quark masses, the theory of QCD depends only on this scale parameter, which determines the properties of the hadrons, e.g. their masses or their magnetic moments. The proton mass can be expressed as a numerical constant, multiplied by the scale parameter. This constant can be calculated in QCD, e.g. by lattice methods. Of course, in reality the proton mass also depends on the quark masses. These contributions can be calculated, using the chiral symmetry. About 20 MeV of the proton mass are due to the u-quarks, about 19 MeV due to the d-quarks, and about 35 MeV are due to the pairs of strange quarks and antiquarks. If the quark masses are set to zero, the mass of the proton will be reduced to about 862 MeV.

The current quark masses do not reflect the SU(3) symmetry. At an energy scale of 1 GeV the u-mass is about 5.4 MeV, the d-mass is about 7.8 MeV, and the s-mass is about 150 MeV. Since the d-mass is larger than the u-mass, the neutron is heavier than the proton.

Feynman predicted in 1974 that in electron–positron annihilation at very high energy the produced quarks would lead to jets of particles, the quark jets. These jets were observed at DESY in 1978. One year later, it was observed at DESY events with three jets. In the annihilation of an electron and a positron, a quark, an antiquark and a gluon were produced. The gluon produced also a particle jet, thus three jets were observed.

Fritzsch and Gell-Mann discussed in 1972 in their paper, which appeared in the proceedings of the Rochester conference in Chicago, that there should exist neutral particles, composed of gluons. Since the gluons are color octets, two or three gluons could form a color singlet hadron, called a glue-meson. One has searched for these particles, but no clear evidence was found. Presumably, these particles mix strongly with neutral quark–antiquark mesons.

6

In 1979, Gell-Mann, Pierre Ramond and Richard Slansky discussed the seesaw mechanism for the neutrino masses. The very small neutrino masses are then related to the masses of the charged leptons and a very heavy Majorana mass for the right-handed neutrino.

After 1980, Gell-Mann got interested in string theories. In 1984, Gell-Mann co-founded the Santa Fe Institute. In this Institute he has now his office.

Gell-Mann had a few very good Ph.D. students. In particular, I like to mention Sidney Coleman, Rodney Crewther, Kenneth Wilson and Kenneth Young. He also wrote a popular book with the title *The Quark and the Jaguar*.

Gell-Mann has worked in physics for more than 60 years, and he continues to publish articles on physics. Since many years he is also interested in linguistics, in birds and in archaeology.

MEMORIES OF MURRAY AND THE QUARK MODEL

G. ZWEIG

26-169, Research Laboratory of Electronics, Massachusetts Institute of Technology,
77 Massachusetts Ave., Cambridge, MA 02139-4307, USA
zweig@mit.edu

Life at Caltech with Murray Gell-Mann in the early 1960's is remembered. Our different paths to quarks, leading to different views of their reality, are described.

Keywords: Quarks; aces; history; Caltech; current algebra.

Prologue: In 1964 Dan Kevles arrived at Caltech from Princeton as a young assistant professor of history, specializing in the history of science. As an undergraduate he had majored in physics. Shortly after his arrival I barged into his office, told him that Elementary Particle Physics was in great flux, tremendously exciting; history was in the making, just waiting for him to record. And much of it involved Richard Feynman and Murray Gell-Mann, whose offices were just 600 feet away!

My excitement was not contagious. Dan lectured me, saying that no one can recognize what is historically important while it is happening. One must wait many years to understand the historical significance of events. What he could have added is that it is convenient for historical figures to be unavailable to contradict historians who document their actions, and sometimes even their motives.[a]

Well, I'm going to risk it. Today I'm going to tell you about the Murray Gell-Mann I saw in action, and a little bit about the history of the quark model. Not only is Murray alive and well, he's in the audience, and will keep me honest. So let's begin.

Early influences: Murray, a belated "Happy Birthday!" I have learned a lot from you, and for that I am truly grateful. We go way back, even further than you realize. In the summer of 1957, after a hard day's work as a counselor at a day camp for children, I came across an article you coauthored in *Scientific American*, which said of elementary particles:[1]

> "At present our level of understanding is about that of Mendeleyev, who discovered only that certain regularities in the properties of the elements existed. What we aim for is the kind of understanding achieved by Pauli,

---

[a]In Dan's defense, when Henry Kissinger asked China's Premier Zhou Enlai to assess the 1789 French revolution, Zhou Enlai is reported to have replied, "It is too early to say."

whose exclusion principle showed why these regularities were there, and by the inventors of quantum mechanics, who made possible exact and detailed predictions about atomic systems."

This article appeared just three months before Sputnik, when it still wasn't fashionable to do physics. At the time I was just starting my junior year at the University of Michigan as a math major, but was thinking of switching to physics when going to grad school. Here was a big green light saying: "Go!"

In my senior year I went in to see my quantum mechanics professor P. V. C. Hough for advice on graduate schools. This was the Hough who would become the Hough of the Hough-Powell bubble chamber digitizer, and the Hough transform in image processing. His comment: "Bethe is at Cornell, where I come from, but he's getting old. There are a couple of young guys at Caltech, Feynman and Gell-Mann, why don't you go there." And I did.

Life at Caltech, the first 3 years: It was wonderful to be at Caltech in the very early 60's. Carl Anderson was the avuncular chairman of the physics department. The theory graduate students included Hung Cheng, Sidney Coleman, Roger Dashen, Jim Hartle, and Ken Wilson, just to name some. Shelly Glashow and Rudy Mössbauer were postdocs, and Yuval Ne'eman and J. J. Sakurai were visitors. And then, of course, there were Murray and Richard Feynman. If that wasn't enough, you could always go across campus and talk with ex-particle-physicist Max Delbrück, who had invented molecular biology, or Linus Pauling, a phenomenologist par excellence.

Money was pouring into particle physics, helped now by Sputnik. Pictures from bubble and spark chambers were just beginning to provide an enormous wealth of information. I still remember driving across LA to my first APS meeting at UCLA. In a cavernous dark half-empty auditorium three speakers, Bogdan Maglić, Bill Walker, and Harold Ticho showed slides demonstrating the existence of the first meson resonances, the ω, ρ, and K^*. APS meetings seemed pretty interesting!

Shortly thereafter, Murray and Yuval Ne'eman independently proposed that these, and other hadronic resonances, be classified according to the representations of SU(3), trumping Lee and Yang who continued to use the representations of G_2.[2] But this is getting ahead of our story.

After my first academic year at Caltech I asked Bob Christy, one of my professors, if I could do theoretical research with him over the summer. In a very disdainful way he replied, "You know nothing. Why don't you go over to the Synchrotron and learn experimental physics. If you do become a theorist later you won't have time to learn what experimental physics is all about." In retrospect, this was great advice.

At the Synchrotron, Alvin Tollestrup was testing his "fast electronics," which would be used to study the nonleptonic decay $K^+ \to \pi^+ + \pi^0 + \gamma$ at the Bevatron in Berkeley. This K particle had other uses. After talking to Alvin, I proposed looking for the violation of time-reversal symmetry in leptonic K-decay, piggybacking on Alvin's experiment. This was to be my thesis problem. Alvin suggested that I talk to Murray to gain a better understanding of the $\Delta I = 1/2$ rule in nonleptonic K-decay, which Alvin's experiment was designed to illuminate.

At this point I remember only one of my meetings with Murray. I had worked out a dynamical mechanism for the suppression of leptonic K decay, which allowed me to predict angular distributions. My first theoretical result! I walked happily into Murray's office, handing him two pieces of paper, one a xerox copy of the published experimental results, and the other the corresponding theoretical angular distributions, which were in good agreement with experiment. Murray looked at the two pieces of paper, looked at me, and said "In our field it is customary to put theory and experiment on the same piece of paper." I was mortified, but the lesson was valuable.

Because I was a graduate student, I got off lightly in my interactions with the faculty. Not so for all. Fred Zachariasen, who had initially suggested Alvin's K-decay experiment, invited one of his collaborators, Marshall Baker, to give a seminar about Marshall's recent work on K-decay. Particle physics seminars took place every Tuesday at 2 o'clock in a very small classroom. As usual, Feynman and Murray sit front row center. Lesser luminaries, postdocs, and graduate students sit in the rows behind them. Murray is wearing his tweed sports coat with tie, while Feynman, dressed more like a graduate student, impatiently taps the floor with his hush puppy shoes. Both of them look oddly out of place, squeezed into drop-leaf chairs, with their paddles out, meant for undergraduates. As Marshall begins, Murray reaches down to his side, picks up a folded newspaper from the floor, unfolds it, snaps it open at eye level, and proceeds to read right in front of Marshall, who is only a yard away. After about a minute, Feynman, who doesn't pay much attention to other people's work, leans over to Murray and asks in his best Far Rockaway accent "Is this guy smart?" Feynman's voice is hushed, but loud enough so that everyone in the room, including the speaker, hears the question. This is not the first time the seminar attendees have witnessed these two in action. They know that if Murray's head nods up and down behind the paper, Feynman will ask questions. If his head rocks back and forth, Feynman won't waste time with questions. This time Murray nods up and down, answering the question for everyone except the speaker. What the seminar attendees didn't know is that Marshall stutters when stressed. Feynman starts questioning, Marshall starts stuttering; the more questions, the longer the stutter. With Feynman's final question, Marshall's stutter goes into an infinite loop, Feynman slams the palm of his hand down on the paddle of his drop-leaf chair, shouts "Goddamn it! I can't get a straight answer out of this guy," and storms out of the classroom, leaving Marshall in full stutter.

The next day I happened to walk by Murray's office. The door was open, and I overheard Fred animatedly asking Murray to give Marshall a $100 honorarium as partial compensation for Feynman's atrocious behavior. Murray seemed sympathetic, but noncommittal.[b]

---

[b]Speakers at Caltech theory seminars never got an honorarium. Murray no longer remembers if this tradition was broken in Marshall's case.

10

I won't describe the next two years of 18-hour days of classes and experimental work. When the smoke cleared, I couldn't find any evidence for the violation of time-reversal symmetry. Faced with the prospect of another two years determining the value of an upper bound, I punted and went to Mexico for a month. Upon returning, I switched to theory, and asked Murray to be my thesis advisor. Despite what you might think from my previous remarks, Murray had been very kind to me, almost fatherly, so he was a natural choice. But Murray said no! He was going to the East Coast on sabbatical, but he "would talk to Dick."

When I went in rather timidly to ask Feynman if he would be my thesis advisor, he responded: "Murray says you're OK, so you must be OK." And then I remembered Murray's nodding up and down at Marshall Baker's seminar. After telling me about life with his thesis advisor, Johnny Wheeler, Feynman said that he wanted to see me from 1:30 in the afternoon till tea time (4:15) every Thursday. I prepared frantically for each meeting, never presenting the same topic twice. This went on for the entire academic year.

How constituent quarks (aces) were discovered:[3] Let me tell you about just one of those meetings, which took place late April 1963. On April 15, *Physical Review Letters* published a paper titled "Existence and Properties of the ϕ Meson."[4] The casual reader of that article, and perhaps even the authors themselves, might have thought this was just a confirmation of the existence of yet another resonance. By then over 25 "credible" meson resonances had been reported. But I thought it remarkable that the ϕ decayed only into $K + \bar{K}$ near threshold, with angular momentum 1, while there was no evidence for the decay into $\rho + \pi$ far above threshold, with angular momentum 0. Phase space arguments greatly favored $\rho + \pi$ over $K + \bar{K}$, but only $K + \bar{K}$ was observed. My calculations showed that the decay into $\rho + \pi$ was suppressed by at least two orders of magnitude. The ϕ was much narrower than expected (see Fig. 1)!

How was this discrepancy to be understood? The authors of the paper noted that there might be a problem, but dismissed the discrepancy. They wrote:

> "The observed rate [for $\phi \to \rho + \pi$] is lower than ... predicted values by one order of magnitude; however the above estimates are uncertain by at least this amount so that this discrepancy need not be disconcerting."

Feynman couldn't be bothered with the discrepancy. He launched into a tirade about how unreliable experiments were, and explained that at the time he proposed the V–A theory for the weak interactions, experiments were against him, and those experiments all turned out to be wrong.[c]

But I couldn't get the suppression of ϕ decay out of my mind. Feynman had taught that "in the strong interactions everything that can possibly happen does,

---

[c]The V–A theory was initially at variance with angular correlations measured in He^6 decay, and the absence of the decay $\pi^- \to e^- + \bar{\nu}$. Later at CERN, Alvin observed this decay at the predicted rate, confirming V–A.

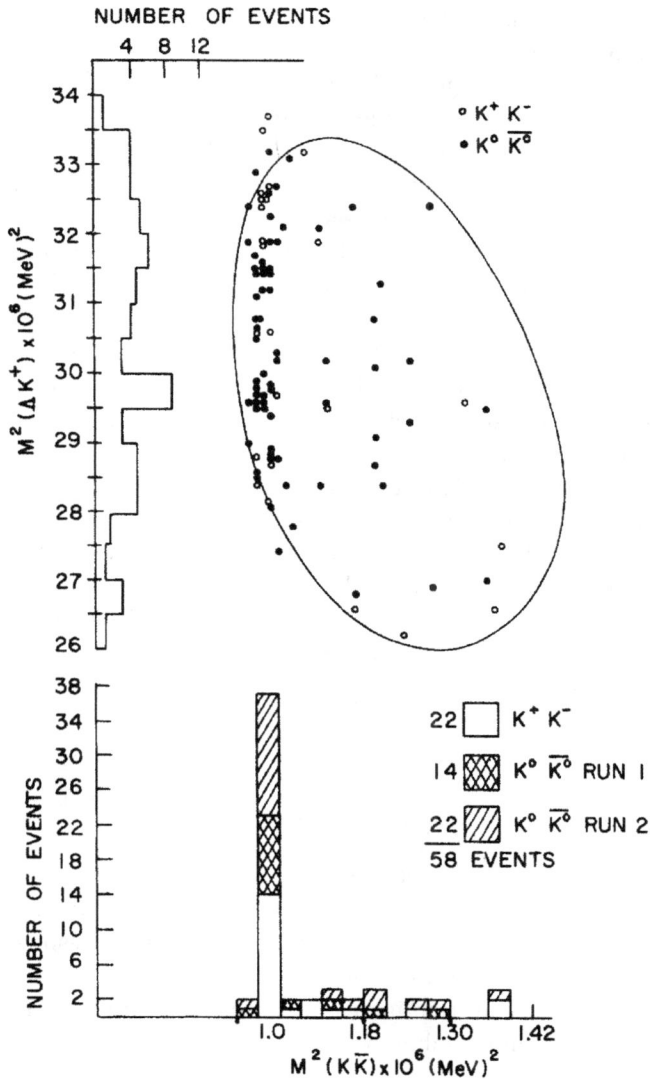

Fig. 1. Dalitz plot taken from "Ref. 4." The expected dominant decay, $\phi \to \rho + \pi$, was not observed. Instead, ϕ decayed into $K + \bar{K}$, even though the K and $\bar{K}$ have angular momentum 1, and all resonant events are at the edge of the Dalitz plot. Reprinted with permission. Copyright 1963 by the American Physical Society.

and with the maximum strength allowed by unitarity."[d] Well here was a strong interaction — a decay — that was not happening with maximal strength. It wasn't happening at all! Current theory said that suppressions exist because of symmetries,

---

[d]This was a different, but more useful, form of Murray's Totalitarian Principle: "Everything which is not forbidden is compulsory."

but in this case there wasn't a symmetry to enforce the suppression. I was convinced that something important must be happening.

In 1949 Fermi and Yang suggested that the pion was not an elementary particle, but rather a bound state of a nucleon and antinucleon.[5] Sakata extended that model to include strangeness, using p, n, and Λ to form both meson and baryon resonances. By 1963 enough was known about hadron dynamics and the baryon resonances to see that these models could not be correct in detail,[e] but the idea that hadrons had constituents fascinated me. I replaced Sakata's constituents with three unknown constituents, p_0, n_0, and Λ_0,[6,7] and called them "aces."[f] The first two aces had strangeness 0, the third, Λ_0, strangeness -1. To avoid problems with the baryon spectrum inherent in the Sakata model, aces were assigned baryon number $1/3$. Fractional baryon number meant fractional charge. The mass splitting between the p_0 and n_0 was assumed to be of electromagnetic origin, and therefore small. The Λ_0 was assumed to be substantially heavier than the other two aces, and responsible for the SU(3) symmetry breaking that occurred in the strong interactions. The ϕ was assumed to consist entirely of $\Lambda_0\bar{\Lambda}_0$, and the ρ and the π to consist only of the *other* two aces and their antiparticles. I didn't want the ϕ to contain any $p_0\bar{p}_0$ or $n_0\bar{n}_0$, since the strong interactions distinguished Λ_0 from p_0 and n_0. Assuming that the squares of meson masses were proportional to the sum of the squares of the masses of their constituents led to two relations among vector meson masses,

$$m_\omega^2 = m_\rho^2 \,,$$

and

$$m_{K^*}^2 = (m_\phi^2 + m_\rho^2)/2 \,.$$

Both relations were remarkably accurate.

What remained was an assumption about dynamics, i.e., an assumption about how mesons decay, expressed in terms of their constituents. I assumed that when a meson $a\bar{a}$ initiated its decay into two other mesons $a\bar{a}' + a'\bar{a}$, the a would separate from the $\bar{a}$, and as the separation increased, a new $a'\bar{a}'$ pair would pop out of the vacuum, also separate, and combine with the now separated a-$\bar{a}$ pair to complete the decay (see Fig. 2),[g]

$$a\bar{a} \to a\bar{a}' + a'\bar{a} \,.$$

---

[e] Indeed, Fermi and Yang had written "Unfortunately we have not succeeded in working out a satisfactory relativistically invariant theory of nucleons among which ... attractive forces act [to form pions]."

[f] There are 4 aces in a deck of cards, so why call them aces? Because in analogy with the 4 leptons known at that time, I though that there should be a fourth constituent. If the τ were known then, I might have called them dice.

[g] The a and $\bar{a}$ were not allowed to separate without the creation of an $a'\bar{a}'$ pair, since aces had fractional charge, and fractionally charged particles were not observed in meson decays. The other possibility, that the a and $\bar{a}$ would "eat each other," was forbidden by fiat.

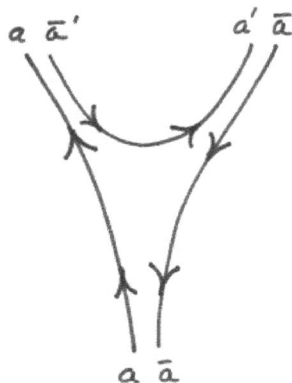

Fig. 2   "Zweig diagram" for the decay of the meson $a\bar{a}$. Murray sometimes called these "twig diagrams," since the English word "twig" is derived from the German word, "zweig," meaning branch.

Since the ϕ only contained Λ_0 and $\bar{\Lambda}_0$, whereas ρ and π only contained n_0, p_0, $\bar{n}_0$, and $\bar{p}_0$, ϕ decay into $\rho + \pi$ was impossible!

The amplitude for any hadronic decay could be computed pictorially. Fig. 3 is an example taken from the original ace paper.[7] These diagrams contained more information than SU(3) provided. "Zweig's rule" not only forbad certain decays, it specified the relative amplitudes of allowed decays.[h] For example, in addition to forbidding $\phi \to \rho + \pi$, the rule determined the F/D ratio for meson-baryon couplings.

Were aces real? Since aces obeyed dynamical rules, it was hard to imagine that they weren't real. Ace-antiace pairs popped out of the vacuum in hadronic decays. Aces and antiaces orbited around one another with angular momentum $\vec{L}$ and total spin $\vec{S}$; the mesons they created had mass that depended on the value of $\vec{L} \cdot \vec{S}$. And the weak leptonic decay of hadrons was attributed to the weak decay of their ace constituents, which were governed by V–A interactions. However, arguing against the reality of aces was the existence of the famous spin $\frac{3}{2}$ Ω^-, which contained 3 identical Λ_0 aces with their spins aligned, violating Pauli's spin-statistics theorem![i]

This "tinker-toy" view of hadron physics that seemed to violate the spin-statistics theorem drove people crazy.[j] When I went in to see Murray to explain my ideas after returning from CERN in the early fall of 1964, he exclaimed "Oh, the concrete quark model. That's for blockheads!" When I explained my reason for the suppression of ϕ decay to Feynman, he became visibly irritated, arguing that

---

[h]Explicit rules for computing decay amplitudes implicit in the graphical calculus are summarized in Appendix 2 of "Ref. 8."

[i]I thought this problem would eventually be solved, and it was, by distinguishing the 3 aces with 3 different colors.

[j]In addition, since aces hadn't been observed, doing physics with aces ignored a fundamental lesson learned from quantum mechanics: "Always work with observables." The "Bootstrap," built on Heisenberg's S matrix of scattering amplitudes, evolved from this maxim.

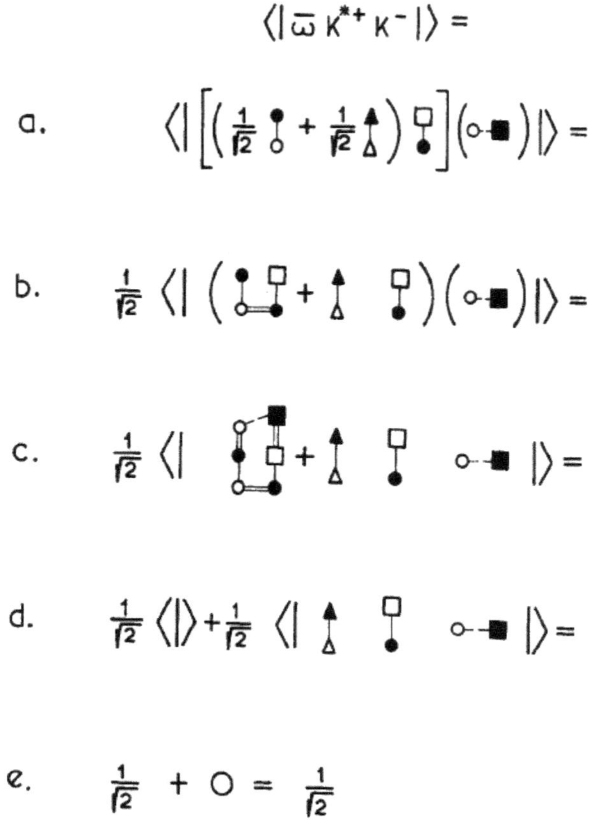

$$\langle | \bar{\omega} \, K^{*+} \, K^- | \rangle =$$

a.

b.

c.

d.

e. $\quad \dfrac{1}{\sqrt{2}} \; + \; O \; = \; \dfrac{1}{\sqrt{2}}$

Fig. 3.   The graphical computation of the $\omega K^{*+} K^-$ coupling constant taken from the original ace paper.[7] Circles, triangles, and squares represent p_0, n_0, and Λ_0, respectively; antiaces are shaded. A meson is formed by tying an ace to an antiace with a spring (straight line). Strong interaction symmetry is broken by making the Λ_0 heavier (larger) than the other two aces. Additional aces, if discovered, were to be represented by pentagons, hexagons, etc. This idiosyncratic graphical calculus did not facilitate the acceptance of aces as constituents of hadrons.

"unitarity mixes all states with the same quantum numbers," making suppression impossible. For example, the ϕ mixes with the ω, which mixes with the $\rho + \pi$, so that ϕ must go to $\rho + \pi$. I was saying that the ω and ϕ mix, but in just such a way as to make the ϕ consist entirely of $\Lambda_0 \bar{\Lambda}_0$, forbidding the decay into $\rho + \pi$. It might seem to have been a bizarre assumption, but I had no alternative.[k] It wasn't until more than a decade later, with the discovery of the exceptionally narrow ψ/J, that people realized that the ϕ and the ψ/J were narrow for similar reasons, and finally accepted the idea that hadrons have constituents with dynamics that obey Zweig's rule.

---

[k]Even today, knowing about QCD, the suppression of ϕ into $\rho + \pi$ is still somewhat mysterious.

Murray's toy field theories: Murray had a completely different view of quarks, using them as fundamental fields in a toy field theory. Murray's use of field theories, from which symmetry relations could be abstracted, first appeared in a 1957 article.[9] The abstract begins with:

"An attempt is made to construct a crude field theory of hyperons and K particles, which are assumed to have spin 1/2 and spin 0, respectively."

Fields in this model correspond to real particles, e.g., the Λ and K, and Murray establishes relations between meson-baryon coupling constants by assuming global symmetry. Most enlightening, however, are the "General Remarks:"

"Supposing that the model we have presented has elements of truth, we may add the following remarks:

(1) The symmetry properties of the model may be correct even though the use of field theory is unjustified. For this reason an analysis purely in terms of the symmetry group of the theory is in order."

Here Murray constructs a field theory that he knows is incorrect in detail, picks properties of the objects in the theory that he believes should also hold in the real theory, and throws away the rest.

Four years later in the "Eightfold Way," Murray proposes that unitary symmetry be used to classify particles, rather than global symmetry, this time using hypothetical particles l and $\bar{L}$ as fundamental fields:[10]

"For the sake of a simple exposition, we begin our discussion of unitary symmetry with 'leptons' [l and $\bar{L}$], although our theory really concerns the baryons and mesons and the strong interactions. The particles we consider here for mathematical purposes do not necessarily have anything to do with real leptons, but there are some suggestive parallels."

After using l and $\bar{L}$ to construct states that transform like real particles, Murray reassures the reader that:

"We shall attach no physical significance to the l and $\bar{L}$ 'particles' out of which we have constructed the baryons. The discussion up to this point is really just a mathematical introduction to the properties of unitary spin."

The "Eightfold Way" was never published in a journal. Ideas from it were distilled, leading to a much more formal paper with a toy field theory based on the Sakata model, and not the model based on l and $\bar{L}$.[11] From Section IV of that paper:

"We generalize the Fermi-Yang description to obtain the symmetrical Sakata model and abstract from it as many physically meaningful relations as possible."

Current quarks, 1964: According to Bob Serber, in the spring of 1963 over lunch at the Columbia faculty club, Serber told Murray about a scheme he had been thinking about in which baryon representations were made from three fundamental representations of SU(3) ($3 \times 3 \times 3$), and meson representations from the fundamental representation and the representation representing the antiparticles of the fundamental representation ($3 \times \bar{3}$).[1] After a moment's calculation Murray found that this would imply that the members of the fundamental representation would have fractional charge, a fact that Serber had not realized. No more was said, but in February of 1964 Murray proposed using the three fractionally charged objects in the fundamental representation as fields from which to construct the currents of a toy field theory.[12]

> "we assign to the triplet t the following properties: spin $\frac{1}{2}$, $z = -\frac{1}{3}$, and baryon number $\frac{1}{3}$. We then refer to the members $u^{\frac{2}{3}}$, $d^{-\frac{1}{3}}$, and $s^{-\frac{1}{3}}$ of the triplet as 'quarks' ... A formal mathematical model based on field theory can be built up for the quarks exactly as for p, n, Λ in the old Sakata model ... All these [current commutation] relations can now be abstracted from the field theory model and used in a dispersion theory treatment."

Finally, Murray ends the paper with the famous lines:

> "It is fun to speculate about the way quarks would behave if they were physical particles of finite mass (instead of purely mathematical entities as they would be in the limit of infinite mass). ... A search for stable quarks of charge $-\frac{1}{3}$ or $+\frac{2}{3}$ and/or stable di-quarks of charge $-\frac{2}{3}$ or $+\frac{1}{3}$ or $+\frac{4}{3}$ at the highest energy accelerators would help to reassure us of the non-existence of real quarks."

Murray's modus operandi is eloquently explained in a paper published five months later:[13]

> "We use the method of *abstraction* from a Lagrangian field theory model. In other words, we construct a mathematical theory of the strongly interacting particles, which may or may not have anything to do with reality, find suitable algebraic relations that hold in the model, postulate their validity, and then throw away the model. We compare this process to a method sometimes employed in French cuisine: a piece of pheasant meat is cooked between two slices of veal, which are then discarded."

Murray's evolving view of quarks: At the end of February 1972, Murray delivered a set of lectures in Schladming Austria titled "Quarks".[14] This is the last record I have showing Murray's views before the "November Revolution" when the ψ/J was discovered, making the existence of real quarks all but obvious. By that time

---

[1]Letter to me from Bob Serber dated July 8, 1980.

Murray spoke of "constituent quarks," but viewed *his* quarks as "current quarks." Murray begins with:

"In these lectures I want to speak about at least two interpretations of the concept of quarks for hadrons and the possible relations between them. First I want to talk about quarks as 'constituent quarks'. These were used especially by G. Zweig (1964) who referred to them as aces.... The whole idea is that hadrons act as if they are made up of quarks, but the quarks do not have to be real. If we use the quark statistics described above, we see that it would be hard to make the quarks real, since the singlet restriction is not one that can be easily applied to real underlying objects;....

There is a second use of quarks, as so-called 'current quarks' which is quite different from their use as constituent quarks;.... In the following discussion of current quarks we attempt to write down properties that may be exact, at least to all orders in the strong interaction, with the weak, electromagnetic and gravitational interactions treated as perturbations....

If quarks are only fictitious there are certain defects and virtues. The main defect would be that we never experimentally discover real ones and thus will never have a quarkonics industry. The virtue is that then there are no basic constituents for hadrons — hadrons act as if they were made up of quarks but no quarks exist — and, therefore, there is no reason for a distinction between the quark and bootstrap picture: they can be just two different descriptions of the same system, like wave mechanics and matrix mechanics. In one case you talk about the bootstrap and when you solve the equations you get something that looks like a quark picture; in the other case you start out with quarks and discover that the dynamics is given by bootstrap dynamics...."[m]

"If we go too far... and try to construct a complete Fock space for quarks and antiquarks on a light-like plane, abstracting the algebraic properties from free quark-theory, we are in danger of ending up with real quarks, and perhaps even with free real quarks as I mentioned before. In our work, we are always between Scylla and Charybdis; we may fail to abstract enough, and miss important physics, or we may abstract too much and end up with fictitious objects in our models turning into real monsters that devour us."

The 1957 article "Elementary Particles," that Murray wrote with Rosenbaum, was viewed as a great success by Jim Flanagan, editor of the *Scientific American*. In late 1971 he flew Frank Bello, an associate editor, to Pasadena to help Murray and me write an article on quarks, the new "elementary particles." Frank

---

[m] However, in order to recover bootstrap dynamics, the algebraic properties of operators abstracted from the free field theory of current quarks would have to be supplemented by additional assumptions about quark dynamics.

and I wrote a draft, but he and Murray completely rewrote it after Frank got back to New York, changing the meaning of constituent quarks. Murray and Frank wrote:[n]

> "As seemed probable from the outset, the quark model may be nothing more than a useful mathematical construct: The known hadrons — including dozens not yet discovered when the model was conceived — behave 'as if' they were composed of quarks. Quarks themselves may have no independent existence."

Murray and I could not agree on the meaning of constituent quarks.[o] When Murray suggested we abandon the article, I agreed.

A tribute from the master: In 1977 Feynman nominated both of us for the Nobel Prize in Physics. When I learned about this relatively recently, I felt great satisfaction. Murray, on the other hand, might think that this is no big deal for him. After all, he already has a Nobel prize, and he presumably gets nominated every year for a second one. But to my knowledge, Feynman never nominated anyone for anything, so I think this is a real tribute, even for Murray. As proof of Feynman's nomination, I offer Fig. 4.[p]

Summary: How can Murray's contributions, described here, be put in some perspective? Causality and CPT symmetry are examples of very general principles that are expected to hold in all particle interactions. By abstracting from toy free field theories, Murray identified certain algebraic relations among operators, e.g., the equal-time current commutation relations, that he postulated as also holding in the strong interactions. Since the matrix elements of these operators were measurable, his postulates were testable, and some were quickly verified to reasonable accuracy.[16,17] These relationships between operators, though limited in scope,[q] could be absolutely true. His genius was to understand that he must find Scylla and Charybdis, and then, like Jason, sail between them.

Science is a social enterprise, and society recognizes individuals that influence the work of others. Murray was concerned with describing reality, making predictions that could be tested experimentally, and providing a theoretical framework that enabled others to expand on his vision. The reason we are here today is because Murray thereby set an agenda for an entire generation of physicists, dominating our field like no other.

Epilogue: Murray's work eventually became less concerned with experiment, and more with theory. I walked into his office one day and asked, "Murray, you're so good at phenomenology, why aren't you doing it?" He replied, "I'm not interested in it any more." I was shocked. It was like Picasso in his prime giving up on painting. It was the end of an era.

---

[n]From a draft Frank sent to Murray on February 25, 1972; Caltech Archives.
[o]Valentine Telegdi provides an independent description of our differing views.[15]
[p]Murray was delighted to hear of Feynman's nomination. He was unaware of it before this talk.
[q]Limited by virtue of their existence in a free field theory.

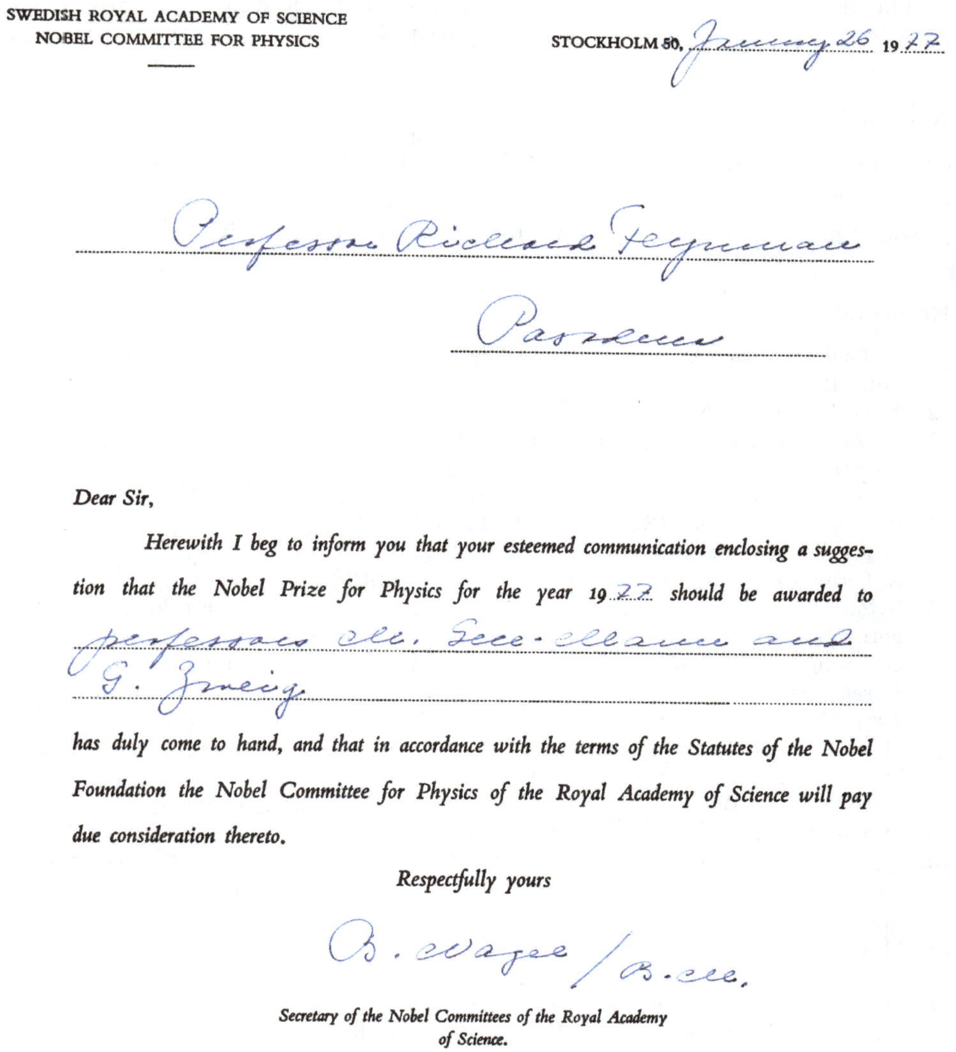

SWEDISH ROYAL ACADEMY OF SCIENCE
NOBEL COMMITTEE FOR PHYSICS
—

STOCKHOLM 50, *January 26* 19 *77*

Professor Richard Feynman

Pasadena

Dear Sir,

     Herewith I beg to inform you that your esteemed communication enclosing a sugges-tion that the Nobel Prize for Physics for the year 19 77 should be awarded to

professors M. Gell-Mann and G. Zweig

has duly come to hand, and that in accordance with the terms of the Statutes of the Nobel Foundation the Nobel Committee for Physics of the Royal Academy of Science will pay due consideration thereto.

     Respectfully yours

B. Nagel / B. ell.

Secretary of the Nobel Committees of the Royal Academy
of Science.

Fig. 4.   Acknowledgement of Feynman's nomination letter by the Nobel Committee for Physics.

    Over the years Murray and I drifted apart. Murray worked on foundations of quantum theory, then complexity and linguistics. I switched to neurobiology, or as Murray put it with a smile, "cutting up cats."

    The historian Dan Kevles — whose office I barged into almost 50 years earlier, asking him to record history in the making — true to his word, went on to research the past and write about George Ellery Hale in the Gilded Age, and Robert Millikan, a founder and first president of Caltech. Eventually Dan did broaden his vision of what historians do. In 1998 he wrote a book about Caltech's then sitting president David Baltimore.[18]

Oh, and whatever happened to aces? They are alive and well! In case you haven't noticed, constituent quarks are really aces in disguise.

Acknowledgments

Charlotte Erwin, Head of Archives and Special Collections at Caltech, and Loma Karklins have been very helpful in locating documents for this talk. Erica Jen has provided invaluable advice as to what to say, and how to say it.

References

1. M. Gell-Mann and E. P. Rosenbaum, Elementary particles, *Scientific American* 72–86 (July 1957).
2. T. D. Lee and C. N. Yang, *Phys. Rev.* **122**, 1954 (1961).
3. G. Zweig, Origins of the quark model, in *Proceedings of the Fourth International Conference on Baryon Resonances*, ed. N. Isgur (University of Toronto, Canada, 1980), pp. 439–479, www-hep2.fzu.cz/~chyla/talks/others/zweig80.pdf.
4. P. L. Connolly *et al.*, *Phys. Rev. Lett.* **10**, 371 (1963), http://link.aps.org/abstract/PRL/v10/p371.
5. E. Fermi and C. N. Yang, *Phys. Rev.* **76**, 1739 (1949).
6. G. Zweig, An SU$_3$ model for strong interaction symmetry and its breaking, CERN Report 8419/TH.401 (January 17, 1964), http://cdsweb.cern.ch/record/352337?ln=en.
7. G. Zweig, An SU$_3$ model for strong interaction symmetry and its breaking II, in *Developments in the Quark Theory of Hadrons, A Reprint Collection, Volume I: 1964–1978*, eds. D. B. Lichtenberg and S. P. Rosen (Hadronic Press, Nonamtum, Massachusetts, 1980), pp. 22–101 [CERN Report 8419/TH.412 (February 21, 1964)], http://cdsweb.cern.ch/record/570209?ln=en.
8. J. Mandula, J. Weyers and G. Zweig, *Ann. Rev. Nucl. Sci.* **20**, 289 (1970).
9. M. Gell-Mann, *Phys. Rev.* **106**, 1296 (1957).
10. M. Gell-Mann, The eightfold way, CIT Synchrotron Laboratory Report CTSL-20 (1961).
11. M. Gell-Mann, *Phys. Rev.* **125**, 1067 (1961).
12. M. Gell-Mann, *Phys. Lett.* **8**, 214 (1964).
13. M. Gell-Mann, *Phys.* **1**, 63 (1964).
14. M. Gell-Mann, *Acta Physica Austriaca* **Suppl. IX**, 733–761 (1972).
15. Interview with Valentine L. Telrgdi, Caltech Archives Oral Histories, March 2002, http://oralhistories.library.caltech.edu/146/.
16. S. L. Adler, *Phys. Rev. Lett.* **14**, 1051 (1965).
17. W. I. Weisberger, *Phys. Rev. Lett.* **14**, 1047 (1965).
18. D. Kevles, *The Baltimore Case: A Trial of Politics, Science, and Character* (W.W. Norton & Co., Inc., New York, 1998).

SOME PROBLEMS IN COLD ATOM RESEARCH*

C. N. YANG

Tsinghua University, Beijing 100084, China

It is a great pleasure to be here today, seeing so many old friends to celebrate the 80th birthday of my longtime friend. I think, probably, of all the people in the audience, I am the one who have known him for the longest period. I met Murray in 1951 when I was at the Institute for Advanced Study in Princeton.

Listening to the talks this morning, I sort of re-experienced the long career I have in elementary particle physics, but I also work in statistical mechanics. We all know Murray has a wide range of interests but probably, few people in this audience know that he has also written good papers in statistical mechanics.

This (Fig. 1) is one of the papers ("Correlation Energy of an Electron Gas at High Density") he wrote with Brueckner — a very good paper. I think he left that year, afterwards, never to return.

In recent years, the field of cold atoms has become very, very hot. The reason is because of the new, incredible technology both in laser technology and microelectronics.

Here [Fig. 2(a)] is a chip manufactured in the laboratory in Amsterdam. The structure of the chip is illustrated here [Fig. 2(b)]. There is a current, I, in red, which generates a cylindrical magnetic field, and there is also a biased field going that way. The two of them could produce together a reading of a small magnetic field, or near-zero magnetic field, but with extremely strong gradient in the neighborhood. That became a one-dimensional trap suspended 90 microns above the chip. As a consequence, you can trap thousands of atoms, either fermions or bosons in that trap. That is becoming an extremely interesting new field of study.

The theory of it is actually very simple, in principle. What we need to study is the Hamiltonian,

$$H = \sum_{i=0}^{N} \left[-\frac{1}{2}\frac{\partial^2}{\partial x_i^2} + \frac{1}{2}x_i^2 \right] + g \sum_{i>j} \delta(x_i - x_j) , \quad g \text{ tunable}, \ e_n = \frac{1}{2} + n . \quad (1)$$

---

*Transcribed by C. H. Oh (National University of Singapore). Unread by C. N. Yang.

PHYSICAL REVIEW   VOLUME 106, NUMBER 2   APRIL 15, 1957

Correlation Energy of an Electron Gas at High Density*

MURRAY GELL-MANN, *Department of Physics, California Institute of Technology, Pasadena, California*

AND

KEITH A. BRUECKNER, *Department of Physics, University of Pennsylvania, Philadelphia, Pennsylvania*
(Received December 14, 1956)

The quantity ϵ_c is defined as the correlation energy per particle of an electron gas expressed in rydbergs. It is a function of the conventional dimensionless parameter r_s, where r_s^{-3} is proportional to the electron density. Here ϵ_c is computed for small values of r_s (high density) and found to be given by $\epsilon_c = A \ln r_s + C + O(r_s)$. The value of A is found to be 0.0622, a result that could be deduced from previous work of Wigner, Macke, and Pines. An exact formula for the constant C is given here for the first time; earlier workers had made only approximate calculations of C. Further, it is shown how the next correction in r_s can be computed. The method is based on summing the most highly divergent terms of the perturbation series under the integral sign to give a convergent result. The summation is performed by a technique similar to Feynman's methods in field theory.

Fig. 1.

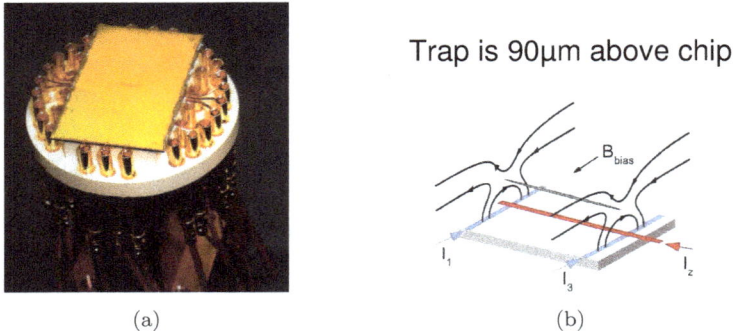

(a)   (b)

Fig. 2.

The first part of the Hamiltonian in the square bracket is the harmonic trap potential, and the last part is a one-dimensional delta function interaction. You can easily prove that in one dimension, the interaction between two atoms is reducible to a term of this form. And in particular, the g can be tuned with the new laser magneto trap and technology. So, this became an extremely interesting theoretical problem.

If you do not have the g term, that is to say one-dimensional trap problem, which can be easily solved. If you do not have the trap, the harmonic trap, that problem is one-dimensional many-body, either bosons or fermions interact with the delta function potential, that problem had been solved in the 1960s by Lieb and by me. So, the present problem is a combination of the two — how do you deal with the trap and the delta function interaction. My colleague Ma and I had solved it and we had recently published it.

Let me explain to you what is the strategy. If you do not have the g term, it is just a one-dimensional harmonic oscillator and the energy of each particle is E_n, which is equal to $\frac{1}{2} + n$ in the right units [Eq. (1)].

Now, let us look at the ground state energy, which is E_0 for this system, for different values of g (Fig. 3). First, let us concentrate on this point, which means

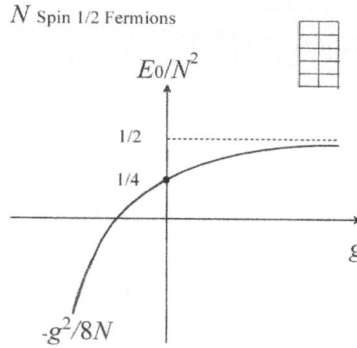

Fig. 3.

there is no delta interaction, this is a system of free fermions. Now for spin one-half fermion, you have various symmetries and it had been proven long time ago by Lieb, that the ground state has this symmetry of space wave function, that is the Young tableau. It is obvious that without the g term, it is just a collection of non-interacting fermions, and therefore for the ground state, you can easily find the energy. Obviously, half of the particles has spin-up and half of the particles with spin-down. Half of the particles with spin-up fill up the space with $n = 0$, $n = 1$, $n = $ so on and so forth. So clearly the total energy is given by $N^2/4$. So, we know that when E_0/N_2 for $g = 0$, is equal to $1/4$:

$$\boxed{g = 0} \quad \begin{matrix} N/2 \text{ spin} \uparrow \\ N/2 \text{ spin} \downarrow \end{matrix} \quad E_0 \cong 2 \sum_0^{N/2} \left(\frac{1}{2} + n \right) \cong \frac{N^2}{4}. \tag{2}$$

Let us look at this graph (Fig. 3), that is for very large g, for N fermions. If you look at the large positive g for N fermions. These N fermions have two kinds of spins, some spin-up and some spin-down but the coupling constant is very large and is positive. That means the wave function has to be zero when the two particles are together. That is true of course if they have the same spin, but for $g = +\infty$, even if they have one spin-up and one spin-down, so the Pauli's exclusion principle does not make it equal to zero, but the delta function would make it equal to zero.

In short, when $g = \infty$, the N particles have to occupy states, free particle states, from $N = 0$, all the way up to N, now up to $N = 2$. Therefore, the total energy is given by this, that is equal to $N^2/2$:

$$\boxed{g = +\infty} \quad N \text{ Fermions} \quad E_0 \cong \sum_0^N \left(n + \frac{1}{2} \right) \cong \frac{N^2}{2}. \tag{3}$$

So, we now should know if we go to this large g limit, $E_0/N^2 = 1/2$, go to $g = 0$, this is equal to $1/4$, which is half of that (Fig. 3). How about here where g is negative?

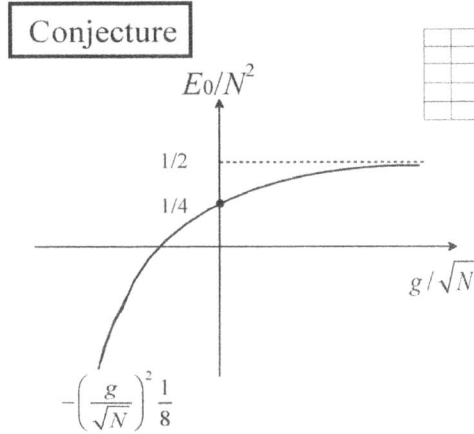

Fig. 4.

For g that is negative and large, the particle has the tendency to form spin up–spin down pairs with very deep attraction, and the pair energy is well-known. This was already done in the 1960s. The pair energy is equal to $-g^2/4$, and you have $N/2$ pairs. So, this is now the total energy:

$$\boxed{g = -\text{large}} \quad \text{pairs of } \uparrow\downarrow \text{ bound states} \quad E_0 \cong \left(\frac{N}{2}\right)\left(-\frac{g^2}{4}\right). \tag{4}$$

The total energy is proportional to Ng^2. So now, if you look at these (Fig. 3), it is very strange, because, here $g = \infty$, it is pretty constant. Here for negative g, there is a factor g^2/N. When Ma and I first looked at it, we realized what this indicates. You should plot it against $g/\sqrt{N}$ (Fig. 4), this is $1/4$, and this is $1/2$, this becomes this. This equals to $\sqrt{N}$, it is a function of this. In other words, we realize that in this plot, this becomes a limit when N goes to infinity. This was the conjecture that we published last year, and now we have proven that conjecture. In fact, we are now able to complete the whole curve.

The strategy of that is as follows. We later proved this conjecture using the Thomas–Fermi method. Thomas–Fermi method is well-known, of course, to every student of physics. It is a very physical idea.

Let us re-examine that. Let us first consider **free particles in harmonic trap with one component**, not spin-up and spin-down, just spin-up with N fermions. Let us consider this problem. Thomas–Fermi method said that in a harmonic trap like this, you consider a small region dx, and you treat the particle in here as if they are free, and you consider the particles within this interval. Then you add the total kinetic energy that would be from this point to here (Fig. 5). The maximum kinetic energy which is this length plus the potential should be equal to this level. You realize that for N fermions, the density at this point should be equal to this. Density as the function of x is a half ellipse (Fig. 6). So this is the Thomas–Fermi

Thomas-Fermi Method
(for one component Fermions)

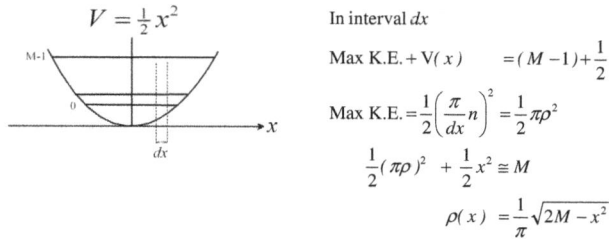

$V = \frac{1}{2}x^2$

In interval dx

Max K.E. $+ V(x)$ $= (M-1) + \frac{1}{2}$

Max K.E. $= \frac{1}{2}\left(\frac{\pi}{dx}n\right)^2 = \frac{1}{2}\pi\rho^2$

$\frac{1}{2}(\pi\rho)^2 + \frac{1}{2}x^2 \cong M$

$\rho(x) = \frac{1}{\pi}\sqrt{2M - x^2}$

Fig. 5.

Limiting density as M→∞

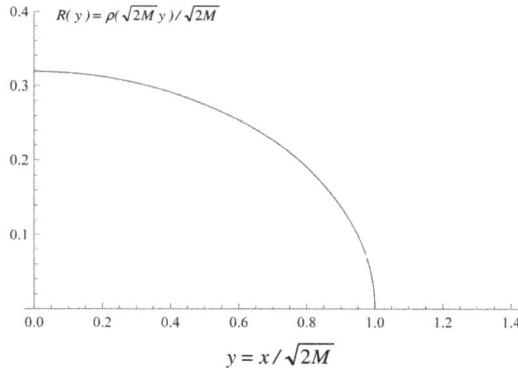

$R(y) = \rho(\sqrt{2M}\,y)/\sqrt{2M}$

$y = x/\sqrt{2M}$

Fig. 6.

result of that problem. But of course, in this case, you can calculate exactly the density distribution because there is no coupling g:

$$\frac{\rho(x)}{\sqrt{2M}} = \frac{1}{\pi}\sqrt{1 - \left(\frac{x}{\sqrt{2M}}\right)^2},$$

$$\frac{x}{\sqrt{2M}} = y, \tag{5}$$

$$R(y) = \frac{\rho(\sqrt{2M}\,y)}{\sqrt{2M}} = \frac{1}{\pi}\sqrt{1 - y^2}.$$

So, the density is just the sum of the M lowest state of particle density, these are the normalized wave functions:

$$\rho(x) = \sum_{0}^{M-1} |\psi_n(x)|^2. \tag{6}$$

If you plot that for $M = 5$, you get this expression (Fig. 7). Let us compare that with Thomas–Fermi, it is in very good agreement (Fig. 8). If we now do it for $M = 15$, the agreement is better (Figs. 9 and 10); if you do it for $M = 55$ particles, the agreement is excellent (Figs. 11 and 12). As the number of particles is increased,

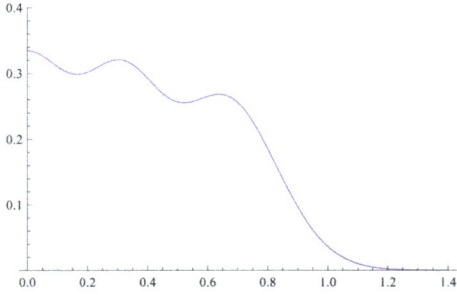

M = 5

Fig. 7.

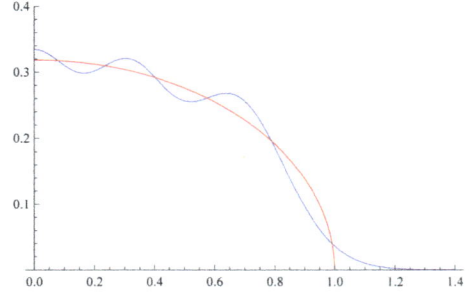

M = 5

Fig. 8.

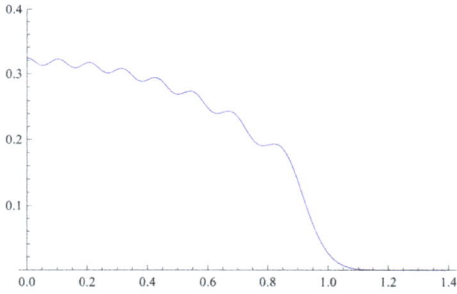

M = 15

Fig. 9.

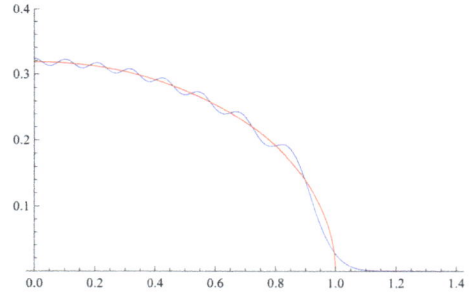

M = 15

Fig. 10.

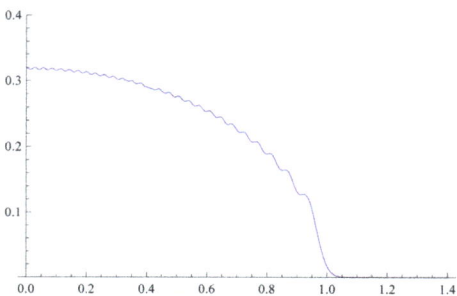

M = 55

Fig. 11.

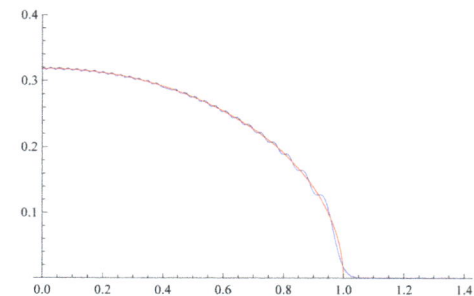

M = 55

Fig. 12.

the Thomas–Fermi method becomes the limit, in fact becomes the limit very rapidly.

So, we conclude that Thomas–Fermi method gives the limit, limiting density as M goes to infinity. Now we applied this idea to the case when you have both the trap and the g. But first, let us ask ourselves why the Thomas–Fermi method gives such a good limit.

As M goes to infinity, almost all particles have very small wavelengths. Therefore, they are classical particles. Therefore, the classical statistical mechanics holds. Therefore, the Thomas–Fermi method gives the correct limit. Once you appreciate this, you realize that the Thomas–Fermi method has a very wide-ranging applicability, in fact, not only in one-dimension, but in two dimensions and in three dimensions. This is the realization which I only reached within the last few months. At first, I thought that the good agreement that I showed you was because it was one-dimensional. We all know that two dimensions and one dimension are very different. Because in one dimension, there is no diffraction. That is a reason why Lieb and I were able to solve the one-dimensional problem without the trap using the Bethe ansatz. The spirit of Bethe ansatz is that there is no diffraction in one dimension. But now, I realize that is irrelevant as far as validity of the Thomas–Fermi method is concerned.

So, let us see how we do it. First, **without the harmonic potential $V(x)$, but with the delta function interaction $\Sigma\delta(x_i - x_j)$**, this problem was solved in 1963 by Lieb and Liniger for bosons, and by me in 1967 for fermions. We used the Bethe's hypothesis of 1931. In one dimension, if two particles p_1, p_2 collided, conservation of momentum gives the first equation [Eq. (7)], conservation of energy gives the second equation [Eq. (8)]:

$$p_1 + p_2 = p_1' + p_2' \,, \tag{7}$$

$$E_1 + E_2 = E_1' + E_2' \,. \tag{8}$$

But when we solved this, you find that there are only two solutions:

$$p_1' = p_1 \,, \quad p_2' = p_2 \,, \tag{9}$$

$$p_1' = p_2 \,, \quad p_2' = p_1 \,. \tag{10}$$

One is this equation [Eq. (9)] which means a straightforward scattering. This [Eq. (10)] means exchange. So, in one dimension, because of conservation of momentum and conservation of energy, you only have reflection and transmission, no diffraction, and that is the reason why Bethe ansatz works in one dimension, and that is also the reason why Bethe ansatz does not work in higher dimensions. Using this, we can solve the problem, in particular for the fermion case. This (slide) says that in N particles, many particles p_1, p_2, p_3, up to p_N, then in each condition, you have just the permutation without change, so all you can do is to permute these momenta, so the analytic problem, the differential equation problems become an algebraic problem. So that is a trick which makes the Bethe ansatz so powerful.

- Lieb & Linnoger 1963: Bosons, spinless.
- Yang 1967: Fermions, spin 1/2.

This piece becomes an algebraic problem, and the algebraic problem eventually resolves into two coupled Fredholm equations for two unknown functions, σ and ρ as a function of p:

$$2\pi\sigma = -\int_{-B}^{B} \frac{2c\sigma(\Lambda')d\Lambda'}{c^2 + (\Lambda - \Lambda')^2} + \int_{-Q}^{Q} \frac{4c\rho dp}{c^2 + 4(p - \Lambda)^2} , \tag{11}$$

$$2\pi\rho = 1 + \int_{-B}^{B} \frac{4c\sigma d\Lambda}{c^2 + 4(p - \Lambda)^2} . \tag{12}$$

If you solve these two equations, you will get all the properties of the ground state for the fermion problem without the harmonic trap.

Now we apply that to the case with the harmonic trap (Fig. 13). And in this region, however, we now have this potential due to the trap, and the kinetic energy part we use the solution due to Bethe ansatz. So, with that, you can solve the problem completely. After that is done, you can show that, in fact, this conjecture was correct (Fig. 14). The Thomas–Fermi method gives this whole curve, and we have now a solution. This is not a sketch, this is the exact calculation.[1-3] The calculation is, of course, based on solution of the Fredholm equations. Of course, we do know how to analytically solve that, but to any degree of accuracy that you want, we can calculate the curve to better accuracy than you have assigned.

All these have appeared in these publication: Refs. 1–3, and in additional publications to come. We are also generalizing it to finite temperature. The trick is because for finite temperature without delta function, that problem has also been solved in the 1960s. And so, we just use that coupled with the Thomas–Fermi

(C) With both $V(x)$ and $\sum \delta(x - x^1)$. T - F method.

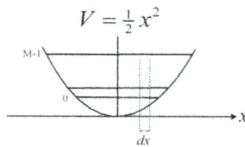

In each dx introduced $\sum \delta$ interaction. Solve the Fredholm equations.

$V = \frac{1}{2}x^2$

Fig. 13.

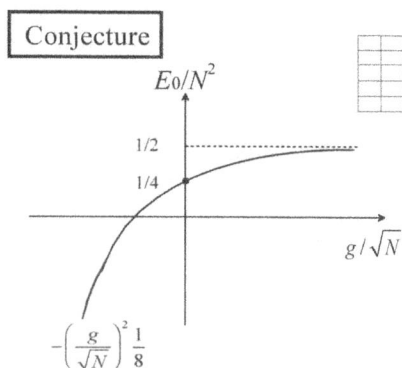

Fig. 14.

method, then we will get the finite temperature solution for the original Hamiltonian problem.

References

1. Z.-Q. Ma and C. N. Yang, *Chin. Phys. Lett.* **26**, 120505 (2009).
2. Z.-Q. Ma and C. N. Yang, *Chin. Phys. Lett.* **26**, 120506 (2009).
3. Z.-Q. Ma and C. N. Yang, *Chin. Phys. Lett.* **27**, 020506 (2010).

Banquet Speech at the Singapore Conference in Honour of Murray Gell-Mann on His 80th Birthday
(February 2010)

C. N. Yang

It was about 60 years ago in 1951 that I first met in Princeton, at the Institute for Advanced Study, our Honoree tonight, Murray Gell-Mann. In the intervening 60 years, man's understanding of the fundamental structure of matter has made great historic advances. At this wonderful Conference, we have the opportunity to review some of these historic advances. It is a little like looking over an old album of memorable photographs in one's lifetime. In many of these photographs, Murray appears either in the foreground, or as the photographer snapping the picture. We know he is pleased with the album.

Anyone who had had contact with Murray cannot fail to be impressed by his catholic interests in many things, by his knowledge, by his humor, but also by his sometimes overbearing self-confidence. Today, at age 80, he is as keen, as impressive and as forward looking, as ever. In another 20 years, when Singapore will become the number one country in the world in per capita income, our host K. K. Phua will organize the 100th birthday celebration for Murray, and I promise I shall come and together we shall examine the new enlarged edition of the album of memorable photographs in the history of 21st as well as 20th century physics.

MURRAY AND THE OMEGA MINUS

NICHOLAS P. SAMIOS

Physics Department, Brookhaven National Laboratory,
Upton, NY 11973, USA

The exciting findings and activities in particle physics in the 50's and 60's will be discussed from an experimentalist's viewpoint. Particular emphasis will be placed on the description of several crucial discoveries (including the omega minus) and on the remarkable insight, guidance, and major contributions of Murray Gell-Mann to the understanding of the symmetry of hadrons which led to the development of the standard model of the strong interactions.

Keywords: Strangeness; eightfold way; SU(3); quarks; Ω^-.

1. Introduction

In this talk I will cover four major contributions by Murray Gell-Mann to high energy particle physics in the period 1953–1964. These involved: (1) the introduction of strangeness by Gell-Mann and Nishijima; (2) the formulation of the eightfold way and SU(3) by Gell-Mann and Ne'eman; (3) the derivation of the mass formula for meson and baryon octets and decuplets by Gell-Mann and Okubo; and (4) the concept of quarks by Gell-Mann and Zweig.

2. Gell-Mann–Nishijima — Strangeness

By 1953, cosmic ray investigations with nuclear emulsions and cloud chambers had unearthed many new particles, the more noteworthy being the so-called V particles. These were the lambda baryon with the observed decay $\Lambda^0 \to p\pi^-$ and the theta meson decaying via $\theta^0 \to \pi^+\pi^-$, both with lifetimes $\approx 10^{-10}$ sec. It had been known from experiments performed at cyclotrons that in the process $\pi^- p \to \pi^- p$ interacted strongly, namely with lifetimes $\simeq 10^{-23}$ sec. As such it was strange that the Λ^0, θ^0 had such long lifetimes. There were many conjectures as to why this was so, but the successful explanation was supplied by Gell-Mann[1-3] and independently by Nishijima.[1-3] This involved assigning a new quantum number, strangeness, to the new particles, a value that was conserved in strong interactions but violated in weak interactions and thus pay a penalty in the rate. As such the Λ^0 was assigned strangeness (-1) and the θ^0 $(+1)$ with the consequence that such particles would be produced in association] as in the reaction $\pi^- p \to \Lambda^0 \theta^0$ which preserves strangeness and proceeds strongly. They proposed the following relationship between the charge, Q, baryon number B, strangeness S and third component of isospin I_3:

$$Q = I_3 + \frac{1}{2}(B + S).$$

In this manner the known Λ^0, Σ^+, Σ^-, Σ^0 were assigned $S = -1$, the Ξ^-, Ξ^0 $S = -2$, the $(\theta \equiv K)$ K^+K^0, S=+1 and the $\bar{K}^0$ K^-, $S = -1$. This classification scheme worked and served to describe the strong and weak interactions of strange particles. One of the consequences of the Gell-Mann–Nishijima formula is the expected preponderance of a negative strangeness baryons. To me this of course strongly suggested the utilization of high energy K^- beams as the preferred strategy for discovering new strange particles. As such, over the following years I advocated and was involved in the construction of a 2 GeV and a 5 GeV K^- beam at BNL.

3. Gell-Mann–Ne'eman — SU(3) and the Eightfold Way

It is informative as well as interesting to review the experimental status of the known elementary particles as tabulated by Barkas and Rosenfeld[4] in 1957 and by Gell-Mann and Rosenfeld[5] in 1961. This shows in Table 1. It was on the basis of this sparse information that Murray,[6–8] and independently Ne'eman,[6–8] formulated the eightfold way and the SU(3) symmetry of particle families in 1961.

Table 1.   Experimental situation.

1957 Barkas–Rosenfeld			
Entries	γ		
	e	μ	v
	π^+	π^-	π^0
	K^+	K^0	$\bar{K}^0 K^-$
	p	N	
	Λ		
	Σ^+	Σ^-	Σ^0
	Ξ^-	Ξ^0	
1961 Gell-Mann–Rosenfeld			
	ρ, ω		
$K^*(895)$			
$\Delta(1238)$		$N(1510)$	$N(1680)$
$Y^*(1385)$		$Y(1405)$	$Y(1815)$

This formulation is shown in Table 2 where several meson octets and the baryon octet are noted with two possibilities for a $Y = -2$ baryon, as a singlet in a decuplet and a triplet in a 27 representation.

Table 2. 1961 Gell-Mann–Neeman Eightfold Way and SU(3).

$$8 \times 8 = 1 + 8 + 8 + 10 + 10^* + 27$$

Mesons

Octet	0^- π K $\bar{K}$	$x(\eta)$
	1^- ρ M (K^*) $\bar{M}$ $(\bar{K}^*)$ ω	

Baryons

Octet (8)	$1/2+$	$\Lambda \Sigma \Xi N$
Decuplet (10)	Quartet	$Y = +1$
	Triplet	$Y = 0$
	Doublet	$Y = -1$
	Singlet	$Y = -2$
(27)	Triplet	$Y = +2$
	Triplet	$Y = -2$

The 1962 Rochester Conference held at CERN was the setting for several major experimental findings and for Murray's dramatic exposition on the unitary symmetry model SU(3). Two groups, we the BNL/Syracuse group[9,10] and UCLA[9,10] presented evidence for an excited cascade resonance, the $\Xi^*(1535)$ a $Y = -1$ doublet. In addition we also reported on the finiding of a meson resonance, the $\varphi(1020)$, a singlet which mainly decayed into two kaons. The data are shown in Fig. 1 where these resonances are clearly evident. In addition, the existence for the $Y_1^*(1385)$ was reenforced by many experiments, and no evidence was found for a K^+p resonance ($Y = +2$). It was at this point that Murray made his famous remark[11] at the end of a theoretical session, during the question period. I quote:

"If we take the unitary symmetry model with baryon and meson octets, with first order violation giving rise to mass differences, we obtain some rules for supermultiplets. Suppose, now we try to incorporate the 3/2–3/2 nucleon resonance into the scheme. The only supermultiplet that does not lead to nonexistent resonances in the K–N channels is the 10 representation, which gives 4 states:

$$I = 3/2, \quad S = 0,$$
$$I = 1, \quad S = -1,$$
$$I = 1/2, \quad S = -2,$$
$$I = 0, \quad S = -3.$$

The mass rule is stronger here and yields equal spacing of these states. Starting with the resonance at 1238 MeV, we may conjecture that the Y^*, at 1385 MeV and the Xi^* at 1535 MeV might belong to this supermultiplet. Certainly they fulfill the requirement of equal spacing. If $J = 3/2+$ is really right for these two cases, then our speculation might have some value and

34

CERN '62

Ξ* (1535)

φ (1020)

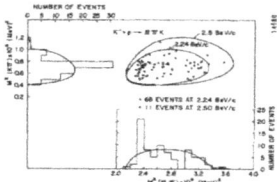

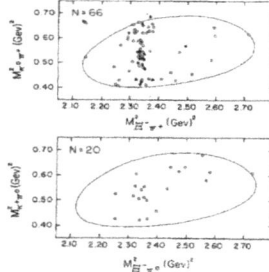

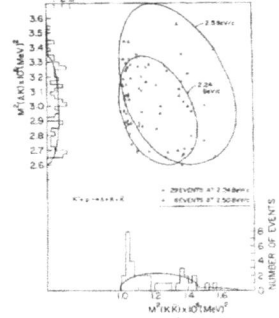

Fig. 8 Dalitz plot of the reaction $K^-p \to \Xi^-\pi^+K^0$. The effective mass distribution for $\Xi^-\pi^0$ and $K^+\pi^-$ are projected on abscissa and ordinate.

Fig. 1 $K^-+p \to \Xi^-+\pi^0+K^0$ at 1.80 GeV/c.

Fig. 6 Dalitz plot for the reaction $\bar{K}^-p \to \Lambda K\bar{K}$.

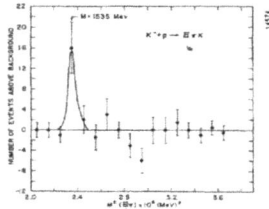

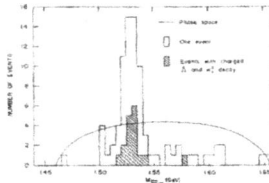

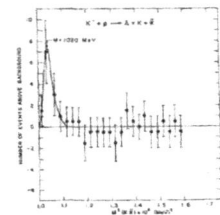

Fig. 9 The effective mass distribution for $\Xi^-\pi$ from the channels $\Xi^-\pi^-K^0$ and $\Xi^-\pi^0K^-$ for those events above phase space.

Fig. 2 $K^-+p \to \Xi^-+\pi^0+K^0$ at 1.80 GeV/c.

Fig. 1.

we should look for the last particle, called, say Ω^- with $S = -3$, $I = 0$. At 1685 MeV, it would be metastable and should decay by weak interactions into $K^- + \Lambda$, and or $\pi^- + \Xi^0$ or $\pi^0 + \Xi^-$. Perhaps it would explain the old Eisenberg event. A beam of K^- with momentum ≥ 3.5 GeV/c could yield Ω^- by means of $K^- + p \to K^+ + K^0 + \Omega^-$."

I spoke to Murray immediately after his remark and decided to emphasize a search for the Ω^- in our proposal for a K^-p experiment at BNL with the newly designed 5 GeV K^- beam and the new 80" bubble chamber. As noted earlier, the Gell-Mann–Nishijima relation strongly suggested that K^- beams were the method of choice for exploring the domain of strange particles. This proved very successful with the 2.3 GeV K^- beam which produced the $\varphi(1020)$ and $\Xi^*(1535)$ and subsequently the spin of the $Y_1^*(1385)$. Prior to the CERN conference, in March 1962 we had proposed to the BNL Program Advisory Committee an experiment for the "Systematic Study of K^-p interactions in the momentum range 3.5–5.0 GeV." It was turned down in June 1962. In light of the recent experimental discoveries and Murray's remark, we updated our proposal emphasizing the success of the SU(3) symmetry scheme in describing the 0^- and 1^- meson octets as well as the unique prediction of the existence of the Ω^- hyperon. The High Energy Advisory Committee approved

Fig. 2.   80" bubble chamber.

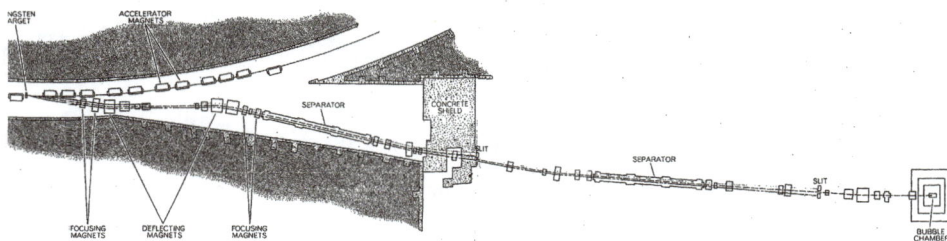

Fig. 3.   Layout of the 5 GeV K^- beam.

this experiment in the 80" bubble chamber with first priority on Oct. 1963. I called this the Gell-Mann Effect.

4. Discovery of Ω^-

The discovery was made in February 1964.[12] Figure 2 shows the 80" bubble chamber in all its complexity and in Fig. 3 is shown the layout of the 5 GeV K^- beam with its numerous magnets and separators. This all became operational in the winter of 1963.

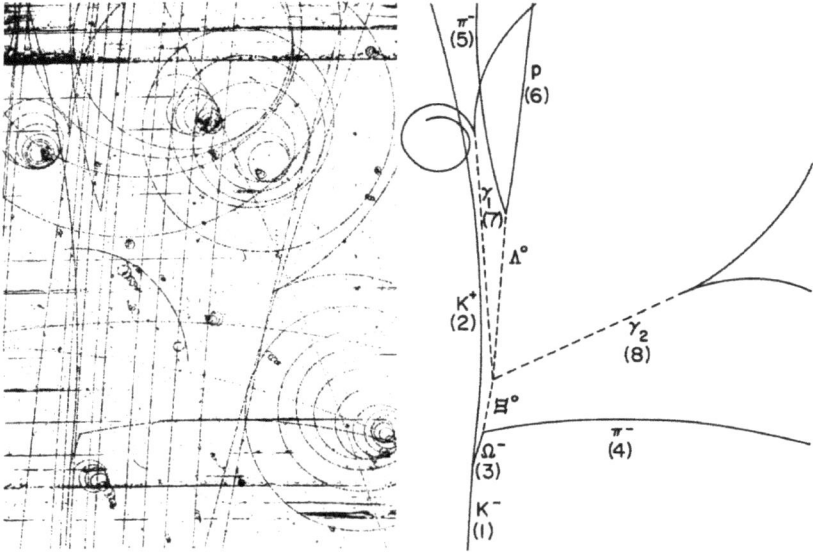

Fig. 4.   The production and decay of the first Ω^- event.

The production and decay of the first Ω^- event is shown in Fig. 4. The observed reaction is:

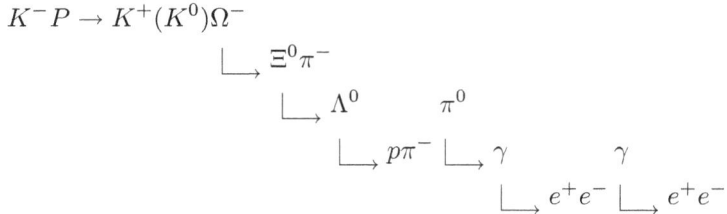

$$K^-P \to K^+(K^0)\Omega^-$$

$$\hookrightarrow \Xi^0\pi^-$$

$$\hookrightarrow \Lambda^0 \qquad \pi^0$$

$$\hookrightarrow p\pi^- \hookrightarrow \gamma \qquad \gamma$$

$$\hookrightarrow e^+e^- \hookrightarrow e^+e^-$$

It is an extremely unusual event in many respects, foremost being the conversion in the liquid hydrogen of both γ's from the π^0 decay into electron positron pairs, a probability $< 1\%$. The calculated line of flight of the Λ^0 decay was inconsistent from coming from the Ω^- decay vertex, and kinematically the unseen particle from the production vertex was that of a K^0. The reconstructed Ξ^0 mass was 1314 MeV (not 1321 as had been established for the Ξ^-) and the mass of the Ω^- was determined to be 1686 ± 12 MeV. In fact this event could have also discovered the Ξ^0. A spectacular event. By 1968 we had collected eight Ω^- events and measured its mass to be 1673.3 ± 1.0 MeV. This is to be compared with the latest Particle Data Group value of 1672.45 ± 0.29 MeV; compiled from over 150,000 Ω^- events, not bad. The Ω^- spin has been determined to be $3/2$ and even its magnetic moment to be -2.024 nuclear magnetons. Ω_c's have been found at ~ 2700 MeV and Ω_b's at ~ 6150 MeV. Quite a family.

5. Gell-Mann–Okubo —Mass Formula

Both Gell-Mann[6–8] and Okubo[13] derived a mass formula between members of a given multiplet. In the case of baryons this resulted in the following.

$$\text{Baryons:} \quad \frac{m_n + m_\Xi}{2} = \frac{3m_\Lambda + m_\Sigma}{4}$$

Octet: 1128.7 MeV   1135.1 MeV   0.6%

Decuplet: Equal mass spacing

$$M(\Omega^-) - m(\Xi^*) = m(\Xi^*) - m(Y^*) = m(Y^*) - m(\Delta)$$

139 MeV   149 MeV   148 MeV

The agreement is rather good, better than 1%. The situation in the case of mesons is a bit more complex. Both a linear and quadratic relation were derived and were considered a success when first presented, although agreement was between 7–25%, in retrospect rather poor.

Mesons:

Octets	Linear: $M_k = \dfrac{3M_\eta + M_\pi}{4}$	Quadratic: $M_k^2 = \dfrac{3M_\eta^2 + M_\pi^2}{4}$

Pseudoscalar:	$\pi(140)$	$K(495)$	$\eta(550)$	7–10%
Vector	$\rho(770)$	$K^*(895)$	$\omega(785)$	14–25%

Nonet:				
	0^-	$\eta'(958)$ mixing angle	$-25, -11$	(1964)
	1^-	$\varphi(1020)$ mixing angle	36,38	(1961)
	2^+	$f^*(1525)$ mixing angle	28,30	(1966)

We subsequently realized that the meson sector consisted of nonets, $1+8$ multijets. We had the good fortune to participate in the discovery of the singlets for the 0^-, the $\eta'(958)$ in 1964; 1^-, the $\varphi(1020)$ in 1961 and the 2^+, $f'(1525)$ in 1968.

The mixing angles for the three nonets vary, with that of the φ corresponding to $38°$, a case of perfect mixing. In the case of the φ we had also measured the decay branching ratio $\varphi \to \frac{\rho\pi}{KK} = 0.35 \pm 0.20$ which we estimated to be at least one order of magnitude smaller than expected. George Zweig calculated it to be two orders of magnitude too small and this was one of the observations that led him to postulate the existence of aces.

6. 1964 Quarks/Aces, Gell-Mann, Zweig

It was soon realized by both Murray[14,15] and George[14,15] that the hadrons may not be elementary but consist of smaller units which they called quarks/aces respectively. They came to this conclusion from different perspectives and arrived at a composition of non-integrally charged quarks, $u(+2/3)$, $d(-1/3)$, $s(-1/3)$, each

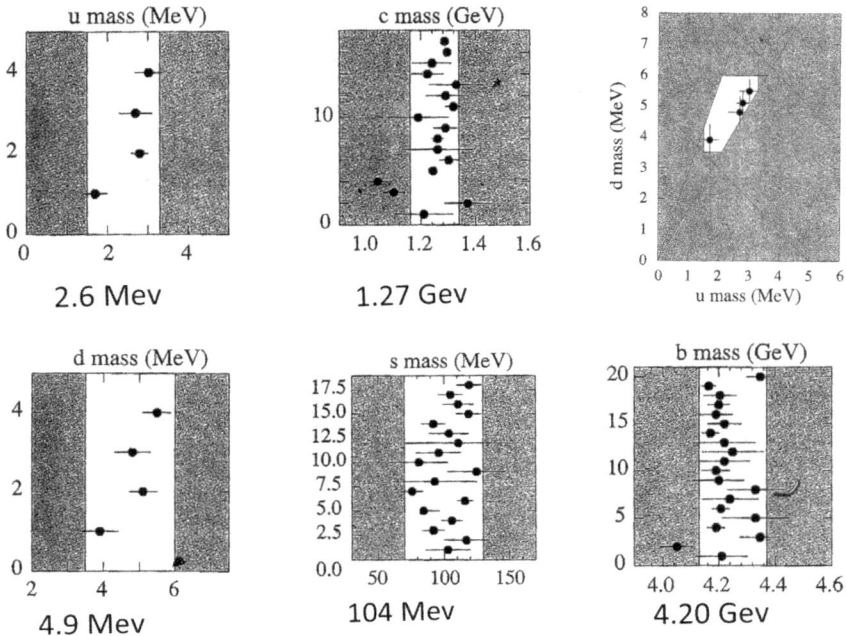

Fig. 5. Quark masses.

with baryon number $1/3$ and u, d with strangeness zero and s with strangeness -1. In this manner

$$\text{Mesons: quark–antiquark pair giving nonets}$$

$$3 \times \bar{3} = 1 + 8$$

and

$$\text{Baryons: 3 quarks yielding singlet, octet and decuplets}$$

$$3 \times 3 \times 3 = 1 + 8 + 8 + 10\,.$$

Several groups claimed to have found free quarks, most notably McCusker and they all turned out to be false. More recently several groups claimed to have found particles belonging to a pentaquark configuration, and, these also appear to be false. We now have quarks and gluons, confinement, asymtotic freedom and QCD. The advent of lattice QCD has allowed for nonperturbative calculations of both quark and hadron masses. Values for the quark masses, as reported in the Particle Data Book are shown in Fig. 5, demonstrating rather good sensitivity and consistency for the various masses. In calculating hadron masses, mesons and baryons, many groups now use, besides, the π, K, the Ω^- as input values. What an irony. I expect that this field of lattice gauge calculations will have a major impact in the future in particle and nuclear physics as the computers become even bigger, pedaflops and the algorithms become more sophisticated.

I close by noting that Murry Gell-Mann has had an enormous influence in particle physics for many, many years. I wish Murray a very happy 80th birthday and also wish him

$$Ei\sigma \quad \Pi o\lambda\lambda\alpha \quad E\text{T}\eta$$

(To Many Years)

Fig. 6.   Looking forward to your 100th Birthday.

References

1. M. Gell-Mann, *Il Nuovo Cimento* **4**, 848 (1956).
2. T. Nakano and K. Nishijima, *Progr. Theor. Phys.* **10**, 581 (1953).
3. K. Nishijima, *Progr. Theor. Phys.* **13**, 285 (1955).
4. W. H. Barkas and A. H. Rosenfeld, *Data for Elementary Particle Physics*, 1957, UCRL 8030 (Particle Data Group, *Review of Particle Properties*, 1968).
5. M. Gell-Mann and A. H. Rosenfeld, *Ann. Rev. Nucl. Sci.* **7**, 407 (1957).
6. M. Gell-Mann, *The Eightfold Way* (1961) [CIT Sup. for Rapid CTSL 20].
7. M. Gell-Mann *Phys. Rev.* **125**, 1067 (1962).
8. Y. Ne'eman *Nucl. Phys.* **26**, 222 (1961).
9. BNL/SYR Collab., K^-p interaction at 2.24 GeV/c, Presented by N. P. Samios, *Effective Mass Distributions*, pp. 279–289.
10. UCLA Group, A resonance in the Ξ_π system at 1.53 GeV, Presented by H. Ticho, pp. 289–290.
11. M. Gell-Mann, Remark, in *Int. Conf. on High Energy Physics*, CERN (1962), p. 805.
12. V. E. Barns *et al.*, *Phys. Rev. Lett.* **12**, 204 (1964).
13. S. Okubo, *Progr. Theor. Phys.* **27**, 949 (1962).
14. M. Gell-Mann, *Phys. Lett.* **4**, 14 (1964).
15. G. Zweig, CERN preprint 8182/Th401 (1962).

FROM Ω^- TO Ω_b

M. KARLINER

Raymond and Beverly Sackler School of Physics and Astronomy,
Tel Aviv University, Tel Aviv, Israel
E-mail: marek@proton.tau.ac.il

I discuss several recent highly accurate theoretical predictions for masses of baryons containing the b quark, especially Ω_b (ssb) very recently reported by CDF. I also point out an approximate effective supersymmetry between heavy quark baryons and mesons and provide predictions for the magnetic moments of Λ_c and Λ_b. Proper treatment of the color-magnetic hyperfine interaction in QCD is crucial for obtaining these results.

Keywords: QCD; hadron masses; heavy quarks; baryons; magnetic moments; effective supersymmetry.

1. Introduction

QCD describes hadrons as valence quarks in a sea of gluons and $\bar{q}q$ pairs. At distances above ~ 1 GeV^{-1} quarks acquire an effective *constituent mass* due to chiral symmetry breaking. A hadron can then be thought of as a bound state of constituent quarks. In the zeroth-order approximation the hadron mass M is then given by the sum of the masses of its constituent quarks m_i,

$$M = \sum_i m_i \,.$$

The binding and kinetic energies are "swallowed" by the constituent quarks masses. The first and most important correction comes from the color hyper-fine (HF) chromo-magnetic interaction,

$$M = \sum_i m_i + V_{i<j}^{HF(QCD)} \,;$$

$$V_{ij}^{HF(QCD)} = v_0 \, (\vec{\lambda}_i \cdot \vec{\lambda}_j) \, \frac{\vec{\sigma}_i \cdot \vec{\sigma}_j}{m_i m_j} \, \langle \psi | \delta(r_i - r_j) | \psi \rangle \tag{1}$$

where v_0 gives the overall strength of the HF interaction, $\vec{\lambda}_{i,j}$ are the $SU(3)$ color matrices, $\sigma_{i,j}$ are the quark spin operators and $|\psi\rangle$ is the hadron wave function. This is a contact spin-spin interaction, analogous to the EM hyperfine interaction, which is a product of the magnetic moments,

$$V_{ij}^{HF(QED)} \propto \vec{\mu}_i \cdot \vec{\mu}_j = e^2 \, \frac{\vec{\sigma}_i \cdot \vec{\sigma}_j}{m_i m_j} \tag{2}$$

in QCD, the $SU(3)_c$ generators take place of the electric charge. From eq. (1) many very accurate results have been obtained for the masses of the ground-state hadrons. Nevertheless, several caveats are in order. First, this is a low-energy phenomenological model, still awaiting a rigorous derivation from QCD. It is far from providing a complete description of the hadronic spectrum, but it provides excellent predictions for mass splittings and magnetic moments. The crucial assumptions of the model are:

(a) HF interaction is considered as a perturbation which does not change the wave function;
(b) effective masses of quarks are the same inside mesons and baryons;
(c) there are no 3-body effects.

2. Quark Masses

As the first example of the application of eq. (1) we can obtain the $m_c - m_s$ quark mass difference from the $\Lambda_c - \Lambda$ baryon mass difference:

$$
\begin{aligned}
M(\Lambda_c) - M(\Lambda) &= \\
&= (m_u + m_d + m_c + V_{ud}^{HF} + V_{uc}^{HF} + V_{dc}^{HF}) \\
&\quad - (m_u + m_d + m_s + V_{ud}^{HF} + V_{us}^{HF} + V_{ds}^{HF}) \\
&= m_c - m_s,
\end{aligned}
\tag{3}
$$

where the light-quark HF interaction terms V_{ud}^{HF} cancel between the two expressions and the HF interaction terms between the heavy and light quarks vanish:

$$
V_{us}^{HF} = V_{ds}^{HF} = V_{uc}^{HF} = V_{dc}^{HF} = 0,
$$

since the u and d light quarks are coupled to a spin-zero diquark and the HF interaction couples to the spin.

Table I below shows the quark mass differences obtained from mesons and baryons [1]. The mass difference between two quarks of different flavors denoted by i and j are seen to have the same value to a good approximation when they are bound to a "spectator" quark of a given flavor.

On the other hand, Table I shows clearly that *constituent quark mass differences depend strongly on the flavor of the spectator quark*. For example, $m_s - m_d \approx 180$ MeV when the spectator is a light quark but the same mass difference is only about 90 MeV when the spectator is a b quark.

Since these are *effective masses*, we should not be surprised that their difference is affected by the environment, but the large size of the shift is quite surprising and its quantitative derivation from QCD is an outstanding challenge for theory.

A second example shows how we can extract the ratio of the constituent quark masses from the ratio of the the hyperfine splittings in the corresponding mesons.

| observable | baryons | | mesons | | | | Δm_{Bar} | Δm_{Mes} |
| | | | $J=1$ | | $J=0$ | | | |
	B_i	B_j	$\mathcal{V}_i$	$\mathcal{V}_j$	$\mathcal{P}_i$	$\mathcal{P}_j$	MeV	MeV
$\langle m_s - m_u \rangle_d$	sud	uud	$s\bar{d}$	$u\bar{d}$	$s\bar{d}$	$u\bar{d}$	177	179
	Λ	N	K^*	ρ	K	π		
$\langle m_s - m_u \rangle_c$			$c\bar{s}$	$c\bar{u}$	$c\bar{s}$	$c\bar{u}$		103
			D_s^*	D_s^*	D_s	D_s		
$\langle m_s - m_u \rangle_b$			$b\bar{s}$	$b\bar{u}$	$b\bar{s}$	$b\bar{u}$		91
			B_s^*	B_s^*	B_s	B_s		
$\langle m_c - m_u \rangle_d$	cud	uud	$c\bar{d}$	$u\bar{d}$	$c\bar{d}$	$u\bar{d}$	1346	1360
	Λ_c	N	D^*	ρ	D	π		
$\langle m_c - m_u \rangle_c$			$c\bar{c}$	$u\bar{c}$	$c\bar{c}$	$u\bar{c}$		1095
			ψ	D^*	η_c	D		
$\langle m_c - m_s \rangle_d$	cud	sud	$c\bar{d}$	$s\bar{d}$	$c\bar{d}$	$s\bar{d}$	1169	1180
	Λ_c	Λ	D^*	K^*	D	K		
$\langle m_c - m_s \rangle_c$			$c\bar{c}$	$s\bar{c}$	$c\bar{c}$	$s\bar{c}$		991
			ψ	D_s^*	η_c	D_s		
$\langle m_b - m_u \rangle_d$	bud	uud	$b\bar{d}$	$u\bar{d}$	$b\bar{d}$	$u\bar{d}$	4685	4700
	Λ_b	N	B^*	ρ	B	π		
$\langle m_b - m_u \rangle_s$			$b\bar{s}$	$u\bar{s}$	$b\bar{s}$	$u\bar{s}$		4613
			B_s^*	K^*	B_s	K		
$\langle m_b - m_s \rangle_d$	bud	sud	$b\bar{d}$	$s\bar{d}$	$b\bar{d}$	$s\bar{d}$	4508	4521
	Λ_b	Λ	B^*	K^*	B	K		
$\langle m_b - m_c \rangle_d$	bud	sud	$b\bar{d}$	$c\bar{d}$	$b\bar{d}$	$c\bar{d}$	3339	3341
	Λ_b	Λ_c	B^*	D^*	B	D		
$\langle m_b - m_c \rangle_s$			$b\bar{s}$	$c\bar{s}$	$b\bar{s}$	$c\bar{s}$		3328
			B_s^*	D_s^*	B_s	D_s		

Table I. Quark mass differences from baryons and mesons.

The hyperfine splitting between K^* and K mesons is given by

$$M(K^*)-M(K) = v_0 \frac{\vec{\lambda}_u \cdot \vec{\lambda}_s}{m_u m_s} \left[(\vec{\sigma}_u \cdot \vec{\sigma}_s)_{K^*} - (\vec{\sigma}_u \cdot \vec{\sigma}_s)_K \right] \langle \psi | \delta(r) | \psi \rangle$$

$$= 4v_0 \frac{\vec{\lambda}_u \cdot \vec{\lambda}_s}{m_u m_s} \langle \psi | \delta(r) | \psi \rangle, \tag{4}$$

and similarly for hyperfine splitting between D^* and D with $s \to c$ everywhere. From (4) and its D analogue we then immediately obtain

$$\frac{M(K^*) - M(K)}{M(D^*) - M(D)} = \frac{4v_0 \dfrac{\vec{\lambda}_u \cdot \vec{\lambda}_s}{m_u m_s} \langle \psi | \delta(r) | \psi \rangle}{4v_0 \dfrac{\vec{\lambda}_u \cdot \vec{\lambda}_c}{m_u m_c} \langle \psi | \delta(r) | \psi \rangle} \approx \frac{m_c}{m_s}. \tag{5}$$

2.1. Color hyperfine splitting in baryons

As an example of hyperfine splitting in baryons, let us now discuss the HF splitting in the Σ (uds) baryons. Σ^* has spin $\frac{3}{2}$, so the u and d quarks must be in a state of relative spin 1. The Σ has isospin 1, so the wave function of u and d is symmetric in flavor. It is also symmetric in space, since in the ground state the quarks are in a relative S-wave. On the other hand, the u-d wave function is antisymmetric in color, since the two quarks must couple to a $\mathbf{3^*}$ of color to neutralize the color of the third quark. The u-d wave function must be antisymmetric in flavor × spin × space × color, so it follows it must be symmetric in spin, i.e. u and d are coupled to spin one. Since u and d are in spin 1 state in both Σ^* and Σ their HF interaction with each other cancels between the two and thus the u-d pair does not contribute to the $\Sigma^* - \Sigma$ HF splitting,

$$M(\Sigma^*) - M(\Sigma) = 6v_0 \frac{\vec{\lambda}_u \cdot \vec{\lambda}_s}{m_u m_s} \langle \psi | \delta(r_{rs}) | \psi \rangle \tag{6}$$

we can then use eqs. (4) and (6) to compare the quark mass ratio obtained from mesons and baryons:

$$\left(\frac{m_c}{m_s} \right)_{Bar} = \frac{M_{\Sigma^*} - M_\Sigma}{M_{\Sigma_c^*} - M_{\Sigma_c}} = 2.84$$

$$\left(\frac{m_c}{m_s} \right)_{Mes} = \frac{M_{K^*} - M_K}{M_{D^*} - M_D} = 2.81 \tag{7}$$

$$\left(\frac{m_c}{m_u} \right)_{Bar} = \frac{M_\Delta - M_p}{M_{\Sigma_c^*} - M_{\Sigma_c}} = 4.36$$

$$\left(\frac{m_c}{m_u} \right)_{Mes} = \frac{M_\rho - M_\pi}{M_{D^*} - M_D} = 4.46. \tag{8}$$

We find the same value from mesons and baryons $\pm 2\%$.

The presence of a fourth flavor gives us the possibility of obtaining a new type of mass relation between mesons and baryons. The $\Sigma - \Lambda$ mass difference is believed to be due to the difference between the $u-d$ and $u-s$ hyperfine interactions. Similarly, the $\Sigma_c - \Lambda_c$ mass difference is believed to be due to the difference between the $u-d$ and $u-c$ hyperfine interactions. We therefore obtain the relation

$$\left(\frac{\frac{1}{m_u^2} - \frac{1}{m_u m_c}}{\frac{1}{m_u^2} - \frac{1}{m_u m_s}} \right)_{Bar} = \frac{M_{\Sigma_c} - M_{\Lambda_c}}{M_\Sigma - M_\Lambda} = 2.16$$

$$\left(\frac{\frac{1}{m_u^2} - \frac{1}{m_u m_c}}{\frac{1}{m_u^2} - \frac{1}{m_u m_s}} \right)_{Mes} = \frac{(M_\rho - M_\pi) - (M_{D^*} - M_D)}{(M_\rho - M_\pi) - (M_{K^*} - M_K)} = 2.10 .$$

(9)

The meson and baryon relations agree to $\pm 3\%$.

We can write down an analogous relation for hadrons containing the b quark instead of the s quark, obtaining the prediction for splitting between Σ_b and Λ_b:

$$\frac{M_{\Sigma_b} - M_{\Lambda_b}}{M_\Sigma - M_\Lambda} = \frac{(M_\rho - M_\pi) - (M_{B^*} - M_B)}{(M_\rho - M_\pi) - (M_{K^*} - M_K)} = 2.51 \qquad (10)$$

yielding $M(\Sigma_b) - M(\Lambda_b) = 194 \, \text{MeV}$ [1, 2].

This splitting was recently measured by CDF [3]. They obtained the masses of the Σ_b^- and Σ_b^+ from the decay $\Sigma_b \to \Lambda_b + \pi$ by measuring the corresponding mass differences in MeV

$$M(\Sigma_b^-) - M(\Lambda_b) = 195.5^{+1.0}_{-1.0} \,(\text{stat.}) \pm 0.1 \,(\text{syst.})$$

$$M(\Sigma_b^+) - M(\Lambda_b) = 188.0^{+2.0}_{-2.3} \,(\text{stat.}) \pm 0.1 \,(\text{syst.})$$

(11)

with isospin-averaged mass difference $M(\Sigma_b) - M(\Lambda_b) = 192$ MeV, as shown in Fig. 1.

There is also the prediction for the spin splittings, good to 5%

$$M(\Sigma_b^*) - M(\Sigma_b) =$$
$$= \frac{M(B^*) - M(B)}{M(K^*) - M(K)} \cdot [M(\Sigma^*) - M(\Sigma)] = 22 \, \text{MeV} \qquad (12)$$

to be compared with 21 MeV from the isospin-average of CDF measurements [3].

The relation (10) is based on the assumption that the qq and $q\bar{q}$ interactions have the same flavor dependence. This automatically follows from the assumption that both hyperfine interactions are inversely proportional to the products of the same quark masses. But all that is needed here is the weaker assumption of same flavor dependence,

$$\frac{V_{hyp}(q_i \bar{q}_j)}{V_{hyp}(q_i \bar{q}_k)} = \frac{V_{hyp}(q_i q_j)}{V_{hyp}(q_i q_k)} \qquad (13)$$

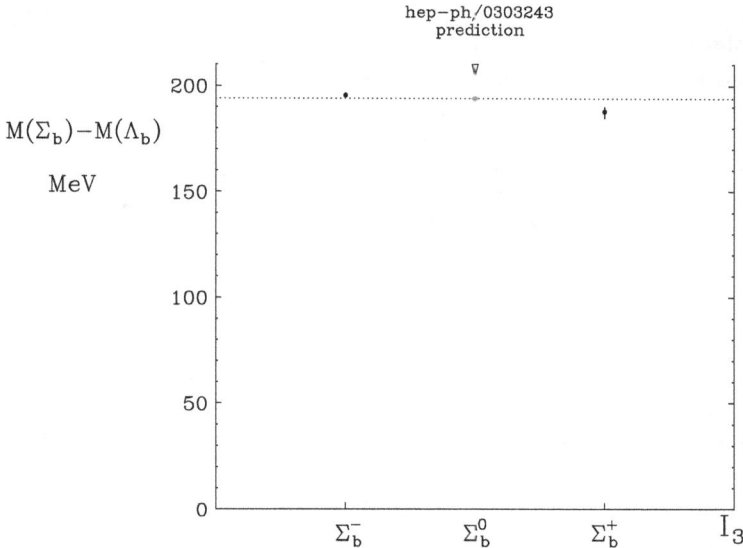

Fig. 1. Experimental results for $\Sigma_b^\pm$ masses [3] vs the theoretical prediction in Ref. [1].

This yields [1]

$$\frac{M_{\Sigma_b} - M_{\Lambda_b}}{(M_\rho - M_\pi) - (M_{B^*} - M_B)} = 0.32 \approx$$

$$\approx \frac{M_{\Sigma_c} - M_{\Lambda_c}}{(M_\rho - M_\pi) - (M_{D^*} - M_D)} = 0.33 \approx \tag{14}$$

$$\approx \frac{M_\Sigma - M_\Lambda}{(M_\rho - M_\pi) - (M_{K^*} - M_K)} = 0.325 \tag{15}$$

The baryon-meson ratios are seen to be independent of the flavor f.

The challenge is to understand how and under what assumptions one can derive from QCD the very simple model of hadronic structure at low energies which leads to such accurate predictions.

3. Effective Meson-Baryon SUSY

Some of the results described above can be understood [2] by observing that in the hadronic spectrum there is an approximate effective supersymmetry between mesons and baryons related by replacing a light antiquark by a light diquark.

This supersymmetry transformation goes beyond the simple constituent quark model. It assumes only a valence quark of flavor i with a model independent structure bound to "light quark brown muck color antitriplet" of model-independent structure carrying the quantum numbers of a light antiquark or a light diquark, cf. Fig. 2. Since it assumes no model for the valence quark, nor the brown muck antitriplet coupled to the valence quark, it holds also for the quark-parton model in which the valence is carried by a current quark and the rest of the hadron is a complicated mixture of quarks and antiquarks.

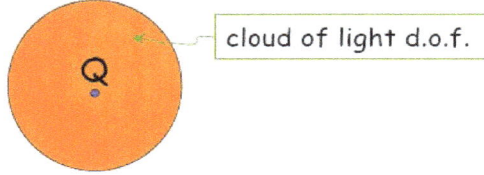

Fig. 2.   A heavy quark coupled to "brown muck" color antitriplet.

This light quark supersymmetry transformation, denoted here by T_{LS}^S, connects a meson denoted by $|\mathcal{M}(\bar{q}Q_i)\rangle$ and a baryon denoted by $|\mathcal{B}([qq]_S Q_i)\rangle$ both containing the same valence quark of some fixed flavor Q_i, $i = (u, s, c, b)$ and a light color-antitriplet "brown muck" state with the flavor and baryon quantum numbers respectively of an antiquark $\bar{q}$ (u or d) and two light quarks coupled to a diquark of spin S.

$$T_{LS}^S|\mathcal{M}(\bar{q}Q_i)\rangle \equiv |\mathcal{B}([qq]_S Q_i)\rangle . \tag{16}$$

The mass difference between the meson and baryon related by this T_{LS}^S transformation has been shown [4] to be independent of the quark flavor i for all four flavors (u, s, c, b) when the contribution of the hyperfine interaction energies is removed. For the two cases of spin-zero [4] $S = 0$ and spin-one $S = 1$ diquarks,

$$M(N) - \tilde{M}(\rho) = 323 \text{ MeV} \approx$$
$$\approx M(\Lambda) - \tilde{M}(K^*) = 321 \text{ MeV} \approx$$
$$\approx M(\Lambda_c) - \tilde{M}(D^*) = 312 \text{ MeV} \approx$$
$$\approx M(\Lambda_b) - \tilde{M}(B^*) = 310 \text{ MeV} \tag{17}$$

$$\tilde{M}(\Delta) - \tilde{M}(\rho) = 517.56 \text{ MeV} \approx$$
$$\approx \tilde{M}(\Sigma) - \tilde{M}(K^*) = 526.43 \text{ MeV} \approx$$
$$\approx \tilde{M}(\Sigma_c) - \tilde{M}(D^*) = 523.95 \text{ MeV} \approx$$
$$\approx \tilde{M}(\Sigma_b) - \tilde{M}(B^*) = 512.45 \text{ MeV} \tag{18}$$

where

$$\tilde{M}(V_i) \equiv \frac{3M_{\mathcal{V}_i} + M_{\mathcal{P}_i}}{4}; \tag{19}$$

are the weighted averages of vector and pseudoscalar meson masses, denoted respectively by $M_{\mathcal{V}_i}$ and $M_{\mathcal{P}_i}$, which cancel their hyperfine contribution, and

$$\tilde{M}(\Sigma_i) \equiv \frac{2M_{\Sigma_i^*} + M_{\Sigma_i}}{3}; \tilde{M}(\Delta) \equiv \frac{2M_\Delta + M_N}{3} \tag{20}$$

are the analogous weighted averages of baryon masses which cancel the hyperfine contribution between the diquark and the additional quark.

4. Magnetic Moments of Heavy Quark Baryons

In Λ, Λ_c and Λ_b baryons the light quarks are coupled to spin zero. Therefore the magnetic moments of these baryons are determined by the magnetic moments of the s, c and b quarks, respectively. The latter are proportional to the chromomagnetic moments which determine the hyperfine splitting in baryon spectra. We can use this fact to predict the Λ_c and Λ_b baryon magnetic moments by relating them to the hyperfine splittings in the same way as given in the original prediction [5] of the Λ magnetic moment,

$$\mu_\Lambda = -\frac{\mu_p}{3} \cdot \frac{M_{\Sigma^*} - M_\Sigma}{M_\Delta - M_N} = -0.61 \,\text{n.m.} \qquad (\text{EXP} = -0.61\,\text{n.m.}). \qquad (21)$$

We obtain

$$\mu_{\Lambda_c} = -2\mu_\Lambda \cdot \frac{M_{\Sigma_c^*} - M_{\Sigma_c}}{M_{\Sigma^*} - M_\Sigma} = 0.43 \;\text{n.m.}; \quad \mu_{\Lambda_b} = \mu_\Lambda \cdot \frac{M_{\Sigma_b^*} - M_{\Sigma_b}}{M_{\Sigma^*} - M_\Sigma} = -0.067\,\text{n.m.}$$

$$(22)$$

We hope these observables can be measured in foreseeable future and view the predictions (22) as a challenge for the experimental community.

5. Testing Confining Potentials Through Meson/Baryon HF Splitting Ratio

The ratio of color hyperfine splitting in mesons and baryons is a sensitive probe of the details of the confining potential. This is because this ratio depends only on the value of the wave function at the origin, which in turn is determined by the confining potential and by the ratio of quark masses, as can be readily seen from eqs. (4) and (6), together with the fact that the color quark-antiquark interaction in mesons is twice as strong as the quark-quark interaction in baryons, $(\vec{\lambda}_u \cdot \vec{\lambda}_s)_{meson} = 2(\vec{\lambda}_u \cdot \vec{\lambda}_s)_{baryon}$. We then have

$$\frac{M(K^*) - M(K)}{M(\Sigma^*) - M(\Sigma)} = \frac{4}{3} \frac{\langle\psi|\delta(\vec{r}_u - \vec{r}_s)|\psi\rangle_{meson}}{\langle\psi|\delta(\vec{r}_u - \vec{r}_s)|\psi\rangle_{baryon}} \qquad (23)$$

and analogous expressions with the s quark replaced by another heavy quark Q. From the experiment we have 3 data points for this ratio, with $Q = s, c, b$. We can then compute the ratio (23) for 5 different representative confining potentials and compare with experiment. The 5 potentials are

- harmonic oscillator
- Coulomb interaction
- linear potential
- linear + Coulomb, i.e. Cornell potential
- logarithmic.

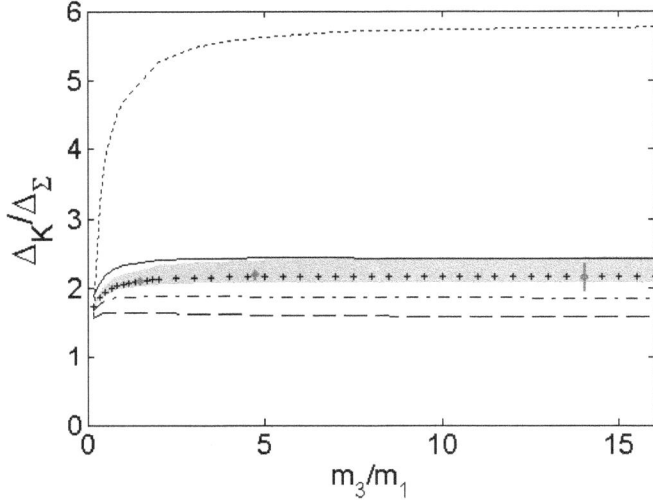

Fig. 3. Ratio of the hyperfine splittings in mesons and baryons, as function of the quark mass ratio. Shaded region: Cornell potential for $0.2 < k < 0.5$; crosses: Cornell, $k = 0.28$; long dashes: harmonic oscillator; short dashes: Coulomb; dot-dash: linear; continuous: logarithmic; thick dots: experimental data.

The results are shown in Fig. 3 and Table II [6].

	Δ_K/Δ_Σ	$\Delta_D/\Delta_{\Sigma_c}$	$\Delta_B/\Delta_{\Sigma_b}$
m_Q/m_q	1.33	4.75	14
EXP	2.08 ± 0.01	2.18 ± 0.08	2.15 ± 0.20
Harmonic	1.65	1.62	1.59
Coulomb	5.07 ± 0.08	5.62 ± 0.02	5.75 ± 0.01
Linear	1.88 ± 0.06	1.88 ± 0.08	1.86 ± 0.08
Log	2.38 ± 0.02	2.43 ± 0.02	2.43 ± 0.01
Cornell (k=0.28)	2.10 ± 0.05	2.16 ± 0.07	2.17 ± 0.08

Table II. Ratio of hyperfine splittings in mesons and baryons, for different potentials.

For all potentials which contain one coupling constant the coupling strength cancels in the meson-baryon ratio. The Cornell potential which is a combination of a Coulomb and linear potential contains two couplings, one of which cancels in the meson-baryon ratio. The remaining coupling is denoted by k. The gray band corresponds to the range of values $0.2 < k < 0.5$ of the Cornell potential. The crosses correspond to $k = 0.28$ which is the value previously used to fit the charmonium data. Clearly the Cornell potential with $k = 0.28$ provides the best fit to the experiment.

6. Predicting the Mass of *b*-Baryons

On top of the already discussed Σ_b with quark content bqq, there are two additional ground-state *b*-baryons, Ξ_b and Ω_b. We will now discuss the theoretical prediction of their masses and compare it with experiment.

6.1. Ξ_b

The Ξ_Q baryons quark content is Qsd or Qsu. They can be obtained from "ordinary" Ξ (ssd or ssu) by replacing one of the s quarks by a heavier quark $Q = c, b$. There is one important difference, however. In the ordinary Ξ, Fermi statistics dictates that two s quarks must couple to spin-1, while in the ground state of Ξ_Q the (sd) and (su) diquarks have spin zero. Consequently, the Ξ_b mass is given by the expression: $\Xi_q = m_q + m_s + m_u - 3v\langle\delta(r_{us})\rangle/m_u m_s$. The Ξ_b mass can thus be predicted using the known Ξ_c baryon mass as a starting point and adding the corrections due to mass differences and HF interactions:

$$\Xi_b = \Xi_c + (m_b - m_c) - 3v \left(\langle\delta(r_{us})\rangle_{\Xi_b} - \langle\delta(r_{us})\rangle_{\Xi_c}\right)/(m_u m_s) \tag{24}$$

Since the Ξ_Q baryon contains a strange quark, and the effective constituent quark masses depend on the spectator quark, the optimal way to estimate the mass difference $(m_b - m_c)$ is from mesons which contain both s and Q quarks:

$$m_b - m_c = \tfrac{1}{4}(3B_s^* + B_s) - \tfrac{1}{4}(3D_s^* + D_s) = 3324.6 \pm 1.4 \ . \tag{25}$$

On the basis of these results we predicted [7] $M(\Xi_b) = 5795 \pm 5$ MeV. Our paper was submitted on June 14, 2007. The next day CDF announced the result [9], $M(\Xi_b) = 5792.9 \pm 2.5 \pm 1.7$ MeV, following up on an earlier D0 measurement, $M(\Xi_b) = 5774 \pm 11 \pm 15$ MeV [8], as shown in Fig. 4 below.

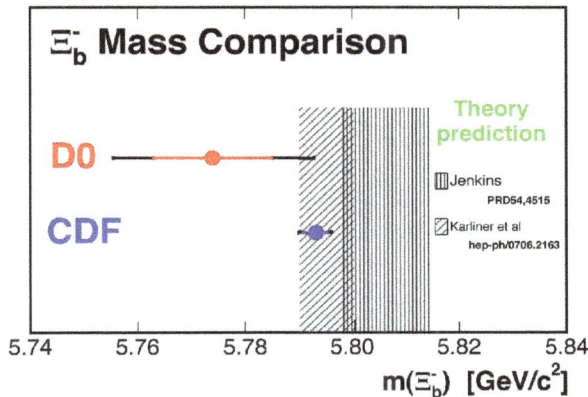

Fig. 4. Ξ_b mass - comparison of theoretical predictions with CDF and X0 data (from a CDF talk by D. Litvinstev).

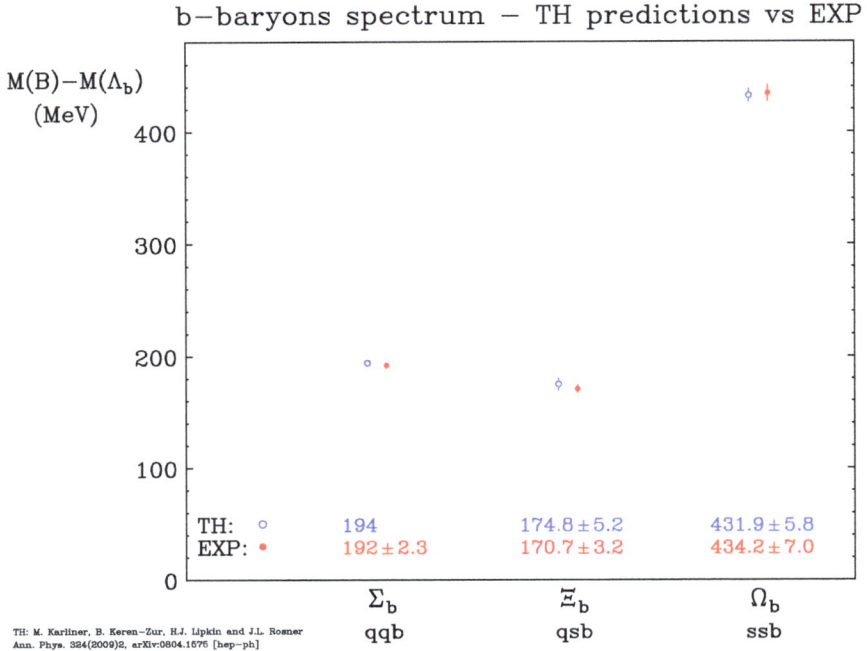

Fig. 5. Masses of b-baryons – comparison of theoretical predictions [7, 10] with experiment.

6.2. Mass of the Ω_b

For the spin-averaged Ω_b mass we have

$$\tfrac{1}{3}(2M(\Omega_b^*) + M(\Omega_b)) = \tfrac{1}{3}(2M(\Omega_c^*) + M(\Omega_c)) + (m_b - m_c)_{B_s - D_s} = 6068.9 \pm 2.4 \text{ MeV}. \tag{26}$$

For the HF splitting we obtain

$$M(\Omega_b^*) - M(\Omega_b) = (M(\Omega_c^*) - M(\Omega_c))\frac{m_c}{m_b}\frac{\langle\delta(r_{bs})\rangle_{\Omega_b}}{\langle\delta(r_{cs})\rangle_{\Omega_c}} = 30.7 \pm 1.3 \text{ MeV} \tag{27}$$

leading to the following predictions:

$$\Omega_b = 6052.1 \pm 5.6 \text{ MeV}; \qquad \Omega_b^* = 6082.8 \pm 5.6 \text{ MeV}. \tag{28}$$

About four months after our prediction (28) for Ω_b mass was published [10], D0 collaboration published the first measurement of Ω_b mass [11]:

$$M(\Omega_b)_{D0} = 6165 \pm 10(stat.) \pm 13(syst.) \text{ MeV}.$$

The deviation from the central value of our prediction was huge, 113 MeV. Understandably, we were very eager to see the CDF result. CDF published their result about nine months later, in May 2009 [12]:

$$M(\Omega_b)_{CDF} = 6054 \pm 6.8(stat.) \pm 0.9(syst.) \text{ MeV}.$$

Quantity	Refs. [13]	Ref. [14]	Value in MeV Ref. [15]	This work	Experiment
$M(\Lambda_b)$	5622	5612	Input	Input	5619.7 ± 1.7
$M(\Sigma_b)$	5805	5833	Input	–	5811.5 ± 2
$M(\Sigma_b^*)$	5834	5858	Input	–	5832.7 ± 2
$M(\Sigma_b^*) - M(\Sigma_b)$	29	25	Input	20.0 ± 0.3	$21.2^{+2.2}_{-2.1}$
$M(\Xi_b)$	5812	5806^a	Input	5790–5800	5792.9 ± 3.0^b
$M(\Xi_b')$	5937	5970^a	5929.7 ± 4.4	5930 ± 5	–
$\Delta M(\Xi^b)^c$	–	–	–	6.4 ± 1.6	–
$M(\Xi_b^*)$	5963	5980^a	5950.3 ± 4.2	5959 ± 4	–
$M(\Xi_b^*) - M(\Xi_b')$	26	10^a	20.6 ± 1.9	29 ± 6	–
$M(\Omega_b)$	6065	6081	6039.1 ± 8.3	6052.1 ± 5.6	6054.4 ± 7^d
$M(\Omega_b^*)$	6088	6102	6058.9 ± 8.1	6082.8 ± 5.6	–
$M(\Omega_b^*) - M(\Omega_b)$	23	21	19.8 ± 3.1	30.7 ± 1.3	–
$M(\Lambda_{b[1/2]}^*)$	5930	5939	–	5929 ± 2	–
$M(\Lambda_{b[3/2]}^*)$	5947	5941	–	5940 ± 2	–
$M(\Xi_{b[1/2]}^*)$	6119	6090	–	6106 ± 4	–
$M(\Xi_{b[3/2]}^*)$	6130	6093	–	6115 ± 4	–

Table III. Comparison of predictions for b baryons with those of some other recent approaches [13–15] and with experiment. Masses quoted are isospin averages unless otherwise noted. Our predictions are those based on the Cornell potential.
[a] Value with configuration mixing taken into account; slightly higher without mixing.
[b] CDF [9] value of $M(\Xi_b^-)$.
[c] $M(bsd) - M(bsu)$.
[d] CDF [12] value of $M(\Omega_b)$.

The central values of theoretical prediction and CDF agree to within 2 MeV, or about 1/3 standard deviation.

Fig. 5 shows a comparison of our predictions for the masses of Σ_b, Ξ_b and Ω_b baryons with the CDF experimental data.

The sign in our prediction

$$M(\Sigma_b^*) - M(\Sigma_b) < M(\Omega_b^*) - M(\Omega_b) \tag{29}$$

appears to be counterintuitive, since the color hyperfine interaction is inversely proportional to the quark mass. The expectation value of the interaction with the same wave function for Σ_b and Ω_b violates our inequality. When wave function effects are included, the inequality is still violated if the potential is linear, but is satisfied in predictions which use the Cornell potential [6]. This reversed inequality is not predicted by other recent approaches [13–15] which all predict an Ω_b splitting smaller than a Σ_b splitting. However the reversed inequality is also seen in the corresponding charm experimental data,

$$M(\Sigma_c^*) - M(\Sigma_c) < M(\Omega_c^*) - M(\Omega_c)$$
$$64.3\pm0.5\text{MeV} \qquad 70.8\pm1.5\text{MeV}. \tag{30}$$

This suggests that the sign of the $SU(3)$ symmetry breaking gives information about the form of the potential. It is of interest to follow this clue theoretically and experimentally.

We have made additional predictions [7, 10] for some excited states of b-baryons. Our results are summarized in Table III.

Acknowledgments

The work described here was done in collaboration with B. Keren-Zur, H.J. Lipkin and J. Rosner. It was supported in part by a grant from the Israel Science Foundation.

References

[1] M. Karliner and H.J. Lipkin,hep-ph/0307243, Phys. Lett. **B575** (2003) 249.
[2] M. Karliner and H. J. Lipkin, Phys. Lett. B **660**, 539 (2008) [arXiv:hep-ph/0611306].
[3] T. Aaltonen *et al.* [CDF Collaboration], Phys. Rev. Lett. **99** (2007) 202001.
[4] M. Karliner and H. J. Lipkin, Phys. Lett. B **650**, 185 (2007) [arXiv:hep-ph/0608004].
[5] A. De Rujula, H. Georgi and S.L. Glashow, Phys. Rev. D12 (1975) 147
[6] B. Keren-Zur, Annals Phys. **323**, 631 (2008) [arXiv:hep-ph/0703011].
[7] M. Karliner, B. Keren-Zur, H. J. Lipkin and J. L. Rosner, arXiv:0706.2163v1 [hep-ph].
[8] V. M. Abazov *et al.* [D0 Collaboration], Phys. Rev. Lett. 99 (2007) 052001.
[9] T. Aaltonen *et al.* [CDF Collaboration], Phys. Rev. Lett. 99 (2007) 052002.
[10] M. Karliner, B. Keren-Zur, H. J. Lipkin and J. L. Rosner, arXiv:0708.4027 [hep-ph] (unpublished) and arXiv:0804.1575 [hep-ph], Annals Phys **324**,2 (2009).
[11] V. M. Abazov *et al.* [D0 Collaboration], Phys. Rev. Lett. **101**, 232002 (2008) [arXiv:0808.4142 [hep-ex]].
[12] T. Aaltonen *et al.* [CDF Collaboration], Phys. Rev. D **80**, 072003 (2009) [arXiv:0905.3123 [hep-ex]].
[13] D. Ebert *et al.*, Phys. Rev. D 72 (2005) 034026; Phys. Lett. B **659** (2008) 612.
[14] W. Roberts and M. Pervin, arXiv:0711.2492 [nucl-th].
[15] E. E. Jenkins, Phys. Rev. D 77 (2008) 034012.

EARLY HISTORY OF QCD AND QUARKS

GABRIEL KARL

Department of Physics, University of Guelph, Guelph, Ontario, N1G 2W1, Canada
gk@physics.uoguelph.ca

Early work on the compositeness of hadrons is reviewed, with emphasis on Quarks and Confinement.

Keywords: QCD; quarks; history.

1. Introduction

A very long time ago, while I was a graduate student,[a] I became aware of group theory[b] and of what was then called Elementary Particle Physics. At that time there was one unusual name mentioned often in connection with new pretty ideas. Later on I met Murray Gell-Mann in Erice, and I was amazed to see that he was not very much older than me. It is a pleasure to speak here in his honour, in such distinguished company, and I thank Harald Fritzsch for the opportunity.

The main subject of this talk is the compositeness of the hadrons and confinement of quarks.

Nuclear Physics is as old as my mother would be today, about 100 years old. About eighty years ago it was proposed that all nuclei are composites of protons and neutrons. At that time it was believed that both protons and neutrons are elementary particles. This view was first challenged by the measurements of magnetic moments which turned out to be (roughly) three nuclear magnetons (proton) and minus two (neutron). This caused consternation for theorists, who rallied with the suggestion that the proton and neutron are not Dirac particles!

Nevertheless the proton and neutron were assumed (by *nuclear* physicists) to be elementary all the way to 1974, when the J/ψ was discovered, a full decade after the proposal of Gell-Mann and Zweig for quarks. The Nuclear community took the

---

[a]I was graduate student of Physical Chemistry at the University of Toronto, from 1961–1964. My supervisor was Professor John C. Polanyi, who won a Nobel Prize in 1986.
[b]The course of Group Theory for Physics, was taught by Professor W. T. Sharp (a student of E. P. Wigner) and G. de B. Robinson (a student of Rev. Alfred Young, of Tableaux fame). Sharp taught SU2 and SU3 including the "eightfold way" and the Gell-Mann–Okubo formula. His presentation was so clear and the ideas so beautiful that this made me change my career. Robinson taught permutation groups and their representations.

view that the strong interaction renormalizes the magnetic moments from the Dirac values $(1, 0)$ to the experimental values $(3, -2)$.

The earliest proposal for composite hadrons,[1] is due to Fermi and Yang, in 1949, in a paper entitled: "Are mesons elementary particles?" — the junior author of this paper spoke earlier in this session. The paper proposed pions are bound states of a nucleon and an antinucleon. This was quite daring since the antiproton (and antineutron) were not yet discovered! And there was some worry prior to experiments (see, e.g. Ref. 2), due to "not Dirac Particles," that antiprotons might not exist. The Fermi–Yang (FY) model was extended by Sakata,[3] to account for strange mesons and baryons. The Sakata and FY models predicted (as noted by Okun[4]) isoscalar mesons (like the eta meson), to be discovered later. In retrospect, the flaw in these models was that two or three baryons were elementary but other baryons and mesons were not. Nevertheless these models played an important role in setting the stage for later ideas of compositeness.

By the early fifties many excited states of the nucleon were observed, in photo-production, elastic scattering, etc. These experiments made it unlikely that ground state nucleons are elementary. Also electron scattering from hydrogen revealed that the proton had an electromagnetic radius of order 10^{-13} cm, larger than the bound on the electron radius.

The compositeness of hadrons turned out to be much more intricate than the early speculations — namely hadrons are composed of quarks and quarks are "confined." This was first seen as a possibility by Murray Gell-Mann. The idea of quarks was suggested by Gell-Mann,[5] Zweig,[6] Neeman, Serber and others. But the deep idea of confinement occurred only to Gell-Mann. He named such quarks "mathematical" (a very unfortunate choice) and described this possibility in a lecture at the 1966 (Rochester) Conference at Berkeley. This proposal was totally rejected by most experts at the time, even by those who accepted quarks. I can only compare this situation with Dirac's proposal of the positron, which received similar unfriendly reception in the community in 1930–1932. The story for positrons is detailed, for example, in the recent excellent biography of Dirac by Graham Farmelo. However Dirac's idea was rescued by Anderson's and Blackett and Ochialini's timely experiments. In contrast, Gell-Mann's idea implied that there would be no discovery of quarks, which turned out also to be the case. By the time experiments showed that there are no free quarks, models of confinement were proposed in the period 1972–1974 (by Mandelstam, Vinciarelli, Ken Wilson, Kogut and Susskind, Nielsen and Olesen and others) and Ken Wilson proposed (as far as I can tell) "confinement" — a descriptive word, which makes this option *sound* more physical. In fact this naming extended the range of what is physical. So it came to be that Murray got no credit for this idea at all. This is not unusual in theoretical physics (in *my* experience), as credit moves somewhat unpredictably, but it was unusual for Gell-Mann. By then however he had received the Nobel Prize which must have provided some consolation.

Gell-Mann's motivation for proposing quarks was to understand why SU3 flavor symmetry was good in hadron physics and baryons are only singlets, octets and decuplets. The paper thanked Serber for stimulating the ideas in 1963 and proposed the exotic name of "quark" with a reference to James Joyce's *Finnegan's Wake* (page 383!, as noted also in the talk by Harald Fritzsch at this conference) where this word is found. The name and idea have survived now for nearly fifty years and probably will outlive most of us and even our grandchildren.

Zweig has spoken this morning and had a chance to explain his motivation, based on Zweig's rule. He also gave a talk in Toronto on his invention, and I'd like to quote from his talk at the Baryon 84 Conference:

> "The reaction of the theoretical physics community to the ace model was generally not benign . . . . The idea that hadrons, citizens of a nuclear democracy, were made of elementary particles with fractional quantum numbers did seem a bit rich. The idea however, was apparently correct."

But there were many who accepted the quark model (if not confinement), and I name here just a few: Dick Dalitz (who was my mentor in Particle Physics) in Oxford and Wally Greenberg in Maryland, Yoichiro Nambu in Chicago, G. Morpurgo in Genoa, Harry Lipkin in Rehovot, Tavkhelidze in Dubna. They all contributed to the acceptance of the quark model, and there were many, many others. Nambu was the earliest to propose color and gauge couplings (with Han). Greenberg saw early on that there were problems with statistics in baryons, and proposed parastatistics — equivalent to color; he also had rather light quarks. Dalitz and Greenberg assigned excited states (baryonic resonances). Morpurgo explained that nonrelativistic quarks were acceptable physically and explained some photon selection rules. This is not an exhaustive list. It is clear that the proposal of the quark model generated a big wave of theoretical endeavour, right from 1964. Nevertheless the quark model was embraced by the nuclear physics community only after the discovery of the J/ψ in 1974.

The reluctance of the nuclear theory community to accept the thought that there are fermionic degrees of freedom which do not have independent existence as free particles is easy to understand. I cannot resist sharing my recollections from a conference in the summer of 1974 (before J/ψ, which appeared in November) in honour of Sir Rudolph Peierls who was retiring from Oxford. I took it upon myself[7] to make a comment on quark confinement (after a talk by Dalitz) by saying there is a "historical parallel" between the explanation of beta decay in the thirties (electrons are not nuclear constituents, but are created in beta decay) and quark confinement (where quarks do not exist outside the nucleon only inside). The two scenarios are each others reverse. The older generation (Bethe, Peierls, Weiskopf , . . .) shook their heads in disapproval while younger folk shook their heads in approval. There was no reply to my comment.

After November 1974 when the J/ψ was observed, the accepted wisdom changed. Also the nice name of "Elementary Particle Physics" became less popular and

56

changed to "High Energy Physics." It is interesting also that somehow no one got full credit for the idea of confinement, even now; the present author feels that Gell-Mann was unlucky in this story.

Another interesting aspect of this story illustrates how flexible is the borderline between mathematical and physical objects. Before the Aharonov–Bohm effect, the electromagnetic potential was thought to be a mathematical object, useful in calculating the fields, which are physical. Not so afterwards. Confined quarks were thought by most to be unphysical, not so after 1974. A nice story on this point is told by Yang[8] on his conversation with E. P. Wigner about the Fermi theory of Beta decay. Wigner credited Fermi with realizing that fermionic fields were physical, even though they were invented by Wigner and Jordan who firmly believed that the fields were mathematical objects.

As noted at the start, most of this story took place a long time ago.[9] Physics was very different, much more fun than now. This is not only because many of us were younger but also because the scale of experiments and theory were much smaller. Now important machines cost billions of dollars, and the theory community is very much larger. But instead of billions of dollars, the physics community then had giants in its ranks. We are celebrating today one of these giants.

I am glad to have participated through this remarkable era, when giants were present in our community.

Finally and more personally, I wish to acknowledge and thank for a conversation with Murray Gell-Mann in 1988 when I needed medical advice, and he gave me a very thorough description of research in North America on chemotherapy for the cancer of the colon.

References

1. E. Fermi and C. N. Yang, *Phys. Rev.* **76**, 1739 (1949).
2. M. Ruderman and R. Finkelstein, *Phys. Rev.* **76**, 1458 (1949).
3. S. Sakata, *Progr. Theor. Phys.* **16**, 686 (1956).
4. L. B. Okun, *Sov. Phys. JETP* **7**, 322 (1958).
5. M. Gell-Mann, *Phys. Lett.* **8**, 214 (1964).
6. G. Zweig, CERN Reports 8182/TH401 and 8419/TH412 (1964), unpublished.
7. G. Karl, in *Rudolf Peierls and Theoretical Physics — Proc. of the Symposium to Mark the Occasion of the Retirement of Professor Sir Rudolph E. Peierls (F.R.S., C.B.E.)*, Oxford, July 11–12, 1974, eds. I. J. R. Aitchison and J. E. Paton (Pergamon Press, Oxford, 1977), pp. 105–106.
8. C. N. Yang, in *Fermi Remembered*, ed. J. W. Cronin (The University of Chicago Press, 2004), p. 241 [I am indebted to Prof. C. N. Yang for this story and reference.]
9. M. Gell-Mann, Quarks, color and QCD, in *The Rise of the Standard Model*, eds. L. Hoddeson, L. Brown, M. Riordan and M. Dresden (Cambridge University Press, 1997), pp. 625–633 [This reference gives M. Gell-Mann's version of this history.]

UNDERSTANDING CONFINEMENT IN QCD: ELEMENTS OF A BIG PICTURE

MIKHAIL SHIFMAN

*William I. Fine Theoretical Physics Institute, University of Minnesota,
Minneapolis, MN 55455, USA*

I give a brief review of advances in the strong interaction theory. This talk was delivered at the Conference in honor of Murray Gell-Mann's 80th birthday, 24–26 February 2010, Singapore.

Keywords: SU(3); quarks; quark confinement; non-Abelian strings.

1. Introduction

In the early 1970s, when quantum chromodynamics (QCD) was born, I was just in the beginning of my career in theoretical physics. My teachers at ITEP[a] tried to convey to me a number of "commandments" which were intended for guidance in my future scientific life. One of them was: always listen to what Gell-Mann says because he has a direct line to God. I always did. Gell-Mann was one of the discoverers of QCD who opened a whole new world. Unlike many recent theoretical constructions, whose relevance to nature is still a big question mark, QCD will stay with us forever. I am happy that I invested so much time and effort in studying QCD. This was a long and exciting journey. Almost 40 years later, I am honored and proud to be invited to this Conference celebrating Professor Gell-Mann's 80th birthday to give a talk on advances in QCD.

I should say that the problem of strong interactions turned out to be extremely difficult (despite the fact that the underlying Lagrangian is firmly established) and the advances slow and painful. This is a usual story with the strong coupling regime: whenever theorists find themselves at strong coupling, they are in trouble. Yang–Mills theories are no exception.

My task today is to outline some contours of the strong interaction theory which gradually emerged from obscurity during these four decades. Yes, the theory is incomplete, but those parts which are already in existence are beautiful, and continue to grow.

---

[a]Institute for Theoretical and Experimental Physics in Moscow.

Fig. 1.   Everybody knows that this is Murray Gell-Mann.

First, to give a general idea of the role which this theory plays in high energy physics (HEP) I would like to chart approximate contours of main areas that are under intense development in the theoretical community of today. To this end I display a symbolic map in Fig. 2. The linkage of the HEP theory map to the earth's geography is arbitrary and does not mean anything. You should pay attention only to interconnections of various areas of HEP. You see that QCD and strongly coupled gauge theories at large, in which Murray Gell-Mann was a trailblazer, occupy a vast area. The advances there and in neighboring areas crucially depend on the exchange of ideas between them. One can say that they feed each other. Of particular importance was a breakthrough impact of supersymmetry of which I will speak later. In the Appendix you can find some additional information regarding the HEP theory map obtained from a HEP world traveller. Unfortunately, the traveller was unable to visit some areas (allegedly, because of clearance issues), namely, that of extremal phenomena (high-energy and high density QCD), nuclear physics, multiverses, and theoretical nonperturbative supersymmetry.

2. QCD

In all processes with hadron participation strong interactions play a role. All matter surrounding us is made of protons and neutrons the most common representatives of the class of hadrons. Even if you do not see them explicitly, they will necessarily show up at a certain stage or in loops. The fundamental Lagrangian governing strong interactions is

$$\mathcal{L} = \sum_f \bar{\psi}_f(i\slashed{D} - m_f)\psi^f - \frac{1}{4}G^a_{\mu\nu}G^{\mu\nu a} \,. \tag{1}$$

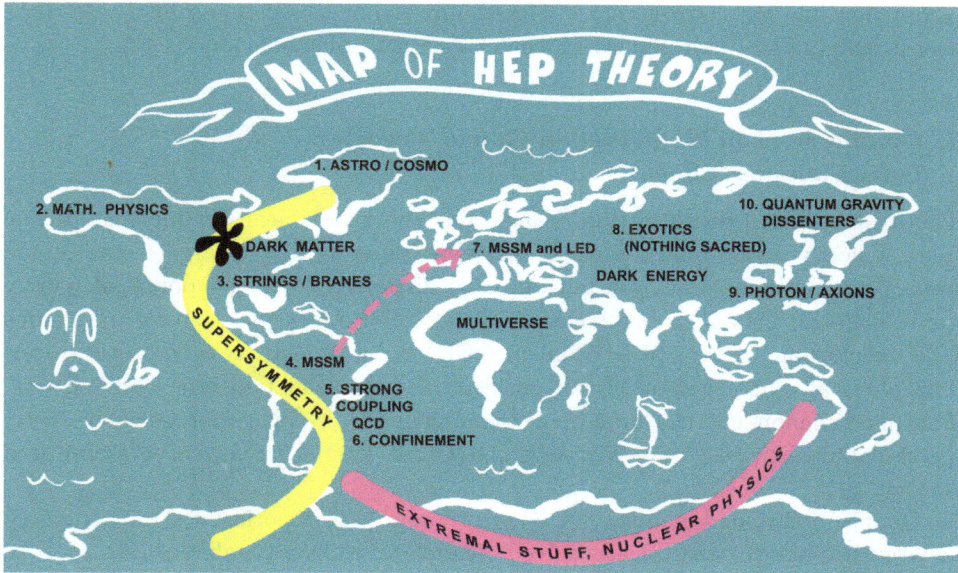

Fig. 2.   Map of the HEP theory.

The first term describes color-triplet quarks and their coupling to color-octet gluons. It ascends to Gell-Mann. The second Yang–Mills term describes the gluon dynamics. Both terms taken together comprise the Lagrangian of quantum chromodynamics (QCD). The very name "quantum chromodynamics" ascends to Gell-Mann too. Much in the same way as the Schrödinger equation codes all of quantum chemistry, the QCD Lagrangian codes all of

- nuclear physics;
- Regge behavior;
- neutron stars;
- chiral physics;
- light & heavy quarkonia;
- glueballs & exotics;
- exclusive & inclusive hadronic scattering at large momentum transfer;
- interplay between strong forces & weak interactions,
- quark-gluon plasma;

and much more.

Although the underlying Lagrangian (1), and asymptotic freedom it implies at short distances,[1,2] are established beyond any doubt, the road from this starting point to theoretical control over the large-distance hadronic world is long and difficult. The journey which started 40 years ago is not yet completed. *En route*, many beautiful theoretical constructions were developed allowing one to understand various corners of the hadronic world. Here I am unable even to list them, let alone discuss in a comprehensible way. Therefore, I will focus only on one — albeit absolutely global — aspect defining the hadronic world: the confinement phenomenon.

3. Confinement in Non-Abelian Gauge Theories: Dual Meissner Effect

The most salient feature of pure Yang–Mills theory is linear confinement. If one takes a heavy probe quark and an antiquark separated by a large distance, the force between them does not fall off with distance, while the potential energy grows linearly. This is the explanation of the empiric fact that quarks and gluons (the microscopic degrees of freedom in QCD) never appear as asymptotic states. The physically observed spectrum consists of color-singlet mesons and baryons. The phenomenon got the name color confinement, or, in a more narrow sense, quark confinement. In the early days of QCD it was also referred to as infrared slavery.

Quantum chromodynamics (QCD), and Yang–Mills theories at strong coupling at large, are not yet analytically solved. Therefore, it is reasonable to ask:

Are there physical phenomena in which interaction energy between two interacting bodies grows with distance at large distances? Do we understand the underlying mechanism?

The answer to these questions is positive. The phenomenon of linearly growing potential was predicted by Abrikosov[3,4] in the superconductors of the second type. The corresponding set up is shown in Fig. 3. In the middle of this figure we see a superconducting sample, with two very long magnets attached to it. The superconducting medium does not tolerate the magnetic field. On the other hand, the flux of the magnetic field must be conserved. Therefore, the magnetic field lines emanating from the N pole of one magnet find their way to the S pole of another magnet, through the medium, by virtue of a flux tube formation. Inside the flux tube the Cooper pair condensate vanishes and superconductivity is destroyed. The flux tube has a fixed tension, implying a constant force between the magnetic poles as long as they are inside the superconducting sample. The phenomenon described above is sometimes referred to as the Meissner effect.

Of course, the Meissner effect of the Abrikosov type occurs in the Abelian theory, QED. The flux tube that forms in this case is Abelian. In Yang–Mills theories we are interested in non-Abelian analogs of the Abrikosov vortices. Moreover, while in the Abrikosov case the flux tube is that of the magnetic field, in QCD and QCD-like theories the confined objects are quarks; therefore, the flux tubes must be "chromoelectric" rather than chromomagnetic. In the mid-1970s Nambu, 't Hooft, and Mandelstam (independently) put forward an idea[5–7] of a "dual Meissner effect" as the underlying mechanism for color confinement. Within their conjecture, in chromoelectric theories "monopoles" condense leading to formation of "non-Abelian flux tubes" between the probe quarks. At this time the Nambu–'t Hooft–Mandelstam paradigm was not even a physical scenario, rather a

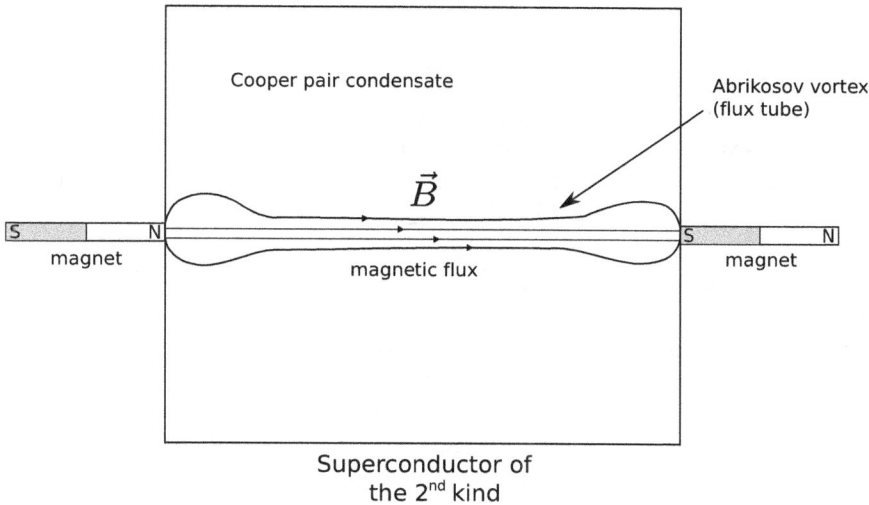

Fig. 3.   The Meissner effect in QED.

dream, since people had no clue as to the main building blocks such as non-Abelian flux tubes. After the Nambu–'t Hooft–Mandelstam conjecture had been formulated many works were published on this subject, to no avail.

A long-awaited breakthrough discovery came 20 years later: the Seiberg–Witten solution[8,9] of $\mathcal{N} = 2$ super-Yang–Mills theory slightly deformed by a superpotential breaking $\mathcal{N} = 2$ down to $\mathcal{N} = 1$. In the $\mathcal{N} = 2$ limit, the theory has a moduli space. If the gauge group is SU(2), on the moduli space, SU(2)$_{\text{gauge}}$ is spontaneously broken down to U(1). Therefore, the theory possesses the 't Hooft–Polyakov monopoles[10,11] in the quasiclassical regime. Of course, in this regime they are very heavy and play no role in dynamics. Using the power of $\mathcal{N} = 2$ supersymmetry, two special points on the moduli space were found[8,9] at strong coupling, (the monopole and dyon points), in which the monopoles (dyons) become massless. In these points the scale of the gauge symmetry breaking

$$SU(2) \to U(1) \tag{2}$$

is determined by Λ, the dynamical scale parameter of $\mathcal{N} = 2$ super-Yang–Mills theory.

All physical states can be classified with regards to the unbroken U(1). It is natural to refer to the U(1) gauge boson as to the photon. In addition to the photon, all its superpartners, being neutral, remain massless at this stage, while all other states, with nonvanishing "electric" charges, acquire masses of the order of Λ. In particular, two gauge bosons corresponding to SU(2)/U(1) — it is natural to call them $W^{\pm}$ — have masses $\sim \Lambda$. All such states are "heavy" and can be integrated out.

In the low-energy limit, near the monopole and dyon points, one deals with electrodynamics of massless monopoles. One can formulate an effective local theory describing interaction of the light states by dualizing the original phton. This is a U(1) gauge theory in which the (magnetically) charged matter fields M, $\tilde{M}$ are those of monopoles while the U(1) gauge field that couples to M, $\tilde{M}$ is *dual* with respect to the photon of the original theory. The $\mathcal{N} = 2$ preserving superpotential has the form $\mathcal{W} = \mathcal{A}M\tilde{M}$, where $\mathcal{A}$ is the $\mathcal{N} = 2$ superpartner of the dual photon/photino fields.

Now, if one switches on a small $\mathcal{N} = 2$ breaking superpotential, the only change in the low-energy theory is the emergence of the extra $m^2\mathcal{A}$ term in the superpotential. Its impact is crucial: it triggers the monopole condensation, $\langle M \rangle = \langle \tilde{M} \rangle = m$, which implies, in turn, that the dual U(1) symmetry is spontaneously broken, and the dual photon acquires a mass $\sim m$. As a consequence, the Abrikosov flux tubes are formed. Viewed inside the dual theory, they carry fluxes of the magnetic field. With regards to the original microscopic theory these are the electric field fluxes.

Thus, Seiberg and Witten demonstrated, for the first time ever, the existence of the dual Meissner effect in a judiciously chosen non-Abelian gauge field theory. If one "injects" a probe (very heavy) quark and antiquark in this theory, a flux tube forms between them, with necessity, leading to linear confinement.

The flux tubes in the Seiberg–Witten solution were investigated in detail in Refs. 12 and 13. These flux tubes are Abelian, and so is confinement caused by their formation. What does that mean? At the scale of distances at which the flux tube is formed (the inverse mass of the Higgsed U(1) photon) the gauge group that is operative is Abelian. In the Seiberg–Witten analysis this is the dual U(1). The off-diagonal (charged) gauge bosons are very heavy in this scale and play no direct role in the flux tube formation and confinement that ensues. Naturally, the spectrum of composite objects in this case turns out to be richer than that in QCD and similar theories with non-Abelian confinement. By non-Abelian confinement I mean such dynamical regime in which at distances of the flux tube formation all gauge bosons are equally important.

Moreover, the string topological stability is based on $\pi_1(\mathrm{U}(1)) = \mathbb{Z}$. Therefore, N strings do not annihilate as they should in $\mathrm{SU}(N)$ QCD-like theories.

The two-stage symmetry breaking pattern, with $\mathrm{SU}(2) \to \mathrm{U}(1)$ occurring at a high scale while the dual $\mathrm{U}(1) \to$ nothing at a much lower scale, has no place in QCD-like theories, as we know from experiment. In such theories, presumably, all non-Abelian gauge degrees of freedom take part in the string formation, and are operative at the scale at which the strings are formed. The strings in the Seiberg–Witten solution are believed to belong to the same universality class as those in QCD-like theories. However, in the limit of large-m deformations, when a non-Abelian regime presumably sets in and non-Abelian strings develop in the model considered by Seiberg and Witten, theoretical control is completely lost. Thus, the status of the statement of the same universality class is conjectural.

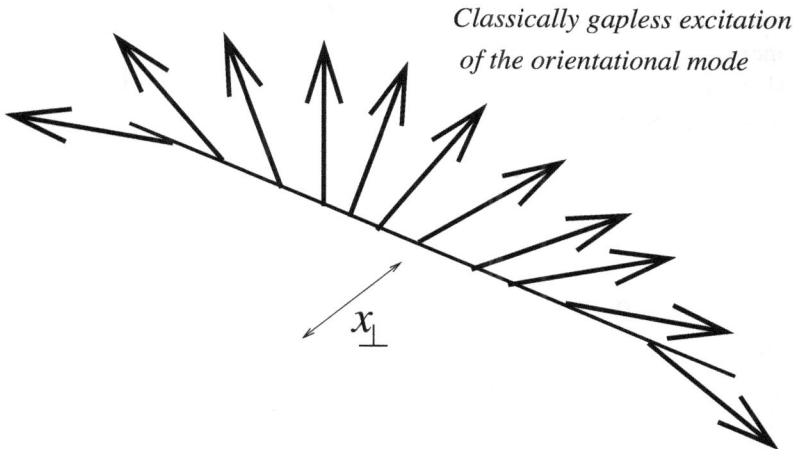

Classically gapless excitation of the orientational mode

$x_{\perp}$

Fig. 4.   Orientational moduli on the string world-sheet.

4. Non-Abelian Strings

In a bid to better understand string-based confinement mechanism in Yang–Mills theories that might be more closely related to QCD people continued searches for models supporting non-Abelian strings. If a model in which non-Abelian strings develop in a fully controllable manner, i.e. at weak coupling, could be found and the passage from Abelian to non-Abelian strings explored, this would provide us with evidence that no phase transition occurs between the two regimes in the Seiberg–Witten solution.

In the technical sense, what does one mean when one speaks of non-Abelian flux tubes? Apparently, the orientation of the magnetic field in the tube interior must be free to strongly fluctuate inside the SU(N) group. There is no such freedom in the Abelian string of the Abrikosov type. In other words, in addition to translational moduli, the theory on the string world-sheet must acquire orientational moduli (Fig. 4).

If one thinks there is a kind of string theory behind QCD confining dynamics, such behavior is natural. Indeed, string theory is formulated in higher dimensions. Bringing it to $D = 4$ requires compactification of some dimensions. If (some of) compact dimensions have isometries, the corresponding sigma model on the 4D string world-sheet will have classically massless internal degrees of freedom. For instance, compactification on S_2 gives rise to CP(1) sigma model on the string world-sheet. Two-dimensional infrared dynamics will then generate a mass gap for these orientational degrees of freedom.

That's why searches for non-Abelian flux tubes and non-Abelian monopoles in the bulk Yang–Mills theories continued, with a decisive breakthrough in 2003–2004.[14,15] By that time the program of finding field-theoretical analogs of all basic constructions of string/D-brane theory was in full swing. BPS domain walls, analogs of D branes, had been identified in supersymmetric Yang–Mills theory.[16] It had been demonstrated that such walls support gauge fields localized on them. BPS saturated string-wall junctions had been constructed.[17] Topological stability of the non-Abelian strings under consideration is due to the fact that

$$\pi_1\left(\frac{\mathrm{SU}(2)\times\mathrm{U}(1)}{Z_2}\right) \to \mathrm{nontrivial}\,. \tag{3}$$

5. Basic Bulk Theory: Setting the Stage

Non-Abelian strings were first found in $\mathcal{N} = 2$ super-Yang–Mills theories with $\mathrm{U}(2)_{\mathrm{gauge}}$ and two matter hypermultiplets.[14,15] The $\mathcal{N} = 2$ vector multiplet consists of the $\mathrm{U}(1)$ gauge field A_μ and the $\mathrm{SU}(2)$ gauge field A_μ^a, (here $a = 1, 2, 3$), and their Weyl fermion superpartners (λ^1, λ^2) and $(\lambda^{1a}, \lambda^{2a})$, plus complex scalar fields a, and a^a. The global $\mathrm{SU}(2)_R$ symmetry inherent to $\mathcal{N} = 2$ models manifests itself through rotations $\lambda^1 \leftrightarrow \lambda^2$.

The quark multiplets consist of the complex scalar fields q^{kA} and $\tilde{q}_{Ak}$ (squarks) and the Weyl fermions ψ^{kA} and $\tilde{\psi}_{Ak}$, all in the fundamental representation of the $\mathrm{SU}(2)$ gauge group ($k = 1, 2$ is the color index while A is the flavor index, $A = 1, 2$). The scalars q^{kA} and $\bar{\tilde{q}}^{kA}$ form a doublet under the action of the global $\mathrm{SU}(2)_R$ group. The quarks and squarks have a $\mathrm{U}(1)$ charge too.

If one introduces a non-vanishing Fayet–Iliopoulos parameter ξ the theory develops isolated quark vacua, in which the gauge symmetry is fully Higgsed, and all elementary excitations are massive. In the general case, two matter mass terms allowed by $\mathcal{N} = 2$ are unequal, $m_1 \neq m_2$. There are free parameters whose interplay determines dynamics of the theory: the Fayet–Iliopoulos parameter ξ, the mass difference Δm and a dynamical scale parameter Λ, an analog of the QCD scale Λ_{QCD} (Fig. 5). Extended supersymmetry guarantees that some crucial dependences are holomorphic, and there is no phase transition.

Both the gauge and flavor symmetries of the model are broken by the squark condensation. All gauge bosons acquire the same masses (which are of the order of inverse string thickness). A global diagonal combination of color and flavor groups, $\mathrm{SU}(2)_{C+F}$, survives the breaking (the subscript $C + F$ means a combination of global color and flavor groups).

While $\mathrm{SU}(2)_{C+F}$ is the symmetry of the vacuum, the flux tube solutions break it spontaneously. This gives rise to orientational moduli on the string world-sheet.

The bulk theory is characterized by three parameters of dimension of mass: ξ, Δm, and Λ. As various parameters vary, the theory under consideration evolves in a very graphic way, see Fig. 5. At $\xi = 0$ but $\Delta m \neq 0$ (and $\Delta m \gg \Lambda$) it presents a very clear-cut example of a model with the standard 't Hooft–Polyakov monopole. This is due to the fact that the relevant part of the bosonic sector is nothing but the Georgi–Glashow model. The monopole is unconfined — the flux tubes are not yet formed.

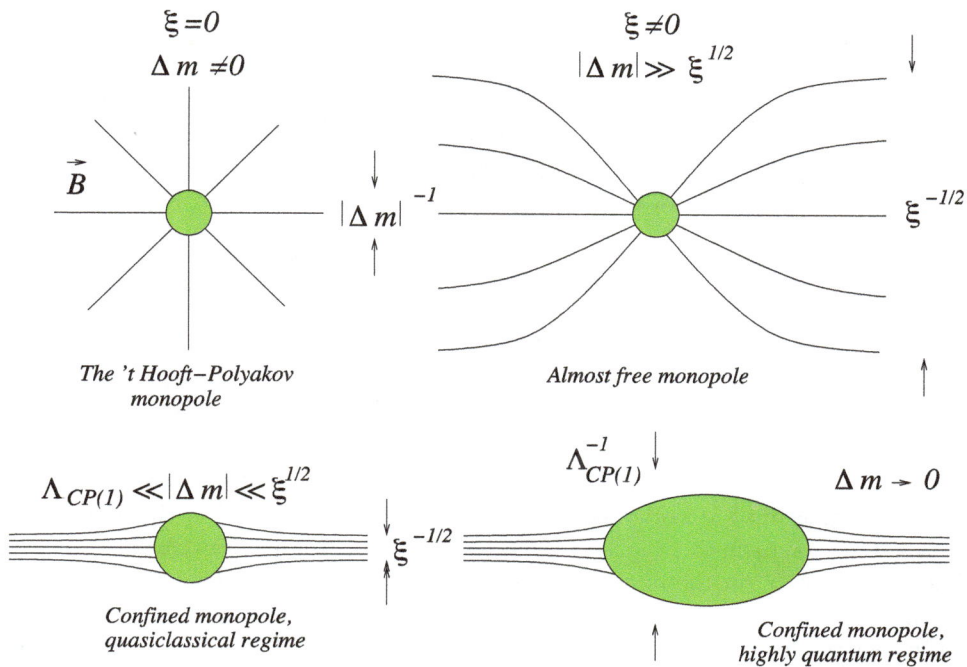

$\xi = 0$
$\Delta m \neq 0$

$\vec{B}$

The 't Hooft–Polyakov
monopole

$\xi \neq 0$
$|\Delta m| \gg \xi^{1/2}$

$|\Delta m|^{-1}$

$\xi^{-1/2}$

Almost free monopole

$\Lambda_{CP(1)} \ll |\Delta m| \ll \xi^{1/2}$

$\xi^{-1/2}$

Confined monopole,
quasiclassical regime

$\Lambda_{CP(1)}^{-1}$

$\Delta m \to 0$

Confined monopole,
highly quantum regime

Fig. 5.   Various regimes for monopoles and strings.

Switching on $\xi \neq 0$ traps the magnetic fields inside the flux tubes, which are weak as long as $\xi \ll \Delta m$. The flux tubes change the shape of the monopole far away from its core, leaving the core essentially intact. Orientation of the chromomagnetic field inside the flux tube is essentially fixed. This is due to the fact that all off-diagonal gauge bosons (W bosons) are heavy in this limit. Thus, the flux tubes supported in this limit are *Abelian*. (They are commonly referred to as the Z_N strings.)

With $|\Delta m|$ decreasing, fluctuations in the orientation of the chromomagnetic field inside the flux tubes grow. Simultaneously, the monopole which no longer resembles the 't Hooft–Polyakov monopole, is seen as a *string junction*.

Finally, in the limit $\Delta m \to 0$ the transformation is complete. A global SU(2) symmetry restores in the bulk. All three gauge bosons have identical masses. Orientational (exact, classically massless) moduli develop on the string world-sheet making it non-Abelian. The string world-sheet theory is CP(1) (CP($N-1$) for generic values of N). Two-dimensional CP($N-1$) models with four supercharges are asymptotically free. They have N distinct vacuum states.

Each vacuum state of the world-sheet CP($N-1$) theory presents a distinct string from the standpoint of the bulk theory. There are N species of such strings; they have degenerate tensions $T_{\text{st}} = 2\pi\xi$. The ANO string tension is N times larger.

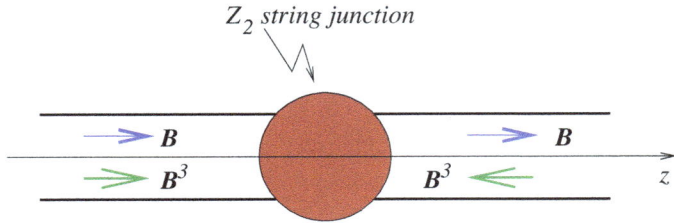

Fig. 6. Z_2 string junction.

Two different strings can form a stable junction. Figure 6 shows this junction in the limit

$$\Lambda_{\mathrm{CP}(1)} \ll |\Delta m| \ll \sqrt{\xi} \qquad (4)$$

corresponding to the lower left corner in Fig. 5. The magnetic fluxes of the U(1) and SU(2) gauge groups are oriented along the z axis. In the limit (4) the SU(2) flux is oriented along the third axis in the internal space. However, as $|\Delta m|$ decreases, fluctuations of B_z^a in the internal space grow, and at $\Delta m \to 0$ it has no particular orientation in SU(2) (the lower right corner of Fig. 5). In the language of the world-sheet theory this phenomenon is due to restoration of the O(3) symmetry in the quantum vacuum of the CP(1) model.

Evolution from the upper right corner in Fig. 5 to the lower right corner is in fact the transformation of the Abelian string into non-Abelian. $\mathcal{N} = 2$ supersymmetry guarantees that it is smooth, with no phase transition.

The junctions of degenerate strings present what remains of the monopoles in this highly quantum regime.[18,19] It is remarkable that, despite the fact we are deep inside the highly quantum regime, holomorphy allows one to exactly calculate the mass of these monopoles. This mass is given by the expectation value of the kink central charge in the world-sheet CP($N-1$) model (including the anomaly term), $M_M \sim N^{-1} \langle R \psi_L^\dagger \psi_R \rangle$.

6. Towards $\mathcal{N} = 1$

The "unwanted" feature of $\mathcal{N} = 2$ theory, making it less similar to QCD than one would desire, is the presence of the adjoint chiral superfields $\mathcal{A}$ and $\mathcal{A}^a$. One can get rid of them making them heavy. To this end we can endow the adjoint superfield with a mass term of the type $\mu \mathcal{A}^2$, through the $\mathcal{N} = 1$ preserving superpotential

$$\mathcal{W} = \frac{\mu}{2}[\mathcal{A}^2 + (\mathcal{A}^a)^2]. \qquad (5)$$

Now, supersymmetry of the bulk model becomes $\mathcal{N} = 1$. At large μ the adjoint fields decouple.

With the deformation superpotential (5) the 1/2 BPS classical flux tube solution stays the same as in the absence of this superpotential.[20] Moreover, the number of the boson and fermion zero modes, which become moduli fields on the string world-sheet does not change either. For the fermion zero modes this statement follows from an index theorem proved in Ref. 21. If the string solution and the number of zero modes remain the same, what can one say about the string world-sheet theory?

The bulk deformation (5) leads to a remarkable, *heterotic* deformation of the CP(1) model on the world-sheet, with $\mathcal{N} = (0,2)$ supersymmetry. The discovery of non-Abelian strings in $\mathcal{N} = 1$ bulk theories is a crucial step on the way to the desired $\mathcal{N} = 0$ theories. Moreover, the heterotically deformed CP(1) model is very rich by itself exhibiting a number of distinct dynamical scenarios unknown previously.

To understand the emergence of $\mathcal{N} = (0,2)$ supersymmetry in the world-sheet Lagrangian recall that $\mathcal{N} = 2$ Yang–Mills theories which support non-Abelian flux tubes have eight supercharges. The flux tube solutions are 1/2 BPS-saturated. Hence, the effective low-energy theory of the moduli fields on the string world-sheet must have four supercharges. The bosonic moduli consist of two groups: two translational moduli $(x_0)_{1,2}$ corresponding to translations in the plane perpendicular to the string axis, and two orientational moduli whose interaction is described by CP(1) (see Fig. 4). The fermion moduli also split in two groups: four supertranslational moduli ζ_L, $\zeta_L^\dagger$, ζ_R, $\zeta_R^\dagger$ plus four superorientational moduli. $\mathcal{N} = 2$ supersymmetry in the bulk and on the world-sheet guarantees that $(x_0)_{1,2}$ and $\zeta_{L,R}$ form a free field theory on the world-sheet completely decoupling from (super)orientational moduli, which in turn form $\mathcal{N} = (2,2)$ supersymmetric CP(1) model.

When one deforms the bulk theory to break $\mathcal{N} = 2$ down to $\mathcal{N} = 1$, one has four supercharges in the bulk and expects two supercharges on the world-sheet. Two out of four supertranslational modes, ζ_R and $\zeta_R^\dagger$, get coupled to two superorientational modes ψ_R and $\psi_R^\dagger$.[23] At the same time, ζ_L and $\zeta_L^\dagger$ remain protected. Thus, the right- and left-moving fermions acquire different interactions; hence, the flux tube becomes heterotic!

This breaks two out of four supercharges on the world-sheet. Edalati and Tong outlined[23] a general structure of the chiral $\mathcal{N} = (0,2)$ generalization of CP(1). Derivation of the heterotic CP(1) model from the bulk theory was carried out in Ref. 22.

7. Heterotic Non-Abelian String

The Lagrangian of the heterotic CP($N-1$) model can be written as[22]

$$L_{\text{heterotic}} = \zeta_R^\dagger i\partial_L \zeta_R + \left[\gamma g_0^2 \zeta_R G_{i\bar{j}}\left(i\partial_L \phi^{\dagger \bar{j}}\right)\psi_R^i + \text{H.c.}\right]$$
$$- g_0^4|\gamma|^2\left(\zeta_R^\dagger \zeta_R\right)\left(G_{i\bar{j}}\psi_L^{\dagger\bar{j}}\psi_L^i\right) + G_{i\bar{j}}\left[\partial_\mu \phi^{\dagger\bar{j}}\partial_\mu \phi^i + i\bar{\psi}^{\bar{j}}\gamma^\mu D_\mu \psi^i\right]$$
$$- \frac{g_0^2}{2}\left(G_{i\bar{j}}\psi_R^{\dagger\bar{j}}\psi_R^i\right)\left(G_{k\bar{m}}\psi_L^{\dagger\bar{m}}\psi_L^k\right)$$
$$+ \frac{g_0^2}{2}\left(1 - 2g_0^2|\gamma|^2\right)\left(G_{i\bar{j}}\psi_R^{\dagger\bar{j}}\psi_L^i\right)\left(G_{k\bar{m}}\psi_L^{\dagger\bar{m}}\psi_R^k\right). \tag{6}$$

The constant γ in Eq. (6) is the parameter which determines the "strength" of the heterotic deformation, and the left-right asymmetry in the fermion sector. It is related to the parameter μ in Eq. (5) (e.g. $\gamma \propto \mu$ at small μ). The third, fourth and fifth lines in Eq. (6) are the same as in the conventional $\mathcal{N} = (2,2)$ CP$(N-1)$ model, except the last coefficient.

Introduction of a seemingly rather insignificant heterotic deformation drastically changes dynamics of the CP(1) model, leading to spontaneous supersymmetry breaking. At small μ (small γ) the field ζ_R represents a massless Goldstino, with the residue $\langle R\psi_R^\dagger \psi_L \rangle$. As well known, a nonvanishing bifermion condensate $\langle R\psi_R^\dagger \psi_L \rangle$ develops in the undeformed model. Thus, the vacuum energy

$$\mathcal{E}_{\text{vac}} = |\gamma|^2 \left| \langle R\psi_R^\dagger \psi_L \rangle \right|^2 \neq 0 \,. \tag{7}$$

Therefore, upersymmetry is spontaneously broken. A nonvanishing $\mathcal{E}_{\text{vac}}$ for arbitrary values of γ in heterotically deformed CP$(N-1)$ models was obtained in[24] from the large-N expansion. Spontaneous breaking of SUSY in heterotic CP$(N-1)$ was anticipated in Ref. 25.

8. Large-N Solution of the Heterotic CP$(N-1)$ Model

To reveal a rich dynamical structure of the heterotically deformed CP$(N-1)$ models it is instructive to add twisted masses which correspond to $\Delta m \neq 0$ introduced above. Moreover, the most convenient choice of the twisted masses is that preserving the Z_N symmetry of the model which exists at $\Delta m = 0$,

$$m_k = m \exp\left(i\frac{2\pi k}{N} \right), \qquad k = 0, 1, 2, \ldots, N-1 \,, \tag{8}$$

where m is a complex parameter setting the scale of the twisted masses. For simplicity I will take it real. Now we have two variable parameters, m and γ, the strength of the heterotic deformation. The breaking vs nonbreaking of the above Z_N determines the phase diagram. This model can be solved at large N using the $1/N$ expansion.[24] I will present here just two plots exhibiting main features of the solution.

Figure 7 displays three distinct regimes and two phase transition lines. Two phases with the spontaneously broken Z_N on the left and on the right are separated by a phase with unbroken Z_N. This latter phase is characterized by a unique vacuum and confinement of all U(1) charged fields ("quarks"). In the broken phases (one of them is at strong coupling) there are N degenerate vacua and no confinement.

Figure 8 shows the vacuum energy density at a fixed value of γ. It demonstrates that supersymmetry is spontaneously broken everywhere except a circle $|m| = \Lambda$ in the Z_N-unbroken phase. The first phase transition occurs at strong coupling (small

69

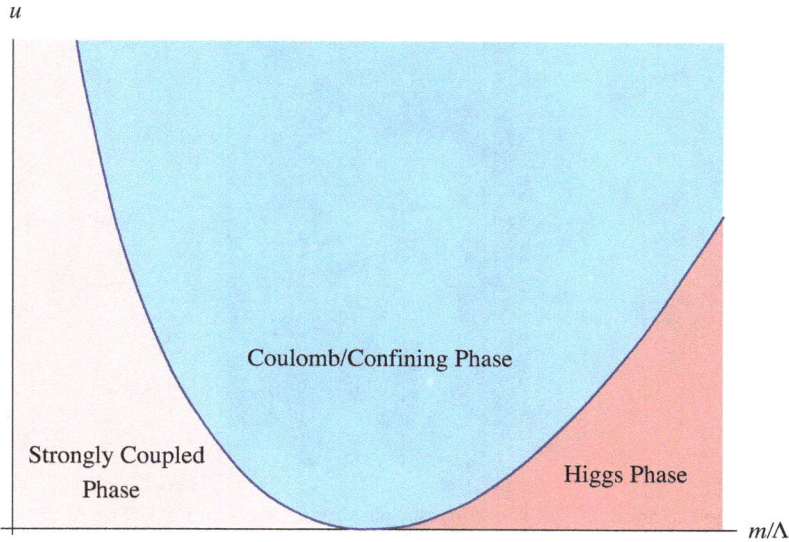

Fig. 7. The phase diagram of the twisted-mass deformed heterotic $CP(N-1)$ theory. The parameter u denotes the amount of deformation and is related to γ.

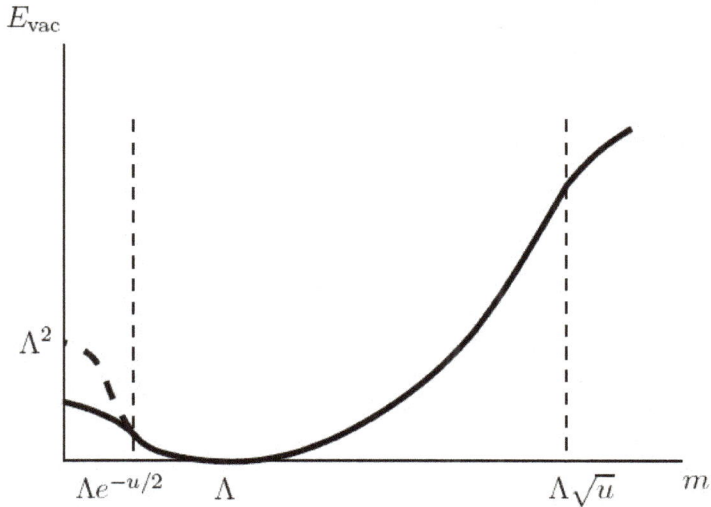

Fig. 8. Vacuum energy density vs m. The dashed line shows an unstable quasivacuum.

$|m|$) while the second phase transition is at weak coupling (large $|m|$). Both phase transitions between the three distinct phases are of the second kind.

One must be able to translate this rich world-sheet-dynamics into (presumably) highly nontrivial statements regarding the bulk theory at strong coupling.

Fig. 9. Everybody knows that this is Murray Gell-Mann. (©Barry Blitt, The Atlantic Monthly, July 2000.)

9. Instead of Conclusion

The progress in understanding dynamics of non-Abelian theories at strong coupling was painfully slow. But what a progress it is! To properly appreciate the scale of achievements, please, look back in the 1970's and compare what was known then about strong interactions to what we know now. Just open old reviews or textbooks devoted to this subject, in parallel with fresh publications. Of course, a pessimist might say that the full analytical theory is still elusive. Will it ever be created? And what does it mean, "the full analytical theory," in the case when we are at strong coupling? The richness of the hadronic world is enormous. Unlike QED we will never be able to analytically calculate all physical observables with arbitrary precision. But do we really need this? To my mind, what is really needed is the completion of the overall qualitative picture of confinement in non-supersymmetric theories, supplemented by a variety of approximate quantitative tools custom-designed to treat particular applications. A large number of such tools are already available.

Acknowledgments

I am very grateful to A. Yung with whom I shared the pleasure of working on the issues discussed in this talk. Generous assistance of Andrey Feldshteyn with cartoons is gratefully acknowledged. This write-up was completed during my stay at the Institut de Physique Théorique, CEA-Saclay, which was made possible due to the support of the Chaires Internacionales de Recherche Blaise Pascal, Fondation de l'Ecole Normale Supérieure. This work is supported in part by the DOE grant DE-FG02-94ER408.

Appendix. Snapshots Made by HEP World Traveller

1. ASTRO / COSMO

Fig. 10. Astroparticle physics and cosmology.

2. MATH. PHYSICS

Fig. 11. Mathematical physics absorbing string theory.

3. STRINGS / BRANES

Fig. 12. The theory of strings and branes.

4

Fig. 13. Supersymmetry-based phenomenology.

5. STRONG COUPLING

Fig. 14. QCD and other gauge theories at strong coupling: from confinement to chiral symmetry breaking and back via supersymmetry.

6. CONFINEMENT

Fig. 15. Deciphering mechanisms of confinement.

72

7. MSSM and LED

Fig. 16. Searching for supersymmetry and/or large extra dimensions "under a lamp post."

7. MSSM and LED

Fig. 17. The search is continued in a different way.

8. EXOTICS (NOTHING SACRED)

Fig. 18. All things "beyond" are highly appreciated. If we only new what exactly to look for . . . .

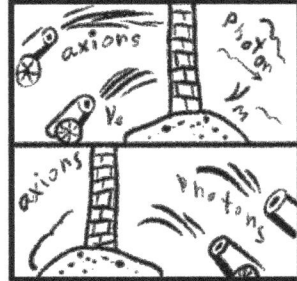

9. PHOTON / AXIONS

Fig. 19. Axion physics.

10. QUANTUM GRAVITY DISSENTERS

Fig. 20. The ever-lasting passion for quantum gravity.

References

1. D. J. Gross and F. Wilczek, *Phys. Rev. Lett.* **30**, 1343 (1973).
2. H. D. Politzer, *Phys. Rev. Lett.* **30**, 1346 (1973).
3. A. Abrikosov, *Sov. Phys. JETP* **32**, 1442 (1957) [Reprinted in *Solitons and Particles*, eds. C. Rebbi and G. Soliani (World Scientific, Singapore, 1984), p. 356].
4. H. Nielsen and P. Olesen, *Nucl. Phys. B* **61**, 45 (1973) [Reprinted in *Solitons and Particles*, eds. C. Rebbi and G. Soliani (World Scientific, Singapore, 1984), p. 365].
5. Y. Nambu, *Phys. Rev. D* **10**, 4262 (1974).
6. G. 't Hooft, Gauge theories with unified weak, electromagnetic and strong interactions, in *Proc. of the E.P.S. Int. Conf. on High Energy Physics*, Palermo, 23–28 June, 1975, ed. A. Zichichi (Editrice Compositori, Bologna, 1976).
7. S. Mandelstam, *Phys. Rep.* **23**, 245 (1976).
8. N. Seiberg and E. Witten, *Nucl. Phys. B* **426**, 19 (1994) [Erratum: *ibid.* **430**, 485 (1994)], arXiv:hep-th/9407087.
9. N. Seiberg and E. Witten, *Nucl. Phys. B* **431**, 484 (1994), arXiv:hep-th/9408099.
10. G. 't Hooft, *Nucl. Phys. B* **79**, 276 (1974).
11. A. M. Polyakov, *JETP Lett.* **20**, 194 (1974).
12. M. R. Douglas and S. H. Shenker, *Nucl. Phys. B* **447**, 271 (1995), arXiv:hep-th/9503163.
13. A. Hanany, M. J. Strassler and A. Zaffaroni, *Nucl. Phys. B* **513**, 87 (1998), arXiv:hep-th/9707244.
14. A. Hanany and D. Tong, *J. High Energy Phys.* **0307**, 037 (2003), arXiv:hep-th/0306150.
15. R. Auzzi, S. Bolognesi, J. Evslin, K. Konishi and A. Yung, *Nucl. Phys. B* **673**, 187 (2003), arXiv:hep-th/0307287.
16. G. R. Dvali and M. A. Shifman, *Phys. Lett. B* **396**, 64 (1997) [Erratum: *ibid.* **407**, 452 (1997)], arXiv:hep-th/9612128.
17. M. Shifman and A. Yung, *Phys. Rev. D* **67**, 125007 (2003), arXiv:hep-th/0212293.
18. M. Shifman and A. Yung, *Phys. Rev. D* **70**, 045004 (2004), arXiv:hep-th/0403149.
19. A. Hanany and D. Tong, *J. High Energy Phys.* **0404**, 066 (2004), arXiv:hep-th/0403158.
20. M. Shifman and A. Yung, *Phys. Rev. D* **72**, 085017 (2005), arXiv:hep-th/0501211.
21. A. Gorsky, M. Shifman and A. Yung, *Phys. Rev. D* **75**, 065032 (2007), arXiv:hep-th/0701040.
22. M. Shifman and A. Yung, *Phys. Rev. D* **77**, 125016 (2008) [Erratum: *ibid.* **79**, 049901 (2009)], arXiv:0803.0158.
23. M. Edalati and D. Tong, *J. High Energy Phys.* **0705**, 005 (2007), arXiv:hep-th/0703045.
24. M. Shifman and A. Yung, *Phys. Rev. D* **77**, 125017 (2008) [Erratum: *ibid.* **81**, 089906 (2010)], arXiv:0803.0698.
25. D. Tong, *J. High Energy Phys.* **0709**, 022 (2007), arXiv:hep-th/0703235.

QCD — MASS AND GAUGE IN A FIELD THEORY.
QCD GLUE MESONS

PETER MINKOWSKI

Albert Einstein Center for Fundamental Physics - ITP,
University of Bern, Germany

Honoring Murray Gell-Manns 80th birthday, Singapore
24.-26. February 2010

1 Meeting you, MGM , while a student at ETH-Zurich

It was a 'strange' feeling and the article(s) I found reading Physical Review papers, with your name as author(s) – Murray Gell-Mann – promised something I should learn more about.

PHYSICAL REVIEW   FEBRUARY 1, 1962

VOLUME 125, NUMBER 3

Symmetries of Baryons and Mesons*

MURRAY GELL-MANN
California Institute of Technology, Pasadena, California
(Received March 27, 1961; revised manuscript received September 20, 1961)

Fig 1 : from reference [1-1961]

It was an intense struggle to eliminate alternatives through the crossroad

- within a renormalizable local field theory base fields starting from quarks must explicitly display all quantum numbers, also color, in a canonical way accompanied by gauge fields, with associated quantum numbers

I would like to present some aspects of gluonic mesons here.

1-1   QCD – the two central anomalies : scale and chiral U1

We face the theoretical abstraction of QCD with $N_{fl} = 6$, representing strong interactions – adaptable to two or three light flavors (u , d , s) of quarks and antiquarks. $\leftrightarrow$

quarks   :   color is counted in $\pi^0 \to \gamma\gamma$ and

$$R = \frac{\sigma\,(e^+e^- \to \text{hadrons})}{\sigma\,(e^+e^- \to \mu^+\mu^-)}$$

spin and flavor are clearly seen
in $q\bar{q}$ and $3q$, $3\bar{q}$ spectroscopy.

$$\mathcal{L} = \left[\bar{q}^{\,c'}_{\dot{\mathcal{B}}\,f} \left\{ \begin{array}{c} \frac{i}{2}\,\overset{\rightarrow}{\partial}_\mu\,\delta_{c'c} \\ -\,v^A_\mu\,\left(\frac{1}{2}\lambda^A\right)_{c'\dot{c}} \\ -\,m_f\,\bar{q}^{\,\dot{c}}_{\dot{\mathcal{A}}\,f}\,q^{\,c}_{\mathcal{A}\,f} \end{array} \right\} \gamma^\mu_{\dot{\mathcal{B}}\mathcal{A}}\,q^{\,c}_{\mathcal{A}\,f} \right] - \frac{1}{4\,g^2}\,F^{\mu\nu\,A}\,F^A_{\mu\nu} + \Delta\mathcal{L} \tag{1}$$

quarks : c' , $c = 1,2,3$ color , $f = 1,\cdots,6$ flavor

$\mathcal{B},\mathcal{A} = 1,\cdots,4$ spin , m_f mass

gauge bosons :

$$F^A_{\mu\nu} = \partial_\nu\,v^A_\mu - \partial_\mu\,v^A_\nu - f_{ABC}\,v^B_\nu\,v^C_\mu$$

$$A,B,C = 1,\cdots,dim\,(\,G = SU3_c\,) = 8$$

Lie algebra labels, $\left[\frac{1}{2}\,\lambda^A\,,\,\frac{1}{2}\,\lambda^B\right] = if_{ABC}\,\frac{1}{2}\,\lambda^C$

$$\tag{2}$$

perturbative rescaling :

$$v^A_\mu = g\,v^A_{\mu\,pert}\,,\,F^A_{\mu\nu} = g\,F^A_{\mu\nu\,pert}$$

Degrees of freedom are seen in jets, in (e.g.) the energy momentum sum rule in deep inelastic scattering but not clearly in spectroscopy.
Completing $\Delta\mathcal{L}$ in Fermi gauges

$$\Delta\mathcal{L} = \left\{ \begin{array}{c} -\frac{1}{2\,\eta\,g^2}\,\left(\partial_\mu\,v^{\mu\,A}\right)^2 \\ +\,\partial^\mu\,\bar{c}^A\,(\,D_\mu\,c\,)^A \end{array} \right\} \;;\; \eta \,:\, \text{gauge parameter}$$

ghost fermion fields : c , $\bar{c}$; $(\,D_\mu\,c\,)^A = \partial_\mu\,c^A - f_{ABD}\,v^B_\mu\,c^D$

gauge fixing constraint : $C^A = \partial_\mu\,v^{\mu\,A}$

$$\tag{3}$$

Gauge boson binary bilocal and adjoint string operators

One goal is, to identify – not just some candidate resonance – gluonic mesons, binary and higher modes, and to relate them to the base quantities within QCD.

$$B_{[\mu_1 \nu_1],[\mu_2 \nu_2]}(x_1, x_2) =$$
$$= F_{[\mu_1 \nu_1]}(x_1 ; A) U(x_1, A ; x_2, B) F_{[\mu_2 \nu_2]}(x_2 ; B) \qquad (4)$$
$$A, B, \cdots = 1, \cdots, 8 ; \text{ no flavor but spin}$$

$F_{[\mu \nu]}(x ; A)$ denote the color octet of field strengths.
The quantity $U(x, A ; y, B)$ in eq. (4) denotes the octet string operator, i. e. the path ordered exponential over a straight line path $\mathcal{C}$ from y to x

$$U(x, A ; y, B) = P \exp \left(\int_y^x \bigg|_{\mathcal{C}} dz^\mu \frac{1}{i} v_\mu(z, D) \mathcal{F}_D \right)_{AB}$$
$$(\mathcal{F}_D)_{AB} = i f_{ADB}$$
$$(5)$$

with the local limit

$$B_{[\mu_1 \nu_1],[\mu_2 \nu_2]}(x_1 = x_2 = x) =$$
$$= (:) F_{[\mu_1 \nu_1]}^A(x) F_{[\mu_2 \nu_2]}^A(x)(:) \qquad (6)$$
$$\text{no flavor but spin}$$

The same procedure involving a triplet string applies to $\bar{q} q$ bilinears

$$B^q_{[\dot{A} f_1, \mathcal{B} f_2]}(x_1, x_2) =$$
$$= \bar{q}^{\dot{c}_1}_{\mathcal{B} f_2}(x_1) U(x_1, c_1 ; x_2, \dot{c}_2) q^c_{\mathcal{A} f_1}(x_2)$$
$$U(x_1, c_1 ; x_2, \dot{c}_2) =$$
$$\qquad (7)$$
$$= P \exp \left(\int_y^x \bigg|_{\mathcal{C}} dz^\mu \frac{1}{i} v_\mu(z, D) \frac{1}{2} \lambda_D \right)_{c_1 \dot{c}_2}$$

flavor *and* spin
with the local limit

$$B^q_{[\dot{\mathcal{B}} f_2, \mathcal{A} f_1]}(x_1 = x_2 = x) = (:) \bar{q}^{\dot{c}}_{\mathcal{B} f_2}(x) q^c_{\mathcal{A} f_1}(x)(:) \qquad (8)$$

The symbols (:) in eqs. 6 and 8 should indicate that normal ordering of regulating the local limits is required and further that such normal ordering is *not* unique , and dependent on quark masses in the case of the $\bar{q} q$ bilinears.
The U1-axial central anomaly involves the local chiral current projections from $B^q_{[\dot{\mathcal{B}} f_2, \mathcal{A} f_1]}(x)$ in eq. 8

$$(j^\pm_\mu)_{f_2 f_1}(x) = B^q_{[\dot{\mathcal{B}} f_2, \mathcal{A} f_1]}(x) (\gamma_\mu \frac{1}{2}(\P \pm \gamma_5))_{\mathcal{B} \dot{\mathcal{A}}}$$
$$= (:) \bar{q}^{\dot{c}}_{f_2} \gamma^\pm_\mu q^c_{f_1}(x)(:) \qquad (9)$$
$$\gamma_5 = \gamma_{5R} = \frac{1}{i} \gamma_0 \gamma_1 \gamma_2 \gamma_3 ; \quad \gamma^\pm_\mu = \gamma_\mu \frac{1}{2}(\P \pm \gamma_5)$$

The equations of motion for the fermion fields are *and superficially imply* (upon $f_1 \leftrightarrow f_2$)

$$\not{p}\, q^{c}_{f_2} = \tfrac{1}{i} \left(\not{p}^{\,c\,\dot{c}'} + \delta^{c\,\dot{c}'} m_{f_2} \right) q^{c'}_{f_2}$$

$$\bar{q}^{\dot{c}}_{f_1}\, \overleftarrow{\not{p}} = \bar{q}^{\dot{c}'}_{f_1} \tfrac{1}{i} \left(-\not{p}^{\,c'\,\dot{c}} - \delta^{c'\,\dot{c}} m_{f_1} \right)$$

$$\partial^{\mu} \left(j^{\pm}_{\mu} \right)_{f_1 f_2} = \tfrac{1}{2i} \begin{pmatrix} (m_{f_2} - m_{f_1})\, S_{f_1 f_2}\, \mp \\ \mp\, (m_{f_2} + m_{f_1})\, P_{f_1 f_2} \end{pmatrix} \qquad (10)$$

$$S_{f_1 f_2} = (:) \bar{q}^{\dot{c}}_{f_1}\, q^{c}_{f_2}\, (:)$$

$$P_{f_1 f_2} = (:) \bar{q}^{\dot{c}}_{f_1}\, \gamma_5\, q^{c}_{f_2}\, (:)$$

; no sums over f_1 , f_2

In eq. 10 m_f denotes the *real*, *nonnegative* quark mass for flavor f.
From eq. 10 the relations for vector and axial vector currents *superficially* follow

$$(j_{\mu})_{f_1 f_2} = (j^{+}_{\mu})_{f_1 f_2} + (j^{-}_{\mu})_{f_1 f_2}$$

$$(j^{5}_{\mu})_{f_1 f_2} = (j^{+}_{\mu})_{f_1 f_2} - (j^{-}_{\mu})_{f_1 f_2}$$

$$\partial^{\mu} (j_{\mu})_{f_1 f_2} = \tfrac{1}{i} (m_{f_2} - m_{f_1})\, S_{f_1 f_2} \qquad (11)$$

$$\partial^{\mu} (j^{5}_{\mu})_{f_1 f_2} = (m_{f_2} + m_{f_1})\, i\, P_{f_1 f_2}$$

As it follows from the original derivation by Adler and Bell and Jackiw [2-1969] in QED, the vecor current Ward identities in eq. 11 can be implemented also in QCD , leaving the axial current ones reduced to the flavor non-singlet case , leaving the U1 axial current divergent anomalous

$$\partial^{\mu} (j_{\mu})_{f_1 f_2} = \tfrac{1}{i} (m_{f_2} - m_{f_1})\, S_{f_1 f_2} \quad \checkmark$$

$$\begin{Bmatrix} j^{5}_{\mu} \\ P \end{Bmatrix}^{NS}_{f_1 f_2} = \begin{Bmatrix} j^{5}_{\mu} \\ P \end{Bmatrix}_{f_1 f_2} - \tfrac{1}{N_{fl}} \delta_{f_1 f_2} \sum_{f} \begin{Bmatrix} j^{5}_{\mu} \\ P \end{Bmatrix}_{f\,f} \qquad (12)$$

and similarly

$$\begin{Bmatrix} j^{5}_{\mu} \\ P \end{Bmatrix}^{S}_{f_1 f_2} = \sum_{f} \begin{Bmatrix} j^{5}_{\mu} \\ P \end{Bmatrix}_{f\,f} \qquad (13)$$

Quark masses and splittings : m_f **and** $\Delta m_f = m_f - \langle m \rangle$

In the subtitle above $\langle m \rangle$ stands for the mean quark mass

$$\langle m \rangle = \tfrac{1}{N_{fl}} \sum_{f} m_f \qquad (14)$$

The identities for vector currents in eqs. 11 and 12 can be extended separating the contributions proportional to Δm_f and $\langle m \rangle$

$$\partial^\mu (j_\mu)_{f_1 f_2} = \tfrac{1}{i} (\Delta m_{f_2} - \Delta m_{f_1}) S_{f_1 f_2} \checkmark$$
$$\partial^\mu (j_\mu^5)_{f_1 f_2}^{NS} = (\Delta m_{f_2} + \Delta m_{f_1}) i P_{f_1 f_2}^{NS} \checkmark$$
$$\partial^\mu (j_\mu^5)_{f_1 f_2}^{S} = 2 \langle m \rangle i P^S \not\checkmark [\longrightarrow + \delta_5] \qquad (15)$$
$$\delta_5 = (2 N_{fl}) \tfrac{1}{32\pi^2} F_{\mu\nu}^A \widetilde{F}^{\mu\nu A} \Big|_{\rightarrow ren.gr.inv}$$
$$\widetilde{F}_{\mu\nu}^A = \tfrac{1}{2} \varepsilon_{\mu\nu\sigma\tau} F^{\mu\nu A}$$

1

The singlet axial current anomaly

We shall return to the question of how the local operator
$ch_2 (F) \equiv \tfrac{1}{32\pi^2} (:) F_{\mu\nu}^A \widetilde{F}^{\mu\nu A} (:)$ is to be normalized and rendered renormalization group invariant [3-1991] . Here we just assume this to have been achieved and denote the U1-axial anomaly, the first of the central two, in its general form
(eq. 15)

$$\left\{ \partial^\mu (j_\mu^5)^S = 2 \langle m \rangle i P^S + \delta_5 \right\} (x)$$
$$\delta_5 = (2 N_{fl}) \tfrac{1}{32\pi^2} (:) F_{\mu\nu}^A \widetilde{F}^{\mu\nu A} (:) \Big|_{\rightarrow ren.gr.inv} \qquad (16)$$

From here it is conceptually clear how the scale- (or trace-) anomaly arises but strictly within QCD . The renormalizability of a field theory in the limit of uncurved space-time gives rise to a local , symmetric and *conserved* energy momentum tensor , implying exact Poincaré invariance

$$\left\{ \vartheta_{\mu\nu} = \vartheta_{\nu\mu} \right\} (x)$$
$$\partial^\nu \vartheta_{\mu\nu} = 0 \qquad (17)$$

In connection with the normal ordering questions it is important to admit in the precise form of the energy momentum tensor a nontrivial vacuum expected value , which
in view of exact Poincaré invariance must be of the form

$$\langle \Omega | \vartheta_{\mu\nu} (x) | \Omega \rangle = \tfrac{1}{4} \eta_{\mu\nu} \tau$$
$$\left\{ \begin{array}{c} \eta_{\mu\nu} = diag (1, -1, -1, -1) \\ \tau \end{array} \right\} \text{ independent of x} \longrightarrow$$

$$\Delta \vartheta_{\mu\nu} (x) = \vartheta_{\mu\nu} (x) - \langle \Omega | \vartheta_{\mu\nu} (x) | \Omega \rangle \times \left\{ \begin{array}{c} \widehat{\P} \\ \text{or } | \Omega \rangle \langle \Omega | \end{array} \right. \qquad (18)$$

with $\partial^\nu \Delta \vartheta_{\mu\nu} (x) = 0$; $\langle \Omega | \Delta \vartheta_{\mu\nu} (x) | \Omega \rangle = 0$

---

[1] δ_5 was – as far as I know from Yair Zarmi and Anthony Hay – introduced by Murray Gell-Mann in lectures $\sim$ 1970 in Hawaii .

In eq. 18 $\mathbf{1}$ denotes the unit operator in the entire Hilbert space of states , while $P_\Omega = | \Omega \rangle \langle \Omega |$ stands for the projector on the ground state .

Furthermore from the two local , *conserved* tensors in eq. 18 only $\Delta \vartheta_{\mu \nu} (x)$ with vanishing vacuum expected value is acceptable as representing the conserved 4 momentum *operators* in the integral form

$$\widehat{P}_\mu = \int_t d^3 x \, \Delta \vartheta_{\mu \, 0} (t , \vec{x}) \tag{19}$$

All these arguments *notwithstanding* to subtract any eventual vacuum expected values of local operators , often put forward as mathematical prerequisites , it is wise *not to do so* in the presence of spontaneous parameters , the dynamical origin of spontaneous symmetry breaking, e.g. chiral symmetries in the limit or neighbourhood of some $m_f \to 0$.

Using the (classical) equations of motion pertaining to the Lagrangean in eqs. 1 - 3

$$(D_\nu F^{\mu\nu})^A = j^{\mu \, A} (\bar{q} , q) ; \quad F \to F_{pert}$$
$$(D_\varrho F^{\mu \nu})^A = \partial_\varrho F^{\mu \nu \, A} - f_{ABD} \, v_\varrho^B \, F^{\mu \nu \, D}$$
$$j_\mu^A (\bar{q} , q) = g \, \bar{q}_{A \, f}^{\dot{c}} \, (\gamma_\mu)_{A B} \, \tfrac{1}{2} \, (\lambda^A)_{c \dot{c}'} \, q_{A \, f}^{c'} \tag{20}$$
$$i \, (\gamma^\mu D_\mu q)_{A \, f}^c = m_f \, q_{A \, f}^c \; \text{and} \; q \to \bar{q} .$$
$$(D_\mu q)_{A \, f}^c = [\, \partial_\mu \, \delta_{c \dot{c}'} + i \, g \, v_\mu^D \, \tfrac{1}{2} \, (\lambda^D)_{c \dot{c}'} \,] \, q_{A \, f}^{c'}$$

the associated form of the energy momentum becomes

$$\vartheta_{\mu \nu}^{(cl)} = \begin{bmatrix} F_{\mu \varrho}^A \, F_\nu^{\varrho \, A} - \tfrac{1}{4} \, \eta_{\mu \nu} \, F_{\sigma \varrho}^A \, F^{\varrho \sigma \, A} + \\ + \tfrac{1}{2} \, \left[\, \bar{q}_f \, \gamma_\mu \, \tfrac{i}{2} \, \overset{\leftrightarrow}{D}_\nu \, q_f + \mu \leftrightarrow \nu \, \right] \end{bmatrix} \tag{21}$$

and using once more the fermion part of the equations of motion the trace of the classical energy momentum tensor becomes [2]

$$\vartheta^\mu{}_\mu^{(cl)} = \sum_f m_f \, S_{f \, f} ; \quad S_{f_1 f_2} = (:) \, \bar{q}_{f_1}^{\dot{c}} \, q_{f_2}^c \, (:) \tag{22}$$

The scale- or trace- anomaly

From the classical soft fermionic contribution to the trace of the energy momentum tensor there is a clear conjecture , also by Murray Gell-Mann , of the anomalous contribution , which subsequently became the scale- or trace- anomaly within QCD

$$\vartheta^\mu{}_\mu = \sum_f m_f \, S_{f \, f} + \delta_0$$
$$\mathcal{E}_0 = - \left(-2 \, \beta (g) / g^3 \right) \left[\tfrac{1}{4} \, (:) \, F_{\mu \nu}^A \, F^{\mu \nu \, A} \, (:) \right]_{\to \, ren.gr.inv} \tag{23}$$

The two central anomalies alongside : scale- or trace- and U1-axial anomaly

---

[2] The classical energy momentum (density-) tensor can be obtained from the coupling to an external gravitational field .

We collect the two anomalous identities in eqs. 23 and 16

$$\left\{ \vartheta^{\mu}_{\ \mu} \quad = \sum_f m_f \, S_{\bar{f}f} + \delta_0 \right\} (x)$$

$$\left\{ \partial^{\mu} \left(j^5_{\ \mu} \right)^S = 2 \langle m \rangle \, i \, P^S + \delta_5 \right\} (x)$$

$$\delta_0 = - \left(-2\,\beta(g)/g^3 \right) \left[\tfrac{1}{4} (:) F^A_{\ \mu\nu} F^{\mu\nu A} (:) \right]_{\to\ ren.gr.inv}$$

$$\delta_5 = \quad (2\,N_{fl}) \, \tfrac{1}{8\pi^2} \left[\tfrac{1}{4} (:) F^A_{\ \mu\nu} \widetilde{F}^{\mu\nu A} (:) \right]_{\to ren.gr.inv} \tag{24}$$

$$-\beta/g^3 = \tfrac{1}{16\pi^2}\, b_0 + O(Y) \; ; \; Y = g^2/(16\,\pi^2$$

$$\beta(g) \quad : \quad \text{Callan-Symanzik rescaling function in QCD}$$

The qualification 'central' for the anomalies in eq. 24 stands for the property that in rendering the square coupling constant and the associated $\vartheta-$ parameter in the gauge boson *renormalized* Lagrangean density x dependent

$$\mathcal{L}_{g.b.} = -\tfrac{1}{g^2} \tfrac{1}{4} (:) F^A_{\ \mu\nu} F^{\mu\nu A} (:) + \vartheta \tfrac{1}{8\pi^2} \tfrac{1}{4} (:) F^A_{\ \mu\nu} \widetilde{F}^{\mu\nu A}$$

$$g^2 \to g^2(x) \; ; \; \vartheta \to \vartheta(x) \tag{25}$$

maintains perturbative renormalizability and acts together with suitable boundary conditions as external sources for the scalar and pseudoscalar local field strength bilinears, rendered renormalization group invariant by absorbing the g dependent part of the rescaling function $\left(\beta(g)/g^3 \right)/\left(\beta(g')/g'^3 \right)_{g'\to 0}$ into the field strength bilinear and likewise normalizing the second Chern character at zero distance $\leftrightarrow \mu = \infty$. The field strength bilinears in brackets without subscripts denote the renormalization group invariant operators.

$$\left[\tfrac{1}{4} (:) F^A_{\ \mu\nu} F^{\mu\nu A} (:) \right] = \left. \frac{\beta(g)/g^3}{\beta(g')/g'^3} \right|_{g'\to 0} \left[\tfrac{1}{4} (:) F^A_{\ \mu\nu} F^{\mu\nu A} (:) \right]_g \tag{26}$$

$$\left[\tfrac{1}{4} (:) F^A_{\ \mu\nu} \widetilde{F}^{\mu\nu A} (:) \right] \; : \; \text{normalized at zero distance}$$

With these renormalizations the central anomalies (eq. 24) take the form [3]

$$\left\{ \vartheta^{\mu}_{\ \mu} \quad = \sum_f m_f \, S_{\bar{f}f} + \delta_0 \right\} (x)$$

$$\left\{ \partial^{\mu} \left(j^5_{\ \mu} \right)^S = 2 \langle m \rangle \, i \, P^S + \delta_5 \right\} (x) \tag{27}$$

$$\delta_0 = \quad - b_0 \, \tfrac{1}{8\pi^2} \quad \left[\tfrac{1}{4} (:) F^A_{\ \mu\nu} F^{\mu\nu A} (:) \right] \; b_0 =$$

$$\delta_5 = (2\,N_{fl}) \, \tfrac{1}{8\pi^2} \left[\tfrac{1}{4} (:) F^A_{\ \mu\nu} \widetilde{F}^{\mu\nu A} (:) \right] \tfrac{1}{3} (33 - 2\,N_{fl})$$

2   Resonances composed of gauge bosons – glueballs here restricting to mainly binaries with $J^{PC} = 0^{++}$, 0^{-+} , 2^{++} [4]

---

[3]   Credit is due to H. Kluberg-Stern and J. B. Zuber segregating gauge variant and invariant twist 4 operators [4-1975] .

[4]   These lowest states we called "the three musketeers" with Wolfgang Ochs .

Color neutral gauge boson binaries can be described by a wave function in a similar way as photon binaries, as originally derived by L.D Landau and C. N. Yang [5-1948] , [6-1950] . We choose a common time and at that time fix gauge potentials and transform to the c.m. frame and spacelike relative momenta

$$v_0^A = 0 \ , \ \partial_m v^{m\,A}(t;\vec{x}) = 0 \ m = 1,2,3 \ \longrightarrow$$
$$\Psi_{mn}^{AB}(t;\vec{k}) = \delta^{AB} \exp(-iMt)\,\psi_{mn}(\vec{k}) \ ; \ k_m \psi_{mn} = k_m \psi_{nm} = 0 \quad (28)$$
$$\psi_{mn}(\vec{k}) = \psi_{nm}(-\vec{k}) \ \text{Bose symmetry}$$

In the wake of excitement after the discovery of the J/ψ $c\bar{c}$ vector-meson in November 1974 [7-1974] , [8-1974] we took a new look at *pure* gluemesons with Harald Fritzsch [9-1975].

d) Phenomenology of Glue Mesons

Thus far no glue mesons have been identified in the hadronic spectrum, although there are several candidates of I ~ 0 -mesons which may consist predominantly of glue. Among those are the broad ϵ-enhancement in the 0^+-wave at ~ 600 MeV, and the E(1420)-meson. Both may be two-gluon resonances.

Fig 2 : Quote from ref. [9-1975]

There exist 3 series with respect to total angular momentum, all with C=+

series	helicity transfer	P	J	candidates
1	0	+	0 ,2,4 $\cdots$	$\rightarrow$ discussed below
2	0	-	0 ,2,4 $\cdots$	η (1405)
3	± 2	+	2 ,3,4 $\cdots$	f2 (1640) f2 (1810) $\cdots$ f2 (2150)

Table 1

We list the local gauge boson bilinear operators, which can connect the ground state to the lowest spin states in the three series in Table 1

$$\langle \Omega | \left[\tfrac{1}{4} (:) F_{\mu\nu}^A F^{\mu\nu\,A} (:) \right] | 0^{++}, p \rangle = f_+ m^2 (0^{++})$$
$$\langle \Omega | \left[\tfrac{1}{4} (:) F_{\mu\nu}^A \widetilde{F}^{\mu\nu\,A} (:) \right] | 0^{-+}, p \rangle = f_- m^2 (0^{-+})$$
$$\langle \Omega | \vartheta_{\mu\nu} (FF) | 2^{++}, p, \varepsilon \rangle = f_2 m^2 (2^{++}) \varepsilon_{\mu\nu}(S_z) \quad (29)$$
$$\varepsilon_{\mu\nu} = \varepsilon_{\nu\mu} \ , \ \varepsilon_\mu^\mu = 0 \ , \ p^\mu \varepsilon_{\mu\nu} = 0 \ , \ \varepsilon^{\mu\nu}\varepsilon_{\mu\nu} = 1 \ , \ p^2 = m^2$$

3   A meson composed of glue among scalar resonances – an unresolved question
partial aspects in elastic $\pi\pi$ scattering and $p(\pi\pi)p$ central production

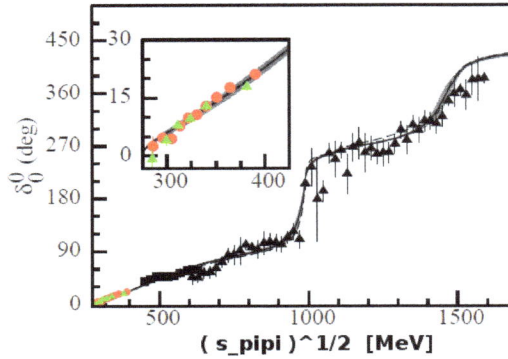

Fig 3 : $\pi\pi$; $I = 0$ phase shifts from [11-2008]

The data points near $\pi\pi$ threshold are the result of the precision measurement *and analysis* of the Na48/2 collaboration [12-2008] (red points) without applying a correction for isospin violations illustrated in figure 4 taken from ref. [12-2008] below and yielding excellent agreement with the determination of the isospin exact extrapolation of the pi pi scattering length by Gilberto Colangelo, Jürg Gasser and Heinrich Leutwyler [13-2001] .

Fig. 3 (from Oller et al. [11-2008]) shall illustrate that the new and precise data from Na48/2 does not pose difficulties to an *interpolation* to the phase shift analysis of the CERN-MUNICH collaboration, as documented in Fig. 5 below.

$$a_0 = 0.220 \pm 0.005_{\text{theo}}$$
$$a_2 = -0.0444 \pm 0.0010_{\text{theo}}$$

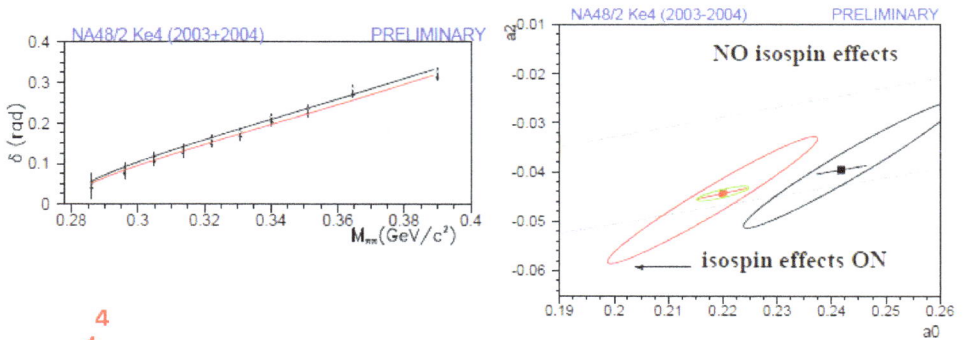

Figure 3: Left: Phase shift (δ) measurements without mass effects (top black line) and with mass effects included (bottom red line). In each case the line corresponds to the 2-parameter fit. Right: Fits of the NA48/2 K_{e4} data in the (a_0, a_2) plane without (black) and with (red) isospin mass effects. The symbols are the result of the one-parameter fit imposing the ChPT constraint. The small (green) ellipse corresponds to the most accurate prediction from ChPT.

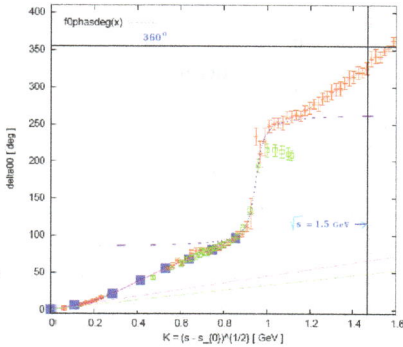

Fig. 5 : The $\pi\pi$, $I = 0$
ideal **elastic s-wave**
from threshold to
$\sim$ 1.625 GeV.

Fig. 5 plots the s-wave phase shifts versus $K = \left(M_{\pi\pi}^2 - 4\,m_\pi^2\right)^{1/2}$.

$\blacksquare$: from ref. [13-2001] :
Colangelo, Gasser and Leutwyler ,
$-$ interpolates $\blacksquare$, $+$: from
ref. [14-2007] : *Na48/2 coll.*
corrected for isospin breaking
$\mathbb{I}$: from ref. [15-1973] :
Protopopescu et al.
$\mathbb{I}$: from ref. [16-1973] *CERN-*
-Munich coll. ; W. Ochs : thesis 1973 .
$--$ $^{--}$: minimal meromorphic parametrization of the
influence of f0(980) .
$-$, $-$ linear approximations :
$\delta 00 = 0.5\,a00\,K \leftrightarrow$
$a00\,m_\pi = 0.22$, $a00\,m_\pi = 0.16$.
ideal in the caption to figure 5 refers to
the limit $e = 0$, $m_d = m_u$.

The rapid phase variation induced by f0(980) defines two fringes, denoted low and high the two regions

low   high

$$0 \leq K \leq \sim 0.9 \text{ GeV} \quad ; \quad \sim 1.0 \text{ GeV} \leq K \leq \sim 1.6 \text{ GeV} \qquad (30)$$
$$2m_\pi \leq \sqrt{s} \leq \sim 0.94 \text{ GeV} ; \sim 1.04 \text{ GeV} \leq \sqrt{s} \leq \sim 1.625 \text{ GeV}$$

The minimal meromorphic parametrization is defined from the complex pole position on the second s - sheet, the K - plane with $\Im K < 0$ $\left(s_0 = 4\,m_\pi^2 \right)$

$$C_R^2 = \left(K_R - \tfrac{1}{2}\,i\,\gamma_R \right)^2 = \mathcal{M}_R^2 - s_0 = \left(M_R - \tfrac{1}{2}\,i\,\Gamma_R \right)^2 - s_0$$
$$S_{mmp}\left(K_R , \gamma_R ; K \right) = \frac{\left| C_R \right|^2 - K^2 + i\,\gamma_R\,K}{\left| C_R \right|^2 - K^2 - i\,\gamma_R\,K} \qquad (31)$$

The minimal meromorphic superposition of N resonances with identical *ideal* quantum numbers
– in any two body channel – corresponds to the multiplication of the individual $S_{mmp}\left(K_{R_\alpha} , \gamma_{R_\alpha} ; K \right)$ factors for resonance R_α ; $\alpha = 1, \cdots , N$ as defined in eq. 31.

$$S_{mmp}^N (K) = \prod_{\alpha=1}^N S_{mmp}\left(K_{R_\alpha} , \gamma_{R_\alpha} ; K \right)$$
$$S = S_{bg}^N S_{mmp}^N ; \quad S_{bg}^N (K) = \eta_{bg}^N (K) \exp\left(2\,i\,\delta_{bg}^N (K) \right) \qquad (32)$$

The background introduced above for $J^{PC} = 0^{++}$, $I = 0$; $\pi\pi \to \pi\pi$ is defined *relative* to S_{mmp}^N given in eq. 32 .

3 a   3 resonances – 1 channel : f0 ($\sim$990) , gb ($\sim$1000) , f0 ($\sim$1500) elastic $\pi\pi$ scattering and p ($\pi\pi$) p central production

In the following we display results using the form of the $\pi\pi$, $I = 0$, 0^{++} elastic amplitude with three resonances as given in eq. 32 with mass and width parameters in MeV

$$R_1 : M = 980 ; \Gamma = 54 \mid R_2 : M = 1000 ; \Gamma = 900 \mid$$
$$R_3 : M = 1510 ; \Gamma = 106 \qquad \textit{To figure 6 :} \qquad (33)$$

Fig 6 : $\pi\pi$; $I = 0$
phase shifts
from [16-1973] , [17-2008]

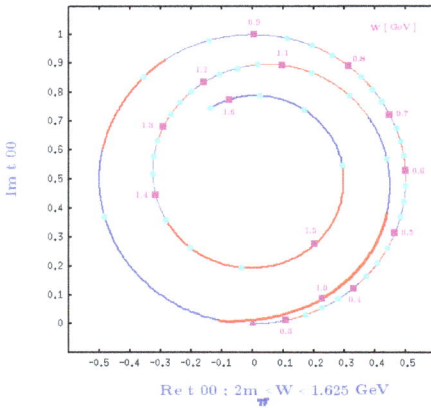

To figure 6 :

— : minimal meromorphic phase
from two resonances gb and
f0(980) with $mf0 = 0.99$ GeV,
same ratio $\Gamma_{f0}/mf0 = 0.055$.

— : background phase added with
same parameters as for —

$\cdots$: for only 2 resonances with
$mf0 = 0.98$GeV ,
$(\Gamma/m)f0 = 0.055$.

— : full phase from three resonances
f0(980) , gb and f0(1500)
(eq. 33) .

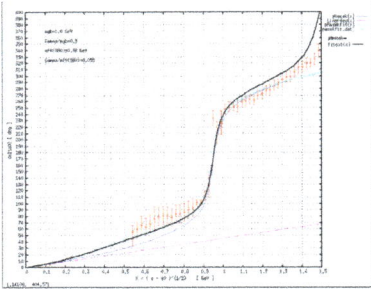

Fig 7 : $\pi\pi$; $I = 0$ Argand
diagram from material used
in ref. [17-2008]

Fig 8 : $\pi\pi$; $I = 0$ Argand
diagram from the CERN-
MUNICH collaboration [16-1973]

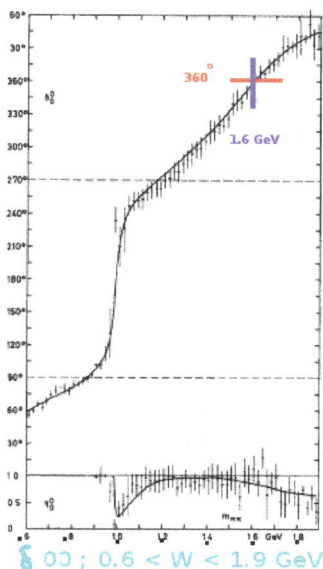

Fig 9

to Fig 9 :
$\pi\pi$; $I = 0$ phase shifts from the
CERN-MUNICH collaboration [16-1973]
a tribute to Wolfgang Ochs

I quote from ref. [18-1986] :
"We conclude by commenting that our exclusive data show interesting behaviour in the $\pi^+\pi^-$ - D-wave, but the production is dominated by the S-wave. Although the S*(980) reveals itself in a most striking manner, there is no evidence* for any new 0^{++} states. The lack of a low-mass scalar glueball candidate poses a problem for some conventional models of the glueball spectrum." [5]

References

[1-1961] M. Gell-Mann, 'Symmetries of Baryons and Mesons', Phys. Rev. 125 (1962) 1067-1084 .

[2-1969] S.L. Adler, 'Axial vector vertex in spinor electrodynamics', Phys.Rev.177 (1969) 2426-2438,
J. Bell and R. Jackiw, 'A PCAC puzzle: $\pi^0 \rightarrow \gamma\gamma$ in the sigma model', Nuovo Cim.A60 (1969) 47-61 .

[3-1991] P. Minkowski, 'Proton spin and quark spin', in Proceedings of the "Workshop on Effective Field Theories of the Standard Model", Dobogókõ, Hungary August 22 - 26, 1991 ; U. Meissner edt., World Scientific Singapore 1992 , BUTP-91-2, Aug 1991. 31pp. .

[4-1975] H. Kluberg-Stern and J. B. Zuber, 'Ward Identities and Some Clues to the Renormalization of Gauge Invariant Operators', Phys.Rev. D12 (1975) 467 ,

---

[5] The results of the AFS collaboration was a guideline to our common work with Wolfgang Ochs [19-1999] .

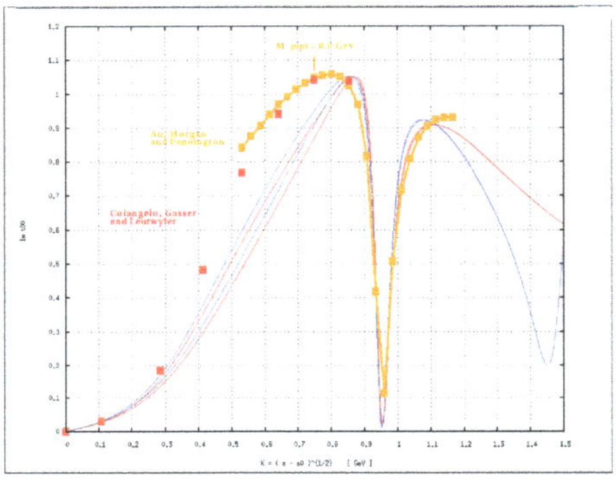

T. Åkesson et al. / Search for glueballs

to Fig 10 :
The imaginary part
$\Im T00 = (W / K) \Im t00$ in Fig 7 ,
of the $\pi\pi$, $I = 0$
ideal elastic s-wave from threshold to
~ 1.526 GeV

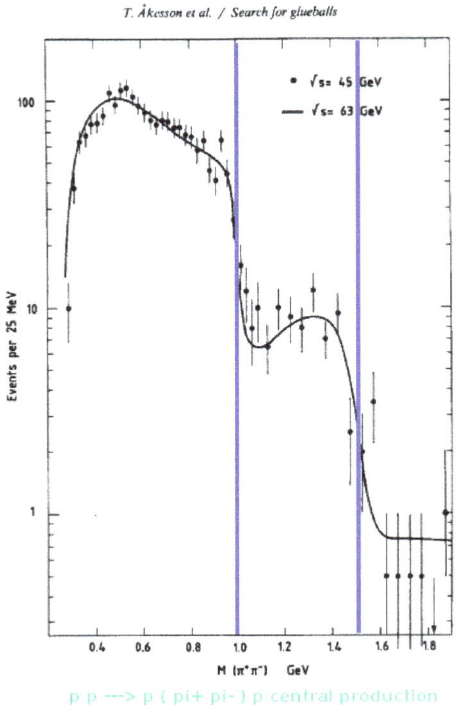

to Fig 11 :
Central production
of pion pairs at the ISR
$(W_{pp} = 63$ GeV$)$
within the invariant mass
range $0.5 < M_{\pi\pi} < 1.9$ GeV
from the Axial Field
Spectrometer collaboration
[18-1986]

'Renormalization of Nonabelian Gauge Theories in a Background Field Gauge.
1. Green Functions', Phys.Rev. D12 (1975) 482 and 'Renormalization of Non-
abelian Gauge Theories in a Background Field Gauge. 2. Gauge Invariant Operators', Phys.Rev.D12 (1975) 3159-3180 .

[5-1948] L. D. Landau, Dokl. Akad.Nauk, USSR 60 (1948) 207-209

[6-1950] C. N. Yang, 'Selection Rules for the Dematerialization of a Particle into
Two Photons', Phys.Rev.77 (1950) 242-245 .

[7-1974] J.E. Augustin et al., 'Discovery of a Narrow Resonance in e^+e^- Annihilation', Phys. Rev. Lett. 33 (1974) 1406-1408 .

[8-1974] J. J. Aubert et al., 'Experimental Observation of a Heavy Particle J', Phys.
Rev. Lett. 33 (1974) 1404-1406 .

Fig 80 : thank you

[9-1975] H. Fritzsch and P. Minkowski, 'Psi Resonances, Gluons and the Zweig Rule', Nuovo Cim.A30 (1975) 393 .

[10-2008] C. Amsler et al. (Particle Data Group), Physics Letters B667, 1 (2008) and 2009 partial update for the 2010 edition .

[11-2008] M. Albaladejo, J.A. Oller and C. Piqueras, 'S-wave meson scattering up to 2-GeV and its spectroscopy', Talk given at Workshop on Scalar Mesons and Related Topics Honoring 70th Birthday of Michael Scadron (SCADRON 70), Lisbon, Portugal, 11-16 Feb 2008, Int.J.Mod.Phys.A24 (2009) 581-585, arXiv:0804.2341 [hep-ph] .

[12-2008] B. Bloch-Deveaux, Na48/2 coll., 'Results from NA48/2 on pi pi scattering lengths measurements in $K^{\pm} \to \pi^+\pi^- e^{\pm}\nu$ and $K^{\pm} \to \pi^0\pi^0\pi^{\pm}$ decays', PoS CONFINEMENT8 (2008) 029 .

[13-2001] C. Colangelo, J. Gasser and H. Leuwyler, 'pi pi scattering', BUTP-01-1, Mar 2001, Nucl.Phys.B603 (2001) 125-179, hep-ph/0103088, *and*
'The Quark condensate from K(e4) decays', BUTP-01-7, Mar 2001. 4pp., Phys.Rev.Lett.86 (2001) 5008-5010, hep-ph/0103063 .

[14-2007] The NA48/2 Collaboration (J.R. Batley et al.) , CERN-PH-EP-2007-035, Oct 2007, 24pp, 'New high statistics measurement of K_{e4} decay form factors and $\pi\pi$ scattering phase shifts' , Eur.Phys.J.C54 (2008) 411-423 .

[15-1973] S.D. Protopopescu, M. Alston-Garnjost, A. Barbaro-Galtieri, Stanley M. Flatte, J.H. Friedman, T.A. Lasinski, G.R. Lynch, M.S. Rabin, F.T. Solmitz, ' $\pi\pi$ partial wave analysis from reactions $\pi^+ p \to \pi^+\pi^-\Delta^{++}$ and $\pi^+ p \to K^+ K^-\Delta^{++}$ at 7.1-GeV/c', Phys.Rev.D7 (1973) 1279 .

[16-1973] B. Hyams, C. Jones, P. Weilhammer, W. Blum, H. Dietl, G. Grayer, W. Koch, E. Lorenz, G. Lütjens, W. Männer, J. Meissburger, W. Ochs, U. Stierlin and F. Wagner, '$\pi\pi$ phase-shift analysis from 600 - 1900 MeV', Nuclear Physics B64 (1973) 134-162 .
W. Ochs, University of Munich thesis 1973, unpublished.

[17-2008] P. Minkowski, 'On concise hypotheses for the interpretation of a wide scalar resonance as gauge boson binary in QCD', Oct 2008, in proc. of 14th Int. Conf. on Quantum Chromodynamics (QCD 08), Montpellier 7-12 July 2008, Nucl.Phys.Proc.Suppl.186 (2009) 302-305, arXiv:0810.0775 [hep-ph].

[18-1986] T. Akesson et al., The Axial Field Spectrometer collaboration, 'A search for glueballs and a study of double Pomeron exchange at the CERN intersecting storage rings', Nuclear Physics B264 (1986) 154-184 .

[19-1999] P. Minkowski and W. Ochs, 'Identification of the glueballs and the scalar meson nonet of lowest mass', BUTP-98-27, MPI-PHT-98-89, Nov 1998. 65pp., Eur.Phys.J.C9 (1999) 283-312, hep-ph/9811518 .

[f1-1973] B. Hyams et al., CERN-Munich collaboration, Nucl. Phys. B64 (1973) 134-162 .

[f2-2008] R. Kaminski, J. Pelaez and F. Yndurain, 'The Pion-pion scattering amplitude. III. Improving the analysis with forward dispersion relations and Roy equations', Phys.Rev.D77 (2008) 054015, arXiv:0710.1150 [hep-ph].

THE QCD COUPLING AND PARTON DISTRIBUTIONS AT HIGH PRECISION

JOHANNES BLÜMLEIN

Deutsches Elektronen-Synchrotron, DESY, Platanenallee 6, D-15738 Zeuthen, Germany
E-mail: Johannes.Bluemlein@desy.de

A survey is given on the present status of the nucleon parton distributions and related precision calculations and precision measurements of the strong coupling constant $c_s(M_Z^2)$. We also discuss the impact of these quantities on precision observables at hadron colliders.

Keywords: Deep-inelastic scattering; strong coupling constant; heavy flavors.

1. Inside Nucleons

The physics of the strong interactions always has been tightly connected to the study of nucleons at shorter and shorter distances. The measurement of the anomalous magnetic moments of the proton[1] and neutron[2] in 1933 and 1939 made clear that nucleons are no elementary particles. During the 1950ies the Hofstadter experiments[3] revealed the charge distributions inside nucleons[4] at scales $Q^2 \simeq 0.5 \cdot M_N^2$. Yet it was unknown how these distributions came about. In 1964 Murray Gell–Mann[5] proposed the quark model, to catalog the plethora of observed baryons and mesons. Independently G. Zweig suggested aces[6] as the building blocks of hadrons. A direct connection to the lepton-nucleon scattering data was not made at that time.

Back in 1954 C.N. Yang and R Mills[7] proposed novel bosonic field theories based on gauge invariance with respect to non-abelian groups. This development went unrelated to strong interactions for a long time. With the advent of the Stanford Linear Accelerator in 1968 the nucleon structure could be resolved at much shorter distances by the MIT-SLAC experiments[8-10] beyond the resonant region $W \geq 2\text{GeV}$ for values Q^2 up to 30 GeV². The remarkable finding by these experiments were that *i*) the structure function $\nu W_2(\nu, Q^2)$ which has been expected to depend on both kinematic variables ν and Q^2 independently, turned out to take the same values for fixed values of $x = Q^2/(2M_N\nu)$ irrespectively of ν and Q^2 at high enough values. This phenomenon is called scaling. *ii*) The ratio of the longitudinal structure function W_L and W_2 turned out to be very small. Bjorken[11] had predicted scaling at asymptotic scales $Q^2, \nu \to \infty$ in 1969. Learning about the SLAC-MIT results R. Feynman very quickly proposed the parton model,[12] which is equivalent to Bjorken's description but based on the observed strict microscopic correlation between Q^2 and $\nu = q.p_i$

$$W(x, Q^2) = \sum_i e_i^2 \int_0^1 dx_i f_i(x_i) \delta \left(\frac{q.p_i}{M^2} - \frac{Q^2}{M^2} \right) , \qquad (1)$$

where e_i and f_i denote the parton's charge and distribution functions. Would the parton model be unique in describing the new data? This has been challenged by other popular formalisms like vector meson dominance.[13] However, they failed to describe the behaviour observed for W_L, which corresponded to that of spin 1/2 partons, according to the calculations by Callan and Gross.[14]

Yang–Mills theories[7] became building blocks of the electro-weak Standard Model,[15] although there renormalizibility had not been proven yet, a conditio sine qua non for a physical theory. The proof was an urgent matter and in 1971 it was achieved both for massless and spontaneously broken Yang-Mills theories, along with designing practical loop computations in this sophisticated theory in an automated way.[16,17] Quantum Chromodynamics (QCD) was proposed as the theory of the strong interactions in 1972 by Gell–Mann and Fritzsch and Leutwyler[18,19] as a renormalized Yang-Mills field theory based on $SU(3)$ gauge interactions.[20] D. Gross, F. Wilczek[21] and D. Politzer[22] studied the asymptotic behaviour of color octet gluon Yang-Mills theory, cf. also Ref. 23, and found asymptotic freedom. This is the essential ingredient, which makes it possible to perform perturbative calculations at large scales in a theory with strong interactions at low scales.

At short distances the nucleon structure functions $F_i(x, Q^2)$ obey the light-cone expansion.[24] At large scales Q^2 the contributions of lowest twist dominate and the representation

$$F_i(x, Q^2) = \sum_j C_i^j(x, Q^2/\mu^2) \otimes f_j(x, \mu^2) \qquad (2)$$

holds. Here $C_i^j(x, Q^2/\mu^2)$ denote the Wilson coefficients and $f_j(x, \mu^2)$ are the parton densities. μ^2 is an arbitrary factorization scale and $\otimes$ denotes the Mellin convolution.

The scale behavior of the nucleon structure functions $F_i(x, Q^2)$ obey renormalization group equations, an important aspect of renormalizable Quantum Field Theories to which Murray Gell–Mann made very essential contributions very early.[25,a] Transforming Eq. (2) to Mellin space one obtains the following Callan-Symanzik[28] equations:

$$\left[\mu \frac{\partial}{\partial \mu} + \beta(g) \frac{\partial}{\partial g} - 2\gamma_\psi(g) \right] F_i(N, Q^2) = 0 \qquad (3)$$

---

[a]It is interesting to note that different approaches to renormalization result into different mathematical structures as shown in Ref. 26. Thus the method by Gell–Mann and Low[25] is in general related to a cocycle, while that by Stückelberg and Petermann[27] relates to a group. I thank A. Petermann for pointing out Ref. 26 to me.

$$\left[\mu\frac{\partial}{\partial\mu} + \beta(g)\frac{\partial}{\partial g} + \gamma_\kappa(N,\mu) - 2\gamma_\psi(g)\right] f_k(N,\mu^2) = 0 \qquad (4)$$

$$\left[\mu\frac{\partial}{\partial\mu} + \beta(g)\frac{\partial}{\partial g} - \gamma_\kappa(N,\mu)\right] C_j^k(N,Q^2/\mu^2) = 0 , \qquad (5)$$

with $\beta(g)$ the QCD β-function and $\gamma_\kappa(N,\mu)$ the anomalous dimensions. Both functions imply the scaling violations of the structure functions.[b] With progressing time the measurement of the deep-inelastic structure functions improved considerably and after 40 years, e.g. the precision of the structure function $F_2^{\text{em}}(x,Q^2)$ reached 1% over a very wide range, cf. Ref. 30. Due to this both the precision measurement of the unpolarized parton distributions and the strong coupling constant $\alpha_s(M_Z^2)$ is possible from these data.

2. Higher Order QCD Corrections to Deep-Inelastic Scattering

On the theory side, the progress in higher order computations, likewise, has been enormous during the same period. The initial 1-loop results[21,22,31] are now widely improved to 3-loop order and even somewhat beyond. This is necessary to comply with the current precision of data. The status of the theory of deep-inelastic scattering is illustrated in the flowchart below. Here the dates indicate the year in which the corresponding correction to the respective quantities has been calculated.

Let me mention the most far reaching results. For the QCD β-function the 4–loop corrections were first computed by Vermaseren et al.[32] in 1997. The unpolarized anomalous dimensions and massless Wilson coefficients are known to 3–loops for a series of moments[33] and in complete form[34–36] since 2004/05. A first moment for the non-singlet+ anomalous dimensions has been computed at 4–loops[37] in 2006 and more moments are in preparation. The heavy flavor Wilson coefficients in the region $Q^2/m^2 = \rho \gg 1$, which is a good approximation in case of $F_2(x,Q^2)$ for $\rho \geq 10$, were computed for a larger number of moments[38] and in complete form for F_L.[39] Currently the computation of the Wilson coefficients at general values of N is underway.[40]

At the level of the leading twist ($\tau = 2$) representation the light cone expansion and the QCD-improved parton model lead to the same results. In the 1980ies it was thought, that the higher order corrections need to be supplemented by different small-x resummations, cf. Refs. 41, 42, to obtain correct results, even in the HERA kinematic region. These perturbative resummations are connected to the problem how, within their approach, perturbative and non-perturbative contributions are clearly separated - an important pre-requisite to apply perturbation theory at all.

---

[b]We remind that Drell and collaborators at the end of the 1960ies[29] were seeking desperately scaling in fermion-meson interactions with loop corrections but ended up with scaling violations in general.

Theory of DIS

'69, QCD: '72   '69 – 72

Parton Model   Light Cone Expansion [f]

Twist 2   Higher Twist

~ 77

polarized   *unpolarized*

Fixed Order PT: QCD   Twist 3 Twist 4 ...

Sum Rules

'60ies - now

α_s Splitting functions   Coefficient functions

$O(\alpha_s^4)$ $O(\alpha_s^3)$ $O(\alpha_s^3)$

'73,'74,'80,'97   '73,'82, '04   '82,'92,'05,'09

Special Kinematics Domain: Small x   '75, '86 '90 - '98   More General View Non-forw. scattering

Resummations ?   Diffractive Scattering   Angular Momentum: q, G

Higher Orders $\Longrightarrow$ New Algorithms Novel Mathematics

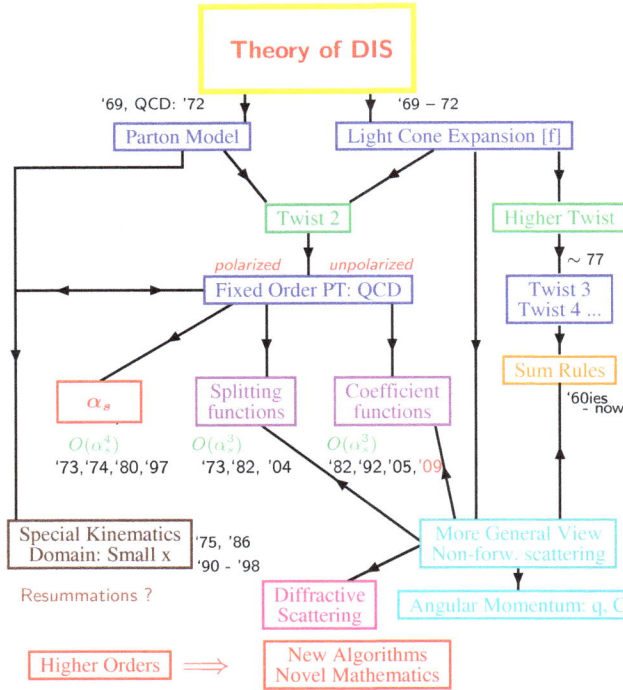

The resummation[41,43] has successfully predicted the so-called 'leading' poles of the QCD anomalous dimensions related to the poles at $N = 1$ in Mellin space, at least up to $O(\alpha_s^3)$. The leading series is related to the scale-invariant limit of QCD. The corresponding resummed anomalous dimension, however, has branch cuts in the complex plane[44] but no poles at all, see also Ref. 45. These singularities are much milder. Phenomenological studies[46] have shown, that subleading effects are as important as the leading ones, since they widely cancel the effect of the former. One estimates that about four complete series of these terms are needed to obtain convergence. Currently the only practical approach relies on the computation of the Wilson-coefficients and anomalous dimensions to high enough order, which includes all the small- and large-x effects automatically. In the latter case, the renormalization group even allows reliable resummations.[47]

Beyond the level of leading twist much less is known on deeply-inelastic structure functions. Most of the results obtained so far concern the 1-loop level, cf. e.g. Ref. 48.[c] Here, the corresponding partonic operator matrix elements depend on several dimensionless invariants x_i, unlike in the case of lowest twist. They cannot be measured individually in the deep-inelastic process, but require ab-initio determinations using reliable non-perturbative methods. For a series of moments,

---

[c]For older references see Ref. 49.

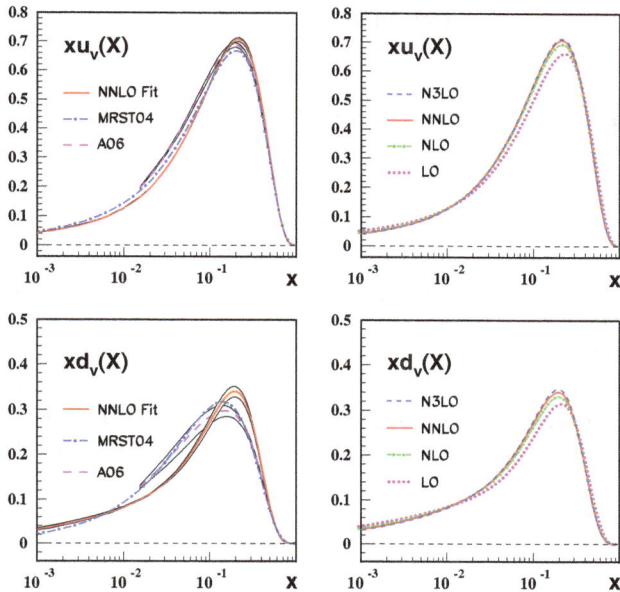

Fig. 1. Left panels: The parton densities xu_v and xd_v at the input scale $Q_0^2 = 4.0$ GeV^2 (solid line) compared to results obtained from NNLO analyses by MRST04 (dashed–dotted line)[55] and A06 (dashed line).[56] The shaded areas represent the fully correlated 1σ statistical error bands. Right panels: Comparison of the same parton densities at different orders in QCD from LO to N^{3}LO Ref. 57.

this may be possible in the future, using lattice techniques. In the polarized case a series of results has been obtained for the twist-3 contributions, cf. e.g. Ref. 50, among them a series of integral relations between different structure functions.[49,51] Finally, deep-inelastic non-forward scattering has been extensively studied during the last two decades,[52] These methods do in principle allow to measure the (quark) angular momentum of nucleons,[53] which is still difficult experimentally. One may apply these techniques to describe inclusive deep-inelastic diffractive scattering, referring to the light cone expansion[54] and proving that the anomalous dimensions are structurally the same as in the forward case.

3. Precision Quark and Gluon Twist-2 Distributions

In the following we describe recent extractions of the unpolarized twist-2 parton densities to 3-loop accuracy and higher.

3.1. Flavor non-singlet analysis

The flavor non-singlet parton distributions obey scalar evolution equations and do not depend on the gluon distribution, which is more difficult to access in deep-inelastic scattering and may cause some systematic uncertainty, in particular determining the strong coupling constant $\alpha_s(M_Z^2)$. Moreover, in the small x region the

QCD evolution leads to moderate changes of the the distributions, unlike in the flavor singlet and gluon-case. Due to this flavor non-singlet analyses are advantageous. One may apply the valence-approximation in the region $x \geq x_0$, $x_0 \sim 0.35, 0.4$ and construct a non-singlet distribution from deuteron and proton data for $x \leq x_0$. To describe the valence quark parton densities, the distribution $x(\bar{d} - \bar{u})(x, Q^2)$ has to be known, which can be measured using Drell-Yan data. Furthermore, the non-singlet $O(\alpha_s^2)$ heavy flavor corrections are applied, which amount to about 1%. The yet unknown 3-loop corrections will be even smaller. To perform a leading twist analysis, kinematic regions with higher twist contributions are cut out in a systematic study, implying the cuts of $Q^2 > 4$ GeV^2, $W^2 > 12.5$ GeV^2, cf. Refs. 57, 58. The results for the parton distributions $x u_v(x, Q^2)$ and $x d_v(x, Q^2)$ are illustrated in Figure 1 and are compared to other determinations. Wile in the case of the $x u_v$ distributions the overall agreement is good, there are still systematic differences in case of the down-valence distribution. We also illustrate the perturbative expansion from LO to N^{3}LO reaching convergence.

Extrapolating the twist-2 QCD fit results into the region 12.5 GeV$^2 > W^2 > 4$ GeV^2 the flavor non-singlet higher twist contributions can be determined empirically.[59] The inclusion of soft resummation terms for the Wilson coefficient beyond the N^{3}LO corrections allows to extract the higher twist terms in the region $x < 0.75$ in a stable way, while lower order analyses overestimate the higher twist contributions. The relative higher twist contributions in the proton and deuteron case turn out to be about of the same size.

3.2. *Combined singlet and non-singlet analysis*

In combined singlet and non-singlet analyses of the deep-inelastic world data at NNLO, cf. Refs. 60–64, one determines also the different sea-quark and gluon densities. This has always to be done together with the measurement of the QCD scale Λ_{QCD} due to strong correlations. To unfold the sea-quark densities one refers to Drell-Yan- and di-muon data as well, through which the distributions $x(\bar{d} - \bar{u})(x, Q^2)$ and $x s(x, Q^2) = x \bar{s}(x, Q^2)$ can be measured individually. Due to the large charm-quark contribution to the deep-inelastic structure functions the description of the heavy flavor contributions is required at the same level of accuracy as for the light partons. Currently it is available to $O(\alpha_s^2)$[65] and the $O(\alpha_s^3)$ corrections are underway.[38,40,66] To obtain an accurate interpolation between low and higher scales Q^2, the so-called BMSN-interpolation is recommended.[67,68] Nowhere in the kinematic region of HERA heavy flavor logs become large to be resummed,[69] i.e. in a very wide kinematic region even the charm quarks cannot be viewed massless. Yet, one may define heavy flavor parton densities in terms of technical quantities[38,67] to some extent ^{d} to evaluate other observables. Here one has always to check to which extent this approximation holds.

---

^{d}At 3-loop order graphs exist with both charm and bottom-quark lines in the operator matrix elements. They do not fall under the paradigm of single parton distributions, despite being universal.

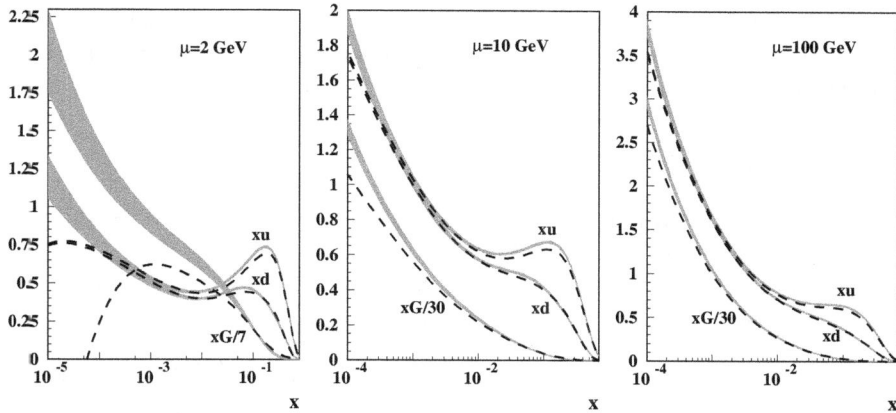

Fig. 2. The light parton densities xu, xd and xG at the scales $\mu^2 = 4, 100, 10000$ GeV^2. The bands denote the parton distributions with 1σ uncertainty of ABKM09.[60] The dashed lines correspond to MSTW08;[64] from Ref. 60.

The present analyses have lead to very precise parton distributions. The NNLO parton distributions determined in Refs. 60, 62 do widely agree within the measured region, with very slight differences in the $x(u + \overline{u})$ and $x(d + \overline{d})$ distributions. In Figure 2 we compare the results of the NNLO fits of Refs. 60, 64 for the light partons. At low scales μ^2 the sea-quark and gluon distributions[64] take lower values than those in Ref. 60, and the NNLO gluon distribution even tends to negative values yielding the largest difference.

Through the evolution the densities get closer. This, however, is partly due to the large value of $\alpha_s(M_Z^2)$, Eq. (14), which leads to a relative acceleration of the evolution compared to Ref. 60. At scales larger than $Q^2 \sim 10^4$ GeV^2, accessible at the LHC, this will lead to a further growing gluon density of Ref. 64 compared to Ref. 60. The precision observables at the LHC will help to constrain the parton distributions further.

4. The Strong Coupling Constant

The QCD parameter Λ_{QCD}, or $\alpha_s(M_Z^2)$, is determined in QCD fits together with the non-perturbative input densities for the different partons at a starting scale Q_0^2. There are tight correlations between the value of $\alpha_s(M_Z^2)$ and some parameters of the parton densities. An important example is the normalization of the gluon density, cf. Refs. 60, 70. In the non-singlet analysis[57] we obtained at NNLO, cf. also Ref. 71,

$$\Lambda_{\mathrm{QCD}}^{N_f=4} = 226 \pm 25 \text{ MeV} , \quad (6)$$

and at N^{3}LO, assigning to the yet unknown 4-loop anomalous dimension a $\pm 100\%$ error,

$$\Lambda_{\mathrm{QCD}}^{N_f=4} = 234 \pm 26 \text{ MeV} . \quad (7)$$

Usually the QCD parameter is expressed in terms of $\alpha_s(M_Z^2)$. In the following we compare the results of recent NNLO and N^{3}LO analyses for the deep-inelastic world data obtained by different groups :

$$\alpha_s(M_Z^2) = 0.1134 \,{}^{+0.0019}_{-0.0021} \quad \text{NNLO} \qquad\qquad [57] \qquad (8)$$

$$\alpha_s(M_Z^2) = 0.1141 \,{}^{+0.0020}_{-0.0022} \quad \text{N}^3\text{LO} \qquad\qquad [57] \qquad (9)$$

$$\alpha_s(M_Z^2) = 0.1135 \pm 0.0014 \quad \text{NNLO, FFS} \qquad [60] \qquad (10)$$

$$\alpha_s(M_Z^2) = 0.1129 \pm 0.0014 \quad \text{NNLO, BSMN} \qquad [60] \qquad (11)$$

$$\alpha_s(M_Z^2) = 0.1124 \pm 0.0020 \quad \text{NNLO, dyn. approach} \qquad [62] \qquad (12)$$

$$\alpha_s(M_Z^2) = 0.1158 \pm 0.0035 \quad \text{NNLO, stand. approach} \qquad [62] \qquad (13)$$

$$\alpha_s(M_Z^2) = 0.1171 \pm 0.0014 \quad \text{NNLO} \qquad\qquad [72] \qquad (14)$$

More recent unpolarized NNLO fits, including the combined HERA data,[30] yield

$$\alpha_s(M_Z^2) = 0.1147 \pm 0.0012 \text{ NNLO} \qquad\qquad [61] \qquad (15)$$

$$\alpha_s(M_Z^2) = 0.1145 \pm 0.0042 \text{ NNLO, preliminary} \quad [73] \qquad (16)$$

Note that the values (8,9) are independent of the gluon distribution. The combined singlet non-singlet analysis, based on rather different data, and being sensitive to both the sea-quark and gluon densities, yield the very similar values (10,11). This analysis has been performed with a different code than used in Ref. 57.

The above values are located below the present weighted average of $\alpha_s(M_Z^2)$ measurements[72] of

$$\alpha_s(M_Z^2) = 0.1184 \pm 0.0007 \,, \qquad\qquad (17)$$

cf. Figure 3. The error given in (17) cannot include the yet unknown relative systematics between the different classes of the same type of measurement.

We would like to mention that recent determinations of $\alpha_s(M_Z^2)$ using event shape moments for high energy e^+e^- annihilation data from PETRA and LEP including power corrections the following values were obtained :

$$\alpha_s(M_Z^2) = 0.1135 \pm 0.0002 \text{ (exp)} \pm 0.005 \,(\Omega_1) \pm 0.0009 \text{ (pert) NNLO } [76] \quad (18)$$

$$\alpha_s(M_Z^2) = 0.1153 \pm 0.0017 \text{ (exp)} \pm 0.0023 \text{ (th)} \qquad\qquad \text{NNLO } [77] \quad (19)$$

Also these measurements of $\alpha_s(M_Z^2)$ yield low values. They show that the results obtained analyzing deep–inelastic data do not form a special case. Also in deep-inelastic scattering off polarized targets $\alpha_s(M_Z^2)$, at NLO, has been measured, however with larger errors, see Refs. 70, 73, 74. The present error on $\alpha_s(M_Z^2)$ at NNLO of ~ 0.0012 is at the margin of the present theory and systematics errors. Different known theoretical uncertainties, cf. also Eqs. (8–11), are of the order of 0.0007, the quoted 1σ error of the world average. The systematics of the different extractions of $\alpha_s(M_Z^2)$ has to be understood in even more detail in the future. The current values of $\alpha_s(M_Z^2)$ obtained from precision deep-inelastic scattering data disfavor the

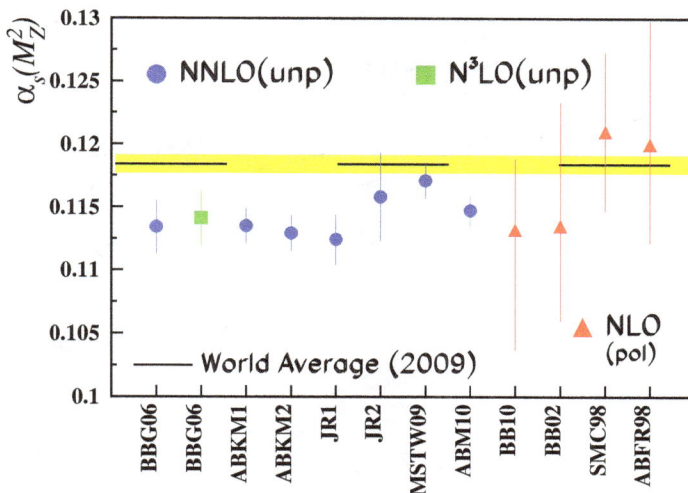

Fig. 3. The strong coupling constant $\alpha_s(M_Z^2)$ from different DIS measurements, at NNLO and N^{3}LO, Eqs. (8–15), and at NLO in the polarized case, cf. Refs. 70, 74. The yellow band describes the weighted average of a wide range of $\alpha_s(M_Z)$ measurements;[72] Ref. 70.

unification of forces, even in the supersymmetric extension of the Standard Model. However, the picture may change soon, with new findings at the LHC.

5. Main Inclusive Cross Sections at Hadron Colliders

The accuracy for the parton distribution functions reached can now be applied to derive precision predictions for inclusive hadronic observables, such as the Drell-Yan cross section, the $W^\pm$, Z-boson, the $t\bar{t}$- and Higgs-boson production cross sections at NNLO. Detailed analyses have been given in Refs. 60–64. For all these quantities at least these corrections are necessary. The Drell-Yan cross section and the $W^\pm$, Z-boson production cross sections are, furthermore, used as 'standard candle' processes to measure the collider luminosity. They have therefore to be known as precisely as possible. As an example we show in Figure 4 predictions on the inclusive Higgs boson production at hadron colliders. Within the present accuracy of the parton distribution functions the predictions[60,64] still show some differences, which are likely related to the different gluon distributions at low scale and the values of $\alpha_s(M_Z^2)$. The predictions agree for $s \sim M_Z$ but differ for higher mass scales.

6. Higher Loop Integrals and Mathematics

The computation of higher order loop integrals is still a difficult task, even in the massless case or in the presence of a single mass scale. At three-loop (and higher) orders this both applies to the zero– and single–scale problems calculated at present. At the one hand, the results expected have a rather simple structure. On the other

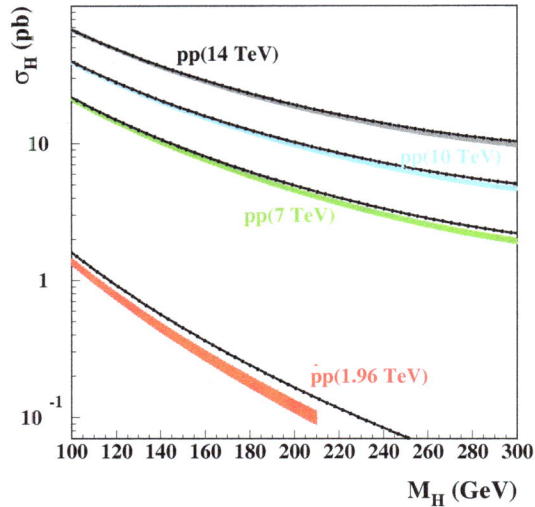

Fig. 4.   Inclusive Higgs boson production cross section at TEVATRON and the LHC. The bands denote the parton distribution uncertainty of ABKM09;[60] the lines correspond to MSTW08;[64] from Ref. 60.

hand, a growing multitude of diagrams which contain more and more difficult structures have to be computed. Obviously, an enormous part of intermediary results simply cancels. It is, however, difficult to let cancel these contributions at a rather early stage of the computation, or to even widely avoid that they occur from the very beginning. This is one of the central problems of all present calculations. Gauss' theorem[75] allows to express Feynman diagrams to a set of master integrals, which finally have to be computed for zero scale quantities. Also in case of single scale quantities, e.g. given by a Mellin variable N, one may obtain similar recursions. This method, simplifying the calculation, may lead to a large number of terms being of higher complexity than those finally appearing in the results. This property seems to be in common with different other approaches, which in the first place appear to simplify the calculation technically, like Mellin-Barnes[76] integrals or multinomial expansion, since they lead to rather elementary Feynman parameter integrals. Another approach, cf. e.g. Refs. 40, 66, 77, 78, consists in evaluating the individual Feynman parameter integrals without applying intermediary simplifying methods, i.e. mapping them directly to the analytic mathematical structure they represent. This method is more demanding but will finally lead us to a deeper understanding of the objects we deal with. For single scale single integrals integrals and one mass, the 2-loop integrals and simpler 3-loop integrals can be represented in terms of generalized hypergeometric functions and extensions thereof, like Appell functions cf. Ref. 79. An important issue in integration is a clear definition of the target space and the knowledge of the relations of its elements. For the Feynman parameter integrals discussed above, shuffle– and Hopf–algebras[80] play a central role, along with

Poincaré iterated integrals over specific alphabets. Feynman integrals are almost always periods.[81] In this context, the simplest structures are multiple zeta values, cf. Ref. 82, nested harmonic sums,[83] harmonic polylogarithms,[84] and generalized harmonic sums.[85–87] Feynman parameter integrals will coin new classes of higher transcendental functions going to higher and higher order, which have to be studied to perform future precision calculations in an efficient way. Their evaluation is intimately connected to modern summation technologies, like SIGMA,[88] and efficient algorithms to establish and solve the associated recurrences of both large order and degree.[40,89] All this will require high performance computer algebra written using highly efficient languages like FORM,[90] nearly 50 years after SCHOONSCHIP was introduced,[17] and the investment of many CPU years, however, at an even more involved level than considered today. Structures of Feynman parameter integrals, e.g. on the level of multiple zeta values, form an interesting recent field in mathematics, cf. Refs. 78, 91, 92, which is also related to irrationality proofs of the basis elements spanning these quantities. Here we face a new era of a tight symbiosis between theoretical physics and modern mathematics, which is regarded to be very essential.

7. Conclusions

Quantum Chromodynamics was a great discovery. With it Murray Gell–Mann completed the revolution of the strong interactions started in the early 1960ies with the introduction of the quarks. During the last 37 years computations grew to a precision of $O(1\%)$ for inclusive quantities, which are described at 3–loop, and partly at 4–loop, level, moving the frontiers of Quantum Field Theory to breathtaking new horizons. The running of the strong coupling constant is understood in great detail, despite in different classes of analyses still values of $\alpha_s(M_Z^2)$ are found which differ by experimental and theoretic systematic effects, being partly yet unknown. To determine $\alpha_s(M_Z^2)$ at the level of its present statistical accuracy of $\sim 1\%$ further studies and even higher order calculations are required for some of the processes. The QCD improved quark-gluon parton model works impressively well at short distances - a clear triumph of Quantum Chromodynamics and proof that quarks and gluons, although being confined, are basic building blocks of matter. Without them the Standard Model would suffer from anomalies[93] and not form a Quantum Field Theory.

During the last two decades the methods of lattice QCD steadily improved, in particular concerning the systematic errors involved. We therefore expect precision computations both on Λ_{QCD} and a series of moments of unpolarized and polarized parton densities in the near future. The results of these calculations ab initio can then be compared to the precision extractions discussed based on precision experimental data and higher order perturbative calculations. A final question concerns the unification of the three forces of the Standard Model, given what we know at low scales at present. Future discoveries, perhaps at the LHC, will lead to a clarification here.

Acknowledgments

I would like to thank the organizers of the conference for invitation. In a conversation with M. Gell–Mann it turned out that we have a second common interest: the physics of the direction of time. Everywhere in physics it runs forward, for all we know.

References

1. R. Frisch and O. Stern, *Z. Phys.* **85**, p. 4 (1933).
2. L. W. Alvarez and F. Bloch, *Phys. Rev.* **57**, 111 (1940).
3. R. Hofstadter Electron scattering and nuclear and nucleon structure. A collection of reprints with an introduction, (New York, Benjamin, 1963), 690 p.
4. D. N. Olson, H. F. Schopper and R. R. Wilson, *Phys. Rev. Lett.* **6**, 286 (1961).
5. M. Gell-Mann, *Phys. Lett.* **8**, 214 (1964).
6. G. Zweig An $SU(3)$ model for the strong interaction symmetry and its breaking, CERN-TH-401, 412 (1964).
7. C.-N. Yang and R. L. Mills, *Phys. Rev.* **96**, 191 (1954).
8. W. K. H. Panofsky Low q^2 electrodynamics, elastic and inelastic electron (and muon) scattering, Proc. 14th International Conference on High-Energy Physics, Vienna, 1968, J. Prentki and J. Steinberger, eds., (CERN, Geneva, 1968), pp. 23.
9. R. E. Taylor Inelastic electron - proton scattering in the deep continuum region, Proc. 4th International Symposium on Electron and Photon Interactions at High Energies, Liverpool, 1969, (Daresbury Laboratory, 1969), eds. D.W. Braben and R.E. Rand, pp. 251.
10. H. W. Kendall, *Rev. Mod. Phys.* **63**, 597 (1991).
    R. E. Taylor, *Rev. Mod. Phys.* **63**, 573 (1991).
    J. I. Friedman, *Rev. Mod. Phys.* **63**, 615 (1991).
11. J. D. Bjorken and E. A. Paschos, *Phys. Rev.* **185**, 1975 (1969).
12. R. P. Feynman The behavior of hadron collisions at extreme energies, Proc. of 3rd International Conference on High Energy Collisions, Stony Brook, 1969, C.N. Yang, J.A. Cole, M. Good, R. Hwa, and J. Lee-Franzini, eds., (Gordon and Breach, New York, 1970), pp. 237.
    R. P. Feynman Photon-hadron interactions, (Benjamin Press, Reading, 1972), 282 p.
13. J. Sakurai Vector-Meson dominance - present status and future prospects, Proc. 4th International Symposium on Electron and Photon Interactions at High Energies, Liverpool, 1969, (Daresbury Laboratory, 1969), eds. D.W. Braben and R.E. Rand, pp. 91.
14. C. G. Callan and D. J. Gross, *Phys. Rev. Lett.* **22**, 156 (1969).
15. S. L. Glashow, *Nucl. Phys.* **22**, 579 (1961).
    S. Weinberg, *Phys. Rev. Lett.* **19**, 1264 (1967).
16. G. 't Hooft, *Nucl. Phys.* **B33**, 173 (1971).
    G. 't Hooft and M. J. G. Veltman, *Nucl. Phys.* **B44**, 189 (1972).
    G. 't Hooft and M. J. G. Veltman, *Nucl. Phys.* **B50**, 318 (1972).
    G. 't Hooft, *Nucl. Phys.* **B61**, 455 (1973).
    G. 't Hooft and M. J. G. Veltman Diagrammar, CERN Yellow Report 73–9 (1973).
    G. 't Hooft and M. J. G. Veltman, *Nucl. Phys.* **B153**, 365 (1979).
17. M. J. G. Veltman SCHOONSCHIP, version 1, Dec. 1963.
18. H. Fritzsch and M. Gell-Mann Current algebra: Quarks and what else?, Proceedings of 16th International Conference on High-Energy Physics, Batavia, Illinois, 6-13 Sep Vol. **2**, J.D. Jackson, A. Roberts, R. Donaldson, eds., pp. 135 (1972), hep-ph/0208010.
19. H. Fritzsch, M. Gell-Mann and H. Leutwyler, *Phys. Lett.* **B47**, 365 (1973).

20. Y. Nambu A Systematics Of Hadrons In Subnuclear Physics, in: Preludes in Theoretical Physics, eds. A. De-Shalit, H. Fehsbach and L. van Hove (North-Holland, Amsterdam, 1966), pp. 133.

21. D. J. Gross and F. Wilczek, *Phys. Rev. Lett.* **30**, 1343 (1973).

22. H. D. Politzer, *Phys. Rev. Lett.* **30**, 1346 (1973).

23. I. B. Khriplovich, *Yad. Fiz.* **10**, 409 (1969).
    G. t'Hooft (1972), unpublished.

24. K. G. Wilson, *Phys. Rev.* **179**, 1499 (1969).
    R. A. Brandt and G. Preparata, *Nucl. Phys.* **B27**, 541 (1972).
    Y. Frishman, *Annals Phys.* **66**, 373 (1971).

25. M. Gell-Mann and F. E. Low, *Phys. Rev.* **95**, 1300 (1954).

26. R. Brunetti, M. Duetsch and K. Fredenhagen (2009), arXiv:0901.2038 [math-ph].

27. E. C. G. Stückelberg and A. Petermann, *Helv. Phys. Acta* **24**, 317 (1951).

28. C. G. Callan, *Phys. Rev.* **D2**, 1541 (1970).
    K. Symanzik, *Commun. Math. Phys.* **18**, 227 (1970).

29. S. D. Drell, D. J. Levy and T.-M. Yan, *Phys. Rev.* **187**, 2159 (1969).
    S. D. Drell, D. J. Levy and T.-M. Yan, *Phys. Rev.* **D1**, 1617 (1970).
    S. D. Drell, D. J. Levy and T.-M. Yan, *Phys. Rev.* **D1**, 1035 (1970).
    T.-M. Yan and S. D. Drell, *Phys. Rev.* **D1**, 2402 (1970).
    S. D. Drell and T.-M. Yan, *Ann. Phys.* **66**, 578 (1971).

30. F. Aaron *et al.*, *JHEP* **1001**, p. 109 (2010).

31. D. J. Gross and F. Wilczek Phys. Rev. **D8**, 3633 (1973).
    D. J. Gross and F. Wilczek Phys. Rev. **D9**, 980 (1974).
    H. Georgi and H. D. Politzer, *Phys. Rev.* **D9**, 416 (1974).
    W. Furmanski and R. Petronzio, *Z. Phys.* **C11**, 293 (1982), and references therein.

32. J. A. M. Vermaseren, S. A. Larin and T. van Ritbergen, *Phys. Lett.* **B405**, 327 (1997).

33. S. A. Larin, T. van Ritbergen and J. A. M. Vermaseren, *Nucl. Phys.* **B427**, 41 (1994).
    S. A. Larin, P. Nogueira, T. van Ritbergen and J. A. M. Vermaseren, *Nucl. Phys.* **B492**, 338 (1997).
    A. Retey and J. A. M. Vermaseren, *Nucl. Phys.* **B604**, 281 (2001).
    J. Blümlein and J. A. M. Vermaseren, *Phys. Lett.* **B606**, 130 (2005).

34. S. Moch, J. A. M. Vermaseren and A. Vogt, *Nucl. Phys.* **B646**, 181 (2002).

35. A. Vogt, S. Moch and J. A. M. Vermaseren, *Nucl. Phys.* **B691**, 129 (2004).

36. J. A. M. Vermaseren, A. Vogt and S. Moch, *Nucl. Phys.* **B724**, 3 (2005).

37. P. A. Baikov and K. G. Chetyrkin, *Nucl. Phys. Proc. Suppl.* **160**, 76 (2006).

38. I. Bierenbaum, J. Blümlein and S. Klein, *Nucl. Phys.* **B820**, 417 (2009).

39. J. Blümlein, A. De Freitas, W. L. van Neerven and S. Klein, *Nucl. Phys.* **B755**, 272 (2006).

40. J. Ablinger, I. Bierenbaum, J. Blümlein *et al.* (2010), arXiv:1007.0375, Nucl. Phys. (Proc. Suppl.) B, in print.

41. V. S. Fadin, E. A. Kuraev and L. N. Lipatov, *Phys. Lett.* **B60**, 50 (1975).

42. L. V. Gribov, E. M. Levin and M. G. Ryskin, *Nucl. Phys.* **B188**, 555 (1981).

43. V. S. Fadin and L. N. Lipatov, *Phys. Lett.* **B429**, 127 (1998).

44. R. Ellis, F. Hautmann and B. Webber, *Phys. Lett.* **B348**, 582 (1995).
    J. Blümlein (1995), hep-ph/9506446.

45. J. Blümlein and W. L. van Neerven, *Phys. Lett.* **B450**, 412 (1999).

46. J. Blümlein and A. Vogt Phys. Lett. **B370**, 149 (1996), hep-ph/9510410; Acta Phys. Polon. **B27**, 1309 (1996), hep-ph/9603450; Phys. Lett. **B386**, 350 (1996), hep-ph/9606254; Phys. Rev. **D58**, 014020 (1998), hep-ph/9712546.
    J. Blümlein, V. Ravindran, W. L. van Neerven and A. Vogt (1998), hep-ph/9806368.

47. G. F. Sterman,Partons, factorization and resummation, hep-ph/9606312, (1995).
    E. Laenen, G. Stavenga and C. D. White, *JHEP* **03**, p. 054 (2009).
48. V. Braun, A. Manashov and J. Rohrwild, *Nucl. Phys.* **B826**, 235 (2010).
49. J. Blümlein and A. Tkabladze, *Nucl. Phys.* **B553**, 427 (1999).
50. J. Kodaira, Y. Yasui and T. Uematsu, *Phys. Lett.* **B344**, 348 (1995).
    B. Geyer, D. Müller and D. Robaschik (1996), hep-ph/9611452.
    J. Kodaira and K. Tanaka, *Nucl. Phys. (Proc. Suppl.) B* **86**, 134 (2000).
    X.-D. Ji, W. Lu, J. Osborne and X.-T. Song, *Phys. Rev.* **D62**, 094016 (2000).
    A. V. Belitsky, X.-D. Ji, W. Lu and J. Osborne, *Phys. Rev.* **D63**, 094012 (2001).
    V. M. Braun, G. Korchemsky and A. Manashov, *Nucl. Phys.* **B597**, 370 (2001).
    V. M. Braun, G. Korchemsky and A. Manashov, *Nucl. Phys.* **B603**, 69 (2001).
    V. Braun, A. Manashov and B. Pirnay, *Phys. Rev.* **D80**, 114002 (2009).
51. J. Blümlein and N. Kochelev, *Nucl. Phys.* **B498**, 285 (1997).
52. A. Belitsky and A. Radyushkin, *Phys. Rep.* **418**, 1 (2005).
53. X.-D. Ji, *Phys. Rev. Lett.* **78**, 610 (1997).
54. J. Blümlein, B. Geyer and D. Robaschik, *Nucl. Phys.* **B560**, 283 (1999).
    J. Blümlein and D. Robaschik, *Phys. Lett.* **B517**, 222 (2001).
    J. Blümlein and D. Robaschik, *Phys. Rev.* **D65**, 096002 (2002).
55. A. Martin, R. Roberts, W. Stirling and R. Thorne, *Phys. Lett.* **B604**, 61 (2004).
56. S. Alekhin, K. Melnikov and F. Petriello, *Phys. Rev.* **D74**, 054033 (2006).
57. J. Blümlein, H. Böttcher and A. Guffanti, *Nucl. Phys.* **B774**, 182 (2007).
58. J. Blümlein, H. Böttcher and A. Guffanti, *Nucl. Phys. Proc. Suppl.* **135**, 152 (2004).
59. J. Blümlein and H. Böttcher, *Phys. Lett.* **B662**, 336 (2008).
60. S. Alekhin, J. Blümlein, S. Klein and S. Moch, *Phys. Rev.* **D81**, 014032 (2009).
61. S. Alekhin, J. Blümlein and S. Moch (2010), arXiv:1007.3657.
62. P. Jimenez-Delgado and E. Reya, *Phys. Rev.* **D79**, 074023 (2009).
63. P. Jimenez-Delgado and E. Reya, *Phys. Rev.* **D80**, 114011 (2009).
64. A. D. Martin, W. J. Stirling, R. S. Thorne and G. Watt, *Eur. Phys. J.* **C63**, 189 (2009).
65. E. Laenen, S. Riemersma, J. Smith and W. L. van Neerven, *Nucl. Phys.* **B392**, 162 (1993); **B392**, 229 (1993).
    S. Riemersma, J. Smith and W. L. van Neerven, *Phys. Lett.* **B347**, 143 (1995).
66. I. Bierenbaum, J. Blümlein, S. Klein and C. Schneider, *Nucl. Phys.* **B803**, 1 (2008).
67. M. Buza, Y. Matiounine, J. Smith and W. L. van Neerven, *Eur. Phys. J.* **C1**, 301 (1998).
68. I. Bierenbaum, J. Blümlein and S. Klein, *Phys. Lett.* **B672**, 401 (2009).
69. M. Glück, E. Reya and M. Stratmann, *Nucl. Phys.* **B422**, 37 (1994).
70. J. Blümlein and H. Böttcher, arXiv:1005.3113, (2010).
71. M. Glück, E. Reya and C. Schuck, *Nucl. Phys.* **B754**, 178 (2006).
72. S. Bethke, *Eur. Phys. J.* **C64**, 689 (2009).
73. J. Blümlein and H. Böttcher, arXiv:1007.2784, (2010).
74. J. Blümlein and H. Böttcher, *Nucl. Phys.* **B636**, 225 (2002).
75. J. Lagrange Nouvelles recherches sur la nature et la propagation du son, Miscellanea Taurinensis, t. II, 1760-61; Oeuvres t. I, p. 263.
    K. G. Chetyrkin, A. L. Kataev and F. V. Tkachov, *Nucl. Phys.* **B174**, 345 (1980).
    S. Laporta, *Int. J. Mod. Phys.* **A15**, 5087 (2000).
76. E. Barnes, *Proc. Lond. Math. Soc.* **6**, 141 (1908).
    E. Barnes, *Quart. J. Math.* **41**, 136 (1910).
    H. Mellin, *Math. Ann.* **68**, 305 (1910).
77. I. Bierenbaum, J. Blümlein and S. Klein, *Nucl. Phys.* **B780**, 40 (2007).

78. F. Brown, *Commun. Math. Phys.* **287**, 925 (2009).
79. W. Bailey Generalized Hypergeometric Series, (Cambridge University Press, Cambridge, 1935), 108 p.
    L. Slater Generalized Hypergeometric Functions, (Cambridge University Press, Cambridge, 1966), 273 p.
80. C. Kreimer, *Adv. Theor. Math. Phys.* **2**, 303 (1998).
    A. Connes and D. Kreimer, *Commun. Math. Phys.* **199**, 203 (1998).
    D. J. Broadhurst and D. Kreimer, *J. Symb. Comput.* **27**, p. 581 (1999).
    S. Weinzierl, *Eur. Phys. J.* **C33**, s871 (2004).
81. C. Bogner and S. Weinzierl, *J. Math. Phys.* **50**, p. 042302 (2009).
    C. Bogner and S. Weinzierl, *Int. J. Mod. Phys.* **A25**, 2585 (2010).
82. J. Blümlein, D. Broadhurst and J. Vermaseren, *Comput. Phys. Commun.* **181**, 582 (2010), and references therein.
83. J. Blümlein and S. Kurth, *Phys. Rev.* **D60**, p. 014018 (1999).
    J. A. M. Vermaseren, *Int. J. Mod. Phys.* **A14**, 2037 (1999).
84. E. Remiddi and J. A. M. Vermaseren, *Int. J. Mod. Phys.* **A15**, 725 (2000).
85. A. Goncharov, *Math. Res. Lett.* **5** *(1998) 497* .
86. S Moch, P. Uwer and S. Weinzierl, *J. Math. Phys.* **43**, 3363 (2002).
87. J. Ablinger, J. Blümlein and C. Schneider (2010), in preparation.
88. C. Schneider *J. Symbolic Comput.* **43**, 611 (2008), arXiv:0808.2543; *Ann. Comb.* **9**, 75 (2005); *J. Differ. Equations Appl.* **11**, 799 (2005); *Ann. Comb.* (2009) to appear, arXiv:0808.2596; Proceedings of the Conference on Motives, Quantum Field Theory, and Pseudodifferential Operators, to appear in the Mathematics Clay Proceedings, (2010); *Sém. Lothar. Combin.* **56**, 1 (2007), Article B56b, Habilitationsschrift JKU Linz (2007) and references therein.
89. J. Blümlein, M. Kauers, S. Klein and C. Schneider, *Comput. Phys. Commun.* **180**, 2143 (2009).
90. J. A. M. Vermaseren New features of FORM, (2000), math-ph/0010025.
91. P. Cartier (2002), Sém. Bourbaki, Mars 2001, 53e année, exp. no. **885**, Asterisque **282** 137–173.
92. V. V. Zudilin, *Uspekhi Mat. Nauk* **56**, 149 (2001).
93. S L. Adler, *Phys. Rev.* **177**, 2426 (1969).
    J. S. Bell and R. Jackiw, *Nuovo Cim.* **A60**, 47 (1969).
    C. Bouchiat, J. Iliopoulos and P. Meyer, *Phys. Lett.* **B38**, 519 (1972).

WHAT WE KNOW AND DON'T KNOW ABOUT THE ORIGIN OF THE SPIN OF THE PROTON

ANTHONY W. THOMAS, ANDREW CASEY and HRAYR H. MATEVOSYAN

CSSM, School of Chemistry and Physics, University of Adelaide,
Adelaide, SA 5005, Australia

The origin of the spin of the proton is one of the most fundamental questions in modern hadron physics. Although tremendous progress has been made since the discovery of the "spin crisis" brought the issue to the fore, much remains to be understood. We carefully review what is known and, especially in the case of lattice QCD, what is not known. We also explain the importance of QCD inspired models in providing a physical picture of proton structure and the connection between those models and what is measured experimentally and on the lattice. We specifically apply these ideas to the issue of quark orbital angular momentum in the proton. We show that the Myhrer–Thomas resolution of the proton spin crisis is remarkably consistent with modern information from lattice QCD.

Keywords: Proton spin; quark spin and orbital angular momentum; gluon angular momentum; quark models; lattice QCD; chiral symmetry.

1. Introduction

As the ground state of the baryon spectrum, the proton is the most fundamental object to be studied in nuclear physics. Its modification in a nuclear medium is critical to the understanding of nuclear structure,[1] nuclear structure functions[2,3] and the equation of state of dense matter.[4] Apart from its mass, which is now reproduced in lattice QCD at the level of a few percent[5,6] — about as accurately as one can hope, given the ambiguities of setting the physical scale in lattice QCD — there are many other properties which have been studied in great detail. For example, the electric and magnetic form factors are accurately measured[7] to several GeV^2 and lattice QCD is meeting that challenge.[8,9] The strange elastic form factors,[10–12] which play a similar role in QCD to the Lamb shift in QED, are beautifully understood in terms of lattice QCD.[13,14] Indeed, this is one case in hadronic physics where the accuracy of the theoretical calculations exceeds that of the state of the art experiments by an order of magnitude. The leading twist deep inelastic (DIS) structure functions have been studied over a tremendous range of momentum transfer and it is in the spin dependent DIS measurements that the modern fascination with the proton spin began.[15] More recently, a whole host of new spin dependent measurements, notably deeply virtual Compton scattering[16–19] (DVCS) and the transverse momentum distributions[20,21] (TMD's) are being studied.

We now know unambiguously that the fraction of the helicity of the proton, colloquially the "fraction of its spin," carried by quarks is around one third.[22,23] The exact value is deduced from the integral of the spin structure function $g_1^p(x)$ after subtracting the axial charges g_A^3 and g_A^8. While the former is known very accurately, the latter is determined using SU(3) symmetry, which recent work suggests could be broken by 20%. Using this most recent determination, the experimental value of the renormalization group invariant proton spin fraction carried by quarks is $g_A^0 = 0.36 \pm 0.03 \pm 0.05$.[24] This leaves 64% of the spin of the proton to be accounted for as orbital angular momentum of the quarks and spin and angular momentum of the gluons — modulo the controversies of defining such a decomposition for gluons.

On the surface, this seems like a very different picture of the proton than we have been used to imagining within popular quark models. On the other hand, using a relativistic quark model that includes the one-gluon-exchange (OGE) hyperfine interaction[25] and also taking into account the pion cloud required by the chiral symmetry of QCD[26,27] seems like a minimal set of requirements for a realistic, modern model. And, as shown by Myhrer, Schreiber and Thomas,[28–30] such a model is completely consistent with the latest experimental data — indeed, the most sophisticated treatment including these effects yields $g_A^0 = 0.42 \pm 0.07$.[24] Furthermore, this minimal chiral quark model naturally leads to the u quarks carrying the majority of the rest of the proton spin as orbital angular momentum.[31]

While the picture just presented is very satisfactory, it is vital that we find ways to independently check it. In particular, one would like to use measurements of DVCS and perhaps TMD's to directly measure the orbital angular momentum on each flavor of quark. Using p and d targets at Hermes[33] and JLab,[32] respectively, it has been possible to derive highly model dependent constraints on J^u and J^d at a scale of order 2 GeV^2. However, the systematic errors associated with this result (through model dependence and possible higher twist terms) are unknown and therefore, while the results are stimulating we can draw no firm conclusions without much more data at considerably higher Q^2.

In comparison with the experimental situation, the results from lattice QCD seem at first glance to be very convincing. The most recent simulation[8] yields values for $J^{u\pm d}$ with quoted errors of typically only a few percent, at a scale of order 4 GeV^2. Moreover, the orbital angular momenta of the u and d quarks deduced from these results appear to be in total contradiction with the model results which we just mentioned, with L^u negative and opposite in sign and almost equal in magnitude to L^d.

A resolution of this problem was proposed almost immediately by Thomas,[31] who showed that the evolution from the relatively low scale historically associated with quark models to 4 GeV^2 for the lattice simulation resolved the qualitative discrepancy. Whether or not this explanation is quantitatively satisfactory depends strongly on the errors that one associates with the lattice data and to a lesser extent on the scale associated with the model. Wakamatsu has taken a very literal intepretation of the quoted lattice errors and made the very strong claim[34] that this rules out the interpetation of Myhrer, Thomas and collaborators. Coincidentally, his calculation appeared to confirm the fascinating consequence of the chiral quark soliton model (CQSM), which he and his collaborators had worked out some years before,[35] namely that the highly nonlinear pion fields within that model lead to a relatively large, negative value for L^{u-d} at the model scale.

Given the major issues which we have outlined, the structure of this work appears naturally. We first address the issue of what constitute realistic errors in the lattice QCD calculations. Then we address the challenges which confront a comparison of quark models with QCD itself. Finally, we summarise the current situation with respect to the spin of the proton in the light of this analysis.

2. Beware the Phrase "Lattice QCD Says"!

Lattice QCD is an extremely powerful technique for deducing quantitative consequences of QCD. There is no other method capable of competing with it at the present time. Its successes are legion, as we already mentioned in the Introduction. In addition to those obvious quantitative successes, the study of hadron properties as a function of quark mass (away from the physical regime) has given us enormous qualitative insights into the way QCD works. For instance, it has established that meson loops are naturally suppressed for quark masses above 40 MeV (pion masses above ~ 0.4 GeV).[36,37] This in turn defines the unphysical region where a simple constituent quark model, without a pion cloud, might be expected to be realistic.

On the other hand, success brings with it many dangers, particularly the natural human tendency to optimism. Realistic physical results in lattice QCD require multiple extrapolations: lattice spacing $a \to 0$; lattice size $L \to \infty$; $m_\pi \to 0.14$ GeV; $m_K \to 0.495$ GeV. Yet there are few cases, indeed none when it comes to hadron form factors or moments, where equally high precision calculations are made for multiple lattice spacings and volumes. One must extrapolate quite long distances to the physical pion mass, with many groups employing formulas appropriate to the so-called power counting regime (PCR), when the actual quark masses lie well outside it. Finally, when it comes to operator matrix elements, the evaluation of disconnected diagrams, quark loops through which, for example, the strange quarks contribute to nucleon form factors, is in its infancy[38] and they are usually not able to be included. It is usually noted that for quantities like J^{u+d}, the calculations are incomplete because of the omission of disconnected terms. On the other hand, there is no agreed way to estimate the corresponding error, and so, for example, plots of the constraint on J^u and J^d, deduced from lattice QCD never reflect the unknown uncertainty from this source. In contrast, when it comes to an iso-vector (nonsinglet) quantity like J^{u-d}, where the u and d disconnected contributions cancel exactly, the lattice simulation is often presented (incorrectly, for the present time at least, as we discuss below) as being very reliable.

We now focus on the uncertainties associated with deriving the quark orbital angular momenta, L^u and L^d, from lattice data on the moments of the generalized parton distributions (GPD's), through which one aims to describe DVCS. We shall see that they are considerably larger than is usually recognised.

2.1. *The nonsinglet combination $J^u - J^d$*

To summarise a very complex calculation in its simplest terms, the total angular momentum carried by each flavor of quark is deduced from the matrix element of the second rank tensor, $\bar{q}\gamma^\mu D^\nu q$, suitably symmetrized and with traces subtracted, evaluated in the limit of zero momentum transfer between nucleon states. The relevant tensor structure at the nucleon level involves γ^μ with coefficient function $A(t)$, and $\sigma^{\mu\alpha}\Delta_\alpha$ with coefficient $B(t)$, where the momentum transfer is Δ and $t = -\Delta^2$. The total angular momentum is $J^q = (A(0)+B(0))/2$, for quark flavor q.[17]

Since Δ vanishes in the limit $t \to 0$, the coefficient in front of $B(0)$ actually vanishes in the limit that we need. Thus one not only has to take the usual limits $a \to 0$ and $L \to \infty$ and $m_\pi \to 0.14$ GeV but also find the limit as $t \to 0$ of $B(t)$, in spite of a vanishing coefficient.

It is important to realize that the physical content of $J^u - J^d \equiv J^{u-d}$ involves *both* the quark orbital angular momentum, L^{u-d}, and the nonsinglet spin combination $g_A^3 = \Delta u - \Delta d$ (divided by two). As we explained earlier, there is great interest in the former, while the latter is very well known from neutron β-decay. However, what is *not* at all well understood is the value implicit in the lattice estimate of J^{u-d}. *That* is what should be subtracted to obtain L^{u-d}. In fact, the direct calculation of g_A^3 is fraught with difficulties in lattice QCD.[39,40] The major problem is the very strong dependence on the lattice volume, with g_A^3 decreasing dramatically as m_π decreases, for fixed L. While this behaviour is not fully understood, it has been reproduced in a chiral quark model.[41] The feedback from the lattice boundary condition on the pion field to the valence quarks certainly plays a role, as does the nonzero contribution from the integral of the gradient of the pion field. The tendency of g_A^3 to lie well below the experimental value, even on quite a large lattice volume, has been emphasized by the RBC Collaboration.[40] Even on a lattice size of 2.7 fm, they reported a reduction by 9% (to $g_A^3 \sim 1.09$) in going from $m_\pi \sim 0.42$ GeV to 0.31 GeV, with the suggestion of a *much* bigger correction at the physical pion mass.

In plotting the effect of finite volume it is usual to show quantities as a function of $m_\pi L$. However, as explained in detail in the context of chiral quark models like the cloudy bag (CBM),[41] the relevant variable is actually $m_\pi(L - 2R)$, with $R \sim 0.8$–1.0 fm the confinement or bag radius, because it is only outside that radius that the pion field begins to decrease! To have even two pion Compton wavelengths outside the bag one therefore needs $L \sim 5.6$ fm for a pion of mass 0.2 GeV. Much of the work on moments of the GPD's has been performed on a lattice of order 2.6 fm and only recently were results reported at a single pion mass on a 3.5 fm lattice. That there is a potentially significant problem is indicated by the fact that the axial form factor calculated even on the largest lattice translates to a size almost a factor of two smaller than the experimental value[8] — a major and unexplained discrepancy.

In addition to the issue of finite size effects on the lattice, one must also make a chiral extrapolation to the physical pion mass. This was first studied by Detmold *et al.* in the context of the CBM and finite range regularization.[42] That approach naturally suppresses pion loops at masses above 0.4 GeV and has produced reliable results for extrapolation of data from beyond the PCR. Traditional chiral perturbation theory, applied to data beyond the PCR, is highly sensitive to the order to which one works and a judicious choice of parameters can lead to agreement with the experimental data for g_A^3. That process is harder to follow when the answer is not known, which is precisely the situation we face when extracting J^{u-d} from GPD data. Not only does one have *no* studies of the finite volume corrections in that case, but one needs to extrapolate in t over a tremendous range. In fact, since one knows that the experimental axial form factor is a 1 GeV dipole, most of the data used is near or beyond the formal radius of convergence and it seems highly unlikely that a linear extrapolation in t over this range could provide an accurate description of the physics.

To summarise:

- For g_A^3 there are well studied finite volume corrections which mean that on the lattice volumes commonly studied g_A^3 may be 10–20% below experiment.
- The uncertainties in the *implicit* calculation of g_A^3 through the appropriate GPD moment are unknown. We simply have no experience to guide us.
- That they may be large is illustrated by: the error of almost a factor of two in the radius of the axial form factor; the fact that one typically extrapolates linearly to $t = 0$, even though the lattice data is near or beyond the formal radius of convergence; the chiral extrapolation uses formulas appropriate to the PCR, whereas the data is not in that regime.

As a naive estimate of the error associated with using a linear function of t, Wang and Thomas recently repeated[43] the chiral analysis of the GPD data in Ref. 44. The effect of replacing the linear fit in t by a dipole was to increase J^{u-d} by 20%. As an admittedly naive estimate of the error on J^{u-d} we combine in quadrature this 20% with the potential finite volume error associated with g_A^3, which we take to be ± 0.15. Using as the central value the latest study of LHPC, this gives $J^{u-d} = +0.23 \pm 0.09$. To extract L^{u-d} from this, we choose the value for the *implicit* axial charge to be $g_A^3|_{\text{latt}} = 1.1$, corresponding to the value found in direct calculations at low pion mass on a lattice of the size used in Ref. 44. This yields $L^{u-d} = -0.32 \pm 0.09$. While, as we admit, this is something of a guess, it is almost certainly a more realistic estimate than found elsewhere. In line with this reasoning, a reasonable procedure for fixing a realistic central value would be to increase J^{u-d} by (half of) the difference between 1.1 and the physical value of $g_A^3 = 1.27$, yielding $J^{u-d} = +0.31 \pm 0.09$.

2.2. *The singlet case* J^{u+d}

To make a reasonable estimate of the error in the lattice simulation for J^{u+d} is even more difficult than for the nonsinglet case we just considered. We know that the chiral extrapolation using the coefficients appropriate to full QCD does effectively build in some disconnected terms. In the case of octet magnetic moments this seemed to be quite accurate. However, we have no experience for angular momentum. For the singlet spin component, we can use the experimental limits on polarized glue in the proton,[45] namely $|\Delta G| < 0.5$, to estimate the maximum contribution it may give through the axial anomaly, which is missing from the lattice simulation in the absence of disconnected terms. For three flavours the contribution to g_A^0 from the U(1) axial anomaly,[46–49] $-3\alpha_s \Delta G/(2\pi)$, is then less than 0.06 in magnitude.

Knowing the possible error in g_A^0 gives no guidance at all with respect to the orbital angular momentum associated with the disconnected diagrams. As a pure guess we take it to be similar in size to the spin contribution. Thus with the central value that given in the latest LHPC study we take $J^{u+d} = 0.23 \pm 0.06$. We emphasize again that this is little more than a guess and indeed it would be remarkable if the total error on J^{u+d} were less than that on J^{u-d}. However, this seems to be a reasonable starting point for a first discussion of what we do and do not know.

To get the flavor singlet orbital angular momentum carried by the quarks we subtract the value of $\Delta u|_{\rm inv} + \Delta d|_{\rm inv}$ deduced using the values of g_A^8 and g_A^0 discussed in the Introduction. Then, combining errors in quadrature, we have $L^{u+d} = +0.04 \pm 0.09$.

2.3. *Conclusions concerning individual quark angular momenta*

To summarise, our best estimates of the values of $J^{u,d}$ and $L^{u,d}$ at the physical pion mass are: $J^u = +0.27 \pm 0.06$; $J^d = -0.04 \pm 0.06$; $L^u = -0.14 \pm 0.07$; $L^d = +0.13 \pm 0.07$. In spite of all the caveats given earlier, in our view these values constitute the most realistic estimate of what we actually know from lattice QCD. Even with these increased errors, this represents a tremendous achievement.

3. Back to the Future

Models play a vital role in physics. While there are cases where exact numerical solutions for observables are feasible, our brains need to grasp issues at an intuitive level. We need to "have a feeling" for the way things work. This is more than mere gratification, novel ideas for new experiments which occasionally change the paradigm in a field are often initiated as a result of a qualitative discussion based on a particular model or a comparison between models. New theoretical ideas are often generated by discussions about conceptual problems associated with a model or a comparison between the expectations in a model and the results of a numerical simulation. In short, an intuitive understanding of phenomena in physics is an essential part of our understanding and models are a vital ingredient in developing that intuition.

The "spin crisis" provides an excellent example of the power of models. The European Muon Collaboration reported a result for the integral of g_1^p. It was a number as meaningful as any other. The excitement it generated came from a comparison with the expectations of the simple models which dominated the thinking at the time.[50] The failure of those simple models led to two decades of important experimental and theoretical work which has dramatically expanded our understanding of proton spin.

As we take stock of those achievements and decide what needs to be done next, it is extremely valuable to see how what we know compares with the more sophisticated models that have been compared with the current data. In the case of g_A^3 and g_A^8, which are the charges associated with partially conserved currents, there is no ambiguity. Chiral quark models may not describe all of the complexities of QCD but for quantities which are renormalization group invariant (RGI) it does not matter. The explanation of the proton spin sum offered by Myhrer and Thomas has recently been updated by Bass and Thomas to ensure that g_A^3 was consistent with experiment. This led to a value for $g_A^8 = 0.46 \pm 0.05$.[24] Although this is some 20% below the SU(3) prediction, many researchers have suggested that SU(3) may be broken by as much as 20% in the baryon sector, so this is not a major surprise.

3.1. *Parton distributions from quark models*

Extracting the predictions of a quark model for quantities which are not RGI presents a greater challenge. For parton distributions, for example, this was first considered in the mid-70's[51,52] and the philosphy has stood the test of time well. At that time the issue was to understand the parton distribution functions, for example the valence and sea distributions. Within constituent quark and bag models the structure of the nucleon is dominated by valence quarks and in particular the fraction of the momentum of the proton carried by its valence quarks is close to one.[53] On the other hand, the data tends to show that at a scale of order 5 GeV^2 the valence quarks carry less than 50% of the proton momentum. QCD evolution provides a natural explanation. That is, within QCD the valence momentum fraction is a monotonically decreasing function of scale, Q^2, and so if one assumes that the model describes QCD at low resolution (i.e. with gluons integrated out or unresolved) one can choose the "model scale" such that it matches the valence momentum fraction at the reference scale, typically 5 GeV^2.

Of course, the model scale is then scheme and order dependent but within a particular scheme and at a particular order, if one matches the valence momentum to data then one can make predictions for all the other moments, singlet as well as nonsinglet. From those moments one can carry out the inverse Mellin transform to construct the parton distribution functions, PDF's, at any scale. Thus the single working hypothesis yields many predictions that can be compared with experiment and more importantly a physical picture that can be used to provide an intuitive understanding of the PDF's. It turns out that the model scale is rather low, typically around 0.5 GeV (a resolution of a few tenths of a Fermi) and this has met some criticism. On the other hand, provided one matches the model to one piece of data (such as the valence quark momentum fraction), in practice the physical predictions are not strongly dependent on the order to which one works. That is, the LO and NLO scales may be different, but provided they are both chosen to match the model to data the resulting structure functions are surprisingly close. Any residual differences represent a fundamental limit to the predictive power of the model.

This technique has been widely applied in testing the predictions of various quark models against data for spin dependent as well as spin independent PDF's,[54] as well as the PDF's of atomic nuclei,[3] where the famous EMC effect provides clear evidence for the modification of the structure of a bound nucleon.[55] One of the best known phenomenological applications of this method is the work of Glück, Reya and collaborators,[56] who showed that essentially all of the data on nucleon PDF's collected at HERA could be very well described by a simple, valence dominated set of PDF's at a very low scale (~ 0.5 GeV) which were then evolved to scales as high as tens of thousands of GeV squared. Those starting distributions bear a striking resemblance to the distributions calculated within models like the MIT bag,[57] the chiral quark soliton model[58] and the NJL model.[54]

Once one has such a link between a model and data, one can begin to ask fundamental questions which may have important consequences. Some of the issues which have been first explored in models whose importance was only realized much later include the asymmetry in the $\bar{u}$ and $\bar{d}$ distributions,[59] the asymmetry between s and $\bar{s}$,[60] the violation of charge symmetry in PDF's[61-63] and the asymmetry between the polarized antiquark distributions $\Delta\bar{u}$ and $\Delta\bar{d}$.[57] To summarise briefly, the first led to a large violation of the Gottfried sum-rule,[64,65] the second and third are vital to the understanding of the NuTeV anomaly[66] and the last to understanding the structure of the nonperturbative vacuum in the proton. All of them have inspired and continue to inspire major experimental programs at the world's leading subatomic physics laboratories.

3.2. *The flavour singlet spin sum*

The comparison between model predictions and data in this case is complicated by the U(1) axial anomaly. In a quark model the operator $J_5^\mu \equiv \sum_q \bar{q}\gamma^\mu\gamma_5 q$ is partially conserved and there is no problem in evaluating it. As Schreiber, Myhrer and Thomas showed more than 20 years ago, the exchange current correction associated with the OGE hyperfine interaction first derived by De Rujula *et al.* and used in almost all quark models, substantially reduces the matrix element of this operator, as does the pion cloud required by chiral symmetry. When combined, something these authors were loathe to do in 1988 for fear of double counting but which we know now from studies of the Δ–N splitting in quenched and full QCD[67] is justified, these corrections reduce the quark spin fraction from around 65% (in a relativistic quark model like the bag) to 0.42 ± 0.07. Within the errors this is in perfectly acceptable agreement with modern data.

In QCD, the renormalization of the operator J_5^μ leads directly to an encounter with the U(1) axial anomaly. Indeed, if one renormalizes the operator in a gauge invariant way the current is no longer partially conserved. At least for N_f light quark flavours, at high enough resolution and modulo the possibility of a topological contribution at $x = 0$,[68] the matrix element of the chiral current and the gauge invariant current differ by the famous term, $-N_f \alpha_s(Q^2) \Delta G(Q^2)/(2\pi)$, which does not vanish as $Q^2 \to \infty$. This discovery originally led most investigators to believe that there might be a large component of polarized glue in the proton which would explain the EMC "spin crisis." After 20 years of sophisticated experimental work we know this is not the case and rather ΔG being of order 4, as needed to resolve the spin crisis, its magnitude is actually less than 0.5. At this level the correction to g_A^0 is less than 0.06 and although one needs to sort this out quantitatively in the future, at present we can simply conclude that polarized gluons play little or no role in resolving the spin crisis and for the time being we can compare quark model predictions directly with the data.

As we have already mentioned, the Myhrer–Thomas explanation is that relativistic quark motion, combined with a OGE exchange current and the pion cloud of the proton very naturally suppresses the quark spin fraction, leading to a current best value of 0.42 ± 0.07, which is in satisfactory agreement with the data, namely $0.36 \pm 0.03 \pm 0.05$. The physics of this success is that all of these mechanisms move quark spin to orbital angular momentum of the quarks. In the first quantitative analysis, Thomas[31] found $L^u = +0.25$ and $L^d = +0.06$, where for the time we do not try to estimate errors in the model predictions and, of course, the labels u and d denote the contributions from both quarks and antiquarks.

That these values differ dramatically from the values deduced from lattice simulations was explained by Thomas in terms of the QCD evolution from the model scale to the scale 4 GeV^2, where the lattice results were reported. Indeed, using LO QCD and starting with a typical model scale $Q_0 \sim 0.4$ GeV, he found $L^u = -0.03$ and $L^d = +0.14$. QCD evolution has reversed the signs of L^u and L^d (as *must* happen on model independent grounds — if $L^u > L^d$ at the model scale — as explained in Ref. 31). Quantitatively, L^d agrees quite well with the values deduced from lattice QCD at the end of the last section, while L^u is a little small in magnitude.

3.3. *Latest analysis*

Since the initial study of the Myhrer–Thomas model there have been several developments. Bass and Thomas have revised the calculation taking into account several well known corrections which raise g_A^3 in the bag model from 1.09 to 1.27. This changes the derived values of $L^{u,d}$ slightly. In addition, we have explored the effects of using NLO evolution as well as small variations in the model scale. We show those results in Fig. 1 and discuss them in detail below.

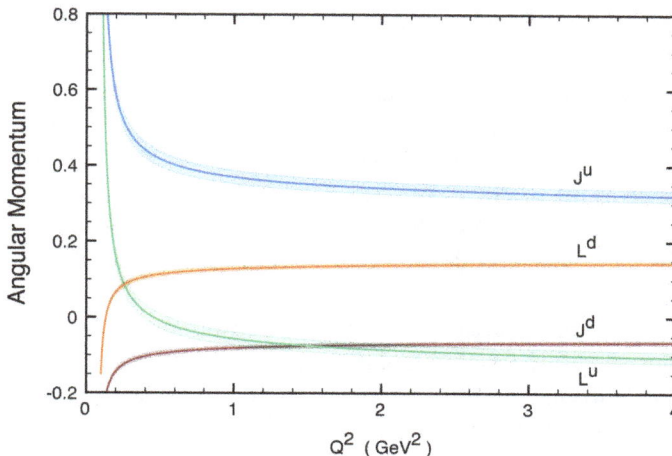

Fig. 1. Calculation of the NLO QCD evolution of J^u, L^d, J^d, L^u, starting with the values found by Bass and Thomas in the Myhrer–Thomas model at a scale $Q_0 = 0.4$ GeV. The error bands show the effect of a variation by ± 0.01 GeV^2 in Q_0^2.

It is a straightforward exercise to show that the Bass–Thomas update of the Myhrer–Thomas analysis of the spin problem yields $J^{u,d,s} = (+0.66, -0.17, +0.01)$, so that $L^{u,d,s} = (+0.23, +0.045, +0.015)$ at the model scale. These are clearly very close to the values given above. In order to compare them with the lattice data we evolve from the model scale to 4 GeV^2 using both LO and NLO evolution with $\Lambda_{\rm QCD}$ fixed to 0.248 GeV. To fix the starting scale we choose it in each case so that the total quark angular momentum at 4 GeV^2 is 0.26. This yields a smaller value for Q_0 in LO than that chosen in Ref. 31, namely 0.32 GeV, while in NLO the value agrees with that chosen by Thomas, namely 0.4 GeV. This is actually reassuring because Thomas took the value of Q_0 that at NLO yields the best fit of the empirical valence PDF's using input distributions calculated in the NJL model.[54]

The resulting values of $L^{u,d}$ are $(-0.12, +0.15)$ and $(-0.13, +0.17)$ at LO and NLO, respectively. Firstly, it is remarkable that these values are really quite close to each other. This strongly suggests that the procedure for relating the predictions in the model to data is meaningful. Secondly, if we compare with the values deduced from the lattice QCD calculations as explained above, namely $(-0.14\pm0.07, +0.18\pm0.07)$, it is obvious that the agreement is excellent. That is, the chiral quark model of Myhrer and Thomas not only resolves the spin crisis but its explanation in terms of the redistribution of spin as quark orbital angular momentum is quantitatively supported by modern lattice simulations. As an illustration of the effect of evolution, in Fig. 1 we show the evolution with Q^2 from the starting scale to 4 GeV^2 at NLO of $J^{u,d}$ and $L^{u,d}$. The error bands show the effect of a change in Q_0^2 by plus or minus 0.01 GeV^2.

4. Discussion and Outlook

We have seen that the errors on the lattice QCD results for *both* J^{u+d} and J^{u-d} are likely to be considerably larger tham the statistical errors typically quoted. Certainly it is not trivial to quantify those errors: for J^{u+d} they are primarily associated with the omission of "disconnected terms"; while for J^{u-d} the primary source of error is likely the implicit value of $\Delta u - \Delta d \equiv g_A^3$, which could be as much as 20% or more below the empirical value. That the simulated axial radius is wrong by almost a factor of two is one clear indication of the potential for trouble in this area.

In the context of such errors, the very literal acceptance by Wakamatsu of the errors typically quoted in lattice QCD papers can be seen to be potentially very misleading. Indeed, the appreciation of those possible errors was the reason Thomas rated the agreement found in Ref. 31 as satisfactory. In contrast, Wakamatsu took the value of $B^{u-d}(0) = 0.289$ and $\langle x \rangle^{u-d} = 0.158$, so that $J^{u-d} = 0.223$ with a very small error. Subtracting the empirical $g_A^3 = 1.27$ he concluded that $L^{u-d} = -0.41$, again with a very small margin of error. In comparison, our analysis of the lattice data in Subsec. 2.2 suggests that $L^{u-d} = -0.32 \pm 0.09$. While these are consistent within the quoted error the conclusions drawn from a comparison with our best NLO result, namely $L^{u-d} = -0.30$, is quite different for these two cases. *If the former is correct our result is excluded. On the other hand, if as we argue strongly is the case, the latter is correct the agreement is excellent.*

This difference becomes even more significant in the light of the prediction of the chiral quark soliton model (χQSM) that at the model scale (in that case taken to be 0.30 GeV^2, although the authors of Ref. 58 note that: "On the whole it looks as though we have determined the distributions at an even lower normalization point than" Glueck *et al.*, namely less than 0.36 GeV^2) $L^{u-d} = -0.330$.[34] If one starts NLO evolution at 4 GeV^2 with $L^{u-d} = -0.41$ it passes through the χQSM value at a scale of order 0.4 GeV^2 and is a little high at the model scale. However, if one starts at our prefered lattice value of -0.32 the evolved value is always above the χQSM prediction and at any reasonable model scale it clearly contradicts that prediction.

In summary, it is our conclusion that once the uncertainties in the current lattice QCD information on quark orbital angular momentum are taken into account, along with the effect of QCD evolution, the Myhrer–Thomas resolution of the spin crisis[30] is in remarkable agreement with the results of lattice QCD. In contrast, the prediction of the χQSM, reported by Wakamatsu and Tsujimoto,[35] appears unlikely to be correct. For the immediate future it will be very important to continue to improve the accuracy of the lattice simulations, taking into account as well as possible the effect of disconnected terms and the potentially strong dependence of J^{u-d} on lattice volume. From the experimental point of view, we look forward to further studies of DVCS and TMD's at higher Q^2, in order to reduce the model dependence of the already promising results obtained at Hermes and JLab. Those experimental efforts will need to be strongly supported by associated theoretical studies.

This is a remarkable period in the study of proton structure, with the day rapidly approaching when we can say that we genuinely understand the origin of the spin of the proton.

Acknowledgments

This work was presented by one of us (A. W. Thomas) at the celebration of Murray Gell-Mann's 80th birthday in Singapore. It was an honour and a pleasure to be able to describe the tremendous progress that has been made in understanding such a fundamental property of the proton in terms of Murray's fundamental quarks and gluons. We would like to acknowledge the efforts of Harald Fritzsch and the staff of Nanyang Technological University in organising this wonderful conference as well as the helpful comments on early drafts of this manuscript from Steven Bass, Robert Edwards, David Richards and Ross Young. This work was supported by the University of Adelaide and by the Australian Research Council through the award of an Australian Laureate Fellowship to A. W. Thomas.

References

1. P. A. M. Guichon, H. H. Matevosyan, N. Sandulescu and A. W. Thomas, *Nucl. Phys. A* **772**, 1 (2006), arXiv:nucl-th/0603044.
2. A. W. Thomas, A. Michels, A. W. Schreiber and P. A. M. Guichon, *Phys. Lett. B* **233**, 43 (1989).
3. I. C. Cloet, W. Bentz and A. W. Thomas, *Phys. Lett. B* **642**, 210 (2006).
4. J. Rikovska-Stone, P. A. M. Guichon, H. H. Matevosyan and A. W. Thomas, *Nucl. Phys. A* **792**, 341 (2007), arXiv:nucl-th/0611030.
5. R. D. Young and A. W. Thomas, *Phys. Rev. D* **81**, 014503 (2010).
6. S. Durr *et al.*, *Science* **322**, 1224 (2008), arXiv:0906.3599.
7. C. E. Hyde and K. de Jager, *Ann. Rev. Nucl. Part. Sci.* **54**, 217 (2004).
8. LHPC Collab. (J. D. Bratt *et al.*), arXiv:1001.3620.
9. H. W. Lin, S. D. Cohen, R. G. Edwards, K. Orginos and D. G. Richards, arXiv:1005.0799.
10. HAPPEX Collab. (A. Acha *et al.*), *Phys. Rev. Lett.* **98**, 032301 (2007), arXiv:nucl-ex/0609002.
11. A4 Collab. (F. E. Maas *et al.*), *Phys. Rev. Lett.* **93**, 022002 (2004).
12. R. D. Young, J. Roche, R. D. Carlini and A. W. Thomas, *Phys. Rev. Lett.* **97**, 102002 (2006), arXiv:nucl-ex/0604010.
13. D. B. Leinweber *et al.*, *Phys. Rev. Lett.* **94**, 212001 (2005), arXiv:hep-lat/0406002.
14. D. B. Leinweber *et al.*, *Phys. Rev. Lett.* **97**, 022001 (2006), arXiv:hep-lat/0601025.
15. European Muon Collab. (J. Ashman *et al.*), *Phys. Lett. B* **206**, 364 (1988).
16. A. V. Belitsky and A. V. Radyushkin, *Phys. Rep.* **418**, 1 (2005).
17. X. D. Ji, *Phys. Rev. D* **55**, 7114 (1997), arXiv:hep-ph/9609381.
18. X. D. Ji, J. Tang and P. Hoodbhoy, *Phys. Rev. Lett.* **76**, 740 (1996), arXiv:hep-ph/9510304.
19. M. Burkardt, *Phys. Rev. D* **72**, 094020 (2005), arXiv:hep-ph/0505189.
20. M. Anselmino *et al.*, *Phys. Rev. D* **75**, 054032 (2007), arXiv:hep-ph/0701006.
21. A. Bacchetta, D. Boer, M. Diehl and P. J. Mulders, *J. High Energy Phys.* **0808**, 023 (2008), arXiv:0803.0227.
22. HERMES Collab. (A. Airapetian *et al.*), *Phys. Rev. D* **75**, 012007 (2007).

23. Compass Collab. (E. S. Ageev *et al.*), *Phys. Lett. B* **647**, 330 (2007).

24. S. D. Bass and A. W. Thomas, *Phys. Lett. B* **684**, 216 (2010), arXiv:0912.1765.

25. T. A. DeGrand *et al.*, *Phys. Rev. D* **12**, 2060 (1975).

26. A. W. Thomas, *Adv. Nucl. Phys.* **13**, 1 (1984).

27. A. W. Thomas, *Progr. Theor. Phys.* **168**, 614 (2007), arXiv:0711.2259.

28. F. Myhrer and A. W. Thomas, *Phys. Rev. D* **38**, 1633 (1988).

29. A. W. Schreiber and A. W. Thomas, *Phys. Lett. B* **215**, 141 (1988).

30. F. Myhrer and A. W. Thomas, *Phys. Lett. B* **663**, 302 (2008).

31. A. W. Thomas, *Phys. Rev. Lett.* **101**, 102003 (2008), arXiv:0803.2775.

32. Jefferson Lab Hall A Collab. (M. Mazouz *et al.*), *Phys. Rev. Lett.* **99**, 242501 (2007), arXiv:0709.0450.

33. HERMES Collab. (A. Airapetian *et al.*), *J. High Energy Phys.* **0806**, 066 (2008), arXiv:0802.2499.

34. M. Wakamatsu, *Eur. Phys. J. A* **44**, 297 (2010), arXiv:0908.0972.

35. M. Wakamatsu and H. Tsujimoto, *Phys. Rev. D* **71**, 074001 (2005).

36. R. D. Young, D. B. Leinweber and A. W. Thomas, *Progr. Part. Nucl. Phys.* **50**, 399 (2003), arXiv:hep-lat/0212031.

37. A. W. Thomas, *Nucl. Phys. B* (*Proc. Suppl.*) **119**, 50 (2003), arXiv:hep-lat/0208023.

38. M. Deka *et al.*, *Phys. Rev. D* **79**, 094502 (2009), arXiv:0811.1779.

39. LHPC Collab. (R. G. Edwards *et al.*), *Phys. Rev. Lett.* **96**, 052001 (2006), arXiv:hep-lat/0510062.

40. RBC+UKQCD Collab. (T. Yamazaki *et al.*), *Phys. Rev. Lett.* **100**, 171602 (2008), arXiv:0801.4016.

41. A. W. Thomas, J. D. Ashley, D. B. Leinweber and R. D. Young, *J. Phys. Conf. Ser.* **9**, 321 (2005), arXiv:hep-lat/0502002.

42. W. Detmold, W. Melnitchouk and A. W. Thomas, *Phys. Rev. D* **66**, 054501 (2002), arXiv:hep-lat/0206001.

43. P. Wang and A. W. Thomas, arXiv:1003.0957.

44. LHPC Collab. (Ph. Hagler *et al.*), *Phys. Rev. D* **77**, 094502 (2008), arXiv:0705.4295.

45. S. Procureur, *Eur. Phys. J. A* **32**, 483 (2007).

46. R. D. Carlitz, J. C. Collins and A. H. Mueller, *Phys. Lett. B* **214**, 229 (1988).

47. G. Altarelli and G. G. Ross, *Phys. Lett. B* **212**, 391 (1988).

48. S. D. Bass, B. L. Ioffe, N. N. Nikolaev and A. W. Thomas, *J. Moscow Phys. Soc.* **1**, 317 (1991).

49. A. V. Efremov, J. Soffer and O. V. Teryaev, *Nucl. Phys. B* **346**, 97 (1990).

50. J. R. Ellis and R. L. Jaffe, *Phys. Rev. D* **9**, 1444 (1974) [Erratum: *ibid.* **10**, 1669 (1974)].

51. A. Le Yaouanc, L. Oliver, O. Pene and J. C. Raynal, *Phys. Rev. D* **12**, 2137 (1975) [Erratum: *ibid.* **13**, 1519 (1976)].

52. R. L. Jaffe, *Phys. Rev. D* **11**, 1953 (1975).

53. A. I. Signal and A. W. Thomas, *Phys. Rev. D* **40**, 2832 (1989).

54. I. C. Cloet, W. Bentz, A. W. Thomas, *Phys. Lett. B* **621**, 246 (2005), arXiv:hep-ph/0504229.

55. D. F. Geesaman, K. Saito and A. W. Thomas, *Ann. Rev. Nucl. Part. Sci.* **45**, 337 (1995).

56. M. Gluck, E. Reya and A. Vogt, *Z. Phys. C* **67**, 433 (1995).

57. A. W. Schreiber, A. I. Signal and A. W. Thomas, *Phys. Rev. D* **44**, 2653 (1991).

58. D. Diakonov, V. Petrov, P. Pobylitsa, M. V. Polyakov and C. Weiss, *Nucl. Phys. B* **480**, 341 (1996), arXiv:hep-ph/9606314.

59. A. W. Thomas, *Phys. Lett. B* **126**, 97 (1983).

60. A. I. Signal and A. W. Thomas, *Phys. Lett. B* **191**, 205 (1987).

61. E. N. Rodionov, A. W. Thomas and J. T. Londergan, *Mod. Phys. Lett. A* **9**, 1799 (1994).

62. E. Sather, *Phys. Lett. B* **274**, 433 (1992).

63. J. T. Londergan, J. C. Peng and A. W. Thomas, to appear in *Rev. Mod. Phys.*, arXiv:0907.2352.

64. W. Melnitchouk, A. W. Thomas and A. I. Signal, *Z. Phys. A* **340**, 85 (1991).

65. BCDMS Collab. (A. C. Benvenuti *et al.*), *Phys. Lett. B* **237**, 592 (1990).

66. W. Bentz, I. C. Cloet, J. T. Londergan and A. W. Thomas, arXiv:0908.3198.

67. R. D. Young, D. B. Leinweber, A. W. Thomas and S. V. Wright, *Phys. Rev. D* **66**, C94507 (2002), arXiv:hep-lat/0205017.

68. S. D. Bass, *Rev. Mod. Phys.* **77**, 1257 (2005), arXiv:hep-ph/0411005.

DETERMINATION OF LIGHT QUARK MASSES IN QCD

C. A. DOMINGUEZ

Centre for Theoretical Physics and Astrophysics,
University of Cape Town, Rondebosch 7700, South Africa,
and Department of Physics, Stellenbosch University,
Stellenbosch 7600, South Africa

The standard procedure to determine (analytically) the values of the quark masses is to relate QCD two-point functions to experimental data in the framework of QCD sum rules. In the case of the light quark sector, the ideal Green function is the pseudoscalar correlator which involves the quark masses as an overall multiplicative factor. For the past thirty years this method has been affected by systematic uncertainties originating in the hadronic resonance sector, thus limiting the accuracy of the results. Recently, a major breakthrough has been made allowing for a considerable reduction of these systematic uncertainties and leading to light quark masses accurate to better than 8%. This procedure will be described in this talk for the up-, down-, strange-quark masses, after a general introduction to the method of QCD sum rules.

1. Introduction

Due to quark and gluon confinement in QCD the (analytical) determination of the values of the quark masses necessarily calls for an approach different from that for ordinary, non confined particles. The ideal approach involves considering a QCD correlation function which on the one hand involves the quark masses and other QCD parameters, and on the other hand it involves a measurable (hadronic) spectral function. Given a framework to relate both representations of this correlator, i.e. the QCD and the hadronic representation, the quark masses would then become a function of QCD parameters, e.g. the strong coupling, some vacuum condensates reflecting confinement, etc., and measurable hadronic parameters. This framework has traditionally been that of the QCD sum rules,[1] to be described in the following sections. Unfortunately, the directly measurable Green functions in the light quark sector are the vector and axial-vector correlators entering τ-decays, and which involve the quark masses as sub-leading contributions. This feature impacts negatively on the accuracy of the results. The pseudoscalar correlator, which involves the quark masses as an overall multiplicative factor, thus making it the ideal object to be used, is not realistically measurable beyond the ground state pseudoscalar meson pole (pion or kaon). In the case of the scalar correlator the situation is worsened by the absence of a ground state pole. The standard approach for close to thirty years in connection with the (pseudo) scalar correlator has been to model the hadronic resonance spectral functions as best as possible. In the early days, radial excitations of the ground state pseudoscalar meson were parametrized in zero-width with model dependent couplings, followed later by finite-width parameterizations, and eventually incorporating sound threshold constraints from chiral perturbation theory.[2-5] At present, the masses and widths of the first two radial excitations of the pion and the kaon are well known experimentally. However, the model dependency is still unavoidable as inelasticity, non-resonant background and resonance interference are realistically impossible to guess.

As a result of this model dependency of the hadronic sector, light quark mass determinations have been historically affected by systematic uncertainties. A major breakthrough has been achieved recently[6,7] by using a successful procedure which reduces considerably the contribution of the unknown resonance sector. This quenching is accompanied by a corresponding enhancement of the better known terms in the correlator, i.e. perturbative QCD and the pseudoscalar meson pole. As a result, the light quark masses can now be determined (analytically) with an unprecedented accuracy of better than 8%, as will be described in this talk.

2. QCD Sum Rules

The method of QCD sum rules, introduced by Shifman, Vainshtein, and Zakharov[8] more than thirty years ago, has become a powerful technique to study hadronic physics in the low energy resonance region by means of QCD.[1] This method is also a complementary tool to numerical simulations of QCD on a lattice.[9] The range of applications has steadily grown over the years and covers the determination of the hadronic spectrum (masses, couplings, and widths), as well as electromagnetic, weak, and strong form factors. QCD sum rules have also been extended to finite temperature and density,[10] thus allowing for a study of chiral symmetry restoration and quark-gluon deconfinement.[11] On the QCD sector, sum rules are employed to extract the values of the quark masses[1] and of the strong coupling at a scale of the τ-lepton mass,[12-14] and they are also a tool to confront QCD predictions with experimental data, e.g. from e^+e^- annihilation and hadronic τ-lepton decays. This method is based on two fundamental pillars:

(i) the operator product expansion of current correlators at short distances, extended beyond perturbation theory to incorporate quark-gluon confinement, and

(ii) Cauchy's theorem in the complex energy (squared) plane, often referred to as quark-hadron duality.

To be more specific, let us consider a typical object in QCD in the form of the two-point function, or current correlator

$$\Pi(q^2) = i \int d^4x e^{iqx} \langle 0|T(J(x)J(0))|0\rangle \,, \tag{1}$$

where the local current $J(x)$ is built from the quark and gluon fields entering the QCD Lagrangian, and it has definite quantum numbers. Equivalently, this current can be written in terms of hadronic fields having the same quantum numbers. The specific choice of current will depend on the application one has in mind. For instance, if one is interested in determining the hadronic properties of the ρ^+-meson, then one would choose the QCD vector isovector current $J_\mu(x) = \bar{d}(x)\gamma_\mu u(x)$, and its hadronic realization in terms of the ρ^+-meson field. If the goal is to determine the values of the light quark masses, then the ideal object would be the correlator involving the axial-vector current divergences $J_5(x)|_j^i = (m_i + m_j)\bar{\psi}^i(x)\gamma_5\psi_j(x)$, with i, j the up, down, or strange quark flavors. The hadronic representation of this correlator contains the pseudoscalar meson (π or K) mass and coupling, followed by its radial excitations and the hadronic continuum. The tool to relate these two representations is Cauchy's theorem in the complex energy (squared) plane, to be discussed shortly.

The QCD correlator, Eq. (1), will contain a perturbative piece (PQCD), computed up to a given loop order in perturbation theory, and a non perturbative part mostly reflecting quark-gluon confinement. The leading order in PQCD is shown in Fig. 1. Since QCD has never been solved analytically, the effects due to confinement can only be introduced by parameterizing quark and gluon propagator corrections effectively in terms of vacuum condensates. This is done as follows.

In the case of the quark propagator

$$S_F(p) = \frac{i}{\not{p} - m} \implies \frac{i}{\not{p} - m + \Sigma(p^2)}, \tag{2}$$

the quark propagator correction $\Sigma(p^2)$ would contain the information on confinement. One expects this correction to peak at and near the quark mass-shell, i.e. for $p \simeq 0$ in the case of light quarks. Effectively, this can be viewed as in Fig. 2, where the (infrared) quarks in the loop have zero momentum and interact strongly with

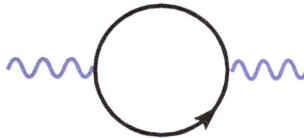

Fig. 1. Leading order PQCD correlator. All values of the four-momentum of the quark in the loop are allowed. The blue wiggly line represents the current of momentum q.

Fig. 2. Quark propagator modification due to (infrared) quarks interacting with the physical QCD vacuum, and involving the quark condensate. Large momentum flows only through the bottom propagator.

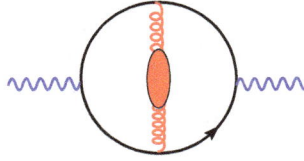

Fig. 3. Gluon propagator modification due to (infrared) gluons interacting with the physical QCD vacuum, and involving the gluon condensate. Large momentum flows only through the quark propagators.

the physical QCD vacuum. This effect is then parameterized in terms of the quark condensate $\langle 0|\bar{q}(0)q(0)|0\rangle$. Similarly, in the case of the gluon propagator one would have

$$D_F(k) = \frac{i}{k^2} \Longrightarrow \frac{i}{k^2 + \Lambda(k^2)} , \tag{3}$$

where the gluon propagator correction will peak at $k \simeq 0$, and the effect of confinement in this case will be parameterized by the gluon condensate $\langle 0|\alpha_s \mathbf{G}^{\mu\nu} \cdot \mathbf{G}_{\mu\nu}|0\rangle$ (see Fig. 3). In addition to the quark and the gluon condensate there is a plethora of higher order condensates entering the OPE of the current correlator at short distances, i.e.

$$\Pi(q^2)|_{\text{QCD}} = C_0 \hat{I} + \sum_{N=0} C_{2N+2}(q^2, \mu^2) \langle 0|\hat{O}_{2N+2}(\mu^2)|0\rangle , \tag{4}$$

where μ^2 is the renormalization scale, and where the Wilson coefficients in this expansion depend on the Lorentz indexes and quantum numbers of $J(x)$ and of the local gauge invariant operators $\hat{O}_N$ built from the quark and gluon fields. These operators are ordered by increasing dimensionality and the Wilson coefficients, calculable in PQCD, fall off by corresponding powers of $-q^2$. Since there are no gauge invariant operators of dimension $d = 2$ involving the quark and gluon fields in QCD, it is normally assumed that the OPE starts at dimension $d = 4$ (with the quark condensate being multiplied by the quark mass). This is supported by results from QCD sum rule analyses of τ-lepton decay data, which show no evidence of $d = 2$ operators.[15–17] The unit operator in Eq. (4) has dimension $d = 0$ and $C_0 \hat{I}$ stands for the purely perturbative contribution. The Wilson coefficients as well as the vacuum condensates depend on the renormalization scale. In the case of the leading $d = 4$ terms in Eq. (4) the μ^2 dependence of the quark mass cancels the corresponding dependence of the quark condensate, so that this contribution is a renormalization group (RG) invariant. Similarly, the gluon condensate is also a RG invariant quantity, hence once determined in some channel these condensates can be used throughout. At dimension $d = 6$ there appears the four-quark condensate, obtained from Fig. 1 at the next to leading order (one gluon exchange) and allowing all four quark lines to interact with the physical vacuum (see Fig. 4). While this condensate has a residual renormalization scale dependence, this is so small that in practice it can be ignored. The four-quark condensate, while relatively small, is

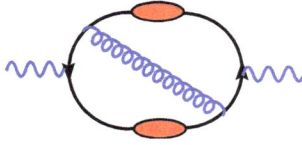

Fig. 4. The four-quark condensate of dimension $d = 6$ in the OPE. This is responsible for the $\rho - a_1$ mass splitting. Large momentum flows only through the gluon propagator.

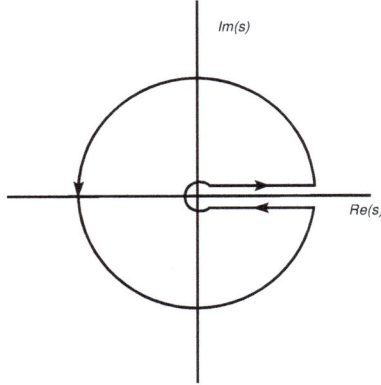

Fig. 5. Integration contour in the complex s-plane. The discontinuity across the real axis brings in the hadronic spectral function, while integration around the circle involves the QCD correlator.

crucial to explain the large $\rho(770)$–$a_1(1260)$ mass splitting. In most applications $-q^2$ is chosen large enough so that the condensates of higher dimension ($d \geq 8$) can be safely ignored. The numerical values of the vacuum condensates cannot be calculated analytically from first principles as this would be tantamount to solving QCD exactly. One exception is that of the quark condensate which enters in the Gell-Mann-Oakes-Renner relation, a QCD low energy theorem following from the global chiral symmetry of the QCD Lagrangian. Otherwise, it is possible to extract values for the leading vacuum condensates using QCD sum rules together with experimental data, e.g. e^+e^- annihilation into hadrons, and hadronic decays of the τ-lepton. Alternatively, as lattice QCD improves in accuracy it should become a valuable source of information on these condensates.

Turning to the hadronic sector, bound states and resonances appear in the complex energy (squared) plane (s-plane) as poles on the real axis, and singularities in the second Riemann sheet. In addition there will be multiple cuts reflecting non-resonant multi-particle production. All these singularities lead to a discontinuity across the positive real axis. Choosing an integration contour as shown in Fig. 5, and given that there are no further singularities in the complex s-plane, Cauchy's theorem leads to the finite energy sum rule (FESR)

$$\int_{\text{sth}}^{s_0} ds \frac{1}{\pi} f(s) \text{Im}\, \Pi(s)|_{\text{HAD}} = -\frac{1}{2\pi i} \oint_{C(|s_0|)} ds f(s) \Pi(s)|_{\text{QCD}}, \qquad (5)$$

where $f(s)$ is an arbitrary (analytic) function, s_{th} is the hadronic threshold, and the finite radius of the circle, s_0, is large enough for QCD and the OPE to be used on the circle. Physical observables determined from FESR should not depend on s_0. In practice, though, this independence is not exact, and there is usually a region of stability where observables are fairly independent of s_0, typically in the range $s_0 \simeq 1$–3 GeV^2. The variation of an observable in the stability region is incorporated into the error of the determination. Equation (5) is the mathematical statement of what is usually referred to as quark-hadron duality. Since QCD is not valid in the time-like region ($s \geq 0$), in principle there is a possibility of problems on the circle near the real axis (duality violations). I shall come back to this issue later. The right hand side of this FESR involves the QCD correlator which is expressed in terms of the OPE as in Eq. (4). The left hand side calls for the hadronic spectral function which is written as

$$\operatorname{Im}\Pi(s)|_{\text{HAD}} = \operatorname{Im}\Pi(s)|_{\text{POLE}} + \operatorname{Im}\Pi(s)|_{\text{RES}} + \operatorname{Im}\Pi(s)|_{\text{PQCD}}\theta(s - s_0), \qquad (6)$$

where the ground state pole, absent in some channels, is followed by the resonances which merge smoothly into the hadronic continuum above some threshold s_0. This continuum is expected to be well represented by PQCD if s_0 is large enough. Due to this, if one were to consider an integration contour in Eq. (5) extending to infinity, the cancellation between the hadronic continuum in the left hand side and the PQCD contribution to the right hand side, would render the sum rule a FESR. The performance of the contour integral in the complex s-plane is discussed in the next section.

3. Finite Energy Sum Rules

The integration in the complex s-plane of the QCD correlator is usually carried out in two different ways, Fixed Order Perturbation Theory (FOPT) and Contour Improved Perturbation Theory (CIPT).[18,19] The first method treats running quark masses and the strong coupling as fixed at a given value of s_0. After integrating all logarithmic terms $(\ln(-s/\mu^2))$ the RG improvement is achieved by setting the renormalization scale to $\mu^2 = -s_0$. In CIPT the RG improvement is performed before integration, thus eliminating logarithmic terms, and the running quark masses and strong coupling are integrated (numerically) around the circle. This requires solving numerically the RGE for the quark masses and the coupling at each point on the circle. The FESR Eq. (5), with $f(s) = 1$, in FOPT can be written as

$$(-)^N C_{2N+2}\langle 0|\hat{O}_{2N+2}|0\rangle = \int_0^{s_0} ds\, s^N \frac{1}{\pi}\operatorname{Im}\Pi(s)|_{\text{HAD}} - s_0^{N+1} M_{2N+2}(s_0), \qquad (7)$$

where the dimensionless PQCD moments $M_{2N+2}(s_0)$ are given by

$$M_{2N+2}(s_0) = \frac{1}{s_0^{(N+1)}} \int_0^{s_0} ds\, s^N \frac{1}{\pi}\operatorname{Im}\Pi(s)|_{\text{PQCD}}, \qquad (8)$$

and $\operatorname{Im}\Pi(s)$ is assumed dimensionless for simplicity.

If the hadronic spectral function is known in some channel from experiment, then $\text{Im}\,\Pi(s)|_{\text{HAD}} \equiv \text{Im}\,\Pi(s)|_{\text{DATA}}$, and Eq. (7) can be used to determine the values of the vacuum condensates. Subsequently, Eq. (7) can be used in a different channel to determine the masses and couplings of the hadrons in that channel. It is important to mention that the correlator $\Pi(q^2)$ is generally not a physical observable. However, this has no effect in FOPT as the unphysical constants in the correlator do not contribute to the integrals. The situation is quite different in CIPT where Eq. (5) cannot be used for unphysical correlators. For instance for a correlator whose physical counterpart is the second derivative (needed to eliminate a first degree polynomial), Cauchy's theorem and the resulting FESR must be written for the second derivative. In this case one has to use the following identity[6,7]

$$\oint_{C(|s_0|)} ds\, g(s)\Pi(s) = \oint_{C(|s_0|)} ds [F(s) - F(s_0)]\Pi''(s), \qquad (9)$$

where

$$F(s) = \int^s ds' \left[\int^{s'} ds'' g(s'') - \int^{s_0} ds'' g(s'') \right], \qquad (10)$$

and $g(s)$ is an arbitrary function. This is easily proved by integrating by parts the right hand side of Eq. (9) and using Eq. (10) to obtain the left hand side. In this case Eq. (5) becomes

$$\int_{\text{sth}}^{s_0} ds\, g(s)\frac{1}{\pi}\text{Im}\,\Pi(s)|_{\text{HAD}} = -\frac{1}{2\pi i} \oint_{C(|s_0|)} ds [F(s) - F(s_0)]\Pi''(s)|_{\text{QCD}}, \qquad (11)$$

which is the master FESR to use in CIPT. The running quark masses and the running strong coupling entering $\Pi''(s)$ are now functions of the integration variable and are not fixed as previously in FOPT. The running coupling obeys the RGE

$$s\frac{da_s(-s)}{ds} = \beta(a_s) = -\sum_{N=0} \beta_N a_s(-s)^{N+2}, \qquad (12)$$

where $a_s \equiv \alpha_s/\pi$, and e.g. for three quark flavors $\beta_0 = 9/4$, $\beta_1 = 4$, $\beta_2 = 3863/384$, $\beta_3 = (421797/54 + 3560\zeta(3))/256$. In the complex s-plane $s = s_0 e^{ix}$ with the angle x defined in the interval $x \in (-\pi, \pi)$. The RGE then becomes

$$\frac{da_s(x)}{dx} = -i\sum_{N=0} \beta_N a_s(x)^{N+2}, \qquad (13)$$

This RGE can be solved numerically at each point on the integration contour of Eq. (11) using e.g. a modified Euler method, providing as input $a_s(x = 0) = a_s(-s_0)$. Next, the RGE for the quark mass is given by

$$\frac{s}{m}\frac{dm(-s)}{ds} = \gamma(a_s) = -\sum_{M=0} \gamma_M a_s^{M+1}, \qquad (14)$$

where e.g. for three quark flavors $\gamma_0 = 1$, $\gamma_1 = 182/48$, $\gamma_2 = [8885/9 - 160\zeta(3)]/64$, $\gamma_3 = [2977517/162 - 148720\zeta(3)/27 + 2160\zeta(4) - 8000\zeta(5)/3]/256$. With the aid of

Eqs. (12) and (13) the above equation can be converted into a differential equation for $m(x)$ and integrated, with the result

$$m(x) = m(0) \exp \left\{ -i \int_0^x dx' \sum_{M=0} \gamma_M [a_s(x')]^{M+1} \right\} , \qquad (15)$$

where the integration constant $m(0)$ can be identified with $m(s_0)$.

4. The Light Pseudoscalar Correlator

I discuss now the use of FESR to determine the values of the QCD light quark masses.[6,7] As mentioned in the Introduction, in this case the ideal current operator in Eq. (1) is the axial-vector current divergence.

$$\psi_5(q^2) = i \int d^4 x e^{iqx} \langle |T(\partial^\mu A_\mu(x)|_i^j , \quad \partial^\nu A_\nu^\dagger(0)|_i^j)| \rangle , \qquad (16)$$

where $\partial^\mu A_\mu(x)|_i^j = (m_i + m_j) : \overline{q_j}(x) i \gamma_5 q_i(x):$ is the divergence of the axial-vector current, and (i, j) are flavour indexes. To simplify the notation we shall use in the sequel $m_i + m_j \equiv m$. The advantage is that the masses appear here as overall multiplicative factors, rather than as sub-leading power corrections like in other correlators, e.g. the vector or axial-vector correlators. The great disadvantage is that there is no direct experimental data beyond the pseudoscalar meson poles, i.e. the hadronic resonance spectral function, $\mathrm{Im}\,\Pi(s)|_{\mathrm{RES}}$ in Eq. (6) is not known experimentally. The only available information is that there are two radial excitations in the non-strange (π) as well as in the strange (K) channel with known masses and widths. This is hardly enough to reconstruct the full spectral function. In fact, inelasticity, non-resonant background, and resonance interference are impossible to guess so that a model is needed for the resonant spectral function. This fact, which introduces a serious systematic uncertainty, has affected all quark mass determinations using QCD sum rules until recently.[6,7] The breakthrough has been to introduce an integration kernel in the FESR tuned to suppress substantially the resonance energy region above the ground state. This kernel is of the form

$$\Delta_5(s) = 1 - a_0 s - a_1 s^2 , \qquad (17)$$

where $\Delta_5(s)$ stands for either $f(s)$ in Eq. (5) for FOPT, or $g(s)$ in Eq. (9) for CIPT. The coefficients are fixed by requiring that $\Delta_5(s)$ vanish at the peak of the two radial excitations, i.e. $\Delta_5(M_1^2) = \Delta_5(M_2^2) = 0$. This has the effect of reducing the resonance contribution to the FESR to a couple of a percent of the ground state contribution, well below the uncertainty due to α_s. Clearly, other more elaborate choices of the integration kernel are possible. It has been found, though, that the simplest form above is optimal in the sense of simplicity and of achieving the goal of reducing considerably the systematic uncertainty from the hadronic resonance sector.

The FESR following from Cauchy's theorem, Eq. (5), takes now the form

$$-\frac{1}{2\pi i} \oint_{C(|s_0|)} ds\psi_5^{\mathrm{QCD}}(s)\Delta_5(s) = 2f_P^2 M_P^4 \Delta_5(M_P^2)$$

$$+ \int_{s_{th}}^{s_0} ds \frac{1}{\pi} \mathrm{Im}\,\psi_5(s)|_{\mathrm{RES}}\Delta_5(s)\,, \qquad (18)$$

where f_P and M_P stand for the pseudoscalar meson pole (π, K) parameters, and this FESR is suitable for FOPT. In the case of CIPT the FESR, Eq. (11), becomes

$$-\frac{1}{2\pi i} \oint_{C(|s_0|)} ds\psi_5''^{QCD}(s)[F(s) - F(s_0)] = 2f_P^2 M_P^4 \Delta_5(M_P^2)$$

$$+ \frac{1}{\pi} \int_{s_{th}}^{s_0} ds\,\mathrm{Im}\,\psi_5(s)|_{\mathrm{RES}}\Delta_5(s)\,, \quad (19)$$

where $F(s)$ is defined in Eq. (10). It should be clear that the distortion introduced by the integration kernel $\Delta_5(s)$, Eq. (17), affects all three terms of the FESR as Cauchy's theorem remains valid as long as the kernel is an analytic function. Hence, the suppression achieved in the hadronic resonance contribution is compensated by corresponding changes in the pseudoscalar meson pole and in the QCD contributions. Since these two terms are reasonably well known, this is a welcome feature.

5. Results

The light-quark pseudoscalar correlator in PQCD is known up to fifth-loop order,[20] with the strong coupling being determined from data on τ-decays[12–14] $\alpha_s(M_\tau^2) = 0.344 \pm 0.009$ which corresponds to a QCD scale in the $\overline{MS}$ scheme of $\Lambda = 365$–397 MeV. The handling of logarithmic quark mass singularities in this correlator requires some care, as explained in Refs. 21 and 22. The leading non-perturbative contributions of dimension $d = 4$ are also known, together with higher order quark mass corrections. Complete expressions of all these contributions to the FESR for both FOPT and CIPT may be found in Refs. 6 and 7. A posteriori, it is found that quark mass terms of order $\mathcal{O}(m^4)$ and higher, as well as vacuum condensates of dimension $d \geq 6$ are negligible on account of the integration kernel, Eq. (17). The hadronic resonance spectral function has been modeled by two Breit-Wigner forms normalized at threshold according to chiral perturbation theory.[2–5] The quark masses were determined with and without this contribution in order to gauge its impact on the final result. It turns out that allowing for a $\pm 30\%$ uncertainty in the hadronic resonance contribution impacts on the quark masses at the level of less than 1%. Assuming the unknown six-loop PQCD contribution to be equal to the known five-loop term also has an impact at the 1% level. The presence of the integration kernel, Eq. (17), thus enhances considerably the importance of the better known contributions to the FESR, i.e. PQCD and the pseudoscalar meson pole. The major sources of uncertainty are then the values of the strong coupling and of the radius of the integration circle s_0. Results for the strange quark mass in FOPT are shown in Fig. 6. The width of the stability region is typical of FESR applications. However, in CIPT as shown in Fig. 7 this region is remarkably wide, with $m_s(2 \text{ GeV})$ being exceptionally stable. This is also the case for the up- and down-quark masses in CIPT as shown in Figs. 8 and 9. While there is no clear cut criterion to establish which integration technique is best, results from this application clearly favour CIPT in terms of the stability region. Taking into account all possible sources of uncertainty, and following a very conservative approach of adding them up, rather than combining them in quadrature, the final results (in the $\overline{MS}$ scheme) are

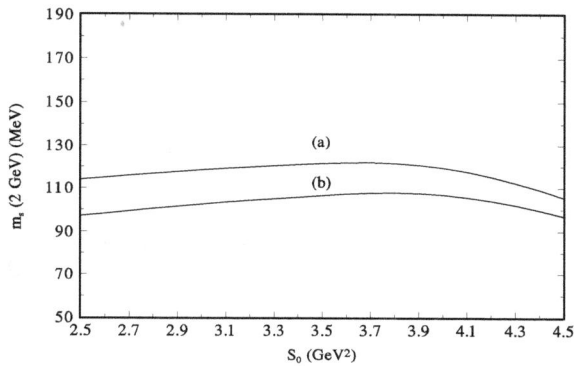

Fig. 6.   Running strange quark mass in FOPT at a scale of 2 GeV as a function of s_0. Curve (a) is for $\Lambda = 330$ MeV, and curve (b) for $\Lambda = 420$ MeV, corresponding, respectively, to $\alpha_s(M_\tau^2) = 0.31$ and $\alpha_\varepsilon(M_\tau^2) = 0.36$.

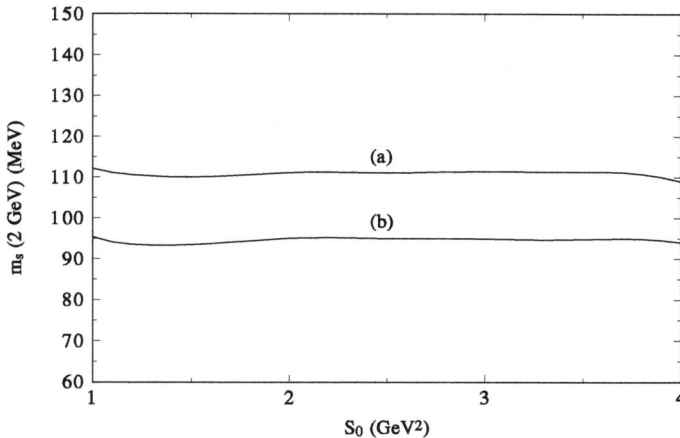

Fig. 7.   Running strange quark mass in CIPT at a scale of 2 GeV as a function of s_0. Curve (a) is for $\Lambda = 330$ MeV, and curve (b) for $\Lambda = 420$ MeV, corresponding, respectively, to $\alpha_s(M_\tau^2) = 0.31$ and $\alpha_\varepsilon(M_\tau^2) = 0.36$.

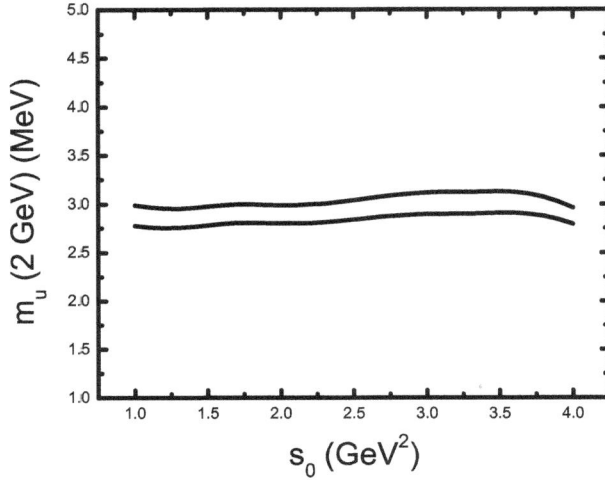

Fig. 8. Up quark mass at 2 GeV as a function of s_0 for $\alpha_s(M_\tau^2) = 0.335(0.353)$, or $\Lambda_{\rm QCD} = 365(397)$ MeV, upper (lower) curve, respectively.

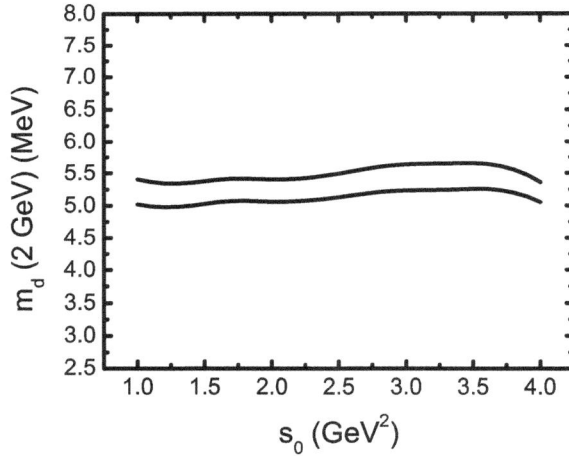

Fig. 9. Down quark mass at 2 GeV as a function of s_0 for $\alpha_s(M_\tau^2) = 0.335(0.353)$, or $\Lambda_{\rm QCD} = 365(397)$ MeV, upper (lower) curve, respectively.

$$m_u(2 \text{ GeV}) = 2.9 \pm 0.2 \text{ MeV},$$

$$m_d(2 \text{ GeV}) = 5.3 \pm 0.4 \text{ MeV},$$

$$m_{ud} \equiv \frac{m_u + m_d}{2} = 4.1 \pm 0.2 \text{ MeV},$$

$$m_s(2 \text{ GeV}) = 102 \pm 8 \text{ MeV}.$$

This is at present the most accurate determination of the light quark masses using QCD sum rules, with the above results in very good agreement with the most recent lattice QCD values.[23-28] With the systematic uncertainties from the hadronic resonance sector under very good control, a future reduction of the errors in the quark masses will mostly rely on more accurate determinations of the strong coupling.

Acknowledgment

This talk is based on work done in collaboration with N. F. Nasrallah, R. H. Röntsch and K. Schilcher. Work supported in part by NRF (South Africa).

References

1. P. Colangelo and A. Khodjamirian, in *At the Frontier of Particle Physics/ Handbook of QCD*, Vol. 3, ed. M. A. Shifman (World Scientific, Singapore, 2001), pp. 1495–1576.
2. C. A. Dominguez, *Z. Phys. C* **26**, 269 (1984).
3. C. A. Dominguez and E. de Rafael, *Annals Phys.* **174**, 372 (1978).
4. C. A. Dominguez, L. Pirovano and K. Schilcher, *Phys. Lett. B* **425**, 193 (1998).
5. J. Bijnens, J. Prades and E. de Rafael, *Phys. Lett. B* **348**, 226 (1995).
6. C. A. Dominguez, N. F. Nasrallah, R. Röntsch and K. Schilcher, *J. High Energy Phys.* **0805**, 020 (2008).
7. C. A. Dominguez, N. F. Nasrallah, R. Röntsch and K. Schilcher, *Phys. Rev. D* **79**, 014009 (2009).
8. M. A. Shifman, A. I. Vainshtein and V. I. Zakharov, *Nucl. Phys. B* **147**, 385 (1979); *Nucl. Phys. B* **147**, 448 (1979).
9. S. Scherer, *Adv. Nucl. Phys.* **27**, 277 (2003).
10. A. I. Bochkarev and M. E. Shaposnikov, *Nucl. Phys. B* **286**, 220 (1986).
11. C. A. Dominguez, M. Loewe, J. C. Rojas and Y. Zhang, *Phys. Rev. D* **81**, 014007 (2010).
12. M. Davier, S. Descotes-Genon, A. Höcker, B. Malaescu and Z. Zhang, *Eur. Phys. J. C* **56**, 305 (2008).
13. P. A. Baikov, K. Chetyrkin and J. H. Kühn, *Phys. Rev. Lett.* **101**, 012002 (2008).
14. A. Pich, arXiv: 1001.0389.
15. C. A. Dominguez, *Phys. Lett. B* **345**, 291 (1995).
16. C. A. Dominguez and K. Schilcher, *Phys. Rev. D* **61**, 114020 (2000).
17. C. A. Dominguez and K. Schilcher, *J. High Energy Phys.* **0701**, 093 (2007).
18. A. A. Pivovarov, *Z. Phys., Particles & Fields, C* **53**, 461 (1992).
19. F. Le Diberder and A. Pich, *Phys. Lett. B* **286**, 147 (1992).
20. P. A. Baikov, K. G. Chetyrkin and J. H. Kuhn, *Phys. Rev. Lett.* **96**, 012003 (2006).
21. M. Jamin and M. Münz, *Z. Phys. C* **66**, 633 (1995).
22. K. G. Chetyrkin, C. A. Dominguez, D. Pirjol and K. Schilcher, *Phys. Rev. D* **51**, 5090 (1995).
23. C. Allton *et al.* (RBC/UKQCD Coll.), *Phys. Rev. D* **78**, 114509 (2008).
24. Y. Nakamura *et al.* (CP-PACS Coll.), *Phys. Rev. D* **78**, 034502 (2008).
25. J. Noaki *et al.* (JLQCD/TWQCD Coll.) *PoS* (**LAT2009**), 096 (2009).
26. B. Blossier *et al.* (ETM Coll.) *J. High Energy Phys.* **0804**, 020 (2008).
27. A. Bazazov *et al.* (MILC Coll.) *PoS* (**CD09**), 007 (2009).
28. A. Bazazov *et al.* (MILC Coll.) *PoS* (**LAT2009**), 077 (2009).

THE ELUSIVE HIGGS BOSON(S)

JOHN F. GUNION

Department of Physics, University of California at Davis,
Davis, CA 95616, USA
jfgunion@ucdavis.edu

We outline the motivations for having a light Higgs boson with fairly SM-like couplings to WW, ZZ but with unusual decays that elude LEP limits and that will also make its detection at the LHC quite challenging. The Next-to-Minimal Supersymmetric Model provides a very appealing model of this type.

Keywords: Higgs; NMSSM; LHC.

1. Introduction

To a large extent the motivations for a light Higgs boson all involve Quantum Loop calculations. In particular: loop corrections to m_W, m_Z, ... needed for precision electroweak (PEW) comparisons; quadratically divergent loop corrections to the Higgs mass leading to the hierarchy problem; loop-derived Renormalization Group Evolution for parameters that, for example, leads to RGE generated Electroweak Symmetry Breaking (EWSB); and so forth. Loop calculations owe a great debt to Murray Gell-Mann and to our conference leader Harald Fritzsch in that the theory of QCD made possible detailed and very successful checks of the RGE evolution of the strong coupling constant, which also plays a crucial role in the apparent coupling constant unification at M_U within the MSSM.

Exposing the nature of electroweak symmetry breaking is goal #1 for the LHC. I will argue that the manner in which the Higgs boson associated with EWSB will be revealed could be very different from conventional expectations. In particular, if we demand that the theory be "ideal" (as specified below), the possibilities are very limited and imply the existence of a Supersymmetric Higgs boson with SM-like WW, ZZ, $f\bar{f}$ couplings *but with unusual decays*.

An "ideal" model should have the following characteristics:

(1) calculable unitarization of $WW \rightarrow WW$ scattering;
(2) excellent agreement with precision electroweak (PEW) data;
(3) consistency with LEP limits;
(4) no hierarchy problem (i.e. the quadratically divergent loop contributions to the Higgs mass should be cut off by a new physics scale of $\mathcal{O}(\text{TeV})$);

(5) coupling constant unification without ad hoc tuning of matter content and/or Lagrangian parameters;

(6) no electroweak fine-tuning (i.e. the value of m_Z is not simply input and/or is not strongly dependent on input global parameters at the GUT, or any other, scale).

With these criteria in mind, the most important points/ingredients leading to a preference for Supersymmetry and, in particular, the Next-to-Minimal Supersymmetric Model, are the following:

- Precision electroweak (PEW) data is beautifully consistent with a light Higgs boson with SM-like couplings to WW, ZZ.
  The best, most "ideal," PEW description is obtained if there is a Higgs that couples to WW, ZZ with SM-like strength and that has $m_h \lesssim 105$ GeV.
- Spin-0 particles have a natural place in chiral SUSY fields.
- Supersymmetry with a supersymmetry breaking scale $\mathcal{O}(\text{TeV})$ is a very beautiful approach to curing the hierarchy problem.
- A supersymmetric model with supersymmetry breaking at the TeV scale and exactly two Higgs doublets gives "dynamical" (i.e. RGE) gauge coupling unification as well as RGE EWSB.
- Minimizing electroweak fine-tuning (i.e. the sensitivity of m_Z to high scale parameters) in Supersymmetric Models implies that the SUSY-breaking scale should be significantly below 1 TeV for which SUSY models generically predict $m_h \lesssim 105$ GeV (for the lightest CP-even SUSY Higgs).
- The LEP limit on an h with SM-like WW, ZZ couplings and SM-like decays or the very similar MSSM-like decays is $m_h > 114$ GeV. To have $m_h < 105$ GeV the h must decay in an elusive way. The last three decay channels below are ones that are not so strongly constrained by LEP.
- The reason why the NMSSM is particularly suitable is that it it is the simplest SUSY model that contains an elusive Higgs decay almost as a matter of course. In particular, it has an extra naturally light ($m_a < 2m_b$) CP-odd Higgs boson, a, for which $B(h \to aa)$ can be large. The latter is natural since the $h \to b\bar{b}$ coupling is so small that $\Gamma(h \to aa) \gg \Gamma(h \to b\bar{b})$ is easily achieved.

Very generally, suppressed $B(h \to b\bar{b})$ is quite "natural" in extended models. The Higgs provides a natural "portal" to new physics of many kinds that could lead to a weak m_h LEP limit. As a result, we really should not count on knowing what the Higgs "looks like."

2. Motivations for Nonstandard Decays — Single H

Let us now consider in more detail the model-independent motivations sketched earlier for there to be a $\lesssim 105$ GeV Higgs with SM-like WW, ZZ couplings but unusual decays. We denote such a Higgs by H in this section.

132

Table 1. LEP m_H Limits for a H with SM-like ZZ coupling, but varying decays.

Mode	SM modes	2τ or $2b$ only	$2j$	$WW^* + ZZ^*$	$\gamma\gamma$	$E\!\!\!/$	$4e, 4\mu, 4\gamma$
Limit (GeV)	114.4	115	113	100.7	117	114	114?

Mode	$4b$	Pure 4τ	any (e.g. $4j$)	$2f + E\!\!\!/$
Limit (GeV)	110	$86 \to \sim 108$[a]	82	90?

[a]Recent preliminary ALEPH limit to be discussed later.

(1) Precision Electroweak data: The latest $\Delta\chi^2$(PEW) vs m_H plot from LEP-EWWG indicates that at 95% CL $m_H < 157$ GeV and the $\Delta\chi^2$ minimum is near 87 GeV when all data are included. However, there is substantial discrepancy between the best m_H value determined from the b and c forward–backward asymmetries and the rest of the data, the latter preferring $m_H \sim 50$ GeV while the former prefers $m_H \sim 500$ GeV. This is reflected in the fact that the SM has a CL for the PEW fit of only 0.14 when all data are included, whereas if the $A_{\text{FB}}^{b,c}$ data is excluded the SM gives a fit with CL near 0.78 and a 95% CL upper on m_H of ~ 105 GeV.[1] In addition, the latest m_W and m_t measurements clearly prefer $m_{h_{\text{SM}}} \lesssim 100$ GeV.

(2) Electroweak Baryogenesis: $m_H \lesssim 105$ GeV is needed for the phase transition to be strong enough.

(3) Largest LEP excess: Perhaps the ideal Higgs should be such as to predict the 2.3σ excess at $M_{b\bar{b}} \sim 98$ GeV seen in the $Z + b\bar{b}$ final state.

The simplest possibility for explaining the excess is to have $m_H \sim 100$ GeV and $B(H \to b\bar{b}) \sim (0.1 - 0.2) \times B(h_{\text{SM}} \to b\bar{b})$ (assuming H has SM ZZ coupling as desired in order to satisfy precision electroweak constraints) with the remaining H decays being to one or more of the channels that are poorly constrained at LEP.

As already noted, one generic way of having a low LEP limit on m_H is to suppress the $H \to b\bar{b}$ branching ratio by having a light a (or h) with $B(H \to aa) > 0.7$ and $m_a < 2m_b$ (to avoid the LEP $Z + 4b$ limit of $m_h > 110$ GeV, i.e. above the ideal range). Then, for $\tan\beta \gtrsim 2$, $a \to \tau^+\tau^-$ is dominant for $2m_\tau < m_a < 2m_b$ while $a \to jj$ is dominant for $m_a < 2m_\tau$. And, as noted, both $H \to 4\tau$ and $H \to 4j$ are more weakly constrained by LEP data than $H \to 2b, 4b$.[2,3]

If a Higgs with SM-like ZZ coupling has mass no larger than 105 GeV, then it is useful to recall the model-independent triviality and global minimum constraints on the scale Λ of new physics. According to the standard analysis, new physics is needed by $\Lambda < 10^4(10^3)$ GeV if $m_H \sim 100$ GeV (~ 50 GeV).

The overall situation can be sketched as in Fig. 1 with the LEP excess in the $b\bar{b}$ channel also preferring $m_H \sim 90$–105 GeV and with baryogenesis preferring $m_H < 105$ GeV.

Finally, it should be noted that the "effective" m_H of order 50 GeV preferred by the PEW data can be achieved by combining loop diagrams involving somewhat

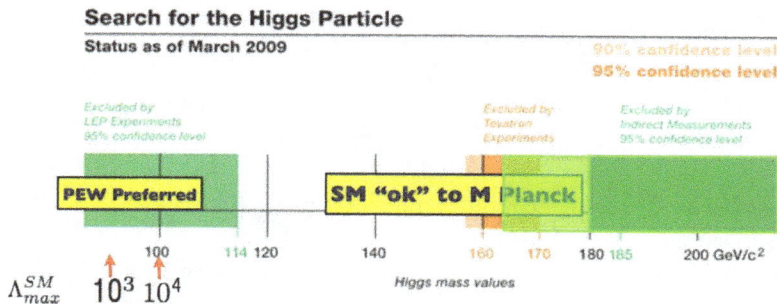

Fig. 1.   Higgs overview.

light SUSY particles (as needed to avoid significant electroweak fine-tuning in any case) with loop corrections associated with an H in the $m_H \sim 90$–105 GeV mass range that has SM-like HZZ and HWW couplings.

3. Why SUSY, in Particular the NMSSM

Now let us turn to the reasons why SUSY also prefers a Higgs with mass $\lesssim 105$ GeV and why the NMSSM is a particularly apt choice of SUSY model in this case. First, SUSY cures the naturalness/hierarchy problem. Second, SUSY + R-parity leads to a dark matter candidate. Thirdly, in the MSSM (with exactly two Higgs doublet superfields), if we assume that all sparticles reside at the $\mathcal{O}(1 \text{ TeV})$ scale and that μ is also $\mathcal{O}(1 \text{ TeV})$, then we get gauge coupling unification and RGE EWSB. However, there is a potential issue of fine-tuning with regard to RGE EWSB. Namely, is it necessary to finetune the GUT scale parameters in order to obtain the correct Z mass? $F \equiv \text{Max}_i \frac{\partial \log m_Z}{\partial \log p_i}$ (p_i = GUT-scale parameter) measures the degree to which GUT parameters must be tuned. Most theorists agree that $F < 10$–20 (10%–5% fine-tuning) is quite acceptable. The analyses of Refs. 2 and 4 show that in almost any SUSY model this requires $m_{\tilde{t}} \lesssim 400$ GeV and a relatively light gluino. The problem with such a low value for $m_{\tilde{t}}$ in the MSSM context is that the lightest CP-even Higgs boson, denoted by h, is then predicted to have mass $\lesssim 100$ GeV and yet is typically SM-like in its decays and therefore excluded for $m_h < 114$ GeV. To get $m_h > 114$ GeV requires $m_{\tilde{t}} > 800$ GeV and then $F > 50$. What is needed is a SUSY model for which the stop mass can be low but for which the resulting light $\lesssim 105$ GeV Higgs is not excluded by LEP.

The NMSSM (see for example Ref. 5) is an extremely attractive choice for achieving this. The scalar component of the extra single superfield $\hat{S}$ contains an additional CP-even Higgs and an additional CP-odd Higgs, leading to three CP-even eigenstates, $h_{1,2,3}$ and two CP-odd eigenstates, $a_{1,2}$. The lightest of the CP-even states, h_1, is quite SM-like and the lightest CP-odd state, a_1, is mainly singlet and has a small mass that is protected by a $U(1)_R$ symmetry. Large $B(h_1 \to a_1 a_1)$ is easy to achieve. We will simplify and denote for the most part $h_1 \to h$ and $a_1 \to a$.

Let us review the many attractive features of the NMSSM:

(1) It solves the μ problem: $W \ni \lambda \hat{S} \hat{H}_u \hat{H}_d$ leads to $\mu_{\text{eff}} = \lambda \langle S \rangle \lesssim 500$ GeV for typical values of λ and typical values of $\langle S \rangle$, where the size of $\langle S \rangle$ is determined by the scale of soft-SUSY-breaking which we have argued is most ideally $\lesssim$ 1 TeV.

(2) It preserves MSSM gauge coupling unification.

(3) It preserves radiative EWSB.

(4) It preserves dark matter (provided that R-parity conservation is invoked).

(5) Like any SUSY model, it solves the quadratic divergence hierarchy problem.

(6) And, $m_h \lesssim 105$ GeV is allowed because of $h \to aa$ decays with $m_a < 2m_b$. As a result, it yields excellent agreement with PEW constraints and it allows minimal fine-tuning F for getting m_Z (i.e. v) correct after evolving all parameters defined at the GUT scale, M_U, down to the m_Z scale.[3] The latter is because $\tilde{t}_1$, $\tilde{t}_2$ can be light (~ 350 GeV is just right). Low F also requires that $m_{\tilde{g}}$ be not too far above 300 GeV. The minimum in F as a function of m_{h_1} is quite sharp. At $\tan \beta = 10$, this minimum is at $m_{h_1} \sim 100$ GeV at $F \sim 6$, rising slowly to $F \sim 10$ by $m_{h_1} \sim 108$ GeV and then very sharply for still higher m_{h_1}.

To repeat, in the MSSM such low stop masses are not acceptable since m_{h^0} would then be below the LEP limit while large enough $m_{\tilde{t}}$ to obtain $m_{h^0} > 114$ GeV would lead to large F, especially if h^0 is SM-like.

(7) An a with large $B(h \to aa)$ and $m_a < 2m_b$ can be achieved without fine-tuning of the A_λ and A_κ soft-SUSY breaking parameters ($V \ni A_\lambda S H_u H_d + \frac{1}{3} A_\kappa S^3$) that control the a properties.[6] When $A_\lambda, A_\kappa \to 0$, the NMSSM has an additional $U(1)_R$ symmetry, in which limit the a is pure singlet and $m_a = 0$. If $U(1)_R$ is exact at M_U, then the the RGE equations imply that the zero GUT-scale values of A_λ and A_κ naturally evolve to low-scale values for which m_a is small and $B(h \to aa)$ has a good chance of being large.

In Ref. 6 we devised a measure G of the A_λ and A_κ tuning needed to achieve $m_a < 2m_B$ and $B(h \to aa) > 0.7$. We found that to achieve small G the a must be largely singlet (e.g. $\sim 10\%$ at amplitude level if $\tan \beta \sim 10$) as expected, but also that m_a cannot be too small; in particular, ~ 7.5 GeV $\lesssim m_a$ (but below $2m_b$) is required for the very smallest G values ($G < 15 - 20$).

Of course, multisinglet extensions of the NMSSM will expand the possibilities. Indeed, typical string models predict a plethora of SM singlet scalar superfields leading to the possibility of many light a's, light h's and light $\tilde{\chi}$'s, to pairs of which the SM-like h with $m_{\tilde{h}} \sim 100$ GeV could decay.

4. Predictions Regarding a Generic Light a and the NMSSM a

It is important to explore the possibilities for detecting the light a and the h in the scenarios discussed above. First, we need to go into more detail regarding the limits on the a that can be obtained from existing data. First, we define a generic

135

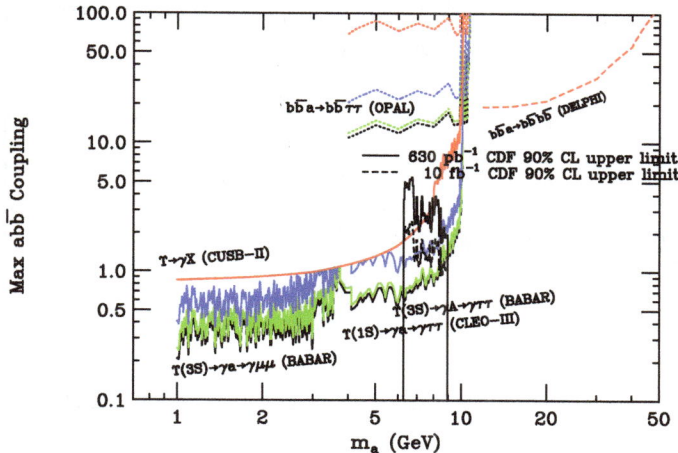

Fig. 2. (Color online) Limits on $C_{ab\bar{b}}$ from Refs. 9 and 7. These limits include recent BaBar $\Upsilon_{3S} \rightarrow \gamma\mu^+\mu^-$ and $\gamma\tau^+\tau^-$ data. Color code: $\tan\beta = 0.5$ (red); $\tan\beta = 1$ (blue); $\tan\beta = 2$ (green); $\tan\beta \geq 3$ (black). The black histograms show $\tan\beta \geq 3$ limits extracted from CDF data as discussed later.

coupling to fermions by

$$\mathcal{L}_{af\bar{f}} \equiv iC_{af\bar{f}} \frac{ig_2 m_f}{2m_W} \bar{f}\gamma_5 fa. \tag{1}$$

In what follows, we will ignore possible large-$\tan\beta$ SUSY corrections.

In the NMSSM context (more generally, in 2HDM(II) models), we can predict the branching ratios of the a as a function of m_a and the ratio of the H_u and H_d Higgs field VEV's: $\tan\beta = \langle H_u \rangle / \langle H_d \rangle$. Especially important, are $B(a \rightarrow \mu^+\mu^-)$ and $B(a \rightarrow \tau^+\tau^-)$. We make special note of a few results. First, for $\tan\beta > 1.5$ and $m_a \lesssim 2m_B$ one finds $B(a \rightarrow \mu^+\mu^-) \sim 0.002$–$0.003$. Second, values of $B(a \rightarrow \tau^+\tau^-)$ at $\tan\beta \gtrsim 2$ are $\gtrsim 0.85$ at moderate m_a but decline to ~ 0.75 for 10 GeV $\lesssim m_a \lesssim 2m_B$, largely as a result of $B(a \rightarrow gg)$ becoming quite significant when m_a is near the $b\bar{b}$ threshold of the b-quark loop contribution to the agg coupling. More details can be found in Refs. 7 and 8. Inputting the relevant branching ratios, one finds that the strongest $|C_{ab\bar{b}}|$ limits derive from BaBar and CLEO data on $\Upsilon(nS) \rightarrow \gamma a$; they appear in Fig. 2 along with some old LEP limits. The most unconstrained region is that with $m_a > 8$ GeV, especially 9 GeV $< m_a < 12$ GeV.

What are the implications in the NMSSM context? The light a mass eigenstate is a mixture of the doublet CP-odd scalar field (that is contained in the MSSM) and the new singlet CP-odd state coming from the S of the NMSSM: $a = \cos\theta_A a_{\text{MSSM}} + \sin\theta_A a_S$. Then, one finds $C_{ab\bar{b}} = \cos\theta_A \tan\beta$, where small $\cos\theta_A$ is expected. The limits on $C_{ab\bar{b}}$ will then imply limits on $\cos\theta_A$ for any given choice of $\tan\beta$. These are shown in Fig. 3.

To see the impact of these limits we can compare scatter plots of points allowed (in G vs $\cos\theta_A$ space) before and after imposing the $\cos\theta_A^{\max}$ limits. This comparison

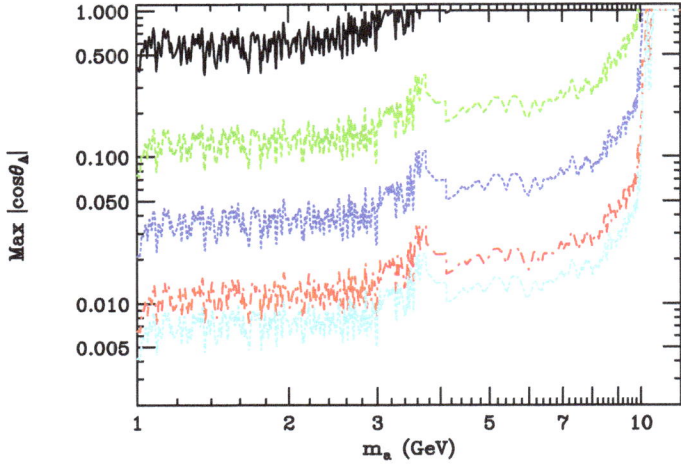

Fig. 3.   The curves are for $\tan\beta = 1$ (upper), 3, 10, 32 and 50 (lower).

appears in Fig. 4. From the figure, we see that many points with low m_{a_1} and large $|\cos\theta_A|$ are eliminated by the $|\cos\theta_A| < \cos\theta_A^{\max}$ requirement, including almost all the $m_{a_1} < 2m_\tau$ (blue) points and a good fraction of the $2m_\tau < m_{a_1} < 7.5$ GeV (red) points. Thus, we have a convergence whereby low "light-a" fine-tuning in the NMSSM *and* direct $\Upsilon_{3S} \to \gamma\mu^+\mu^-$ limits *both* single out the small $|\cos\theta_A|$ and $m_a > 7.5$ GeV part of parameter space. LHC studies of the h and a should (and have) focused on this case.

In fact, results from ALEPH[10] further shift the focus to high m_a in the NMSSM context. ALEPH examines $e^+e^- \to Zh$ with $h \to aa$ and $a \to \tau^+\tau^-$ and places

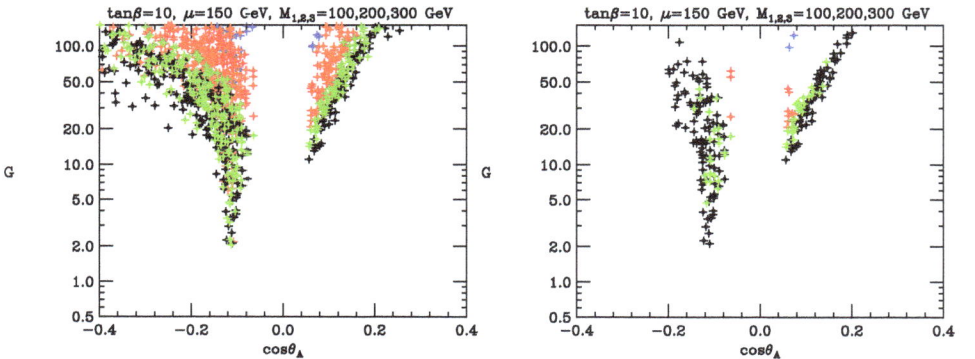

Fig. 4.   (Color online) Light-a_1 fine-tuning measure G before and after imposing $|\cos\theta_A| \leq \cos\theta_A^{\max}$. Color code: $m_a < 2m_\tau$ (blue); $2m_\tau < m_a < 7.5$ GeV (red); 7.5 GeV $< m_a < 8.8$ GeV (green); 8.8 GeV $< m_a < 2m_B$ GeV (black).

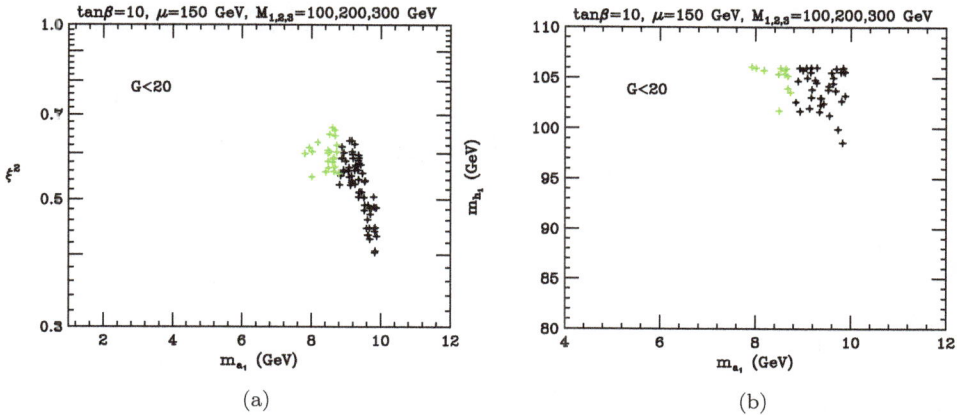

Fig. 5. (a) ξ^2 vs m_{a_1} for points with $|\cos\theta_A| < \cos\theta_A^{\max}$. (b) scatter plot of points in the left-hand plot that survive the ALEPH limits. Plotted points are taken from a $\tan\beta = 10$ fixed-μ scan.

limits on

$$\xi^2 \equiv \frac{\sigma(Zh)}{\sigma(Zh_{\text{SM}})} B(h \to aa)[B(a \to \tau^+\tau^-)]^2 . \qquad (2)$$

The limits on ξ^2 are fairly strong for $m_h \lesssim 100$ GeV, but weaken rapidly as one increases m_h up to the upper limit of interest to us of $m_h \sim 105$ GeV. This latter remark is particularly relevant for $\tan\beta \geq 10$ since $m_{h_1} \sim 105$ GeV can be achieved with low F and G for $m_a = 10$ GeV. In contrast, for $\tan\beta = 3$ the $a_1 \to \tau^+\tau^-$ branching ratio is still large but for the stop masses that give small electroweak fine-tuning one finds $m_{h_1} \lesssim 100$ GeV. Thus, no $\tan\beta = 3$ points survive the new ALEPH limits (although they would have survived the "expected" ALEPH limits — the actual ALEPH limits disagree with expectations from the SM by nearly 2.5σ). A plot of ξ^2 vs m_{a_1} in the case of $\tan\beta = 10$ appears in Fig. 5(a), and the points that survive the ALEPH limits are plotted in m_{h_1}–m_{a_1} space in Fig. 5(b). Note that since only h_1 has significant ZZ, WW coupling and since m_{h_1} of the surviving points is at most just above 105 GeV, agreement with PEW observables will be excellent for these scenarios.

As an important side remark, one should note in Fig. 4 that there is a strict lower bound on $|\cos\theta_A|$ arising from the need for $B(h_1 \to a_1 a_1) > 0.7$. This implies that if one can place sufficiently strong limits on $|\cos\theta_A|$ then one could rule out the ideal scenarios in which $m_{h_1} < 105$ GeV can be consistent with LEP results by virtue of $h_1 \to a_1 a_1$ decays being dominant. An important question is whether the implied minimum doublet component of the a_1 will allow its direct detection at the LHC. We turn to this issue shortly.

For $\tan\beta < 3$ it quickly becomes easier to escape the ξ^2 limits. This is illustrated by repeating the plots of Fig. 5 for $\tan\beta = 2$ and $\tan\beta = 1.7$ in Figs. 6 and 7. For these figures we have imposed the "ideal" requirement that $m_{\text{eff}} \leq 105$ GeV,

tanβ=2.0, μ=150 GeV M_{SUSY}=300 GeV, A_t=−300 GeV

tanβ=2.0, μ=150 GeV M_{SUSY}=300 GeV, A_t=−300 GeV

(a)   (b)

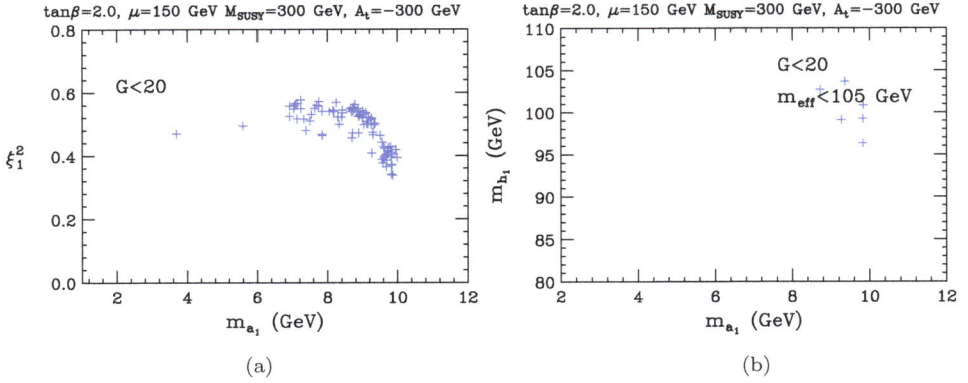

Fig. 6.   (a) ξ_1^2 for h_1 vs m_{a_1} for points with $|\cos\theta_A| < \cos\theta_A^{max}$. (b) m_{h_1} vs m_{a_1} scatter plot of points in the left-hand plot that survive the ALEPH limits. Plotted points are taken from a $\tan\beta = 2$ fixed-μ scan.

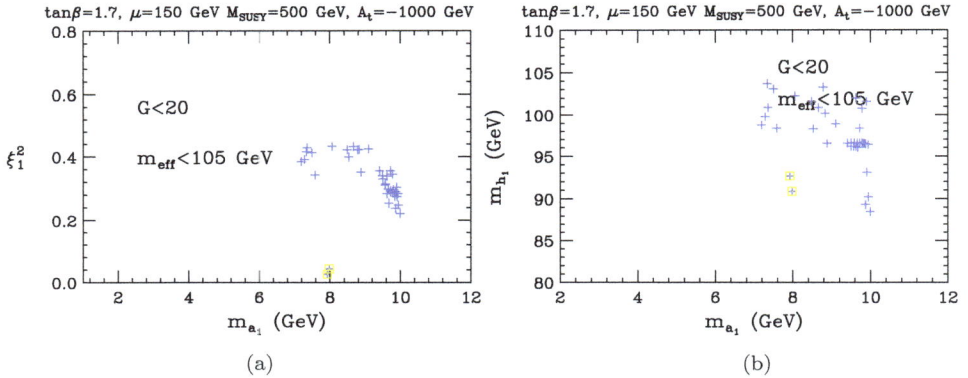

tanβ=1.7, μ=150 GeV M_{SUSY}=500 GeV, A_t=−1000 GeV

tanβ=1.7, μ=150 GeV M_{SUSY}=500 GeV, A_t=−1000 GeV

(a)   (b)

Fig. 7.   (a) ξ_1^2 for h_1 vs m_{a_1} for points with $|\cos\theta_A| < \cos\theta_A^{max}$. (b) m_{h_1} vs m_{a_1} scatter plot of points in the left-hand plot — in this case, all left-hand plot points survive the ALEPH constraints. Plotted points are taken from a $\tan\beta = 1.7$ fixed-μ scan.

where $\ln m_{eff} = \sum_i [g_{ZZh_i}^2/g_{ZZh_{SM}}^2] \ln m_i$ characterizes to good approximation the net effect of the CP-even Higgs bosons on PEW observables. The reason that low-$\tan\beta$ points escape the new ALEPH limits more easily is that $B(a_1 \to jj)$ increases and $B(a_1 \to \tau^+\tau^-)$ decreases as $\tan\beta$ decreases. The Higgs is increasingly "buried" by having $h_1 \to a_1 a_1 \to 4j$ decays dominate.

5. Direct Discovery of the a at a Hadron Collider

At a hadron collider, one studies $gg \to a \to \mu^+\mu^-$ and tries to reduce the heavy flavor background by isolation cuts on the muons. At the Tevatron, one can use results from a related CDF search for narrow resonances to obtain limits on $|C_{ab\bar{b}}|$ in the 6.3 GeV $\lesssim m_a \lesssim$ 9 GeV range. The actual CDF analysis was done for an

$L = 630$ pb^{-1} data set. We have statistically extrapolated to $L = 10$ fb^{-1} to obtain higher integrated luminosity expectations. The $L = 630$ pb^{-1} and extrapolated $L = 10$ fb^{-1} Tevatron limits are shown by the black histograms in Fig. 2 (which appeared earlier). Comparing to the limits from $\Upsilon \to \gamma a$ channels obtained by BaBar and others, one sees that with a data sample of order $L = 10$ fb^{-1} the CDF limits would compete with BaBar limits for $\tan\beta \geq 3$ and $m_a \gtrsim 9$ GeV, approaching the $C_{ab\bar{b}} = \tan\beta\cos\theta_A \sim 1$ level that impacts the most preferred NMSSM scenarios.

For $m_a > 9$ GeV (above their narrow resonance analysis window) one can extract implicit limits assuming no 3σ excesses in any of the $M_{\mu^+\mu^-}$ bins they plot in this range. In the region $M_{\mu^+\mu^-} \sim 10$ GeV (but away from $\Upsilon(nS)$ peaks), this implicit limit is $|C_{ab\bar{b}}| \lesssim 2$. If m_a is degenerate with the $\Upsilon(1S)$ peak the limit is of order $|C_{ab\bar{b}}| < 3.5$. Once one crosses the b quark threshold, the limits rise quickly to $|C_{ab\bar{b}}| \lesssim 5\text{-}6$. However, all these are well below the $|C_{ab\bar{b}}| \sim 30$ level for which a light a might have explained Δa_μ if $C_{ab\bar{b}} \gtrsim 32$.[9] Thus, one can finally conclude that Δa_μ cannot be due to a light a.

One can hope that the LHC can place even stronger limits on a light a than the Tevatron. There have been studies of the Upsilon, Drell–Yan and heavy-flavor backgrounds by CMS and ATLAS, but only ATLAS has presented public results — see Fig. 8. In the figure, one sees that the Drell–Yan background is much smaller than the heavy flavor background, even after muon isolation cuts.

To see if a light a is observable above this background one needs to know the cross section for the a at a give LHC energy and the efficiency for a events given the cuts used in obtaining the plotted background. A recent Monte Carlo study gives $\varepsilon_{\text{ATLAS}} = 0.1$. We write $\varepsilon_{\text{ATLAS}} = 0.1r$. After adding in maximal estimates for additional heavy flavor backgrounds not simulated in obtaining Fig. 8, using ATLAS

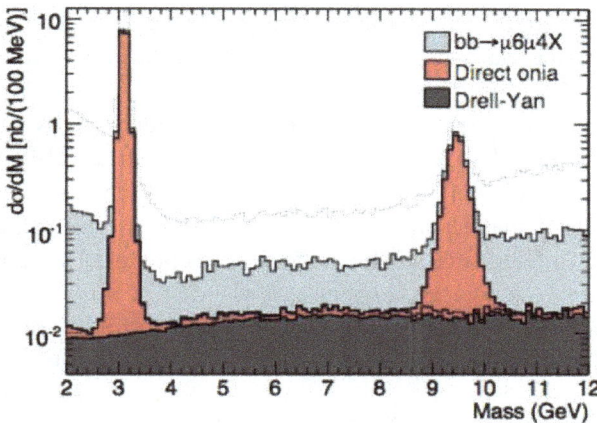

Fig. 8. ATLAS dimuon spectrum prediction after corrections for acceptance and efficiencies.[11] Only the $b\bar{b}$ heavy flavor background was simulated for this plot.

Table 2. Luminosities (fb^{-1}) needed for $S/\sqrt{B} = 5$ if $C_{ab\bar{b}} = 0.1$ and $\tan\beta = 10$.

Case	$m_a = 8$ GeV	$m_a = M_{\Upsilon_{1S}}$	$m_a \lesssim 2m_B$
ATLAS LHC7	$17/r^2$	$63/r^2$	$9/r^2$
ATLAS LHC10	$13/r^2$	$47/r^2$	$7/r^2$
ATLAS LHC14	$10/r^2$	$38/r^2$	$5.4/r^2$

resolutions for a very narrow resonance, and taking $\tan\beta = 10$ and $\cos\theta_A = 0.1$ (typical of the $\tan\beta = 10$ models with the smallest G), we obtain Table 2 which gives the L's required to obtain a $S/\sqrt{B} = 5$ signal for various LHC energies (7, 10, and 14 TeV).

For $r = 1$, the required L's away from the Upsilon resonance may be achieved in a year or two of LHC operation after the 2012 shutdown. However, smaller $\cos\theta_A$ values are possible at $\tan\beta = 10$ and for lower $\tan\beta$ values $C_{ab\bar{b}} = \cos\theta_A \tan\beta$ can be quite a bit smaller than the value of unity assumed in Table 2. The L required for $S/\sqrt{B} = 5$ increases rapidly as $|C_{ab\bar{b}}|$ falls below 1. In short, the prospects for direct discovery of the a of the NMSSM ideal Higgs scenarios at the LHC are significant, but there is no guarantee. Of course, it is very possible that more refined techniques with $r > 1$ and/or smaller background will be developed by ATLAS and CMS that would further improve sensitivity to the a.

6. Detecting the Light h of the NMSSM

I will focus first on possibilities at a hadron collider, especially the LHC. There are basically two distinct cases to discuss. First, once $\tan\beta \geq 3$ the branching ratios of an a with $2m_\tau < m_a < 2m_B$ are essentially $\tan\beta$-independent and $B(a \to \tau^+\tau^-)$ is fairly large, of order 0.89 at its maximum, declining to ~ 0.75 just below the $B\bar{B}$ threshold. In this case, the final state with the largest LHC rate would be $h \to aa \to 4\tau$, with a possibly useful rate for $h \to aa \to \tau^+\tau^-\mu^+\mu^-$ given that $B(a \to \mu^+\mu^-)$ is of order 0.003. Meanwhile, the "standard" LHC search channels would all be suppressed relative to SM and MSSM expectations by roughly a factor of 10. The implication of this latter is that all these standard search modes would fail, including in particular the $h \to \gamma\gamma$ channel that is the most important in the SM and MSSM for a Higgs mass below 125 GeV (as favored by PEW).

The second case, to which we turn later, is the case of $\tan\beta \leq 2$. In this region, $B(a \to \tau^+\tau^-)$ and $B(a \to \mu^+\mu^-)$ both decline rapidly while $B(a \to 2j)$ increases. This is an even more difficult situation for LHC detection of the h.

Let us first focus on the $\tan\beta \geq 3$ case. (As we have discussed earlier, scenarios that clearly escape the new ALEPH limits are found for $\tan\beta \gtrsim 10$.) The new LHC production/decay channels in which a Higgs signal might be detectable then include:

(1) $gg \rightarrow h \rightarrow aa \rightarrow 4\tau$ and $2\tau + \mu^+\mu^-$.

There is an actual D0 analysis[12] of this mode using $L \sim 4$ fb^{-1} of data. There are even small $\sim 1\sigma$ excesses for $m_a \sim 4$ GeV and $10-11$ GeV consistent with the predicted signal. About $L \sim 40$ fb^{-1} would be needed for a 3σ signal. Some estimates for this channel at the LHC have appeared in Ref. 13. My own estimate of the net useful cross-section at $\sqrt{s} = 14$ TeV (more or less consistent with that in Ref. 13) is

$$\sigma(gg \rightarrow h)B(h \rightarrow aa)[2B(a \rightarrow \mu^+\mu^-)B(a \rightarrow \tau^+\tau^-)]\epsilon \sim 3-6 \text{ fb}. \quad (3)$$

Backgrounds are small, so perhaps $10-20$ events in a single $\mu^+\mu^-$ bin would be convincing. This would imply that $L = 4$ fb^{-1} might reveal a signal. As an aside, note that if $m_a < 2m_\tau$, then $B(a \rightarrow \mu^+\mu^-) \sim 0.025$–$0.1$ (depending on m_a and $\tan\beta$) and

$$\sigma(gg \rightarrow h)B(h \rightarrow aa)[B(a \rightarrow \mu^+\mu^-]^2\epsilon > (50 \text{ fb}) \times \epsilon.$$

If $\epsilon > 0.02$ (as seems likely) then one finds $\sigma_{\text{eff}} > 1$ fb. This $\mu^+\mu^- + \mu^+\mu^-$ final state should be background free (two $\mu^+\mu^-$ pairs in the same narrow mass bin). Even though $m_a < 2m_\tau$ NMSSM scenarios are strongly disfavored by BaBar data, this channel should be pursued at the LHC with the expectation that one could eliminate $m_a < 2m_\tau$ ideal Higgs scenarios once and for all.

(2) $WW \rightarrow h \rightarrow aa \rightarrow \tau^+\tau^- + \tau^+\tau^-$.

The key will be to tag relevant events using spectator quarks and require very little activity in the central region by keeping only events with 4 or 6 tracks. As far as the signal alone is concerned this channel looks moderately promising,[14,15] but definitive results for the background are not yet available.

(3) $t\bar{t}h \rightarrow t\bar{t}aa \rightarrow t\bar{t} + \tau^+\tau^- + \tau^+\tau^-$.

A proper study has not been performed as of yet. Would isolated tracks/leptons from the τ's make this easier than the $t\bar{t}h \rightarrow t\bar{t}b\bar{b}$ channel that appears to be so challenging because of QCD backgrounds?

(4) $W, Z + h \rightarrow W, Z + aa \rightarrow W, Z + \tau^+\tau^- + \tau^+\tau^-$.

Leptons from W, Z decay and isolated tracks/leptons from the τ's could very possibly provide a clean signal. So far, a detailed study of signal and background is not available.

(5) $\tilde{\chi}_2^0 \rightarrow h\tilde{\chi}_1^0$ with $h \rightarrow aa \rightarrow 4\tau$.

Again, there is no detailed study as of yet, but one should recall that the $\tilde{\chi}_2^0 \rightarrow h\tilde{\chi}_1^0$ channel provides a viable signal in the MSSM for some parameter choices when $h \rightarrow b\bar{b}$ decays are dominant.

(6) Last, but definitely not least: diffractive production $pp \rightarrow pph \rightarrow ppX$.

Using the precise measurements of the 4-momenta of the final-state protons, the mass M_X can be reconstructed with roughly a 1–2 GeV resolution, potentially revealing a Higgs peak, independent of the decay of the Higgs. The event is quiet so that the tracks from the τ's appear in a relatively clean environment, allowing track counting and associated cuts.

142

Signal significances from Ref. 16 approach 3 to 5 sigma depending on triggering assumptions and luminosity. The triggering assumptions for this level of signal were very conservative. It is possible that we can increase our rates by about a factor of 2 to 3 using additional triggering techniques.

Let us now turn to NMSSM ideal Higgs scenarios in the case of $\tan\beta \leq 2$. As noted earlier, as $\tan\beta$ decreases the $a \to \tau^+\tau^-$ and $a \to \mu^+\mu^-$ branching ratios become smaller and the $a \to jj$ channels become more dominant. The backgrounds in these latter channels are much larger. The Higgs begins to be "buried" under the QCD background. A quick review of possibly useful channels leads one to focus on:

(1) $gg \to h \to aa \to \mu^+\mu^- X$.

If a single a tag is ok then the effective useful cross section ($\sqrt{s} = 14$ TeV) is

$$\sigma(gg \to h)B(h \to aa)[2 \times B(a \to \mu^+\mu^-]\epsilon > (70\text{ fb}) \times \epsilon$$

for $B(a \to \mu^+\mu^-) > 0.001$ (as applies for $\tan\beta > 1$ and $m_a < 2m_\tau$). If $\epsilon > 0.02$ (seems likely) then $\sigma_{eff} > 1.4$ fb. The background is probably significant, but maybe not too large after zeroing in on the a peak in the $\mu^+\mu^-$ channel. If we suppose that 50 events would suffice to pick out the signal, this would imply that only $L = 30$ fb^{-1} would be needed. It seems that this approach should be pursued.

Of course, if SUSY is sufficiently light to predict $m_h < 105$ GeV (recall this is needed to avoid electroweak fine-tuning) then we will have a plethora of SUSY signals. We could encounter a situation at the LHC where supersymmetric particles are easily seen in relatively early data but detection of the Higgs is impossible or requires very large L. In this case, there is still an indirect way of checking that something like a light h with SM-like couplings to WW, ZZ is present. Namely, one should find that $WW \to WW$ scattering is perturbative for WW com energies of 1 TeV and above. Such a check is relatively certain to require integrated luminosity at $\sqrt{s} = 14$ TeV of at least 100 fb^{-1} and more probably substantially more than that.

Of course, Higgs detection in the case of difficult h decays is vastly simpler at a linear e^+e^- collider, for which I use the generic shorthand ILC. At such a facility and for the planned $\sqrt{s} \geq 350$ GeV energy and $L > 500$ fb^{-1}, $e^+e^- \to ZX$ is guaranteed to reveal the Higgs peak in M_X (independently of how the h decays since one can determine M_X using the incoming beam momenta and the measured Z momentum in the $Z \to e^+e^-$ and $Z \to \mu^+\mu^-$ channels) just as LEP might have. Unfortunately, it is now clear that > 10 years will pass before an ILC might begin operation.

7. Other Related "Nightmare" Scenarios

As noted earlier, the NMSSM is just the simplest extension of the MSSM obtained by adding just one singlet chiral superfield. String models with more than one additional singlet are easily constructed. All of the singlet CP-odd and CP-even scalars will mix with the corresponding MSSM doublet scalar fields and create a series of CP-even and CP-odd Higgs eigenstates, h_i and a_k. As one adds more singlets, detecting any one of the h_i eigenstates is likely to become increasingly difficult. First, each h_i will have reduced ZZ, WW coupling. Second, it can be arranged that these eigenstates decay in complex ways that are only weakly constrained by LEP data. In particular, $h_i \to a_j a_k$ decays will generically be present and can easily be dominant in precise analogy to the one-singlet NMSSM case we have described in detail above. A further bonus of such a complicated spectrum of states is that it is very easy to construct a model in which the effective PEW Higgs mass defined earlier is such that $m_{\mathrm{eff}} < 105$ GeV. Even $m_{\mathrm{eff}} \sim 50$ GeV can be achieved for an appropriate spectrum of m_i and g_{ZZh_i} choices without violating LEP constraints.

In addition, if the spectrum of h_i states is sufficiently dense that the resonances are separated by less than or order of the experimental resolutions in the relevant detection channels, then one cannot simply do bump-hunting. Rather, it would be necessary to ascertain the presence of an enhanced smooth continuum in the mass reconstructed in a given final state. This is the "worst case" scenario envisioned long ago in Ref. 17. Detection of such an enhanced continuum would be incredibly difficult at a hadron collider. The only possibility that comes to mind is to use the $pp \to pph_i$ process in which m_i can be reconstructed independently of the decays of the h_i. With enough events one could possibly detect an excess in the M_X spectrum of some $pp \to ppX$ channels. However, cross-sections are low and reliable isolation of this class of events requires low event overlap rates, implying that the large L would have to be accumulated without going to the highest instantaneous luminosities.

In contrast, an ILC with $\sqrt{s} \gtrsim 250$ GeV and sufficient integrated L is guaranteed to detect the h_i by measuring the M_X spectrum in the process $e^+e^- \to ZX$. For example, an isolated h_i peak will be observable at $M_X \sim m_i \sim 90$–100 GeV no matter how the h_i decays so long as $g_{ZZh_i}^2 \gtrsim 0.05g_{ZZh_{\mathrm{SM}}}^2$ for $L \gtrsim 100$ fb^{-1}. As shown in Ref. 17, for $L \gtrsim 500$ fb^{-1} the ILC could even detect the continuum enhancement in M_X of a series of *overlapping* Higgs bosons with spectrum such that $m_{\mathrm{eff}} < 150$ GeV.

Of course, models in which there are multiple Higgs decaying in difficult to detect manners have been proliferating. Even in the context of supersymmetry, if one is willing to allow R-parity violation (which of course means that the LSP could not be a dark matter candidate) Higgs decays such as $h \to \tilde{\chi}_1^0 + \tilde{\chi}_1^0 \to 3j + 3j$ (for baryonic R-parity violation) become a generic possibility and are obviously very difficult to detect at a hadron collider. The Higgs would be completely buried under the QCD background.

8. Conclusions

There is ample opportunity for the CP-even Higgs boson(s) that carry the ZZ, WW coupling to be extremely elusive at a hadron collider. Such elusiveness is more or less required in order to escape LEP limits when these Higgses are sufficiently light to give an ideal description of precision electroweak constraints and no electroweak fine-tuning. A light spectrum for the CP-even Higgs bosons also improves the chances that electroweak baryogenesis is viable. The NMSSM and its extensions allow for elusive Higgs to Higgs-pair decays and do so in the context of low-scale supersymmetry with all its benefits: a cure of the hierarchy problem; coupling constant unification; RGE electroweak symmetry breaking; .... While supersymmetry should be discovered early in these scenarios, the Higgs might never be found without constructing a high energy, high luminosity e^+e^- (or $\mu^+\mu^-$) collider. If the LHC has not discovered a SM-like Higgs from the SM or MSSM after its second run beginning in 2012, we must take these elusive-Higgs scenarios seriously.

Acknowledgments

I would like to thank the National Technological University of Singapore for their exceptional hospitality and support and H. Fritzsch for his kind invitation to talk at this gathering in honor of Murray Gell-Mann. I wish Murray many more years of productivity and good health.

References

1. M. S. Chanowitz, arXiv:0806.0890.
2. R. Dermisek and J. F. Gunion, *Phys. Rev. Lett.* **95**, 041801 (2005).
3. R. Dermisek and J. F. Gunion, *Phys. Rev. D* **73**, 111701 (2006).
4. R. Dermisek and J. F. Gunion, *Phys. Rev. D* **77**, 015013 (2008).
5. J. R. Ellis, J. F. Gunion, H. E. Haber, L. Roszkowski and F. Zwirner, *Phys. Rev. D* **39**, 844 (1989).
6. R. Dermisek and J. F. Gunion, *Phys. Rev. D* **75**, 075019 (2007).
7. R. Dermisek and J. F. Gunion, arXiv:0911.2460.
8. R. Dermisek and J. F. Gunion, arXiv:1002.1971.
9. J. F. Gunion, *J. High Energy Phys.* **08**, 032 (2009).
10. ALEPH Collab., arXiv:1003.0705.
11. D. D. Price, arXiv:0808.3367.
12. V. M. Abazov *et al.*, *Phys. Rev. Lett.* **103**, 061801 (2009).
13. M. Lisanti and J. G. Wacker, *Phys. Rev. D* **79**, 115006 (2009).
14. A. Belyaev *et al.*, arXiv:0805.3505.
15. J. F. Gunion and T. Tait, private communication.
16. J. R. Forshaw, J. F. Gunion, L. Hodgkinson, A. Papaefstathiou and A. D. Pilkington, *J. High Energy Phys.* **04**, 090 (2008).
17. J. R. Espinosa and J. F. Gunion, *Phys. Rev. Lett.* **82**, 1084 (1999).

PROSPECTS FOR NEW PHYSICS AT THE LHC*

JOHN ELLIS

Theory Division, Physics Department, CERN, CH-1211 Geneva 23, Switzerland
John.Ellis@cern.ch

High-energy collisions at the LHC are now starting. The new physics agenda of the LHC is reviewed, with emphasis on the hunt for the Higgs boson (or whatever replaces it) and supersymmetry. In particular, the prospects for discovering new physics in the 2010-2011 run are discussed.

Keywords: LHC; Higgs boson; supersymmetry; extra dimensions.

1. Preamble

Back in the dawn of prehistory, when I was a student, Murray Gell-Mann was an inspiration to me. His work dominated the particle physics landscape that I entered then: strangeness, $V - A$ theory, the eightfold way, quarks, current algebra, the renormalization group, and so much more, and laid the basis for the developments that have occurred since. These fundamental contributions are now so embedded in the fabric of particle physics that perhaps we sometimes forget to remember them and celebrate their originator with all the respect he deserves. It is therefore a pleasure for me to participate in this meeting honouring Murray and his achievements, and a privilege to be given the opportunity to speak here about the next chapter in particle physics that is now unfolding.

2. Open Questions Beyond the Standard Model

There is a standard list of open questions beyond the Standard Model of particle physics. What is the origin of particle masses, and are they due to a single elementary Higgs boson, or to something else? Why are there so many types of matter particles, and is the answer to this question related to the origin of matter in the Universe? What is the nature of the cold dark matter that makes up some 80% of the matter in the Universe? How to unify the fundamental forces? How to construct a quantum theory of gravity?

The LHC may be able to address all these questions. One of its main motivations has been to solve the mass problem, and its experiments should tell us definitively

---

*CERN-PH-TH/2010-074.

whether or not there exists a Higgs boson resembling that in the Standard Model. There is a large class of models in which cold dark matter is composed of particles that were in thermal equilibrium in the early Universe, in which case they should weigh $\sim$ 1 TeV, and be produced at the LHC. One example of such a theory is supersymmetry, which would also assist in the unification of the fundamental forces. Measuring the masses of supersymmetric particles, if they exist, would be a great way of testing predictions based on such theories. Supersymmetry and extra dimensions are key aspects of string theory, the only promising candidate for a consistent quantum theory of gravity, which could be tested in very novel ways if the LHC produces microscopic black holes. What are the prospects that the LHC might cast light on these enticing scenarios?

3. Hunt for the Higgs

In the Standard Model, particles acquire their masses from couplings to a universal scalar field whose associated quantum, called the Higgs boson, has become the 'Holy Grail' of particle physics. Direct searches for the Higgs boson by experiments at the LEP accelerator established the lower bound $m_h > 114.4$ GeV.[1] Precision electroweak data are sensitive to m_h through quantum corrections, and yield the preferred range[2]

$$m_h = 87^{+35}_{-26} \text{ GeV}. \tag{1}$$

The 95% confidence-level upper limit on the mass of the Higgs boson is 157 GeV if only the precision electroweak data are used, or 186 GeV if the LEP direct lower limit is included. Recently, the Tevatron experiments CDF and D0 have excluded a range of heavier masses:[3]

$$162 \text{ GeV} < m_h < 166 \text{ GeV}. \tag{2}$$

A combined fit to all the data, shown in Fig. 1, yields the asymmetric estimate[4]

$$m_h = 116.4^{+15.6}_{-1.3} \text{ GeV} \tag{3}$$

at the 68% confidence level.

If the Higgs mass is large, so is the Higgs quartic self-coupling λ, and renormalization-group effects cause it to blow up at some relatively low scale Λ, as seen in Fig. 2, heralding the appearance of new non-perturbative physics. On the other hand, if m_h is small, negative renormalization by the large t-quark Yukawa coupling drives $\lambda < 0$, leading to an instability in the electroweak vacuum, unless new physics such as supersymmetry intervenes.[5] Only a narrow range of $m_h \in (130, 180)$ GeV is compatible with the survival of the Standard Model at all scales up to the Planck mass. This would be the 'maximal conceivable disaster' scenario for the LHC: a single Standard Model Higgs boson and nothing else! The precision electroweak data favour small values of m_h, and the combination with the Tevatron exclusion (2) excludes the blow-up scenario at the 99% confidence level.[6] The unstable-vacuum scenario is preferred, but the 'disaster' scenario is not even disfavoured at the 1-σ level.

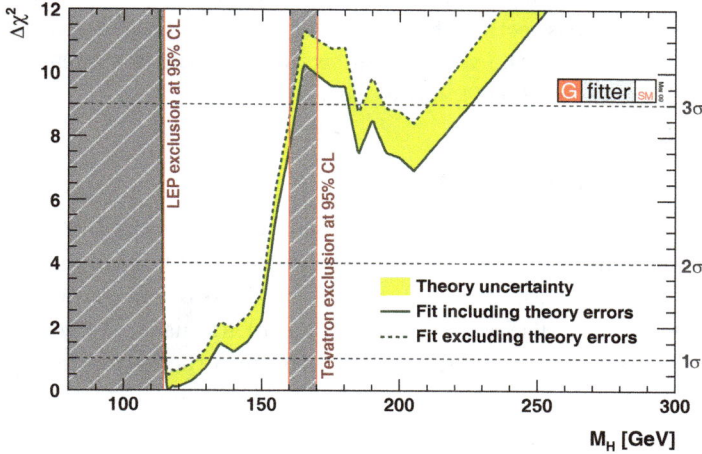

Fig. 1.   The χ^2 function for the Standard Model as a function of the Higgs mass, combining[4] the precision electroweak data[2] with the LEP[1] and Tevatron[3] exclusions.

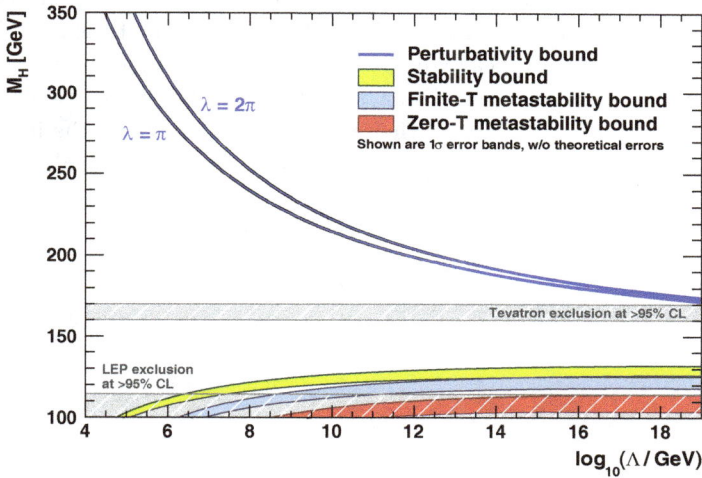

Fig. 2.   If the Standard Model Higgs boson weighs more than ~ 180 GeV, the Higgs self-coupling blows up at some scale Λ below the Planck scale, inducing new non-perturbative physics. If it weighs less than ~ 130 GeV, our current electroweak vacuum is unstable. The data summarized in Fig. 1 disfavour the blow-up scenario at the 99% confidence level.[6]

4. The LHC Physics Haystack

The standard list of the primary physics objectives of the LHC includes the search for this Higgs boson (or whatever replaces it), the nature of dark matter, the primordial plasma that filled the Universe when it was less than a microsecond old, and matter–antimatter asymmetry. It is worth remembering that the cross sections for producing interesting heavy new particles at the LHC are typically $\mathcal{O}(1/\text{TeV})^2$

148

and possibly with additional factors $\sim \alpha^2$, much less than the total cross section $\mathcal{O}(1/m_\pi)^2$. Typical cross sections for producing the Higgs boson or supersymmetric particles are $\sim 10^{-12}$ of the total cross section: looking for them will resemble searching for a needle in 100,000 haystacks!

5. The Search for the Higgs Boson

For this reason, the Higgs boson will not appear as soon as the LHC produces high-energy collisions. The left panel of Fig. 3 shows the amount of luminosity required at 14 TeV in the centre of mass either to exclude the Higgs boson at the 95% confidence level or to claim a 5-σ discovery, as a function of its mass. We see that a couple of hundred inverse picobarns could get the LHC into the exclusion business, while several inverse femtobarns would be needed to guarantee detection at any mass. There is an intermediate range of masses $m_h \in (150, 500)$ GeV where detection via $h \to WW$, ZZ decays is relatively easy, but in the preferred low-mass range (3) these decay modes are less important and Higgs detection becomes more difficult.

The stakes in the Higgs search are high, since it is key to many puzzles in cosmology as well as particle physics. How is electroweak symmetry broken? Is there such a thing as an elementary scalar field? What is the fate of the Standard Model at high scales? Did mass appear when the Universe was a picosecond old through an electroweak phase transition? Did CP-violating Higgs interactions help create the matter in the Universe? Did another elementary scalar field, the inflaton, cause (near-)exponential expansion of the early Universe, and hence make the Universe

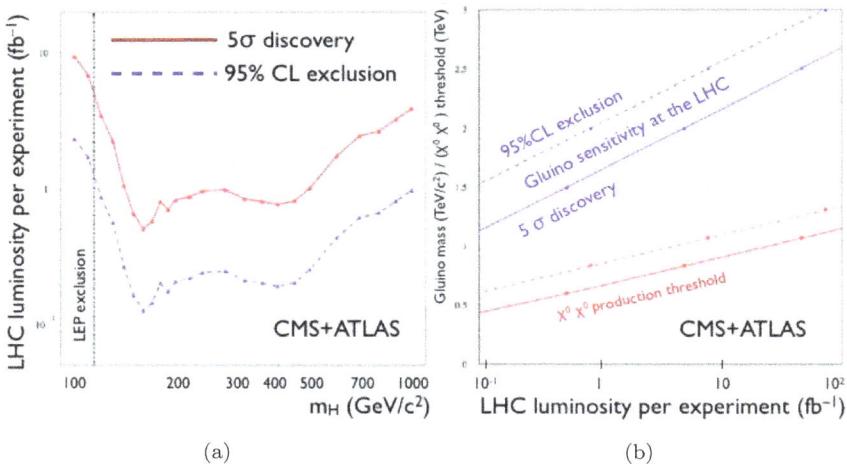

(a) (b)

Fig. 3. (a) The sensitivity of the LHC running at 14 TeV for 95% confidence level exclusion of the Standard Model Higgs boson (lower, dash-dotted curve) and 5-σ discovery (upper, solid curve). (b) In the constrained MSSM, the mass of the gluino (which may be excluded or discovered at the LHC, upper lines) is correlated with the threshold for producing supersymmetric particles at a linear e^+e^- collider (lower lines).[7]

so big and old? The typical scale of vacuum (dark) energy in the Higgs potential is some 60 orders of magnitude larger than the measured value: why is there so little dark energy? Discovering the Higgs boson, or proving that it does not exist, may not answer all these questions, but it may be our best experimental probe of them.

The most exciting Higgs scenario for the LHC may be that it is proven *not* to exist. That would really shake us smug theorists out of our torpor! The best alternative known to me would be to break electroweak symmetry by boundary conditions in extra dimensions (see, for example, Ref. 8, and references therein). Finding evidence for extra dimensions might be easier to explain to a lay audience than an ill-named God particle.

6. The Search for Supersymmetry

There are many motivations for supersymmetry: its beauty, it would render the hierarchy of mass scales more natural, it predicts a light Higgs boson, it stabilizes the electroweak vacuum, it facilitates grand unification, and it is apparently needed for the consistency of string theory. Here I focus on the fact that supersymmetry could provide the dark matter required by astrophysics and cosmology.[9]

In many supersymmetric models, there is a multiplicatively-conserved quantum number called R-parity, that may be represented as $R = (-1)^{2S-L+3B}$, where S is spin, L is lepton number, and B is baryon number. It is easy to verify that known particles have $R = +1$, whereas their putative supersymmetric partners, differing in spin by $1/2$, would have $R = -1$. The conservation of R parity would imply that sparticles are produced in pairs, that heavier sparticles decay into lighter ones, and that the lightest supersymmetric particle (LSP) is stable, because it has no legal decay mode. Hence, it should still be around after being produced early in the Big Bang, and could provide the needed dark matter. Presumably, the LSP is some neutral, weakly-interacting particle, otherwise it would have bound to ordinary matter and been detected by now. In this case, the favoured signature for supersymmetry at colliders is missing transverse energy carried away by the invisible dark matter particles.

There are important constraints on supersymmetry due to the absence of sparticles at LEP and the Tevatron, the LEP lower limit on m_h and the consistency of b-quark decays with the Standard Model. Some hint of new physics at the TeV scale may be provided by the measurement[10] of the anomalous magnetic moment of the muon, $g_\mu - 2$, that could be explained by supersymmetry, although there are still uncertainties in the Standard Model calculation of $g_\mu - 2$.[11] The measured density of dark matter, $0.097 < \Omega_{\mathrm{DM}} h^2 < 0.122$, provides a very tight constraint on some combination of supersymmetric model parameters, if the LSP provides the dark matter. The interplay of these constraints is shown in Fig. 4 for one particularly simple supersymmetric model with universal supersymmetry-breaking parameters $m_{1/2}$ and m_0 assumed at the GUT scale, the CMSSM.[12]

7. Where is Supersymmetry?

We have recently made a global supersymmetric fit using a frequentist approach to analyze the precision electroweak data, the LEP Higgs mass limit, the cold dark matter density, b-decay data and (optionally) $g_\mu - 2$. We combined the likelihood functions from these different observables to construct a global likelihood that can be used to infer preferred regions of the supersymmetric parameter space.[13,14]

We see in Fig. 5 that the preferred regions of the $(m_0, m_{1/2})$ planes in both the CMSSM and a model with common non-universal supersymmetry-breaking contributions to the Higgs masses (NUHM1) correspond to relatively low masses where the relic LSP density is brought into the WMAP range by coannihilations with light sleptons, particularly the lighter stau. The 'focus-point' region at large m_0 is disfavoured, principally but not exclusively by $g_\mu - 2$. If one drops this constraint, considerably larger ranges of m_0 and $m_{1/2}$ would be allowed, though small values are slightly preferred by other data.[13,14]

Figure 6 compares the preferred regions of these $(m_0, m_{1/2})$ planes with the 5-σ discovery reach of the LHC with given amounts of integrated luminosity at certain centre-of-mass energies. The 5-σ discovery reach of the LHC with 1/fb of luminosity at 7 TeV in the centre of mass would probably include all the 68% confidence-level regions in Fig. 6.

If supersymmetric particles are this light, they would contribute significantly to the electroweak radiative corrections that test the Standard Model at the quantum level. Using low-energy precision measurements as input, one may use radiative corrections to predict high-energy observables such as m_t, M_W and m_h. Within the Standard Model, the predictions of m_t and M_W have been very successful, and one would hope that their success would not be undermined in supersymmetric extensions of the Standard Model. On the one hand, supersymmetry predicts a restricted range for m_h, tending to reduce an uncertainty in the calculations of the radiative corrections. On the other hand, light sparticles could themselves contribute significantly to the radiative corrections.

As seen in Fig. 7, the net effect of supersymmetry in the context of the global fits to the CMSSM and NUHM1 introduced above is to *reduce* the predicted ranges of m_t and M_W.[15] The predictions are in both cases in good agreement with the measured values: indeed, the supersymmetric predictions for M_W are even slightly better than those in the Standard Model.

8. Extra Dimensions?

Another fashionable scenario for physics beyond the Standard Model is the appearance of large extra dimensions. Their possible existence was suggested decades ago, and given extra momentum by string theory, which appears to require their appearance at some distance scale. However, our current understanding of string theory does not offer any firm guidance as to their possible sizes. It has been suggested that, if they are large, they might solve or at least alleviate the mass

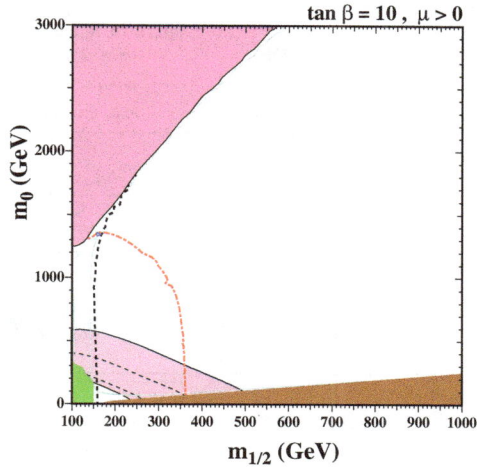

Fig. 4. The $(m_{1/2}, m_0)$ plane of the CMSSM for sample values of the other supersymmetric model parameters, showing the different theoretical, phenomenological, experimental and cosmological constraints. There is no consistent electroweak vacuum in the dark pink shaded region at large m_0, the LSP would be charged in the brown shaded region at small m_0, $b \rightarrow s\gamma$ excludes the green shaded region, LEP excludes the regions to the left of the dashed black and red lines by unsuccessful chargino and Higgs searches, respectively, and $g_\mu - 2$ favours the paler shaded pink region. The LSP would have the appropriate cosmological density in the narrow turquoise strip close to the boundaries of the allowed region.[12]

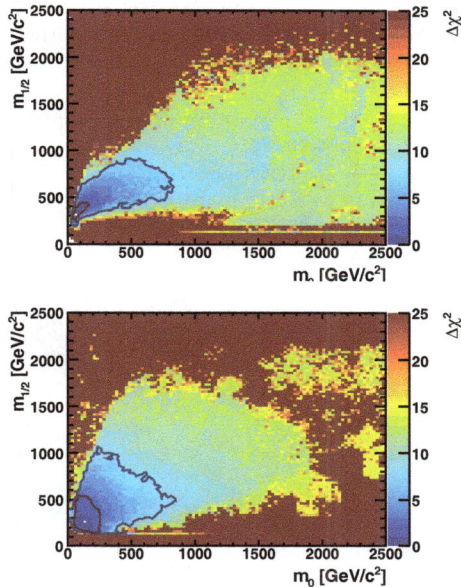

Fig. 5. The preferred regions in the $(m_0, m_{1/2})$ planes of the CMSSM (upper panel) and the NUHM1 (lower panel), as found in a frequentist analysis.[14] The best-fit points are shown as white points, and the 68% and 95% confidence-level contours are shown as solid black lines.

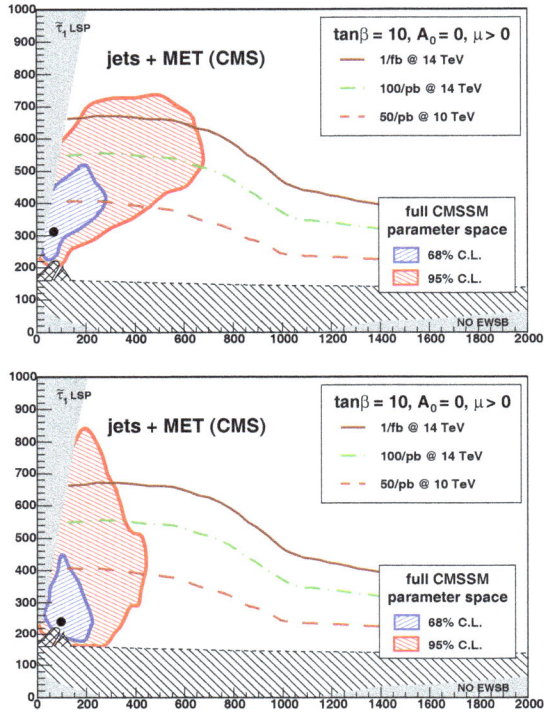

Fig. 6. The preferred regions in the $(m_0, m_{1/2})$ planes of the CMSSM (upper panel) and the NUHM1 (lower panel), compared with the estimated discovery sensitivity of the LHC with different amounts of luminosity and centre-of-mass energy.[13]

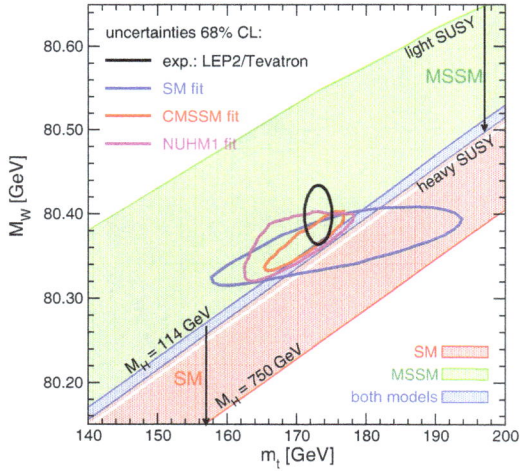

Fig. 7. Predictions for m_t and M_W on the basis of low-energy precision data in the Standard Model (blue curve), the CMSSM (red curve) and the NUHM1 (violet curve) compared with the experimental measurements at LEP and the Tevatron (black ellipse).[15]

hierarchy problem. Perhaps I am showing my age, but I find large extra dimensions less attractive than supersymmetry. Nevertheless, the final arbiters will be experiments, particularly at the LHC.

They may find Kaluza–Klein excitations of Standard Model particles, or they may discover missing energy leaking into extra dimensions, or they may discover that gravity becomes strong at the TeV scale and LHC collisions produce microscopic black holes. This last speculation has sparked concerns that have no scientific basis. The same theories that predict microscopic black holes also predict that they decay via Hawking radiation, with interesting grey-body factors that would be fascinating tests of string theory or any other quantum theory of gravity. Even if they were stable, the continued existence of the Earth and other celestial objects tells us to fear nothing from the LHC.[16,17] Do I hear the sound of other axes being ground?

9. The Restart of the LHC

Following the first start of the LHC on September 10th, 2008 and the electrical fault nine days later that laid the LHC low, there was jubilation on November 20th, 2009 when the LHC was restarted. This was redoubled 3 days later when the first collisions were observed at 900 GeV in the centre of mass. Soon afterwards, the beams were successfully ramped up to 1.18 TeV and the highest-energy human-made collisions were seen.

Remarkably quickly, the LHC experiments were able to reconstruct the decays of known particles and remeasure their masses. Two-photon decays of the π^0 and η were quickly seen, but seeing the Higgs will take a while, as discussed above: no Higgs yet! Both ATLAS and CMS have seen multi-jet events, and have shown distributions of the missing transverse energy. So far, these agree only too well with Monte Carlo simulations: no supersymmetry yet! Multi-jet events occur at rates compatible with QCD: no Hawking-decaying black holes yet!

10. What will the Future Bring?

While I was writing up this talk, the LHC produced its first collisions at 3.5 TeV per beam on March 30th, 2010, and Fig. 8 shows some of the first events observed by the ATLAS and CMS experiments. The default LHC operating scenario is to collide at 3.5 TeV per beam until the end of 2011, aiming to accumulate 1/fb of integrated luminosity. Depending on the LHC running experience during this period, the LHC beam energy may be increased slightly during this period. As illustrated in Fig. 9, this should enable the LHC to at least equal the Tevatron sensitivity to an intermediate-mass Higgs boson weighing between 150 and 180 GeV. Moreover, it would give the LHC a reach for new physics such as supersymmetry that would extend beyond the Tevatron. Indeed, by discovering or excluding the gluino, this first physics run of the LHC may already be able to tell us whether supersymmetric are light enough to be produced with a 500 GeV linear e^+e^- collider.[7]

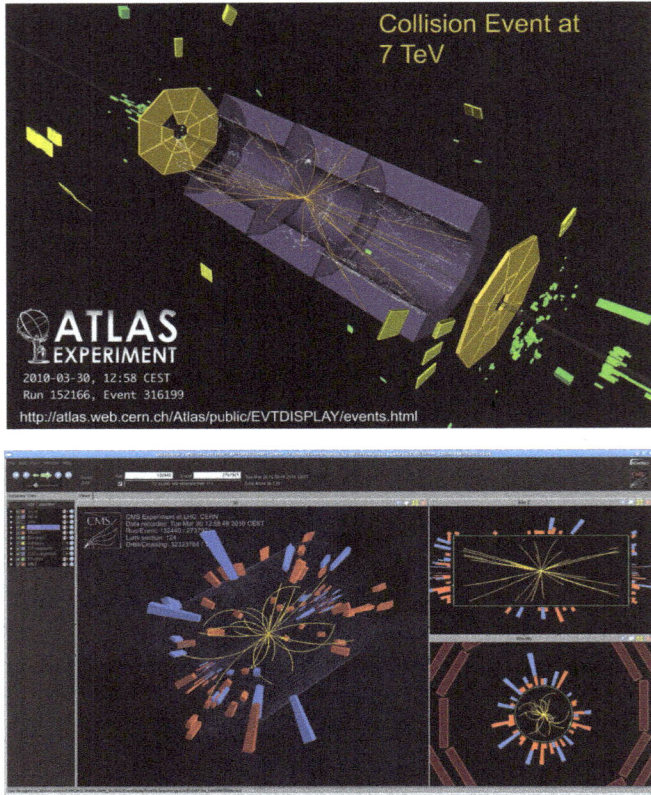

Fig. 8.   First 7-TeV collisions from ATLAS (upper panel)[18] and CMS (lower panel).[19]

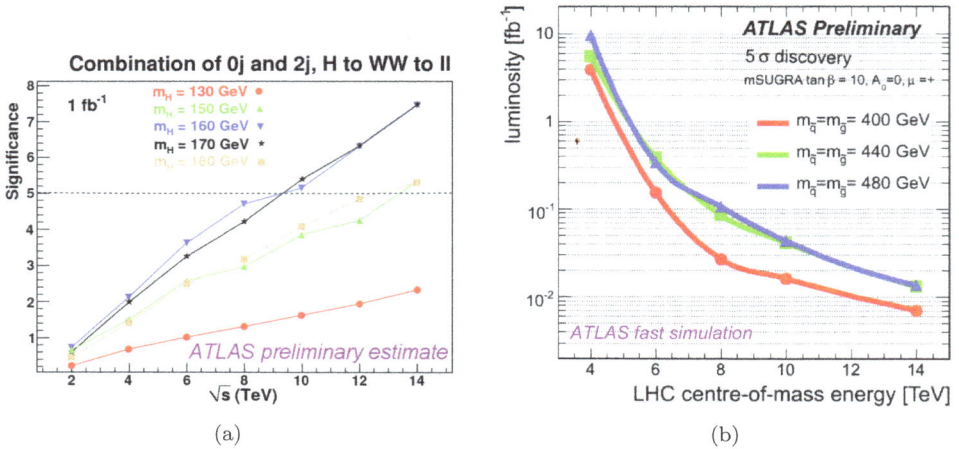

(a)   (b)

Fig. 9.   (a) The ATLAS signal sensitivity for a Standard Model Higgs boson with 1/fb integrated luminosity at various centre-of-mass energies. (b) The sensitivity of ATLAS for 5-σ discovery of supersymmetric particles at reduced centre-of-mass energies.[20]

A long shutdown is then planned for consolidation of the LHC and its injectors, including the LHC magnet interconnects and training of the dipole magnets, making it possible to run the LHC at or close to its design energy of 7 TeV per beam. Further in the future, there will be at least major upgrade of the LHC, incorporating Linac4 and new interaction-region insertions. The scope of possible further upgrades (an SPL? a higher extraction energy for the PS booster? replacement of the PS? new collision insertions? crab cavities?) are still under discussion, and will be decided only in the light of operational experience with the LHC.

11. A Conversation with Mrs Thatcher

Back in 1982, while she was Prime Minister of the UK, Mrs. Thatcher visited CERN and was introduced to British physicists. When she was told I was a theoretical physicist, she asked me "What do you do?" I explained that my job was to think of things for the experiments to look for, and hope they find something different. "Wouldn't it be better if they found what you predicted?" asked Mrs. T., who always liked *her* ideas to be vindicated. My response was that, in that case, we would not be learning anything really new. Likewise, I sincerely hope that the LHC will be remembered in the history of physics for something *not* described in this talk.

Acknowledgments

It is a pleasure to thank Harald Fritzsch for his kind invitation to this enjoyable event, and K. K. Phua and his team for their generous hospitality. Above all, it is a pleasure to thank Murray for his inspiration over the years.

References

1. ALEPH, DELPHI, L3 and OPAL Collab. LEP Working Group for Higgs boson searches (R. Barate *et al.*), *Phys. Lett. B* **565**, 61 (2003), arXiv:hep-ex/0306033.
2. ALEPH, CDF, D0, DELPHI, L3, OPAL and SLD Collab., LEP Electroweak Working Group, Tevatron Electroweak Working Group and SLD electroweak and heavy flavour groups, arXiv:0911.2604.
3. CDF and D0 Collab. (T. Aaltonen *et al.*), *Phys. Rev. Lett.* **104**, 061802 (2010), arXiv:1001.4162.
4. H. Flacher, M. Goebel, J. Haller, A. Hocker, K. Moenig and J. Stelzer, *Eur. Phys. J. C* **60**, 543 (2009), arXiv:0811.0009.
5. J. R. Ellis and D. Ross, *Phys. Lett. B* **506**, 331 (2001), arXiv:hep-ph/0012067.
6. J. Ellis, J. R. Espinosa, G. F. Giudice, A. Hoecker and A. Riotto, *Phys. Lett. B* **679**, 369 (2009), arXiv:0906.0954.
7. J.-J. Blaising, A. De Roeck, J. Ellis, F. Gianotti, P. Janot, G. Rolandi and D. Schlatter, http://council-strategygroup.web.cern.ch/council-strategygroup/BB2/contributions/Blaising2.pdf.
8. C. Grojean, New theories for the Fermi scale, arXiv:0910.4976.
9. J. R. Ellis, J. S. Hagelin, D. V. Nanopoulos, K. A. Olive and M. Srednicki, *Nucl. Phys. B* **238**, 453 (1984).

156

10. G. W. Bennett *et al.*, *Phys. Rev. D* **73**, 072003 (2006), arXiv:hep-ex/0602035.
11. M. Davier, A. Hoecker, B. Malaescu, C. Z. Yuan and Z. Zhang, arXiv:0908.4300.
12. J. Ellis and K. A. Olive, Supersymmetric dark matter candidates, in *Particle Dark Matter: Observations, Models and Searches*, ed. G. Bertone (Cambridge University Press, 2010), Chap. 8, pp. 142–163, arXiv:1001.3651.
13. O. Buchmueller *et al.*, *J. High Energy Phys.* **0809**, 117 (2008), arXiv:0808.4128.
14. O. Buchmueller *et al.*, *Eur. Phys. J. C* **64**, 391 (2009), arXiv:0907.5568.
15. O. Buchmueller *et al.*, *Phys. Rev. D* **81**, 035009 (2010), arXiv:0912.1036.
16. S. B. Giddings and M. L. Mangano, *Phys. Rev. D* **78**, 035009 (2008), arXiv:0806.3381.
17. J. R. Ellis, G. Giudice, M. L. Mangano, I. Tkachev and U. Wiedemann, *J. Phys. G* **35**, 115004 (2008), arXiv:0806.3414.
18. ATLAS Collab., http://atlas.web.cern.ch/Atlas/public/EVTDISPLAY/events.html.
19. CMS Collab., http://cms.web.cern.ch/cms/News/e-commentary/cms-e-commentary10.htm.
20. ATLAS Collab. (G. Aad *et al.*), Expected performance of the ATLAS experiment — detector, trigger and physics, arXiv:0901.0512.

MURRAY GELL-MANN AND THE LAST FRONTIER OF LHC PHYSICS: THE QGCW PROJECT

A. ZICHICHI

INFN and University of Bologna, Italy
CERN CH-1211, Geneva 23, Switzerland
Enrico Fermi Centre, Rome, Italy
and
Ettore Majorana Foundation and Centre for Scientific Culture, Erice, Italy

Keywords: QGCW (Quark–Gluon–Colored-World); Gell-Mann.

1. Introduction

This paper is my personal testimony of the role played, in those experimental and technological activities where I have been directly involved, by some of Murray Gell-Mann original ideas which go from the isospin $\frac{1}{2}$ for the θ-meson to SU(3) flavor, to SU(3) color, to complexity. The starting point is the θ-meson having isospin $\frac{1}{2}$ [1-3] and the $(\theta-\bar{\theta})$ mixing in 1955;[4,5] it continued in 1960[6,a] with the parameter ε, proposed in order not to spoil the universality of the Fermi-coupling. In the same paper[6] Gell-Mann and Lévy proposed the σ-model whose consequences ended up with the prediction of the top-quark. One year later, 1961, the eightfold way was elaborated[8-10] and in 1964 "a schematic model of baryons and mesons" was proposed having SU(3)-flavor[11-13] with "quarks" as building blocks. In 1968, the $(\eta-\eta')$ mesonic mixing[14] was investigated where from, the vector meson (ω, ϕ) mixing problem came out. The existence of the QCD "color"[15-17] is the basis of the present QGCW (Quark–Gluon–Colored-World) project, which is the first step. The next one being linked to the ELN collider with 300 km ring and total energy in the PeV range. And finally, complexity in 1995.[18,19] If complexity exists at the fundamental level, then for the years to come, what should be discovered at the frontier of our knowledge is in the field of totally unexpected events.

---

[a]In a footnote of this paper the authors suggest that a parameter can be associated with the "strange" currents in order not to spoil the universality of the Fermi-coupling. This parameter gave rise, three years later, to the "Cabibbo angle."[7]

2. The $(\theta-\bar{\theta})$ Mixing in 1955 and Its Consequences

The first reason to be grateful to Murray is the establishment of the Ettore Majorana Foundation and Centre for Scientific Culture in Erice (EMFCSC). I was the youngest member of the Italian delegation at the 1955 International Conference in Particle Physics where the most powerful group of experimental physics, led by Patrick M. S. Blackett, was expected to present their most recent cosmic rays results obtained at Jungfraujoch, Switzerland.

The report was presented by a fellow of the Blackett group and the chairman of the session was the famous Buthler, co-discoverer with Rochester of the first experimental evidence for the existence of two totally unexpected particles, called by Blackett, V^0, since they appeared in the cloud chamber as inverted V's. These two V^0's were in fact a baryon Λ^0 and a meson θ^0. No one knew the reason for their existence. In order to describe their production and decay properties, Murray proposed the existence of a new quantum number called *strangeness*, conserved at production and violated at decay. The great news presented by the Blackett group[20] was a series of observations where the following reaction was needed to explain the results:

$$\theta^0 + N \to \Lambda + N'.$$

A young fellow, at the end of the presentation, stood up and said: "I am sorry but I would like to point out that this reaction violates 'strangeness-conservation'." Silence in the over-crowded lecture hall. The chairman saw another young fellow who wanted to say something; his English was broken, but apparently he was defending the Blackett group results. So he was invited by the chairman to go to the blackboard and using the chalk he explained that in the "strangeness theory" proposed by Gell-Mann[1-3] the θ-meson isospin was $\frac{1}{2}$, therefore the θ neutral component had to be with positive and negative strangeness. Furthermore, according to a very recent paper by Gell-Mann and Pais,[4] the θ^0-meson, produced with strangeness $+1$, becomes a mixture of strangeness $+1$ and -1; thus the production of Λ^0 could indeed be via the reaction

$$\bar{\theta}^0 + N \to \Lambda^0 + N'.$$

The young fellow was me and this is how I became the pupil of the great Blackett, whose cloud chamber was a powerful instrument producing very interesting pictures. In two of them I found two examples of pair production of heavy mesons $(K^0, \bar{K}^0)$ and $(K^+, \bar{K}^0)$,[21] thus proving the correctness of the Gell-Mann strangeness model.[1-5] Blackett was very happy that old pictures had finally found the right interpretation and contributed to the understanding of the "strange" world started with the V^0-particles.

Let me add a few details on these very exciting years.

The V^0-particles also gave rise to the $(\theta-\tau)$ problem,[22,23] which culminated in the discovery of the breaking of the symmetry operators C and P. The discovery of the noninvariance of these symmetry operators was suggested in 1956 in a detailed analysis of all weak processes by T. D. Lee and C. N. Yang;[24] the first experimental

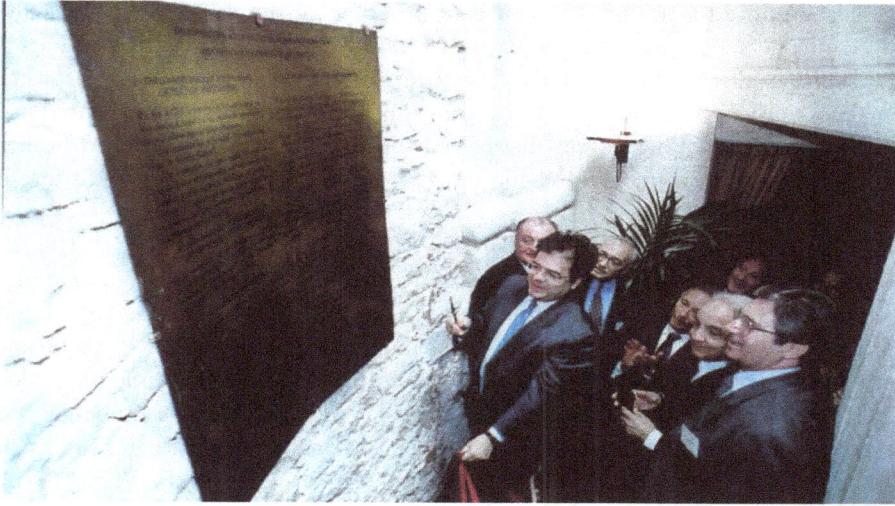

Fig. 1. The Minister of Home Affairs, Dr. Enzo Bianco, unveils the bronze reproduction of the document which established the constitution of the Ettore Majorana Foundation and Centre for Scientific Culture (EMFCSC) signed by J. S. Bell, P. M. S. Blackett, I. I. Rabi, V. F. Weisskopf and A Zichichi at CERN on May 8, 1962. In his inaugural address (Erice, May 8, 2000) H. E. the Minister Enzo Bianco recalled how in 1964 he was named one of the 100 "best students" in Italy in an EMFCSC competition. At the extreme right is Enzo Iarocci, President of the INFN (1998–2004).

evidence was provided by C. S. Wu and collaborators one year later.[25–27] This is how the θ-mesons became K-mesons.

Seven years after my arrival in the Blackett group, in 1962, the document establishing the existence of the Majorana Centre in Erice (EMFCSC) was signed by Blackett, Bell, Rabi, Weisskopf and myself at CERN, Geneva. Murray, thanks to his isospin $\frac{1}{2}$ for the θ-meson and to the $(\theta, \bar{\theta})$ mixing, contributed to the foundation of the Erice Centre, despite his signature not being in the 1962 document.

In Fig. 1 there is a photo which was taken at the celebration by the Minister of Home Affairs of the Italian Government of the bronze reproduction dedicated to the founding paper of the Ettore Majorana Centre (EMFCSC).

Let me go back to the "strangeness mixing." This mixing predicts the existence of two mesons, θ_1^0 and θ_2^0, on the basis of the validity of C invariance in weak interactions. The discovery of $\theta_2^0 \to 3\pi$ by Lederman[28] was interpreted as a proof that C invariance holds in weak interactions. With the discovery of C and P breaking, the $(\theta - \tau)$ mesons became, as mentioned before, a unique particle, the K-meson, which splits into two components, K_1^0 and K_2^0, each one thought to be an eigenstate of the symmetry operator CP proposed by Landau[29] to replace the two broken P and C invariances.

In 1956, Lee, Oehme and Yang (LOY), before parity violation was experimentally proven by C. S. Wu, pointed out that the existence of K_2^0 could not be taken as a proof of C invariance, nor as a proof of CP invariance;[30] LOY showed that strangeness mixing does not imply C invariance. In fact, even if CP is not valid, there would still be a long-lived neutral K-meson and, in order to prove that strangeness mixing is or is not CP invariant, other experiments had to be done in K decay physics. In 1964, it was discovered that CP invariance is indeed broken[31] and this is why the two neutral K-mesons (K_1^0, K_2^0), became (K_S^0, K_L^0), as foreseen by LOY in 1957.[30]

Let me quote an amusing detail of this great discovery, started with the $(\theta, \bar{\theta})$ mixing. The experiment[31] was not planned to search for the 2π decay mode of the K_2^0 meson. The aim of the experiment was to check the anomalous regeneration in hydrogen, previously reported by R. Adair *et al.*[32] (and found[31] to be more than an order of magnitude lower). The search for the 2π decay mode of the long-lived K_2^0 was proposed by us at CERN, but was rejected because the neutral beam in the PS experimental hall had already been allocated to another group's programme. On the other hand, we were already engaged with the PAPLEP (Proton AntiProton Annihilation into LEpton Pairs) experiment to search for the production of the third lepton through the $(e\mu)$ final state produced in $(\bar{p}p)$ annihilation,[33] using the CERN-PS beam which was next to the neutral beam we wanted for the $K_2^0 \to 2\pi$ search. I was told by the CERN research director of that time to "give other people the chance," when trying to convince him that the existence of the long-lived K_2^0 was not a proof of CP invariance as shown by LOY in 1957,[30] therefore the search for the $K_2^0 \to 2\pi$ decay mode, violating CP invariance, was not in contradiction with the existence of the long-lived K_2^0 meson. It would have been too much to give two PS beams to the same group, he told me later. Moreover, we were not proposing to check the anomalous regeneration in hydrogen (a proposal considered very interesting by many CERN theorists). Our aim was to follow the deep theoretical remark by LOY and check if CP was really valid in K_2^0 decay.

The flavor mixing problem and its CP invariance or noninvariance, is extremely topical today with many experiments being planned in order to understand the basic distinction between flavor mixing and CP invariance, for all flavors. How and why the quark flavors (u, c, t) and (d, s, b) mix and why this mixing is linked to the breaking of CP has no theoretical understanding, so far. All we can do is to measure the various flavor mixings and CP breakings.

Flavor mixing, started by Murray in 1955,[1–5] appears to be active also in the lepton sector as discovered by M. Koshiba *et al.*[34] and now being experimentally investigated over the world (see for example the proceedings of the Erice School of the last three years 2006, 2008, 2009).

Another chain of consequences originated by the existence of the V^0-particles was the proliferation of mesons and baryons with two branches: "statics" and "dynamics." The "static" proliferation gave rise, first to the eightfold way of Gell-Mann and Ne'eman,[8–10] and then to the "flavor" global symmetry SU(3)$_f$ based on the existence of three quark flavors: u, d, s.[11–13] SU(3)$_f$ contributed to open the way towards $SU(3)_c$. It is in fact the notion that two baryons

$$\Omega^- \quad \text{and} \quad \Delta^{++}_{3/2\,3/2}$$

had to be fermions, but appeared to be perfectly symmetric in their quark composition,[35-37] that prompted the idea of the existence of a new intrinsic quantum number.[38,39]

This chain of consequences, started with Murray's strangeness, led to the discovery of Quantum Chromodynamics (QCD)[15-17] — the fundamental forces acting among quarks and gluons. This force was affected by a theoretical trouble: confinement, since from QCD it is not possible to predict it. And here comes another experimental game: to see if in a violent collisions the proton breaks into its pieces.

In fact, DIS (Deep Inelastic Scattering) between electrons and protons revealed in 1968 at SLAC[b] a totally unexpected phenomenon: only part of the proton, called "parton" by Feynman, was involved in the interaction. The rest of the proton was totally inactive. If at high energy the proton behaves as if its pieces were "free" and therefore noninteracting among themselves, then in a high energy collision two protons should break up into their constituents, for example into the "quarks" of Gell-Mann.

The discovery of scaling at SLAC[40] prompted the CERN's implementation[41-43] of a sophisticated experimental setup intended to establish if fractionally charged particles were "freely" produced at the highest energy (pp) collisions (using the ISR collider). No quarks were observed by us at ISR thus establishing a firm contradiction: at high energy the pieces of the protons were losing their coupling (this is the meaning of scaling) but no quarks were observed at ISR.[41-43]

Scaling was finally understood as a consequence of the non-Abelian nature of the force acting between the constituents of a proton (or a neutron). Consequently, the nonexistence of quarks, searched for at ISR in violent collisions,[41-43] was understood in terms of the low energy behavior of this new force. Thus, "asymptotic freedom" and "confinement" came in the construction of the Standard Model, with QCD as the third fundamental force of Nature, to be added to the other two forces: electroweak and gravitational.

It is really incredible that the same mathematics, SU(3), first used to describe the proliferation of mesons and baryons, with $SU(3)_f$, became the basic structure of the third fundamental force of Nature with $SU(3)_c$.

It is the non-Abelian property of QCD that allowed all of us, after many decades, to finally understand the origin of nuclear isospin and of $SU(3)_f$. It is the gluon–gluon interaction that guarantees the "flavor" independence of α_3 as shown in the three-gluon diagram (Fig. 2).

---

[b]The first report on "scaling" was presented by J. I. Friedman at the *14th International Conference on High Energy Physics*, in Vienna, 28 August–5 September 1968. The report was presented as paper No. 563 but not published in the Conference Proceedings. For a detailed reconstruction of the events see Ref. 40.

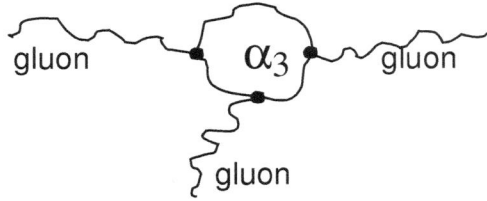

Fig. 2.

3. From the Proliferation of Mesons and Baryons to the Effective-Energy, E^{had}, in All Interactions, Regardless of the Difference Between the Interacting Particles

The physics of strong interactions was characterized by two classes of phenomena, one of "static" nature, the other of "dynamic" nature. Both were affected by proliferation in the most fundamental component of this physics: its elementary particles.

The proliferation in the static sector of the strong interaction was the huge number of mesons and baryons.[c] This multitude of states was reduced by an order of magnitude through the octets and decuplets of Gell-Mann and Ne'eman $SU(3)_f$.[8-10]

The proliferation in the dynamic sector was the multitude of final states produced by pairs of interacting particles, in strong, electromagnetic and weak processes:

Strong	EM	Weak
πp	γp	νp
Kp	ep	
pp	μp	
pn	$e^+ e^-$	
$\bar{p}p$		

The properties measured in the multihadronic final states were, for example,

(i) the average charged multiplicity $\langle n_{\text{ch}} \rangle$;
(ii) the fractional energy distribution $d\sigma/dx_i$;
(iii) the transverse momentum distribution $d\sigma/dp_{t_i}$; etc.

All these quantities appeared to depend on the nature of the interacting particles.

---

[c]An example of the proliferation in the meson resonances is Ref. 44. A record of the evolution in the multitude of baryonic and mesonic states can be found in the proceedings of the Erice Subnuclear Physics Schools.

How can it be that a unique fundamental force, QCD, acting among its two very simple basic components, quarks and gluons, produces such a variety of final states as those observed when a pair of particles interact?

It is the introduction of the Effective Energy which allowed us to put all the different final states on the same basis.

The results are the universality features measured in all multihadronic final states, no matter what is the pair of interacting particles in the initial state. The universality features are a QCD nonperturbative effect.

The first and basic step in this long "nonperturbative" QCD trip is the introduction of the Effective Energy.

This new quantity came about by studying (pp) interactions at the CERN-ISR where it was proven that the set of final states produced at the ISR nominal energy of 62 GeV consisted of a sum of final states, each one having a different Effective Energy, called E^{had}, as shown in Figs. 3 and 4.

These final states have properties like those produced in (e^+e^-) annihilation provided that $(\sqrt{s})e^+e^-$ is equal to E^{had}. The "nominal" ISR energy, $(\sqrt{s})_{pp}$, had to be corrected event by event in order to find the correspondent Effective Energy.

In order to check the validity of the Effective Energy down to the lowest value, the ISR collider was used at its lowest nominal energy: $(\sqrt{s})_{pp} = 30$ GeV.

This allowed a set of very low Effective Energies to be obtained using purely hadronic interactions.

It was shown (Fig. 4) that the momenta of the secondary particles produced in (pp) collisions had the same distribution as those produced in (e^+e^-) annihilation, at the same Effective Energy, i.e. $E^{\text{had}} \equiv (\sqrt{s})e^+e^-$.

Another universality feature, the average charged multiplicity, $\langle n_{\text{ch}} \rangle$, measured in (pp) and in (e^+e^-) interactions, is reported in Fig. 5.

With the advent of LHC we predict for $\langle n_{\text{ch}} \rangle$ the values shown in Fig. 6.

4. The QCD Color, the QGCW Project and Complexity

With the advent of the LHC supercollider at CERN we propose to study the properties of the "new world" which should be produced in a collision between heavy nuclei $(_{208}\text{Pb}^{82+})$ at the maximum energy so far available in our planet, i.e. 1150 TeV $(1.15 \times 10^{15}\text{eV})$.

The new world is the Quark–Gluon–Colored-World (QGCW).

We avoid in purpose to call it "quark–gluon plasma" since, in the extremely high energy collision between heavy ions, many QCD open-color-states should be produced. The number of these QCD open-color-states is by far higher than the number of baryons and mesons so far known, since these baryons and mesons have to obey the condition of being QCD-colorless. We want to search for specific effects due to the fact that the colorless condition is avoided.

In principle, many different phase transitions could take place and a vast variety of complex systems should show up. The properties of this new world should open unprecedented horizons in understanding the logic of Nature. How to study the new

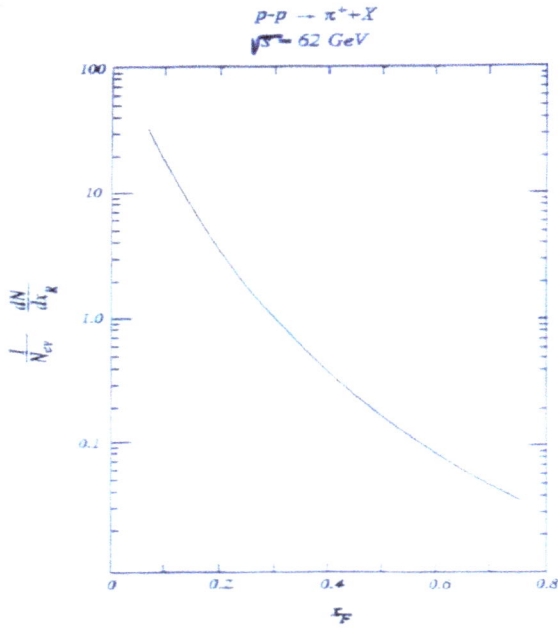

Fig. 3. Momentum distribution of charged pions produced in pp collisions at the ISR with the nominal energy $(\sqrt{s})_{pp} = 62$ GeV.

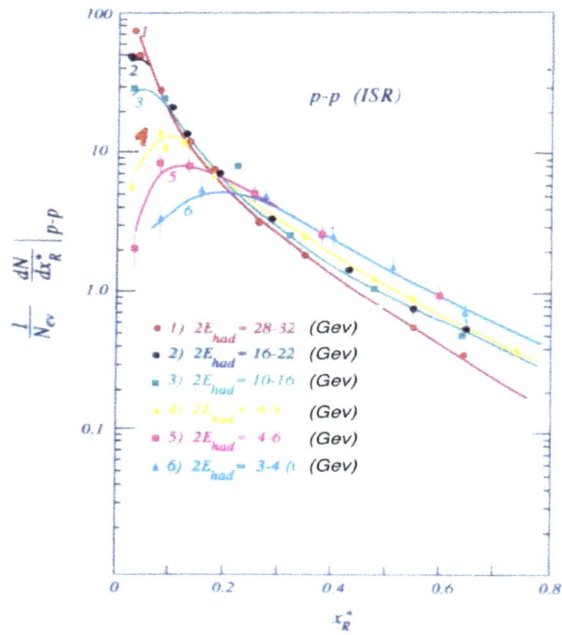

Fig. 4. The momentum spectrum in Fig. 3 is the sum of all spectra in Fig. 4, each one with a fixed E^{had}.

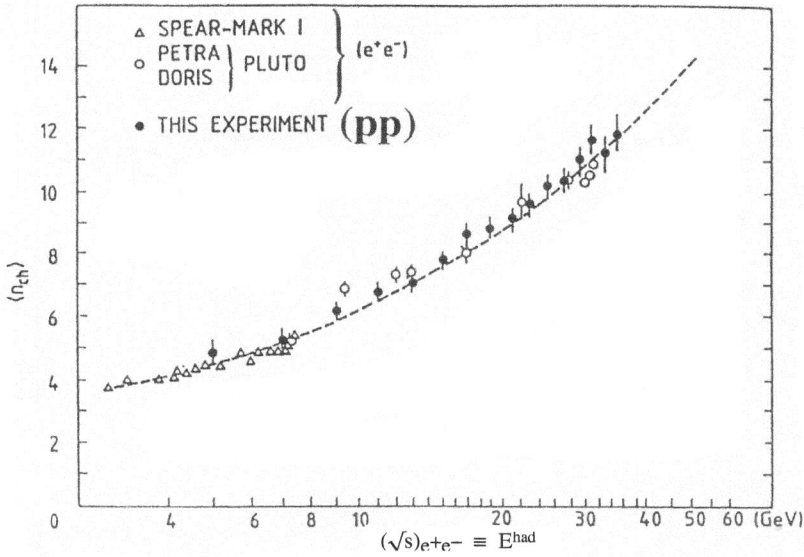

Fig. 5. Direct comparison between (pp) and (e^+e^-) processes. Mean charged multiplicity as measured in e^+e^- using $(\sqrt{s})_{e^+e^-}$ and pp, using the Effective Energy, E^{had}, not the nominal $(\sqrt{s})_{pp}$.

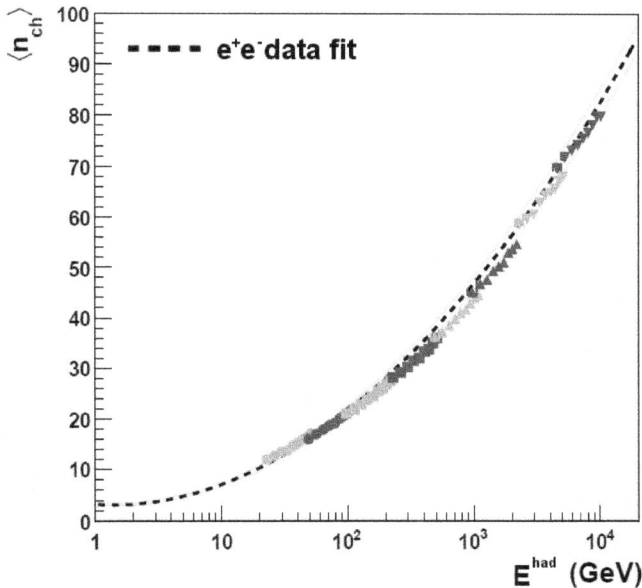

Fig. 6. Mean charged multiplicity $\langle n_{\mathrm{ch}} \rangle$ predicted for (pp) events as function of the Effective Energy, E^{had}. The nominal (pp) center-of-mass energy $(\sqrt{s})_{pp}$ goes from 70 to 14000 GeV. The dotted line is our fit to the e^+e^- data vs $(\sqrt{s})_{e^+e^-}$ whose highest values (from 91 to 209 GeV) are from LEP. The Effective Energies in (pp) events correspond to $(\sqrt{s})_{e^+e^-}$ in (e^+e^-) annihilation.

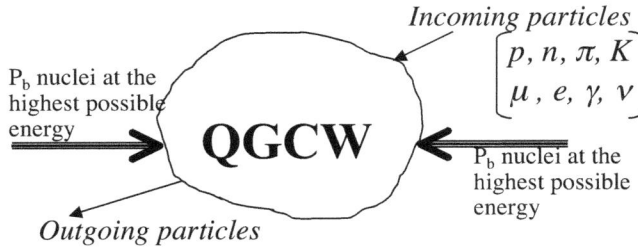

Fig. 7.

world is illustrated in Fig. 7 where beams of known particles $(p, n, \pi, K, \mu, e, \gamma, \nu)$ bombard the QGCW-volume and a special set of detectors allows one to measure the properties of the outgoing particles.

The QGCW project is based on $(P_b–P_b)$ nuclei which collide at the highest possible energy, starting with LHC, as a first step. Inside the QGCW-volume there are free quarks and antiquarks, an enormous number of gluons and all open-QCD-color-states allowed by $SU(3)_c$. If we bombard this QGCW-volume with particles having EM interactions (photons and charged leptons), these particles should have some difficulty in going through the QGCW. But if we bombard the same QGCW-volume with protons, neutrons, pions or any other hadron, these particles should have enormous difficulties to go through. For this to become real physics an effective R&D work covering the following two fields is necessary: one is the accelerator technology and the other is the detectors technology. Furthermore, our detectors should be as sophisticated as possible in order to allow totally unexpected effects to be detected, if Nature follows the logic of complexity at the fundamental level.

The most recent contribution by Gell-Mann which attracted my attention is complexity.[18,19] Here my concern was to find out the experimental evidences in order for complexity to exist.

Since these experimental evidences exist at the level of elementary particle physics,[45–47] the conclusion is that, when we construct instruments to detect what should come out from LHC, we should be prepared to see something which is totally unexpected, as was the case with photographic plates (Becquerel, from radioactivity to weak forces), with Wilson cloud chamber (from strange particles to $SU(3)_f$ and $SU(3)_c$), just to quote two examples. At present, we want to build extremely sophisticated detectors to study the world imagined by Gell-Mann to be with open QCD color states, but we should be prepared for something that is totally unexpected.

References

1. M. Gell-Mann, On the classification of particles, preprint, unpublished, August (1953).
2. M. Gell-Mann, *Phys. Rev.* **92**, 833 (1953).
3. M. Gell-Mann and A. Pais, *Proc. Glasgow Conf. on Mesons and Nuclear Physics* (1954), p. 342.

4. M. Gell-Mann and A. Pais, *Phys. Rev.* **97**, 1387 (1955).

5. M. Gell-Mann, *Suppl. Nuovo Cimento IV*, Serie **X**, n. 2 (1956).

6. M. Gell-Mann and M. Lévy, *Nuovo Cimento* **16**, 705 (1960).

7. N. Cabibbo, *Phys. Rev. Lett.* **10**, 531 (1963).

8. M. Gell-Mann, The eightfold way — A theory of strong-interaction symmetry, California Institute Technology Synchrotron Lab. Report 20 (1961).

9. Y. Ne'eman, *Nucl. Phys.* **26**, 222 (1961).

10. M. Gell-Mann and Y. Ne'eman, *The Eightfold Way* (W.A. Benjamin Inc., 1964).

11. M. Gell-Mann, *Phys. Lett.* **8**, 214 (1964).

12. G. Zweig, Fractionally charged particles and SU_6, CERN Report TH 401 (1964).

13. A. Zichichi (ed.), Erice lecture 1964, in *Symmetries in Elementary Particle Physics* (Academic Press, 1965).

14. M. Gell-Mann, R. J. Oakes and B. Renner, *Phys. Rev.* **175**, 2195 (1968).

15. H. Fritzsch and M. Gell-Mann, Current algebra: Quarks and what else?, in *Proc. XVI Int. Conf. on High Energy Physics*, Chicago, Vol. 2 (1972).

16. W. A. Bardeen, H. Fritzsch and M. Gell-Mann, Light cone current algebra, pion decay and electron-positron annihilation, in *Proc. of the Meeting on Conformal Invariance*, Frascati, 1972 [Printed in *Scale and Conformal Symmetry in Hadron Physics* (John Wiley and Sons, Inc., 1973)].

17. H. Fritzsch, M. Gell-Mann and H. Leutwyler, *Phys. Lett. B* **47**, 365 (1973).

18. M. Gell-Mann, *Complexity* **1**, 16 (1995).

19. M. Gell-Mann and S. Lloyd, *Complexity* **2**, 44 (1996).

20. G. D. James, On the production of V particles, in *Proc. 1955 Int. Conf. on Particle Physics*, Nuovo Cimento Supplemento, Vol. 4 (1956), p. 325.

21. W. A. Cooper, H. Filthuth, J. A. Newth, G. Petrucci, R. A. Salmeron and A. Zichichi, *Nuovo Cimento* **5**, 1388 (1957).

22. R. H. Dalitz, *Phil. Mag.* **44**, 1068 (1953).

23. R. H. Dalitz, *Proc. Phys. Soc. A* **69**, 527 (1956).

24. T. D. Lee and C. N. Yang, *Phys. Rev.* **104**, 254 (1956).

25. C. S. Wu, E. Ambler, R. W. Hayward and D. D. Hoppes, *Phys. Rev.* **105**, 1413 (1957).

26. R. Garwin, L. Lederman and M. Weinrich, *Phys. Rev.* **105**, 1415 (1957).

27. J. J. Friedman and V. L. Telegdi, *Phys. Rev.* **105**, 1681 (1957).

28. K. Lande, E. T. Booth, J. Impeduglia, L. M. Lederman and W. Chinowski, *Phys. Rev.* **103**, 1901 (1956).

29. L. D. Landau, *Zh. Éksp. Teor. Fiz.* **32**, 405 (1957).

30. T. D. Lee, R. Oehme and C. N. Yang, *Phys. Rev.* **106**, 340 (1957).

31. J. Christenson, J. W. Cronin, V. L. Fitch and R. Turlay, *Phys. Rev. Lett.* **13**, 138 (1964).

32. R. Adair, W. Chinowsky, R. Crittenden, L. B. Leipuner, B. Musgrave and F. T. Shively, *Phys. Rev.* **132**, 2285 (1963).

33. C. S. Wu, T. D. Lee, N. Cabibbo, V. F. Weisskopf, S. C. C. Ting, C. Villi, M. Conversi, A. Petermann, B. H. Wiik and G. Wolf, *The Origin of the Third Family*, eds. O. Barnabei, L. Maiani, R. A. Ricci and F. Roversi Monaco (Rome, 1997), (World Scientific, 1998).

34. M. Koshiba, Kamiokande and super-Kamiokande, in *From the Planck Length to the Hubble Radius — Proc. of the 1998-Erice Subnuclear Physics School*, Vol. 36 (World Scientific, 2000).

35. H. J. Lipkin, Particle physics for nuclear physicists, in *Physique Nucléaire, Les-Houches 1968*, eds. C. DeWitt and V. Gillet (Gordon and Breach, 1969), p. 585.

36. H. J. Lipkin, *Phys. Lett. B* **45**, 267 (1973).

37. Y. Nambu, A systematics of hadrons in subnuclear physics, in *Preludes in Theoretical Physics*, eds. A. de-Shalit, H. Feshbach and L. Van Hove (North-Holland, 1966), p. 133.

38. O. W. Greenberg, *Phys. Rev. Lett.* **13**, 598 (1964).

39. M. Y. Han and Y. Nambu, *Phys. Rev. B* **139**, 1006 (1965).

40. J. I. Friedman, Deep inelastic scattering evidence for the reality of quarks, in *History of Original Ideas and Basic Discoveries in Particle Physics*, eds. H. B. Newman and T. Ypsilantis (Plenum Press, 1994), p. 725.

41. T. Massam and A. Zichichi, Quark Search at the ISR, CERN preprint, June 1968.

42. M. Basile, G. Cara Romeo, L. Cifarelli, P. Giusti, T. Massam, F. Palmonari, G. Valenti and A. Zichichi, *Nuovo Cimento A* **40**, 41 (1977).

43. M. Basile, G. Cara Romeo, L. Cifarelli, A. Contin, G. D'Alì, P. Giusti, T. Massam, F. Palmonari, G. Sartorelli, G. Valenti and A. Zichichi, *Nuovo Cimento A* **45**, 281 (1978).

44. R. H. Dalitz and A. Zichichi (eds.), Editrice compositori, in *Proc. of the EPS Conference*, Bologna, 1971.

45. A. Zichichi, *Complexity at the Fundamental Level: Consequences for LHC*, Desy, Hamburg, November 2005.

46. A. Zichichi, *Proc. 44th Course of the International School of Subnuclear Physics*, Erice, Italy, September 2006 (World Scientific, 2008), Vol. 44, p. 269.

47. A. Zichichi, Complexity exists at the fundamental level, in *42nd Course of the International School of Subnuclear Physics*, Erice, Italy, September 2004 (World Scientific, 2007), Vol. 42, p. 251.

SOME LESSONS FROM SIXTY YEARS OF THEORIZING

MURRAY GELL-MANN

Santa Fe Institute, USA

It gives me great pleasure that so many distinguished colleagues, many of you old friends, have gathered here for a conference associated with my near-centennial, my reaching the age of fourscore years. It is a pleasure to see all of you and to listen to what you have to say.

I have been asked to make some remarks too, and I thought hard about what would be appropriate. The result was a decision not to present a paper on some of the research that I'm carrying out, but rather to mention a few of the things I have learned about doing research in theoretical science, things I wish someone had explained to me around 60 years ago.

The one I should like to emphasize particularly has to do with the frequently encountered need to go against certain received ideas. Sometimes these ideas are taken for granted all over the world and sometimes they prevail only in some broad region or in certain institutions. Often they have a negative character and they amount to prohibitions of thinking along certain lines. Now we know that most challenges to scientific orthodoxy are wrong and many are crank. Now and then, however, the only way to make progress is to defy one of those prohibitions that are uncritically accepted without good reason.

Some years ago I was paid by a large corporation to appear in a very brief television advertisement. I didn't have to praise the company or even mention it. All I had to do was encourage people to ask, "Why not?"

The TV clip was judged successful and the company ran it for a second year. According to the rules of Actors' Equity they had to pay my fee a second time and also give me a second year's membership in the Screen Actors' Guild. They also invited me to visit their headquarters and give a talk to the employees there and also, via their intranet, to employees all over the world. The subject was, of course, asking "Why not?"

I gave some examples from the sciences of cases where it was important to challenge certain received ideas, and finally I remarked that although I knew little about business I assumed that in business as well as in scientific research there was usually (although not always) a damned good reason why not. I said that surely in business there are certain things to which attention must always be paid, such as profit and loss and legal and ethical considerations. It turned out that those were the right concerns to stress, because the company was Enron. Fortunately, I was not paid in stock.

Most of us remember how until very recently gastric ulcers were attributed mostly to stress, for example stress caused by a tyrannical boss. Hollywood moguls were quoted as boasting, "I don't get ulcers. I give ulcers." Occasionally a physician got good results treating ulcers with antibiotics, but that work was generally ignored because everyone "knew" that the primary cause was stress. Finally, a brave Australian doctor gave himself ulcers by swallowing a preparation of the bacterium *H. pylori*, and the received idea was eventually defeated.

Another familiar medical example is that of Semmelweis in mid-nineteenth century Budapest. He saw large numbers of mothers and newborn babies dying of puerperal fever. He correctly attributed that to the failure of doctors and other health workers to wash their hands properly. Not only was Semmelweis running afoul of a received idea, that careful washing had nothing to do with preventing the spread of disease. He was also insulting the medical profession. Think how many mothers and babies died unnecessarily because his correct observation was ignored. Years later Lister finally gained acceptance for the idea of cleanliness in surgical and obstetric practice, washing everything in sight with carbolic acid. He became Lord Lister.

From archaeology we can take the example of non-calendrical Maya glyphs. The calendrical ones, including those referring to the apparent motions of the sun, the moon, and Venus, were deciphered around a century ago, but the other glyphs were not. What were they? The chief expert on Maya archaeology, Sir J. Eric S. Thompson, proclaimed that they were not writing, and it was dangerous to disagree. In fact, writing was the correct answer, but it took many years for it to be deciphered. An important role was played by Yuri Knorozov in Soviet Russia, a scholar who lived outside the sphere of influence of Thompson.

I noted just now that sometimes a negative received idea is restricted to a region. For example, it was mainly in the U.S. that geologists were forbidden to take continental drift seriously. Shortly after I joined the Caltech faculty in 1955, I was entering the lunchroom at the Athenaeum, our faculty club, when I spotted a round table full of geologists. I was invited to join them, but only if I promised not to mention continental drift, which they "knew" was wrong. In fact, if I remember correctly, out of all of them, it was only Heinz Lowenstam, an immigrant from Europe, who believed in it. The physicists at Caltech invited Patrick Blackett and Teddy Bullard to give colloquia at which they presented strong evidence in favor of continental drift. The geologists attended but most of them left shaking their heads at the naivete of the physicists. They were committing a logical error. The various models that had been put forward to explain continental drift were wrong, as our geologists knew. But that didn't mean that the whole idea of continental drift was wrong. When sea floor spreading and plate tectonics came along, American geologists had to change their minds (except, I am told, at Harvard, where drift was denied for another decade or so).

Another lunch table at the Athenaeum held the cosmologists and nuclear physicists working on nucleogenesis in stars. They had been led to that research by the ideas of Fred Hoyle on continuous creation as opposed to an evolutionary cosmology. Continuous creation had been suggested by Herman Bondi and Tommy Gold to resolve the problem of the earth's being apparently older than the universe. As soon as that puzzle was solved, by recalibrating the yardstick for galactic distances, continuous creation should have been shelved. Instead, it was taken up by Fred and then by many others. It ruled out the genesis of nuclei in the early universe since there was no early universe. That led to the excellent work on nucleogenesis in stars, but it also led to another logical error. That a weird idea led to good work on something else didn't make the weird idea itself correct. The problem showed up when Willy Fowler, Fred, and the Burbidges couldn't account for the very light elements. I pointed out to them, as many others must have done, that Alpher and Gamow (with no help from Bethe, although Gamow put Bethe's name on the paper just for fun) had shown how the light elements would have been created in the early universe. "Ah, but Hoyle has taught us that there was no early universe," they replied. I should note that many years later Willy went to the trouble of acknowledging graciously that he should have listened. Here the "continuous creationists" were wrongly challenging a correct idea (evolutionary cosmology), while I and others were challenging the Caltech orthodoxy on the subject.

When Einstein proposed special relativity in 1905, he was suggesting that mechanics was covariant under the same symmetry group as Maxwell's equations for electromagnetism. The transformation formulae were, of course, those of the Lorentz group, which was already known, but Einstein went further and dared to take on the received idea of absolute time and space, arguing that there is only relative time and space.

We need hardly bring up the obvious examples of Copernicus and Galileo challenging the heliocentric idea.

Now the cases we've been reviewing are very important ones. It may be hard for us to conclude that a similar situation is occurring in our own much more modest work, which we would not attempt to compare with that of Einstein or Copernicus or Galileo. But I believe it is important, when faced with a theoretical puzzle at any level, to look at the negative received ideas that are relevant and see whether they are actually justified.

On a modest level, I have run into received ideas in presenting my own research. For example, in working out the notion of strangeness, I had to contradict the received idea that half-integral spin implied half-integral isospin and integral spin implied integral isospin. I was able to see that there was no reason for that association, even though it was widely believed.

Those of us who liked quarks got into trouble with at least three received ideas:

(i) the neutron and proton are elementary and not made of simpler things,

(ii) elementary particles do not have fractional electric charges in units of the proton charge, and

(iii) elementary particles are not confined inside objects like the neutron and proton, unable to emerge singly or to be used in industry.

We had to check that there was really no need for any of these prohibitions. That is the point I am trying to emphasize.

Nowadays, I do some work with a team of bright, mostly Russian linguists, who study distant relationships among human languages, ones that involve bigger (and older) groupings than generally accepted families like Indo-European or Uralic or Austronesian. Virtually all such work is rejected by most "mainstream" historical linguists in Western Europe and the U.S. They impose a huge burden of proof on anyone who suggests a distant genealogical relationship, preferring to resort to borrowing of words or to chance or to onomatopeia to explain similarities in basic vocabulary, no matter how unlikely these alternatives may appear in a given case. Are we dealing here with another of those doomed received ideas? I suspect we are.

So much for challenging unnecessary prohibitions. Another lesson I have learned is that it is sometimes very important to distinguish ideas that are important for to today's problems from those that will be useful in the context of tomorrow's deeper problems. I can think of at least two cases from my own research.

One has to do with the introduction of charm for the hadrons and, correspondingly, two different neutrinos for the electron and the muon. That makes two weak doublets. Why was I slow in embracing charm? I knew that two different neutrinos gave the best explanation of the absence of muon disintegration into an electron and a photon. I knew that a weak angle of around 15 degrees between the strangeness-changing and the strangeness-preserving parts of the hadronic weak current worked well, again reinforcing the idea of weak doublets. What was the matter? Why was I slow in embracing charm?

I wanted the electric charge operator to be a generator of a simple or at least semisimple Lie group — no U(1) factors. That meant the sum of the electric charges of the elementary particles had to be zero. For the quarks, that worked fine without charm: $2/3 -1/3 -1/3$ is indeed zero. Adding another $2/3$ would ruin it. Bt I was stupidly thinking in terms of the immediate problem of finding the right hadron theory, not the long-range problem of a theory of hadrons and leptons together. There, with the aid of color, charm works beautifully: $3(2/3 - 1/3 + 2/31/3) - 1 - 1 = 0$. If we adjoin the top and bottom quarks and the tau lepton with its neutrino, we still get zero.

Let's look at another occasion when I should have distinguished ideas to be used today from ones that might be useful tomorrow. When Harald and I thought of a Yang–Mills theory of color SU(3) coupled to quarks (what we later called QCD), I talked about it at the 1972 Rochester meeting in Chicago, with David Gross (another QCD pioneer) in the chair. I was quite enthusiastic, but in the written version of the talk I was much more tentative. The main reason is that I was learning about string theory and I wondered whether some kind of colored strings might be involved somehow. Now string theory may be useful, but not in formulating the dynamics of hadrons in terms of quarks and gluons (even though in QCD the stretched gluon bonds can behave somewhat like strings). Strings were mainly an idea for the morrow!

The third and final lesson I would like to draw from my experience in theoretical research is this: doubt and messiness are inevitable. We should tolerate them, even embrace them.

I read the new biography of Dirac by Graham Farmelo when it came out some months ago. Despite a few errors, it is an excellent book. (I was delighted, by the way, to find that two of the anecdotes in it came from me.) I was fascinated by the accounts of the great theoretical physicists of the twenties, thirties, and forties of the last century, many of whom I got to know later. What struck me particularly was the messiness of the process of figuring out quantum mechanics, atomic and nuclear physics, and quantum electrodynamics. Those great scientists had doubts and changed their minds and changed them back again. They were at times uncertain about which path to follow. Sometimes they were simply wrong.

One of the anecdotes that Farmelo got from me was about my asking Dirac why he didn't simply predict the positron right away, using his idea of "holes" in a sea of negative energy electrons. (Instead, he tried at one time to make his "holes" into protons, with a mysterious factor of almost two thousand in the mass.) His answer to me was "Cowardice, pure cowardice."

Now those hesitations were just the kinds of things that my colleagues and I experienced during the next era in elementary particle theory. For some reason I hadn't drawn the obvious conclusion that those distinguished predecessors were just as confused as we were or nearly so.

I suffered for many decades from the belief that when hesitating between alternatives I had to choose the correct one. Lyova Okun once quoted to me advice he had received from an older colleague: "Publish your idea along with the objections to it." I would now add "Publish the two contradictory ideas along with their consequences and choose later." Apparently the messiness of the process is inevitable. Instead of suffering while trying to make it perfectly clean and neat, why not embrace the messiness and enjoy it?

NEW GAUGE SYMMETRY IN GRAVITY AND THE EVANESCENT ROLE OF TORSION

H. KLEINERT

Institut für Theoretische Physik, Freie Universität Berlin,
Arnimallee 14, D14195 Berlin, Germany
ICRANeT, Piazzale della Republica 1, 10 -65122, Pescara, Italy

If the Einstein–Hilbert action $\mathcal{L}_{\mathrm{EH}} \propto R$ is re-expressed in Riemann-Cartan spacetime using the gauge fields of translations, the vierbein field h_μ^α, and the gauge field of local Lorentz transformations, the spin connection $A_{\mu\alpha}{}^\beta$, there exists a new gauge symmetry which permits reshuffling the torsion, partially or totally, into the Cartan curvature term of the Einstein tensor, and back, via a *new multivalued gauge transformation*. Torsion can be chosen at will by an arbitrary gauge fixing functional. There exist many equivalent ways of specifying the theory, for instance Einstein's traditional way where $\mathcal{L}_{\mathrm{EH}}$ is expressed completely in terms of the metric $g_{\mu\nu} = h_\mu^\alpha h_{\alpha\nu}$, and the torsion is zero, or Einstein's teleparallel formulation, where $\mathcal{L}_{\mathrm{EH}}$ is expressed in terms of the torsion tensor, or an infinity of intermediate ways. As far as the gravitational field in the far-zone of a celestial object is concerned, matter composed of spinning particles can be replaced by matter with only orbital angular momentum, without changing the long-distance forces, no matter which of the various new gauge representations is used.

PACS numbers: 98.80.Cq, 98.80. Hw, 04.20.Jb, 04.50+h.

1. In theoretical physics it often happens that a mathematical structure has a simple extension for which a natural phenomenon is waiting to be discovered. The most prominent example is the existence of a negative square root of the relativistic mass shell relation $p_0 = \sqrt{\mathbf{p}^2 + m^2}$ which led Dirac to postulate the existence of a positron, discovered in 1932 by Carl Anderson.[1] Sometimes, this rule does not seem to work initially, only to find out later that nature has chosen an unexpected way to make it work after all. Here the best example is the existence of a solution of the above energy-momentum relation for negative m^2, which was interpreted by some theoreticians as the signal for the existence of a particle faster than light. Such particles were never found. A simple physical realization appeared, however, with the discovery of the Ginzburg–Landau field theory of phase transitions and its quantum versions (now referred to as Higgs field theory). Since there are always *interactions*, a negative parameter m^2 destabilizes the field fluctuations. The fields move to a new ground state, around which they fluctuate with *positive* m^2. The situation is completely analogous to what happens in any building if ω^2 of one of its eigenfrequencies turns negative. The building collapses until the debris settles in a ruin, and that has only positive ω^2's. The collapse of an interacting field system with negative m^2 is observed as a phase transition to a state with stable fluctuations and positive m^2.

2. For many years, theorists have been wondering why Einstein's theory of gravity represents such a perfect geometrization of the gravitational forces.[2] Since the work of Cartan in 1922 it is known that the Riemannian spacetime, in which the celestial objects move, has a "natural" extension to *Riemann–Cartan* spacetime. This possesses a further geometric property called *torsion*. Why is there no trace of it in the movements of planets? Einstein himself has asked this question and discussed it in letters with Cartan.[3] He set up a theory of *teleparallelism* which explains gravity by a theory in Riemann–Cartan spacetime, in which the total curvature tensor vanishes identically. The Einstein–Hilbert action is then equal to a combination of scalars formed from torsion tensors,[4] and torsion forces provide us merely with an alternative way of describing gravitational forces, as emphasized in Refs. 5 and 6.

3. Yet another extension of Einstein's theory to Riemann–Cartan spacetime was advanced since 1959.[7-9] It has the appealing feature that it can be rewritten as a gauge theory invariant under local Poincaré transformations, i.e. both local translations and local Lorentz transformations, thus bringing it to a similar form as the gauge theories of weak, electromagnetic, and strong interactions. This gauge theory treats torsion as an *independent* field which couples only to the intrinsic spin of the elementary particles in a celestial body. Unfortunately, however, such an approach has several unsatisfactory features. First, the theory is meant to be classical, but the spin carries a power of $\hbar$ which vanishes in the classical limit. So there is really no classical source of torsion. Indeed, if torsion couples to spin with the gravitational coupling strength, the factor $\hbar$ implies that it cannot play any sizable role in the forces between celestial bodies. For example, even if the earth consisted only of polarized atoms, its intrinsic spin would be 10^{-15} times smaller than the rotational spin around the axis.

Moreover, there exist severe conceptual problems. One was emphasized in Ref. 11. As long as we do not know precisely the truly *elementary particles*, and it is doubtful that we ever will, many particles are described by effective fields, and it is impossible to specify whether the spin of those fields is caused by orbital motion or by the intrinsic spins of more elementary constituents. As an example, the spin-one field of a ρ-meson contains a wave function of two spinless pions in a p-wave, which do not couple to torsion. But it also contains two spin-$\frac{1}{1}$ quarks in an s-wave which would couple. Another problem is that if torsion couples to all spins, the photon becomes massive. In order to avoid this, the authors advocating this approach postulate that the photon is is an exception, and is not coupled to torsion. However, this contradicts the fact that roughly one percent of a photon is a virtual ρ-meson, which is strongly coupled to baryons. These, in turn, are supposed to be coupled to torsion (see Fig. 1), so the photon would become massive after all.

Thus the existence of an independent torsion field is highly dubious, and we may ask ourselves, whether the description of gravity in Riemann-Cartan spacetime

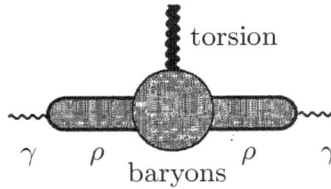

Fig. 1.   Diagram for mass generation of photon. It couples via a ρ-meson to baryonic matter which would be coupled torsion if $q \neq 1$.

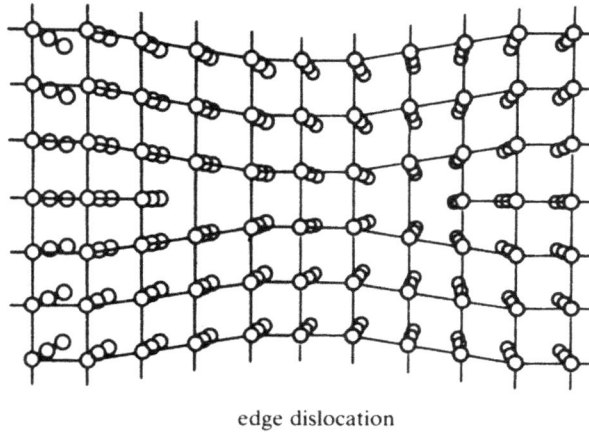

edge dislocation

Fig. 2.   Formation of a dislocation line (of the edge type) by a Volterra process. The Burgers vector **b** characterizes the missing layer. There exist two more types where **b** points in orthogonal directions.

proposed in Refs. 7–9 has really a chance of being true, or whether nature doesn't have a deeper reason for avoiding the above problems. It is the purpose of this note to answer this question affirmatively. Inspiration comes from a simple model of gravity, a "world crystal" with defects,[9,12,13] whose lattice constant is of the order of a Planck length. Some consequences of such a world crystal were pointed out in a recent study of black holes in such a scenario.[14]

4. We begin by showing that in the absence of matter, a world crystal is a model for Einstein's theory with a new type of extra gauge symmetry in which zero torsion is merely a special gauge. A completely equivalent gauge is the absence of Cartan curvature, which is found in Einstein's teleparallel universe. Before presenting the argument, recall that a crystal can have two different types of topological line-like defects,[8,9] which in a four-dimensional world crystal are world surfaces (which may be the objects of string theory).

First, there are translational defects called *dislocations* (Fig. 2). These are produced by a cutting process due to Volterra: a single-atom layer is removed from the crystal, allowing the remaining atoms to relax to equilibrium under the elastic

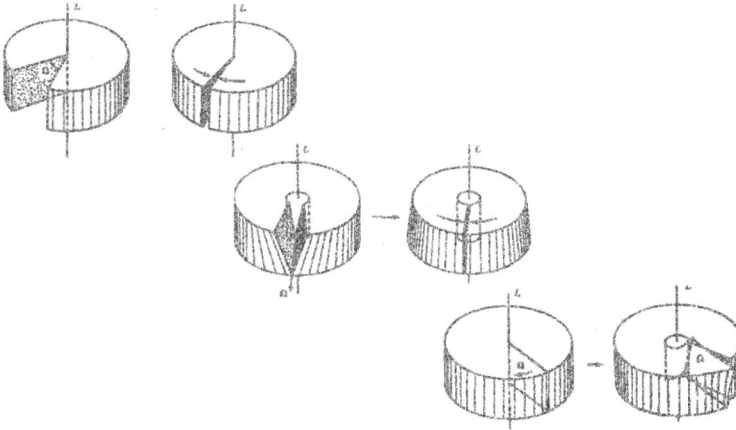

Fig. 3. Three different possibilities of constructing disclinations: wedge, splay, and twist disclinations. They are characterized by the Frank vector $\mathbf{\Omega}$.

forces. A second type of topological defects is of the rotation type, the so-called *disclinations* (Fig. 3). They arise by removing an entire wedge from the crystal and re-gluing the free surfaces.

The defects imply a failure of derivatives to commute in front of the displacement field $u_i(\mathbf{x})$. In three dimensions, the dislocation density is given by the tensor

$$\alpha_{ij}(\mathbf{x}) = \epsilon_{ikl}\nabla_k\nabla_l u_j(\mathbf{x}) \,. \tag{1}$$

If $\omega_i \equiv \frac{1}{2}\epsilon_{ijk}[\nabla_j u_k(\mathbf{x}) - \nabla_k u_j(\mathbf{x})]$ denotes the local rotation field, the disclination density is defined by

$$\theta_{ij}(\mathbf{x}) = \epsilon_{ikl}\nabla_k\nabla_l\omega_j(\mathbf{x}) \,. \tag{2}$$

The defect densities satisfy the conservation laws

$$\nabla_i\theta_{ij} = 0 \,, \quad \nabla_i\alpha_{ij} = -\epsilon_{jkl}\theta_{kl} \,. \tag{3}$$

These are fulfilled as Bianchi identities if we express $\theta_{ij}(\mathbf{x})$, $\alpha_{ij}(\mathbf{x})$ in terms of plastic gauge fields β_{kl}^p, ϕ_{lj}^p, setting $\theta_{ij} = \epsilon_{ikl}\nabla_k\phi_{lj}^p$, $\alpha_{il} = \epsilon_{ijk}\nabla_j\beta_{kl}^p + \delta_{il}\phi_{kk}^p - \phi_{li}^p$. The defect densities are invariant under the gauge transformations $\beta_{kl}^p \to \beta_{kl}^p + \nabla_k u_l^p - \epsilon_{klr}\omega_r^p$, $\phi_{li}^p \to \phi_{li}^p + \partial_l\omega_i^p$, where $\omega_i^p \equiv \frac{1}{2}\epsilon_{ijk}\nabla_j u_k^p$. Thus $h_{ij} \equiv \beta_{ij}^p + \epsilon_{ijk}\omega_k^p$ and $A_{ijk} \equiv \phi_{ij}^p\epsilon_{jkl}$ are *translational* and *rotational* defect gauge fields in the crystal.[10]

The Volterra processes can be represented mathematically by multivalued transformations from an Euclidean crystal with coordinates $\bar{x}^a$ to a crystal with defects and coordinates x^μ, as illustrated in Figs. 4 and 5 for two-dimensional crystals.

For an edge dislocation the mapping is $\bar{x}^1 = x^1$, $\bar{x}^2 = x^2 + (\frac{b}{2\pi})\phi(x)$, where $\phi(x) \equiv (1/2\pi)\arctan(x^2/x^1)$. Initially, this function has a cut from the origin towards left infinity. In a second step, the cut is removed and the multivalued version of the arctan is taken. This makes $\phi(\mathbf{x})$ the Green function of the commutator

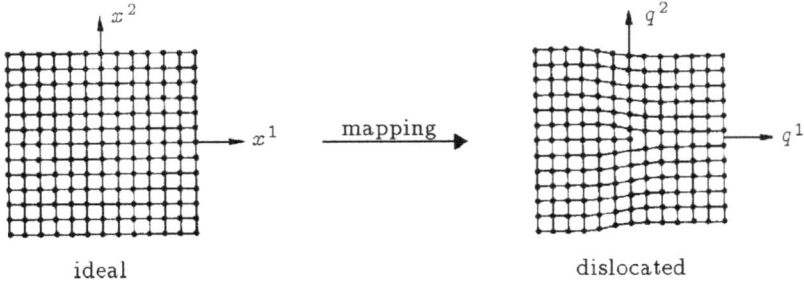

Fig. 4. Multivalued mapping of the perfect crystal to an edge dislocation with a Burgers vector **b** pointing in the 2-direction.

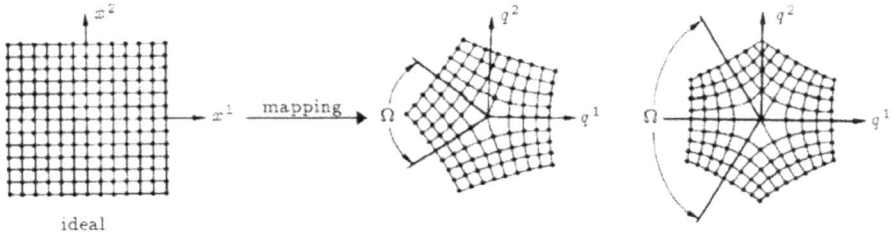

Fig. 5. Multivalued mapping of the perfect crystal to a wedge disclination of Frank vector Ω in the third direction.

$[\partial_1, \partial_2]$: $(\partial_1\partial_2 - \partial_2\partial_1)\phi(x) = \delta^{(2)}(\mathbf{x})$. For a wedge disclination, the mapping is $d\bar{x}^i = \delta^i_\mu[x^\mu + (\frac{\Omega}{2\pi})\varepsilon^\mu_\nu x^\nu \phi(x)]$.

A combination of the two

$$\eta_{ij}(\mathbf{x}) \equiv \theta_{ij}(\mathbf{x}) - \frac{1}{2}\nabla_m[\epsilon_{\min}\alpha_{jn}(\mathbf{x}) + \{ij\} + \epsilon_{ijn}\alpha_{mn}] \tag{4}$$

forms the *defect tensor*

$$\eta_{ij}(\mathbf{x}) \equiv \epsilon_{ikl}\epsilon_{jmn}\nabla_k\nabla_m u^p_{ln}(\mathbf{x}), \quad u^p_{ln} \equiv \frac{1}{2}(\beta^p_{ln} + \beta^p_{nl}). \tag{5}$$

It is a symmetric tensor due to the conservation laws (3), and represents the Einstein tensor $G_{ij} \equiv R_{ij} - \frac{1}{2}g_{ij}R^k_k$ of the geometry of the world crystal.

The expressions can easily be defined on a simple-cubic world lattice if we replace ∇_i by lattice derivatives, as shown in Refs. 8 and 9. There it is also shown that, in three spacetime dimensions, the disclination density $\theta_{ij}(\mathbf{x})$ represents the Einstein tensor G^C_{ij} associated with the *Cartan curvature tensor* $R^{C\ \ l}_{ijk}$ of the Riemann–Cartan geometry of the world crystal. The relation is

$$G^C_{ji}(\mathbf{x}) = \epsilon_{ikl}\nabla_k\nabla_l\omega_j(\mathbf{x}) = \theta_{ij}(\mathbf{x}). \tag{6}$$

The dislocation density $\alpha_{ij}(\mathbf{x})$ represents the torsion $S_{lkj} = \frac{1}{2}(\Gamma_{lkj} - \Gamma_{klj})$ of the Riemann–Cartan geometry. Here the relation is

$$\alpha_{ij} = \epsilon_{ikl}S_{lkj}. \tag{7}$$

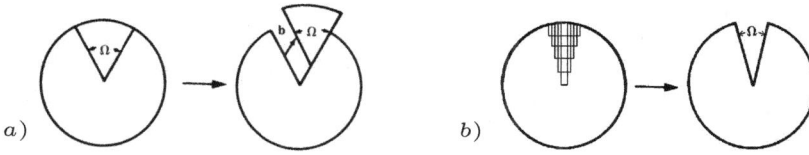

Fig. 6. Equivalence between a) dislocation and pair of disclination lines, b) disclination and stack of dislocation lines.

5. The standard form of a defect with Burgers vector b_l and Frank vector Ω_q has a displacement field

$$u_l(\mathbf{x}) = -\delta(\mathbf{x}, V)[b_l + \epsilon_{lqr}\Omega_q(x_r - \bar{x}_r)] \,, \tag{8}$$

where ϵ_{lqr} is the antisymmetric unit tensor, $\bar{x}_r$ the axis of rotation of the disclination part, and $\delta(\mathbf{x}; V)$ is the delta function on the volume V, i.e. in three dimensions:

$$\delta(\mathbf{x}; V) = \int_V d^3x' \delta^{(3)}(\mathbf{x} - \mathbf{x}') \,. \tag{9}$$

Its derivative is the delta function on the Volterra surface S of V:

$$-\boldsymbol{\nabla}\delta(\mathbf{x}; V) = \boldsymbol{\delta}(\mathbf{x}; S) = \int_S d\mathbf{S}' \delta^{(3)}(\mathbf{x} - \mathbf{x}') \,. \tag{10}$$

For the new gauge symmetry, the crucial observation is that as a simple consequence of (8), a dislocation line in the world crystal can either be obtained by a Volterra process of cutting out a thin slice of material of thickness $\mathbf{b}$, or alternatively by cutting out a wedge of Frank vector $\boldsymbol{\Omega}$, and reinserting it at distance $\mathbf{b}$ from the cut. Thus the dislocation line is indistinguishable from a pair of disclination lines with opposite Frank vector Ω whose axes of rotation are separated by a distance $\mathbf{b}$ [Fig. 6(a)]. Conversely, a disclination line is equivalent to a stack of dislocation lines with fixed Burgers vector $\mathbf{b}$ [Fig. 6(b)].

Analytically, this is most easily seen in the two-dimensional version of the relation (4):

$$\eta_{33} = \theta_{33} + \epsilon_{3mn}\nabla_m\alpha_{3n} \,. \tag{11}$$

Each term is invariant under the plastic gauge transformations $\beta_{kl}^p \to \beta_{kl}^p + \nabla_k u_l^p - \epsilon_{kl}\omega_3^p$, $\phi_l^p \to \phi_l^p + \partial_l\omega_3^p$. The general defect has a displacement field

$$u_l = -\delta(V_2)[b_l - \Omega\epsilon_{3lr}(x_r - \bar{x}_r)] \,. \tag{12}$$

The first term is a dislocation, the second term a disclination. According to Fig. 6, the latter can be read as a superposition of dislocations with the same Burgers vector $\tilde{b}_l = -\int_{\bar{x}}^x dx_r' \Omega\epsilon_{3lr}$. The former may be viewed as a dipole of disclinations: $-\bar{\nabla}_l[-\frac{1}{2}b_m\epsilon_{3km}]\epsilon_{3kr}(x_r - \bar{x}_r)$.

6. Let us now derive the emerging gravitational forces in the world crystal. Consider the partition function, at unit temperature, of the world crystal which we take to be three-dimensional, for simplicity:

$$Z = \sum_{n_{ij}(\mathbf{x})} \prod_{\mathbf{x},i} \left[\frac{du_i(\mathbf{x})}{a} \right] e^{-H} . \tag{13}$$

In linear elasticity, the energy depends quadratically only on the difference between the elastic and the plastic strain tensors $u_{ij} = \frac{1}{2}(\nabla_i u_j + \nabla_j u_i)$ and u_{ij}^p, and reads on the lattice

$$H = \frac{\mu}{4} \sum_{\mathbf{x}} \sum_{i<j} [\nabla_i u_j(\mathbf{x}) + \nabla_j u_i(\mathbf{x}) - n_{ij}(\mathbf{x})]^2 . \tag{14}$$

Here μ is the elastic constant,[16] and the integer numbers n_{ij} of are the lattice versions of $2u_{ij}^p$ in Eq. (5). This partition function explains for low temperature the correct classical specific heat. If the temperature is increased, it reaches a point where the configuration entropy of the defects wins over the Boltzmann factors of their energy, and the world crystal melts.

We have shown in Ref. 9 that in order to arrive at the proper Newton forces at long distances we have to insert one more derivative in the lattice action and start out with what is called the *floppy world crystal* where

$$H = \mu \sum_{\mathbf{x}} \sum_{i<j} \{\nabla_k[\nabla_i u_j(\mathbf{x}) + \nabla_j u_i(\mathbf{x}) - n_{ij}(\mathbf{x})]\}^2 . \tag{15}$$

The partition function depends on the defect configuration only via the defect tensor formed from n_{ij}, i.e. on η_{ij}. It is a functional of this tensor which can be expanded into powers of η_i^i, $\eta_{ij}\eta^{ij}$, $\eta_{ij}\eta^{jk}\eta_k^i$, $\eta_{ij}\eta^{ij}\eta_k^k$, ... The expansion coefficient are proportional to powers of the Planck length, for each η_{ij} two powers. Since η_{ij} is the defect representation of the Einstein tensor G_{ij}, the partition function defines a gravitational action which is a power series of the Einstein tensor. To leading (second) order in the Planck length is is proportional to the scalar $G = G_i^i = -R$.

Note that the gravitational action arises in this model from the *entropy* of the fluctuations[13] in the same way as rubber elasticity in polymer physics.[17]

The defect tensor $\eta_{ij} \hat{=} G_{ij}$ can be decomposed into an Cartan part and a torsion part as in Eq. (4). From the equivalence of defects illustrated in Fig. 6 it is now obvious that we can re-express the action, which contains only to defect tensor $\eta_{ij} \hat{=} G_{ij}$, completely in terms of the dislocation density α_{ij}, i.e. in terms of the torsion tensor S_{lkj} via Eq. (7). The Cartan curvature tensor is then identically zero, showing that Einstein's teleparallel formulation of gravity is completely equivalent to the original Einstein theory. Alternatively, we may make the torsion vanish identically, and recover the original Einstein theory.

In addition, there exists an infinite number of intermediate formulations of the theory with both Riemann-Cartan curvature and torsion in some well-defined mixture.

7. Generalizing the defect relations (4) and (11) to $D \geq 4$ spacetime dimensions and allowing for large deviations from Euclidean space, we find[18]

$$G_{\mu\nu} = G_{\mu\nu}^{\mathrm{C}} - \frac{1}{2} D^{*\lambda}(S_{\mu\nu,\lambda} - S_{\nu\lambda,\mu} + S_{\lambda\mu,\nu}) \tag{16}$$

where $G_{\mu\nu}$ is the Einstein tensor and $G_{\mu\nu}^{\mathrm{C}}$ its Cartan version, while $S_{\mu\kappa}{}^{,\tau}$ is the Palatini tensor related to the torsion field $S_{\mu\kappa}{}^{\tau}$ by

$$\frac{1}{2} S_{\mu\kappa}{}^{,\tau} \equiv S_{\mu\kappa}{}^{\tau} + \delta_{\mu}{}^{\tau} S_{\kappa\lambda}{}^{\lambda} - \delta_{\kappa}{}^{\tau} S_{\mu\lambda}{}^{\lambda} . \tag{17}$$

The symbol D_{μ} denotes the covariant derivative defined by $D_{\mu}v_{\nu} \equiv \partial_{\mu}v_{\nu} - \Gamma_{\mu\nu}{}^{\lambda}v_{\lambda}$, $D_{\mu}v^{\lambda} \equiv \partial_{\mu}v^{\lambda} + \Gamma_{\mu\nu}{}^{\lambda}v^{\nu}$, and $D_{\mu}^{*} \equiv D_{\mu} + 2S_{\mu\kappa}{}^{\kappa}$. The defect conservation laws (3) read

$$D_{\mu}^{*}G_{\lambda}^{\mathrm{C}\mu} + 2S^{\nu\lambda\kappa}G_{\kappa\nu}^{\mathrm{C}} - \frac{1}{2}S^{\nu\kappa,\mu}R_{\lambda\mu\nu\kappa}^{\mathrm{C}} = 0 \,, \tag{18}$$

$$D^{*\mu}S_{\lambda\kappa,\mu} = G_{\lambda\kappa}^{\mathrm{C}} - G_{\kappa\lambda}^{\mathrm{C}} . \tag{19}$$

They are Bianchi identities ensuring the single-valuedness of observables, connection $\Gamma_{\mu\nu}{}^{\lambda}$ and metric $g_{\mu\nu}$, via the integrability conditions $[\partial_{\sigma}, \partial_{\tau}]\Gamma_{\mu\nu}{}^{\lambda} = 0$ and $[\partial_{\sigma}, \partial_{\tau}]g_{\mu\nu} = 0$.

In a four-dimensional Riemann–Cartan spacetime, the geometry is described by the direct generalizations of *translational* and *rotational* defect gauge fields h_{ij} and A_{ijk}, which are here the vierbein field $h^{\alpha}{}_{\mu}$, and the spin connection $A_{\mu\alpha}{}^{\beta}$. The square of the former is the metric $g_{\mu\nu} = h^{\alpha}{}_{\mu}h_{\alpha\nu}$. The latter is defined by the covariant derivative $D_{\lambda}h_{\beta}{}^{\mu} = \partial_{\lambda}h_{\beta}{}^{\mu} - A_{\lambda\beta}{}^{\gamma}h_{\gamma}{}^{\mu} + \Gamma_{\lambda\nu}{}^{\mu}h_{\beta}{}^{\nu} \equiv D_{\lambda}^{L}h_{\beta}{}^{\mu} + \Gamma_{\lambda\nu}{}^{\mu}h_{\beta}{}^{\nu}$. The field strength of $A_{\mu\alpha}{}^{\beta} \equiv (A_{\mu})_{\alpha}{}^{\beta}$

$$F_{\mu\nu\beta}{}^{\gamma} \equiv \{\partial_{\mu}A_{\nu} - \partial_{\nu}A_{\mu} - [A_{\mu}, A_{\nu}]\}_{\lambda}{}^{\kappa} \,, \tag{20}$$

determines the Cartan curvature $R_{\mu\nu\lambda}^{\mathrm{C}}{}^{\kappa} \equiv h_{\lambda}^{\beta}h_{\gamma}{}^{\kappa}F_{\mu\nu\beta}{}^{\gamma}$. The field strength of $h^{\gamma}{}_{\nu}$ is the torsion:

$$S_{\alpha\beta}{}^{\gamma} \equiv \frac{1}{2}h_{\alpha}{}^{\mu}h_{\beta}{}^{\nu}[D_{\mu}^{L}h^{\gamma}{}_{\nu} - (\mu \leftrightarrow \nu)] . \tag{21}$$

The relations (16), (18), and (19) follow from this.

8. The theory is gauge invariant under local Lorentz transformations as a direct consequence of the fact that the metric can alternatively be written as

$$g_{\mu\nu} = h^{\gamma}{}_{\mu}\Lambda^{a}{}_{\gamma}\Lambda_{a}{}^{\beta}h_{\beta\nu} \,, \tag{22}$$

where $\Lambda_{a}{}^{\beta}$ is an *arbitrary local* Lorentz transformation, and that the Einstein–Hilbert Lagrangian $\mathcal{L}_{\mathrm{EH}} = -(1/2\kappa)R$ is *independent* of $\Lambda^{a}{}_{\alpha}$. The extra $\Lambda_{a}{}^{\beta}$ transforms the gauge field $A_{\mu\alpha}{}^{\beta}$ as

$$A_{\mu\alpha}{}^{\beta} \to A_{\mu\alpha}{}^{\beta} + \Delta A_{\mu\alpha}{}^{\beta} \,, \quad \Delta A_{\mu\alpha}{}^{\beta} \equiv \Lambda_{a}{}^{\beta}\partial_{\mu}\Lambda^{a}{}_{\alpha} . \tag{23}$$

At this point are ready to introduce the *new* gauge invariance announced in the title: we allow $\Lambda_{a}{}^{\beta}$ in Eq. (22) to be a *multivalued Lorentz transformation*. This is *not* integrable, so that $\Delta A_{\mu\alpha}{}^{\beta}$ is a *nontrivial* gauge field. Indeed, the rotational field strength $F_{\mu\nu\alpha}{}^{\gamma}$ can be expressed as $F_{\mu\nu\alpha}{}^{\gamma} \equiv \Lambda_a{}^{\gamma}[\partial_\mu, \partial_\nu]\Lambda^a{}_\alpha \neq 0$ and yields a nonzero Cartan curvature $R^{\mathrm{C}}_{\mu\nu\lambda}{}^{\kappa} \neq 0$. The important observation is that a *multivalued* $\Lambda^a{}_\alpha$ is able to *change the geometry*.[19] The right-hand side of (16) is *independent* of the vector field $A_{\mu\alpha}{}^{\beta}$. This allows us to shuffle torsion into Cartan curvature and back, fully or partially, by complete analogy with the defect transformations in two-dimensional crystals in Fig. 6. We may choose for $A_{\mu\alpha}{}^{\beta}$ any function we like. For example we may choose it to make the torsion vanish, and $A_{\mu\alpha}{}^{\beta}$ reduces to the usual spin connection of Einstein's gravity, the well-known combination of the objects of anholonomity

$$\Omega_{\mu\nu}{}^{\lambda} = \frac{1}{2}[h_\alpha{}^{\lambda}\partial_\mu h^\alpha{}_\nu - (\mu \leftrightarrow \nu)]. \tag{24}$$

In the opposite extreme $A_{\mu\alpha}{}^{\beta} = 0$, the Cartan curvature is zero, spacetime is teleparallel, and the Lagrangian is equal to the combination of torsion tensors:

$$\mathcal{L}_S = -\frac{1}{2\kappa}(-4D_\mu S^\mu + S_{\mu\nu\lambda}S^{\mu\nu\lambda} + 2S_{\mu\nu\lambda}S^{\mu\lambda\nu} - 4S^\mu S_\mu), \tag{25}$$

where $S_\mu \equiv S_{\mu\nu}{}^{\nu}$.

In any of the new gauges, the correct gravitational field equations are derived by extremizing the Einstein–Hilbert action

$$\mathcal{A}_{\mathrm{EH}} = -\frac{1}{2\kappa}\int dx\sqrt{g}R^{\mathrm{C}} + \int dx\sqrt{g}\mathcal{L}_S + \mathcal{A}_{\mathrm{GF}}, \tag{26}$$

where $\mathcal{A}_{\mathrm{GF}}$ is a functional of $h^\alpha{}_\mu$ and $A_{\mu\alpha}{}^{\beta}$ fixing some convenient gauge. For $\mathcal{A}_{\mathrm{GF}} = \delta[A_{\mu\alpha}{}^{\beta}]$ this leads to the teleparallel theory, and for $\mathcal{A}_{\mathrm{GF}} = \delta[S_{\alpha,\beta,\gamma}[h^\alpha{}_\mu, A_{\mu\alpha}{}^{\beta}]]$ we re-obtain Einstein's original theory.

9. Adding matter fields of masses m to the Einstein Lagrangian, and varying with respect to $h^\alpha{}_\mu$, we find in the zero-torsion gauge the Einstein equation

$$G_{\mu\nu} = \kappa T_{\mu\nu}, \tag{27}$$

where $T_{\mu\nu}$ is the sum over the symmetric energy-momentum tensors of all matter fields. Each contains the canonical energy-momentum tensor ${}^{\mathrm{m}}\Theta_{\mu\nu}$ and the spin current densities ${}^{\mathrm{m}}\Sigma_{\mu\nu},{}^{\lambda}$ in the combination due to Belinfante,[20]

$${}^{\mathrm{m}}T_{\kappa\nu} = {}^{\mathrm{m}}\Theta_{\kappa\nu} - \frac{1}{2}D^{*\mu}\left({}^{\mathrm{m}}\Sigma_{\kappa\nu,\mu} - {}^{\mathrm{m}}\Sigma_{\nu\mu,\kappa} + {}^{\mathrm{m}}\Sigma_{\mu\kappa,\nu}\right), \tag{28}$$

which is the matter analog of the defect relation (16).

The new gauge invariance of (16) has the physical consequence that the external gravitational field in the far-zone of a celestial body does not care whether angular momentum comes from rotation of matter or from internal spins. The off-diagonal elements of the metric in the far-zone, and thus the Lense-Thirring effect measured in Ref. 15, depend only on the total angular momentum $J^{\lambda\mu} = \int d^3x(x^\lambda T^{\mu 0} - x^\mu T^{\lambda 0})$, which by the Belinfante relation (28) is the sum of orbital angular momentum $L^{\lambda\mu} = \int d^3x(x^\lambda\Theta^{\mu 0} - x^\mu\Theta^{\lambda 0})$ and spin $S^{\lambda\mu} = \int d^3x\Sigma^{\lambda\mu,0}$. A star consisting of polarized matter has the same external gravitational field in the far-zone as a star rotating with the corresponding orbital angular momentum. This is the *universality of orbital momentum and intrinsic angular momentum* in gravitational physics observed in Ref. 11.

Since torsion is merely a *new-gauge degree* of freedom in describing a gravitational field, it cannot be detected experimentally, not even by spinning particles. A field with arbitrary spin may be coupled to gravity via the covariant derivative $D_\mu \equiv \partial_\mu \mathbf{1} + \frac{i}{2} A_{\mu\alpha}{}^\beta \Sigma^\alpha{}_\beta$, where $\Sigma^\alpha{}_\beta$ are the generators of the Lorentz group, in the Dirac case $\Sigma_{\alpha\beta} = \frac{i}{4}[\gamma^\alpha, \gamma_\beta]$. But since the torsion is a tensor, we may equally well use an infinity of alternative covariant derivatives $D_\mu^q \equiv \partial_\mu \mathbf{1} + \frac{i}{2} A_{\mu\alpha}^q{}^\beta \Sigma^\alpha{}_\beta$, where $A_{\mu\alpha}^q{}^\beta \equiv A_{\mu\alpha}{}^\beta - gK_{\mu\alpha}{}^\beta$, and $K_{\mu\alpha\beta} = h_\alpha{}^\nu h_\beta{}^\lambda K_{\mu\nu\lambda} \equiv h_\alpha{}^\nu h_\beta{}^\lambda (S_{\mu\nu\lambda} - S_{\nu\lambda\mu} + S_{\lambda\mu\nu})$. Any coupling constant q is permitted by covariance. In order to see which q is physically correct we come back to the above-discussed photon mass problem, and consider the covariant electromagnetic field tensor $F_{\mu\nu}^q \equiv D_\mu^q A_\nu - D_\nu^q A_\mu$. Working out the covariant derivative we find $\partial_\mu A_\nu - \partial_\nu A_\mu - 2(1-q) S_{\mu\nu}{}^\lambda A_\lambda$, which shows that Maxwell Lagrangian $-\frac{1}{4} F_{\mu\nu}^q F^{q\mu\nu}$ acquires a mass term, unless we fix the coupling strength has the value $q = 1$.

For this value of q, a little algebra[8,9] shows that the torsion drops out from the gauge field $A_{\mu\alpha}^q{}^\beta$. This reduces to the good-old Fock–Ivanenko spin connection that has been used in Einstein gravity without torsion:

$$A_{\mu\alpha}^1{}^\beta = \bar{A}_{\mu\alpha}{}^\beta = h_\alpha{}^\nu h^{\beta\lambda} (\Omega_{\mu\nu\lambda} - \Omega_{\nu\lambda\mu} + \Omega_{\lambda\mu\nu}). \tag{29}$$

Having ensured that the photon does not couple to torsion, we must also prevent all all other spinning baryonic matter to do so, to avois giving a mass to the photons via virtual processes of the type discussed above an illustrated in Fig. 1.

10. How about the motion of a spinless point particle in the infinitely many different descriptions of the same theory? Since the metric $g_{\mu\nu} = h^\gamma{}_\mu \Lambda^a{}_\gamma \Lambda_a{}^\beta h_{\beta\nu}$ is independent of the local Lorentz transformations $\Lambda^a{}_\alpha$, and the action $\mathcal{A} = -mc \int ds = -mc \int (g_{\mu\nu} dx^\mu dx^\nu)^{1/2}$ depends only on $g_{\mu\nu}$, the trajectories are geodesics for all $\Lambda_a{}^\beta$. The same result can of course be obtained my integrating the local conservation law of the total energy-momentum tensor $T_{\mu\nu}$ along a thin world-tube.

A spinning particle "sees" the gauge field of Lorentz transformations $A_{\mu\alpha}^q{}^\beta$, but it does so only via the $q = 1$-version (29). This contains only the vierbein fields, not the torsion, and is invariant under the multivalued version of the gauge transformation (23). Hence the motion of a spinning particle is blind the torsion field, which can therefore not be detected by any experiment.

11. What we have done can be understood better by a simple analogy. Insted of Einstein's theory, we consider a model of a real field ρ with an Euclidean Lagrangian $\mathcal{L} = (\partial_\mu \rho)^2 - \rho^2 + \rho^4$ and a partition function $Z = \int \mathcal{D}\psi \mathcal{D}\psi^* e^{-\int dx \mathcal{L}}$. The field ρ is the analog of the metric $g_{\mu\nu}$. We may now trivially introduce an extra gauge structure by re-expressing the Lagrangian in terms of a complex field $\psi = e^{i\theta}\rho$ and a gauge field A_μ as $\bar{\mathcal{L}} = |(\partial_\mu - iA_\mu)\psi|^2 - |\psi|^2 + |\psi|^4$. Now we form the partition function $\bar{Z} = \int \mathcal{D}\psi \mathcal{D}\psi^* \mathcal{D}A_\mu \Phi e^{-\int dx \bar{\mathcal{L}}}$, where Φ is an arbitrary gauge-fixing functional multiplied by the associated Faddev–Popov determinant. The new $\bar{Z}$ is completely equivalent to the original Z. Obviously there is no way of observing A_μ. The partition function Z plays the role of Einstein's theory, whereas $\bar{Z}$ gives its reformulation in terms of a gauge field, which does not change the physical content of the theory. The decomposition $\rho = \psi^*\psi = (\rho e^{-i\theta})(e^{i\theta}\rho)$ is the analog of the decomposition (22).

12. Higher gradient terms in elastic energy of the world crystal will generate an extra action $\mathcal{A}_{A_\mu}$ of the gauge field $A_{\mu\alpha}{}^\beta$.[21] This would, in general, violate the new symmetry discussed above and give torsion a life of its own. However, as long as the gravitational effects of spinning constituents in celestial bodies are suppressed with respect to that of the orbital angular momenta by many orders of magnitude, there is not much sense in conjecturing explicit forms of $\mathcal{A}_{A_\mu}$, unless we want to compete with string theory in setting up an ultimate *theory of everything* as a substitute of religion.

13. In summary, we have shown that if the Einstein–Hilbert Lagrangian is expressed in terms of the translational and rotational gauge fields $h^\alpha{}_\mu$ and $A_{\mu\alpha}{}^\beta$, the Cartan curvature can be converted to torsion and back, totally or partially, by a *new type of multivalued gauge transformation* in Riemann–Cartan spacetime, a *hypergauge transformation*. In this general formulation, Einstein's original theory is obtained by going to the zero-torsion hypergauge, while his teleparallel theory is in the hypergauge in which the Cartan curvature tensor vanishes. But any intermediate choice of the field $A_{\mu\alpha}{}^\beta$ is also allowed.

Acknowledgments

I thank F. W. Hehl, J. G. Pereira, and especially Jan Zaanen for a critical reading of the manuscript.

References

1. C. D. Anderson, *Phys. Rev.* **43**, 49 (1933).
2. See the letter exchanges in Physics Today, **60**, 10, 16 (2006) following S. Weinberg's article in Vol. **58**, 31 (2005).
3. R. Debever (ed.), *Elie Cartan–Albert Einstein, Letters on Absolute Parallelism 1929-1932* (Princeton University Press, Princeton, 1979).
4. K. Hayashi and T. Shirafuji, *Phys. Rev. D* **19**, 3524 (1979).
5. H. I. Arcos, and J. G. Pereira, (gr-qc/0408096v2); *Int. J. Mod. Phys. D* **13**, 2193 (2004), (gr-qc/0501017v1); V. C. de Andrade, H.I. Arcos, and J. G. Pereira, (gr-qc/0412034).
6. V. C. de Andrade and J. G. Pereira, *Int. J. Mod. Phys. D* **8**, 141 (1999), (gr-qc/9708051).
7. R. Utiyama, *Phys. Rev.* **101**, 1597 (1956); T. W. B. Kibble, *J. Math. Phys.* **2**, 212 (1961); D.W. Sciama, *Rev. Mod. Phys.* **36**, 463 (1964); F.W. Hehl, P. von der Heyde G. D. Kerlick and J. M. Nester, *Rev. Mod. Phys.* **48**, 393 (1976); F. W. Hehl, J. D. McCrea, E. W. Mielke and Y. Neemann, *Phys. Rep.* **258**, 1 (1995); R. T. Hammond, *Rep. Prog. Phys.* **65**, 599 (2002); W.-T. Ni, *Rep. Prog. Phys.* **73**, 056901 (2010), arXiv:0912.5057.

8. H. Kleinert, *Gauge Fields in Condensed Matter*, Vol. II, Stresses and Defects, World Scientific, Singapore 1989, pp. 744–1443 (`www.physik.fu-berlin.de/~kleinert/b2`).

9. H. Kleinert, *Multivalued Fields in Condensed Matter, Electromagnetism, and Gravitation* (World Scientific, Singapore, 2008), pp. 1-497 (`www.physik.fu-berlin.de/~kleinert/b11`).

10. E. Kröner, in *The Physics of Defects*, eds. R. Balian *et al.* (North-Holland, Amsterdam, 1981), p. 264.

11. H. Kleinert, *Gen. Rel. Grav.* **32**, 1271 (2000) (`physik.fu-berlin.de/~kleinert/271/271j.pdf`).

12. H. Kleinert, *Ann. d. Physik* **44**, 117 (1987) (`http://physik.fu-berlin.de/~kleinert/172/172.pdf`).

13. The entropy origin of the stiffenss of spacetime has recently been emphasized by E. P. Verlinde, arXiv:1001.0785. It has previously been used to generate the stiffness of strings: H. Kleinert, Dynamical Generation of String Tension and Stiffness, *Phys. Lett. B* **211**, 151 (1988); Membrane Stiffness from v.d. Waals forces, *Phys. Lett. A* **136**, 253 (1989). This is of course just another formulation of good-old Sakharov's idea. See A. D. Sakharov, *Dokl. Akad. Nauk SSSR* **170**, 70 (1967) [Soviet Physics-Doklady **12**, 1040 (1968)]. See also H. J. Schmidt, *Gen. Rel. Grav.* **32**, 361 (2000) (`www.springerlink.com/content/t51570769p123410/fulltext.pdf`).

14. P. Jizba, H. Kleinert and F. Scardigli, *Uncertainty Relation on World Crystal and its Applications to Micro Black Holes*, (arXiv:0912.2253).

15. See `einstein.stanford.edu`.

16. We ignore here the second elastic constant since it is irrelevant to the argument.

17. See Chapter 15 in H. Kleinert, *Path Integrals in Quantum Mechanics, Statistics, Polymer Physics, and Financial Markets* (World Scientific, Singapore, 2009) (`www.physik.fu-berlin.de/~kleinert/b5`).

18. B. A. Bilby, R. Bullough and E. Smith, *Proc. Roy. Soc. London, A* **231**, 263 (1955); K. Kondo, in *Proceedings of the II Japan National Congress on Applied Mechanics* (Tokyo, 1952), publ. in *RAAG Memoirs of the Unified Study of Basic Problems in Engeneering and Science by Means of Geometry*, Vol. 3, 148, ed. K. Kondo (Gakujutsu Bunken Fukyu-Kai, 1962).

19. The new freedom brought about by multivalued gauge transformations in many areas of physics is explained in the textbook.[9] For instance, we can *derive* the physical laws with magnetism from those without it, in particlular the minimal coupling law. Similarly, we can *derive* the physical laws in curved space from those in flat space.

20. F. J. Belinfante, *Physica* **6**, 887 (1939). For more details see Sec. 17.7 in the textbook.[9]

21. See p. 1453 in Ref. 8 (`physik.fu-berlin.de/~kleinert/b1/gifs/v1-1453s.html`).

NEUTRINO MASS AND GRAND UNIFICATION OF FLAVOR

R. N. MOHAPATRA

Maryland Center for Fundamental Physics, University of Maryland,
Maryland, MD-20742, USA
rmohapat@umd.edu

The problem of understanding quark mass and mixing hierarchies has been an outstanding problem of particle physics for a long time. The discovery of neutrino masses in the past decade, exhibiting mixing and mass patterns so very different from the quark sector has added an extra dimension to this puzzle. This is specially difficult to understand within the framework of conventional grand unified theories which are supposed to unify the quarks and leptons at short distance scales. In the paper, I discuss a recent proposal by Dutta, Mimura and this author that appears to provide a promising way to resolve this puzzle. After stating the ansatz, we show how it can be realized within a SO(10) grand unification framework. Just as Gell-Mann's suggestion of SU(3) symmetry as a way to understand the hadronic flavor puzzle of the sixties led to the foundation of modern particle physics, one could hope that a satisfactory resolution of the current quark-lepton flavor problem would provide fundamental insight into the nature of physics beyond the standard model.

Keywords: SO(10); type II seesaw; rank one.

1. Introduction

The quark masses as well as their mixings exhibit a hierarchical pattern i.e. for masses $m_{u,d} \ll m_{c,s} \ll m_{t,b}$; for mixing angles $V_{ub} \ll V_{cb} \ll V_{cd} \ll V_{ud,cs,tb}$. This is known as the flavor puzzle for quarks. Unravelling this puzzle has long been recognized as a challenge for physics beyond the standard model.[1] In the 1960's, a puzzle of similar nature was the focus of attention of many when particle physicists tried to understand why there were different baryons and mesons with masses close to each other. This puzzle, the flavor puzzle of the sixties was solved in a seminal paper by Prof. Gell-Mann- the famous "Eight-fold Way" paper, that practically the first read for every graduate student aspiring to be a particle theorist in the sixties. In this paper, he proposed that there is an underlying symmetry for hadrons interactions, the SU(3) symmetry which is responsible for the closeness of observed mass and quantum number patterns. This led to the so called Gell-Mann-Okubo mass formula, which was extremely successful in understanding the baryon and meson spectra and it culminated with the discovery of the Ω^- meson by N. Samios et al. This proposal of Gell-Mann unleashed an idea that had a profound impact on particle physics: it led to the concepts of quarks as the constituents of hadrons which forms the foundation of modern particle physics. It subsequently led to the birth of Quantum Chromodynamics as the theory of strong forces etc.

In the modern day particle physics of quarks and leptons, the hope has always been that solving the flavor puzzle may have similar ground breaking implication for physics beyond the standard model.

The puzzle of quark flavors was already mysterious but it got more so after the discovery of neutrino masses and mixings in 1998 and subsequent years. Unlike the quark mixings, lepton mixings do not exhibit a hierarchical pattern (i.e. neutrino mixings between generations denoted by θ_{ij} are given by $\theta_{23}^l \sim 45°$ and $\theta_{12}^l \simeq 35°$ as against $\theta_{23}^q \sim 2.5°$ and $\theta_{12}^q \sim 13°$) and the neutrino masses also do not exhibit as strong a hierarchy as quarks or charged leptons i.e. for a normal hierarchy for neutrinos, $m_2/m_3 \simeq V_{us} \ll m_\mu/m_\tau$. In fact the lepton mixing matrix (the PMNS matrix) appears to very closely resemble the following pattern: called the tri-bi-maximal mixing matrix:[2]

$$U_{\text{TB}} = \begin{pmatrix} \sqrt{\frac{2}{3}} & \sqrt{\frac{1}{3}} & 0 \\ -\sqrt{\frac{1}{6}} & \sqrt{\frac{1}{3}} & -\sqrt{\frac{1}{2}} \\ -\sqrt{\frac{1}{6}} & \sqrt{\frac{1}{3}} & \sqrt{\frac{1}{2}} \end{pmatrix}. \tag{1}$$

On top of this, neutrinos being electrically neutral particles could be their own anti-particles, the so-called Majorana fermions. In fact, the most popular way to understand the small neutrino masses seems to be the seesaw mechanism,[3] which predicts that the neutrinos are their own anti-particles. It could be that the different mixing pattern for leptons is related to this feature. Our proposal below has this as one of its ingredients.

The problem of quark-lepton masses and mixings becomes specially puzzling in grand unified theories where the quarks and leptons unify at a very high scale. One would naively expect that when quarks and leptons unify, their masses and mixings would exhibit a similar pattern. In fact the seesaw mechanism also suggests that the scale of neutrino mass is about 10^{14} GeV which is very close to the conventional grand unification scale $\sim 10^{16}$ GeV or so. This then raises the fundamental question of how we understand the diverse pattern of quarks and leptons in a seesaw motivated grand unified theory. Cracking this code may provide a hint of some really new exciting underlying physics.

In this note, I describe a recent ansatz proposed by Dutta and Mimura and this author which promises a new way to have a unified understanding of quark lepton flavor[4] and its realization in SO(10) grand unified theories .

As a prelude to this discussion, let us realize that one way to understand the quark mass hierarchy is to start with a rank one mass matrix for up, down quarks and charged leptons in the leading order i.e.

$$M_{u,d,l} = m_{t,b,\tau} \begin{pmatrix} 0 & 0 & 0 \\ 0 & 0 & 0 \\ 0 & 0 & 1 \end{pmatrix} \tag{2}$$

and consider corrections coming from non-leading operators. Second clue for our ansatz is the observation that small quark mixings are an indication that one could write the up and down quark mass matrices as a sum a "big" matrix plus a small matrix with the "big" matrix part for both sectors being proportional to each other so that in the leading order the CKM angles vanish e.g.

$$M_{u,d} = M_{u,d}^0 + \delta_{u,d} \tag{3}$$

with $M_d^0 = rM_u^0$ being the "big" matrix and $\delta_{u,d}$ being the smaller part. As just mentioned, the proportionality of the large parts of the mass matrices guarantees that the mixing angles will necessarily be small since the diagonalizing matrix for the large parts are "parallel" or "aligned" and the nontrivial CKM matrix represents the "small" misalignment between the two matrices determined by the "smaller" parts of the mass matrix.

We can therefore now state our ansatz[4] which consists of two parts:

(1) The quark and lepton mass matrices have the following general feature:

$$\begin{aligned} M_u &= M_0 + \delta_u; \\ M_d &= rM_0 + \delta_d; \\ M_l &= rM_0 + \delta_l; \\ M_\nu &= fv_L \end{aligned} \tag{4}$$

(2) M_0 has rank one.

Note that by a choice of the lepton basis, we can make f diagonal without loss of generality. It is then clear from the first part of the ansatz (item one) that for an "anarchic" form of the matrices M_0 and δ as long as $\delta_{ij} \ll M_0$, the lepton mixing angles are large whereas the quark mixing angles are small. The second rank one property than guarantees that quark and charged lepton masses are hierarchical whereas since f matrix is arbitrary, any hierarchy in the neutrino sector is likely to be milder. Incidentally, rank one property to understand mass hierarchy has been used in the past; see for instance Ref. 5.

2. Gauge Group Required to Implement the Ansatz

The question that arises next is how to implement our ansatz with a gauge model framework. To implement the first part, it is important to notice that the up and down quark mass matrices must be proportional to each other. Such relations do not emerge from the standard model since the u_R and d_R fields are separate fields and their Yukawa couplings responsible for the up and down quark mass matrices are therefore independent of each other. The situation however changes once we expand the gauge group to the left-right symmetric group $SU(2)_L \times SU(2)_R \times U(1)_{B-L}$ group since the (u_R, d_R) form a doublet of the $SU(2)_R$ group and thus relate the up and down Yukawa matrices.[6] Since quark-lepton unifications arises naturally within grand unification theories, the obvious group to consider is the SO(10) group as we do in the next section, although the basic conditions of the ansatz could also be realized with less predictive power within the $SU(2)_L \times SU(2)_R \times SU(4)_c$ partial unification groups.

3. SO(10) Realization of Flavor Unification Ansatz

As is well known, in SO(10) models, the matter fermions belong to **16**-dim. spinor representations. To get fermion masses, we will consider SO(10) models with **10**, **126** plus possibly another **10** or **120** Higgs fields where fermion masses are generated by renormalizable Yukawa couplings[7] only and where type II seesaw[8] is responsible for neutrino masses.[9] To implement our idea, we require that one of the **10** Yukawa couplings is the dominant one contributing to up, down and charged lepton masses and has rank one with other smaller couplings providing neutrino masses as well as most of the quark lepton flavor hierarchy. We postpone the discussion of how to get rank one till later. Let us see if this model does indeed give us our ansatz. It is well known that in these models,[7] we have the following form for all fermion masses:

$$Y_u = h + r_2 f + r_3 h', \tag{5}$$
$$Y_d = r_1(h + f + h'),$$
$$Y_e = r_1(h - 3f + c_e h'),$$
$$Y_{\nu D} = h - 3r_2 f + c_\nu h',$$

where Y_a are mass matrices divided by the electro-weak vev v_{wk} and r_i and $c_{e,\nu}$ are the mixing parameters which relate the $H_{u,d}$ to the doublets in the various GUT multiplets. More precisely, the matrices h, f and h' in Y_a are multiplied by the Higgs mixing parameters when they appear in the fermion mass matrices.

Furthermore, we use the type II seesaw formula[8] for getting neutrino masses gives[9]

$$\mathcal{M}_\nu = f v_L. \tag{6}$$

Note that f is the same coupling matrix that appears in the charged fermion masses in Eq. (5), up to factors from the Higgs mixings and the Clebsch-Gordan coefficients. This helps us to connect the neutrino parameters to the quark-sector parameters. The equations (5) and (6) are the key equations in our unified approach to addressing the flavor problem and obviously satisfy our flavor unification ansatz.

4. Implementing Rank One Strategy

The rank one Yukawa coupling with **10** Higgs field generates the features of flavor hierarchy, and rank 1 matrices can often appear in various ways (flavor symmetry, discrete symmetry, and string models). In this section, we give an SO(10) model, where the rank one ansatz used in our discussion of flavor emerges from extra vector like spinors above the GUT scale as well as a discrete symmetry.

When the direct couplings of chiral fermions with a Higgs field are forbidden by the chosen discrete symmetry, and the effective Yukawa couplings are generated by propagating vector-like matter fields, the rank of the effective Yukawa matrix depends on the number of the vector-like fields. Actually, when there are only one pair of vector-like matter fields as a flavor singlet, the effective Yukawa matrix is rank 1.

To illustrate this in a warm-up example, we consider a model which has one extra vector-like pair of matter fields to start with with mass slightly above the GUT scale contributing to the **10** coupling (denoted by $\psi_V \equiv \mathbf{16}_V \oplus \bar{\psi}_V \equiv \overline{\mathbf{16}}_V$) and three gauge singlet fields Y_a. We add a Z_4 discrete symmetry to the model under which the fields $\psi_a \to i\psi_a$, and $Y_a \to -iY_a$. The **10**-Higgs field H is invariant under this symmetry. The gauge invariant Yukawa superpotential under this assumption is given by

$$W = \psi_V H \lambda \psi_V + M_V \psi_V \bar{\psi}_V + \bar{\psi}_V \sum_a Y_a \psi_a. \tag{7}$$

When we give vevs $\langle Y_a \rangle \neq 0$, ψ_V and ψ_a are mixed. The heavy vector-like fields, $\bar{\psi}_V$ and a linear combination of ψ_V and ψ_a (i.e. $M_V \psi_V + \sum_a Y_a \psi_a$), and the effective operator below its scale and at the GUT scale is given by

$$\mathcal{L}_{eff} = \frac{\lambda}{M_V^2 + \sum_a Y_a^2} \left[\sum_a Y_a \psi_a \right] H \left[\sum_b Y_b \psi_b \right]. \tag{8}$$

This gives rise to a rank one h coupling. We note that it does not contradict the $O(1)$ top Yukawa coupling, when $M_V^2 \sim \sum_a Y_a^2$ (or $M_V^2 < \sum_a Y_a^2$).

If we let the $\overline{\mathbf{126}}$ Higgs field transform like -1 under Z_4, it can induce the f coupling with rank three. Our final model given below builds on this but differs in details e.g. it has got two vectorlike spinor multiplets instead of one etc.

5. Making the Model Predictive

Simply using the above ansatz in the context of an SO(10) model with mass relations in Eq. (5), turns out to reproduce the qualitative features of the quark and lepton spectra quite well. For example in the context of a two generation model (involving the second and third generation), this simple ansatz predicts $V_{cb} \simeq (m_s/m_b + e^{i\sigma} m_c/m_t) \cot\theta$, where θ is the atmospheric mixing angle. This relation is in rough agreement with observations to leading order. In addition, we have at GUT scale $m_b \sim m_\tau$ as well as $m - \mu \sim -3m_s$ also in rough agreement with observations.

Encouraged by these results, we can be more ambitious and start using this ansatz in combination of other ideas to make as many predictions as possible. To this end, we note that in the limit vanishing Yukawa couplings, the standard model has $[U(3)]^5$ global symmetry. It is therefore quite possible that in the final understanding of flavor, a subgroup of this large symmetry does play an important role,[10] specially subgroups which have three dimensional representation to fit three generations. In order to exploit this observation, one may replace all Yukawa couplings by flavon fields which transform as three dimensional representations of a subgroup of $[SU(3)]^5$ and consider the minima of the flavon theory in flavon space as determining the values of the Yukawa couplings. It turns out that there are nontrivial examples where this program is realized. In the second paper of Ref. 4, we presented an S_4 subgroup example. Below I briefly recapitulate this example.

6. The $SO(10) \times S_4 \times Z_n$ Model of Flavor

Recall that the S_4 group is a 24 element group describing permutations of four distinct objects and has five irreducible representations with dimensions $\mathbf{3_1} \oplus \mathbf{3_2} \oplus \mathbf{2} \oplus \mathbf{1_2} \oplus \mathbf{1_1}$. The distinction between the representations with subscripts 1 and 2 is that the later change sign under the transformation of group elements involving the odd number of permutations of S_4.

We assign the three families of **16**-dim. matter fermions ψ to $\mathbf{3_2}$-dim. representation of S_4 and the Higgs field H, $\bar{\Delta}$ and H' to $\mathbf{1_1}, \mathbf{1_2}$, and $\mathbf{1_1}$ reps, respectively. We then choose three $SO(10)$ singlet flavons ϕ_i transforming as $\mathbf{3_2}, \mathbf{3_1}, \mathbf{3_2}$ reps of S_4 and one gauge and S_4 singlet fields s_1, s_2 transforming as $\mathbf{1_2}$ and $\mathbf{1_1}$ respectively. We further assume that at a scale slightly above the GUT scale, there are two S_4 singlet vectorlike pairs of $\mathbf{16} \oplus \overline{\mathbf{16}}$ fields denoted by ψ_V and $\bar{\psi}_V$. In order to get the desired Yukawa couplings naturally from this high scale theory, we supplement the S_4 group by an Z_n group with all the above fields belonging to representations given in the table below. The fields and representations to generate the desired Yukawa couplings. $\omega = e^{i\frac{2\pi}{n}}$. $\alpha = 2 + a - b$. In addition to the fields below, there are two gauge and S_4 singlet $\mathbf{1_2}, \mathbf{1_1}$ fields with Z_n quantum numbers ω^a and ω^b respectively.

	ψ	H	$\bar{\Delta}$	H'	ϕ_1	$\bar{\phi}_1$	ϕ_2	$\bar{\phi}_2$	ϕ_3	$\bar{\phi}_3$	ψ_{V1}	$\bar{\psi}_{V1}$	ψ_{V2}	$\bar{\psi}_{V2}$
$SO(10)$	16	10	$\overline{126}$	10	1	1	1	1	1	1	16	$\overline{16}$	16	$\overline{16}$
S_4	$\mathbf{3_2}$	$\mathbf{1_1}$	$\mathbf{1_2}$	$\mathbf{1_1}$	$\mathbf{3_2}$	$\mathbf{3_2}$	$\mathbf{3_1}$	$\mathbf{3_1}$	$\mathbf{3_2}$	$\mathbf{3_2}$	$\mathbf{1_1}$	$\mathbf{1_1}$	$\mathbf{1_2}$	$\mathbf{1_2}$
Z_n	1	ω^{-4}	ω^{-2-a}	ω^{-1}	ω^2	ω^{-2}	ω	ω^{-1}	ω^α	$\omega^{-\alpha}$	ω^2	ω^{-2}	ω	ω^{-1}

The most general high scale Yukawa superpotential involving matter fields invariant under this symmetry is given by

$$W = (\phi_1 \psi)\bar{\psi}_{V1} + \psi_{V1}\psi_{V1}H + M_1\bar{\psi}_{V1}\psi_{V1} \qquad (9)$$
$$+ (\phi_2 \psi)\bar{\psi}_{V2} + \frac{1}{M_P}s_1\psi_{V2}\psi_{V2}\bar{\Delta} + M_2\bar{\psi}_{V2}\psi_{V2}$$
$$+ \frac{1}{M_P^2}s_2(\phi_3\psi\psi)\bar{\Delta} + \frac{1}{M_P}(\phi_2\psi\psi)H',$$

where the brackets stand for the S_4 singlet contraction of flavor index. The singlet field s_i can have large vev as follows: consider its Z_n charge to be such that the only polynomial term involving the s_i in the superpotential has the form $s_i^{k_i}/M_P^{k_i-3}$ (in order to describe the essential potential, we ignore a possible $s_1^{\ell_1}s_2^{\ell_2}$ term). The dominant part of the potential in the presence of SUSY breaking has the form:

$$V(s_i) = -m_{s_i}^2 |s_i|^2 + k \frac{s_i^{2k_i-2}}{M_P^{2k_i-6}} + \cdots . \tag{10}$$

Minimizing this leads to $\langle s_i \rangle \sim [m_{S_i}^2 M_P^{2k_i-6}]^{\frac{1}{2k_i-4}}$, which is above GUT scale for larger values of the integer k_i (which in turn is determined by the Z_n symmetry charge of s_i). One could also have large vevs for s_1, s_2 by using anomalous $U(1)$ charges for them using D-terms to break the $U(1)$ symmetry.

The effective theory below the scales $M_{1,2}$ and $\langle s_i \rangle$ of the vector-like pair masses and the s_i-vevs respectively is given by

$$W = (\phi_1\psi)(\phi_1\psi)H + (\phi_2\psi)(\phi_2\psi)\bar{\Delta} + (\phi_3\psi\psi)\bar{\Delta} + (\phi_2\psi\psi)H', \tag{11}$$

where we have omitted the dimensional coupling constants to make it simple for the purpose of writing. The discrete symmetries prevent ϕ^2/M^2 corrections to these terms. So our predictions based on this effective superpotential do not receive large corrections. We note that the non-renormalizable terms in Eq.(9) can also be obtained from renormalizable couplings if we introduce further S_4-triplet vectorlike fields. Here, however we use only S_4-singlet vectorlike fields to get rank 1 contribution to h and f Yukawa couplings and that is why we need the non-renormalizable terms to be present in Eq.(9.

In order to get fermion masses, we have to find the alignment[12] of the vevs of the flavon fields $\phi_{1,2,3}$. We show below that the following choice of vevs are among the minima of the flavon superpotential provided the couplings of mixed terms between different ϕ_i's are small compared to other couplings:

$$\phi_1 = \begin{pmatrix} 0 \\ 0 \\ 1 \end{pmatrix}, \quad \phi_2 = \begin{pmatrix} 0 \\ -1 \\ 1 \end{pmatrix}, \quad \phi_3 = \begin{pmatrix} 1 \\ 1 \\ 1 \end{pmatrix}. \tag{12}$$

Clearly, there are other vacua for the flavon model that we do not choose. What is however nontrivial is that the alignments are along quantized directions. This is a consequence of supersymmetry combined with discrete symmetries in the theory. Given these vev, we find from Eq. (11) that the Yukawa coupling matrices h, f, h' have the form:

$$h \propto \begin{pmatrix} 0\,0\,0 \\ 0\,0\,0 \\ 0\,0\,1 \end{pmatrix}, \tag{13}$$

$$f \propto \begin{pmatrix} 0 & 0 & 0 \\ 0 & 1 & -1 \\ 0 & -1 & 1 \end{pmatrix} + \lambda \begin{pmatrix} 0\,1\,1 \\ 1\,0\,1 \\ 1\,1\,0 \end{pmatrix}, \tag{14}$$

$$h' \propto \begin{pmatrix} 0 & 1 & -1 \\ 1 & 0 & 0 \\ -1 & 0 & 0 \end{pmatrix}, \tag{15}$$

and the charged fermion mass matrices can then be inferred. The neutrino mass matrix in this basis has the form:

$$\mathcal{M}_\nu = \begin{pmatrix} 0 & c & c \\ c & a & c-a \\ c & c-a & a \end{pmatrix}, \qquad (16)$$

where $c/a = \lambda \ll 1$. It is diagonalized by the tri-bi-maximal matrix

$$U_{\text{TB}} = \begin{pmatrix} \sqrt{\frac{2}{3}} & \sqrt{\frac{1}{3}} & 0 \\ -\sqrt{\frac{1}{6}} & \sqrt{\frac{1}{3}} & -\sqrt{\frac{1}{2}} \\ -\sqrt{\frac{1}{6}} & \sqrt{\frac{1}{3}} & \sqrt{\frac{1}{2}} \end{pmatrix}. \qquad (17)$$

This is however not the full PMNS matrix which will receive small corrections from diagonalization of the charged lepton matrix, which not only make small contributions to the θ_{atm} and $\theta_\odot$ but also generate a small θ_{13}.

The neutrino masses are given by $m_{\nu 3} = 2a - c$; $m_{\nu 2} = 2c$ and $m_{\nu 1} = -c$. To fit observations, we require $\lambda = c/a \simeq \sqrt{\Delta m_\odot^2/\Delta m_{\text{atm}}^2} \sim 0.2$, which fixes the neutrino masses $m_{\nu 3} \simeq 0.05$ eV, $m_{\nu 2} \simeq 0.01$ eV, and $m_{\nu 1} \simeq 0.005$ eV. We will see below that λ is also the Cabibbo angle substantiating our claim that neutrino mass ratio and Cabibbo angle are related.

For the charged lepton, up and down quark mass matrices, we have:

$$M_\ell = \frac{r_1}{\tan\beta} \begin{pmatrix} 0 & -3m_1 + \delta & -3m_1 - \delta \\ -3m_1 + \delta & -3m_0 & 3m_0 - 3m_1 \\ -3m_1 - \delta & 3m_0 - 3m_1 & -3m_0 + M \end{pmatrix}, \qquad (18)$$

$$M_d = \frac{r_1}{\tan\beta} \begin{pmatrix} 0 & m_1 + \delta & m_1 - \delta \\ m_1 + \delta & m_0 & -m_0 + m_1 \\ m_1 - \delta & -m_0 + m_1 & m_0 + M \end{pmatrix},$$

$$M_u = \begin{pmatrix} 0 & r_2 m_1 + r_3\delta & r_2 m_1 - r_3\delta \\ r_2 m_1 + r_3\delta & r_2 m_0 & -r_2 m_0 + r_3 m_1 \\ r_2 m_1 + r_3\delta & -r_2 m_0 + r_3 m_1 & r_2 m_0 + M \end{pmatrix},$$

where $\tan\beta$ is a ratio of $H_{u,d}$ vevs. Note that $m_1/m_0 = \lambda \sim 0.2$ and of course $m_0 \ll M$. A quick examination of these mass matrices leads to several immediate conclusions:

(1) The model predicts that at GUT scale $m_b \simeq m_\tau$.
(2) Since $(M_d)_{11} \to 0$, we get $V_{us} \simeq \sqrt{m_d/m_s}$.
(3) The empirically satisfied relation $m_\mu m_e \simeq m_s m_d$ can be obtained by the choice of parameters $-3m_1 + \delta = (m_1 + \delta)e^{i\sigma}$, where σ is a phase. Solving this equation, we find that $\delta = m_1(1 + i\cot\sigma/2)$. We obtain $V_{us} \simeq (1 - r_3/r_2)\delta/m_0$, thereby relating Cabibbo angle to the neutrino mass ratio $m_\odot/m_{\text{atm}} \simeq \lambda$.

(4) $m_\mu \sim -3m_s$.

(5) The leptonic mixing angle to diagonalize M_ℓ is related to quark mixing $\theta^l_{12} \sim \frac{1}{3}V_{us}$, which leads to a prediction for $\sin\theta_{13} \equiv U_{e3} \sim \frac{V_{us}}{3\sqrt{2}} \simeq 0.05$.[11]

(6) $V_{cb} \sim \dfrac{m_s}{m_b}\cot\theta_{\mathrm{atm}}$.

(7) The masses of up and charm quarks are given by the parameters $r_{2,3}$ and are therefore not predictions of the model.

(8) CP violation in quark sector can put in by making the parameters h' complex.

(9) The model predicts a small amplitude for neutrino-less double beta decay from light neutrino mass: $m_{\nu_{ee}} \sim c\sin\theta^l_{12} \simeq 0.3$ meV.

The first four relations are fairly well satisfied by observations; the fifth prediction (i.e. that for U_{e3}) can be tested in upcoming reactor and long baseline experiments. Note that the deviation from tri-bi-maximal mixing pattern coming from the charged lepton mass diagonalization could be thought of as a small perturbation of the neutrino mass matrix except that we predict the form of the perturbation from symmetry considerations. The sixth prediction gives a smaller value for V_{cb} (0.02 as against observed GUT scale value of 0.03) if one uses GUT scale extrapolated value of the known b mass. However, in the MSSM there are threshold corrections to the $b - s$ quark mass mixing from gluino and wino exchange one-loop diagrams; by choosing this contribution, one could obtain the desired V_{cb}.

Note that in this model, the top quark Yukawa coupling at GUT scale arises from an effective higher dimensional operator. We have showed the effective operator in Eq. (11) by expanding ϕ/M. The more precise form for the top Yukawa coupling is $\phi^2/(M_1^2 + \phi^2)h_{\psi_V\psi_V H}$, where $h_{\psi_V\psi_V H}$ is a coupling of $\psi_V\psi_V H$ term, and ϕ is the vev of ϕ_1 multiplied by $\phi_1\psi\bar\psi_V$ coupling. This is simply because the low energy third generation field is a linear combination of the form $\cos\alpha\,\psi_3 - \sin\alpha\,\psi_V$ with the mixing angle $\sin\alpha \simeq \phi/\sqrt{M_1^2 + \phi^2}$.

Therefore, in general, there is no gross contradiction to the fact that the top Yukawa coupling is order 1. However, in our case, if ϕ/M_1 becomes close to 1, the atmospheric mixing shifts from the maximal angle. Given the error in the determination of the atmospheric mixing angle, this is consistent with data and as this measurement sharpens, this is going to provide a test of this particular model. The desired smallness of the effective f and h' couplings however are more naturally obtained due to the presence of the Planck mass in the denominator. In order to make the f-coupling dominate over the h', we have to choose a small coupling for the H' Higgs field in Eq. (4). Similarly the λ term in Eq. (9) is assumed to be small compared to the coefficient of the first matrix.

Thus within these set of assumptions, this model is in good phenomenological agreement with observations. In a more complete theory, these assumptions need to be addressed. We however find it remarkable that despite these shortcomings, the model provides a very useful unification strategy of the diverse quark-lepton mixing patterns.

7. Vev Alignment as Minima of Flavon Theories

In this section, we give examples of how the minima of flavon theories can determine the Yukawa couplings of the fermions and lead to predictive flavor models. We discuss the specific case of the S_4 model at hand. This mechanism is of course applicable to any general group.

We start our discussion by giving some simple examples and discussing the flavon alignment as a prelude to the more realistic example. First thing to note is that $\mathbf{3_1}^3$ is invariant under S_4, but $\mathbf{3_2}^3$ is not. Denoting $\phi = (x, y, z)$, we see that in the first case, the singlet of $\phi^3 = xyz$. The superpotential for a $\mathbf{3_1}$ flavon field ϕ can therefore be written as

$$W = \frac{1}{2}m\phi^2 - \lambda\phi^3 = \frac{1}{2}m(x^2 + y^2 + z^2) - \lambda xyz. \tag{19}$$

The solution of F-flat vacua ($\phi \neq 0$) are

$$\phi = \frac{m}{\lambda}\{(1,1,1) \text{ or } (1,-1,-1) \text{ or } (-1,1,-1) \text{ or } (-1,-1,1)\}. \tag{20}$$

These vacua break S_4 down to S_3 and in the process determine the Yukawa couplings.

On the other hand, when $\mathbf{3_2}$ flavon is used (or the cubic term is forbidden by a discrete symmetry), quartic term involving the triplet is crucial for the F-flat vacua. The invariant quartic term ϕ^4 gives two linear combinations of the form $x^4 + y^4 + z^4$ and $x^2y^2 + y^2z^2 + z^2x^2$. This is because they have to be symmetric homogenous terms and invariant under the Klein's group, which is π rotation around the x, y, z axes.

Thus, the superpotential term for $\mathbf{3_2}$ field ϕ is

$$W = \frac{1}{2}m\phi^2 - \frac{\kappa^{(1)}}{M}(\phi^4)_1 - \frac{\kappa^{(2)}}{M}(\phi^4)_2 \tag{21}$$

$$= \frac{1}{2}(x^2 + y^2 + z^2) - \frac{\kappa^{(1)}}{4M}(x^4 + y^4 + z^4) - \frac{\kappa^{(2)}}{2M}(x^2y^2 + y^2z^2 + z^2x^2).$$

The nontrivial F-flat vacua ($\phi \neq 0$) are

$$\phi = \sqrt{\frac{mM}{\kappa^{(1)}}}\,\vec{a}, \quad \sqrt{\frac{mM}{\kappa^{(1)} + 2\kappa^{(2)}}}\,\vec{b}, \quad \sqrt{\frac{mM}{\kappa^{(1)} + \kappa^{(2)}}}\,\vec{c}, \tag{22}$$

where $\vec{a} = (0,0,\pm 1)$, $(0,\pm 1,0)$, $(\pm 1,0,0)$, $\vec{b} = (\pm 1,\pm 1,\pm 1)$, and $\vec{c} = (0,\pm 1,\pm 1)$, $(\pm 1,\pm 1,0)$, $(\pm 1,0,\pm 1)$. We note that these vectors correspond to the axes of the regular hexahedron. The vacua break S_4 down to Z_4, Z_3, and Z_2, respectively. More importantly, the vacuum states in Eq. (12) used in the analysis of fermion masses in the previous section are a subset of the above vacua.

Note that if we add a ϕ^4 term to the superpotential involving the $\mathbf{3_1}$ flavon field, $\vec{a}$ vacuum is possible, in addition to the original $\vec{b}$ vacua. However, $\vec{c}$ vacuum is absent.

8. Comments

A complete understanding of flavor is clearly a very ambitious task. Our proposal should be considered as a simple beginning towards a final theory. It should be noted that even though we have considered on SO(10) group, our general unification ansatz (without as much predictivity) in $SU(2)_L \times SU(2)_R \times SU(4)_c$ theories as well and perhaps other groups such as E_6. Similarly, one should explore other flavor models.

A second point of importance is that while we have kept only leading order terms, one should clearly consider higher order corrections to our predictions systematically. In the above mode, we have checked next order corrections and found them to be absent due to the discrete symmetries.

A final source of corrections could come from anomalies in the discrete symmetries, although we expect them to be small.[13]

9. Conclusion

In summary, I have discussed a recently proposed ansatz that has the potential to provide a unified description of the diverse quark and lepton flavor. This could provide the first opening into a very difficult problem of particle physics- the problem of flavor. A simple realization of this ansatz is shown to occur within a grand unified SO(10) model with type II seesaw describing the neutrino masses. The successes of that model are that it seems to provide an understanding of several observed quark-lepton mass relations such as bottom-tau mass unification, strange quark-muon mass ratio (1/3) etc. and predicts a value for $\theta_{13} \sim 0.05$ and atmospheric mixing angle different from the maximal value. The model like most grand unified theories of neutrinos predicts a normal hierarchy and observation of inverted hierarchy will therefore rule out this model (as well as most grand unified theories). Under certain reasonable approximations, this also seems to explain why $m_{solar}/m_{atm} \sim \theta_C$. Both the predictions given above (θ_{13} and θ_{atm}) could be used to test the model in the upcoming long baseline neutrino experiments. It also predicts a value of 0.3 meV for the neutrinoless double beta decay experiments.

Acknowledgments

I would like to thank Harald Fritzsch for inviting me to speak at this stimulating conference honoring Prof. M. Gell-Mann. This work is supported by the US National Science Foundation under grant No. PHY-0652363. The author is grateful to Y. Mimura and B. Dutta in whose collaboration this work was done and to Michael Ratz for discussions on anomalous discrete symmetries.

References

1. S. Weinberg, Trans. New York Acad. Sci. **38**, 185 (1977); F. Wilczek and A. Zee, Phys. Lett. B **70**, 418 (1977) [Erratum-ibid. **72B**, 504 (1978)]; H. Fritzsch, Phys. Lett. B **73**, 317 (1978).

2. P. F. Harrison, D. H. Perkins and W. G. Scott, Phys. Lett. B **530**, 167 (2002) [hep-ph/0202074]; X. G. He and A. Zee, Mod. Phys. Lett. A **22**, 2107 (2007) [arXiv:hep-ph/0702133]; Z. z. Xing, Phys. Lett. B **533**, 85 (2002) [hep-ph/0204049].

3. P. Minkowski, Phys. Lett. B **67** (1977) 421; T. Yanagida in *Workshop on Unified Theories, KEK Report 79-18*, p. 95, 1979; M. Gell-Mann, P. Ramond and R. Slansky, *Supergravity*, p. 315. Amsterdam: North Holland, 1979; S. L. Glashow, *1979 Cargese Summer Institute on Quarks and Leptons*, p. 687. New York: Plenum, 1980; R. N. Mohapatra and G. Senjanovic, Phys. Rev. Lett, **44** (1980) 912.

4. B. Dutta, Y. Mimura and R. N. Mohapatra, Phys. Rev. D **80**, 095021 (2009); JHEP **1005**, 034 (2010).

5. B. S. Balakrishna, A. L. Kagan and R. N. Mohapatra, Phys. Lett. B **205**, 345 (1988); K. S. Babu and R. N. Mohapatra, Phys. Rev. Lett. **64**, 2747 (1990); B. A. Dobrescu and P. J. Fox, JHEP **0808**, 100 (2008); L. Ferretti, S. F. King and A. Romanino, JHEP **0611**, 078 (2006).

6. K. S. Babu, B. Dutta and R. N. Mohapatra, Phys. Rev. D **60**, 095004 (1999).

7. K. S. Babu and R. N. Mohapatra, Phys. Rev. Lett. **70**, 2845 (1993).

8. G. Lazarides, Q. Shafi and C. Wetterich, Nucl. Phys. B **181**, 287 (1981); J. Schechter and J. W. F. Valle, Phys. Rev. D **22**, 2227 (1980); R. N. Mohapatra and G. Senjanovic, Phys. Rev. D **23**, 165 (1981).

9. B. Bajc, G. Senjanovic and F. Vissani, Phys. Rev. Lett. **90**, 051802 (2003).

10. G. Altarelli and F. Feruglio, Nucl. Phys. B **741**, 215 (2006) [hep-ph/0512103]; G. Altarelli, F. Feruglio and C. Hagedorn, JHEP **0803**, 052 (2008) [arXiv:0802.0090 [hep-ph]]; F. Feruglio, C. Hagedorn and L. Merlo, JHEP **1003**, 084 (2010) [arXiv:0910.4058 [hep-ph]]; S. Morisi and E. Peinado, Phys. Rev. D **80**, 113011 (2009) [arXiv:0910.4389 [hep-ph]]; M. C. Chen and S. F. King, JHEP **0906**, 072 (2009) [arXiv:0903.0125 [hep-ph]]; Y. Cai and H. B. Yu, Phys. Rev. D **74**, 115005 (2006) [hep-ph/0608022]; F. Bazzocchi and S. Morisi, Phys. Rev. D **80**, 096005 (2009) [arXiv:0811.0345 [hep-ph]]; H. Ishimori, Y. Shimizu and M. Tanimoto, Prog. Theor. Phys. **121**, 769 (2009) [arXiv:0812.5031 [hep-ph]]; F. Bazzocchi, L. Merlo and S. Morisi, Phys. Rev. D **80**, 053003 (2009) [arXiv:0902.2849 [hep-ph]]; G. Altarelli, F. Feruglio and L. Merlo, JHEP **0905**, 020 (2009) [arXiv:0903.1940 [hep-ph]]; W. Grimus, L. Lavoura and P. O. Ludl, J. Phys. G **36**, 115007 (2009) [arXiv:0906.2689 [hep-ph]]; G. J. Ding, Nucl. Phys. B **827**, 82 (2010) [arXiv:0909.2210 [hep-ph]]; Y. Daikoku and H. Okada, arXiv:0910.3370 [hep-ph]; M. K. Parida, Phys. Rev. D **78**, 053004 (2008) [arXiv:0804.4571 [hep-ph]]; C. Luhn, S. Nasri and P. Ramond, Phys. Lett. B **652**, 27 (2007) [arXiv:0706.2341 [hep-ph]]; C. Hagedorn, M. A. Schmidt and A. Y. Smirnov, Phys. Rev. D **79**, 036002 (2009) [arXiv:0811.2955 [hep-ph]]; F. Bazzocchi and I. de Medeiros Varzielas, Phys. Rev. D **79**, 093001 (2009) [arXiv:0902.3250 [hep-ph]]; A. Adulpravitchai, A. Blum and C. Hagedorn, JHEP **0903**, 046 (2009) [arXiv:0812.3799 [hep-ph]]; S. F. King and C. Luhn, Nucl. Phys. B **820**, 269 (2009) [arXiv:0905.1686 [hep-ph]]; M. C. Chen and K. T. Mahanthappa, Phys. Lett. B **681**, 444 (2009) [arXiv:0904.1721 [hep-ph]]; arXiv:0910.5467 [hep-ph]; For some earlier models, see K. S. Babu, E. Ma and J. W. F. Valle, Phys. Lett. B **552**, 207 (2003) [hep-ph/0206292]; K. S. Babu and X. G. He, hep-ph/0507217; R. N. Mohapatra, S. Nasri and H. B. Yu, Phys. Lett. B **639**, 318 (2006) [hep-ph/0605020].
For recent reviews, see G. Altarelli and F. Feruglio, arXiv:1002.0211 [hep-ph]; H. Ishimori, T. Kobayashi, H. Ohki, H. Okada, Y. Shimizu and M. Tanimoto, arXiv:1003.3552 [hep-th].

11. J. Ferrandis and S. Pakvasa, Phys. Lett. B **603**, 184 (2004) [hep-ph/0409204]; S. F. King, JHEP **0508**, 105 (2005) [hep-ph/0506297]; B. Dutta and Y. Mimura, Phys.

Lett. B **633**, 761 (2006) [hep-ph/0512171]; M. C. Chen and K. T. Mahanthappa, Phys. Lett. B **652**, 34 (2007).

12. S. F. King and G. G. Ross, Phys. Lett. B **520**, 243 (2001) [hep-ph/0108112]; S. F. King, JHEP **0508**, 105 (2005) [hep-ph/0506297]; I. de Medeiros Varzielas and G. G. Ross, Nucl. Phys. B **733**, 31 (2006) [hep-ph/0507176]; I. de Medeiros Varzielas, S. F. King and G. G. Ross, Phys. Lett. B **644**, 153 (2007) [hep-ph/0512313]; Phys. Lett. B **648**, 201 (2007) [hep-ph/0607045]. S. F. King and C. Luhn, Nucl. Phys. B **832**, 414 (2010); C. Hagedorn, S. F. King and C. Luhn, JHEP **1006**, 048 (2010).

13. M. Ratz, private communications. R. Kappl, H. P. Nilles, S. Ramos-Sanchez, M. Ratz, K. Schmidt-Hoberg and P. K. S. Vaudrevange, Phys. Rev. Lett. **102**, 121602 (2009)

LHC AND THE SEESAW MECHANISM

GORAN SENJANOVIĆ

International Centre for Theoretical Physics, 34100 Trieste, Italy

It is often said that neutrino mass is a window to a new physics beyond the standard model (SM). This is certainly true if neutrinos are Majorana particles since the SM with Majorana neutrino mass is not a complete theory. The classical text-book test of neutrino Majorana mass, the neutrino-less double beta decay depends on the completion, and thus cannot probe neutrino mass. As pointed out already more than twenty five years ago, the colliders such as Tevatron or LHC offer a hope of probing directly the origin of neutrino Majorana mass through lepton number violating production of like sign lepton pairs. I discuss this in the context of $L - R$ symmetric theories, which led originally to the seesaw mechanism. A W_R gauge boson with a mass in a few TeV region could easily dominate neutrino-less double beta decay, and its discovery at LHC would have spectacular signatures of parity restoration and lepton number violation. At the end I give an example of a predictive $SU(5)$ grand unified theory that results in a hybrid type I and III seesaw with a light fermion triplet below TeV scale. This theory can be tested at LHC.

1. Foreword

It gives me both great pleasure and honour to make part of the celebration of Murray Gell-Manns 80th Birthday. I first came across of his name in 1965 at the tender age of fifteen when I heard of mysterious quarks. This is the age when you discover girls or better, you hope to be discovered by them, and at that age I already knew I wanted to become a theoretical physicist and do particle physics. And all of a sudden, due to Gell-Mann and quarks, an otherwise boring topic became full of charm and promise and working and a great hit with the girls. For this many of us in my generation will be forever grateful to him.

I was introduced to James Joyce in the same year, but I never managed to fall in love with him as I did with the quarks. Sadly enough, an attempt to read Finnegan's Wake stopped long before I could discover the origin of the name quarks. And the irony of life would bring me to Trieste where Joyce spent a lot of time and is a kind of hero.

My first physical contact with Gell-Mann came in 1979 when he gave a summary talk at a big conference at CALTECH. The conference was clearly devoted to Feynman's 60th birthday, a great event for a young postdoc like me, and in his talk Gell-Mann not only mentioned my work with Rabi Mohapatra on the strong CP problem but also perfectly pronounced my last name, the first time ever in the

USA. I felt rather proud and only later I noticed that he pronounced perfectly every name, Italian, Chinese, French, you name it. It was not me being important, simply a polyglot on stage.

Now, when you work in our field it is impossible not to work on things not related to Gell-Mann. I have though one important thing in common with him: the seesaw mechanism discovered independently by a number of us. It is natural then that I speak of seesaw here. I will focus on what is the crucial question in my opinion: how to probe directly the seesaw, or better to say how to probe the origin of neutrino mass and its Majorana character. The answer is lepton number violation at colliders such as LHC, as Wai-Yee Keung and I suggested almost thirty years ago. The idea is completely analogous to a neutrino-less double beta decay: one can produce two electrons out of 'nothing' if neutrinos are Majorana particles.

2. Introduction

We know that neutrinos are massive but light.[1] If we wish to account for tiny neutrino masses with only the Standard Model (SM) degrees of freedom, we need Weinberg's[2] $d = 5$ effective operator

$$\mathcal{L} = Y_{ij} \frac{L_i H H L_j}{M},$$

where L_i stands for left-handed leptonic doublets and H for the usual Higgs doublet (with a vev v). This in turn produces neutrino Majorana mass matrix

$$M_\nu = Y \frac{v^2}{M}.$$

The non-renormalizable nature of the above operator signals the appearence of new physics through the mass scale M. The main consequence is the $\Delta L = 2$ violation of lepton number through

- neutrino-less double beta decay $\beta\beta 0\nu$[3]
- same sign charged lepton pairs in colliders.

While the neutrino-less double beta decay is a text-book probe of Majorana neutrino mass, the like sign lepton pair production, although suggested already long time ago,[4] has only recently received wide attention. In what follows I argue that this process may be our best bet in probing directly the origin of neutrino mass.

If M is huge, there is no hope of direct observation of new physics. It is often said that large M is more natural, for then Yukawas do not have to be small. However, small Yukawas are natural in a sense of being protected by chiral symmetries and anyway most of the SM Yukawas are small. Furthermore, large ratios of mass scales need fine-tuning, so there is nothing more natural about large M.

In order to get a window to new physics, we need a renormalizable theory of the above effective operator. In the minimal scenario of adding just one new type of

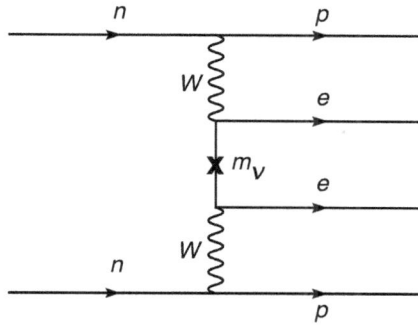

Fig. 1.   Neutrino-less double beta decay through the neutrino Majorana mass.

particles, there are only three different ways of producing it through the exchange of heavy

I) fermion singlet (1C , 1W , Y = 0), called right-handed neutrino; type I seesaw,[5]
II) bosonic weak triplet (1C , 3W , Y = 2); the type II seesaw,[6]
III) fermion weak triplet (1C , 3W , Y = 0); called type III seesaw[9]

where C stands for color and W for $SU(2)$ weak quantum numbers.

It is easy to see that all three types of seesaw lead to one and the same $d = 5$ operator above, and in a sense the effective theory is more useful unless we have a theory of seesaw.

The Majorana neutrino masses lead to lepton number violation (LNV) as in $\beta\beta0\nu$, in Fig.1. This probes neutrino mass in the range 0.1 - 1 eV.

However, in general m_ν is not directly connected to $\beta\beta0\nu$ decay. While it does produce it, the inverse is not true. $\beta\beta0\nu$ decay does not imply the measure of neutrino mass, since it depends on the completion of the SM needed for the $d = 5$ neutrino mass. An example is provided by the $L - R$ symmetric theory discussed in the next section. This is the theory that led originally to the seesaw mechanism, and as such deserves attention. As we will see, if the scale of parity restoration is in the few TeV region, the theory offers a rich LHC phenomenology and a plethora of lepton flavor violating (LFV) processes.

3. Left-Right Symmetry and the Origin of Neutrino Mass

$L - R$ symmetric theories[10] are based on the $SU(2)_L \times SU(2)_R \times U(1)$ gauge group augmented by parity or charge conjugation. They were originally suggested to account for the ad-hoc asymmetry of the SM, with an idea that LR symmetry is spontaneously broken.

Then:

• W_L implies W_R,
• ν_L implies ν_R, with m_{ν_R} of order M_R through the breaking of $L - R$ symmetry,
• Type I seesaw: connects neutrino mass to the scale of parity restoration.

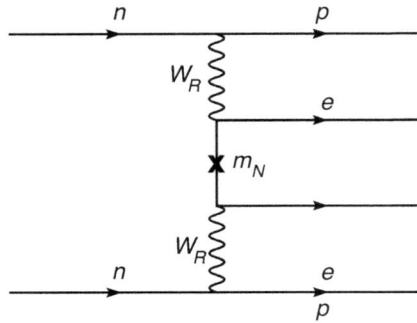

Fig. 2. Neutrino-less double beta decay induced by the right-handed gauge boson and right-handed neutrino.

These facts lead immediately to the new contribution to the neutrino-less double beta decay mentioned above, see Fig. 2. With W_R in the TeV region and the right-handed neutrino mass m_N in the 100 GeV -TeV region, this contribution can easily dominate over the left-handed one. Neutrino mass can even go to zero (vanishing Dirac Yukawa) while keeping the W_R contribution finite.

• Colliders: produce W_R through Drell-Yan as in Fig. 3.

Once the right-handed gauge boson is produced, it will decay into a right-handed neutrino and a charged lepton. The right-handed neutrino, being a Majorana particle, decays equally often into charged leptons or anti-leptons and jets. This often confuses people for naively one argues that the production of a wrong sign lepton must be suppressed by the mass of the right handed neutrino. True, but so does the production of the right sign lepton in its decay; this is the usual time dilation. It is enough that N is heavy enough as to decay into a lepton and two jets, and then the above claim must be true. In turn one has exciting events of same sign lepton

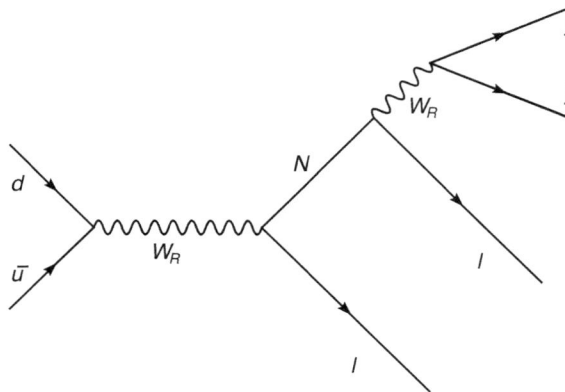

Fig. 3. The production of W_R and the subsequent decay into same sign leptons and two jets through the Majorana character of the right-handed neutrino.

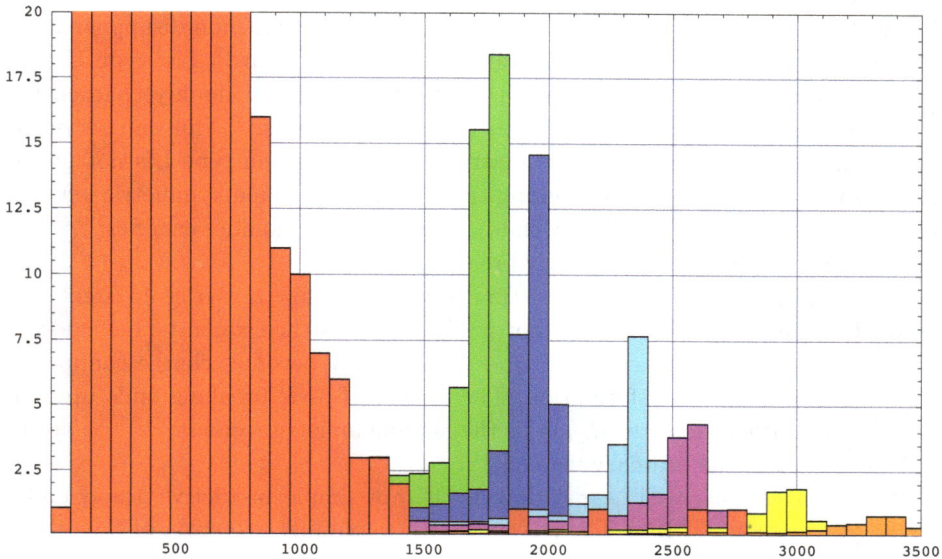

Fig. 4. The number of events as a function of energy (GeV) for L = 8fb^{-1} (courtesy of F. Nesti) where M$_R$ (TeV) is taken to be: 1.8; 2, 0; 2.4; 2.6; 3, 0; 3.4.

pairs and two jets, as a clear signature of lepton number violation. This is a collider analog of neutrino-less double beta decay, and it allows for the determination of W_R mass as shown in the Fig. 4.

This offers

a) direct test of parity restoration through a discovery of W_R,
b) direct test of lepton number violation through a Majorana nature of ν_R,
c) determination of W_R and N masses.

A detailed study[11] concludes an easy probe of W_R up to 3.5 TeV and ν_R in 100 - 1000 GeV for integrated luminosity of 30 fb^{-1}. It needs a study of flavor dependence, i.e. connection with LFV. There have been claims of $M_{W_R} \gtrsim 4$ TeV[12] (or even $M_{W_R} \gtrsim 10$ TeV[13]) in the minimal theory, from CP violating observables. It turns out though that they are not so severe,[14] and that they depend crucially on the choice of LR symmetry.

Recall that $L - R$ symmetry can be P as in the original works or C as it happens in $SC(10)$. The authors of Refs. 12, 13 use P, but it can be shown[14] that in the case of C, the freedom in CP phases leaves only the CP conserving limit $M_{W_R} \gtrsim 2.4$ TeV. This allows for both W_R and the accompanying neutral gauge boson Z_R to see seen at LHC. It was also argued in Ref. 14 that even in the case of P, the actual limit is lower, $M_{W_R} \gtrsim 4$ TeV, leaving hope for for observing at least W_R.

It is worth noting that the same signatures can be studied in the SM with ν_R,[15] but it requires miraculous cancellations of large Dirac Yukawa couplings in order to keep neutrino masses small. When a protection symmetry is called for, one ends up effectively with lepton number conservation and the phenomenon disappears.[16]

The $L - R$ theory possesses naturally also type II seesaw.[7] The type II offers another potentially interesting signature: pair production of doubly charged Higgses which decay into same sign lepton (anti lepton) pairs.[17] This can serve as a determination of the neutrino mass matrix in the case when type I is not present or very small.[18] It is worth commenting that the minimal supersymmetric left-right symmetric model[19] predicts doubly charged scalars at the collider energies[19-21] even for large scale of left-right symmetry breaking.

This is all very nice, but the question is whether a low $L - R$ scale is expected or not. It is perfectly allowed, but not predicted. This theory can be embedded in $SO(10)$ grand unified theory, where $L - R$ symmetry becomes charge conjugation and is a finite gauge transformation. The scale of $L - R$ breaking ends up being high though, either close to M_{GUT} in the supersymmetric version,[22,23] or around 10^{10} GeV or so in the ordinary version.[24]

We are faced then with a question: is there a simple predictive grand unified theory with seesaw at LHC? The answer is yes, a minimal extension of the original Georgi-Glashow theory[25], with an addition of an adjoint fermion representation.[26]

4. Minimal Non-Supersymmetric $SU(5)$

The minimal $SU(5)$ theory consists of: $24_H + 5_H$ Higgs multiplets, where 24_H is used to break the original symmetry to the SM one, and 5_H completes the symmetry breaking; and the three generation of quarks and leptons $3(10_F + \bar{5}_F)$. The theory fails for two reasons:

- gauge couplings do not unify
  α_2 and α_3 meet at about 10^{16} GeV (similar as in the MSSM), but α_1 meets α_2 too early, at $\approx 10^{13}$ GeV
- neutrinos remains massless as in the SM.

The $d = 5$ Weinberg operator for neutrino mass we started with is not enough: neutrino mass comes out too small ($\lesssim 10^{-4} eV$) since the cut-off scale M must be at least as large as M_{GUT} due to $SU(5)$ symmetry. In any case, one must first make sure that the theory is consistent and the gauge couplings unify.

A simple extension cures both problems: add just one extra fermionic 24_F.[26] This requires higher dimensional operators just as in the minimal theory, but can be made renormalizable as usual by adding extra 45_H scalar.[27]

Under $SU(3)_C \times SU(2)_W \times U(1)_Y$ the adjoint is decomposed as: $24_F = (1,1)_0 + (1,3)_0 + (8,1)_0 + (3,2)_{-5/6} + (\bar{3},2)_{5/6}$. The unification works as follows: triplet fermion (like wino in MSSM) slows down α_2 coupling without affecting α_1. In order that they meet above 10^{15} GeV for the sake of proton's stability, the triplet must be light, with a mass below TeV. Then in turn α_3 must be slowed down, which is achieved with an intermediate scale mass for the color octet in 24_F around 10^7 GeV or so.

For a practitioner of supersymmetry, the theory behaves effectively as the MSSM with a light wino, gluino heavy (10^7GeV), no Higgsino, no sfermions (they are irrelevant for unification being complete representations). This shows how splitting supersymmetry[28] opens a Pandora's box of possibilities for unification. The great success of low energy supersymmetry was precisely the prediction of gauge coupling unification,[29–32] ten years before the LEP confirmation of its prediction $\sin^2 \theta_W = .23$. In 1981 when it was thought that $\sin^2 \theta_W = .21$, this required asking for a heavy top quark, with $m_t \simeq 200$ GeV.[32] Unlike the case of supersymmetry, where the scale was fixed by a desire for the naturalness of the Higgs mass, and then unification predicted, in this case the $SU(5)$ structure demands unification which in turn fixes the masses of the new particles in 24_F. The price is the fine-tuning of these masses, but a great virtue is the tightness of the theory: the low mass of the fermion triplet (and other masses) is a true phenomenological prediction not tied to a nice but imprecise notion of naturalness.

With the notation singlet $S = (1,1)_0$, triplet $T = (1,3)_0$, it is evident that we have mixed Type I and Type III seesaw

$$(M_\nu)^{ij} = v^2 \left(\frac{y_T^i y_T^j}{m_T} + \frac{y_S^i y_S^j}{m_S} \right)$$

An immediate consequence is one massless neutrino. Thus one cannot have four generations in this theory, for then all four neutrinos would be light which the Z decay width does not allow. Since the triplet may be out of LHC reach, seeing the fourth generation would serve an important test of a theory; it would simply rule it out.

4.1. *T at LHC*

We saw that unification predicts the mass of the fermion triplet below TeV, and thus it becomes accessible to the colliders such as Tevatron and LHC. It can be produced through gauge interactions (Drell-Yan)

$$pp \rightarrow W^\pm + X \rightarrow T^\pm T^0 + X$$
$$pp \rightarrow (Z \,\text{or}\, \gamma) + X \rightarrow T^+ T^- + X$$

with the cross section for the T pair production in Fig. 5.

The best channel is like-sign dileptons + jets

$$BR(T^\pm T^0 \rightarrow l_i^\pm l_j^\pm + 4\text{jets}) \approx \frac{1}{20} \times \frac{|y_T^i|^2 |y_T^j|^2}{(\sum_k |y_T^k|^2)^2}$$

Same couplings y_T^i contribute to ν mass matrix and T decays, so that T decays can serve to probe the neutrino mass matrix[33] and the nature of the hierarchy of neutrino masses.

With proper cuts SM backgrounds appear under control.[34] With integrated luminosity of 10 fb^{-1} one could find the fermionic triplet T for M_T up to about 400 GeV.

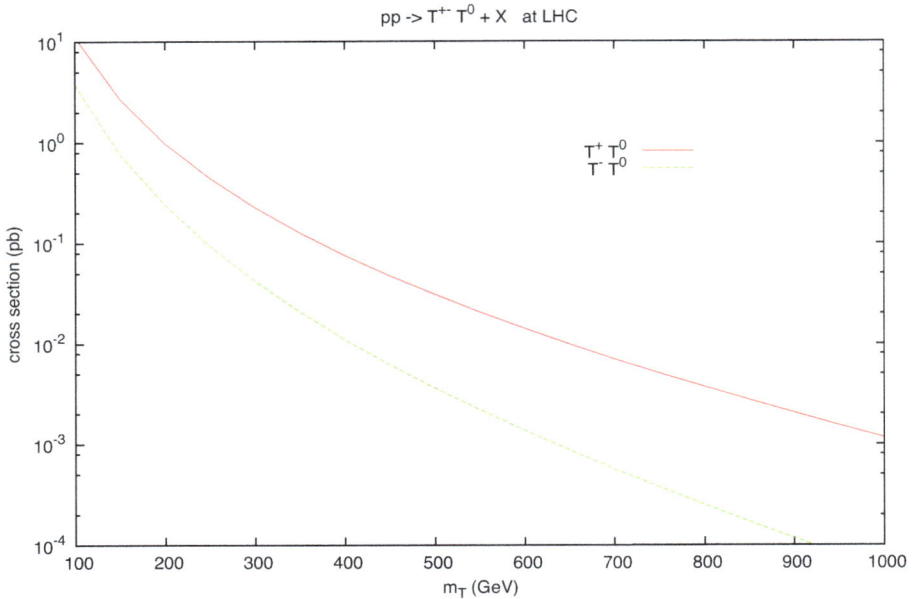

Fig. 5.   Cross section for the T pair production at LHC.

The light triplet fermion also plays an important role in lepton flavor violation, especially in $\mu \to e$ conversion in nuclei, which is induced at the tree level and could be observed even for a triplet out of LHC reach.[35]

Before concluding, it should be mentioned that one can also add a 15-dimensional scalar as an alternative of curing the minimal $SU(5)$ theory. This leads instead to the type II seesaw with possibly light lepto-quarks and its own interesting phenomenology.[37]

5. Summary and Outlook

I discussed here an experimental probe of Majorana neutrino mass origin, both at colliders through the production of the same sign dileptons, and a neutrinoless double beta decay. A classical example is provided by the $L - R$ symmetric theory that predicts the existence of right-handed neutrinos and leads to the seesaw mechanism. A TeV scale $L - R$ symmetry, as discussed here, would have spectacular signatures at LHC, with a possible discovery of W_R and ν_R. This offers a possibility of observing parity restoration and the Majorana nature of neutrinos. It is important to search for an underlying theory that predicts it naturally. For a recent attempt, see Ref. 38.

I have provided next an explicit example of a predictive grand unified theory: ordinary minimal $SU(5)$ with extra fermionic adjoint. A weak fermionic triplet is predicted in the TeV range (type III seesaw) whose decay is connected with neutrino mass. This offers good chances for discovery at LHC with integrated luminosity of 10 fb^{-1} for M_T up to about 400 GeV.

One can also simply study the minimalist scheme of pure seesaw in the connection with the colliders. The type II and III are naturally rather exciting from the experimental point of view, for the new states can be easily produced through the gauge couplings. In the case of the type I it becomes a long shot, since the Dirac Yukawas must be large and the smallness of neutrino mass is then attributed to the cancellations. Strictly speaking that should not be called the seesaw whose name was meant to indicate a natural smallness of neutrino mass after the heavy states are integrated out.

In summary, I argued here that in spite the smallness of neutrino masses, the hope of probing their origin at LHC is not just wishful thinking. Small Yukawa couplings are as natural as the large ones, and the low scale seesaw is perfectly realistic, and even likely in the context of the $SU(5)$ grand unified theory. There are other possible ways of having TeV scale seesaw, as e.g. with mirror leptons[39] and in the case of dynamical symmetry breaking.[40]

Acknowledgments

I wish to thank Harald Fritzsch and other organizers of the Gell-Mann Fest for an excellent conference and a warm hospitality. I am deeply grateful to my collaborators on the topics covered above Abdesslam Arhrib, Charan Aulakh, Borut Bajc, Dilip Ghosh, Tao Han, GuiYu Huang, Alessio Maiezza, Alejandra Melfo, Rabi Mohapatra, Miha Nemevšek, Fabrizio Nesti, Ivica Puljak, Andrija Rašin (Andrija, vrati se, sve ti je oprosteno), and Francesco Vissani. I acknowledge with great pleasure my old collaboration with Wai-Yee Keung on our original work on lepton number violation at colliders. This work was partially supported by the EU FP6 Marie Curie Research and Training Network "UniverseNet" (MRTN-CT-2006-035863).

References

1. For a review and references on neutrino masses and mixings, see A. Strumia and F. Vissani, arXiv:hep-ph/0606054.
   R. N. Mohapatra and A. Y. Smirnov, Ann. Rev. Nucl. Part. Sci. **56**, 569 (2006) [arXiv:hep-ph/0603118].
2. S. Weinberg, Phys. Rev. Lett. **43**, 1566 (1979).
3. G. Racah, Nuovo Cim. **14**, 322 (1937)
   W. H. Furry, Phys. Rev. **56**, 1184 (1939).
   For recent reviews and references, see e.g., the lectures in the Course CLXX of the International School of Physics "Enrico Fermi", Varenna, Italy, June 2008. Published in Measurements of Neutrino Mass, Volume 170 International School of Physics Enrico Fermi Edited by: F. Ferroni, F. Vissani and C. Brofferio, September 2009.
4. W. Y. Keung and G. Senjanović, Phys. Rev. Lett. **50**, 1427 (1983).
5. P. Minkowski, Phys. Lett. B **67** (1977) 421.
   T. Yanagida, proceedings of the *Workshop on Unified Theories and Baryon Number in the Universe*, Tsukuba, 1979, eds. A. Sawada, A. Sugamoto, KEK Report No. 79-18, Tsukuba.

S. Glashow, in *Quarks and Leptons, Cargèse 1979*, eds. M. Lévy. et al., (Plenum, 1980, New York).

M. Gell-Mann, P. Ramond, R. Slansky, proceedings of the *Supergravity Stony Brook Workshop*, New York, 1979, eds. P. Van Niewenhuizen, D. Freeman (North-Holland, Amsterdam).

R. Mohapatra, G. Senjanović, Phys.Rev.Lett. **44** (1980) 912

6. M. Magg and C. Wetterich, Phys. Lett. B **94** (1980) 61.
7. R. N. Mohapatra and G. Senjanović, Phys. Rev. D **23** (1981) 165.
8. G. Lazarides, Q. Shafi and C. Wetterich, Nucl. Phys. B **181** (1981) 287.
9. R. Foot, H. Lew, X. G. He and G. C. Joshi, Z. Phys. C **44** (1989) 441.
    See also E. Ma and D. P. Roy, Nucl. Phys. B **644**, 290 (2002) [arXiv:hep-ph/0206150].
10. J. C. Pati and A. Salam, Phys. Rev. D **10** (1974) 275.
    R. N. Mohapatra and J. C. Pati, Phys. Rev. D **11** (1975) 2558.
    G. Senjanović and R. N. Mohapatra, Phys. Rev. D **12** (1975) 1502.
    G. Senjanović, Nucl. Phys. B **153** (1979) 334.
11. A. Ferrari *et al.*, Phys. Rev. D **62**, 013001 (2000).
    S. N. Gninenko, M. M. Kirsanov, N. V. Krasnikov and V. A. Matveev, Phys. Atom. Nucl. **70**, 441 (2007).
    See also a recent talk on the ATLAS study by V. Bansal, arXiv:0910.2215[hep-ex].
12. Y. Zhang, H. An, X. Ji and R. N. Mohapatra, Nucl. Phys. B **802**, 247 (2008) [arXiv:0712.4218 [hep-ph]].
13. F. Xu, H. An and X. Ji, arXiv:0910.2265 [hep-ph].
14. A. Maiezza, M. Nemevsek, F. Nesti and G. Senjanovic, arXiv:1005.5160 [Unknown].
15. A. Datta and A. Pilaftsis, Phys. Lett. B **278**, 162 (1992).
    A. Datta, M. Guchait and D. P. Roy, Phys. Rev. D **47**, 961 (1993) [arXiv:hep-ph/9208228].
    T. Han and B. Zhang, Phys. Rev. Lett. **97**, 171804 (2006) [arXiv:hep-ph/0604064].
    F. del Aguila, J. A. Aguilar-Saavedra and R. Pittau, JHEP **0710**, 047 (2007) [arXiv:hep-ph/0703261].
16. J. Kersten and A. Y. Smirnov, Phys. Rev. D **76**, 073005 (2007) [arXiv:0705.3221 [hep-ph]].
17. A. G. Akeroyd and M. Aoki, Phys. Rev. D **72**, 035011 (2005) [arXiv:hep-ph/0506176].
    G. Azuelos, K. Benslama and J. Ferland, J. Phys. G **32** (2006) 73 [arXiv:hep-ph/0503096].
    T. Han, B. Mukhopadhyaya, Z. Si and K. Wang, Phys. Rev. D **76**, 075013 (2007) [arXiv:0706.0441 [hep-ph]].
    A. G. Akeroyd, M. Aoki and H. Sugiyama, Phys. Rev. D **77**, 075010 (2008) [arXiv:0712.4019 [hep-ph]].
    P. Fileviez Pérez, T. Han, G. y. Huang, T. Li and K. Wang, Phys. Rev. D **78**, 015018 (2008) [arXiv:0805.3536 [hep-ph]].
18. M. Kadastik, M. Raidal and L. Rebane, Phys. Rev. D **77**, 115023 (2008) [arXiv:0712.3912 [hep-ph]].
    J. Garayoa and T. Schwetz, JHEP **0803** (2008) 009 [arXiv:0712.1453 [hep-ph]].
    P. Fileviez Pérez, T. Han, G. Y. Huang, T. Li and K. Wang, Phys. Rev. D **78**, 071301 (2008)
    [arXiv:0803.3450 [hep-ph]].
19. C. S. Aulakh, A. Melfo and G. Senjanović, Phys. Rev. D **57**, 4174 (1998) [arXiv:hep-ph/9707256].
20. Z. Chacko and R. N. Mohapatra, Phys. Rev. D **58**, 015003 (1998) [arXiv:hep-ph/9712359].

21. C. S. Aulakh, A. Melfo, A. Rašin and G. Senjanović, Phys. Rev. D **58**, 115007 (1998) [arXiv:hep-ph/9712551].
22. An example of a renormalizable $SO(10)$ models with large Higgs representations (smaller representations give M_R even closer to M_{GUT})
    C. S. Aulakh, B. Bajc, A. Melfo, A. Rašin and G. Senjanović, Nucl. Phys. B **597**, 89 (2001) [arXiv:hep-ph/0004031].
    M. Drees and J. M. Kim, arXiv:0810.1875 [hep-ph].
23. Minimal supersymmetric $SO(10)$ with large Higgs representations
    C.S. Aulakh, R.N. Mohapatra, Phys. Rev. D **28** (1983) 217.
    T. E. Clark, T. K. Kuo and N. Nakagawa, Phys. Lett. B **115** (1982) 26.
    K. S. Babu and R. N. Mohapatra, Phys. Rev. Lett. **70** (1993) 2845 [arXiv:hep-ph/9209215].
    D. G. Lee and R. N. Mohapatra, Phys. Rev. D **51**, 1353 (1995) [arXiv:hep-ph/9406328].
    L. Lavoura, Phys. Rev. D **48** (1993) 5440 [arXiv:hep-ph/9306297].
    E. Brahmachari and R. N. Mohapatra, Phys. Rev. D **58** (1998) 015001 [arXiv:hep-ph/9710371].
    K. Matsuda, Y. Koide and T. Fukuyama, Phys. Rev. D **64** (2001) 053015.
    T. Fukuyama and N. Okada, JHEP **0211** (2002) 011.
    B. Bajc, G. Senjanović and F. Vissani, Phys. Rev. Lett. **90**, 051802 (2003) [arXiv:hep-ph/0210207].
    B. Bajc, G. Senjanović and F. Vissani, Phys. Rev. D **70** (2004) 093002 [arXiv:hep-ph/0402140].
    H. S. Goh, R. N. Mohapatra and S. P. Ng, Phys. Lett. B **570**, 215 (2003) [arXiv:hep-ph/0303055].
    C. S. Aulakh, B. Bajc, A. Melfo, G. Senjanović and F. Vissani, Phys. Lett. B **588**, 196 (2004) [arXiv:hep-ph/0306242].
    H. S. Goh, R. N. Mohapatra and S. P. Ng, Phys. Rev. D **68**, 115008 (2003) [arXiv:hep-ph/0308197].
    B. Bajc, A. Melfo, G. Senjanović and F. Vissani, Phys. Rev. D **70** (2004) 035007 [arXiv:hep-ph/0402122].
    C. S. Aulakh and A. Girdhar, Nucl. Phys. B **711**, 275 (2005) [arXiv:hep-ph/0405074].
    T. Fukuyama, A. Ilakovac, T. Kikuchi, S. Meljanac and N. Okada, J. Math. Phys. **46**, 033505 (2005) [arXiv:hep-ph/0405300].
    T. Fukuyama, A. Ilakovac, T. Kikuchi, S. Meljanac and N. Okada, Phys. Rev. D **72**, 051701 (2005) [arXiv:hep-ph/0412348].
    C. S. Aulakh, Phys. Rev. D **72** (2005) 051702 [arXiv:hep-ph/0501025].
    C. S. Aulakh and S. K. Garg, Nucl. Phys. B **757** (2006) 47 [arXiv:hep-ph/0512224].
    B. Bajc, A. Melfo, G. Senjanović and F. Vissani, Phys. Lett. B **634**, 272 (2006) [arXiv:hep-ph/0511352].
    S. Bertolini, T. Schwetz and M. Malinsky, Phys. Rev. D **73**, 115012 (2006) [arXiv:hep-ph/0605006].
    For an interesting version with heavy sfermions, see B. Bajc, I. Dorsner and M. Nemevšsek, JHEP **0811**, 007 (2008) [arXiv:0809.1069 [hep-ph]].
    For a review see G. Senjanović, Talk given at SEESAW25: International Conference on the Seesaw Mechanism and the Neutrino Mass, Paris, France, 10-11 Jun 2004 (Published in *Paris 2004, Seesaw 25* 45-64) [arXiv:hep-ph/0501244].
    C. S. Aulakh, arXiv:hep-ph/0506291.
    G. Senjanović "Theory of neutrino masses and mixings", lectures in the Course CLXX of the International School of Physics "Enrico Fermi", Varenna, Italy, June 2008.

Published in Measurements of Neutrino Mass, Volume 170 International School of Physics Enrico Fermi Edited by: F. Ferroni, F. Vissani and C. Brofferio, September 2009

24. For a recent study see B. Bajc, A. Melfo, G. Senjanović and F. Vissani, Phys. Rev. D **73**, 055001 (2006) [arXiv:hep-ph/0510139].
    S. Bertolini, L. Di Luzio and M. Malinsky, arXiv:0903.4049 [hep-ph].
    See also, N. G. Deshpande, E. Keith and P. B. Pal, Phys. Rev. D **47**, 2892 (1993) [arXiv:hep-ph/9211232].
    D. Chang, R. N. Mohapatra, J. Gipson, R. E. Marshak and M. K. Parida, Phys. Rev. D **31**, 1718 (1985).

25. H. Georgi and S. L. Glashow, Phys. Rev. Lett. **32**, 438 (1974).

26. B. Bajc and G. Senjanović, JHEP **0708**, 014 (2007) [arXiv:hep-ph/0612029].

27. P. Fileviez Pérez, Phys. Lett. B **654**, 189 (2007) [arXiv:hep-ph/0702287].

28. N. Arkani-Hamed and S. Dimopoulos, JHEP **0506** (2005) 073 [arXiv:hep-th/0405159].
    G. F. Giudice and A. Romanino, Nucl. Phys. B **699** (2004) 65 [Erratum-ibid. B **706** (2005) 65] [arXiv:hep-ph/0406088].

29. S. Dimopoulos, S. Raby, F. Wilczek, Phys. Rev. D **24** (1981) 1681.

30. L.E. Ibáñez, G.G. Ross, Phys. Lett. B **105** (1981) 439.

31. M.B. Einhorn, D.R. Jones, Nucl. Phys. B **196** (1982) 475.

32. W. J. Marciano and G. Senjanović, Phys. Rev. D **25**, 3092 (1982).

33. B. Bajc, M. Nemevšek and G. Senjanović, Phys. Rev. D **76**, 055011 (2007) [arXiv:hep-ph/0703080].

34. R. Franceschini, T. Hambye and A. Strumia, Phys. Rev. D **78**, 033002 (2008) [arXiv:0805.1613 [hep-ph]].
    F. del Aguila and J. A. Aguilar-Saavedra, arXiv:0808.2468 [hep-ph].
    F. del Aguila and J. A. Aguilar-Saavedra, arXiv:0809.2096 [hep-ph]. These papers contain a lot useful references.
    A. Arhrib, B. Bajc, D. K. Ghosh, T. Han, G. Y. Huang, I. Puljak and G. Senjanović, arXiv:0904.2390 [hep-ph].
    T. Li and X. G. He, arXiv:0907.4193 [hep-ph].

35. See e.g. A. Abada, C. Biggio, F. Bonnet, M. B. Gavela and T. Hambye, Phys. Rev. D **78** (2008) 033007 [arXiv:0803.0481 [hep-ph]].
    X. G. He and S. Oh, JHEP **0909**, 027 (2009) [arXiv:0902.4082 [hep-ph]].

36. J. F. Kamenik and M. Nemevšek, JHEP **0811** (2008) 007 arXiv:0908.3451 [hep-ph].

37. I. Dorsner, P. Fileviez Pérez and R. Gonzalez Felipe, Nucl. Phys. B **747**, 312 (2006) [arXiv:hep-ph/0512068].
    I. Dorsner and P. Fileviez Pérez, Nucl. Phys. B **723**, 53 (2005) [arXiv:hep-ph/0504276].

38. P. S. B. Dev and R. N. Mohapatra, arXiv:0910.3924 [hep-ph].

39. P. Q. Hung, Phys. Lett. B **649**, 275 (2007) [arXiv:hep-ph/0612004].

40. See for example, T. Appelquist and R. Shrock, Phys. Lett. B **548**, 204 (2002) [arXiv:hep-ph/0204141].

NEUTRINO MIXING, LEPTONIC CP VIOLATION, THE SEE-SAW MECHANISM AND BEYOND

S. T. PETCOV

SISSA/INFN, Trieste, 34136 Italy
IPMU, University of Tokyo, Tokyo, Japan

The phenomenology of 3-neutrino mixing and of the related Dirac and Majorana leptonic CP violation is reviewed. The leptogenesis scenario of generation of the baryon asymmetry of the Universe, which is based on the see-saw mechanism of neutrino mass generation, is considered. The results showing that the CP violation necessary for the generation of the baryon asymmetry of the Universe in leptogenesis can be due exclusively to the Dirac and/or Majorana CP-violating phase(s) in the neutrino mixing matrix U are briefly reviewed.

Keywords: Neutrino mixing; Dirac and Majorana leptonic CP violation; see-saw mechanism; leptogenesis.

1. Introduction

It is both an honor and a pleasure to speak at this Conference, organized in honor of Prof. Murray Gell-Mann's 80th birthday. My talk will be devoted to aspects of neutrino physics. The most famous contribution of Prof. Gell-Mann to neutrino physics is, without any doubt, the idea of the see-saw mechanism of neutrino mass generation, which he developed together with P. Ramond and R. Slansky.[1] As is well-known, the see-saw mechanism provides a natural explanation of the smallness of neutrino masses. A very appealing feature of the see-saw scenario is that, through the leptogenesis theory,[2,3] it connects the generation of neutrino masses to the generation of the matter-antimatter (or baryon) asymmetry of the Universe. In this talk we will review this remarkable connection. We will also recall an episode in which Prof. Gell-Mann played an instrumental role in stimulating a theoretical investigation[4] that lead to the realisation of the presence of Majorana CP-violation phases in the neutrino mixing matrix when the massive neutrinos are Majorana particles. These Majorana phases, in particular, can provide the CP violation necessary for the generation of the baryon asymmetry of the Universe within the leptogenesis scenario.[5,6]

2. The Three Neutrino Mixing: Current Status

The experiments with solar, atmospheric, reactor and accelerator neutrinos[7] have provided compelling evidences for flavour neutrino oscillations[8,9] - transitions in flight between the different flavour neutrinos ν_e, ν_μ, ν_τ (antineutrinos $\bar{\nu}_e$, $\bar{\nu}_\mu$, $\bar{\nu}_\tau$), caused by nonzero neutrino masses and neutrino mixing. Thus, the existence of oscillations of the solar ν_e, atmospheric ν_μ and $\bar{\nu}_\mu$, accelerator ν_μ (at $L \sim 250$; 730 km, L being the distance traveled by the neutrinos) and reactor $\bar{\nu}_e$ (at $L \sim 180$ km), is firmly established. The data imply the presence of mixing in the weak charged lepton current:

$$\mathcal{L}_{CC} = -\frac{g}{\sqrt{2}} \sum_{l=e,\mu,\tau} \overline{l_L}(x)\,\gamma_\alpha \nu_{lL}(x)\,W^{\alpha\dagger}(x) + h.c.\,, \quad \nu_{lL} = \sum_{j=1}^{n} U_{lj}\nu_{jL}\,, \qquad (1)$$

where $\nu_{lL}(x)$ are the flavour neutrino fields, $\nu_{jL}(x)$ is the left-handed (LH) component of the field of the neutrino ν_j having a mass m_j, and U is a unitary matrix - the Pontecorvo-Maki-Nakagawa-Sakata (PMNS) neutrino mixing matrix,[8–10] $U \equiv U_{PMNS}$. All compelling neutrino oscillation data can be described assuming 3-neutrino mixing in vacuum, $n = 3$. The number of massive neutrinos n can, in general, be bigger than 3 if, e.g. there exist right-handed (RH) sterile neutrinos[10] and they mix with the LH flavour neutrinos. It follows from the current data that at least 3 of the neutrinos ν_j, say ν_1, ν_2, ν_3, must be light, $m_{1,2,3} \lesssim 1$ eV, and must have different masses, $m_1 \neq m_2 \neq m_3$. At present there are no compelling experimental evidences for the existence of more than 3 light neutrinos.

In the case of 3 light neutrinos, the neutrino mixing matrix U can be parametrised by 3 angles and, depending on whether the massive neutrinos ν_j are Dirac or Majorana particles, by 1 or 3 CP violation (CPV) phases:[4]

$$U = VP\,, \quad P = \text{diag}(1, e^{i\frac{\alpha_{21}}{2}}, e^{i\frac{\alpha_{31}}{2}})\,, \qquad (2)$$

where $\alpha_{21,31}$ are two Majorana CPV phases and V is a CKM-like matrix,

$$V = \begin{pmatrix} c_{12}c_{13} & s_{12}c_{13} & s_{13}e^{-i\delta} \\ -s_{12}c_{23} - c_{12}s_{23}s_{13}e^{i\delta} & c_{12}c_{23} - s_{12}s_{23}s_{13}e^{i\delta} & s_{23}c_{13} \\ s_{12}s_{23} - c_{12}c_{23}s_{13}e^{i\delta} & -c_{12}s_{23} - s_{12}c_{23}s_{13}e^{i\delta} & c_{23}c_{13} \end{pmatrix}. \qquad (3)$$

In Eq. (3), $c_{ij} = \cos\theta_{ij}$, $s_{ij} = \sin\theta_{ij}$, the angles $\theta_{ij} = [0, \pi/2]$, and $\delta = [0, 2\pi]$ is the Dirac CPV phase. Thus, in the case of massive Dirac neutrinos, the neutrino mixing matrix U is similar, in what concerns the number of mixing angles and CPV phases, to the CKM quark mixing matrix. The presence of two additional physical CPV phases in U if ν_j are Majorana particles is a consequence of the special properties of the latter (see, e.g. Refs. 4, 11). On the basis of the existing neutrino data it is impossible to determine whether the massive neutrinos are Dirac or Majorana fermions.

The neutrino oscillation probabilities depend on the neutrino energy, E, the source-detector distance L, on the elements of U and, for relativistic neutrinos used in all neutrino experiments performed so far, on the neutrino mass squared differences $\Delta m_{ij}^2 \equiv (m_i^2 - m_j^2)$, $i \neq j$ (see, e.g. Ref. 11). In the case of 3-neutrino mixing there are only two independent Δm_{ij}^2, say $\Delta m_{21}^2 \neq 0$ and $\Delta m_{31}^2 \neq 0$.

The numbering of the neutrinos ν_j is arbitrary. We will employ the widely used convention which allows to associate θ_{13} with the smallest mixing angle in the PMNS matrix, and θ_{12}, $\Delta m_{21}^2 > 0$, and θ_{23}, Δm_{31}^2, with the parameters which drive the solar (ν_e) and the dominant atmospheric ν_μ and $\bar{\nu}_\mu$ oscillations, respectively. In this convention $m_1 < m_2$, $0 < \Delta m_{21}^2 < |\Delta m_{31}^2|$, and, depending on $\mathrm{sgn}(\Delta m_{31}^2)$, we have either $m_3 < m_1$ or $m_3 > m_2$. The existing data allow us to determine $\Delta m_\odot^2 \equiv \Delta m_\odot^2$, θ_{12}, and $|\Delta m_{31}^2| \equiv |\Delta m_A^2|$, θ_{23}, with a relatively good precision[12] and to obtain stringent limits on the angle θ_{13}.[13] The best fit values and the 3σ allowed ranges of Δm_{21}^2, s_{12}^2, $|\Delta m_{31(32)}^2|$ and s_{23}^2 read:[12]

$$(\Delta m_{21}^2)_{\mathrm{BF}} = 7.65 \times 10^{-5} \ \mathrm{eV}^2, \quad \Delta m_{21}^2 = (7.05 - 8.34) \times 10^{-5} \ \mathrm{eV}^2 \quad (4)$$

$$(\sin^2 \theta_{12})_{\mathrm{BF}} = 0.304, \quad 0.25 \le \sin^2 \theta_{12} \le 0.37, \quad (5)$$

$$(|\Delta m_{31}^2|)_{\mathrm{BF}} = 2.40 \times 10^{-3} \ \mathrm{eV}^2, \quad |\Delta m_{31}^2| = (2.07 - 2.75) \times 10^{-3} \ \mathrm{eV}^2 \quad (6)$$

$$(\sin^2 \theta_{23})_{\mathrm{BF}} = 0.5, \quad 0.36 \le \sin^2 \theta_{23} \le 0.67. \quad (7)$$

Thus, we have $\Delta m_{21}^2 / |\Delta m_{31}^2| \cong 0.03$, and $|\Delta m_{31}^2| = |\Delta m_{32}^2 - \Delta m_{21}^2| \cong |\Delta m_{32}^2|$. Maximal solar neutrino mixing, i.e. $\theta_{12} = \pi/4$, is ruled out at more than 6σ by the data. Correspondingly, one has $\cos 2\theta_{12} \ge 0.26$ (at 99.73% C.L.). A combined 3-neutrino oscillation analysis of the global data gives:[12]

$$\sin^2 \theta_{13} < 0.035 \ (0.056) \quad \mathrm{at} \ 90\% \ (99.73\%) \ \mathrm{C.L.} \quad (8)$$

These results imply that $\theta_{23} \cong \pi/4$, $\theta_{12} \cong \pi/5.4$ and that $\theta_{13} < \pi/13$. Correspondingly, the pattern of neutrino mixing is drastically different from the pattern of quark mixing.

At present no experimental information on the Dirac and Majorana CPV phases in the neutrino mixing matrix is available. Thus, the status of CP symmetry in the lepton sector is unknown. If $\theta_{13} \ne 0$, the Dirac phase δ can generate CPV effects in neutrino oscillations,[4,14] i.e. a difference between the probabilities of $\nu_l \to \nu_{l'}$ and $\bar{\nu}_l \to \bar{\nu}_{l'}$ oscillations in vacuum: $P(\nu_l \to \nu_{l'}) \ne P(\bar{\nu}_l \to \bar{\nu}_{l'})$, $l \ne l' = e, \mu, \tau$. The magnitude of the CPV effects is determined[15] by the rephasing invariant J_{CP} associated with the Dirac CPV phase in U. It is analogous to the rephasing invariant associated with the Dirac CPV phase in the CKM quark mixing matrix.[16] In the "standard" parametrisation of the PMNS matrix, Eqs. (2) - (3), we have:

$$J_{CP} \equiv \mathrm{Im} \left(U_{\mu 3} U_{e3}^* U_{e2} U_{\mu 2}^* \right) = \frac{1}{8} \cos \theta_{13} \sin 2\theta_{12} \sin 2\theta_{23} \sin 2\theta_{13} \sin \delta. \quad (9)$$

Thus, the size of CP violation effects in neutrino oscillations depends on the magnitude of the presently unknown values of the "small" angle θ_{13} and the Dirac phase δ. The currently reached sensitivity on $\sin^2 \theta_{13}$ is planned to be improved by a factor of (5-10) in three oscillation experiments with reactor $\bar{\nu}_e$ - Double CHOOZ, Daya Bay and RENO,[17-19] which are under preparation at present. Data on θ_{13} and on the Dirac phase δ will be obtained in the long baseline neutrino oscillation experiments T2K, NOνA, etc. (see, e.g. Ref. 20). Testing the possibility of Dirac CP violation in the lepton sector is one of the major goals of the next generation of neutrino oscillation experiments. It should be noted, however, that these experiments cannot give information on the absolute scale of neutrino masses. They are insensitive also to the nature - Dirac or Majorana, of massive neutrinos ν_j and, correspondingly, to the Majorana CPV phases present in U.[4,21]

If ν_j are Majorana fermions, getting experimental information about the Majorana CPV phases in U will be remarkably difficult.[22–26] The phases $\alpha_{21,31}$ can affect significantly the predictions for the rates of the (LFV) decays $\mu \rightarrow e + \gamma$, $\tau \rightarrow \mu + \gamma$, etc. in a large class of supersymmetric theories incorporating the seesaw mechanism.[27] As we will discuss, the Majorana phase(s) in the PMNS matrix can play the role of the leptogenesis CPV parameter(s) at the origin of the baryon asymmetry of the Universe.

The existing data do not allow one to determine the sign of $\Delta m_A^2 = \Delta m_{31(2)}^2$. In the case of 3-neutrino mixing, the two possible signs of $\Delta m_{31(2)}^2$ correspond to two types of neutrino mass spectrum. In the convention of numbering the neutrinos ν_j employed by us, the two spectra read:
i) spectrum with normal ordering (NO): $m_1 < m_2 < m_3$, $\Delta m_A^2 = \Delta m_{31}^2 > 0$, $\Delta m_\odot^2 \equiv \Delta m_{21}^2 > 0$, $m_{2(3)} = (m_1^2 + \Delta m_{21(31)}^2)^{\frac{1}{2}}$;
ii) spectrum with inverted ordering (IO): $m_3 < m_1 < m_2$, $\Delta m_A^2 = \Delta m_{32}^2 < 0$, $\Delta m_\odot^2 \equiv \Delta m_{21}^2 > 0$, $m_2 = (m_3^2 + \Delta m_{23}^2)^{\frac{1}{2}}$, $m_1 = (m_3^2 + \Delta m_{23}^2 - \Delta m_{21}^2)^{\frac{1}{2}}$. Depending on the values of the lightest neutrino mass, $\min(m_j)$, the neutrino mass spectrum can also be: *a) Normal Hierarchical (NH):* $m_1 \ll m_2 < m_3$, $m_2 \cong (\Delta m_\odot^2)^{\frac{1}{2}}$, $m_3 \cong |\Delta m_A^2|^{\frac{1}{2}}$; or *b) Inverted Hierarchical (IH):* $m_3 \ll m_1 < m_2$, with $m_{1,2} \cong |\Delta m_A^2|^{\frac{1}{2}} \sim 0.05$ eV; or *c) Quasi-Degenerate (QD):* $m_1 \cong m_2 \cong m_3 \cong m_0$, $m_j^2 \gg |\Delta m_A^2|$, $m_0 \gtrsim 0.10$ eV.
All three types of spectrum are compatible with the existing constraints on the absolute scale of neutrino masses m_j. Determining the type of neutrino mass spectrum is one of the main goals of the future experiments in the field of neutrino physics[a] (see, e.g. Ref. 20).

Information about the absolute neutrino mass scale (or about $\min(m_j)$) can be obtained, e.g. by measuring the spectrum of electrons near the end point in ^{3}H β-decay experiments[30] and from cosmological and astrophysical data. The most stringent upper bounds on the $\bar{\nu}_e$ mass were obtained in the Troitzk[31] and Mainz[32] experiments:

$$m_{\bar{\nu}_e} < 2.3 \text{ eV} \quad \text{at } 95\% \text{ C.L.} \tag{10}$$

We have $m_{\bar{\nu}_e} \cong m_{1,2,3}$ in the case of QD spectrum. The KATRIN experiment[32] is planned to reach sensitivity of $m_{\bar{\nu}_e} \sim 0.20$ eV, i.e. it will probe the region of the QD spectrum.

The Cosmic Microwave Background (CMB) data of the WMAP experiment, combined with supernovae data and data on galaxy clustering can be used to obtain

---

[a]For a brief discussion of experiments which can provide data on the type of neutrino mass spectrum see, e.g. Ref. 28; for some specific proposals see Ref. 29.

an upper limit on the sum of neutrinos masses[33] $\sum_j m_j \lesssim 0.68$ eV, 95% C.L. A more conservative estimate of the uncertainties in the astrophysical data leads to a somewhat weaker constraint (see, e.g. Ref. 34): $\sum_j m_j \lesssim 1.7$ eV, 95% C.L.

It follows from these data that neutrino masses are much smaller than the masses of charged leptons and quarks. If we take as an indicative upper limit $m_j \lesssim 0.5$ eV, we have $m_j/m_{l,q} \lesssim 10^{-6}$, $l = e, \mu, \tau$, $q = d, s, b, u, c, t$. It is natural to suppose that the remarkable smallness of neutrino masses is related to the existence of a new fundamental mass scale in particle physics, and thus, to new physics beyond that predicted by the Standard Model.

3. The See-Saw Mechanism

A natural explanation of the smallness of neutrino masses is provided by the see-saw mechanism of neutrino mass generation.[1,35] An integral part of the simplest version of this mechanism - the so-called "type I see-saw", are the $SU(2)_L$ singlet RH neutrinos ν_{lR}. The latter are assumed to possess a Majorana mass term as well as Yukawa type coupling with the Standard Model lepton and Higgs doublets $\psi_{lL}(x)$ and $\Phi(x)$, respectively, $(\psi_{lL}(x))^T = (\nu_{lL}^T(x) \; l_L^T(x))$, $l = e, \mu, \tau$, $(\Phi(x))^T = (\Phi^{(0)} \; \Phi^{(-)})$. In the basis in which the Majorana mass matrix of RH neutrinos is diagonal we have:

$$\mathcal{L}_{Y,M}(x) = -(\lambda_{kl} \, \overline{N_{kR}}(x) \, \Phi^\dagger(x) \, \psi_{lL}(x) + \text{h.c.}) - \frac{1}{2} \, M_k \, \overline{N_k}(x) \, N_k(x), \qquad (11)$$

where λ_{lk} is the matrix of neutrino Yukawa couplings and $N_k(x)$ is the heavy (RH) Majorana neutrino field possessing a mass $M_k > 0$. The fields $N_k(x)$ satisfy the Majorana condition $C\overline{N_k}^T(x) = \rho_k N_k(x)$, where C is the charge conjugation matrix and ρ_k is a phase. When the electroweak symmetry is broken spontaneously, the neutral component of the Higgs doublet develops non-zero vacuum expectation value $v = 174$ GeV and the neutrino Yukawa coupling generates a Dirac mass term: $m_{kl}^D \, \overline{N_{kR}}(x) \, \nu_{lL}(x) + \text{h.c.}$, with $m^D = v\lambda$. In the case when the elements of m^D are much smaller than M_k, $|m_{jl}^D| \ll M_k$, $j, k = 1, 2, 3$, $l = e, \mu, \tau$, the interplay between the Dirac mass term and the Majorana mass term of the heavy singlets N_k generates an effective Majorana mass (term) for the LH flavour neutrinos:[1,35]

$$(m_\nu)_{l'l} \cong (m^D)_{l'k}^T \, M_k^{-1} \, m_{kl}^D. \qquad (12)$$

In grand unified theories, m^D is typically of the order of the charged fermion masses. In $SO(10)$ theories,[1] for instance, m^D coincides with the up-quark mass matrix. Taking indicatively $m_\nu \sim 0.05$ eV, $m^D \sim 100$ GeV, one finds $M_k \sim 2 \times 10^{14}$ GeV, which is close to the scale of unification of electroweak and strong interactions, $M_{GUT} \cong 2 \times 10^{16}$ GeV. In GUT theories with RH neutrinos one finds that indeed the heavy singlets N_k naturally obtain masses which are by few to several orders of magnitude smaller than M_{GUT}.

One of the characteristic predictions of the see-saw mechanism is that both the light and heavy neutrinos ν_j and N_k are Majorana particles. The Majorana nature of the light neutrinos can be revealed in the neutrinoless double beta $((\beta\beta)_{0\nu}-)$ decay experiments (see, e.g. Refs. 11, 36).

4. The Nature of Massive Neutrinos and $(\beta\beta)_{0\nu}$-Decay

Establishing whether the neutrinos with definite mass ν_j are Dirac fermions possessing distinct antiparticles, or Majorana fermions, i.e., spin $1/2$ particles that are identical with their antiparticles, is of fundamental importance for understanding the origin of ν-masses and mixing and the underlying symmetries of particle interactions (see, e.g. Ref. 37). We recall that the neutrinos ν_j will be Dirac fermions if the particle interactions conserve some additive lepton number, e.g. the total lepton charge $L = L_e + L_\mu + L_\tau$. If no lepton charge is conserved, the neutrinos ν_j will be Majorana fermions. As we have seen, the massive neutrinos ν_j are predicted to be of Majorana nature by the see-saw mechanism.[1,35] The observed patterns of neutrino mixing and of neutrino mass squared differences can be related to Majorana massive neutrinos and the existence of an *approximate* symmetry in the lepton sector corresponding, e.g. to the conservation of the *non-standard* lepton charge $L' = L_e - L_\mu - L_\tau$.[38] Determining the nature (Dirac or Majorana) of massive neutrinos ν_j is one of the fundamental and most challenging problems in the future studies of neutrino mixing.

Extensive studies have shown that the only feasible experiments having the potential of establishing that ν_j are Majorana fermions at present are the $(\beta\beta)_{0\nu}$-decay experiments searching for the process $(A, Z) \rightarrow (A, Z+2) + e^- + e^-$ (for reviews see, e.g. Refs. 11, 36, 39). The observation of $(\beta\beta)_{0\nu}$-decay and the measurement of the corresponding half-life with sufficient accuracy, would not only be a proof that the total lepton charge is not conserved, but might provide also a unique information on the i) type of ν_j mass spectrum[40]), ii) Majorana phases in U,[22,41] and iii) absolute scale of neutrino masses (for details see, e.g. Refs. 22–25, 36, 39).

Under the assumptions of 3-ν mixing, of massive neutrinos ν_j being Majorana particles, and of $(\beta\beta)_{0\nu}$-decay generated *only by the (V-A) charged current weak interaction via the exchange of the three Majorana neutrinos ν_j* having masses $m_j \lesssim$ few MeV, the $(\beta\beta)_{0\nu}$-decay amplitude has the form (see, e.g. Ref. 22): $A(\beta\beta)_{0\nu} \cong$ $<m> M$, where M is the corresponding nuclear matrix element (NME) which does not depend on the neutrino mixing parameters, and $<m>$ is the effective Majorana mass in $(\beta\beta)_{0\nu}$-decay,

$$|<m>| = \left| m_1 |U_{e1}|^2 + m_2 |U_{e2}|^2 e^{i\alpha_{21}} + m_3 |U_{e3}|^2 e^{i(\alpha_{31} - 2\delta)} \right|, \qquad (13)$$

$|U_{e1}| = c_{12}c_{13}$, $|U_{e2}| = s_{12}c_{13}$, $|U_{e3}| = s_{13}$. In the case of CP-invariance one has $2\delta = 0, 2\pi$ and[42] $\eta_{21} \equiv e^{i\alpha_{21}} = \pm 1$, $\eta_{31} \equiv e^{i\alpha_{31}} = \pm 1$, $\eta_{21(31)}$ being the relative CP-parity of Majorana neutrinos $\nu_{2(3)}$ and ν_1.

The problem of obtaining the allowed values of $|<m>|$ using the data on the neutrino oscillation parameters, and, more generally, of the physics potential of $(\beta\beta)_{0\nu}$-decay experiments, was first studied in Ref. 43 and subsequently in a large

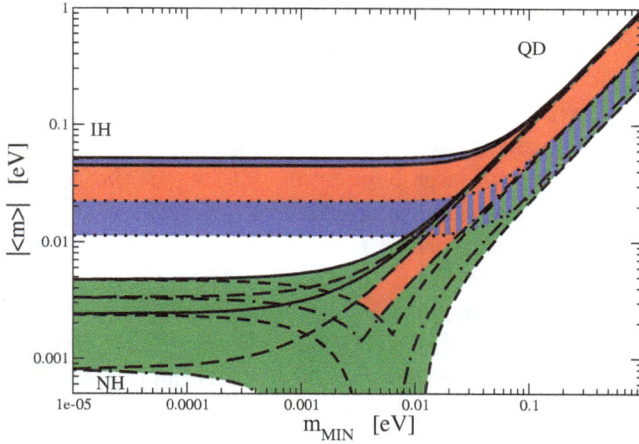

Fig. 1. (Color online) The value of $|<m>|$ as a function of $m_{\text{MIN}} \equiv min(m_j)$, obtained using the 95% C.L. allowed ranges of $\Delta m_\odot^2$, $|\Delta m_A^2|$, $\sin^2\theta_{12}$ and $\sin^2\theta_{13}$. The regions shown in red/grey correspond to violation of CP-symmetry. (From Ref. 28.)

number of papers.[b] Detailed analyses were performed more recently in Refs. 25, 28, 44. The main features of the predictions for $|<m>|$ are Refs. 22, 23, 40 (Fig. 1):

i) for NH spectrum,

$|<m>| \cong |(\Delta m_\odot^2)^{\frac{1}{2}} s_{12}^2 + (\Delta m_A^2)^{\frac{1}{2}} s_{13}^2 e^{-i(\alpha_{31}-\alpha_{21}-2\delta)}| \lesssim 0.006$ eV;

ii) for IH spectrum, $|<m>| \cong \sqrt{|\Delta m_A^2|}(1 - \sin^2 2\theta_{12} \sin^2 \frac{\alpha_{21}}{2})^{\frac{1}{2}}$, thus $|\Delta m_A^2|^{\frac{1}{2}} \cos 2\theta_{12} \lesssim |<m>| \lesssim |\Delta m_A^2|^{\frac{1}{2}}$, or 0.015 eV $\lesssim |<m>| \lesssim 0.055$ eV, the limits corresponding to the CP-conserving values of $\alpha_{21} = 0; \pi$;

iii) for QD spectrum,

$|<m>| \cong m_0 (1 - \sin^2 2\theta_{12} \sin^2 \frac{\alpha_{12}}{2})^{\frac{1}{2}}$, $m_0 \gtrsim |<m>| \gtrsim m_0 \cos 2\theta_{12} \gtrsim 0.03$ eV, with $m_0 \gtrsim 0.1$ eV, $m_0 < 2.3$ eV[32] or $m_0 \lesssim 0.5$ eV.[34]

For IH (QD) spectrum we have: $\sin^2(\alpha_{21}/2) \cong (1 - |<m>|^2/\tilde{m}^2)/\sin^2 2\theta_{21}$, $\tilde{m}^2 = |\Delta m_A^2| (m_0^2)$. Thus, a measurement of $|<m>|$ (and m_0 for QD spectrum) can allow to determine α_{21}.

The experimental searches for $(\beta\beta)_{0\nu}$-decay have a long history.[36] The best sensitivity was achieved in Heidelberg-Moscow ^{76}Ge experiment:[45] $|<m>| < (0.35 - 1.05)$ eV (90% C.L.), where a factor of 3 uncertainty in the relevant NME (see, e.g. Ref. 46) is taken into account. The IGEX collaboration has obtained:[47] $|<m>| < (0.33 - 1.35)$ eV (90% C.L.). A positive signal at $\sim 3\sigma$, corresponding to $|<m>| = (0.1 - 0.9)$ eV, is claimed to be observed in Ref. 48, while a more recent analysis reports[49] evidence at 6σ for $(\beta\beta)_{0\nu}$-decay with $|<m>| = 0.32 \pm 0.03$ eV. Two experiments, NEMO3[50] (with ^{100}Mo) and CUORICINO[51] (with ^{130}Te), designed to reach sensitivity to $|<m>| \sim (0.2 - 0.3)$ eV, obtained the limits: $|<m>| < (0.61 - 1.26)$ eV [50] and $|<m>| < (0.16 - 0.68)$ eV [51] (90% C.L.), where

---

[b]Extensive list of references on the subject is given in Ref. 39.

estimated uncertainties in the NME are accounted for. Most importantly, a number of projects[52] aim at sensitivity to $|<m>|$ $\sim$(0.01–0.05) eV: CUORE (^{130}Te), GERDA (^{76}Ge), SuperNEMO (^{100}Mo), EXO (^{136}Xe), MAJORANA (^{76}Ge), etc. These experiments will probe values of $|<m>|$ corresponding to the IH and QD spectra and test the positive result claimed in Ref. 48.

The existence of robust lower bounds on $|<m>|$ in the cases of IH and QD spectra,[40] which lie within the range of sensitivity of the future $(\beta\beta)_{0\nu}$-decay experiments, is one of the most important features of the predictions of $|<m>|$. These minimal values are given by $|\Delta m^2_A| \cos 2\theta_{21} \cong 0.015$ eV and $m_0 \cos 2\theta_{21} \cong 0.03$ eV. As Fig. 1 indicates, $|<m>|$ cannot exceed ~ 6 meV for the NH spectrum. Thus, $\max(|<m>|)$ in the case of NH spectrum is considerably smaller than $\min(|<m>|)$ for the IH and QD spectra. This opens the possibility of obtaining information about the type of ν-mass spectrum from a measurement of $|<m>| \neq 0$.[40] A positive result in the future $(\beta\beta)_{0\nu}$-decay experiments with $|<m>| > 0.01$ eV would imply, in particular, that the NH spectrum is strongly disfavored (if not excluded). Prospective experimental errors in the values of the oscillation parameters, in $|<m>|$, etc. and the uncertainty in the relevant NME, can weaken somewhat but do not invalidate these results.[c,24,25,53]

As Fig. 1 indicates, a measurement of $|<m>| \gtrsim 0.01$ eV would[23] either determine a relatively narrow interval of possible values of the lightest ν-mass $\min(m_j) \equiv m_{\text{MIN}}$, or would establish an upper limit on m_{MIN}.

The possibility of establishing CP violation in the lepton sector due to the Majorana CPV phases was found to be very challenging:[23–25] it requires quite accurate measurements of $|<m>|$ (and of m_0 for QD spectrum), and holds only for a limited range of values of the relevant parameters. More specifically,[24,25] establishing at 2σ Majorana CP violation in the case of QD spectrum requires for $\sin^2 \theta_{12}$=0.31, in particular, a relative error on the measured values of $|<m>|$ and m_0 smaller than 15%, an uncertainty $F \lesssim 1.5$ in the value of $|<m>|$ due to an imprecise knowledge of the corresponding NME, and a value of the relevant Majorana phase α_{21} typically within the ranges of $\sim (\pi/4 - 3\pi/4)$ and $\sim (5\pi/4 - 7\pi/4)$.

The knowledge of NME with sufficiently small uncertainty is crucial for obtaining quantitative information on the ν-mixing parameters from a measurement of $(\beta\beta)_{0\nu}$-decay half-life.

5. Baryon Asymmetry from the CPV Phases in U_{PMNS}

We will discuss next briefly the interesting possibility[5,6] that the CP violation necessary for the generation of the baryon asymmetry of the Universe, Y_B, in the leptogenesis scenario can be due exclusively to the Majorana (and/or Dirac) CPV phases in the PMNS matrix, and thus can be directly related to the low energy leptonic CP violation (e.g. in neutrino oscillations, etc.). We recall that leptogenesis[2] is

---

[c]Encouraging results, in what regards the problem of calculation of the NME, were reported in Ref. 46. A possible test of the NME calculations is discussed in Ref. 54.

a simple mechanism which allows to explain the observed baryon asymmetry of the Universe. The simplest scheme in which this mechanism can be implemented is the type I see-saw model. In its minimal version it includes the Standard Model (SM) plus two or three right-handed (RH) heavy Majorana neutrinos, N_k. Thermal leptogenesis (see, e.g. Ref. 55) can take place, e.g. in the case of hierarchical spectrum of the heavy neutrino masses, $M_1 \ll M_2 \ll M_3$. The lepton asymmetry is produced in out-of-equilibrium lepton number and CP nonconserving decays of the lightest RH Majorana neutrino, N_1, mediated by the neutrino Yukawa couplings, λ. The lepton asymmetry is converted into a baryon asymmetry by $(B - L)$-conserving but $(B + L)$-violating sphaleron interactions[3] which exist within the SM. In grand unified theories the heavy neutrino masses fall typically in the range of $\sim (10^8 - 10^{14})$ GeV. This range coincides with the range of values of M_k, required for a successful thermal leptogenesis.[55]

The Majorana mass matrix of the LH flavour neutrinos, generated by the see-saw mechanism, Eq. (12), can be written as:

$$m_\nu = v^2 \lambda^T M^{-1} \lambda = U^* m U^\dagger, \quad U \equiv U_{PMNS}, \tag{14}$$

where $M \equiv \mathrm{Diag}(M_1, M_2, M_3)$, $m \equiv \mathrm{Diag}(m_1, m_2, m_3)$, $M_k > 0$, $m_j \geq 0$. In our further discussion it is convenient to use the "orthogonal parametrisation" of the matrix of neutrino Yukawa couplings:[56]

$$\lambda = v^{-1} \sqrt{M} R \sqrt{m} U^\dagger, \quad R R^T = R^T R = 1, \tag{15}$$

where R is, in general, a complex matrix.

The realization that the CP violation necessary for the generation of the baryon asymmetry of the Universe can be due exclusively to the CPV phases in the PMNS matrix, is related to the progress in the understanding of the importance of lepton flavour effects in leptogenesis[57,58] (for earlier discussion see Ref. 59). In the case of hierarchical heavy neutrinos N_k, $M_1 \ll M_2 \ll M_3$, the flavour effects in leptogenesis can be significant for[57,58] 10^8 GeV $\lesssim M_1 \lesssim (0.5 - 1.0) \times 10^{12}$ GeV. If the requisite lepton asymmetry is produced in this regime, the CP violation necessary for successful leptogenesis can be provided entirely by the CPV phases in U_{PMNS}.[5]

Indeed, suppose that the mass of N_1 lies in the interval of interest, 10^9 GeV $\lesssim M_1 \lesssim 10^{12}$ GeV. The CP violation necessary for the generation of the baryon asymmetry Y_B in "flavoured" leptogenesis can arise both from the "low energy" neutrino mixing matrix U and/or from the "high energy" part of the matrix of neutrino Yukawa couplings λ - the matrix R, which can mediate CP-violating phenomena only at some high energy scale. The matrix R does not affect the "low" energy neutrino mixing phenomenology. Suppose further that the matrix R has real and/or purely imaginary CP-conserving elements: we are interested in the case when the CP violation necessary for leptogenesis is due exclusively to the CPV phases in U_{PMNS}. Under these assumptions, Y_B generated via leptogenesis can be written as[57,53] $|Y_B| \cong 3 \times 10^{-3} |\epsilon_\tau \eta|$, where ϵ_τ is the CPV asymmetry in the τ flavour

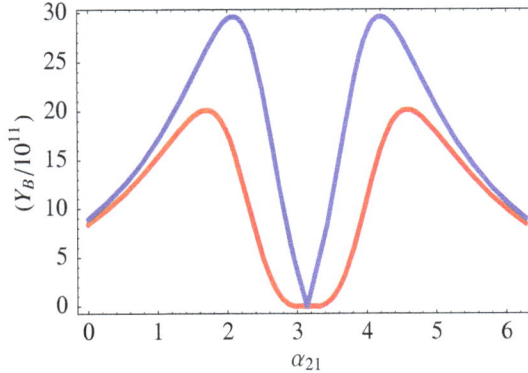

Fig. 2. (Color online) The asymmetry $|Y_B|$ versus the Majorana phase $\alpha_{21} = [0, 2\pi]$, for IH spectrum, purely imaginary $R_{11}R_{12} = i\kappa|R_{11}R_{12}|$, $\kappa = -1$, $|R_{11}|^2 - |R_{12}|^2 = 1$, $M_1 = 2 \times 10^{11}$ GeV, $\delta = 0$ and $s_{13} = 0$ (0.2) - blue/black (red/grey) line. (From Ref. 5.)

(lepton charge) produced in N_1-decays,[d]

$$\epsilon_\tau = -\frac{3M_1}{16\pi v^2}\frac{\text{Im}(\sum_{jk} m_j^{1/2} m_k^{3/2} U_{\tau j}^* U_{\tau k} R_{1j} R_{1k})}{\sum_i m_i |R_{1i}|^2},$$

η is the efficiency factor,[57] $|\eta| \cong |\eta(0.71\widetilde{m}_2) - \eta(0.66\widetilde{m}_\tau)|$, $\widetilde{m}_{2,\tau}$ being the wash-out mass parameters, $\widetilde{m}_2 = \widetilde{m}_e + \widetilde{m}_\mu$, $\widetilde{m}_l = |\sum_j m_j\ R_{1j}\ U_{lj}^*|^2$. Approximate analytic expression for $\eta(\widetilde{m})$ is given in Refs. 57, 58.

Consider the specific example of IH spectrum, $m_3 \ll m_{1,2} \cong |\Delta m_A^2|^{\frac{1}{2}}$. Under the simplifying assumptions of $m_3 \cong 0$ and $R_{13} \cong 0$ (N_3 decoupling), leptogenesis can be successful for $M_1 \lesssim 10^{12}$ GeV only if $R_{11}R_{12}$ is not real,[5] so we consider the case of CP-conserving purely imaginary $R_{11}R_{12} = i\kappa|R_{11}R_{12}|$, $\kappa = \pm 1$. The requisite CP violation can be due to the i) Majorana phase α_{21}, and/or ii) Dirac phase δ, in U. If we set $s_{13} = 0$, the maximum of $|Y_B|$ for, e.g. $\kappa = -1$, is reached for[5] $|R_{11}|^2 \cong 1.4$ ($|R_{12}|^2 = |R_{11}|^2 - 1 = 0.4$), and $\alpha \cong 2\pi/3; 4\pi/3$, and at the maximum $|Y_B| \cong 1.5 \times 10^{-12}(|\Delta m_A^2|^{\frac{1}{2}}/0.05\text{ eV})(M_1/10^9\text{ GeV})$ (Fig. 2). The observed baryon asymmetry, $|Y_B| \cong (8.0-9.2) \times 10^{-11}$, can be reproduced if $M_1 \gtrsim 5.3 \times 10^{10}$ GeV. As was shown in Ref. 5, we can have successful leptogenesis in the case of IH spectrum also if the source of CP-violation is the Dirac phase δ in U. In this case values of $|\sin\theta_{13}\sin\delta| \gtrsim 0.02$, or $|J_{CP}| \gtrsim 4.6 \times 10^{-3}$, are required. Values of $|J_{CP}|$ as small as 4.6×10^{-3}, can be probed in neutrino oscillation experiments at neutrino factories.[20]

Similar results can be obtained[5] for NH light neutrino mass spectrum, as well as in the case of QD in mass heavy Majorana neutrinos.

The interplay in "flavoured" leptogenesis between contributions in Y_B due to the "low energy" and "high energy" CP violation, originating from the PMNS matrix U and the R-matrix, respectively, was investigated in Ref. 6. It was found, in

---

[d]We have given the expression for Y_B normalised to the entropy density, see, e.g. Ref. 5.

particular, that under certain conditions which can be tested in low energy neutrino experiments (IH spectrum, $(-\sin\theta_{13}\cos\delta) \gtrsim 0.1$), the "high energy" contribution in Y_B due to R-matrix, can be so strongly suppressed that it would play practically no role in the generation of baryon asymmetry compatible with the observations. One would have successful leptogenesis in this case only if the requisite CP violation is provided by the Majorana and/or Dirac phases in U_{PMNS}.

6. Instead of Conclusions: Murray and Majorana Neutrinos

I met Prof. Gell-Mann first in December of 1979 at CERN, where I was a post-doctoral fellow. This had important implications for me: our work with S.M. Bilenky and J. Hosek, Ref. 4, in which it was understood that in the case of massive Majorana neutrinos the neutrino mixing matrix contains additional CP violating phases, which are usually called "Majorana phases", was inspired by the discussion I had with Murray after the talk on the see-saw mechanism he gave at the Theory Division. I had a few questions which arose during the talk, but I was a young post-doc at the beginning of my scientific career, while Murray was so famous for his fundamental contributions to particle physics that I was not sure he would be willing to spare even few minutes for a discussion with me. Nevertheless, with much hesitation I approached him with my questions after his talk. It turned out, Murray was accessible, the questions I had interested him and I had very stimulating discussion with him. In essence, I was curious to learn from Murray what was known about the properties of massive Majorana neutrinos. Murray was frank and did not try to obscure the fact that certain aspects of the phenomenology of neutrino mixing with massive Majorana neutrinos were still not well understood. I visited JINR in Dubna at the beginning of 1980 and with S.M. Bilenky and J. Hosek we tried to figure out whether there is any difference between the mixing of massive Dirac and of massive Majorana neutrinos and if there was, how one would observe it. Soon we realised that in the case of massive Majorana neutrinos there are additional CP-violating phases in the PMNS matrix; one could have CP violation even in the mixing of only 2 massive Majorana neutrinos. In contrast, the mixing of 2 massive Dirac neutrinos, as is well known, conserves the CP symmetry. The results of our study appeared in May of 1980. Later similar conclusions were reached by other authors.

This turned out to be just the beginning of my research "involvement" with Majorana neutrinos and, more generally, with Majorana particles, which still continues today. The rather brief discussion with Murray in 1979 arose my curiosity and later fascination with the subtleties of the physics of the Majorana fermions (light and heavy neutrinos, the neutralinos in SUSY theories, etc.), of their mixing and phenomenology, which continued all through my scientific career and is still fresh today.

Happy Anniversary, Prof. Gell-Mann! My best wishes for personal happiness and many professional successes.

Acknowledgments

This work was supported in part by the Italian INFN program on "Fisica Astroparticellare". Partial support from the Organising Committee of the Conference is acknowledged with gratefulness.

References

1. M. Gell-Mann, P. Ramond, R. Slansky, *Proceedings of the Supergravity Stony Brook Workshop*, New York 1979, eds. P. Van Nieuwenhuizen, D. Freedman.
2. M. Fukugita and T. Yanagida, *Phys. Lett. B* **174**, 45 (1986).
3. V.A. Kuzmin *et al.*, *Phys. Lett. B* **155**, 36 (1985).
4. S.M. Bilenky, J. Hosek and S.T. Petcov, *Phys. Lett. B* **94**, 495 (1980).
5. S. Pascoli *et al.*, hep-ph/0609125; *Nucl. Phys. B* **739**, 208 (2006).
6. E. Molinaro, S.T. Petcov, *Phys. Lett. B* **671**, 60 (2009); arXiv:0803.4120.
7. See, e.g. the review articles by K. Nakamura and B. Kayser in C. Amsler *et al.*, *Phys. Lett. B* **667**, 1 (2008), which include also extensive lists of references.
8. B. Pontecorvo, *Zh. Eksp. Teor. Fiz.* **33**, 549 (1957) and **34**, 247 (1958).
9. Z. Maki, M. Nakagawa and S. Sakata, *Prog. Theor. Phys.* **28**, 870 (1962).
10. B. Pontecorvo, *Zh. Eksp. Teor. Fiz.* **53**, 1717 (1967).
11. S. M. Bilenky and S. T. Petcov, *Rev. Mod. Phys.* **59**, 671 (1987).
12. T. Schwetz, M. Tórtola and J.W. F. Valle, arXiv:0808.2016.
13. M. Apollonio *et al.*, *Phys. Lett. B* **466**, 415 (1999).
14. N. Cabibbo, *Phys. Lett. B* **72**, 333 (1978).
15. P.I. Krastev and S.T. Petcov, *Phys. Lett. B* **205**, 84 (1988).
16. C. Jarlskog, *Z. Phys. C* **29**, 491 (1985).
17. F. Ardellier *et al.* [Double Chooz Collaboration], hep-ex/0606025.
18. See, e.g. the Daya Bay homepage http://dayawane.ihep.ac.cn/.
19. J.K. Ahn *et al.* [RENO Collaboration], arXiv:1003.1391.
20. A. Bandyopadhyay *et al.*, *Rept. Prog. Phys.* **72**, 106201 (2009).
21. P. Langacker *et al.*, *Nucl. Phys. B* **282**, 589 (1987).
22. S.M. Bilenky, S. Pascoli and S.T. Petcov, *Phys. Rev.* D**64**, 053010 (2001).
23. S. Pascoli, S.T. Petcov and L. Wolfenstein, *Phys. Lett. B* **524**, 319 (2002).
24. S. Pascoli, S.T. Petcov and W. Rodejohann *Phys. Lett. B* **549**, 177 (2002).
25. S. Pascoli, S.T. Petcov and T. Schwetz, *Nucl. Phys. B* **734**, 24 (2006).
26. A. De Gouvea, B. Kayser, R. Mohapatra, *Phys. Rev. D* **67**, 053004 (2003).
27. S. Pascoli *et al.*, *Phys. Lett. B* **564**, 241 (2003).
28. S. Pascoli and S.T. Petcov *Phys. Rev. D* **77**, 113003 (2008).
29. J. Bernabéu *et al.*, *Nucl. Phys. B* **669**, 255 (2003); S.T. Petcov, T. Schwetz, hep-ph/0511277; S.T. Petcov, M. Piai, *Phys. Lett. B* **533**, 94 (2002).
30. E. Fermi, *Nuovo Cim.* **11**, 1 (1934).
31. V. Lobashev *et al.*, *Nucl. Phys. A* **719**, 153c (2003).
32. K. Eitel *et al.*, *Nucl. Phys. Proc. Suppl.* **143**, 197 (2005).
33. D.N. Spergel *et al.*, *Astrophys. J. Supp.* **170**, 377 (2007).
34. M. Fukugita *et al.*, *Phys. Rev. D* **74**, 0027302 (2006).
35. P. Minkowski, *Phys. Lett. B* **67**, 421 (1977); T. Yanagida, in *Unified Theories and Baryon Number in theUniverse*, Tsukuba, Japan 1979, eds. A. Sawada, A. Sugamoto; R. N. Mohapatra, G. Senjanovic, *Phys. Rev. Lett.* **44**, 912 (1980).
36. C. Aalseth *et al.*, hep-ph/0412300.
37. R. Mohapatra *et al.*, *Rep. Prog. Phys.* **70**, 1757 (2007).

38. S.T. Petcov, *Phys. Lett. B* **110**, 245 (1982); P.H. Frampton, S.T. Petcov and W. Rodejohann, *Nucl. Phys. B* **687**, 31 (2004).

39. S.T. Petcov, *New J. Phys.* **6**, 109 (2004); *Physica Scripta* **T121**, 94 (2005).

40. S. Pascoli, S.T. Petcov, *Phys. Lett. B* **544**, 239 (2002); hep-ph/0310003.

41. S.M. Bilenky *et al.*, *Phys. Rev. D* **56**, 4432 (1996).

42. L. Wolfenstein, *Phys. Lett. B* **107**, 77 (1981); S.M. Bilenky *et al.*, *Nucl. Phys. B* **247**, 61 (1984); B. Kayser, *Phys. Rev. D* **30**, 1023 (1984).

43. S.T. Petcov and A.Yu. Smirnov, *Phys. Lett. B* **322**, 109 (1994).

44. M. Lindner, A. Merle, W. Rodejohann,*Phys. Rev. D* **73**, 053005 (2006).

45. H.V. Klapdor-Kleingrothaus *et al.*, *Nucl. Phys. Proc. Suppl.* **100**, 309 (2001).

46. V.A. Rodin *et al. Phys. Rev. C* **68**, 044302 (2003) and nucl-th/0503063; E. Caurier *et al.*, arXiv:0709.2137 and arXiv:0709.0277.

47. C.E. Aalseth *et al.*, *Phys. Atomic Nuclei* **63**, 1225 (2000).

48. H.V. Klapdor-Kleingrothaus *et al.*, *Phys. Lett. B* **586**, 198 (2004)

49. H.V. Klapdor-Kleingrothaus *et al.*, *Mod. Phys. Lett. A*, **21** 1547, (2006).

50. A. Barabash *et al.*, *J. Phys. Conf. Ser.* **173**, 012008 (2009).

51. C. Amaboldi *et al.*, *Phys. Rev. C* **78**, 035502 (2008).

52. F. Avignone, *Nucl. Phys. Proc. Suppl.* **143**, 233 (2005).

53. S. Pascoli, S.T. Petcov and W. Rodejohann, *Phys. Lett. B* **558**, 141 (2003); H. Murayama and C. Peña-Garay, *Phys. Rev. D* **69**, 031301 (2004).

54. S.M. Bilenky and S.T. Petcov, hep-ph/0405237.

55. W. Buchmuller *et al.*, *Annals Phys.* **315**, 305 (2005).

56. J. A. Casas and A. Ibarra, *Nucl. Phys. B* **618**, 171 (2001).

57. A. Abada *et al.*, *JCAP* **0604**, 004 (2006); *JHEP* **0609**, 010 (2006).

58. E. Nardi *et al.*, *JHEP* **0601**, 164 (2006).

59. R. Barbieri *et al.*, *Nucl. Phys. B* **575**, 61 (2000).

SOME RECENT PROGRESS IN AdS/CFT

JOHN H. SCHWARZ

Department of Physics, California Institute of Technology,
Pasadena, CA 91125, USA
jhs@theory.caltech.edu
http://theory.caltech.edu/~jhs/

Much of modern string theory research concerns AdS/CFT duality, or more generally, gauge/gravity duality. The main subjects are

- Testing and understanding such dualities by exploring how they work for systems with a lot of supersymmetry;
- Constructing and exploring approximate string theory duals of QCD;
- Applying gauge/gravity duality to other areas of physics such as condensed matter and nuclear physics.

I will briefly discuss the first topic.

Keywords: Superstring theory; M-theory; AdS/CFT duality; integrable models.

1. Personal Remarks

Since this is a conference in honor of Murray Gell-Mann on the occasion of his 80th birthday, it seems appropriate to begin by making some personal remarks. Murray Gell-Mann has been a very important influence in my physics career. This is a good opportunity for me to briefly reminisce about this.

In the fall of 1963, the beginning of my second year of graduate study at UC Berkeley, a new physics building (Birge Hall) opened. At the time Berkeley (as well as various other institutions) was trying to enhance its efforts in theoretical particle physics by hiring Murray Gell-Mann. Murray was already a professor in Caltech, but Berkeley hoped that he might be enticed into moving. A prime corner office in Birge Hall, with a great view of the San Francisco Bay and the Berkeley campus, was selected to be Murray's. However, he hadn't yet decided whether to accept the Berkeley offer, and it was time to occupy the building. The office could not be left vacant. If another Professor were assigned to it, it would be awkward to ask him to give it up for Murray. Therefore, it was decided to assign it to graduate students, who would be much easier to dislodge. As a result, that was the office that David Gross and I shared for the next three years. In this way Murray impacted my life before I even met him! Of course, in this period I studied the eight-fold way (known nowadays as SU(3) flavor symmetry) with great interest. So he was also influencing me scientifically.

After graduating from Berkeley I spent six years in Princeton, the last three as an Assistant Professor. In 1972 it was time for me to leave Princeton. The job market at the time was absolutely terrible. There had been enormous expansion of science faculties for more than a decade following the launch of Sputnik, but suddenly it came to a screeching halt and there were almost no jobs. Many good people were driven from the field at that time. My survival entailed an element of luck. In 1971 Neveu and I discovered a string theory, which we called the "dual pion model." This, together with Ramond's work on fermionic strings, led to what is now known as superstring theory. This work was motivated by the desire to describe hadron physics – the application to gravity and unification came later. String theory (called "dual resonance theory" in those days) had a couple hundred enthusiastic devotees, but it was still a relatively small, and somewhat isolated, segment of the particle theory community. In 1971–72 Murray was spending a sabbatical year at CERN, which had a strong group of string theorists. Some of them, such as Brink, Olive, and Scherk, were very interested in supersymmetric string theory and contributed to its development. Even though Murray was not working on this himself, he learned of these developments, and decided that this research could be important. As a consequence, he arranged for me to be offered a senior research position at Caltech, which I was delighted to accept.

During my first couple years at Caltech, Murray collaborated with Fritzsch and Minkowski on the development of QCD, and the standard model quickly fell into place. I followed all this closely, but I continued to work on string theory. Murray made funds available to me to bring collaborators to Caltech for extended visits. This facilitated my collaborations with Brink and Scherk, and (much later) with Green. String theory fell out of favor once it was realized that QCD is the right theory of hadrons. However, during Scherk's visit in 1974 we realized that string theory could be used for gravity and unification instead, and this converts several of its shortcomings into advantages. This change in direction is what convinced me that it would be worthwhile to continue pursuing the subject. By then, the community had little interest in string theory, and it took ten years for this proposal to gain traction. Murray, however, understood that it could be very important, and so he continued to support me. I recall him saying that as a committed environmentalist he recognized the importance of protecting endangered species, and I represented one of them. I have always felt that he has exceptional judgment in these matters.

In January 1989, nine months before his actual birthday, I organized a celebration of Murray's 60th birthday at Caltech. There were two days of lectures. The first day was devoted to physics and the second day to a variety of other subjects in which Murray was interested. A few years later Murray left Caltech and moved to Santa Fe, where he was a founding member of the Santa Fe Institute. I have missed seeing him on a regular basis, a privilege that I had for about 20 years.

2. Review of Some Basic Facts

Let me now turn to gauge/gravity duality.[a] In Maldacena's original paper,[2] he proposed three maximally supersymmetric examples of AdS/CFT duality. A basic indication that the dualities (or equivalences) are plausible is that the symmetries match. In each case, there is a supergroup, which describes the isometries of the string theory or M-theory background geometry. The same supergroup appears as the superconformal symmetry group of the dual quantum field theory. Also, the string theory or M-theory solution has N units of flux threading the sphere factor in the geometry. In fact, the background configuration corresponds to the near-horizon geometry of N coincident branes, each of which contributes one unit of flux. The dual conformal field theory, which also depends on the integer N, is the low energy world-volume theory on the branes.

- **M2-brane Duality:** M-theory on $AdS_4 \times S^7$ is dual to a superconformal field theory (SCFT) in three dimensions. The superconformal symmetry is described by the supergroup is $OSp(8|4)$.
- **D3-brane Duality:** Type IIB superstring theory on $AdS_5 \times S^5$ is dual to a SCFT in four dimensions, specifically $\mathcal{N} = 4$ super Yang–Mills (SYM) theory. The superconformal symmetry in this case is $PSU(2,2|4)$.
- **M5-brane Duality:** M theory on $AdS_7 \times S^4$ is dual to a SCFT in six dimensions. The superconformal symmetry in this case is $OSp(6,2|4)$.

2.1. *The type IIB/$\mathcal{N} = 4$ SYM example*

This by far the most studied, and best understood, example. The N units of flux ($\int_{S^5} F_5 \approx N$) in the superstring solution correspond to the gauge group $SU(N)$ in the $\mathcal{N} = 4$ super Yang–Mills theory.[3] The gauge theory has a well-known large-N topological ('t Hooft) expansion.[4] The expansion is in powers of $1/N$ for large N at fixed λ, where the 't Hooft parameter is

$$\lambda = g_{\text{YM}}^2 N. \tag{1}$$

This expansion corresponds to the loop expansion of the string theory. One also identifies

$$R^2/\alpha' \approx \sqrt{\lambda} \quad \text{and} \quad g_{\text{s}} \approx \lambda/N, \tag{2}$$

where R is the radius of the S^5 and the AdS_5. g_{s} is the string coupling constant determined by the value of the dilaton field, which is a massless scalar mode of the string.

2.2. *The type IIA/ABJM example*

There has been significant progress in the last few years in understanding the M2-brane duality. The suggestion[5] that the three-dimensional SCFT should be Chern–

---

[a]The remainder of this manuscript is very similar to one that I wrote for Shifman's 60th birthday.[1]

Simons gauge theory was implemented for maximal supersymmetry ($\mathcal{N} = 8$) by Bagger and Lambert[6] and by Gustavsson.[7] However, their construction only works for the gauge group $SO(4)$, and it does not provide the desired dual to M-theory on $AdS_4 \times S^7$.

The correct construction was eventually obtained by Aharony, Bergman, Jafferis, and Maldacena (ABJM).[8] One key step in their work was to consider a more general problem: M-theory on $AdS_4 \times S^7/\mathbb{Z}_k$, with N units of flux. This gives 3/4 maximal supersymmetry for $k > 2$. Thus, the dual gauge theory is an $\mathcal{N} = 6$ superconformal Chern–Simons theory in three dimensions. The appropriate gauge group turns out to be $U(N)_k \times U(N)_{-k}$, where the subscripts are the levels of the Chern–Simons terms. The ABJM theory also contains bifundamental scalar and spinor fields. This theory has a topological large-N expansion, just like the usual ones in four dimensions, for which the 't Hooft parameter that is held constant in the limit is

$$\lambda = N/k. \tag{3}$$

The only unusual feature is that the 't Hooft parameter is rational. The extension of the supersymmetry from $\mathcal{N} = 6$ to $\mathcal{N} = 8$ for $k = 1, 2$ is a nontrivial property of the quantum theory.

The orbifold $S^7/\mathbb{Z}_k$ can be described as a circle bundle over a CP^3 base. The circle has radius R/k, where R is the S^7 radius. When $k^5 \gg N$, there is a weakly coupled type IIA superstring interpretation with string coupling constant

$$g_{\mathrm{s}} \approx (N/k^5)^{1/4}. \tag{4}$$

One then obtains the correspondences

$$R^2/\alpha' \approx \sqrt{\lambda} \quad \text{and} \quad g_{\mathrm{s}} \approx \lambda^{5/4}/N, \tag{5}$$

which is very similar to the previous duality. This type IIA duality has 3/4 as much supersymmetry as the type IIB duality, and it is somewhat more complicated.

2.3. AdS energies and conformal dimensions

The geometry of Anti de Sitter space is usually described in Poincaré coordinates, which describes all of the spacetime that is within the light-cone of a given observer, but does not cover the entire spacetime. For the purpose of defining energies that correspond to the dimensions of conformal operators, one needs to use different coordinates, called global coordinates, that cover the entire spacetime. The metric of AdS_{p+2} in global coordinates is

$$ds^2[AdS_{p+2}] = d\rho^2 - \cosh^2 \rho \, dt^2 + \sinh^2 \rho \, ds^2[S^p]. \tag{6}$$

Here, $ds^2[S^p]$ denote the metric of a unit p-dimensional sphere. Actually, AdS/CFT duality requires taking the covering space of AdS, which means that the global time coordinate t runs from $-\infty$ to $+\infty$.

Witten[9] and Gubser, Klebanov, Polyakov[10] gave a prescription for relating n-point correlation functions in the gauge theory to corresponding quantities in the string theory. In the case of two-point functions, the duality relates the energy E_A of a string state $|A\rangle$ (defined with respect to the global time coordinate t),

$$H_{\text{string}}|A\rangle = E_A|A\rangle, \tag{7}$$

to the conformal dimensions Δ_A of the corresponding gauge-invariant local operator $\mathcal{O}_A$ for which

$$\langle \mathcal{O}_A(x)\mathcal{O}_B(y)\rangle \approx \frac{\delta_{AB}}{|x-y|^{2\Delta_A}}. \tag{8}$$

Specifically, the duality requires that

$$\Delta_A(\lambda, 1/N) = E_A(R^2/\alpha', g_{\text{s}}). \tag{9}$$

The 't Hooft expansion of the dimension of $\mathcal{O}_A$ is

$$\Delta_A(\lambda, 1/N) = \Delta_A^{(0)} + \sum_{g=0}^{\infty} \frac{1}{N^{2g}} \sum_{l=1}^{\infty} \lambda^l \Delta_{l,g}. \tag{10}$$

$\Delta_A^{(0)}$ is the classical dimension, and the rest is called the anomalous dimension.

Almost all studies have focused on the planar approximation, (genus $g = 0$), which is dual to free string theory. This restriction may make the problem fully tractable, but it is still very challenging. After all, it would be an extraordinary achievement to solve an interacting four-dimensional quantum field theory even in the planar approximation.

2.4. *Approaches to testing the dualities*

Given that it is not possible to completely solve any of these theories, the question arises how best to test and explore the workings of AdS/CFT duality. The most obvious things—matching symmetries and the dimensions of chiral primary operators—have been done long ago. One wants to dig deeper. One approach is to match, as much as possible, energies and dimensions of fields/operators that are not protected by supersymmetry. It should be noted, however, that a complete test of the duality would also require matching three-point correlators, since a conformal field theory is completely characterized by its two-point and three-point functions. There has been much less progress on this front.

One approach that has been quite successful is the following. First, identify tractable examples of classical solutions of the string world-sheet theory. Next, examine the spectrum of small excitations about these solutions and compute their energies E_A. Finally, identify the corresponding class of operators in the dual gauge theory and compute their dimensions Δ_A in the planar approximation. Then compare to E_A. One subtlety in this analysis is that this comparison requires an extrapolation from large λ, where the classical world sheet theory is valid, to small λ, where the gauge theory can be studied perturbatively. Thus, one needs to identify examples in which this is possible. As we will see, in practice this has conjectural aspects.

A variant of the preceding procedure is to compare equations that determine E_A and Δ_A rather than the solutions. Approaches based on integrability and algebraic curves try to obtain equations of "Bethe type" on both sides and to match them. This is a very active area of research, but I will not be able to review it here. One important issue is that it is much easier to study the world-sheet theory when the range of σ is infinite (rather than a circle). In other words, the string itself is infinite, rather than a loop. In the gauge theory analysis this corresponds to the thermodynamic limit of the Bethe equations arising from a spin-chain analysis. There has been progress recently in extending the integrability techniques to the compact case.[11] However, the story is quite technical, and I don't think it is completely settled.

3. Classical String Solutions

For the reasons outlined above, we want to identify classical string solutions in the $AdS_5 \times S^5$ background that can be used to test the duality. The discussion that follows largely follows an excellent review article by Plefka.[12] Other useful reviews include.[13,14]

The bosonic part of the string world-sheet action has six cyclic coordinates:

$$(t, \varphi_1, \varphi_2; \phi_1, \phi_2, \phi_3), \tag{11}$$

where the first three coordinates pertain to AdS_5 and the second three to S^5. Specifically, we parametrize S^5 as follows:

$$ds^2(S^5) = d\gamma^2 + \cos^2 \gamma \, d\phi_3^2 + \sin^2 \gamma \, ds^2(S^3), \tag{12}$$

where

$$ds^2(S^3) = d\psi^2 + \cos^2 \psi \, d\phi_1^2 + \sin^2 \psi \, d\phi_2^2. \tag{13}$$

Associated to these cyclic coordinates one has conserved charges

$$(E, S_1, S_2; J_1, J_2, J_3). \tag{14}$$

E is the energy and the other five charges are angular momenta.

One much-studied class of string solutions involves a line up the center of AdS_5, described by $\rho = 0$ and $t = \kappa\tau$, where κ is a constant and τ is the world-sheet time coordinate. These configurations have $S_1 = S_2 = 0$.

3.1. Point-particle solutions

The simplest solution is a point particle (collapsed string) encircling the sphere. In addition to $\rho = 0$ and $t = \kappa\tau$, this is described by

$$\gamma = \pi/2, \quad \phi_1 = \kappa\tau, \quad \psi = 0. \tag{15}$$

This has $J_2 = J_3 = 0$.

The quantum excitations of this solution have energies that can be expanded in powers of $1/J$ for large $J = J_1$, where

$$\kappa = J/\sqrt{\lambda} \tag{16}$$

is held fixed. This is equivalent to the BMN analysis of strings in a plane-wave background.[15] One obtains

$$E - J \approx E_2(\kappa) + \frac{1}{J}E_4(\kappa) + \ldots \tag{17}$$

The exact BMN result is

$$E_2 = \sum_{n=-\infty}^{\infty} \sqrt{n^2 + \kappa^2} N_n, \tag{18}$$

where $N_n = \sum_{i=1}^{8} \alpha_n^{i\dagger} \alpha_n^i$ + fermions is expressed in terms of ordinary oscillators

$$[\alpha_m^i, \alpha_n^{j\dagger}] = \delta^{ij} \delta_{mn}. \tag{19}$$

The level matching condition is $\sum n N_n = 0$.

The BMN paper proposed a scaling rule, known as BMN scaling, which predicts agreement with the anomalous dimensions of operators in the dual gauge theory, even though one calculation is valid for large λ and the other for small λ. In other words, their scaling hypothesis, if valid, would justify the extrapolation from small λ to large λ. In fact, it turns out that E_2 agrees perfectly, but agreement for E_4 breaks down at three loops.[16] This is not a problem for AdS/CFT duality, only for the BMN scaling conjecture.[b]

3.2. *Spinning string solutions*

A class of interesting generalizations of the preceding solution describes circular or folded strings that are extended on the $S^3 \subset S^5$. These have $t = \kappa\tau$, $\rho = 0$, and $\gamma = \pi/2$, as before. But now one takes

$$\phi_1 = \omega_1 \tau, \quad \phi_2 = \omega_2 \tau, \quad \psi = \psi(\sigma). \tag{20}$$

For these choices, the string equation of motion gives

$$\psi'' + \omega_{21}^2 \sin\psi \cos\psi = 0, \tag{21}$$

where $\omega_{21}^2 = \omega_2^2 - \omega_1^2$. This is the well-known pendulum equation.

This equation has a first integral

$$\psi' = \omega_{21}\sqrt{q - \sin^2\psi}, \quad q = (\kappa^2 - \omega_1^2)/\omega_{21}^2. \tag{22}$$

The solution for $q < 1$, which involves the elliptic integrals $E(q)$ and $K(q)$, describes a folded string. It corresponds to a pendulum that oscillates back and forth. The

---

[b]Perhaps it would be more fair to say that the BMN scaling conjecture was made for the plane-wave limit only, which corresponds to E_2; what fails is an attempt to generalize the scaling conjecture beyond that.

solution for $q > 1$, which involves the elliptic integrals $E(q^{-1})$ and $K(q^{-1})$, describes a circular string. It corresponds to a pendulum that goes round and round. In the classical limit, the energy has the form

$$E = \sqrt{\lambda} F(J_1/\sqrt{\lambda}, J_2/\sqrt{\lambda}). \tag{23}$$

3.3. Dual gauge theory analysis

This string theory result can be extrapolated to small λ and compared to the dual gauge theory. The operators that carry J_1, J_2 charges have the form

$$\mathcal{O}_\alpha^{J_1,J_2} = \text{Tr}\left(Z^{J_1} W^{J_2}\right) + \dots \tag{24}$$

where Z and W are complex scalar fields in the adjoint of $SU(N)$. The additional terms denoted by dots involve different orderings of the Zs and Ws. Such a trace can be viewed as a ring configuration of an $S = 1/2$ quantum spin chain, where W corresponds to spin up and Z corresponds to spin down.

The conformal dimensions of operators $\mathcal{O}_\alpha^{J_1,J_2}(x)$ with these charges are eigenvalues of the dilatation operator

$$\mathcal{D}\mathcal{O}_\alpha^{J_1,J_2}(x) = \sum_\beta D_{\alpha\beta}\mathcal{O}_\beta^{J_1,J_2}(x). \tag{25}$$

In the planar one-loop approximation the equations are precisely those of a ferromagnetic Heisenberg spin chain, which is a well-known integrable system, whose Hamiltonian is proportional to

$$\mathcal{H} = \sum_{i=1}^{J} \left(\frac{1}{4} - \vec{\sigma}_i \cdot \vec{\sigma}_{i+1}\right). \tag{26}$$

This can be solved using Bethe ansatz techniques, thereby obtaining conformal dimensions that can be compared (successfully) with energies of the corresponding string solutions. Higher-order terms, which correspond to more complicated spin-chain Hamiltonians, have also been studied.

3.4. Strings spinning in AdS

Another interesting class of classical string solutions are ones in which the string position is extended in the AdS space and a point moving on the sphere. The first example of this type is the straight folded string rotating in $AdS_3 \subset AdS_5$.[17] One finds that for large S

$$E = 2\Gamma(\lambda) \log S + O(S^0), \tag{27}$$

where

$$\Gamma(\lambda) = \frac{\sqrt{\lambda}}{2\pi} + O(\lambda^0) \quad \text{for} \quad \lambda \gg 1. \tag{28}$$

The dual gauge theory operators are

$$\mathrm{Tr}(D_+^{s_1} Z\, D_+^{s_2} Z) \quad s_1 + s_2 = S. \tag{29}$$

Their anomalous dimensions take the same form as the energy with

$$\Gamma(\lambda) = \frac{\lambda}{4\pi^2} + O(\lambda^2) \quad \text{for} \quad \lambda \ll 1. \tag{30}$$

In order to compare these, one needs a procedure to extrapolate between small and large λ. In fact, an exact formula for the *cusp anomalous dimension* $\Gamma(\lambda)$ has been deduced using the assumption of exact integrability.[18] It passes all tests and is likely to be correct.

The generalization of this duality to twist J operators, which have the form

$$\mathrm{Tr}(D_+^{s_1} Z\, D_+^{s_2} Z \ldots D_+^{s_J} Z) \quad \text{where} \quad \sum s_l = S, \tag{31}$$

has been explored by Dorey and Losi.[19] They computed the corresponding conformal dimensions using an $SL(2)$ spin chain model. For large S the correspond classical string solutions are *spiky strings* with J cusps. The duality predictions are verified to the extent that they have been explored.

4. Conclusion

There has been a lot of progress in testing AdS/CFT in various special cases for maximally supersymmetric theories. Much of this progress has exploited the integrability of the string world-sheet theory on the one hand and the integrability of various spin-chain models that arise in studies of the dual gauge theory in the planar approximation on the other hand. More recently, there has also been very interesting work exploring analogous constructions for the M2-brane duality following the discovery of the ABJM theory. Much less is known about the M5-brane theory, though there has been significant progress when two of the dimensions wrap a Riemann surface.[20,21] The superconformal theory on flat M5-branes is strongly coupled, and it does not appear to have a Lagrangian description. Moreover, it seems to involve tensionless strings, which is a poorly understood subject.

It has been fun traveling half way around the world to the amazing country of Singapore in order to celebrate my long-time friend and colleague Murray Gell-Mann. I am looking forward to his 100th birthday celebration.

Acknowledgments

This work was supported in part by the U.S. Dept. of Energy under Grant No. DE-FG03-92-ER40701.

References

1. J. H. Schwarz, "Recent Progress in AdS/CFT," Int. J. Mod. Phys. **A25**, 310-318 (2010). [arXiv:0907.4972 [hep-th]].

2. J. M. Maldacena, "The Large N Limit of Superconformal Field Theories and Super-gravity," Adv. Theor. Math. Phys. **2**, 231 (1998) [Int. J. Theor. Phys. **38**, 1113 (1999)] [arXiv:hep-th/9711200].

3. L. Brink, J. H. Schwarz and J. Scherk, "Supersymmetric Yang–Mills Theories," Nucl. Phys. B **121**, 77 (1977).

4. G. 't Hooft, "A Planar Diagram Theory for Strong Interactions," Nucl. Phys. B **72**, 461 (1974).

5. J. H. Schwarz, "Superconformal Chern–Simons Theories," JHEP **0411**, 078 (2004) [arXiv:hep-th/0411077].

6. J. Bagger and N. Lambert, "Gauge Symmetry and Supersymmetry of Multiple M2-Branes," Phys. Rev. D **77**, 065008 (2008) [arXiv:0711.0955 [hep-th]].

7. A. Gustavsson, "Algebraic Structures on Parallel M2-Branes," Nucl. Phys. B **811**, 66 (2009) [arXiv:0709.1260 [hep-th]].

8. O. Aharony, O. Bergman, D. L. Jafferis and J. M. Maldacena, "$N = 6$ Superconformal Chern–Simons-Matter Theories, M2-Branes and Their Gravity Duals," JHEP **0810**, 091 (2008) [arXiv:0806.1218 [hep-th]].

9. E. Witten, "Anti de Sitter Space and Holography," Adv. Theor. Math. Phys. **2**, 253 (1998) [arXiv:hep-th/9802150].

10. S. S. Gubser, I. R. Klebanov and A. M. Polyakov, "Gauge Theory Correlators from Non-critical String Theory," Phys. Lett. B **428**, 105 (1998) [arXiv:hep-th/9802109].

11. N. Gromov, V. Kazakov and P. Vieira, "Exact AdS/CFT Spectrum: Konishi Dimension at any Coupling," arXiv:0906.4240 [hep-th].

12. J. Plefka, "Spinning Strings and Integrable Spin Chains in the AdS/CFT Correspondence," Living Rev. Rel. **8**, 9 (2005) [arXiv:hep-th/0507136].

13. A. A. Tseytlin, "Spinning Strings and AdS/CFT Duality," arXiv:hep-th/0311139.

14. C. Kristjansen, M. Staudacher, and A. Tseytlin, eds., *Special Issue on Integrability and the AdS/CFT Correspondence*, J. Phys. A **42**, 250301 (2009).

15. D. E. Berenstein, J. M. Maldacena and H. S. Nastase, "Strings in Flat Space and pp Waves from $N = 4$ Super Yang Mills," JHEP **0204**, 013 (2002) [arXiv:hep-th/0202021].

16. C. G. Callan, H. K. Lee, T. McLoughlin, J. H. Schwarz, I. Swanson and X. Wu, "Quantizing String Theory in AdS(5) x S**5: Beyond the pp-Wave," Nucl. Phys. B **673**, 3 (2003) [arXiv:hep-th/0307032].

17. S. S. Gubser, I. R. Klebanov and A. M. Polyakov, "A Semi-Classical Limit of the Gauge/String Correspondence," Nucl. Phys. B **636**, 99 (2002) [arXiv:hep-th/0204051].

18. N. Beisert, B. Eden and M. Staudacher, "Transcendentality and Crossing," J. Stat. Mech. **0701**, P021 (2007) [arXiv:hep-th/0610251].

19. N. Dorey and M. Losi, "Spiky Strings and Spin Chains," arXiv:0812.1704 [hep-th].

20. D. Gaiotto, "$N = 2$ Dualities," arXiv:0904.2715 [hep-th].

21. D. Gaiotto and J. M. Maldacena, "The Gravity Duals of $N = 2$ Superconformal Field Theories," arXiv:0904.4466 [hep-th].

ASPECTS OF STRING PHENOMENOLOGY

I. ANTONIADIS*

Department of Physics, CERN – Theory Division,
1211 Geneva 23, Switzerland
ignatios.antoniadis@cern.ch

Lowering the string scale in the TeV region provides a theoretical framework for solving the mass hierarchy problem and unifying all interactions. The apparent weakness of gravity can then be accounted by the existence of large internal dimensions, in the sub-millimeter region, and transverse to a braneworld where our universe must be confined. I review the main properties of this scenario and its implications for observations at both particle colliders, and in nonaccelerator gravity experiments.

1. Introduction

During the last few decades, physics beyond the Standard Model (SM) was guided from the problem of mass hierarchy. This can be formulated as the question of why gravity appears to us so weak compared to the other three known fundamental interactions corresponding to the electromagnetic, weak and strong nuclear forces. Indeed, gravitational interactions are suppressed by a very high energy scale, the Planck mass $M_P \sim 10^{19}$ GeV, associated to a length $l_P \sim 10^{-35}$ m, where they are expected to become important. In a quantum theory, the hierarchy implies a severe fine tuning of the fundamental parameters in more than 30 decimal places in order to keep the masses of elementary particles at their observed values. The reason is that quantum radiative corrections to all masses generated by the Higgs vacuum expectation value (VEV) are proportional to the ultraviolet cutoff which in the presence of gravity is fixed by the Planck mass. As a result, all masses are "attracted" to become about 10^{16} times heavier than their observed values.

Besides compositeness, there are two main theories that have been proposed and studied extensively during the last years, corresponding to different approaches of dealing with the mass hierarchy problem. (1) Low energy supersymmetry with all superparticle masses in the TeV region. Indeed, in the limit of exact supersymmetry, quadratically divergent corrections to the Higgs self-energy are exactly cancelled, while in the softly broken case, they are cutoff by the supersymmetry breaking mass splittings. (2) TeV scale strings, in which quadratic divergences are cutoff by the string scale and low energy supersymmetry is not needed. Both ideas are

---

*On leave from CPHT (UMR CNRS 7644) Ecole Polytechnique, F-91128 Palaiseau.

experimentally testable at high-energy particle colliders and in particular at LHC. Below, I discuss their implementation in string theory.

2. Strings and Extra Dimensions

The appropriate and most convenient framework for low energy supersymmetry and grand unification is the perturbative heterotic string. Indeed, in this theory, gravity and gauge interactions have the same origin, as massless modes of the closed heterotic string, and they are unified at the string scale M_s. As a result, the Planck mass M_P is predicted to be proportional to M_s:

$$M_P = \frac{M_s}{g}, \tag{1}$$

where g is the gauge coupling. In the simplest constructions all gauge couplings are the same at the string scale, given by the four-dimensional (4D) string coupling, and thus no grand unified group is needed for unification. In our conventions $\alpha_{\text{GUT}} = g^2 \simeq 0.04$, leading to a discrepancy between the string and grand unification scale M_{GUT} by almost two orders of magnitude. Explaining this gap introduces in general new parameters or a new scale, and the predictive power is essentially lost. This is the main defect of this framework, which remains though an open and interesting possibility.[a]

The other other perturbative framework that has been studied extensively in the more recent years is type I string theory with D-branes. Unlike in the heterotic string, gauge and gravitational interactions have now different origin. The latter are described again by closed strings, while the former emerge as excitations of open strings with endpoints confined on D-branes.[2] This leads to a braneworld description of our universe, which should be localized on a hypersurface, i.e. a membrane extended in p spatial dimensions, called p-brane (see Fig. 1). Closed strings propagate in all nine dimensions of string theory: in those extended along the p-brane, called parallel, as well as in the transverse ones. On the contrary, open strings are attached on the p-brane. Obviously, our p-brane world must have at least the three known dimensions of space. But it may contain more: the extra $d_\parallel = p - 3$ parallel dimensions must have a finite size, in order to be unobservable at present energies, and can be as large as $\text{TeV}^{-1} \sim 10^{-18}$ m.[3] On the other hand, transverse dimensions interact with us only gravitationally and experimental bounds are much weaker: their size should be less than about 0.1 mm.[4] In the following, I review the main properties and experimental signatures of low string scale models.[5,6,b]

2.1. *Framework of low scale strings*

In type I theory, the different origin of gauge and gravitational interactions implies that the relation between the Planck and string scales is not linear as (1) of the

---

[a] For a review see Ref. 1, and references therein.
[b] For a review see Ref. 7, and references therein.

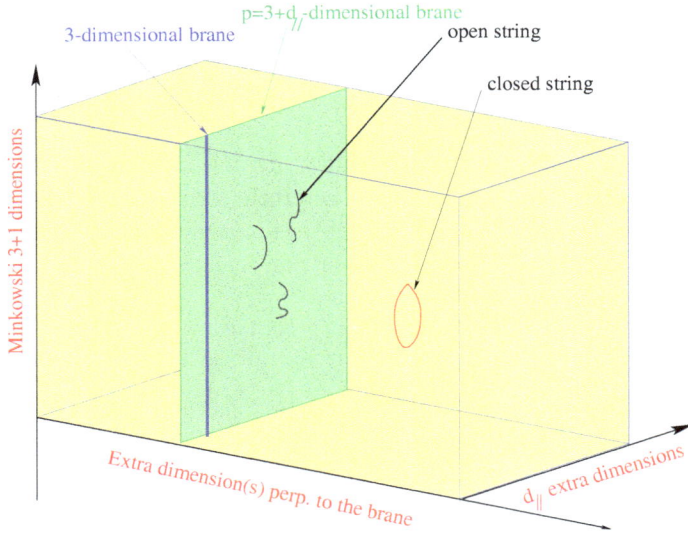

Fig. 1. (Color online) In type I string framework, our Universe contains, besides our three spatial dimensions (denoted by the blue line), some extra dimensions ($d_\parallel = p - 3$) parallel to our world p-brane (green plane) where endpoints of open strings are confined, as well as some transverse dimensions (yellow space) where only gravity (described by closed strings) propagate.

heterotic string. The requirement that string theory should be weakly coupled, constrain the size of all parallel dimensions to be of order of the string length, while transverse dimensions remain unrestricted. Assuming an isotropic transverse space of $n = 9 - p$ compact dimensions of common radius $R_\perp$, one finds

$$M_P^2 = \frac{1}{g^4} M_s^{2+n} R_\perp^n \,, \quad g_s \simeq g^2 \,, \tag{2}$$

where g_s is the string coupling. It follows that the type I string scale can be chosen hierarchically smaller than the Planck mass[5,6,8] at the expense of introducing extra large transverse dimensions felt only by gravity, while keeping the string coupling small.[5,6] The weakness of 4D gravity compared to gauge interactions (ratio M_W/M_P) is then attributed to the largeness of the transverse space $R_\perp$ compared to the string length $l_s = M_s^{-1}$.

An important property of these models is that gravity becomes effectively $(4 + n)$-dimensional with a strength comparable to those of gauge interactions at the string scale. The first relation of Eq. (2) can be understood as a consequence of the $(4 + n)$-dimensional Gauss's law for gravity, with

$$M_*^{(4+n)} = \frac{M_s^{2+n}}{g^4} \tag{3}$$

the effective scale of gravity in $4 + n$ dimensions. Taking $M_s \simeq 1$ TeV, one finds a size for the extra dimensions $R_\perp$ varying from 10^8 km, 0.1 mm, down to a Fermi for

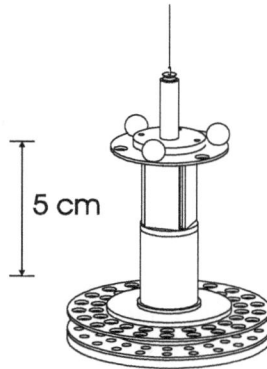

Fig. 2.   Torsion pendulum that tested Newton's law at 55 μm.

$n = 1, 2$, or 6 large dimensions, respectively. This shows that while $n = 1$ is excluded, $n \geq 2$ is allowed by present experimental bounds on gravitational forces.[4,9-13] Thus, in these models, gravity appears to us very weak at macroscopic scales because its intensity is spread in the "hidden" extra dimensions. At distances shorter than $R_\perp$, it should deviate from Newton's law, which may be possible to explore in laboratory experiments (see Fig. 2).

The main experimental implications of TeV scale strings in particle accelerators are of three types, in correspondence with the three different sectors that are generally present: (i) new compactified parallel dimensions, (ii) new extra large transverse dimensions and low scale quantum gravity, and (iii) genuine string and quantum gravity effects. On the other hand, there exist interesting implications in non accelerator table-top experiments due to the exchange of gravitons or other possible states living in the bulk.

3. Experimental Implications in Accelerators

3.1. *World-brane extra dimensions*

In this case $RM_s \gtrsim 1$, and the associated compactification scale $R_\parallel^{-1}$ would be the first scale of new physics that should be found increasing the beam energy.[3,14-16] The main consequence is the existence of Kaluza–Klein (KK) excitations for all SM particles that propagate along the extra parallel dimensions. Their masses are given by

$$M_m^2 = M_0^2 + \frac{m^2}{R_\parallel^2}; \quad m = 0, \pm 1, \pm 2, \ldots, \tag{4}$$

where we used $d_\parallel = 1$, and M_0 is the higher dimensional mass. The zero-mode $m = 0$ is identified with the 4D state, while the higher modes have the same quantum numbers with the lowest one, except for their mass given in (4). There are two types of experimental signatures of such dimensions:[14,17-23] (i) virtual exchange of

KK excitations, leading to deviations in cross-sections compared to the SM prediction, that can be used to extract bounds on the compactification scale; (ii) direct production of KK modes.

On general grounds, there can be two different kinds of models with qualitatively different signatures depending on the localization properties of matter fermion fields. If the latter are localized in 3D brane intersections, they do not have excitations and KK momentum is not conserved because of the breaking of translation invariance in the extra dimension(s). KK modes of gauge bosons are then singly produced giving rise to generally strong bounds on the compactification scale and new resonances that can be observed in experiments. Otherwise, they can be produced only in pairs due to the KK momentum conservation, making the bounds weaker but the resonances difficult to observe.

In addition to virtual effects, KK excitations can be produced on-shell at LHC as new resonances[17–22] (see Fig. 3). There are two different channels, neutral Drell–Yan processes $pp \to l^+l^-X$ and the charged channel $l^\pm\nu$, corresponding to the production of the KK modes $\gamma^{(1)}$, $Z^{(1)}$ and $W_\pm^{(1)}$, respectively. The discovery limits are about 6 TeV, while the exclusion bounds 15 TeV.

On the other hand, if all SM particles propagate in the extra dimension (called universal), KK modes can only be produced in pairs and the lower bound on the compactification scale becomes weaker, of order of 300–500 GeV. Moreover, no resonances can be observed at LHC, so that this scenario appears very similar to low energy supersymmetry. In fact, KK parity can even play the role of R-parity, implying that the lightest KK mode is stable and can be a dark matter candidate in analogy to the LSP.[24]

3.2. *Extra large transverse dimensions*

The main experimental signal is gravitational radiation in the bulk from any physical process on the worldbrane. In fact, the very existence of branes breaks translation invariance in the transverse dimensions and gravitons can be emitted from the brane into the bulk. During a collision of center of mass energy $\sqrt{s}$, there are $\sim (\sqrt{s}R_\perp)^n$ KK excitations of gravitons with tiny masses, that can be emitted. Each of these states looks from the 4D point of view as a massive, quasistable, extremely weakly coupled (s/M_P^2 suppressed) particle that escapes from the detector. The total effect is a missing-energy cross-section roughly of order

$$\frac{(\sqrt{s}R_\perp)^n}{M_P^2} \sim \frac{1}{s}\left(\frac{\sqrt{s}}{M_s}\right)^{n+2}. \tag{5}$$

Explicit computation of these effects leads to the bounds given in Table 1.

Figure 4 shows the cross-section for graviton emission in the bulk, corresponding to the process $pp \to$ jet $+$ graviton at LHC, together with the SM background.[25–32] There is a particular energy and angular distribution of the produced gravitons that arise from the distribution in mass of KK states of spin-2. This can be contrasted

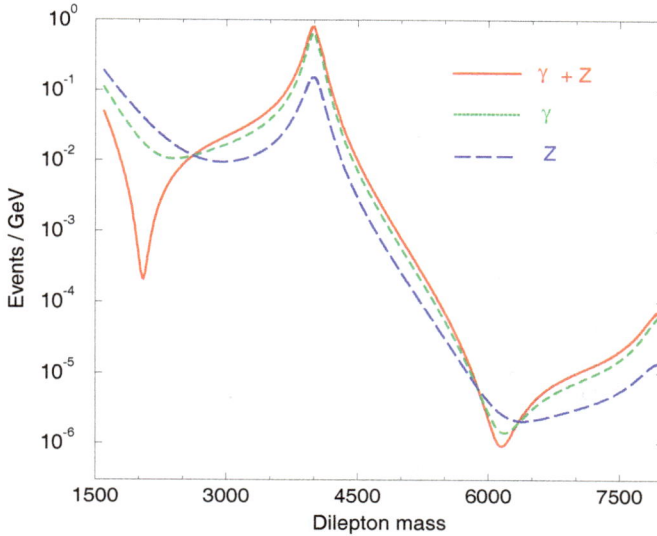

Fig. 3. Production of the first KK modes of the photon and of the Z boson at LHC, decaying to electron–positron pairs. The number of expected events is plotted as a function of the energy of the pair in GeV.

Table 1. Limits on $R_\perp$ in mm.

Experiment	$n = 2$	$n = 4$	$n = 6$
Collider bounds			
LEP 2	5×10^{-1}	2×10^{-8}	7×10^{-11}
Tevatron	5×10^{-1}	10^{-8}	4×10^{-11}
LHC	4×10^{-3}	6×10^{-10}	3×10^{-12}
NLC	10^{-2}	10^{-9}	6×10^{-12}
Present noncollider bounds			
SN1987A	3×10^{-4}	10^{-8}	6×10^{-10}
COMPTEL	5×10^{-5}	—	—

to other sources of missing energy and might be a smoking gun for the extra dimensional nature of such a signal.

In Table 1, there are also included astrophysical and cosmological bounds. Astrophysical bounds[33–35] arise from the requirement that the radiation of gravitons should not carry on too much of the gravitational binding energy released during core collapse of supernovae. The best cosmological bound[36,37] is obtained from requiring that decay of bulk gravitons to photons do not generate a spike in the energy spectrum of the photon background measured by the COMPTEL

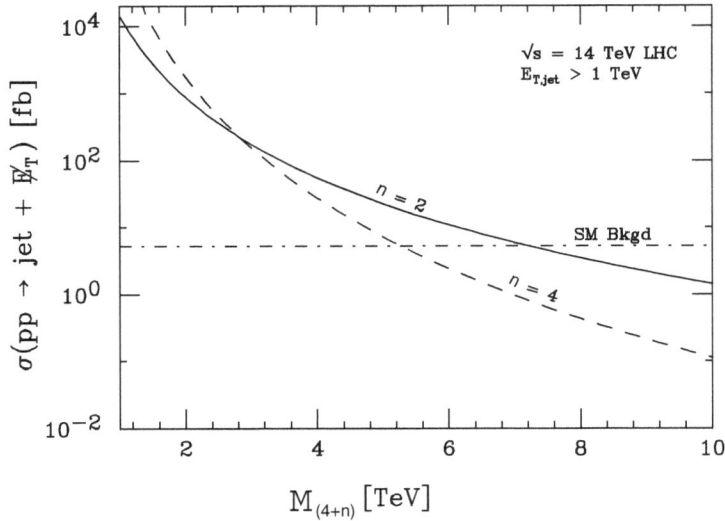

Fig. 4. Missing energy due to graviton emission at LHC, as a function of the higher-dimensional gravity scale M_*, produced together with a hadronic jet. The expected cross-section is shown for $n = 2$ and $n = 4$ extra dimensions, together with the SM background.

instrument. Bulk gravitons are expected to be produced just before nucleosynthesis due to thermal radiation from the brane. The limits assume that the temperature was at most 1 MeV as nucleosynthesis begins, and become stronger if temperature is increased.

3.3. String effects

At low energies, the interaction of light (string) states is described by an effective field theory. Their exchange generates in particular four-fermion operators that can be used to extract independent bounds on the string scale. In analogy with the bounds on longitudinal extra dimensions, there are two cases depending on the localization properties of matter fermions. If they come from open strings with both ends on the same stack of branes, exchange of massive open string modes gives rise to dimension eight effective operators, involving four fermions and two space–time derivatives.[38–42] The corresponding bounds on the string scale are then around 500 GeV. On the other hand, if matter fermions are localized on non-trivial brane intersections, one obtains dimension six four-fermion operators and the bounds become stronger: $M_s \gtrsim$ 2–3 TeV.[7,42] At energies higher than the string scale, new spectacular phenomena are expected to occur, related to string physics and quantum gravity effects, such as possible micro-black hole production.[43–47] Particle accelerators would then become the best tools for studying quantum gravity and string theory.

Direct production of string resonances in hadron colliders leads generically to a universal deviation from Standard Model in jet distribution.[48] In particular, the

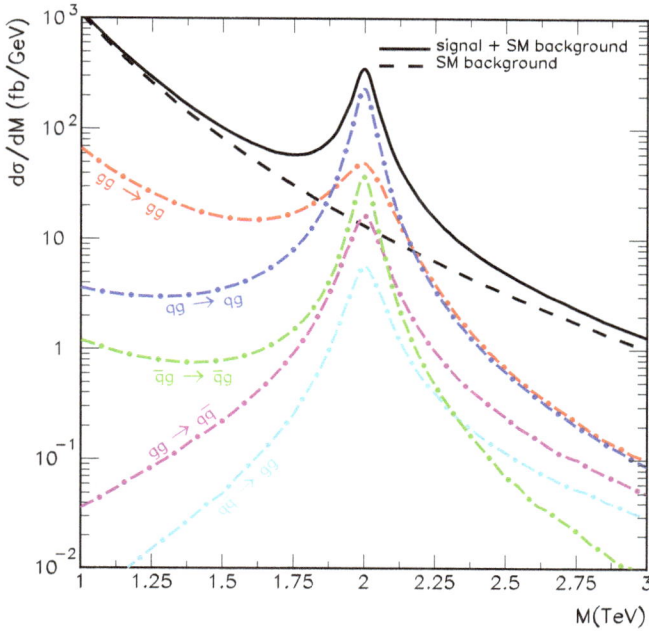

Fig. 5. Production of the first Regge excitations at LHC in the dijet channel, for $M_s = 2$ TeV. The cross-section is plotted as a function of the dijet invariant mass M.

first Regge excitation of the gluon has spin-2 and a width an order of magnitude lower than the string scale, leading to a characteristic peak in dijet production; similarly, the first excitations of quarks have spin-3/2. The dijet cross-section is shown in Fig. 5 for LHC energies. The reason for the universal behavior is that tree N-point open superstring amplitudes involving at most two fermions and gluons are completely model independent from the details of the compactification, including the number of supersymmetries that are left unbroken in four dimensions (even if all are broken). Such tree-level amplitudes do not receive contributions from KK, string winding or closed string graviton modes, but are given as a universal sum over exchanges of open string Regge excitations lying in Regge trajectories with masses

$$M_n^2 = M_s^2 n; \quad n = 0, 1, \ldots \qquad (6)$$

and maximal spin $n + 1$.

The relevant partonic cross-sections for dijet production, involving at most two quarks are $|\mathcal{M}(gg \to gg)|^2$, $|\mathcal{M}(gg \to q\bar{q})|^2$, $|\mathcal{M}(q\bar{q} \to gg)|^2$ and $|\mathcal{M}(qg \to qg)|^2$, where g and q denote gluons and quarks, respectively. They can be obtained from the first two, up to crossing symmetries, at the full (tree) string level and they are model independent.[49] The low energy expansion of these amplitudes reproduce the usual QCD expressions, while higher order terms describe the string corrections due to the exchange of Regge excitations. Besides these amplitudes, there are also those involving four quarks, such as $|\mathcal{M}(q\bar{q} \rightarrow q\bar{q})|^2$ and $|\mathcal{M}(qq \rightarrow qq)|^2$. These are model dependent, because the details of the compactification do not decouple. The reason is that they involve four vertices containing twist fields that describe quark states arising from open strings stretched in brane intersections. However, taking into account QCD color factors, their contribution is suppressed because parton luminosities in proton–proton collisions at TeV energies favor at least one gluon in the initial state. As a result, the dominant contribution comes from the model independent cross-sections described above, leading to the effect of Fig. 5.

We finish this section with some comments related to the possible micro-black hole production. Independently on the unresolved issue of the convergence of string perturbation theory in the kinematic region relevant to micro-black hole formation, there is a simple argument showing that at least within the perturbative TeV string framework, the energy threshold for black hole production is far above the LHC reach. Indeed, a string size black hole has a horizon radius $r_H \sim 1$ in string units, while the d-dimensional Newton's constant behaves as $G_N \sim g_s^2$. It follows that the mass of a d-dimensional black hole is[50]

$$M_{\mathrm{BH}} \sim \frac{r_H^{d/2-1}}{G_N} \simeq \frac{1}{g_s^2} \,. \tag{7}$$

Thus, for a weakly coupled theory, this energy threshold is much higher than the string and the higher dimensional Planck scales M_s and M_* of Eq. (3). Comparing this energy threshold with the mass of Regge excitations (6), one finds $n \sim 1/g_s^4$ which is actually compatible with the relation one obtains by identifying the black hole entropy $S_{\mathrm{BH}} \sim 1/G_N \sim 1/g_s^2$ with the perturbative string entropy $S_{\mathrm{string}} \sim \sqrt{n}$. Using now relation (2), and the value of the Standard Model gauge couplings $g_s \simeq g^2 \sim 0.1$, one finds that the energy threshold M_{BH} of micro-black hole production is about four orders of magnitude higher than the string scale, implying that one would produce 10^4 string states before reaching M_{BH}.

4. Supersymmetry in the Bulk and Short Range Forces

4.1. *Sub-millimeter forces*

Besides the spectacular predictions in accelerators, there are also modifications of gravitation in the sub-millimeter range, which can be tested in "table-top" experiments that measure gravity at short distances. There are three categories of such predictions:

(i) Deviations from the Newton's law $1/r^2$ behavior to $1/r^{2+n}$, which can be observable for $n = 2$ large transverse dimensions of sub-millimeter size.
(ii) New scalar forces in the sub-millimeter range, related to the mechanism of supersymmetry breaking, and mediated by light scalar fields φ with masses[5,6,51,52]

$$m_\varphi \simeq \frac{m_{\mathrm{susy}}^2}{M_P} \simeq 10^{-4} - 10^{-6} \text{ eV} \,, \tag{8}$$

for a supersymmetry breaking scale $m_{\text{susy}} \simeq$ 1–10 TeV. They correspond
to Compton wavelengths of 1 mm to 10 μm. m_{susy} can be either $1/R_{\parallel}$ if
supersymmetry is broken by compactification,[51,52] or the string scale if it is
broken "maximally" on our worldbrane.[5,6] A universal attractive scalar force
is mediated by the radion modulus $\varphi \equiv M_P \ln R$, with R the radius of the longitudinal or transverse dimension(s). For $n = 2$, there may be an enhancement
factor of the radion mass by $\ln R_{\perp} M_s \simeq 30$ decreasing its wavelength by an
order of magnitude.[53]

The coupling of the radius modulus to matter relative to gravity can be easily
computed and is given by

$$\sqrt{\alpha_{\varphi}} = \frac{1}{M} \frac{\partial M}{\partial \varphi}; \quad \alpha_{\varphi} = \begin{cases} \dfrac{\partial \ln \Lambda_{\text{QCD}}}{\partial \ln R} \simeq \dfrac{1}{3} & \text{for } R_{\parallel}, \\[2ex] \dfrac{2n}{n+2} = 1 - 1.5 & \text{for } R_{\perp}, \end{cases} \tag{9}$$

where M denotes a generic physical mass. Such a force can be tested in micro-
gravity experiments and should be contrasted with the change of Newton's
law due the presence of extra dimensions that is observable only for $n =$
2.[4,9–13] The resulting bounds from an analysis of the radion effects are:[54]
$M_* \gtrsim 6$ TeV.

(iii) Nonuniversal repulsive forces much stronger than gravity, mediated by possible
abelian gauge fields in the bulk.[33,55] Such fields acquire tiny masses of the order
of M_s^2/M_P, as in (8), due to brane localized anomalies.[55] Although their gauge
coupling is infinitesimally small, $g_A \sim M_s/M_P \simeq 10^{-16}$, it is still bigger that
the gravitational coupling E/M_P for typical energies $E \sim 1$ GeV, and the
strength of the new force would be 10^6–10^8 stronger than gravity.

In Fig. 6 we depict the actual information from previous, present and upcoming
experiments.[9–13,53] The solid lines indicate the present limits from the experiments
indicated. The excluded regions lie above these solid lines. Measuring gravitational
strength forces at short distances is challenging. The horizontal lines correspond to
theoretical predictions, in particular for the graviton in the case $n = 2$ and for the
radion in the transverse case. These limits are compared to those obtained from
particle accelerator experiments in Table 1.

5. Standard Model on D-branes

The gauge group closest to the Standard Model one can easily obtain with D-
branes is U(3) × U(2) × U(1). The first factor arises from three coincident "color"
D-branes. An open string with one end on them is a triplet under SU(3) and carries the same U(1) charge for all three components. Thus, the U(1) factor of U(3)
has to be identified with *gauged* baryon number. Similarly, U(2) arises from two
coincident "weak" D-branes and the corresponding abelian factor is identified with

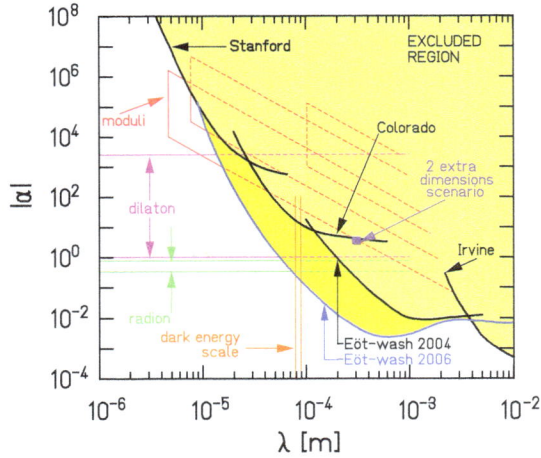

Fig. 6. (Color online) Present limits on new short-range forces (yellow regions), as a function of their range λ and their strength relative to gravity α. The limits are compared to new forces mediated by the graviton in the case of two large extra dimensions, and by the radion.

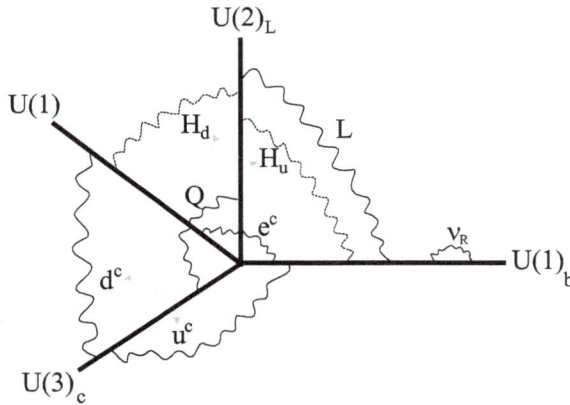

Fig. 7. A minimal Standard Model embedding on D-branes.

gauged weak-doublet number. Finally, an extra U(1) D-brane is necessary in order to accommodate the Standard Model without breaking the baryon number.[56,57]

It turns out that there are two possible ways of embedding the Standard Model particle spectrum on these stacks of branes,[56,57] which are shown pictorially in Fig. 7. The quark doublet Q corresponds necessarily to a massless excitation of an open string with its two ends on the two different collections of branes (color and weak). As seen from the figure, a fourth brane stack is needed for a complete embedding, which is chosen to be a $U(1)_b$ extended in the bulk. This is welcome since one can accommodate right handed neutrinos as open string states on the bulk with sufficiently small Yukawa couplings suppressed by the large volume of the bulk.[58-60] The two models are obtained by an exchange of the up and down antiquarks, u^c and d^c, which correspond to open strings with one end on the color branes and the other either on the U(1) brane, or on the U(1)$_b$ in the bulk. The lepton doublet L arises from an open string stretched between the weak branes and U(1)$_b$, while the antilepton l^c corresponds to a string with one end on the U(1) brane and the other in the bulk. For completeness, we also show the two possible Higgs states H_u and H_d that are both necessary in order to give tree-level masses to all quarks and leptons of the heaviest generation.

Acknowledgments

Work supported in part by the European Commission under the ERC Advanced Grant 226371 and the contract PITN-GA-2009-237920, and in part by the CNRS grant GRC APIC PICS 3747.

References

1. K. R. Dienes, *Phys. Rep.* **287**, 447 (1997), arXiv:hep-th/9602045.
2. C. Angelantonj and A. Sagnotti, *Phys. Rep.* **371**, 1 (2002) [Erratum: *ibid.* **376**, 339 (2003)], arXiv:hep-th/0204089.
3. I. Antoniadis, *Phys. Lett. B* **246**, 377 (1990).
4. D. J. Kapner, T. S. Cook, E. G. Adelberger, J. H. Gundlach, B. R. Heckel, C. D. Hoyle and H. E. Swanson, *Phys. Rev. Lett.* **98**, 021101 (2007).
5. N. Arkani-Hamed, S. Dimopoulos and G. R. Dvali, *Phys. Lett. B* **429**, 263 (1998), arXiv:hep-ph/9803315.
6. I. Antoniadis, N. Arkani-Hamed, S. Dimopoulos and G. R. Dvali, *Phys. Lett. B* **436**, 257 (1998), arXiv:hep-ph/9804398.
7. I. Antoniadis, Prepared for *NATO Advanced Study Institute and EC Summer School on Progress in String, Field and Particle Theory*, Cargese, Corsica, France, 2002.
8. J. D. Lykken, *Phys. Rev. D* **54**, 3693 (1996), arXiv:hep-th/9603133.
9. J. C. Long and J. C. Price, *Comptes Rendus Physique* **4**, 337 (2003).
10. R. S. Decca, D. Lopez, H. B. Chan, E. Fischbach, D. E. Krause and C. R. Jamell, *Phys. Rev. Lett.* **94**, 240401 (2005).
11. R. S. Decca *et al.*, arXiv:0706.3283.
12. S. J. Smullin, A. A. Geraci, D. M. Weld, J. Chiaverini, S. Holmes and A. Kapitulnik, arXiv:hep-ph/0508204.
13. H. Abele, S. Haeßler and A. Westphal, in *271th WE-Heraeus-Seminar*, Bad Honnef, 2002.
14. I. Antoniadis and K. Benakli, *Phys. Lett. B* **326**, 69 (1994).
15. K. R. Dienes, E. Dudas and T. Gherghetta, *Phys. Lett. B* **436**, 55 (1998), arXiv:hep-ph/9803466.
16. K. R. Dienes, E. Dudas and T. Gherghetta, *Nucl. Phys. B* **537**, 47 (1999), arXiv:hep-ph/9806292.
17. I. Antoniadis, K. Benakli and M. Quirós, *Phys. Lett. B* **331**, 313 (1994).
18. I. Antoniadis, K. Benakli and M. Quirós, *Phys. Lett. B* **460**, 176 (1999).
19. P. Nath, Y. Yamada and M. Yamaguchi, *Phys. Lett. B* **466**, 100 (1999).

20. T. G. Rizzo and J. D. Wells, *Phys. Rev. D* **61**, 016007 (2000).
21. T. G. Rizzo, *Phys. Rev. D* **61**, 055005 (2000).
22. A. De Rujula, A. Donini, M. B. Gavela and S. Rigolin, *Phys. Lett. B* **482**, 195 (2000).
23. E. Accomando, I. Antoniadis and K. Benakli, *Nucl. Phys. B* **579**, 3 (2000).
24. G. Servant and T. M. P. Tait, *Nucl. Phys. B* **650**, 391 (2003).
25. G. F. Giudice, R. Rattazzi and J. D. Wells, *Nucl. Phys. B* **544**, 3 (1999).
26. E. A. Mirabelli, M. Perelstein and M. E. Peskin, *Phys. Rev. Lett.* **82**, 2236 (1999).
27. T. Han, J. D. Lykken and R. Zhang, *Phys. Rev. D* **59**, 105006 (1999).
28. K. Cheung and W.-Y. Keung, *Phys. Rev. D* **60**, 112003 (1999).
29. C. Balázs *et al.*, *Phys. Rev. Lett.* **83**, 2112 (1999).
30. L3 Collab. (M. Acciarri *et al.*), *Phys. Lett. B* **464**, 135 (1999).
31. L3 Collab. (M. Acciarri *et al.*), *Phys. Lett. B* **470**, 281 (1999).
32. J. L. Hewett, *Phys. Rev. Lett.* **82**, 4765 (1999).
33. N. Arkani-Hamed, S. Dimopoulos and G. Dvali, *Phys. Rev. D* **59**, 086004 (1999).
34. S. Cullen and M. Perelstein, *Phys. Rev. Lett.* **83**, 268 (1999).
35. V. Barger, T. Han, C. Kao and R. J. Zhang, *Phys. Lett. B* **461**, 34 (1999).
36. K. Benakli and S. Davidson, *Phys. Rev. D* **60**, 025004 (1999).
37. L. J. Hall and D. Smith, *Phys. Rev. D* **60**, 085008 (1999).
38. E. Dudas and J. Mourad, *Nucl. Phys. B* **575**, 3 (2000), arXiv:hep-th/9911019.
39. S. Cullen, M. Perelstein and M. E. Peskin, *Phys. Rev. D* **62**, 055012 (2000).
40. D. Bourilkov, *Phys. Rev. D* **62**, 076005 (2000).
41. L3 Collab. (M. Acciarri *et al.*), *Phys. Lett. B* **489**, 81 (2000).
42. I. Antoniadis, K. Benakli and A. Laugier, *J. High Energy Phys.* **0105**, 044 (2001).
43. P. C. Argyres, S. Dimopoulos and J. March-Russell, *Phys. Lett. B* **441**, 96 (1998), arXiv:hep-th/9808138.
44. T. Banks and W. Fischler, arXiv:hep-th/9906038.
45. S. B. Giddings and S. Thomas, *Phys. Rev. D* **65**, 056010 (2002), arXiv:hep-ph/0106219.
46. S. Dimopoulos and G. Landsberg, *Phys. Rev. Lett.* **87**, 161602 (2001), arXiv:hep-ph/0106295.
47. P. Meade and L. Randall, arXiv:0708.3017.
48. L. A. Anchordoqui, H. Goldberg, D. Lust, S. Nawata, S. Stieberger and T. R. Taylor, *Phys. Rev. Lett.* **101**, 241803 (2008), arXiv:0808.0497.
49. D. Lust, S. Stieberger and T. R. Taylor, *Nucl. Phys. B* **808**, 1 (2009), arXiv:0807.3333.
50. G. T. Horowitz and J. Polchinski, *Phys. Rev. D* **55**, 6189 (1997), arXiv:hep-th/9612146.
51. I. Antoniadis, S. Dimopoulos and G. Dvali, *Nucl. Phys. B* **516**, 70 (1998).
52. S. Ferrara, C. Kounnas and F. Zwirner, *Nucl. Phys. B* **429**, 589 (1994).
53. I. Antoniadis, K. Benakli, A. Laugier and T. Maillard, *Nucl. Phys. B* **662**, 40 (2003), arXiv:hep-ph/0211409.
54. E. G. Adelberger, B. R. Heckel, S. Hoedl, C. D. Hoyle, D. J. Kapner and A. Upadhye, *Phys. Rev. Lett.* **98**, 131104 (2007).
55. I. Antoniadis, E. Kiritsis and J. Rizos, *Nucl. Phys. B* **637**, 92 (2002).
56. I. Antoniadis, E. Kiritsis and T. N. Tomaras, *Phys. Lett. B* **486**, 186 (2000).
57. I. Antoniadis, E. Kiritsis, J. Rizos and T. N. Tomaras, *Nucl. Phys. B* **660**, 81 (2003).
58. K. R. Dienes, E. Dudas and T. Gherghetta, *Nucl. Phys. B* **557**, 25 (1999), arXiv:hep-ph/9811428.
59. N. Arkani-Hamed, S. Dimopoulos, G. R. Dvali and J. March-Russell, *Phys. Rev. D* **65**, 024032 (2002), arXiv:hep-ph/9811448.
60. G. R. Dvali and A. Y. Smirnov, *Nucl. Phys. B* **563**, 63 (1999).

STRING CORRECTIONS TO QCD

DIETER LÜST

Arnold-Sommerfeld-Center for Theoretical Physics,
Ludwig-Maximilians-Universität München,
Theresienstraße 37, D-80333 München, Germany
and
Max-Planck-Institut für Physik,
Föhringer Ring 6, D-80805 München, Germany
dieter.luest@lmu.de, luest@mppmu.mpg.de

In this article we are considering coorections to strong interactions which are due to cclored string Regge excitations. In case of a low string scale within the TeV region these higher spin excitations of quarks and gluons will lead to spectacular, universal dijet signatures at the LHC, which are true for a large class of models in the string landscape.

1. Introduction

It is a great honor and pleasure for me to contribute to conference and the proceedings in Honor of Professor Murray Gell-Mann's 80th. Birthday. The name "Murray Gell-Mann" is intrinsically tied to the quarks as basic constituents of matter. After his famous eightfold way,[1] Gell-Mann[2] and also Zweig[3] proposed the quarks to explain the symmetry structure inside the spectrum of hadrons. Some ten years later in 1973, the QCD lagrangian with spin 1/2 quarks as color triplet fields together with spin 1 color octet gluons of $SU(3)_c$ was formulated by Fritzsch, Gell-Mann and Leutwyler.[4] The figure1 shows Murray together with Harald Fritzsch and Bill Bardeen, another person, who largely contributed during the early days of QCD.

Sc far only color triplet quarks and color octet gluons were seen in nature as colored objects. Nevertheless it is an interesting question, whether there exist other, more heavy particles with color? So far the experimental evidence for this is lacking, however interesting theoretical models were proposed that predict new colored objects. For example, in Grand Unification massive color triplets with spin 1 are needed to complete the known gauge multiplets of the Standard Model into the adjoint representation of a bigger gauge group. These particles can then be indirectly observed by proton decay. In the supersymmetric extension of the Standard Model colored spin 0 squarks as well as spin 1/2 color triplet gluinos are needed to fill out complete supermultiplets. One hopes to find direct evidence of these colored superparticles in the LHC collider experiment, in case their masses are in the TeV region.

Fig. 1. At Harald Fritzsch Symposium, Munich June 2008 (picture thanks to Hagen Kleinert and his wife).

String theory was originally invented as a model (the dual resonance model) for describing the spectrum and the S-matrix of hadrons. The famous Veneziano amplitudes[5] contains as poles an infinite number of string excitations, where the hadronic particles essentially follow the Regge trajectories of vibrating strings,

$$j = j_0 + \alpha' M^2 , \tag{1}$$

with the spin j and α' the Regge slope parameter that determines the fundamental string mass scale $M^2_{\text{string}} = \alpha'^{-1}$. In case of the dual resonance model the string scale has to be chosen of order of the relevant hadronic mass scale, i.e. $M_{\text{string}} = \mathcal{O}(1\,\text{GeV})$. However, due to several difficulties within the dual resonance model, and as a result of the advent of QCD as a field theoretical description of strong interactions, string theory (the dual resonance model) as a model for hadrons was left aside.

In 1974 a radical change of paradigm in string theory took place, when Scherk and Schwarz[6] proposed string theory as a fundamental theory for quantum gravity. The massless spin 2 closed string excitation has all properties of the graviton particle in the quantized version of general relativity. As pointed out by Scherk and Schwarz, this observation seemingly implies that the string scale now has to be identified with the fundamental scale of gravity, the Planck scale of about Planck scale of about $M_{\text{Planck}} \simeq 10^{19}$ GeV. If this is indeed the case, then the masses of all string excitations are as high as Planck scale of about M_{Planck} and hence not accessible for direct production and discovery at present days' terrestrial accelerators like the LHC. Later, during the first and second superstring revolutions, it became clear that the 10-dimensional superstring theory is consistent and free of quantum anomalies, and that M-theory will eventually provide a unique framework for quantum gravity and ten-dimensional gauge interactions.

The situation drastically changes when compactifying the ten-diemnsional superstring to lower, i.e. four space-time dimensions in order to make contact with the Standard Model of particles physics. Two aspects which arise during compactification are especially important. First, the uniqueness of string theory gets vastly destroyed, since there exist a huge number of lower-dimensional vacua of the unique ten-dimensional string. Their number apparently is so huge that the corresponding framework is called string landscape, in analogy to condensed matter physics, where one also deals with a landscape of energetically different vacua of the of the theory. In string theory, basically each point (vacuum) in the landscape describes an universe of its own, where most of these universes have rather different properties compared to our Standard Model or also compared to standard cosmology.

Let us be more precise what we actually mean by the string landscape (for a review see Ref. 7). It is defined to be the space of all possible solutions of the string equations of motion. In ten space-time dimensions, there exist just five different formulations of string theory (two heterotic strings, type I type IIA and type IIB superstrings). Exploring several kind of duality symmetries, it is conjectured that all these string theories can be unified into M-theory, where also 11-dimensional supergravity is included. However the number of lower-dimensional string solutions, i.e. lower dimensional string ground states, which are obtained after compactification, is enormous. This fact became clear already in 1986 constructing heterotic strings in four dimensions,[8–10] and within the covariant lattice construction,[9] the number of possible four-dimensional string ground states was estimated to be of order 10^{1500}. More recently, the number of discrete flux vacua of an effective supergravity potential for type II compactifications on a generic Calabi-Yau manifold was shown to be of order 10^{500}.[11,12] To deal with such a huge number of possibilities, certain strategies are required in order to proceed within the top-down approach. One possible and legitimate approach is given by the investigation of the statistical properties of the string landscape. I.e. one has to determine by statistical methods what is the fraction of string vacua with good phenomenological properties. Possible statistical correlations resp. anti-correlations would be especially worth to be discovered, like e.g. between the number of families and the rank of the low-energy gauge groups, because they could provide a step towards verifying or resp. falsifying string theory. Investigating the statistics of certain orientifold compactifications on orbifolds, it was found that there exist in total about 10^{28} cosnistent D-brane models.[13–16] Only roughly 10^6 of them possess Standard Model-like particle spectra with gauge group $SU(3)_c \times SU(2)_L \times U(1)_Y$ and three generations of quarks and leptons (without massless color exotics). Therefore, in this class of models, the likelihood to find the Standard Model in the sample of all string compactifications with D-branes is of order 10^{-22}.

The existence of the string landscape and also even before the picture of cosmic inflation[17] inspired several people to consider a multiverse of vacua, where one can also compute transition amplitudes among different universes. multiverse with a huge number of bubbles, with each being filled by one of the vacua of the landscape. The population of all possible bubbles in the universe is possible in the context of eternal inflation, where transitions between different bubbles due to quantum tunneling processes are going to happen. Therefore, different universes appear as different phases of an extremely complex and perhaps even chaotic system. However still the problem of calculability and predictability in the landscape of multiverses is a great challenge. Eventually, the multiverse picture is likely to be merged with the anthropic principle[18] (see also Ref. 19). In this context I like to quote Murray Gell-Mann (from M. Duff, 1987): "If we really live in a multiverse, Physics will have been reduced to an environmental science like Botany".

The second new aspect that arises after compactification, is much more in favor of predictability in the landscape, and may open a door towards model independent experimental discovery of string theory: after compactification, the string scale M_{string} is not necessarily any more of the order of the Planck mass, but it is a free parameter, which can be as low as the energy scale of the Standard Model, i.e. of order of a few TeV. This is particularly true in D-brane compactifications (for a review see Ref. 20), where the Standard Model is living on a lower dimensional brane, that might be partly embedded into the internal, compact six-dimensional space that is used for compactification. In this brane world scenario the elementary particles, such as quarks, leptons, gluons, photons, weak bosons and the Higgs particles arise as open string excitations, whose ends are attached to the world volumes of the intersecting D-branes. So in contrast to the old dual resonance model, now the colored quarks and gluons are elementary open strings. It immediately follows that there exist higher spin Regge excitations of the quarks and of the gluons (of all Standard Model fields). In fact, the first massive gluonic resonances can be shown to have spin 0,1, and 2, whereas the first quarks resonances have spin 1/2 and spin 3/2, respectively. These lowest excitations will be found exactly at mass M_{string}, and they are followed by the infinite tower of higher Regge excitations. So in case $M_{\text{string}} \simeq \mathcal{O}(1 \text{ TeV})$, these universal string Regge excitations should be easily found at the LHC.

Therefore, for low string scale D-brane models, there is a very clear picture, how string theory meets collider physics, and how model independent predictions are possible: at low energies around 1 GeV, QCD is of course valid. Here the hadrons build the first tower of Regge states, corresponding to the QCD string picture (which can be made more precise using the AdS/CFT correspondence). QCD processes will be the background for the new stringy physics to show up at higher energies. In fact, at energies around 1 TeV or higher, stringy corrections due to new colored Regge modes will become important. Their production and subsequent decay will then lead the discovery of these universal heavy string excitations. Eventually there will be a full tower of elementary open string Regge modes. Both type of Regge excitations, first the hadronic fields and second the elementary open stings, will interact with each other, whereas these interactions will suppressed by certain powers of the string scale M_{string}. These effective, non-renormalizable interactions may lead to rare FCNC processes, which will imply additional model dependent constraints on the string scale M_{string}. However the main aim of the paper will be to show that the production cross sections of gluons and quarks at the LHC into massive string excitations can be computed in a completely universal, model independent way, allowing for universal string predictions in case the string scale is low.[21] The corresponding tree level string cross sections are independent from in the internal geometry and hence independent from the particular location of the model in the string landscape. This observation nullifies in some sense the string landscape problem at the LHC.

2. String Corrections to QCD: Possible Low Energy (LHC) Signatures of Intersecting D-Brane Models

There is some good reason to believe that the resolution of the hierarchy problem lies in new physics around the TeV mass scale. The LHC collider at CERN is designed to discover new physics precisely in this energy range, hopefully giving important clues about the nature of dark matter and perhaps at the same time about the solution of the hierarchy problem. In fact, there are at least three, not necessarily mutually exclusive scenarios, offered as solutions of the hierarchy problem:

- Low energy supersymmetry at around 1 TeV.
- New strong dynamics at around 1 TeV (technicolor, little Higgs models, etc).
- Large extra dimensions and a low scale for (quantum) gravity at around 1 TeV.

Here we discuss some universal features of the large extra dimensions scenario[22,23] relevant for its possible discovery at the LHC. In this scenario, the gravitational and gauge interactions are unified at around 1 TeV, and the observed weakness of gravity at lower energies is due to the existence of large extra dimensions. Gravitons may scatter into the extra space and by this the gravitational coupling constant is decreased to its observed value. Extra dimensions arise naturally in string theory. Hence, one obvious question is how to embed the above scenario into string theory. Then the next important question is what are the possible signatures of large extra dimensions and low gravity in string theory, and how to detect them at the LHC. Large extra dimensions can appear in string theory in case that the intrinsic scale of the string excitations, called the string mass M_{string} is very low, namely at the order of TeV. In this case a whole tower of infinite string excitations will open up at around 1 TeV, where the new particles essentially follow the well known Regge trajectories of vibrating strings with masses

$$M_n^2 = M_{\text{string}}^2 \; n. \tag{2}$$

The production of string Regge excitations will lead to new contributions to standard model scattering processes, like QCD jets or scattering of quarks into leptons or gauge bosons, which can be measurable at LHC in case the string scale is low.[21,24] Second there are the KK and winding excitations along the small internal dimensions, i.e. KK and winding excitations of the SM fields. Their masses depend on the internal volumes, and they should be also near the string scale M_{string}.

For those amplitudes involving four gauge bosons or two gauge bosons and two matter fermions, the amplitudes do not depend on the geometry of the underlying Calabi-Yau spaces.[21] This model independence still also holds for the four–fermion

matter amplitudes, but only w.r.t. their dependence on the four-dimensional kinematical variables s, t, u. On the other hand, the four–fermion amplitudes do depend
on the internal Calabi-Yau geometry and topology. Concretely, the four–fermion
amplitudes in general depend on the Calabi-Yau intersection numbers, and also on
the rational instanton numbers of the Calabi-Yau space.

The general structure of a four point amplitude of four open string states is as
follows (higher point amplitudes were computed in Ref. 25). Let Φ^i, $i = 1, 2, 3, 4$,
represent gauge bosons, quarks of leptons of the standard model realized on three
or more stacks of intersecting D-branes. The corresponding string vertex operators
V_{Φ^i} are constructed from the fields of the underlying superconformal field theory
(SCFT) and contain explicit (group-theoretical) Chan-Paton factors. In order to
obtain the scattering amplitudes, the vertices are inserted at the boundary of a disk
world-sheet, and the following SCFT correlation function is evaluated:

$$
\mathcal{A}(\Phi^1, \Phi^2, \Phi^3, \Phi^4) = \sum_{\pi \in S_4/Z_2} V_{CKG}^{-1} \int_{\mathcal{I}_\pi} \left(\prod_{k=1}^4 dz_k \right) \langle V_{\Phi^1}(z_1) \, V_{\Phi^2}(z_2) \, V_{\Phi^3}(z_3) \, V_{\Phi^4}(z_4) \rangle \ .
$$

(3)

Here, the sum runs over all six cyclic inequivalent orderings π of the four vertex
operators along the boundary of the disk. Each permutation π gives rise to an in-
tegration region $\mathcal{I}_\pi = \{z \in R \,|\, z_{\pi(1)} < z_{\pi(2)} < z_{\pi(3)} < z_{\pi(4)}\}$. The group-theoretical
factor is determined by the trace of the product of individual Chan-Paton factors,
ordered in the same way as the vertex positions. The disk boundary contains four
segments which may be associated to as many as four different stacks of D-branes,
since each vertex of a field originating from a D-brane intersection connects two
stacks. Thus the Chan-Paton factor may actually contain as many a four traces,
all in the fundamental representations of gauge groups associated to the respective
stacks. However, purely partonic amplitudes for the scattering of quarks and gluons
involve no more than three stacks. Then, after performing all Wick contractions in
eq.(3) the correlators become basic, and generically for each partial amplitude the
integral may be reduced to the Euler Beta function:

$$
B(s, u) = \int_0^1 x^{s-1} (1 - x)^{u-1} = \frac{\Gamma(s) \, \Gamma(u)}{\Gamma(s + u)} = \frac{1}{s} + \frac{1}{u} - \frac{\pi^2}{6} (s + u) + \mathcal{O}(\alpha'^2) \ . \quad (4)
$$

Due to the extended nature of strings, the world–sheet string amplitudes are
generically non–trivial functions in α' in addition to the usual dependence on the
kinematic invariants and degrees of freedom of the external states. In the effective
field theory description this α'–dependence gives rise to a series of infinite many resonance channels due to Regge excitations and/or new contact interactions. Generically, as we already saw, tree–level string amplitudes involving four gluons or amplitudes with two gluons and two fermions are described by the Euler Beta function
depending on the kinematic invariants $s = (k_1 + k_2)^2$, $t = (k_1 - k_3)^2$, $u = (k_1 - k_4)^2$,
with $s + t + u = 0$ and k_i the four external momenta. The whole amplitudes
$A(k_1, k_2, k_3, k_4; \alpha')$ may be understood as an infinite sum over s–channel poles with

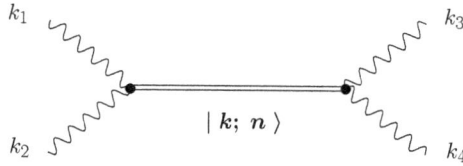

Fig. 2. Exchange of an infinite tower of Regge excitations in open string scattering processes.

intermediate string states $|k;n\rangle$ exchanged, as it can be seen in the figure 2: After neglecting kinematical factors the string amplitude $A(k_1,k_2,k_3,k_4;\alpha')$ assumes the form

$$A(k_1,k_2,k_3,k_4;\alpha') \sim -\frac{\Gamma(-\alpha's)\,\Gamma(1-\alpha'u)}{\Gamma(-\alpha's-\alpha'u)} = \sum_{n=0}^{\infty}\frac{\gamma(n)}{s-M_n^2} \qquad (5)$$

as an infinite sum over s–channel poles at the masses

$$M_n^2 = M_{\text{string}}^2\, n \qquad (6)$$

of the string Regge excitations. In eq.(5) the residues $\gamma(n)$ are determined by the three–point coupling of the intermediate states $|k;n\rangle$ to the external particles and given by

$$\gamma(n) = \frac{t}{n!}\frac{\Gamma(-u\alpha'+n)}{\Gamma(-u\alpha')} = \frac{t}{n!}\prod_{j=1}^{n}[-u\alpha'-1+j] \sim (-\alpha'\,u)^n \quad, \qquad (7)$$

with $n+1$ being the highest possible spin of the state $|k;n\rangle$.

Another way of looking at the expression (5) appears, when we express each term in the sum as a power series expansion in α':

$$A(k_1,k_2,k_3,k_4;\alpha') \sim \underbrace{\frac{t}{s}}_{n=0} - \underbrace{\frac{\pi^2}{6}\,tu\,\alpha'^2+\dots}_{n\neq0}\,. \qquad (8)$$

In this form, the massless state $n=0$ gives rise to a field–theory contribution ($\alpha'=0$), while at the order α'^2 all massive states $n\neq0$ sum up to a finite term. The $n=0$ term in (8) describes the field–theory contribution to the scattering diagram, e.g. the exchange of a massless gluon. On the other hand, the term at the order α'^2 describes a new string contact interaction as a result of summing up all heavy string states. For the four gluon superstring amplitude the first string contact interaction is given by $\alpha'^2\,g_{Dp}^{-2}\,\mathrm{tr}F^4$, which represent a correction to YM theory, as shown in figure 3.

2.1. Four gluon scattering amplitude

Let us start with the open string tree level scattering of four gauge bosons on the disk. The gauge bosons are open strings with ends on same brane, for gluons say

254

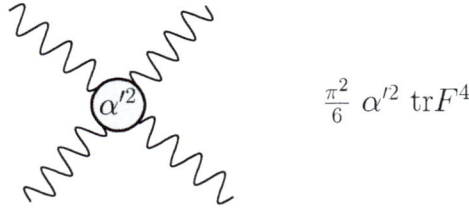

$$\frac{\pi^2}{6} \; \alpha'^2 \; \mathrm{tr} F^4$$

Fig. 3.   Order $(\alpha')^2$ contact interaction in the scattering of four open string gluons.

the QCD stack a. The gauge boson vertex operator in the (-1)-ghost picture reads

$$V_{A^a}^{(-1)}(z,\xi,k) = g_A [T^a]_{\alpha_2}^{\alpha_1} \; e^{-\phi(z)} \; \xi^\mu \; \psi_\mu(z) \; e^{ik_\rho X^\rho(z)} \;, \qquad (9)$$

while in the zero–ghost picture we have:

$$V_{A^a}^{(0)}(z,\xi,k) = \frac{g_A}{(2\alpha')^{1/2}} [T^a]_{\alpha_2}^{\alpha_1} \xi_\mu \; [\; i\partial X^\mu(z) + 2\alpha' \; (k\psi) \; \psi^\mu(z) \;] \; e^{ik_\rho X^\rho(z)} \;. \qquad (10)$$

where ξ^μ is the polarization vector. The vertex must be inserted on the segment of disk boundary on stack a, with the indices α_1 and α_2 describing the two string ends.

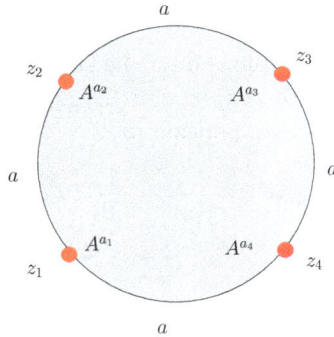

Fig. 4.   The four gauge boson disk diagram.

Four-gluon amplitudes have been known for many years. The corresponding string disk diagram is shown in the figure 4.: The complete amplitude can be generated from the maximally helicity violating MHV amplitudes.[26,27] Averaging over helicities and colors of the incident partons and summed over helicities and colors of the outgoing particles, we obtain for gluon scattering SU(3)

$$|\mathcal{M}(gg \to gg)|^2 = \left(\frac{1}{s^2} + \frac{1}{t^2} + \frac{1}{u^2} \right) \left[\frac{9}{4} \left(s^2 V_s^2 + t^2 V_t^2 + u^2 V_u^2 \right) - \frac{1}{3} \left(s V_s + t V_t + u V_u \right)^2 \right]$$

$$(11)$$

In the D-brane models under consideration, the ordinary $SU(3)$ color gauge symmetry is extended to $U(3)$, so that the open strings terminating on the stack of "cclor" branes contain an additional $U(1)$ gauge boson C. Replacing one gluon by the $U(1)$ color singlet gauge boson component C in the QCD stack a, there is also a non-vanishing string amplitude with one photon or one Z-boson ($C = \gamma, Z$), since the photon or the Z-boson always has an admixture of this $U(1)$ gauge group:

$$|\mathcal{M}(gg \to gC)|^2 = \frac{5}{6}Q_C^2\left(\frac{1}{s^2} + \frac{1}{t^2} + \frac{1}{u^2}\right)(sV_s + tV_t + uV_u)^2 . \tag{12}$$

In the zero-slop field theory limit $\alpha' \to 0$ the functions $V_s, V_t, V_u \to 1$, and the four gauge boson amplitudes get contributions only from the exchange of SM fields. In this limit the string amplitudes approach the known results true in the SM Note that the $\mathcal{M}(gg \to gC) \to 0$, as required in the tree level SM. Note that the four gauge boson amplitude is completely model independent, there are no KK-particles being exchanged in the s-channel.

2.2. *Two gluon, two quark scattering amplitudes*

We now consider the following correlation function between two gauge bosons and two matter fermions:

$$\langle V_{A^x}^{(0)}(z_1, \xi_1, k_1)\, V_{A^y}^{(-1)}(z_2, \xi_2, k_2)\, V_{\psi_{\beta_3}^{\alpha_3}}^{(-1/2)}(z_3, u_3, k_3)\, V_{\bar{\psi}_{\alpha_4}^{\beta_4}}^{(-1/2)}(z_4, \bar{u}_4, k_4)\rangle . \tag{13}$$

The fermion vertex operators are boundary changing operators, being inserted at the intersection of brane stack a and b. Specifically, the chiral fermion vertex operators of the quarks and leptons are:

$$V_{\psi_\beta^\alpha}^{(-1/2)}(z, u, k) = g_\psi [T^\alpha_\beta]_{\alpha_1}^{\beta_1} e^{-\phi(z)/2}\, u^\lambda S_\lambda(z)\, \Xi^{a \cap b}(z)\, e^{ik_\rho X^\rho(z)} ,$$

$$V_{\bar{\psi}_\alpha^\beta}^{(-1/2)}(z, \bar{u}, k) = g_\psi [T^\beta_\alpha]_{\beta_1}^{\alpha_1} e^{-\phi(z)/2}\, \bar{u}_{\dot{\lambda}} S^{\dot{\lambda}}(z)\, \overline{\Xi}^{a \cap b}(z)\, e^{ik_\rho X^\rho(z)} . \tag{14}$$

These vertices connect two segments of disk boundary, associated to stacks a and b, with the indices α_1 and β_1 representing the string ends on the respective stacks. The internal field $\Xi^{a \cap b}$ of conformal dimension $3/8$ is the fermionic boundary changing operator. In the intersecting D-brane models, the intersections are characterized by angles θ_{ba}. Then $\Xi^{a \cap b}$ can be expressed in terms of bosonic and fermionic twist fields σ and s:

$$\Xi^{a \cap b} = \prod_{j=1}^{3} \sigma_{\theta_{ba}^j}\, s_{\theta_{ba}^j} , \qquad \overline{\Xi}^{a \cap b} = \prod_{j=1}^{3} \sigma_{-\theta_{ba}^j}\, s_{-\theta_{ba}^j} . \tag{15}$$

The spin fields

$$s_{\theta^j} = e^{i(\theta^j - \frac{1}{2})H^j} , \qquad s_{-\theta^j} = e^{-i(\theta^j - \frac{1}{2})H^j} \tag{16}$$

have conformal dimension $h_s = \frac{1}{2}(\theta^j - \frac{1}{2})^2$ and twist the internal part of the Ramond ground state spinor. The field σ_θ has conformal dimension $h_\sigma = \frac{1}{2}\theta^j(1 - \theta^j)$ and

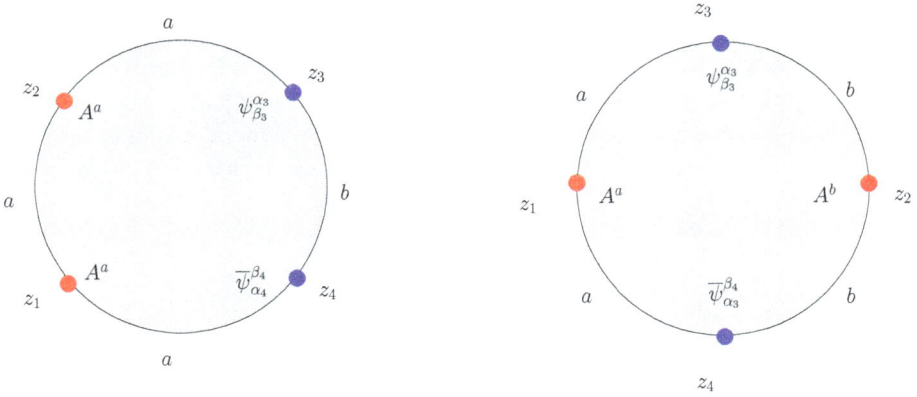

Fig. 5.   The two gauge boson - two fermion disk diagrams.

produces discontinuities in the boundary conditions of the internal complex bosonic Neveu–Schwarz coordinates Z^j.

The fact that fermions originate from the same pair of stacks, say a and b is forced upon us by the conservation of twist charges, in a similar way as their opposite helicities are forced by the internal charge conservation. It follows that both gauge bosons must be associated either to one of these stacks, say $(x, y) = (a_1, a_2)$, or one of them is associated to a while the other to b, say $(x, y) = (a, b)$. The corresponding disk diagrams are shown in the figure 5. Using these informations one obtains the following tree level (squared) amplitudes for two gauge boson, two fermion scattering processes,[21]

$$|\mathcal{M}(gg \to q\bar{q})|^2 = \frac{t^2 + u^2}{s^2}\left[\frac{1}{6}\frac{1}{ut}(tV_t + uV_u)^2 - \frac{3}{8}V_tV_u\right] \qquad (17)$$

and

$$|\mathcal{M}(gq \to gq)|^2 = \frac{s^2 + u^2}{t^2}\left[V_sV_u - \frac{4}{9}\frac{1}{su}(sV_s + uV_u)^2\right] \qquad (18)$$

Again, in the s-channel there can be only the exchange of heavy Regge states and no KK-states. Hence also these two amplitudes are completely independent from the internal geometry.

2.3. Dijet signals for lowest mass strings at the LHC

In this section we will determine the contribution from the exchange of excited, heavy Regge states to dijet processes at the LHC.[28,29] The first Regge excitations of the gluon (g) and quarks (q) will be denoted by g^*, q^*, respectively. The first excitation of the C will be denoted by C^*. In the following we isolate the contribution to the partonic cross section from the first resonant state. Note that far below the string threshold, at partonic center of mass energies $\sqrt{s} \ll M_s$, the form factor $V(s, t, u) \approx 1 - \frac{\pi^2}{6} su/M_s^4$ and therefore the contributions of Regge excitations are strongly suppressed. The s-channel pole terms of the average square amplitudes contributing to dijet production at the LHC can be obtained from the general formulae given in in the previous subsection. However, for phenomenological purposes, the poles need to be softened to a Breit-Wigner form by obtaining and utilizing the correct *total* widths of the resonances.[30] After this is done, the contributions of the various channels are as follows:

$$|\mathcal{M}(gg \to gg)|^2$$

$$= \frac{19}{12} \frac{g^4}{M_s^4} \left\{ W_{g^*}^{gg \to gg} \left[\frac{M_s^8}{(s - M_s^2)^2 + (\Gamma_{g^*}^{J=0} M_s)^2} + \frac{t^4 + u^4}{(s - M_s^2)^2 + (\Gamma_{g^*}^{J=2} M_s)^2} \right] \right.$$

$$\left. + W_{C^*}^{gg \to gg} \left[\frac{M_s^8}{(s - M_s^2)^2 + (\Gamma_{C^*}^{J=0} M_s)^2} + \frac{t^4 + u^4}{(s - M_s^2)^2 + (\Gamma_{C^*}^{J=2} M_s)^2} \right] \right\}, \quad (19)$$

$$|\mathcal{M}(gg \to q\bar{q})|^2 = \frac{7}{24} \frac{g^4}{M_s^4} N_f \left[W_{g^*}^{gg \to q\bar{q}} \frac{ut(u^2 + t^2)}{(s - M_s^2)^2 + (\Gamma_{g^*}^{J=2} M_s)^2} \right.$$

$$\left. + W_{C^*}^{gg \to q\bar{q}} \frac{ut(u^2 + t^2)}{(s - M_s^2)^2 + (\Gamma_{C^*}^{J=2} M_s)^2} \right] \quad (20)$$

$$|\mathcal{M}(q\bar{q} \to gg)|^2 = \frac{56}{27} \frac{g^4}{M_s^4} \left[W_{g^*}^{q\bar{q} \to gg} \frac{ut(u^2 + t^2)}{(s - M_s^2)^2 + (\Gamma_{g^*}^{J=2} M_s)^2} \right.$$

$$\left. + W_{C^*}^{q\bar{q} \to gg} \frac{ut(u^2 + t^2)}{(s - M_s^2)^2 + (\Gamma_{C^*}^{J=2} M_s)^2} \right], \quad (21)$$

$$|\mathcal{M}(qg \to qg)|^2$$

$$= -\frac{4}{9} \frac{g^4}{M_s^2} \left[\frac{M_s^4 u}{(s - M_s^2)^2 + (\Gamma_{q^*}^{J=1/2} M_s)^2} + \frac{u^3}{(s - M_s^2)^2 + (\Gamma_{q^*}^{J=3/2} M_s)^2} \right], \quad (22)$$

where g is the QCD coupling constant ($\alpha_{\text{QCD}} = \frac{g^2}{4\pi} \approx 0.1$) and $\Gamma_{g^*}^{J=0} = 75 \, (M_s/\text{TeV})$ GeV, $\Gamma_{C^*}^{J=0} = 150 \, (M_s/\text{TeV})$ GeV, $\Gamma_{g^*}^{J=2} = 45 \, (M_s/\text{TeV})$ GeV, $\Gamma_{C^*}^{J=2} = 75 \, (M_s/\text{TeV})$ GeV, $\Gamma_{q^*}^{J=1/2} = \Gamma_{q^*}^{J=3/2} = 37 \, (M_s/\text{TeV})$ GeV are the total decay widths for intermediate states g^*, C^*, and q^* (with angular momentum J).[30] The associated weights of these intermediate states are given in terms of the probabilities for the various entrance and exit channels

$$W_{g^*}^{gg \to gg} = \frac{(\Gamma_{g^* \to gg})^2}{(\Gamma_{g^* \to gg})^2 + (\Gamma_{C^* \to gg})^2} = 0.09, \quad (23)$$

$$W_{C^*}^{gg \to gg} = \frac{(\Gamma_{C^* \to gg})^2}{(\Gamma_{g^* \to gg})^2 + (\Gamma_{C^* \to gg})^2} = 0.91, \quad (24)$$

258

$$W_{g^*}^{gg \to q\bar{q}} = W_{g^*}^{q\bar{q} \to gg} = \frac{\Gamma_{g^* \to gg}\,\Gamma_{g^* \to q\bar{q}}}{\Gamma_{g^* \to gg}\,\Gamma_{g^* \to q\bar{q}} + \Gamma_{C^* \to gg}\,\Gamma_{C^* \to q\bar{q}}} = 0.24\,, \qquad (25)$$

$$W_{C^*}^{gg \to q\bar{q}} = W_{C^*}^{q\bar{q} \to gg} = \frac{\Gamma_{C^* \to gg}\,\Gamma_{C^* \to q\bar{q}}}{\Gamma_{g^* \to gg}\,\Gamma_{g^* \to q\bar{q}} + \Gamma_{C^* \to gg}\,\Gamma_{C^* \to q\bar{q}}} = 0.76\,. \qquad (26)$$

Superscripts $J = 2$ are understood to be inserted on all the Γ's in Eqs.(23), (24), (25), (26). Equation (19) reflects the fact that weights for $J = 0$ and $J = 2$ are the same.[30] In what follows we set the number of flavors $N_f = 6$.

In figure 6 we show a representative plot of the invariant mass spectrum, for $M_s = 2$ TeV, detailing the contribution of each subprocess. The QCD background has been calculated at the partonic level from the same processes as designated for the signal, with the addition of $qq \to qq$ and $q\bar{q} \to q\bar{q}$. Our calculation, making use of the CTEQ6D parton distribution functions[31] agrees with that presented in Ref. 32. Finally we estimate (at the parton level) the LHC discovery reach, namely one may calculate a signal-to-noise ratio, with the signal rate estimated in the invariant mass window $[M_s - 2\Gamma, M_s + 2\Gamma]$. The noise is defined as the square root of the number of background events in the same dijet mass interval for the same integrated luminosity. The top two and bottom curves in the next figure show the behavior of the signal-to-noise (S/N) ratio as a function of the string scale for three integrated luminosities (100 fb^{-1}, 30 fb^{-1} and 100 pb^{-1}) at the LHC. It is remarkable that within 1-2 years of data collection, *string scales as large as 6.8 TeV are open to discovery at the $\geq 5\sigma$ level.* For 30 fb^{-1}, the presence of a resonant state with mass as large as 5.7 TeV can provide a signal of convincing significance

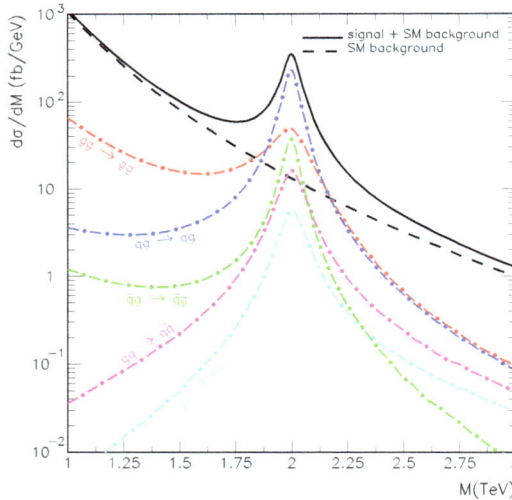

Fig. 6.   Dijet cross sections.

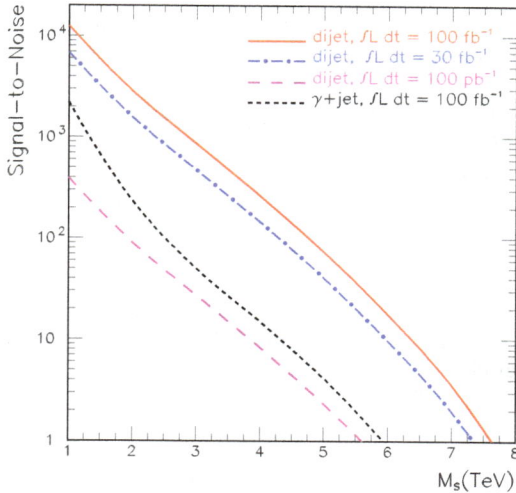

Fig. 7.   Signal to noise ratio.

($S/N \geq 13$). The bottom curve in figure 7, corresponding to data collected in a very early run of 100 pb^{-1}, shows that a resonant mass as large as 4.0 TeV can be observed with 10σ significance! Once more, we stress that these results contain no unknown parameters. They depend only on the D-brane construct for the standard model, and *are independent of compactification details.*

3. Conclusions

The discovery of new colored string Regge excitations at the LHC would provide direct evidence that quantum gravity is starting to operate already in the TeV region. This would have far reaching consequences for our standing about fundamental interactions and the structure of space and time. What has begun in the sixties with Murray Gell-Mann's discovery of the $SU(3)$ flavor symmetry of hadrons and ten years later finding the $SU(3)$ color group of quarks and gluons, would end in a much larger, in fact infinite dimensional symmetry structure at the TeV scale, in case the fundamental scale of strings and quantum gravity really happens to be so low.

Alles Gute zum achtzigsten Geburtstag, Murray Gell-Mann!

Acknowledgments

I like to thank Harald Fritzsch for inviting me to this very inspiring meeting. I am very much indebted to my collaborators on the work presented here: L. Anchordoqui, H. Goldberg, S. Nawata, O. Schlotterer, S. Stieberger and T. Taylor.

References

1. M. Gell-Mann, "The Eightfold Way: A Theory Of Strong Interaction Symmetry,"
2. M. Gell-Mann, "A Schematic Model Of Baryons And Mesons," Phys. Lett. **8**, 214 (1964).
3. G. Zweig, "An SU(3) Model For Strong Interaction Symmetry And Its Breaking,"
4. H. Fritzsch, M. Gell-Mann and H. Leutwyler, "Advantages Of The Color Octet Gluon Picture," Phys. Lett. B **47**, 365 (1973).
5. G. Veneziano, "Construction of a crossing - symmetric, Regge behaved amplitude for linearly rising trajectories," Nuovo Cim. A **57**, 190 (1968).
6. J. Scherk and J. H. Schwarz, "Dual Models And The Geometry Of Space-Time," Phys. Lett. B **52**, 347 (1974).
7. D. Lüst, "Seeing through the String Landscape - a String Hunter's Companion in Particle Physics and Cosmology," JHEP **0903**, 149 (2009) [arXiv:0904.4601 [hep-th]].
8. H. Kawai, D. C. Lewellen and S. H. H. Tye, "Construction of Four-Dimensional Fermionic String Models," Phys. Rev. Lett. **57**, 1832 (1986) [Erratum-ibid. **58**, 429 (1987)].
9. W. Lerche, D. Lüst and A. N. Schellekens, "Chiral Four-Dimensional Heterotic Strings from Selfdual Lattices," Nucl. Phys. B **287**, 477 (1987).
10. I. Antoniadis, C. P. Bachas and C. Kounnas, "Four-Dimensional Superstrings," Nucl. Phys. B **289**, 87 (1987).
11. R. Bousso and J. Polchinski, "Quantization of four-form fluxes and dynamical neutralization of the cosmological constant," JHEP **0006**, 006 (2000) [arXiv:hep-th/0004134].
12. M. R. Douglas, "The statistics of string/M theory vacua," JHEP **0305**, 046 (2003) [arXiv:hep-th/0303194].
13. R. Blumenhagen, F. Gmeiner, G. Honecker, D. Lüst and T. Weigand, "The statistics of supersymmetric D-brane models," Nucl. Phys. B **713**, 83 (2005) [arXiv:hep-th/0411173].
14. F. Gmeiner, D. Lüst and M. Stein, "Statistics of intersecting D-brane models on T^6/Z_6," JHEP **0705**, 018 (2007) [arXiv:hep-th/0703011].
15. F. Gmeiner, R. Blumenhagen, G. Honecker, D. Lüst and T. Weigand, "One in a billion: MSSM-like D-brane statistics," JHEP **0601**, 004 (2006) [arXiv:hep-th/0510170].
16. F. Gmeiner and G. Honecker, "Millions of Standard Models on Z6-prime?," JHEP **0807**, 052 (2008) [arXiv:0806.3039 [hep-th]].
17. A. Linde, "Inflationary Cosmology," Lect. Notes Phys. **738**, 1 (2008) [arXiv:0705.0164 [hep-th]].
18. L. Susskind, "The anthropic landscape of string theory," arXiv:hep-th/0302219.
19. A. N. Schellekens, "The Emperor's Last Clothes?," Rept. Prog. Phys. **71**, 072201 (2008) [arXiv:0807.3249 [physics.pop-ph]].
20. R. Blumenhagen, B. Körs, D. Lüst and S. Stieberger, "Four-dimensional String Compactifications with D-Branes, Orientifolds and Fluxes," Phys. Rept. **445**, 1 (2007) [arXiv:hep-th/0610327].
21. D. Lüst, S. Stieberger and T. R. Taylor, "The LHC String Hunter's Companion," Nucl. Phys. B **808**, 1 (2009) [arXiv:0807.3333 [hep-th]].
22. N. Arkani-Hamed, S. Dimopoulos and G. R. Dvali, "The hierarchy problem and new dimensions at a millimeter," Phys. Lett. B **429**, 263 (1998) [arXiv:hep-ph/9803315].
23. I. Antoniadis, N. Arkani-Hamed, S. Dimopoulos and G. R. Dvali, "New dimensions at a millimeter to a Fermi and superstrings at a TeV," Phys. Lett. B **436**, 257 (1998) [arXiv:hep-ph/9804398].
24. S. Cullen, M. Perelstein and M. E. Peskin, "TeV strings and collider probes of large extra dimensions," Phys. Rev. D **62**, 055012 (2000) [arXiv:hep-ph/0001166].

25. D. Lüst, O. Schlotterer, S. Stieberger and T. R. Taylor, "The LHC String Hunter's Companion (II): Five-Particle Amplitudes and Universal Properties," Nucl. Phys. B **828**, 139 (2010) [arXiv:0908.0409 [hep-th]].
26. S. Stieberger and T. R. Taylor, "Amplitude for N-gluon superstring scattering," Phys. Rev. Lett. **97**, 211601 (2006) [arXiv:hep-th/0607184].
27. S. Stieberger and T. R. Taylor, "Multi-gluon scattering in open superstring theory," Phys. Rev. D **74**, 126007 (2006) [arXiv:hep-th/0609175].
28. L. A. Anchordoqui, H. Goldberg, D. Lüst, S. Nawata, S. Stieberger and T. R. Taylor, "Dijet signals for low mass strings at the LHC," arXiv:0808.0497 [hep-ph].
29. L. A. Anchordoqui, H. Goldberg, D. Lüst, S. Nawata, S. Stieberger and T. R. Taylor, "LHC Phenomenology for String Hunters," arXiv:0904.3547 [hep-ph].
30. L. A. Anchordoqui, H. Goldberg and T. R. Taylor, "Decay widths of lowest massive Regge excitations of open strings," Phys. Lett. B **668**, 373 (2008) [arXiv:0806.3420 [hep-ph]].
31. J. Pumplin, D. R. Stump, J. Huston, H. L. Lai, P. M. Nadolsky and W. K. Tung, "New generation of parton distributions with uncertainties from global QCD analysis," JHEP **0207**, 012 (2002) [arXiv:hep-ph/0201195].
32. A. Bhatti *et al.*, "CMS search plans and sensitivity to new physics with dijets," J. Phys. G **36**, 015004 (2009) [arXiv:0807.4961 [hep-ex]].

MAXIMAL SUPERSYMMETRY AND EXCEPTIONAL GROUPS

LARS BRINK

Department of Fundamental Physics, Chalmers University of Technology,
Göteborg, Sweden
lars.brink@chalmers.se
http://fy.chalmers.se/~tfelb/

The article is a tribute to my old mentor, collaborator and friend Murray Gell-Mann. In it I describe work by Pierre Ramond, Sung-Soo Kim and myself where we describe the $\mathcal{N} = 8$ Supergravity in the light-cone formalism. We show how the Cremmer-Julia $E_{7(7)}$ non-linear symmetry is implemented and how the full supermultiplet is a representation of the $E_{7(7)}$ symmetry. I also show how the $E_{7(7)}$ symmetry is a key to understand the higher order couplings in the theory and is very useful when we discuss possible counterterms for this theory.

Keywords: Maximal supersymmetry; supergravity; exceptional symmetries.

1. Introduction

Murray Gell-Mann was the leading scientist in particle physics for some twenty years, a very long time in such a competitive subject. When I started as a young graduate student I found his name behind almost all the developments in the field that I studied and when I eventually saw him first at a Nobelsymposium here in Sweden and then at CERN, where I was a fellow in the beginning of the 70's, he made such an impression on me that his stature got even bigger. In 1976 I came to Caltech and when I met him my body was shaking. I was so nervous. After some time we got closer and we started to collaborate. This was my most lucky time in life. All through since then we have been very good friends and our discussion have been on-going even if they sometimes have been interrupted for a year or two. We always made a point to start again exactly where we ended the year before. Murray taught me many things, especially the beauty of symmetries. It has since then been a leading star in my scientific life to follow the symmetries and try to implement as big a symmetry as possible. The possible breaking of the symmetry can come later.

Almost thirty years ago I started to study supersymmetric theories in the light-cone gauge/frame formulation. The starting point was the $\mathcal{N} = 4$ SuperYang-Mills theory.[1] In order to implement as much supersymmetry as possible we[2] found that if we choose the light-cone gauge $A^+ = \frac{1}{\sqrt{2}}(A^0 + A^3) = 0$ for the vector fields and choose the light-cone direction x^+ as the evolution parameter we could solve for A^-. We could also solve for half the spinors leaving a theory with only the physical degrees of freedom. (The unphysical degrees of freedom satisfies algebraic equations when x^+ is the "time" and x^- "space".) They could then be assembled to a superfield and the action could be written in terms of this superfield. With this formalism we could prove that the $\mathcal{N} = 4$ is perturbatively finite.[3] We[4] also constructed the $\mathcal{N} = 8$ Supergravity[5] this way but failed to construct higher order couplings than the three-point ones.

Some years ago Pierre Ramond and I took up this programme again in the light of the modern developments. The maximally supersymmetric $\mathcal{N} = 8$ Supergravity and $\mathcal{N} = 4$ SuperYang-Mills play a very important rôle in modern theory. In the standard descriptions they look quite different and are naturally related to eleven- and ten-dimensional theories, respectively. In the light-cone frame description, however, they are described in a remarkably similar way hinting at a deep relation between them. In four dimension, they are the only two (except possibly for some higher-spin theories) that are described by one chiral *constrained* light-cone superfield which captures *all* their physical degrees of freedom.[2] Also, tree level Supergravity amplitudes are related to the square of Yang-Mills amplitudes,[6,7] and the light-cone Hamiltonian of both theories can be written as a *positive definite* quadratic form in their superfields.[8,9] Some even suggest that the ultraviolet finiteness of $\mathcal{N} = 4$ SuperYang-Mills[3] might extend to $\mathcal{N} = 8$ Supergravity.[10,11] There are important structural differences though. $\mathcal{N} = 8$ Supergravity, unlike $\mathcal{N} = 4$ SuperYang-Mills, is not Superconformal invariant. Instead it has the on-shell, non-linear Cremmer-Julia, $E_{7(7)}$ duality symmetry.[5] It was therefore natural to ask if this symmetry can be exploited to bring simplicty to the quartic and higher-order interactions of $\mathcal{N} = 8$ Supergravity. This was done in a paper that we wrote some time ago,[12] which I will describe here.

There are two approaches that one can follow to find the Hamiltonian or the action in this formalism. On the one hand we can follow the scheme mentioned above by fixing the light-cone gauge and eliminate all unphysical degrees of freedom. The other one is to look for a (non-linear) representation of the SuperPoincaré based on the superfield. In our work we have had to use both methods as I will indicate later.

2. $\mathcal{N} = 8$ Supergravity in Light-Cone Superspace

The 256 physical degrees of freedom of $\mathcal{N} = 8$ Supergravity form *one* constrained chiral superfield in the superspace spanned by eight Grassmann variables, θ^m and their complex conjugates $\bar{\theta}_m$ $(m = 1, ..., 8)$, on which $SU(8)$ acts linearly. We introduce the chiral derivatives

$$d^m \equiv -\frac{\partial}{\partial \bar{\theta}_m} - \frac{i}{\sqrt{2}}\theta^m \partial^+ , \quad \bar{d}_m \equiv \frac{\partial}{\partial \theta^m} + \frac{i}{\sqrt{2}}\bar{\theta}_m \partial^+ , \tag{1}$$

which satisfy canonical anticommutation relations

$$\{d^m , \bar{d}_n\} = -i\sqrt{2}\delta^m{}_n \partial^+ . \tag{2}$$

The physical degrees of freedom of $\mathcal{N} = 8$ Supergravity, the spin-2 graviton h and $\bar{h}$; eight spin-$\frac{3}{2}$ gravitinos, ψ^m and $\bar{\psi}_m$, twenty-eight vector fields

$$\overline{B}_{mn} \equiv \frac{1}{\sqrt{2}} \left(B^1_{mn} + i B^2_{mn} \right) ,$$

and their conjugates, fifty-six gauginos $\overline{\chi}_{mnp}$ and χ^{mnp}, and finally seventy real scalars $\overline{D}_{mnpq}$ appear in one superfield

$$
\begin{aligned}
\varphi(y) = {} & \frac{1}{\partial^{+2}} h(y) + i\theta^m \frac{1}{\partial^{+2}} \bar{\psi}_m(y) + i\theta^{mn} \frac{1}{\partial^+} \overline{B}_{mn}(y) \\
& - \theta^{mnp} \frac{1}{\partial^+} \overline{\chi}_{mnp}(y) - \theta^{mnpq} \overline{D}_{mnpq}(y) + i\tilde{\theta}_{mnp} \chi^{mnp}(y) \\
& + i\tilde{\theta}_{mn} \partial^+ B^{mn}(y) + \tilde{\theta}_m \partial^+ \psi^m(y) + 4\tilde{\theta} \partial^{+2} \bar{h}(y) ,
\end{aligned}
\tag{3}
$$

where the bar denotes complex conjugation, and

$$\theta^{a_1 a_2 \ldots a_n} = \frac{1}{n!} \theta^{a_1} \theta^{a_2} \cdots \theta^{a_n} , \qquad \tilde{\theta}_{a_1 a_2 \ldots a_n} = \epsilon_{a_1 a_2 \ldots a_n b_1 b_2 \ldots b_{(8-n)}} \theta^{b_1 b_2 \cdots b_{(8-n)}} .$$

The arguments of the fields are the chiral coordinates

$$y = \left(x, \bar{x}, x^+, y^- \equiv x^- - \frac{i}{\sqrt{2}} \theta^m \bar{\theta}_m \right) , \qquad x = \frac{1}{\sqrt{2}} (x_1 + ix_2) ,$$

so that φ and its complex conjugate $\overline{\varphi}$ satisfy the chiral constraints

$$d^m \varphi = 0, \qquad \bar{d}_m \overline{\varphi} = 0 , \tag{4}$$

The complex chiral superfield is related to its complex conjugate by the *inside-out constraint*

$$\varphi = \frac{1}{4\,\partial^{+4}} d^1 d^2 \cdots d^8 \, \overline{\varphi} , \tag{5}$$

in accordance with the duality condition of D^{mnpq}.

$$D^{mnpq} = \frac{1}{4!} \epsilon^{mnpqrstu} \overline{D}_{rstu}.$$

On the light-cone, the eight kinematical supersymmetries (the spectrum-generating part of the symmetry) are linearly represented by the operators q^m and $\bar{q}_m$. (We use the light-cone notation for the generators of the SuperPoincaré algebra, dividing the spinors into two two-components spinors, which are then linearly combined into complex anticommuting ones.)

$$q^m = -\frac{\partial}{\partial\bar{\theta}_m} + \frac{i}{\sqrt{2}} \theta^m \partial^+ , \qquad \bar{q}_m = \frac{\partial}{\partial\theta^m} - \frac{i}{\sqrt{2}} \bar{\theta}_m \partial^+ , \tag{6}$$

which also satisfy anticommutation relation

$$\{ q^m , \bar{q}_n \} = i\sqrt{2}\, \delta^m{}_n \, \partial^+ , \tag{7}$$

and anticommute with the chiral derivatives. Hence, their linear action on the chiral superfield

$$\delta_s \varphi(y) = \bar{\epsilon}_m q^m \varphi(y) , \tag{8}$$

where $\bar{\epsilon}_m$ is the parameter of the supersymmetry transformation, preserves chirality. The kinematical supersymmetry transformations of the physical fields are then

$$\delta_s h \;=\; 0 \,, \qquad \delta_s \overline{h} \;=\; -\,i\frac{\sqrt{2}}{4}\,\bar{\epsilon}_m\,\psi^m \,,$$

$$\delta_s \psi^m \;=\; 2\sqrt{2}\,\bar{\epsilon}_n \partial^+ B^{mn} \,, \qquad \delta_s \overline{\psi}_m \;=\; -\,\sqrt{2}\,\bar{\epsilon}_m\,\partial^+ h \,,$$

$$\delta_s B^{mn} \;=\; -\,3i\sqrt{2}\,\bar{\epsilon}_p\,\chi^{mnp} \,, \qquad \delta_s \overline{B}_{mn} \;=\; -\,2i\sqrt{2}\,\bar{\epsilon}_{[m}\overline{\psi}_{n]} \,,$$

$$\delta_s \chi^{lmn} \;=\; -\,\frac{\sqrt{2}}{3!}\,\bar{\epsilon}_k\,\partial^+ D^{klmn} \,, \qquad \delta_s \overline{\chi}_{mnp} \;=\; -\,3\sqrt{2}\,\epsilon_{[p}\,\partial^+ \overline{B}_{mn]} \,,$$

and finally

$$\delta_s \overline{D}_{klmn} \;=\; -\,4i\sqrt{2}\,\bar{\epsilon}_{[n}\,\overline{\chi}_{klm]} \,.$$

The quadratic operators

$$T^i{}_j \;=\; -\,\frac{i}{\sqrt{2}\,\partial^+}\left(q^i\bar{q}_j - \frac{1}{8}\delta^i{}_j q^k \bar{q}_k\right) \,, \qquad (9)$$

which satisfy the $SU(8)$ algebra

$$[\,T^i{}_j\,,\,T^k{}_l\,] \;=\; \delta^k{}_j\,T^i{}_l - \delta^i{}_l\,T^k{}_j \,,$$

also act linearly on the chiral superfield

$$\delta_{SU_8}\,\varphi(y) \;=\; \omega^j{}_i\,T^i{}_j\,\varphi(y) \,.$$

They constitute the R-symmetry of the theory. Similarly we have found the linear part of the representation of the remaining generators of the SuperPoincaré algebra. All generators with a minus-component are generators that take the field forward in the light-cone time and they get non-linear contributions in the interacting theory. All the three-point couplings were found already in the paper.[4] These *dynamical* generators include $P^- = H$, the Hamiltonian, where we also used the masslessness condtion $P^2 = 0$ and the other part of the supersymmetry Q_-. In our work on the $\mathcal{N} = 4$ SuperYang-Mills theory[8] we found that maximally supersymmetric theories have the unique property that

$$H \;=\; \frac{1}{2\sqrt{2}} \int d^4x\, d^4\theta\, d^4\bar{\theta}\,\, \overline{\mathcal{W}} \cdot \mathcal{W} \,, \qquad (10)$$

where

$$W = \delta_{Q_-}\,\varphi \qquad (11)$$

and $\cdot$ represents a power of $\frac{1}{\partial^+}$ to make the integral dimensionally correct. This fact was used for the $\mathcal{N} = 8$ Supergravity Theory in ref. (9) to find the four-point coupling by a trial and error method. It was found to consist of some 96 terms with no obvious relations among them. It was clear that we needed some further insight.

3. The $E_{7(7)}$ Symmetry in the Light-Cone Formulation

The $E_{7(7)}$ symmetry has been quite instrumental in the construction and the interpretation of the covariant formulation of the $\mathcal{N} = 8$ Supergravity Theory.[5] In this formulation it is a symmetry at the level of the equations of motion and it only affects the vector and the scalar particles. In our light-cone formulation it must act on the whole superfield. One could try to find a non-linear representation on the whole superfield but the method we eventually found to work was to go back to the original works[5,13] and implement the light-cone gauge condition on the vector field and then eliminate all unphysical degrees of freedom by solving their algebraic equations of motion. We could so eliminate time-derivatives in the action by non-linear field redefinitions and read off the $E_{7(7)}/SU(8)$ transformations of the new vector and scalar fields. However, we then found that these transformations do not close properly. All the fields have to transform as I alluded to above. We could include the other fields of the theory by demanding that the $E_{7(7)}/SU(8)$ transformations commute with the kinematical supersymmetries, that is

$$[\delta_q, \delta_{E/SU_8}]\varphi = 0 . \tag{12}$$

In this way we could read off the transformations for all the fields and check that the transformations close properly. Including the inhomogeneous term, the $E_{7(7)}/SU(8)$ transformation could be written in a compact way by introducing a coherent state-like representation

$$\delta_{E/SU_8}\varphi = -\frac{2}{\kappa}\theta^{ijkl}\overline{\Xi}_{ijkl} + \frac{\kappa}{4!}\Xi^{ijkl}\left(\frac{\partial}{\partial\eta}\right)_{ijkl}\frac{1}{\partial^{+2}}\left(e^{\eta\hat{\bar{d}}\partial^{+3}}\varphi\, e^{-\eta\hat{\bar{d}}\partial^{+3}}\varphi\right)\Big|_{\eta=0} + \mathcal{O}(\kappa^3) , \tag{13}$$

where

$$\eta\hat{\bar{d}} = \eta^m \frac{\bar{d}_m}{\partial^+}, \quad \text{and} \quad \left(\frac{\partial}{\partial\eta}\right)_{ijkl} \equiv \frac{\partial}{\partial\eta^i}\frac{\partial}{\partial\eta^j}\frac{\partial}{\partial\eta^k}\frac{\partial}{\partial\eta^l} .$$

We note that these $E_{7(7)}/SU(8)$ transformations do close properly to an $SU(8)$ transformation on the superfield

$$[\delta_{E/SU_8\, 1}, \delta_{E/SU_8\, 2}]\varphi = \delta_{SU(8)}\varphi .$$

It is chiral by construction $d^n\delta_{E/SU_8}\varphi = 0$, with the power of the first inverse derivative set by comparing with the graviton transformation. Hence, all physical fields, including the graviton transform under $E_{7(7)}$ and can be read off from this equation. I have left out all details of these calculations and I refer the reader to the paper[12] for all the details.

We can now extend the method to the dynamical supersymmetries, and determine the form of the interactions implied by the $E_{7(7)}$ symmetry.

3.1. *Superspace action*

The $\mathcal{N} = 8$ Supergravity action in superspace was first obtained in Ref. 14 and its LC_2 form is derived in Ref. 4 to order κ, using algebraic consistency and simplified further in Ref. 15. It is remarkably simple:

$$S = -\frac{1}{64} \int d^4x \int d^8\theta \, d^8\bar{\theta} \left\{ -\overline{\varphi} \frac{\Box}{\partial^{+4}} \varphi - 2\kappa \left(\frac{1}{\partial^{+2}} \overline{\varphi} \, \overline{\partial} \varphi \, \overline{\partial} \varphi + c.c. \right) + \mathcal{O}(\kappa^2) \right\} , \tag{14}$$

where $\Box \equiv 2 \left(\partial\bar{\partial} - \partial^+\partial^- \right)$. The light-cone superfield Hamiltonian density is then written as

$$\mathcal{H} = 2\overline{\varphi} \frac{\partial\bar{\partial}}{\partial^{+4}} \varphi + 2\kappa \left(\frac{1}{\partial^{+2}} \overline{\varphi} \, \overline{\partial} \varphi \, \overline{\partial} \varphi + c.c. \right) + \mathcal{O}(\kappa^2) . \tag{15}$$

It can be derived from the action of the dynamical supersymmetries on the chiral superfield

$$\delta_s^{dyn} \varphi = \delta_s^{dyn\,(0)} \varphi + \delta_s^{dyn\,(1)} \varphi + \delta_s^{dyn\,(2)} \varphi + \mathcal{O}(\kappa^3) , \tag{16}$$

$$= \epsilon^m \left\{ \frac{\partial}{\partial^+} \bar{q}_m \varphi + \kappa \frac{1}{\partial^+} \left(\bar{\partial} \, \bar{d}_m \, \varphi \partial^{+2} \varphi - \partial^+ \bar{d}_m \, \varphi \partial^+ \, \bar{\partial} \varphi \right) + \mathcal{O}(\kappa^2) \right\} .$$

We now require that the $E_{7(7)}/SU(8)$ commutes with the dynamical supersymmetries. Let me introduce the notation $\delta_{E_{7(7)}/SU(8)} = \boldsymbol{\delta}$.

$$[\boldsymbol{\delta}, \delta_s^{dyn}]\varphi = 0 . \tag{17}$$

This commutativity is valid only on the chiral superfield. For example, $[\boldsymbol{\delta}_1, \delta_s]\boldsymbol{\delta}_2\varphi \neq 0$, due to the non-linearity of the $E_{7(7)}$ transformation. This helps us understand how the Jacobi identity

$$([\boldsymbol{\delta}_1, [\boldsymbol{\delta}_2, \delta_s]] + [\boldsymbol{\delta}_2, [\delta_s, \boldsymbol{\delta}_1]] + [\delta_s, [\boldsymbol{\delta}_1, \boldsymbol{\delta}_2]]) \, \varphi = 0 ,$$

is algebraically consistent. In the last term the commutator of the two $E_{7(7)}/SU(8)$ transformations, $[\boldsymbol{\delta}_1, \boldsymbol{\delta}_2]$, yields an $SU(8)$ under which the supersymmetry transforms. This is precisely compensated by contributions from the first two terms.

Although the dynamical supersymmetry to order κ is already known, we rederive $\delta_s^{dyn\,(1)} \varphi$ from the commutativity between the dynamical supersymmetries and $E_{7(7)}/SU(8)$ transformations.

The inhomogeneous $E_{7(7)}$ transformations link interaction terms with different order in κ. To zeroth order, one finds

$$[\boldsymbol{\delta}^{(-1)}, \delta_s^{dyn\,(1)}]\varphi = \boldsymbol{\delta}^{(-1)} \delta_s^{dyn\,(1)}\varphi = 0 , \tag{18}$$

since $\delta_s^{dyn\,(1)} \boldsymbol{\delta}^{(-1)}\varphi = 0$. To find $\delta_s^{dyn\,(1)}\varphi$ that satisfies both the above equation and the SuperPoincaré algebra, one may start with a general form that satisfies all the commutation relations with the kinematical SuperPoincaré generators (the forms of the kinematical SuperPoincaré generators can be found in Ref. 16),

$$\delta_s^{dyn\,(1)}\varphi \propto \frac{\partial}{\partial a} \frac{\partial}{\partial b} \frac{1}{\partial^{+(m+n+1)}} \left(e^{a\hat{\bar{\partial}}} e^{b\,\epsilon\hat{\bar{q}}} \partial^{+(2+m)} \varphi \, e^{-a\hat{\bar{\partial}}} e^{-b\,\epsilon\hat{\bar{q}}} \partial^{+(2+n)} \varphi \right) \Big|_{a=b=0} ,$$

where $\hat{\bar{\partial}} = \frac{\bar{\partial}}{\partial^+}$, $\epsilon\hat{\bar{q}} = \epsilon^m \frac{\bar{q}_m}{\partial^+}$. It is not difficult to see that this form with non-negative m, n satisfies (18). The number of powers of ∂^+ can be determined by checking the commutation relation between two dynamical generators δ_{p-} (Hamiltonian variation which is derived from the supersymmetry algebra) and δ_{j-} (the boost which can also be obtained through $[\delta_{j-}, \delta_{\bar{q}}]\varphi = \delta_s^{dyn}\varphi$), yielding that the commutator between δ_{j-} and δ_{p-} vanishes only when $m = n = 0$, which leads to the the same form as (16) written in a coherent-like form

$$\delta_s^{dyn\,(1)}\varphi = \frac{\kappa}{2}\frac{\partial}{\partial a}\frac{\partial}{\partial b}\frac{1}{\partial^+}\left[e^{a\hat{\bar{\partial}}}e^{b\,\epsilon\hat{\bar{q}}}\partial^{+2}\varphi\, e^{-a\hat{\bar{\partial}}}e^{-b\,\epsilon\hat{\bar{q}}}\partial^{+2}\varphi\right]\Big|_{a=b=0}.$$

It is worth noting that this is the solution that has the least number of powers of ∂^+ in the denominator, and thus the least "non-local".

The same reasoning can be applied to higher orders in κ. To order κ, we find that commutativity

$$[\boldsymbol{\delta}^{(-1)}, \delta_s^{dyn\,(2)}]\varphi + [\boldsymbol{\delta}^{(1)}, \delta_s^{dyn\,(0)}]\varphi = 0 \tag{19}$$

requires

$$\boldsymbol{\delta}^{(-1)}\delta_s^{dyn\,(2)}\varphi \tag{20}$$

$$= \frac{\kappa}{4!}\Xi^{ijkl}\frac{1}{\partial^{+3}}\left[-\bar{d}_{ijkl}\frac{\partial}{\partial^+}\varphi\partial^{+3}\epsilon\bar{q}\varphi + 4\bar{d}_{ijk}\partial\varphi\bar{d}_l\partial^{+2}\epsilon\bar{q}\varphi - 3\bar{d}_{ij}\partial\partial^+\varphi\bar{d}_{kl}\partial^+\epsilon\bar{q}\varphi\right.$$

$$-\bar{d}_{ijkl}\frac{\epsilon\bar{q}}{\partial^+}\varphi\partial\partial^{+3}\varphi + 4\bar{d}_{ijk}\epsilon\bar{q}\varphi\,\bar{d}_l\partial\partial^{+2}\varphi - 3\bar{d}_{ij}\partial^+\epsilon\bar{q}\varphi\,\bar{d}_{kl}\partial\partial^+\varphi$$

$$+ \bar{d}_{ijkl}\frac{\partial}{\partial^{+2}}\epsilon\bar{q}\,\varphi\,\partial^{+4}\varphi - 4\bar{d}_{ijk}\frac{\partial}{\partial^+}\epsilon\bar{q}\,\varphi\,\bar{d}_l\partial^{+3}\varphi + 3\bar{d}_{ij}\partial\epsilon\bar{q}\,\varphi\,\bar{d}_{kl}\partial^{+2}\varphi$$

$$\left.+ \bar{d}_{ijkl}\,\varphi\,\partial\partial^{+2}\epsilon\bar{q}\,\varphi - 4\bar{d}_{ijk}\partial^+\varphi\,\bar{d}_l\partial\partial^+\epsilon\bar{q}\,\varphi + 3\bar{d}_{ij}\partial^{+2}\varphi\,\bar{d}_{kl}\partial\epsilon\bar{q}\varphi\right],$$

where $\epsilon\bar{q}$ denotes $\epsilon^m\bar{q}_m$, which can be written in a simpler form by rewriting it in terms of a coherent state-like form:

$$\boldsymbol{\delta}^{(-1)}\delta_s^{dyn\,(2)}\varphi \tag{21}$$

$$= \frac{\kappa}{2\cdot 4!}\Xi^{ijkl}\frac{\partial}{\partial a}\frac{\partial}{\partial b}\left(\frac{\partial}{\partial\eta}\right)_{ijkl}\frac{1}{\partial^{+3}}\left[e^{a\hat{\bar{\partial}}}e^{b\epsilon\hat{\bar{q}}}e^{\eta\hat{\bar{d}}}\partial^{+4}\varphi\, e^{-a\hat{\bar{\partial}}}e^{-b\epsilon\hat{\bar{q}}}e^{-\eta\hat{\bar{d}}}\partial^{+4}\varphi\right]\Big|_{a=b=\eta=0}.$$

To find $\delta_s^{dyn(2)}\varphi$ that satisfies (20), consider the chiral combination

$$Z_{mnpq} \equiv \left(\frac{\partial}{\partial\xi}\right)_{mnpq}\left(e^{\xi\hat{\bar{d}}}\partial^{+4}\varphi e^{-\xi\hat{\bar{d}}}\partial^{+4}\varphi\right)\Big|_{\xi=0}, \tag{22}$$

$$= \bar{d}_{mnpq}\varphi\,\partial^{+4}\varphi - 4\bar{d}_{mnp}\partial^+\varphi\,\bar{d}_q\partial^{+3}\varphi + 3\bar{d}_{mn}\partial^{+2}\varphi\,\bar{d}_{pq}\partial^{+2}\varphi.$$

The inhomogeneous $E_{7(7)}$ transformation of

$$Z^{ijkl} \equiv \frac{1}{4!}\epsilon^{ijklmnpq}Z_{mnpq},$$

has the simple form

$$\delta^{(-1)} Z^{ijkl} = \frac{1}{4!} \epsilon^{ijklmnpq} \bar{d}_{mnpq} \delta^{(-1)} \varphi \, \partial^{+4} \varphi = \frac{2}{\kappa} \Xi^{ijkl} \, \partial^{+4} \varphi , \qquad (23)$$

which leads to the solution

$$\delta_s^{dyn(2)} \varphi = \frac{\kappa^2}{2 \cdot 4!} \frac{\partial}{\partial a} \frac{\partial}{\partial b} \left(\frac{\partial}{\partial \eta} \right)_{ijkl} \frac{1}{\partial^{+4}} \left(e^{a\hat{\partial} + b\epsilon\hat{\bar{q}} + \eta\hat{d}} \partial^{+5} \varphi e^{-a\hat{\partial} - b\epsilon\hat{\bar{q}} - \eta\hat{d}} Z^{ijkl} \right) \Bigg|_{a=b=\eta=0} , \qquad (24)$$

where we have fixed the ambiguity discussed earlier by choosing the expression with the least number of ∂^+ in the denominator. This coherent state-like form is very efficient; Written out explicitly $\delta_s^{dyn\,(2)} \varphi$ consists of 60 terms.

The dynamical supersymmetry is then written in terms of the coherent state-like form.

$$\delta_s^{dyn} \varphi = \frac{\partial}{\partial a} \frac{\partial}{\partial b} \Bigg\{ e^{a\hat{\partial}} e^{b\,\epsilon\hat{\bar{q}}} \partial^+ \varphi + \frac{\kappa}{2} \frac{1}{\partial^+} \left(e^{a\hat{\partial} + b\,\epsilon\hat{\bar{q}}} \partial^{+2} \varphi e^{-a\hat{\partial} - b\,\epsilon\hat{\bar{q}}} \partial^{+2} \varphi \right)$$

$$+ \frac{\kappa^2}{2 \cdot 4!} \left(\frac{\partial}{\partial \eta} \right)_{ijkl} \frac{1}{\partial^{+4}} \left(e^{a\hat{\partial} + b\,\epsilon\hat{\bar{q}} + \eta\hat{d}} \partial^{+5} \varphi \, e^{-a\hat{\partial} - b\,\epsilon\hat{\bar{q}} - \eta\hat{d}} Z^{ijkl} \right)$$

$$+ \mathcal{O}(\kappa^3) \Bigg\} \Bigg|_{a=b=\eta=0} . \qquad (25)$$

We now use the fact, as Ananth et al[9] have shown, that the $\mathcal{N} = 8$ supergravity light-cone Hamiltonian can be written as a quadratic form (to order κ^2),

$$\mathcal{H} = \frac{1}{4\sqrt{2}} \left(\mathcal{W}_m , \mathcal{W}_m \right) \equiv \frac{2i}{4\sqrt{2}} \int d^8\theta \, d^8\bar{\theta} \, d^4x \, \overline{\mathcal{W}}_m \frac{1}{\partial^{+3}} \mathcal{W}_m ,$$

where the fermionic superfield $\mathcal{W}_m$ is the dynamical supersymmetry variation of φ

$$\delta_s^{dyn} \varphi \equiv \epsilon^m \, \mathcal{W}_m ,$$

with

$$\mathcal{W}_m = \mathcal{W}_m^{(0)} + \mathcal{W}_m^{(1)} + \mathcal{W}_m^{(2)} + \cdots .$$

Up to order κ, the Hamiltonian is simply

$$\mathcal{H} = \frac{1}{4\sqrt{2}} \left[\left(\mathcal{W}_m^{(0)} , \mathcal{W}_m^{(0)} \right) + \left(\mathcal{W}_m^{(0)} , \mathcal{W}_m^{(1)} \right) + \left(\mathcal{W}_m^{(1)} , \mathcal{W}_m^{(0)} \right) \right] , \qquad (26)$$

while the Hamiltonian of order κ^2 consists of three parts:

$$\mathcal{H}^{\kappa^2} = \frac{1}{4\sqrt{2}} \left[\left(\mathcal{W}_m^{(1)} , \mathcal{W}_m^{(1)} \right) + \left(\mathcal{W}_m^{(0)} , \mathcal{W}_m^{(2)} \right) + \left(\mathcal{W}_m^{(2)} , \mathcal{W}_m^{(0)} \right) \right] , \qquad (27)$$

where the first part was computed by Ananth et al[9]

$$\left(\mathcal{W}_m^{(1)} , \mathcal{W}_m^{(1)} \right) = i \frac{\kappa^2}{2} \frac{\partial}{\partial a} \frac{\partial}{\partial b} \frac{\partial}{\partial r} \frac{\partial}{\partial s} \int d^8\theta \, d^8\bar{\theta} \, d^4x \qquad (28)$$

$$\frac{1}{\partial^{+5}} \left(e^{a\hat{\partial} + b\hat{q}^m} \partial^{+2} \overline{\varphi} e^{-a\hat{\partial} - b\hat{q}^m} \partial^{+2} \overline{\varphi} \right) \left(e^{r\hat{\partial} + s\hat{\bar{q}}_m} \partial^{+2} \varphi e^{-r\hat{\partial} - s\hat{\bar{q}}_m} \partial^{+2} \varphi \right) \Bigg|_{a=b=r=s=0} ,$$

and the second and third parts are complex conjugate of each other. It suffices to consider

$$\left(\mathcal{W}_m^{(0)}, \mathcal{W}_m^{(2)}\right) = i\frac{\kappa^2}{4!}\frac{\partial}{\partial a}\frac{\partial}{\partial b}\left(\frac{\partial}{\partial \eta}\right)_{ijkl}\int d^8\theta\, d^8\bar{\theta}\, d^4x \tag{29}$$

$$\frac{\bar{\partial}}{\partial^+}q^m\overline{\varphi}\frac{1}{\partial^{+7}}\left(e^{a\hat{\partial}+b\,\hat{\bar{q}}_m+\eta\hat{\bar{d}}}\partial^{+5}\varphi\; e^{-a\hat{\partial}-b\,\hat{\bar{q}}_m-\eta\hat{\bar{d}}}Z^{ijkl}\right)\Bigg|_{a=b=\eta=0}.$$

Integration by parts with respect to $\bar{d}$'s and use of the inside-out constraint (5) allow for an efficient rearrangement of terms to yield the final expression

$$\left(\mathcal{W}_m^{(0)}, \mathcal{W}_m^{(2)}\right) \tag{30}$$

$$= -i\frac{\kappa^2}{4!}\frac{\partial}{\partial a}\frac{\partial}{\partial b}\int d^8\theta\, d^8\bar{\theta}\, d^4x\, \frac{\bar{\partial}}{\partial^{+4}}q^m d^{ijkl}\overline{\varphi}\left(e^{a\hat{\partial}+b\hat{\bar{q}}_m}\partial^+\overline{\varphi}\, e^{-a\hat{\partial}-b\hat{\bar{q}}_m}\frac{1}{\partial^{+4}}Z_{ijkl}\right)\Bigg|_{a=b=0}.$$

Therefore, the Hamiltonian to order κ^2 is written as

$$\mathcal{H}^{\kappa^2} = i\frac{\kappa^2}{4\sqrt{2}}\int d^8\theta\, d^8\bar{\theta}\, d^4x\, \frac{\partial}{\partial a}\frac{\partial}{\partial b} \tag{31}$$

$$\left\{\frac{1}{2}\frac{\partial}{\partial r}\frac{\partial}{\partial s}\frac{1}{\partial^{+5}}\left(e^{a\hat{\partial}+b\hat{q}}\partial^{+2}\overline{\varphi}e^{-a\hat{\partial}-b\hat{q}}\partial^{+2}\overline{\varphi}\right)\left(e^{r\hat{\bar{\partial}}+s\hat{\bar{q}}}\partial^{+2}\varphi e^{-r\hat{\bar{\partial}}-s\hat{\bar{q}}}\partial^{+2}\varphi\right)\right.$$

$$\left.-\left[\frac{1}{4!}\frac{\bar{\partial}}{\partial^{+4}}q^m d^{ijkl}\overline{\varphi}\left(e^{a\hat{\partial}+b\hat{\bar{q}}_m}\partial^+\overline{\varphi}e^{-a\hat{\partial}-b\hat{\bar{q}}_m}\frac{1}{\partial^{+4}}Z_{ijkl}\right)+c.c.\right]\right\}\Bigg|_{a=b=r=s=0}.$$

to be compared with the 96 terms of Ananth et al[9]!

4. Higher-Point Terms and Possible Counterterms

Our technique with coherent-state like expressions has given us hope that we could find also higher-point terms in the expansion in the coupling constant κ. In principle we need to find these terms for all dynamical generators as well as for the $E_{7(7)}$ transformations. The key calculations are extensions of (19). There is a good chance that we would find systematics enough to find the higher order terms for both $E_{7(7)}$ symmetry and well as for the dynamical supersymmetry transformations. We have not succeeded in doing this so far and there is one problem that we find more urgent and interesting. That is the following.

In the construction of the $\mathcal{N}=4$ SuperYang-Mills theory we could easily prove that the interaction terms in the dynamical generators are unique since the coupling constant is dimensionless and the algebra gives a unique choice for the number of derivatives in the expressions. For a gravity theory the situation is different. The coupling constant κ is dimensionfull and there is a possibility that a higher order of κ could be compensated by more derivatives in the expressions for the generators. In fact it can be shown in ordinary gravity that a possible two-loop counterterm exists,[17] which of course is a well-known result. Taking over this expression to $\mathcal{N}=8$

Supergravity by finding a superspace expression that contains the gravity result to order θ^0, we found that it should be of the form

$$\delta_{P-}\varphi \sim \bar{\varphi}\bar{\varphi} \tag{32}$$

This expression which contains six derivatives is evidently not satisfying the chiral constraint and is hence not consistent. The result that there is no three-point counterterm in supergravity has been known since the first year of supergravity.[18]

Hence the first possible counterterm must be looked for at the four-point level. Remember the condition (19). It says that the kinetic term leads uniquely to a four-point coupling with the same number of derivatives as the kinetic one. If there is another possible four-point coupling it must hence satisfy

$$[\, \boldsymbol{\delta}^{(-1)}, \, \delta_{s,ct}^{dyn\,(2)}\,]\varphi \;=\; 0, \tag{33}$$

where $\delta_{s,ct}^{dyn\,(2)}$ represents a possible counterterm for the dynamical supersymmetry, ie. a term with an even number higher than one of derivatives. This term must then transform correctly under the remaining generators of the SuperPoincaré algebra, and the dynamical ones have to be given higher counterterms too.

We can use the expression (24) as the starting point extending it with exponentials also in $\hat{\bar{\partial}}$ demanding one more $\hat{\partial}$ than the number of $\hat{\bar{\partial}}$'s. We know that it satisfies the correct commutation relations with the kinematic generators. What remains to be performed are the commutations with the dynamical generators extended to the same order in derivatives and κ. This work is in progress.[17] If we could not find any counterterms, that would be a strong indication that the theory is finite in the perturbation series. If we find a possible one, only explicit calculations of the coefficient in front would tell us if the theory is finite or not to that order.

References

1. L. Brink, J. H. Schwarz and J. Scherk, "Supersymmetric Yang-Mills Theories," Nucl. Phys. B **121**, 77 (1977); F. Gliozzi, J. Scherk and D. I. Olive, "Supersymmetry, Supergravity Theories And The Dual Spinor Model," Nucl. Phys. B **122**, 253 (1977).
2. L. Brink, O. Lindgren and B. E. W. Nilsson, "N=4 Yang-Mills Theory On The Light Cone," Nucl. Phys. B **212**, 401 (1983).
3. L. Brink, O. Lindgren and B. E. W. Nilsson, "The Ultraviolet Finiteness Of The N=4 Yang-Mills Theory," Phys. Lett. B **123**, 323 (1983).
4. A. K. H. Bengtsson, I. Bengtsson and L. Brink, "Cubic Interaction Terms For Arbitrarily Extended Supermultiplets," Nucl. Phys. B **227**, 41 (1983).
5. E. Cremmer and B. Julia, "The N=8 Supergravity Theory. 1. The Lagrangian," Phys. Lett. B **80**, 48 (1978); "The SO(8) Supergravity," Nucl. Phys. B **159**, 141 (1979).
6. F. A. Berends, W. T. Giele and H. Kuijf, "On relations between multi - gluon and multigraviton scattering," Phys. Lett. B **211**, 91 (1988).
7. H. Kawai, D. C. Lewellen and S. H. H. Tye, "A Relation Between Tree Amplitudes Of Closed And Open Strings," Nucl. Phys. B **269**, 1 (1986).
8. S. Ananth, L. Brink, S. -S. Kim and P. Ramond, "Non-linear realization of $PSU(2,2|4)$ on the light-cone," Nucl. Phys. B **722**, 166 (2005)

9. S. Ananth, L. Brink, R. Heise and H. G. Svendsen, "The N=8 Supergravity Hamiltonian as a Quadratic Form," Nucl. Phys. B **753**, 195 (2006)

10. Z. Bern, L. J. Dixon, D. C. Dunbar, M. Perelstein and J. S. Rozowsky, "On the relationship between Yang-Mills theory and gravity and its implication for ultraviolet divergences," Nucl. Phys. B **530**, 401 (1998); Z. Bern, J. J. Carrasco, L. J. Dixon, H. Johansson, D. A. Kosower and R. Roiban, "Three-Loop Superfiniteness of N=8 Supergravity," Phys. Rev. Lett. **98**, 161303 (2007)

11. M. B. Green, J. G. Russo and P. Vanhove, "Non-renormalisation conditions in type II string theory and maximal JHEP **0702**, 099 (2007); "Ultraviolet properties of maximal supergravity," Phys. Rev. Lett. **98**, 131602 (2007)

12. L. Brink, S. -S. Kim and P. Ramond, "$E_{7(7)}$ on the Light Cone," JHEP **0806**, 034 (2008) [AIP Conf. Proc. **1078**, 447 (2009)] [arXiv:0801.2993 [hep-th]].

13. B. de Wit and H. Nicolai, "N=8 Supergravity," Nucl. Phys. B **208**, 323 (1982).

14. L. Brink and P. S. Howe, "The N=8 Supergravity In Superspace," Phys. Lett. B **88**, 268 (1979).

15. S. Ananth, L. Brink and P. Ramond, "Eleven-dimensional supergravity in light-cone superspace," JHEP **0505**, 003 (2005).

16. A. K. H. Bengtsson, I. Bengtsson and L. Brink, "Cubic Interaction Terms For Arbitrary Spin," Nucl. Phys. B **227**, 31 (1983); A. K. H. Bengtsson, I. Bengtsson and L. Brink, "Cubic Interaction Terms For Arbitrarily Extended Supermultiplets," Nucl. Phys. B **227**, 41 (1983).

17. L. Brink, S. -S. Kim, to be published.

18. S. Deser, J. H. Kay and K. S. Stelle, "Renormalizability Properties Of Supergravity," Phys. Rev. Lett. **38**, 527 (1977).

GAUGE/GRAVITY DUALITY AND SOME APPLICATIONS

SPENTA R. WADIA

Tata Institute of Fundamental Research,
Homi Bhabha Road, Colaba, Mumbai 400 005, India
wadia@theory.tifr.res.in

We discuss the AdS/CFT correspondence in which space-time emerges from an interacting theory of D-branes and open strings. These ideas have a historical continuity with QCD which is an interacting theory of quarks and gluons. In particular we review the classic case of D3 branes and the non-conformal D1 brane system. We outline by some illustrative examples the calculations that are enabled in a strongly coupled gauge theory by correspondence with dynamical horizons in semi-classical gravity in one higher dimension. We also discuss implications of the gauge fluid/gravity correspondence for the information paradox of black hole physics.

1. Introduction

A look at the history of elementary particles reveals that a majority of the constituents of the Standard Model, were conceived by theory before they were experimentally established. This includes quarks that were predicted by Murray Gell-Mann[1] and George Zweig[2] to explain hadron spectra. In this note we want to discuss the elementary constituents of space-time and their interactions from which emerges a theory of gravitation.

We hope that the theoretical conceptions we dwell upon in this note contribute to a quantum theory of gravity that enables a description of phenomena like the formation and evaporation process of a black hole and basic issues in cosmology like the origin and fate of the universe. Besides being a framework to address these difficult questions, 'string theoretic' methods seem to have applications to various problems in gauge theories, fluid dynamics and condensed matter physics. This note is written with the hope that the ideas and methods of string theory are accessible to a large community of physicists.

In order to pose the question better we first consider the more familiar setting of quantum chromo-dynamics (QCD) where the quarks interact via color gluons. This theory accounts for the spectrum of hadrons, their interactions and properties of nuclei. In particular in the limit of long wave lengths the chiral non-linear sigma model[3] (another invention of Gell-Mann with Levy) describes the interactions of pions. This theory is characterized by a dimensional coupling, the pion coupling constant f_π^{-1}. In 3+1 dimensions f_π has the dimensions of $(\text{mass})^2$ and if we generalize the QCD gauge group from $SU(3)$ to $SU(N)$, then $f_\pi \sim N$. Hence in the limit of large N the pions are weakly coupled and the theory has soliton solutions with 'baryon number'. The mass of the baryon is proportional to N, the number of quark constituents.[a] In modern terminology one would say that the chiral model is 'emergent' from an underlying theory of more elementary constituents and their interactions. The phenomenon of 'emergence' occurs in complex systems in many areas of science, and also social sciences. Gell-Mann has also contributed to this area.[6]

Over the last 25 years one question that has occupied theoretical physicists is: In what sense is gravity an emergent phenomenon? What are the fundamental constituents of 'space-time' and their interactions. The question is quite akin to that of the emergence of the chiral model from QCD. If gravity is the analogue of the chiral model, where the gravitational (Newton) coupling is dimensional and akin to the 'pion coupling constant', what is the QCD analogue for gravity, from which gravity is emergent?

Perturbative string theory gave the first hint that gravity may be derived from a more microscopic theory because its spectrum contains a massless spin 2 excitation.[7,8] However real progress towards answering this question came about with the discovery of D-branes.[9]

2. D-Branes the Building Blocks of String Theory

A D-p brane (in the simplest geometrical configuration) is a domain wall of dimension p, where $0 \leq p \leq 9$. It is characterized by a charge and it couples to a $(p + 1)$ form abelian gauge field $A^{(p+1)}$, e.g. a D0 brane couples to a 1-form gauge field $A_\mu^{(1)}$, a D1 brane couples to a 2-form gauge field $A_{\mu\nu}^{(2)}$ etc. The D-p brane has a brane tension T_p which is its mass per unit volume. The crucial point is that $T_p \propto 1/g_s$. This dependence on the coupling constant (instead of g_s^{-2}) is peculiar to string theory. It has a very important consequence. A quick estimate of the gravitational field of a D-p brane gives, $G_N^{(10)} T_p \sim g_s^2/g_s \sim g_s$. Hence as $g_s \to 0$, the gravitational field goes to zero! If we stack N D-p branes on top of each other then the gravitational field of the stack $\sim N g_s$. A useful limit to study is to hold $g_s N = \lambda$ fixed, as $g_s \to 0$ and $N \to \infty$. In this limit when $\lambda \gg 1$ the stack of branes can source a solution of supergravity. On the other hand when $\lambda \ll 1$ there is a better description of the stack of D-branes in terms of open strings. A stack of D-branes interacts by the exchange of open strings very much like quarks which interact by the exchange of gluons. Fig. 1a illustrates the self-interaction of a D2-brane by the emission and absorption of an open string and Fig. 1b illustrates the interaction of 2 D2-branes by the exchange of an open string. In the infra-red limit only the lowest mode of the open string contributes and hence the stack of N D-branes can be equivalently described as a familiar $SU(N)$ non-abelian gauge theory in $p + 1$ dim.

---

[a]This theory, which emerges from QCD, had phenomenological antecedents in the theory of superconductivity in the work of Nambu and Jona-Lasinio.[4,5]

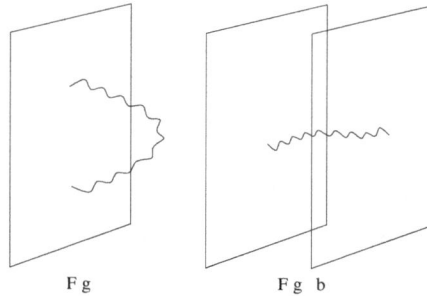

F g   F g b

Fig. 1.

3. Statistical Mechanics of D-brane Systems and Black Hole Thermodynamics

One of the earliest applications of the idea that D-branes are the basic building blocks of 'string theory' (and hence of a theory of gravity) was to account for the entropy and dynamics of certain near extremal black holes in $4 + 1$ dim. As is well known, Strominger and Vafa[10] in a landmark paper showed that the Benkenstein-Hawking entropy of these black holes is equal to the Boltzmann entropy calculated from the micro-states of a system of D1 and D5 branes

$$S_{BH} = \frac{A_h}{4G_N} = k_B ln\Omega = S_{\text{Boltzmann}}$$

This result established that black hole entropy can be obtained from the statistical mechanics of the collective states of the brane system and it provided a macroscopic basis of the first law of thermodynamics, $dS_{BH} = TdM$, where M is the mass of the black hole. Hawking radiation can be accounted for from the averaged scattering amplitude of external particles and the micro-states.[11]

4. D3 Branes and the AdS/CFT Correspondence[12–16]

We now discuss the dynamics of a large number N, of D3 branes. A D3 brane is a 3+1 dim. object. A stack of N D3 branes interacts by the exchange of open strings. In the long wavelength limit ($\ell_s \to 0$, ℓ_s is the string length), only the massless modes of the open string are relevant. These correspond to 4 gauge fields A_μ, 6 scalar fields ϕ^I ($I = 1, \cdots, 6$) (corresponding to the fact that the brane extends in 6 transverse dimensions) and their supersymmetric partners. These massless degrees of freedom are described by $\mathcal{N} = 4$, $SU(N)$ Yang-Mills theory in 3+1 dim. This is a maximally supersymmetric, conformally invariant superconformal field theory in 3+1 dimensions. The coupling constant of this gauge theory g_{YM}, is simply related to the string coupling $g_s = g_{YM}^2$. The 'tHooft coupling is $\lambda = g_s N$ and the theory admits a systematic expansion in $1/N$, for fixed λ. Further as $\ell_s \to 0$ the coupling of the D3 branes to gravitons also vanishes, and hence we are left with the $\mathcal{N} = 4$ SYM theory and free gravitons.

On the other hand when $\lambda \gg 1$, various operators of the large N gauge theory source a supergravity fluctuations in 10-dimension e.g. the energy-momentum tensor $T_{\mu\nu}$ sources the gravitational field in one higher dimension. The supergravity fields include the metric, two scalars, two 2-form potentials, and a 4-form potential whose field strength F_5 is self-dual and proportional to the volume form of S^5. The fact that there are N D3 branes is expressed as $\int_{S^5} F_5 = N$. There are also fermionic fields required by supersymmetry. It is instructive to write down the supergravity metric:

$$ds^2 = H^{-1/2}(-dt^2 + d\vec{x} \cdot d\vec{x}) + H^{1/2}(dr^2 + r^2 d\Omega_5^2)$$

$$H = \left(1 + \frac{R^4}{r^4}\right), \quad \left(\frac{R}{\ell_s}\right)^4 = 4\pi g_s N \tag{1}$$

Since $|g_{00}| = H^{-1/2}$ the energy depends on the 5th coordinate r. In fact the energy at r is related to the energy at $r = \infty$ (where $g_{00} = 1$) by $E_\infty = \sqrt{|g_{00}|}E_r$. As $r \to 0$ (the near horizon limit), $E_\infty = \frac{r}{R}E_r$ and this says that E_∞ is red-shifted as $r \to 0$. We can allow for an arbitrary excitation energy in string units (i.e. arbitrary $E_r \ell_s$) as $r \to 0$ and $\ell_s \to 0$, by holding a mass scale 'U' fixed:

$$\frac{E_\infty}{\ell_s E_r} \sim \frac{r}{\ell_s^2} = U \tag{2}$$

Note that in this limit the gravitons in the asymptotically flat region also decouple from the near horizon region. This is the famous near horizon limit of Maldacena[13] and in this limit the metric (2) becomes

$$ds^2 = \ell_s^2 \left[\frac{U^2}{\sqrt{4\pi\lambda}}\left(-dt^2 + d\vec{x} \cdot d\vec{x}\right) + 4\sqrt{4\pi\lambda}\frac{dU^2}{U^2} + \sqrt{4\pi\lambda}d\Omega_5^2\right] \tag{3}$$

This is locally the metric of $\text{AdS}_5 \times \text{S}^5$. AdS_5 is the anti-de Sitter space in 5 dim. This space has a boundary at $U \to \infty$, which is conformally equivalent to 3+1 dim. Minkowski space-time.

The AdS/CFT conjecture

The conjecture of Maldacena is that $\mathcal{N} = 4$, $SU(N)$ super Yang-Mills theory in 3+1 dim. is dual to type IIB string theory with $\text{AdS}_5 \times S^5$ boundary conditions.

The gauge/gravity parameters are related as $g_{YM}^2 = g_s$ and $R/\ell_s = (4\pi g_{YM}^2 N)^{1/4}$. It is natural to consider the $SU(N)$ gauge theory living on the boundary of AdS_5. The gauge theory is conformally invariant and its global exact symmetry $SO(2,4) \times SO(6)$, is also an isometry of $\text{AdS}_5 \times \text{S}^5$. The metric (3) has a "horizon" at $U = 0$ where $g_{tt} = 0$. It admits an extension to the full AdS_5 geometry which has a globally defined time like killing vector. The boundary of this space is conformal to $S^3 \times R^1$ and the gauge theory on the boudary is well-defined in the IR since S_3 is compact.

The AdS/CFT conjecture is difficult to test because at $\lambda \ll 1$ the gauge theory is perturbatively calculable but the dual string theory is defined in $\text{AdS}_5 \times S^5$ with $R \ll \ell_s$. On the other hand for $\lambda \gg 1$, the gauge theory is strongly coupled and hard to calculate. In this regime $R \gg \ell_s$ and the string theory can be approximated by supergravity in a derivative expansion in ℓ_s/R. It turns out that for large N and large λ, D-branes source supergravity fields $\leq$ spin 2. The gravitational coupling is given by

$$G_N \sim g_s^2 \sim \frac{\lambda^2}{N^2} \ll 1$$

Note the analogy with the constituent formula $f_\pi^{-1} \sim \frac{1}{N}$. The region $\lambda \sim 1$ is most intractable as we can study neither the gauge theory nor the string theory in a reliable way. *However since the conjecture can be verified for supersymmetric states on both sides of the duality, one assumes that the duality is true in general and then uses it to derive interesting consequences for both the gauge theory and the dual string theory (which includes quantum gravity).*

Interpretation of the radial direction of AdS

Before we discuss the duality further we would like to explain the significance of the extra dimension 'r'. Let us recast the AdS_5 metric by a redefinition: $\frac{r}{R} = e^{-\phi}$

$$ds^2 = e^{-2\phi}\left(-dt^2 + d\bar{x} \cdot d\bar{x}\right) + R^2(d\phi)^2 + R^2 d\Omega_5^2 \qquad (4)$$

The boundary in these coordinates is situated at $\phi = -\infty$. Now this metric has a scaling symmetry. For $\alpha > 0$, $\phi \to \phi + \log \alpha$, $t \to \alpha t$ and $\bar{x} \to \alpha \bar{x}$, leaves the metric invariant. From this it is clear that the additional dimension 'Re^ϕ' represents a length scale in the boundary space-time: $\phi \to -\infty$ corresponds to $\alpha \to 0$ which represents a localization or short distances in the boundary coordinates $(\bar{x}, t)$, while $\phi \to +\infty$ represents long distances on the boundary. ϕ is reminiscent of the Liouville or conformal mode of non-critical string theory, where the idea of the emergence of a space-time dim. from string theory was first seen.[17]

The AdS/CFT correspondence clearly indicates that gravity is an emergent phenomenon. What this means is that all gravitational phenomena can be calculated in terms of the correlators of the energy-momentum tensor of the gauge theory, whose microscopic constituents are D-branes interacting via open strings.

5. Black Holes and AdS/CFT

The $\mathcal{N} = 4$, super Yang-Mills theory defined on $S^3 \times R^1$ can be considered at finite temperature if we work with euclidean time and compactify it to be a circle of radius $\beta = 1/T$, where T is the temperature of the gauge theory. We have to supply boundary conditions which are periodic for bosonic fields and are anti-periodic for fermions. These boundary conditions break the $\mathcal{N} = 4$ supersymmetry, and the conformal symmetry. However the AdS/CFT conjecture continues to hold and we will discuss the relationship of the thermal gauge theory with the physics of black holes in AdS.

As we have mentioned, in the limit of large N (i.e. $G_N \ll 1$) and large λ (i.e. $R \gg \ell_s$), the string theory is well approximated by supergravity, and we can imagine considering the Euclidean string theory partition function as a path integral over all metrics which are asymptotic to AdS$_5$ space-time. (For the moment we ignore S^5).

The saddle points are given by the solutions to Einstein's equations in 5-dim. with a negative cosmological constant

$$R_{ij} + \frac{4}{R^2} g_{ij} = 0 \tag{5}$$

As was found by Hawking and Page, a long time ago, there are only two spherically symmetric metrics which satisfy these equations with AdS$_5$ boundary conditions: AdS$_5$ itself and a black hole solution.

It was shown in Ref. [15] that the 'deconfinement' phase of the gauge theory corresponds to the presence of a large black hole in AdS. The temperature of the black hole is the temperature of the deconfinement phase. The AdS/CFT correspondence says that the equilibrium thermal properties of the gauge theory in the regime when $\lambda \to \infty$ are the same as those of the black hole. This correspondence enables us to make precise quantitative statements about the gauge theory at strong coupling ($\lambda \gg 1$), using the fact that on the AdS side the calculation in gravity is semi-classical.

We list a few exact results of thermodynamics of the gauge theory at strong coupling.[15]

(i) the temperature at which the first order confinement-deconfinement transition occurs:

$$T_c = \frac{3}{2\pi R_{S^3}} \tag{6}$$

where R_{S^3} is the radius of S^3.

(ii) the free energy for $T > T_c$

$$F(T) = -N^2 \frac{\pi^2}{8} T^4$$

Here we see a typical use of the AdS/CFT correspondence calculations in the strongly coupled gauge theory ($\lambda \gg 1$) which can be done using the correspondence by using semi-classical gravity since $G_N \sim \frac{1}{N^2} \ll 1$ and $\frac{R}{\ell_s} \sim \lambda^{1/4} \gg 1$.

Conformal fluid dynamics and dynamical horizons

We have seen that the thermodynamics of the strongly coupled gauge theory in the limit of large N and large λ is calculable, in the AdS/CFT correspondence, using the thermodynamic properties of a large black hole in AdS$_5$ with horizon $r_h \gg R$. Similar results hold for a black brane, except that in this case the gauge theory in $R^3 \times S^1$ is always in the deconfinement phase since $T_c = 0$ if $R_{S^3} = \infty$, by Eq. (6).

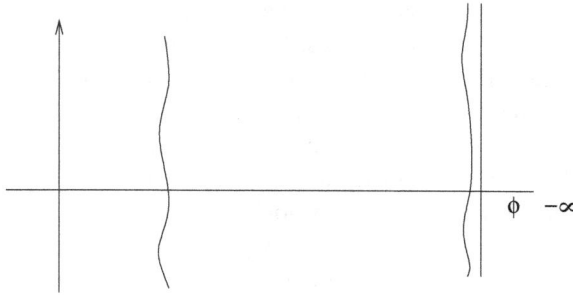

Fig. 2.

We now discuss how this correspondence can be generalized to real time dynamics in this gauge theory when both N and λ are large.

Let us generalize black brane (hole) thermodynamics to fluid dynamics. In conformal fluid dynamics the system is in local thermodynamic equilibrium over a length scale L so that $L \gg \frac{1}{T}$. In the bulk theory in one higher dim. this corresponds to a horizon that is a slowly varying (see Fig. 2) function of the boundary co-ordinates $(\vec{x}, t)$.

$$r_h \to r_h + \delta r_h(\vec{x}, t)$$

$$T \to T + \delta T(\vec{x}, t)$$

$$\frac{1}{T} \frac{\partial}{\partial x^\mu} \frac{\delta T}{T} \sim \frac{1}{LT} \ll 1$$

The ripples on the horizon of a black brane at the linerized level are analysed in terms of quasi-normal modes with a complex frequencies $\omega = \omega_R + i\omega_I$, $\omega_I \propto T$, where T is the temperature of the non-fluctuating brane. The complex frequency arises because of the presence of a horizon when we impose only 'in-falling' boundary conditions. For the dual gauge theory the quasi-normal mode spectrum implies the dissipation of a small disturbance of the fluid in a characteristic time. This is the qualitative reasoning behind the calculation of 'transport coefficients' of the gauge theory like viscosity, thermal and heat conductivity which can be done using semi-classical gravity and the Kubo formula for retarded Green's functions of the corresponding conserved currents. This important step was taken by Policastro, Son and Starinets.[18]

While linear response theory enables us to calculate transport coefficients of fluid dynamics, we now briefly discuss non-linear fluid dynamics and gravity, and indicate a remarkable connection between the (relativistic) Navier-Stokes equations of fluid dynamics and the long wavelength oscillations of the horizon of a black brane which is described by Einstein's equations of general relativity with a negative cosmological constant.

On general physical grounds a local quantum field theory at very high density can be approximated by fluid dynamics. In a conformal field theory in $3+1$ dim. we expect the energy density $\epsilon \propto T^4$, where T is the local temperature of the fluid. Hence fluid dynamics is a good approximation for length scales $L \gg 1/T$. The dynamical variables of relativistic fluid dynamics are the four velocities: $u_\mu(x)$ $(u_\mu u^\mu = -1)$, and the densities of local conserved currents. The conserved currents are expressed as local functions of the velocities, charge densities and their derivatives. The equations of motion are given by the conservation laws. An example is the conserved energy-momentum tensor of a charge neutral conformal fluid:

$$T^{\mu\nu} = (\epsilon + P)u^\mu u^\nu + P\eta^{\mu\nu} - \eta\left(P^{\mu\alpha}P^{\nu\beta}(\partial_\alpha u_\beta + \partial_\beta u_\alpha) - \frac{1}{3}P^{\mu\nu}\partial_\alpha u^\alpha\right) + \cdots \quad (7)$$

where ϵ is the energy density, P the pressure, η is the shear viscocity and $P^{\mu\nu} = u^\mu u^\nu + \eta^{\mu\nu}$. These are functions of the local temperature. Since the fluid dynamics is conformally invariant (inheriting this property from the parent field theory) we have $\eta_{\mu\nu}T^{\mu\nu} = 0$ which implies $\epsilon = 3P$. Since the speed of sound in the fluid is given by $v_s^2 = \dfrac{\partial P}{\partial \epsilon}$, $v_s = \dfrac{1}{\sqrt{3}}$ or re-instating units $v_s = \dfrac{c}{\sqrt{3}}$, where c is the speed of light in vacuum. The pressure and the viscosity are then determined in terms of temperature from the microscopic theory. In this case conformal symmetry and the dimensionality of space-time tells us that $P \sim T^4$ and $\eta \sim T^3$. However the numerical coefficients need a microscopic calculation. The Navier-Stokes equations are given by (7) and

$$\partial_\mu T^{\mu\nu} = 0 \quad (8)$$

The conformal field theory of interest to us is a gauge theory and a gauge theory expressed in a fixed gauge or in terms of manifestly gauge invariant variables is not a local theory. In spite of this (7) seems to be a reasonable assumption and the local derivative expansion in (7) can be justified using the AdS/CFT correspondence.

We now briefly indicate that the eqns.(7), (8) can be deduced systematically from black brane dynamics.[19] Einstein's equation (5) admits a boosted black-brane solution

$$ds^2 = -2u_\mu dx^\mu dv - r^2 f(br)u_\mu u_\nu dx^\mu dx^\nu + r^2 P_{\mu\nu} dx^\mu dx^\nu \quad (9)$$

where v, r, x^μ are in-going Eddington-Finkelstein coordinates and

$$f(r) = 1 - \frac{1}{r^4}$$

$$u^v = \frac{1}{\sqrt{1-\beta_i^2}}, \quad u^i = \frac{\beta^i}{\sqrt{1-\beta_i^2}} \quad (10)$$

where the temperature $T = 1/\pi b$ and the velocities β_i are all constants. This 4-parameter solution can be obtained from the solution with $\beta^i = 0$ and $b = 1$ by a boost and a scale transformation. The key idea is to make b and β^i slowly varying functions of the brane volume i.e. of the co-ordinates x^μ. One can then develop a perturbative non-singular solution of (5) as an expansion in powers of $1/LT$. Einstein's equations are satisfied provided the velocities and pressure that characterise (9) satisfy the Navier-Stokes eqns. The pressure P and viscosity η can be exactly calculated to be[19,20]

$$P = (\pi T)^4 \text{ and } \eta = 2(\pi T)^3 \tag{11}$$

Using the thermodynamic relation $dP = sdT$ we get the entropy density to be $s = 4\pi^4 T^3$ and hence obtain the famous equation of Policastro, Son and Starinets,

$$\frac{\eta}{s} = \frac{1}{4\pi} \tag{12}$$

which is a relation between viscosity of the fluid and the entropy density. Strongly coupled fluid behaves more like a liquid than a gas. Systematic higher order corrections to (7) can also be worked out.

The experiments at RHIC seem to support very rapid thermalization and a strongly coupled quark-gluon plasma with very low viscosity coefficient, $\frac{\eta}{s} \gtrsim \frac{1}{4\pi}$.

The fluid dynamics/gravity correspondence can also be used to study non-equilibrium processes like thermalization which are dual to black hole formation in the gravity theory. An important result in this study is that the thermalization time is more rapid than the expected value $\propto \frac{1}{T}$ where T is the temperature.[21]

Another important result is the connection between the area theorems of general relativity and the positivity of entropy in fluid dynamics.[23]

The fluid/gravity correspondence is firmly established for a $3+1$ dim. conformal fluid dynamics which is dual to gravity in AdS$_5$ space-time. A similar connection holds for $2+1$ dim. fluids and AdS$_4$ space-time. We shall discuss the case of non-conformal fluid dynamics in $1+1$ dim. separately. A special (asymmetric) scaling limit of the relativistic Navier-Stokes equations, where we send $v_s = \frac{c}{\sqrt{3}} \to \infty$ leads to the standard non-relativistic Navier-Stokes equations for an incompressible fluid.[22] *In summary we have a truly remarkable relationship between two famous equations of physics viz. Einstein's equations of general relativity and the Navier-Stokes equations.*

Finally it is hoped that the AdS/CFT correspondence lends new insights to the age old problem of turbulence in fluids. Towards this goal the AdS/CFT correspondence has also been established for forced fluids, where the 'stirring' term is provided by an external metric and dilaton field.[24]

6. Non-Conformal Fluid Dynamics in $1+1$ Dim. from Gravity[25,26]

The famous Policastro, Son and Starinets result (12) is indeed a cornerstone of the gauge/gravity duality. It was originally derived in the context of conformal fluid dynamics. However one suspects that the conjectured bound $\frac{\eta}{s} \geq \frac{1}{4\pi}$ may be more generally valid. We present a summary of a project of the fluid dynamics description, via the gauge/gravity duality for the case of N D1 branes at finite temperature T. The gauge theory describing the collective excitations of this system is a $1+1$ dim.

$SU(N)$ gauge theory with 16 supersymmetries. Note that this gauge theory is not conformally invariant. At high temperatures we expect the theory to have a fluid dynamics description, in terms of a 2-velocity u^μ and stress tensor

$$T^{\mu\nu} = (\epsilon + P)u^\mu u^\nu + P\eta^{\mu\nu} - \xi P^{\mu\nu}\partial_\lambda u^\lambda$$

Note that $\eta_{\mu\nu}T^{\mu\nu} = -3\xi\partial_\lambda u^\lambda$, where ξ is the bulk viscosity. The dual gravity description corresponds to 2 regimes. For $\sqrt{\lambda}N^{-2/3} \ll T \ll \sqrt{\lambda}$, the gravity dual is a classical solution corresponding to a non-external D1 brane. For $\sqrt{\lambda}N^{-1} \ll T \ll \sqrt{\lambda}N^{-2/3}$ the gravity solution corresponds to a fundamental string. Here $\lambda = g_{YM}^2 N$.

For both regimes we find the following exact answers for the strongly coupled fluid dynamics. There is exactly one gauge invariant quasi-normal mode with dispersion:

$$\omega = \frac{q}{\sqrt{2}} - \frac{i}{8\pi T}q^2 \tag{13}$$

The linearized fluid dynamics equations lead to the dispersion relation:

$$\omega = v_s q - \frac{i\xi}{2(\epsilon + P)}q^2 \tag{14}$$

$v_s^2 = \frac{\partial P}{\partial \epsilon}$ is the velocity of the sound mode. Using $v_s^2 = \frac{1}{2}$ and the relation $\epsilon + P = Ts$, we once more arrive at

$$\frac{\xi}{s} = \frac{1}{4\pi} \tag{15}$$

It is worth pointing out that (15) is valid even if we work with the geometry of D1 branes at cones over Sasaki-Einstein manifolds. Here the corresponding gauge theory is different from the gauge theory with 16 supercharges that we mentioned before. Using similar techniques we have also studied the case of the $SU(N)$ gauge theory in $1 + 1$ dim. with finite R-charge density. The dual supergravity solution is that of a non-extremal D1 brane spinning along one of the Cartan directions of $SO(8)$ which reflects the isometry of S^7 present in the near horizon geometry. In this case, besides energy transport, there is also charge transport. The transport coefficients like electrical and heat conductivity can be calculated, and the Weidemann-Franz law can be verified. Once again (14) is valid.

7. A New Term in Fluid Dynamics

The fluid dynamics of a charged fluid is described by the conserved stress tensor $T_{\mu\nu}$ and charged current J_μ. The constituent equations are (to leading order in the derivative expansion)

$$T_{\mu\nu} = P(\eta_{\mu\nu} + 4u_\mu u_\nu) - 2\eta\sigma_{\mu\nu} + \cdots \tag{16}$$

$$J_\mu = nu_\mu - DP_\mu^\nu D_\nu n. \tag{17}$$

where n is the charge density. However in the study of the charged black brane dual to a fluid at temperature T and chemical potential μ, a new term was discovered in the charged current

$$J_\mu = nu_\mu - DP_\mu^\nu D_\nu n + \zeta\ell_\mu \qquad (18)$$

$\ell_\mu = \epsilon_{\mu\nu\rho\sigma}u_\nu\omega_{\rho\sigma}$, $\omega_{\rho\sigma} = \partial_\rho u_\sigma - \partial_\sigma u_\rho$ (the vorticity). The appearance of the new voriticity induced current in (18) is directly related to the presence of the Chern-Simons term in the Einstein-Maxwell lagrangian in the dual gravity description.[27,28]

In a remarkable paper Son and Surowka[29] showed that the vorticity dependent term in (18) always arises in a relativistic fluid dynamics in which there is an anomalous axial $U(1)$ current: $\partial_\mu J_\mu^A = -\frac{1}{8}CF_{\mu\nu}F_{\rho\sigma}\epsilon_{\mu\nu\rho\sigma}$. They showed on general thermodynamic grounds that

$$\zeta = C\left(\mu^2 - \frac{2}{3}\frac{\mu^3 n}{\epsilon + P}\right)$$

where ϵ and P are the energy density and pressure, and μ is the chemical potential. If $C = 0$ one recovers the result (17) of Landau and Lifshitz. This new term may be relevant in understanding bubbles of strong parity violation observed at RHIC and generally in the description of rotating charged fluids.

8. Implications of the Gauge Fluid/Gravity Correspondence for the Information Paradox of Black Hole Physics

The Navier-Stokes equations imply dissipation and violate time reversal invariance. The scale of this violation is set by η/ρ (η is the viscosity and ρ is the density) which has the dim. of length (in units where the speed of light $c = 1$). There is no paradox here with the fact that the underlying theory is non-dissipative and time reversal invariant, because we know that the Navier-Stokes equations are not a valid description of the system for length scales $\ll \eta/\rho$, where the micro-states should be taken into account. *An immediate important implication of this fact via the AdS/CFT correspondence is that there will always be information loss in a semi-classical treatment of black holes in general relativity.* This fact raises an important question: while we understand that information loss in fluid dynamics because we know the underlying constitutent gauge theory, a similar level of understanding does not exist on the string/gravity side, because we as yet do not know the exact equations for all values of the string coupling.

9. Concluding Remarks

In this note we have reviewed the emergence, via the AdS/CFT correspondence, of a quantum theory of gravity from an interacting theory of D-branes. Besides giving a precise definition of quantum gravity in terms of non-abelian gauge theory, this correspondence turns out to be a very useful tool to calculate properties of strongly coupled gauge theories using semi-classical gravity. The correspondence of dynamical horizons and the fluid dynamics limit of the gauge theory enables calculation of transport coefficients like viscosity and conductivity. We also indicated that dissipation in fluid dynamics implies that in semi-classical gravity there will always be 'information loss'.

We conclude with a brief mention of other applications of the AdS/CFT correspondence to various problems in physics.

Condensed Matter:

The AdS/CFT correspondence offers a tool to explore many questions in strongly coupled condensed matter systems in the vicinity of a quantum critical point. It enables calculation of transport properties, non-fermi liquid behavior, quantum oscillations and properties of fermi-surfaces etc.[30-33] There is a puzzling aspect in the application of semi-classical gravity to condensed matter systems: what determines the smallness of the gravitational coupling, which in the gauge theory goes as N^{-2}?

Another interesting development is bulk superconductivity i.e. the presence of a charged scalar condensate in a black hole geometry.[34] This has interesting implications for superfluidity in the quantum field theory on the boundary.

QCD and Gauge theories:

The AdS/CFT correspondence is a powerful tool to calculate multi-gluon scattering amplitudes in $\mathcal{N} = 4$ gauge theories in 3+1 dims. This is done by relating the amplitude to the calculation of polygonal Wilson lines in a momentum space version of AdS_5.[35,36]

Even though the basic theory of the quark-gluon plasma is QCD, calculations in $\mathcal{N} = 4$ gauge theories do indicate qualitative agreement with RHIC observations. We have already remarked that the observed value of $\frac{\eta}{s}$ in (11) is in qualitative agreement with RHIC data. Another calculation of interest is that of jet quenching which corresponds to computing the drag force exerted by a trailing string attached to a quark on the boundary, in the presence of a AdS black hole.[37]

The AdS/CFT correspondence has also yielded a geometric understanding of the phenomenon of chiral symmetry breaking.[38,39]

Singularities in Quantum Gravity

The AdS/CFT correspondence provides a way to discuss the quantum resolution of the singularities of classical general relativity. One strategy would be to study the resolution of singularities that occur in the gauge theory in the $N \to \infty$ limit. Among these are singularities corresponding to transitions of order greater than two[40,41] which admit a resolution in a double scaling limit. The difficult part here is the construction of the map between the gauge theory simgularity and the gravitational singularity. A proposal in the context of the Horowitz-Polchinski cross-over was made in Ref. 42.

Acknowledgments

I would like to thank K.K. Phua and Belal Baaquie for their warm hospitality and Gautam Mandal for a critical reading of the draft and very useful discussions.

References

1. M. Gell-Mann, *A Schematic Model of Baryons and Mesons.* Hydrodynamics from the D1-brane, *Phys. Lett.* **8** 214-215 (1964).
2. G. Zweig, *An SU(3) Model for Strong Interaction Symmetry and its Breaking*, CERN Report 8419/TH.401 (January 17, 1964).
3. M. Gell-Mann and M. Levy, *The axial vector current in beta decay, Nuovo Cim.* **16** 705 (1964).
4. Y. Nambu and G. Jona-Lasinio, *Dynamical Model Of Elementary Particles Based On An Analogy With Superconductivity. Ii, Phys. Rev.* **124** 246-254 (1961).
5. A. Dhar, R. Shankar and S.R. Wadia, *Nambu-Jona-Lasinio Type Effective Lagrangian. 2. Anomalies and Nonlinear Lagrangian of Low-Energy, Large N QCD, Phys. Rev.* **D31** 3256 (1985).
6. M. Gell-Mann, *The Quark and the Jaguar: Adventures in the Simple and the Complex*, (1994).
7. T. Yoneya, *Quantum Gravity and the zero slope limit of the generalized Virasoro model, Nuovo. Cim. Lett.* **8** 951 (1973).
8. J. Sherk and J. Schwarz, *Dual Models for non-hadrons, Nucl. Phys.* **B81** 118 (1974).
9. J. Polchinski, *Dirichlet Branes and Ramond-Ramond Charges, Phys. Rev. Lett.* **75** 4724-4727 (1995).
10. A. Strominger, C. Vafa, *Microscopic Origin of the Bekenstein-Hawking Entropy, Phys.Lett.* **B379** 99-104 (1996).
11. J. David, G. Mandal and S.R. Wadia, *Microscopic Formulation of Black Holes in String Theory, Physics Reports* **369** 551-679 (2002).
12. J. Maldacena, *The large N limit of superconformal field theories and supergravity, Adv. Theo. Math. Phys.* **2** 231 (1997).
13. S. Gubser, I. Klebanov and A. Polyakov, *Gauge theory correlators from non-critical string theory, Phys. Lett.* **B428** 105-114 (1998).
14. E. Witten, *Anti-de Sitter Space and Holography, Adv. Theor. Math. Phys.* **2** 253-291 (1998).
15. E. Witten, *Anti-de Sitter Space, Thermal Phase Transition, and Confinement in Gauge Theories, Adv. Theor. Math. Phys.* **2** 505-532 (1998).
16. O. Aharony, S. Gubser, J. Maldacena, H. Ooguri, Y. Oz, *Large N field theories, string theory and gravity, Physics Reports* **323** 183-386 (2000).
17. S.R. Das, S. Naik and S.R. Wadia, *Quantization of the Liouville Mode and String Theory, Mod. Phys. Lett.* **A4** 1033 (1989); S.R. Das, A. Dhar and S.R. Wadia, *Critical Behavior In Two-Dimensional Quantum Gravity And Equations Of Motion Of The String, Mod. Phys. Lett.* **A5** 799 (1990).
18. D.T. Son and A.O. Starinets, *Viscosity, Black Holes and Quantum Field Theory, Ann. Rev. Nucl. Part. Sci.* **57** 95-118 (2007).
19. S. Bhattacharya, V. Hubeny, S. Minwalla and M. Rangamani, *Nonlinear Fluid Dynamics from Gravity, JHEP* **045** 0802 (2008).
20. R. Baier, P. Romatschke, D.T. Son, A.O. Starinets and M.A. Stephanov, *Relativistic Viscous Hydrodynamics, Conformal Invariance, and Holography, JHEP* **100** 0804 (2008).

21. S. Bhattacharyya and S. Minwalla, *Weak Field Black Hole Formation in Asymptotically AdS Spacetimes*, *JHEP* **034** 0909 (2009), e-Print: arXiv:0904.0464 [hep-th].

22. S. Bhattacharya, S. Minwalla and S.R. Wadia, *The incompressible Non-Relativistic Navier-Stokes equation from Gravity*, arXiv; 0810: 1545 [hep-th].

23. S. Bhattacharyya, V.E. Hubeny, R. Loganayagam, G. Mandal, S. Minwalla, T. Morita, M. Rangamani and H.S. Reall, *Local Fluid Dynamical Entropy from Gravity*, *JHEP* **055** 0806 (2008), e-Print: arXiv:0803.2526 [hep-th].

24. S. Bhattacharyya, R. Loganayagam, S. Minwalla, S. Nampuri, S.P. Trivedi and S.R. Wadia, *Forced Fluid Dynamics from Gravity*, e-Print: arXiv:0806.0006 [hep-th].

25. J.R. David, M. Mahato and S.R. Wadia, *Hydrodynamics from the D1-brane*, *JHEP* **042** 0904 (2009), e-Print: arXiv:0901.2013 [hep-th].

26. J.R. David, M. Mahato, S. Thakur and S.R. Wadia, *Hydrodynamics of R-charged D1-brane* (in preparation).

27. N. Banerjee, J. Bhattacharya, S. Bhattacharyya, S. Dutta, R. Loganayagam and P. Surowka, *Hydrodynamics from charged black branes*, e-Print: arXiv:0809.2596 [hep-th].

28. J. Erdmenger, M. Haack, M. Kaminski and A. Yarom, *Fluid dynamics of R-charged black holes*, *JHEP* **055** 0901 (2009), e-Print: arXiv:0809.2488 [hep-th].

29. D.T. Son and P. Surowka, *Hydrodynamics with Triangle Anomalies*, *Phys. Rev. Lett.* **103** 191601 (2009), e-Print: arXiv:0906.5044 [hep-th].

30. S. Sachdev, *Condensed matter and AdS/CFT*, e-Print: arXiv:1002.2947 [hep-th].

31. S.A. Hartnoll, *Lectures on holographic methods for condensed matter physics*, *Class. Quant. Grav.* **26** 224002 (2009), e-Print: arXiv:0903.3246 [hep-th].

32. H. Liu, J. McGreevy and D. Vegh, *Non-Fermi liquids from holography*, e-Print: arXiv:0903.2477 [hep-th].

33. C.P. Herzog (Princeton U.), *Lectures on Holographic Superfluidity and Superconductivity*, *J. Phys.* **A42** 343001 (2009), e-Print: arXiv:0904.1975 [hep-th].

34. S.S. Gubser, *Breaking an Abelian gauge symmetry near a black hole horizon*, *Phys. Rev.* **D78** 065034 (2008), e-Print: arXiv:0801.2977 [hep-th].

35. L.F. Alday and J. Maldacena, *Lectures on scattering amplitudes via AdS/CFT*, *AIP Conf. Proc.* **1031** 43-60 (2008).

36. L.F. Alday, B. Eden, G.P. Korchemsky, J. Maldacena and E. Sokatchev, *From correlation functions to Wilson loops*, e-Print: arXiv:1007.3243 [hep-th].

37. S.S. Gubser, *Using string theory to study the quark-gluon plasma: Progress and perils*, *Nucl. Phys.* **A830** 657C-664C (2009), arXiv:0907.4808 [hep-th].

38. T. Sakai and S. Sugimoto, *Low energy hadron physics in holographic QCD*, *Prog. Theor. Phys.* **113** 843-882 (2005), e-Print: hep-th/0412141.

39. A. Dhar and P. Nag, *Intersecting branes and Nambu-Jona-Lasinio model*, *Phys. Rev.* **D79** 125013 (2009), e-Print: arXiv:0901.4942 [hep-th].

40. D.J. Gross and E. Witten, *Possible Third Order Phase Transition in the Large N Lattice Gauge Theory*, *Phys. Rev.* **D21** 446-453 (1980).

41. S.R. Wadia, *N = infinity phase transition in a class of exactly soluble model lattice gauge theories*, *Phys. Lett.* **B93** 403 (1980).

42. L. Alvarez-Gaume, P. Basu, M. Marino and S.R. Wadia, *Blackhole/String Transition for the Small Schwarzschild Blackhole of AdS(5) × S^5 and Critical Unitary Matrix Models*, *Eur. Phys. J.* **C48** 647-665 (2006), e-Print: hep-th/0605041

QUANTUM MECHANICS AND FIELD THEORY WITH
MOMENTUM DEFINED ON AN ANTI-DE-SITTER SPACE

MYRON BANDER

Department of Physics and Astronomy,
University of California,
Irvine, CA 92717, USA
mbander@uci.edu

Relativistic dynamics with energy and momentum restricted to an anti-de-Sitter space is presented. Coordinate operators conjugate to such momenta are introduced. Definition of functions of these operators, their differentiation and integration, all necessary for the development of dynamics is presented. The resulting algebra differs from the standard Heisenberg one, notably in that the space-time coordinates do not commute among each other. The resulting time variable is discrete and the limit to continuous time presents difficulties. A parallel approach, in which an overlap function, between position and momentum states, is obtained from solutions of wave equations on this curved space are also investigated. This approach, likewise, has problems in the that high energy behavior of these overlap functions precludes a space-time definition of action functionals.

1. Introduction

Dynamics on space-time manifolds more general than flat Minkowski space leads to restrictions on the corresponding momentum space. For example, placing coordinate space on a periodic lattice forces momenta to a hyper-torus $S_1 \times \cdots \times S_1$. In general such a construction breaks Lorentz symmetry. In this work we will pursue an opposite approach. We consider energy and momenta to be defined on a space whose isometries include the Lorentz group and in turn investigate the properties of the corresponding position operators. Specifically, we consider energy and momenta defined on an anti-de-Sitter (AdS) space.[1] The full isometry group is $O(2,3)$ which, manifestly, contains the $O(1,3)$ Lorentz group. The group $O(2,3)$ replaces the Poincaré group, the isometry group of Minkowski space. We loose translation invariance in return for invariance under four additional boost-like transformations.

The problem now becomes one of identifying the corresponding coordinates. Two approaches are pursued, both based on relations between time-space and energy-momentum in Minkowski space; both approaches have problems preventing further development. In the first approach we note that in flat space the coordinates $(t, \vec{x})$ are operators that translate momenta; as in going to AdS space we lose translation invariance we relate the position operators to the four, aforementioned boosts. The resulting commutation relations among the eight position and momentum operators differ from the Heisenberg algebra, especially in that the position operators do not commute among themselves. Functions of such operators can be introduced and we can define differentiation of these. Although it is not obvious how to introduce integration over functions of noncommuting operators, it can be done and the resulting integrals have desired properties. With differentiation and integration procedures in place we are able to define an action integral for a dynamical system. Some of the consequences of this formulation are:

- A lower limit on the localizabilty of wave packets,
- An upper limit on possible masses of particles.
- Time, instead of being continuous, is discrete.

In the limit where the curvature of the AdS space goes to zero we recover Minkowski dynamics, with one exception. With n being the discrete time, there are states whose time evolution approaches $(-1)^n \exp(-\imath Et)$ and thus do not have a reasonable continuum limit. *A resolution of this problem is lacking.*

In the second approach states labeled by definite time and position are defined via the overlap function $\langle t, \vec{x} | p_0, \vec{p} \rangle$. In flat space this function, $\exp(-ip_\mu x^\mu)$ is a solution of the wave equation on momentum space with the coordinates labeling the solutions. We try the same approach of for momenta in AdS space. The problem that arises in this approach is that high energy and momentum behavior of these overlap functions precludes the definition of a position space action. We cannot even obtain a position space wave equation corresponding to $p_\mu p^\mu - m^2 = 0$. Again this problem is unresolved.

Restricting discussion to nonrelativistic dynamics, in which the energy (and time) are treated separatley results in only the spacial momenta being treated as operators. Discussion of this and a proof of the limit of the size of wave packets is presented in the Appendix.

2. Geometry of Anti-de-Sitter Space

We consider four dimensional energy-momentum $(p_0, \vec{p})$ on a an anti-de-Sitter (AdS) hyper-surface embedded in a flat five dimensional Minkowski space $(p_\tau, p_0, \vec{p})$ subject to the constraint

$$p_\tau^2 + p_0^2 - \vec{p} \cdot \vec{p} = M^2 \,. \tag{1}$$

It is convenient to describe this surface using coordinates $\vec{p}$ and ω related to p_τ and to p_0 by

$$p_\tau = \sqrt{\vec{p} \cdot \vec{p} + M^2} \cos \omega \,,$$
$$p_0 = \sqrt{\vec{p} \cdot \vec{p} + M^2} \sin \omega \,. \tag{2}$$

With these coordinates the invariant energy-momentum volume on this AdS space takes a simple form

$$d^5 p \, \delta(p^2 - M^2) = d\vec{p} \, d\omega \,; \tag{3}$$

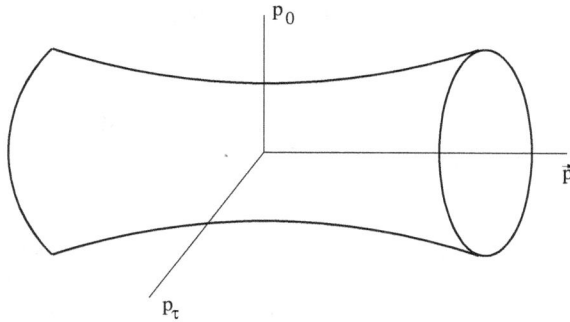

Fig. 1. AdS space embedded in 5-d Minkowski space. he line marked $\vec{p}$ represents three dimensional momenta.

This parametrization makes it is easy to note that energy-momentum restricted to this curved space places a limit on the mass of any state

$$m^2 = p_0^2 - \vec{p} \cdot \vec{p} = M^2 \sin^2 \omega - (\vec{p} \cdot \vec{p}) \cos^2 \omega \,, \tag{4}$$

and thus is always less than M^2. There is no bound on the energy or momenta, only on the mass of any state.

3. Space Coordinates

Placing energy and momenta on an AdS manifold raises the question of how to introduce space and time operators. We do this in analogy with a procedure valid for the ordinary situation of energy-momenta in a flat Minkowski space. In that case the full isometry group of the energy-momentum manifold (not necessarily of any dynamical equations) is the Poincaré group consisting of the Lorentz transformations and momentum translations generated by

$$x^\mu = -i \frac{\partial}{\partial p_\mu} \,; \tag{5}$$

in the present case the full isometry group is the five dimensional anti-de-Sitter group consisting, in addition to Lorentz transformations of the $(p_0, \vec{p})$ subspace, the four Lorentz transformations connecting p_τ with $(p_0, \vec{p})$

$$K^\mu = \sqrt{p_\tau}[-i \frac{\partial}{\partial p_\mu}]\sqrt{p_\tau} \,. \tag{6}$$

Using Eq.(1) to write p_τ in terms of $\vec{p}$ and p_0, it is straightforward to check that the $SO(3,1)$ Lorentz operators $M^{\mu\nu} = i(p^\mu \frac{\partial}{\partial p\nu} - p^\nu \frac{\partial}{\partial p_\mu})$ and the K^μ's generate the desired $O(2,3)$ AdS group with the commutation relations

$$[K^\mu, K^\nu] = -iM^{\mu\nu} \,. \tag{7}$$

The minus sign in the above is crucial as it distinguishes this set of operators as forming the algebra of the aforementioned $O(2,3)$ group, rather than the $O(1,4)$ group that a plus sign would have yielded.

In analogy with Eq.(5) we postulate the following space-time operators,

$$X^\mu = \frac{K^\mu}{M} \,.$$

(8)

Again replacing p_τ by $(M^2 + \vec{p} \cdot \vec{p} - p_0^2)^{\frac{1}{2}}$, we obtain the following coordinate operators:

$$X^\mu = \frac{1}{M}(M^2 + \vec{p} \cdot \vec{p} - p_0^2)^{\frac{1}{4}} \left(-i\frac{\partial}{\partial p_\mu}\right)(M^2 + \vec{p} \cdot \vec{p} - p_0^2))^{\frac{1}{4}} \,.$$

(9)

In the limit $M \to \infty$ these position operators go over to the usual ones in Eq.(5). It is amusing to note that the identifications in Eq.(8) together with the commutation relations in Eq.(7) reproduce the spacial noncommuting quantum mechanics originally introduced by Snyder[2] in 1946.

4. Hilbert Space and Modified Heisenberg Algebra

We shall be working primarily in the Hilbert space consisting of eigenstates of the operators p_0 and $\vec{p}$, labeled as $|p_0, \vec{p}\rangle$; note: p_0 and $\vec{p}$ are not constrained by any mass shell condition. Using the parametrization of Eq.(2) the inner product of these states is

$$\langle p_0', \vec{p}' | p_0, \vec{p} \rangle = \delta(\omega' - \omega)\delta(\vec{p} - \vec{p}') \,.$$

(10)

The Heisenberg algebra of momenta and the coordinate operators defined in Eq.(9) is modified from the usual one to

$$[p_\mu, X_\nu] = ig_{\mu\nu}\frac{p_\tau}{M} \,,$$

$$[X_\mu, X_\nu] = -i\frac{M_{\mu\nu}}{M^2} \,.$$

(11)

Again, in the limit $M \to \infty$ we recover the usual commutation relations. As mentioned earlier, the space-space commutator is the one discussed in Ref. [2].

5. Functions of Position Operators and Differentiation of These

A function $f(X_\mu)$ of the operators introduced in Eq.(8) corresponding to one of the ordinary position operators, $f(x_\mu)$ can be obtained by assuming they have the same Fourier transforms. Namely

$$f(x_\mu) = \int d^4q \tilde{f}(q)e^{iq \cdot x}$$

leads to the suggestion that we define the corresponding $f(X)$ as

$$f(X_\mu) = \int d^4q \tilde{f}(q)e^{iq \cdot X} \,.$$

(12)

However, for technical reasons to which we shall soon return (see discussion towards the end of Sect.(6)), we will modify Eq.(12); we first introduce a vector Q_μ related to q_μ,

$$Q_\mu = \frac{q_\mu}{M}\arcsin\left(\frac{q}{M}\right) \,,$$

(13)

The above definition is valid for q timelike; with the arcsin going over to an arcsinh when q is spacelike. We note that for small q/M $Q_\mu \to q_\mu$. With this definition of Q_μ change Eq.(12) to

$$f(X_\mu) = \int d^4 q \tilde{f}(q) e^{iQ \cdot X} . \tag{14}$$

The derivative of $f(X)$ is, as expected, defined as as

$$\frac{\partial f(X)}{\partial X_\mu} = -i[p_\mu, f(X)] . \tag{15}$$

6. Integration of Functions of the Operators X

For many purposes, both in quantum mechanics and in field theory we need to define an "integral" over the operator X_μ. Primarily, we want to be able to define an action whose variation will yield appropriate equations of motion. With this in mind we will abstract from the definition of ordinary integration the steps needed to carry over this procedure to functions, as defined previously, of the noncommuting specie-time coordinates, X_μ. For ordinary functions we can use the Fourier transforms, $\tilde{f}_i(q_\mu)$ of $f_i(x_\mu)$ to obtain the integral of

$$\int d^4 x f_1(x) \cdots f_N(x) = (2\pi)^4 \int d^4 q_1 \cdots d^4 q_N \tilde{f}_1(q_1) \cdots \tilde{f}_N(q_N \delta^4(q_1 + \cdots q_n) . \tag{16}$$

At this point we are left with the problem of finding the analog of the δ function in the above valid for our coordinates. Noting that the position operator acts as a translation operator on momentum states, $\exp(iq \cdot x)|p\rangle = |p + q\rangle$ allows us to represent the delta function in Eq. (16) as

$$\delta^4(q_1 + \cdots + q_N) = \langle p_0, \vec{p}|e^{i(q_1 + \cdots + q_N) \cdot x}|p_0, \vec{p}\rangle , \tag{17}$$

where $|p_0, \vec{p}\rangle$ is any state. Carrying this over to the representation of functions of X_μ as given in Eq.(14) yields

$$\text{``} \int d^4 X \text{''} f_1(X) \cdots f_N(X) = (2\pi)^4 \int d^4 q_1 \cdots d^4 q_N \tilde{f}_1(q_1) \cdots \tilde{f}_N(q_N) \times$$
$$\langle p_0, \vec{p}|e^{iQ_1 \cdot X} \cdots e^{iQ_N \cdot X}|p_0, \vec{p}\rangle . \tag{18}$$

As mentioned, the $(p_0, \vec{p})$ can refer to any state; for most calculations it is convenient to take the above matrix elements in the state $|0, \vec{0}\rangle$. the AdS symmetry insures that this definition is independent of the choice of the state $(p_0, \vec{p})$.

Some properties of this integration prescription are:

(i) With derivatives defined by Eq.(15) we find

$$\text{``} \int d^4 X \text{''} \partial_\mu f(X_\mu) = 0 .$$

(ii)

$$\text{``} \int d^4 X \text{''} e^{iq^1 \cdot X} e^{-iq^2 \cdot X} = \delta(q_\mu^1 - q_\mu^2) .$$

Had we used q_μ instead of Q_μ, Eq.(13), in the definition of $f(X)$, Eq. (14), the right hand side of item(ii), above, would have been multiplied by $q/[M\sin(q/M))$.

7. Translation Invariance, or Lack Thereof

The modified Heisenberg algebra, Eq. (11) precludes having a unitary operator shifting the position operator X. Using the momentum operator produces

$$e^{ip\cdot a} X_\mu e^{-ip\cdot a} = X_\mu + a_\mu \times \frac{p_\tau}{M} \,. \tag{19}$$

This should come as no surprise as the isometry group of our space is the de Sitter group consisting of Lorentz transformations and the four boosts K_μ, Eq. (6, involving the τ direction, and not the Poincaré group consisting of Lorentz transformations and translations.

The fact that the different components of the position operators do not commute puts a limit on localizing wave packets. It is straightforward to show, see Appendix A, that the expectation value of $X_1^2 + X_2^2 + X_3^2$ in any packet must exceed $1/M^2$.

Integrals of products of more than two fields require the evaluation of matrix elements of the form $\langle 0, \vec{0} \| exp(iQ_1 \cdot X) \cdots \| exp(iQ_N \cdot X) |0, \vec{0}\rangle$; the results are complicated and no closed expression is available. The order of the exponentials cannot, in general, be reversed; this is another indication of the noncommutativity of the operators X_μ.

8. Discreteness of Time

We may diagonalize one of the space-time coordinates and we choose it to be X_0. In the $\omega, \vec{p}$ parametrization, Eq. (2), it takes a simple form

$$X_0 = \frac{-i}{M} \frac{\partial}{\partial \omega} \,; \tag{20}$$

the eigenvalues of this time variable are discrete, $t = n/M$, with n integer. It is characteristic of noncommuting space-time coordinates to result in a discrete time.[3,4]

In the $M \to \infty$ limit time goes over to a continuum limit. With time discrete we expect the energy interval to be finite for a fixed $\vec{p}$ and indeed Eq. (2) shows that $|E| \leq \sqrt{\vec{p} \cdot \vec{p} + M^2}$. Parametrizing ω as $\omega = E/M$ leads to, in the large M limit, the identification $p_0 = E$. Subsequently we will encounter problems with this interpretation.

9. Field Theory

We shall try a naive procedure to set up a field theory where the fields $\phi(X)$ are functions of the operators X_μ by postulating action functionals for these fields. For a free field with mass μ action is taken to be

$$S_F[\phi(X)] = \text{``} \int d^4X\text{''} \left\{ -[p_\nu, \phi^\dagger(X)][p^\nu, \phi(X)] - \mu^2\phi^\dagger(X)\phi(X) \right\}$$

$$= \langle 0\vec{0}| - [p_\nu, \phi^\dagger(X)][p^\nu, \phi(X)] - \mu^2\phi^\dagger(X)\phi(X)|0, \vec{0}\rangle . \tag{21}$$

For $\phi(X)$ of the form $\phi(X) = \int d^3p\tilde{\phi}(q)\exp[iQ \cdot X]$ we readily obtain the mass shell condition $q^2 - \mu^2 = 0$ (q and Q are related by Eq. (13)) Parametrizing q_μ as in Eq(2), namely, $q_0 = \sqrt{\vec{q} \cdot \vec{q} + M^2}\sin\omega$, this mass condition translates to

$$\sin\omega = \pm\frac{\sqrt{\vec{p} \cdot \vec{p} + \mu^2}}{X} \tag{22}$$

For M large we obtain four solutions

$$\omega = \pm\sqrt{\vec{p} \cdot \vec{p} + \mu^2}/M , \tag{23}$$

$$\omega = \pm[\pi - \sqrt{\vec{p} \cdot \vec{p} + \mu^2}/M] .$$

As the time evolution of the states is

$$|S; t = n\rangle = |S; t = 0\rangle e^{i\omega n} , \tag{24}$$

we have to interpret these four solutions. The $\pm$ indicates the usual positive and negative frequencies; the first set goes over to the usual time dependence $\exp(-iEt)$ while the second one has a discrete time propagation of the form $(-1)^n\exp(-iEt)$ which has no smooth continuum limit. *An interpretation of this behavior is lacking at present and this is one of the problems we mentioned in the Introduction.*

10. Space-time Coordinates

In order to resolve the problem brought up at the end of the last section and for possible computational simplifications we shall try a different method of introducing coordinates appropriate to momenta on an AdS space. Again we shall use analogies with such procedures in flat Minkowski space as a guide. This time, however, rather then studying space-time operators we will look for states that correspond, in the $M \to \infty$ limit, to the usual ones, $|x_0, \vec{x}\rangle$. As the X_μ's do not commute, we cannot look for simultaneous eigenstates of these. For flat energy-momentum we can relate momentum and position eigenstates by the overlap

$$\langle x_0, \vec{x}|p_0, \vec{p}\rangle = \frac{1}{(2\pi)^2}e^{-i\vec{p}\cdot\vec{x}} . \tag{25}$$

The states $|x_0, \vec{x}\rangle$ are eigenstates of the commuting operators x_μ. As mentioned earlier, in the present situation with the operators X_μ not commuting we cannot define the analogous state $|X_0, \vec{X}\rangle$. The question we will address in this section is whether we can still obtain a version of the right hand side of Eq. (25) and thus define a state $|X_0, \vec{X}\rangle$ as $\int d^4p|p_0, \vec{p}\rangle\langle p_0, \vec{p}||X_0, \vec{X}\rangle$.

Eq. (25) or its spherical coordinate version,

$$\langle x_0; x, \theta, \phi | p_0; p, l, m \rangle = e^{-ip_0x_0} \frac{1}{\sqrt{2\pi}} j_l(pr) Y_{l,m}(\theta, \phi) \,, \tag{26}$$

is an eigenfunction of the wave equation on momentum space with (x_0, x, l, m) or (x_0, x, θ, ϕ) labeling the solutions,

$$[\partial_{p_0}\partial_{p_0} - \partial_{p_i}\partial_{p_i}] e^{-ip_0x_0} j_l(px) Y_{lm}(\theta, \phi) = (-x_0^2 + x^2) e^{-ip_0x_0} j_l(px) Y_{lm}(\theta, \phi) \,. \tag{27}$$

We are thus led to look for eigenfunctions of

$$g^{\mu\nu} \frac{\partial}{\partial p_\mu} \frac{\partial}{\partial p_\nu} = \bar{K}_\mu K_\mu - M_{\mu\nu} M^{\mu\nu} \,. \tag{28}$$

$g^{\mu\nu}$ is the AdS metric, $ds^2 = (\vec{p} \cdot \vec{p} + M^2) d\omega^2 - d\vec{p} \cdot d\vec{p}$; The right hand side in Eq. (28) is a Casimir operator for the anti-de-Sitter group.

Explicitly this wave equation ,written using the energy-momentum coordinates $\vec{q} = \vec{p}/M$ and ω, is

$$-g^{\mu\nu} \frac{\partial}{\partial p_\mu} \frac{\partial}{\partial p_\nu} = \frac{\partial}{\partial q}(1 + q^2)\frac{\partial}{\partial q} + (\frac{2}{q} + 4q)\frac{\partial}{\partial q} - \frac{\vec{L}^2}{q^2} - \frac{1}{1 + q^2}\frac{\partial^2}{\partial \omega^2}$$

$$= M^2 x^2 \,. \tag{29}$$

with solutions

$$Z_{\lambda, l, m:n}(q, \hat{q}; \omega) = B^{l,n}_{-\frac{1}{2}+i\lambda}(iq) Y_{l,m}(\hat{q}) e^{-in\omega} \,, \tag{30}$$

where the functions $B^{l,n}_{-\frac{1}{2}+i\lambda}(iq)$ are related to the Gegenbauer polynomials[5] and the parameter $\lambda = M^2 x \cdot x$ can be real, implying space-like x or equal to iN, with $N \le (n-1)$ integer for time-like x.

Summarizing these results we shall try to obtain a local field theory for states $|x_\mu\rangle$ related to the momentum ones by

$$\langle n; x, l, m | \omega, \vec{p} \rangle = Z_{\lambda, l, m:n}(Mq, \hat{q}; \omega) \,. \tag{31}$$

In the limit $M \to \infty$ Eq. (31) approaches Eq. (25).

10.1. *Problem*

We would like to investigate analog of the Klein-Gordon or the free field equation in the states defined by Eq. (31),

$$\langle n(1q'; x', l', m' | \left[(\vec{p} \cdot \vec{p} + M^2) \sin^2 \omega - (\vec{p} \cdot \vec{p}] \right] |x_0; x, l, m \rangle \,. \tag{32}$$

The problem is that due to the large q behavior of the functions $B^{l,n}_{-\frac{1}{2}+i\lambda}(iq)$, Eq. (30), these matrix elements do not converge, not even to Dirac delta functions or to their derivatives. The easiest way to see this problem is to look in detail at the case of (1 + 1) dimensions where Eq. (29) takes on a simpler form

$$-g^{\mu\nu} \frac{\partial}{\partial p_\mu} \frac{\partial}{\partial p_\nu}[(1+1)\mathrm{dim}] = \frac{\partial}{\partial q}(1 + q^2)\frac{\partial}{\partial q} - \frac{1}{1 + q^2}\frac{\partial^2}{\partial \omega^2} \,. \tag{33}$$

The eigenfunctions of the above are Legendre functions[5] of imaginary argument; smooth behavior at $q = 0$ restricts them to the form

$$\langle n; x|\omega; Mq\rangle = \begin{cases} P^n_{-\frac{1}{2}+i\lambda}(iq)\,; M^2 x^2 = \lambda^2 - \frac{1}{4};\, \text{x space-like}, \\ Q^n_{-\frac{1}{2}-N}(iq)\, N \le (n-1)\,; M^2 x^2 = N^2 + \frac{1}{4};\, \text{x timelike}. \end{cases} \quad (34)$$

As the large q behavior of $P^n_{-\frac{1}{2}+i\lambda}(iq)$ is $q^{-\frac{1}{2}+i\lambda}$, the matrix elements of q, q^2, etc. do not converge. Again, the resolution of this problem is unclear.

Acknowledgments

This article is based on a talk at a symposium celebrating the 80th birthday of Professor Murray Gell-Mann held at Nanyang Technical University, Sinapore, 24th 26th February, 2010. I wish to thank Professor Harald Fritzsch for organizing this meeting and for inviting me to give this talk.

Appendix A. Nonrelativistic (Three Dimensional) Quantum Mechanics

A simpler application of the ideas discussed in this article can be used to study the case where only the three momenta are placed on a de Sitter space, which in this case may be viewed as a surface embedded in a (3+1) dimensional Minkowski space with coordinates $(p_\tau, \vec{p})$ subject to the constraint

$$p_\tau^2 - \vec{p} \cdot \vec{p} = M^2\,; \quad (A.1)$$

The energy coordinate, p_0 ranges over the full interval, $-\infty \le p_0 \le \infty$. The operators conjugate to $p_0, \vec{p}$ are t and, using eq. (6) and (8) as a guide,

$$X_i = \frac{1}{M}(\vec{p} \cdot \vec{p} + M^2)^{\frac{1}{4}}\left(i\frac{\partial}{\partial p_i}\right)(\vec{p} \cdot \vec{p} + M^2)^{\frac{1}{4}}\,. \quad (A.2)$$

This time The Heisenberg algebra is modified to

$$[p_i, X_j] = -i\delta_{ij}\frac{p_\tau}{M}\,.$$

$$\quad (A.3)$$

$$[X_i, X_j] = -i\frac{M_{ij}}{M}\,,$$

where M_{ij} is the angular momentum. Again the minus sign in front of the M_{ij} distinguishes this algebra as that of the (1,3) Lorentz group rather than the SO(4) rotation group.

As expected, the noncommutativity of the X's prevents a localization of wave packets. The extent to which a packet may be localized is controlled by the eigenvalues of X^2 and a lower bound on such eigenvalues may be obtained by noting that the $SO(1,3)$ Casimir operator $\mathcal{K}^2 - J^2$ equals $\rho^2 - j_0^2 + 1$[6] for representations labeled by (ρ, j_0), with real $\rho \ge 0$, and with all angular momenta in the representation having values greater than j_0. As $X^2 = \left(\mathcal{K}^2 - J^2 + J^2\right)/M^2$, its eigenvalues are $\left[\rho^2 + 1 - j_0^2 + j(j+1)\right]/M^2$, with $j \geq j_0$; thus we find that $X^2 \geq 1/M^2$ and wave packets cannot be localized to better than $1/M$.

Quantum mechanics for one particle in an external potential $V(X)$ involves the operator eigenvalue equation

$$E = \frac{1}{2m}\vec{p} \cdot \vec{p} + V(X), \tag{A.4}$$

while the analogous two body problem requires a bit more care. Due to noncummutativity of the coordinate components we cannot follow the usual procedure and change coordinates from $X^{(1)}$ and $X^{(2)}$ to relative and center of mass ones; these have to be introduced from the beginning. With the usual definitions of relative and center of mass coordinates, $\vec{p}_{\mathrm{rel}} = (m^{(2)}\vec{p}^{(1)} - m^{(1)}\vec{p}^{(2)})/(m^{(1)} + m^{(2)})$, $\vec{x}_{\mathrm{rel}} = \vec{x}^{(1)} - \vec{x}^{(2)}$ and $\vec{p}_{\mathrm{cm}} = \vec{p}^{(1)} + \vec{p}^{(2)}$, $\vec{x}_{\mathrm{cm}} = (m^{(1)}\vec{x}^{(1)} + m^{(2)}\vec{x}^{(2)})/(m^{(1)} + m^{(2)})$, we define

$$X_i^{\mathrm{rel}} = \frac{i}{M}\sqrt{\vec{p}_{\mathrm{rel}} \cdot \vec{p}_{\mathrm{rel}} + M^2}\left(\frac{\partial}{\partial p_i^{(1)}} - \frac{\partial}{\partial p_i^{(2)}}\right)\sqrt{\vec{p}_{\mathrm{rel}} \cdot \vec{p}_{\mathrm{rel}} + M^2}, \tag{A.5}$$

$$X_i^{\mathrm{cm}} = \frac{i}{M(m^{(1)} + m^{(2)})}\sqrt{\vec{p}_{\mathrm{cm}} \cdot \vec{p}_{\mathrm{cm}} + M^2}\left(m^{(1)}\frac{\partial}{\partial p_i^{(1)}} + m^{(2)}\frac{\partial}{\partial p_i^{(2)}}\right)\sqrt{\vec{p}_{\mathrm{cm}} \cdot \vec{p}_{\mathrm{cm}} + M^2}.$$

A direct computation shows that these relative and center of mass variables commute and the coordinates within each class obey the commutation relations of eq. (A.3) and have the desired limit for large M. From the start we would formulate a two body problem as

$$H = \frac{p^{(1)2}}{2m^{(1)}} + \frac{p^{(2)2}}{2m^{(2)}} + V(X^{\mathrm{rel}}). \tag{A.6}$$

The use of these relative coordinated may be extended to many body situations.

References

1. Part of the material presented here is based on M. Bander, Phys. Rev. D **75**, 105010 (2007) [arXiv:hep-th/0701253].
2. H. S. Snyder, Phys. Rev. **71**, (1947) 38.
3. A. P. Balachandran, T. R. Govindarajan, A. G. Martins and P. Teotonio-Sobrinho, JHEP **0411**, 068 (2004) [arXiv:hep-th/0410067].
4. M. Chaichian, A. Demichev, P. Presnajder and A. Tureanu, Eur. Phys. J. C **20**, 767 (2001) [arXiv:hep-th/0007156].
5. I.S.Gradshteyn and I.M.Ryzyk Table of Integrals,Series and Products (ed. A.Jeffrey), 6th ed. (Academic Press (San Diego), 2000).
6. I. M. Gelfand, R. A. Minlos, and Z. Ya. Shapiro, Representations of the Rotation and Lorentz Group and their Applications (Pergamon Press, Inc., Oxford, 1963), translated by G. Cummins and T. Boddington from Russian (Fizmatgiz, Moscow, 1958).

GAUGE SYMMETRY IN PHASE SPACE CONSEQUENCES FOR PHYSICS AND SPACETIME

ITZHAK BARS

Department of Physics and Astronomy, University of Southern California,
Los Angeles, CA 90089-0484, USA

Gell-Mann: Anything which is not forbidden is compulsory!
This paper is dedicated to Murray Gell-Mann for his 80th birthday

Position and momentum enter at the same level of importance in the formulation of classical or quantum mechanics. This is reflected in the invariance of Poisson brackets or quantum commutators under canonical transformations, which I regard as a global symmetry. A gauge symmetry can be defined in phase space (X^M, P_M) that imposes equivalence of momentum and position for every motion at every instant of the worldline. One of the consequences of this gauge symmetry is a new formulation of physics in spacetime. Instead of one time there must be two, while phenomena described by one-time physics in 3+1 dimensions appear as various "shadows" of the same phenomena that occur in 4+2 dimensions with one extra space and one extra time dimensions (more generally, d+2). The 2T-physics formulation leads to a unification of 1T-physics systems not suspected before and there are new correct predictions from 2T-physics that 1T-physics is unable to make on its own systematically. Additional data related to the predictions, that provides information about the properties of the extra 1-space and extra 1-time dimensions, can be gathered by observers stuck in 3+1 dimensions. This is the probe for investigating indirectly the extra 1+1 dimensions which are neither small nor hidden. This 2T formalism that originated in 1998 has been extended in recent years from the worldline to field theory in d+2 dimensions. This includes 2T field theories that yield 1T field theories for the Standard Model and General Relativity as shadows of their counterparts in 4+2 dimensions. Problems of ghosts and causality in a 2T space-time are resolved automatically by the gauge symmetry, while a higher unification of 1T field theories is obtained. In this lecture the approach will be described at an elementary worldline level, and the current status of 2T-physics will be summarized.

1. Some Consequences of the Gauge Symmetry

Gauge symmetry in phase space is an unfamiliar concept. What motivates it? In my own work the motivation emerged from my observation in 1995 that the 11-dimensional extended supersymmetry (SUSY) in M-theory is really a 12-dimensional SUSY with an $SO(10,2)$ symmetry, indicating the possibility of a 12D spacetime with two-time (2T) signature[1] (see also F-theory in 12D that developed soon afterwards,[2] and S-theory in 11+2 dimensions[1]). However, if taken seriously, a

298

theory in such a spacetime would be riddled with problems of ghosts and causality. If such a spacetime is more than a mathematical accident in M- or F- or S- theory, one would have to understand how to construct a physical theory that overcomes these problems. In 1995 the challenge seemed to be worthwhile in its own right, in addition to possibly providing a guide for constructing M- or F- or S- theory and explaining the origin of the $SO(10,2)$ or $SO(11,2)$ symmetry.

Yesterday, Murray Gell-Mann talked about "Some Lessons from 60 Years of Theorizing". One of his main messages was that from time to time we should question some of the "received ideas". Usually, he said, there are good reasons for why they exist, but they are sometimes wrong. He gave examples in many areas of "received ideas" which turned out to be wrong, including some famous ones such as the Earth being the center of the universe. Then he elaborated on how he overcame some of those received ideas in his own research, leading to some of his major successes that we are celebrating in this conference. Such refreshing words from our energetic honoree lifted my spirits and reaffirmed my admiration for his intellect and his seminal work.

Well, that spacetime has only one time coordinate is one of those received ideas. A major argument in favor of this one is that apparently insurmountable problems with ghosts and causality prevent additional timelike dimensions. I questioned this received idea in 1995 and three years later, in 1998, found how to overcome it. The key is the gauge symmetry in phase space[3,4] that I will discuss in this talk.

The resolution of similar problems in one-time theories taught us over the past century that the solution to the ghost problem associated with the first time dimension is to have some carefully constructed gauge symmetries. Gauge symmetries in general relativity, Maxwell or Yang-Mills theories, as well as string theory are essential to remove ghosts, thus providing a physically sensible theory. A gauge symmetry has a dual role. On the one hand it is the very reason for the existence of the fundamental forces while dictating the form of fundamental equations of physics, on the other hand it removes ghosts. It was evident to me that, to remove the ghost and causality problems, that are the stumbling blocks in a theory with two times, a much stronger gauge symmetry was needed. Furthermore, if such a thing existed it would lead to some powerful constraints on the fundamental formulation of physics. This could also be a guiding principle for constructing M-theory.

Before I describe the phase space gauge symmetry based on symplectic transformations $Sp(2,R)$ let me highlight some of its important consequences.

- The $Sp(2,R)$ gauge symmetry requires that all physics be reformulated in 4+2 dimensions (more generally d+2). 2T is a consequence, not an input. Thus, for phase space $\left(X^M, P_M\right)$, and all fields $A_M\left(X\right), G_{MN}\left(X\right)$, etc., 2T signature is required by the symmetry, not just permitted. The underlying $Sp(2,R)$ leads to greater gauge symmetry and constraints that remove all ghosts or causality problems. I called this 2T-physics.

- All 1T physics for which we have experimental evidence so far, at all known scales of energy or distance, fits into 2T-physics. The gauge invariant sector of 2T-physics in 4+2 dimensions, namely the ghost free physical sector, becomes effectively a one-time (1T) theory with an effective 3+1 dimensions. There remains no Kaluza-Klein type degrees of freedom at all. But the outcome is not the same as the 1T formulation of physics. Finding again 3+1 within 4+2 is not a zero sum game, because there are many ways in which 3+1 *phase space* is embedded in 4+2 *phase space*, leading to many emergent "times" and corresponding "Hamiltonians" within 4+2. I call the emergent 3+1 spacetimes and dynamical systems "shadows" of the "substance" in 4+2. This leads systematically to a large number of correct predictions by 2T-physics, in the form of hidden relations between dynamical systems and hidden symmetries in 3+1 dimensions, that the standard 1T formulation of physics (1T-physics) is not equipped to predict but can only verify. The new information in 3+1 provided by the systematic predictions from 4+2 (more generally d+2) is the main new content of 2T-physics.

- 2T-physics was initially formulated as a theory for particles moving on worldlines. In recent years the formulation was successfully extended to 2T field theory, which includes the Standard Model (SM)[8] and General Relativity (GR)[9,10] as 2T field theories in 4+2 dimensions. These are consistent with their 1T counterparts in 3+1 dimensions. In fact the usual SM and GR emerge as one of the shadows from 4+2, namely the "conformal shadow". The status of further developments of the 2T approach, including SUSY, higher dimensions, string theory, will be summarized at the end. Suffice it to say that 2T-physics agrees with 1T-physics but it goes beyond by its potential to make new testable predictions, that 1T-physics misses and, which so far are consistent with known data. This additional information, namely the hidden symmetries and the systematic relationships among the emergent multiple shadows, provides a probe for discovering indirectly the properties of the extra 1+1 dimensions.

2. Phase Space Gauge Symmetry $Sp(2, R)$

A clue for the fundamental principle is a *position$\leftrightarrow$momentum* global symmetry in classical or quantum mechanics. Specifically, position and momentum appear at the same level of importance in specifying boundary conditions or in reporting the results of any measurement. More importantly the formulation of classical mechanics in terms of Poisson brackets $\{X^M, P_N\} = \delta^M_N$ is invariant under all infinitesimal canonical transformations, $\delta_\varepsilon X^M = \frac{\partial \varepsilon(X,P)}{\partial P_M}$, $\delta_\varepsilon P_M = -\frac{\partial \varepsilon(X,P)}{\partial X^M}$, since $\delta_\varepsilon \{X^M, P_N\} = 0$ for any $\varepsilon(X, P)$. A quantum ordered version of the same symmetry holds for the fundamental quantum commutators $[X^M, P_N] = i\hbar \delta^M_N$, since $\delta_\varepsilon [X^M, P_N] = 0$. The symmetry under infinitesimal classical canonical transformations is also the symmetry of the *first term* of any action in the first order formalism $S = \int d\tau (\dot{X}^M P_M - \cdots)$ since one gets a total derivative for

$$\delta_\varepsilon \left(\dot{X}^M P_M \right) = \frac{d}{d\tau} \left(P_M \frac{\partial \varepsilon(X, P)}{\partial X^M} - \varepsilon(X, P) \right). \tag{1}$$

This symmetry is spoiled for the action when a specific Hamiltonian $H(X, P)$ is inserted in the action as part of the "$\cdots$". However a specific Hamiltonian focusses on a specific dynamical system rather than the general formalism. We learned in special and general relativity that the notion of time and the corresponding Hamiltonian are dependent on the observer. Einstein showed us how to detach the formulation of fundamental laws from the perspective of observers by requiring equivalence of all perspectives in all spacetime frames. My idea was to take this equivalence notion one step further to all perspectives in phase space (X^M, P_M), not only perspectives in spacetime X^M, by requiring a gauge symmetry in phase space.

In this approach the Hamiltonian (and the associated time) would be regarded as an emergent concept that depends on some perspective from the point of view of phase space. Hence, I ignored the Hamiltonian and instead focused on requiring an action principle with a local symmetry in phase space.

To begin, the canonical transformation above should be regarded as a *global* symmetry on the worldline since the $\varepsilon(X, P)$ of canonical transformations depends on the proper time τ only through the "fields" $X(\tau), P(\tau)$. So any infinitesimal parameters included in $\varepsilon(X, P)$ are global parameters. To have a local symmetry on the worldline one needs a symmetry with parameters that depend arbitrarily on the worldline parameter τ, through additional τ dependent parameters, leading to $\varepsilon(X(\tau), P(\tau), \tau)$ local on the worldline. It turns out that there is a limit on how large the symmetry can be because the system may turn out to be trivial if constrained by too much local symmetry. What worked is an $\mathrm{Sp}(2, R)$ local symmetry[a] formulated as follows.

Introduce the three generators of $\mathrm{Sp}(2, R)$ as a symmetric 2×2 tensor $Q_{ij}(X, P)$, namely $Q_{11}(X, P), Q_{22}(X, P)$ and $Q_{12}(X, P) \equiv Q_{21}(X, P)$ and require that they form the Lie algebra of $\mathrm{Sp}(2, R)$ under Poisson brackets. I will then require local symmetry on the worldline with arbitrary local prameters $\omega^{ij}(\tau)$ that define $\varepsilon(X, P, \tau) = \frac{1}{2}\omega^{ij}(\tau) Q_{ij}(X, P)$. An action invariant under this local transformation can be constructed by introducing the $\mathrm{Sp}(2, R)$ gauge potentials on the worldline $A^{ij}(\tau)$

$$S = \int d\tau \left(\dot{X}^M P_M - \frac{1}{2} A^{ij}(\tau) Q_{ij}(X(\tau), P(\tau)) \right). \tag{2}$$

It can be verified that the action is invariant under the local transformations of the matter and gauge degrees of freedom $\delta_\varepsilon X^M = \frac{1}{2}\omega^{ij}(\tau) \frac{\partial Q_{ij}(X,P)}{\partial P_M}$, $\delta_\varepsilon P_M = -\frac{1}{2}\omega^{ij}(\tau) \frac{\partial Q_{ij}(X,P)}{\partial X^M}$, $\delta_\varepsilon A^{ij} = D_\tau \omega^{ij} \equiv \partial_\tau \omega^{ij} + [A, \omega]^{ij}$ (summed indiced are contracted with the antisymmetric $\mathrm{Sp}(2, R)$ metric ε_{ij}), provided the Q_{ij} form the Lie algebra under Poisson brackets. It is possible to generalize this action by adding a term of the form $S' = -\int d\tau U(X(\tau), P(\tau))$ provided $U(X, P)$ is invariant under $\mathrm{Sp}(2, R)$, namely $\{Q_{ij}, U\} = 0$.

---

[a]This is for a spinless particle. For particles with spin and/or supersymmetry the symmetry group is larger, but it must include $\mathrm{Sp}(2, R)$ as a subgroup in a special way.[16−3,20]

The equivalence principle in phase space I outlined suggests that one should consider all possible $Q_{ij}(X, P)$ that satisfy $Sp(2, R)$ to recover all possible physical systems for a spinless particle (for particles with spin see footnote a). I found an infinite number of $Q_{ij}(X, P)$ that form $Sp(2, R)$ and I classified them up to canonical transformations.[4,5] I now consider some examples.

An example of the $Q_{ij}(X, P)$ that satisfy $Sp(2, R)$ is

$$\text{Example: } Q_{11} = X \cdot X, \quad Q_{22} = P \cdot P, \quad Q_{12} = X \cdot P. \tag{3}$$

These special Q_{ij} are constructed by using a dot product $X \cdot X = X^M X^N \eta_{MN}$ where the signature of the flat metric η_{MN} in target space is not specified à priori. The $Sp(2, R)$ invariants that satisfy $\{Q_{ij}, U\} = 0$ are all possible functions $U\left(L^{MN}\right)$ of the angular momentum generators $L^{MN} = X^M P^N - X^N P^M$. For this example the $Sp(2, R)$ transformation defined above through Poisson brackets amounts to a *local* linear transformation on $\left(X^M, P^M\right)$ such that these behave like a doublet for each M, as follows[3]

$$\begin{pmatrix} X'^M(\tau) \\ P'^M(\tau) \end{pmatrix} = \begin{pmatrix} a(\tau) & b(\tau) \\ c(\tau) & d(\tau) \end{pmatrix} \begin{pmatrix} X^M(\tau) \\ P^M(\tau) \end{pmatrix}, \quad ad - bc = 1, \tag{4}$$

where for the infinitesimal transformation $a(\tau) = 1 + \omega^{12}(\tau) + \cdots$, $b(\tau) = \omega^{22}(\tau) + \cdots$, $c(\tau) = -\omega^{11}(\tau) + \cdots$, $d(\tau) = 1 - \omega^{12}(\tau) + \cdots$. Furthermore, the action above can be rewritten in terms of usual Yang-Mills type covariant derivatives appropriate for the doublets (X, P).[3]

One of the consequences of the general action (2) is the equation of motion for the gauge field A^{ij} which acts like a Lagrange multiplier. This requires that the $Sp(2, R)$ charges should vanish

$$Q_{ij}(X, P) = 0. \tag{5}$$

The meaning of this equation is that only the $Sp(2, R)$ gauge invariant subspace of phase space is physical. Hence, only gauge invariant motion is allowed. The solution space for these $Sp(2, R)$ conditions are called "shadows". I will show some examples of shadows in the next section.

It turns out that nontrivial solutions to $Q_{ij} = 0$ exist only if the target space has 2 times, no less and no more. To see why, consider the example in Eq.(3). If the metric η_{MN} is Euclidean (0T) then the only solution is trivial $X^M = P^M = 0$. If the metric η_{MN} is Minkowski (1T) then a solution is possible only if X^M, P^M are lightlike and parallel, which means the angular momentum vanishes $L^{MN} = 0$. This is trivial because it does not describe even a free particle. To have non-trivial solutions one must have a metric η_{MN} with two times (2T) or more. With 2T it turns out there is just enough gauge symmetry to remove the ghosts, but with three or more times the $Sp(2, R)$ gauge symmetry is insufficient to remove ghosts. Hence there must be two times, no less and no more. I have shown that 2T is an outcome, not an input, since it is demanded by the gauge symmetry and nontrivial physical content.

For the model of 2T-physics based on the Q_{ij} in Eq.(3) there is an automatic *global* SO(d, 2) symmetry. This SO(d, 2) is the symmetry of the dot products and has generators L^{MN} that are Sp(2, R) gauge invariant $\{Q_{ij}, L^{MN}\} = 0$, with $L^{MN} = X^M P^N - X^N P^M$. The action may be modified by an additional Sp(2, R) gauge invariant term of the form $S' = -\int d\tau U(L(\tau))$, where $U(L)$ is an arbitrary function of the L^{MN}, which could break the global SO(d, 2) symmetry partially or fully. The inclusion of U does not change the essential point that the Q_{ij} must vanish, leading to the same 1T shadows (see next section), and that spacetime is $d + 2$ at the fundamental level, even if the global SO(d, 2) symmetry is broken.

I am often asked if it is possible to have more times by enlarging the gauge symmetry beyond Sp(2, R). My answer is that it is unlikely, but I don't have a theorem so far. This is based on the following considerations. First, it is certainly possible to write a gauge invariant action identical in form to (2) for any Lie algebra whose generators $Q_a(X, P)$ close (assuming these can be constructed in phase space). The issue is whether the gauge invariance condition $Q_a(X, P) = 0$ has nontrivial content and also if the emerging shadows are ghost free. In all attempts so far, with specific examples $Q_a(X, P)$ for spinless particles, we have found that, such a scheme based on noncompact groups, either leads to trivial content for the solutions of $Q_a = 0$, or the emergent spaces (i.e. shadows) have ghosts because all the timelike dimensions could not be removed from X^M, P_M. One remarkable exception that has worked so well is Sp(2, R), which seems to indicate that 2T-physics may be special.[15]

The consequences of the local worldline Sp(2, R) symmetry for local field theory (fields that depend only on X^M)[6–14] and an extension of these concepts to field theory in phase space (fields that depend on both X and P)[5] have been developed, but there will not be sufficient time to discuss them in this talk. They will be described only briefly at the end of this paper.

3. Shadows

In this section I concentrate on the *2T free particle in flat spacetime* described by the Q_{ij} in Eq.(3). To obtain the 1T shadows I will make two gauge choices and solve the two constraints $X^2 = 0$ and $X \cdot P = 0$. This fixes two components of X^M and two components of P^M in terms of the remaining independent degrees of freedom, thus reducing the theory from $d + 2$ dimensions to various shadows in d dimensions. There will remain still one gauge symmetry and one unsolved constraint that can remove the ghosts in the remaining timelike degree of freedom in the shadow.

To perform these steps it is useful to define a lightcone type basis $X^M = (X^{+'}, X^{-'}, X^\mu)$ so that the flat metric in $d + 2$ dimensions is expressed as $ds^2 = -2dX^{+'}dX^{-'} + dX^\mu dX^\mu \eta_{\mu\nu}$ with $\eta_{\mu\nu}$ the Minkowski metric in d dimensions including 1 time.

As the first example of a shadow, consider the following solution[3] which I call the "*conformal shadow*"

$$X^{+'}(\tau) = 1,\ X^{-'} = \tfrac{1}{2}x^2(\tau),\quad X^\mu(\tau) \equiv x^\mu(\tau)$$
$$P^{+'}(\tau) = 0,\ P^{-'} = x(\tau) \cdot p(\tau),\ P^\mu(\tau) \equiv p^\mu(\tau) \quad,\ p^2 = 0. \tag{6}$$

Here the two gauges are $X^{+'}(\tau) = 1$ and $P^{+'}(\tau) = 0$ for all τ. The solution of the constraint $X^2 = 0$ yields $X^{-'}$ and the solution of the constraint $X \cdot P = 0$ yields $P^{-'}$ as given above. The remaining degrees of freedom which were named as x^μ, p^μ are still subject to the constraint $P^2 = -2P^{+'}P^{-'} + P^\mu P_\mu = 0$ which takes the form $p^2 = 0$. This phase space $(x^\mu(\tau), p^\mu(\tau))$ describes the free massless 1T relativistic particle in d dimensions. This is confirmed by inserting the gauge choices into the original action, yielding $S = \int d\tau \left(\dot{x}^\mu p_\mu - \tfrac{1}{2}A^{22}p^2 \right)$, which is the action for the free massless relativistic particle. One can also start from the equations of motion for $X^M(\tau), P^M(\tau)$, insert the gauge fixed configuration above, and obtain the equations of motion of the free massless relativistic particle.

The original action had an SO$(d,2)$ global symmetry. The global symmetry of the action does not disappear since the action is gauge invariant. However, it becomes hard to notice the symmetry in terms of the remaining degrees of freedom because it takes a non-linear form. The generators $L^{MN} = X^M P^N - X^N P^M$ were also gauge invariant, so they can be expressed in terms of the remaining degrees of freedom by inserting the configuration in Eq.(6). This gives the following components of L^{MN} in their shadow form

$$L^{+'-'} = x \cdot p,\quad L^{\mu\nu} = x^\mu p^\nu - x^\nu p^\mu,$$
$$L^{+'\mu} = p^\mu,\quad L^{-'\mu} = \tfrac{1}{2}x^2 p^\mu - x \cdot p x^\mu. \tag{7}$$

This is recognized as the generators of the conformal group SO$(d,2)$.[b] Their action on the massless degrees of freedom x^μ, p^μ is given by computing their Poisson brackets $\delta_\varepsilon x^\mu = \tfrac{1}{2}\varepsilon_{MN}\left[L^{MN}, x^\mu\right]$ and similarly for $\delta_\varepsilon p^\mu$. Using this, one can check that these are indeed generators of symmetry for the action for the massless particle.[3,22] This is expected automatically since both S and L^{MN} are gauge invariants and one already knew that S was invariant under SO$(d,2)$.

A second example is the massive relativistic particle given by the following shadow configuration (a different looking form in Ref. 22 is gauge equivalent)

$$X^{+'} = \tfrac{1+a}{2a},\ X^{-'} = \tfrac{x^2 a}{1+a},\ X^\mu \equiv x^\mu(\tau)\quad a \equiv \left(1 + \tfrac{m^2 x^2}{(x \cdot p)^2}\right)^{1/2}$$
$$P^{+'} = \tfrac{-m^2}{2ax \cdot p},\ P^{-'} = ax \cdot p,\ P^\mu \equiv p^\mu(\tau)\quad,\ p^2 + m^2 = 0 \tag{8}$$

---

[b]Dirac[31] was the first to use a 6 dimensional space to describe conformal symmetry SO$(4,2)$. His approach, which was further developed[32−38] had faded away when I discovered my approach, *as a gauge symmetry of phase space*, without being aware of Dirac's different formalism or reasoning. We now know that Dirac's work is automatically part of 2T-physics, since it coincides with my "conformal shadow" for the special case of Q_{ij} in (3). But this is just an example of a particular shadow within the larger scope of 2T-physics.

The remaining constraint in this case $P^2 = 0$ takes the form $p^2 + m^2 = 0$, which says that the shadow phase space $(x^\mu(\tau), p^\mu(\tau))$ now corresponds to the 1T massive relativistic particle. As before, this can be confirmed by computing both the action and the equations of motion. Now we get a surprize not noticed before in 1T-physics. The 2T approach leads us to expect that the action for the massive relativistic particle $S = \int d\tau \left(\dot{x}^\mu p_\mu - \frac{1}{2}A^{22}\left(p^2 + m^2\right)\right)$, which is a gauge fixed form of the original action (2), should be invariant under $SO(d,2)$, but no one suggested this before in 1T-physics. To find out how to construct the symmetry generators in this case (the massive analog of (7))[22] all one needs to do is insert the shadow phase space of Eq.(8) into $L^{MN} = X^M P^N - X^N P^M$.

Even more surprizing (for 1T-physics) in this regard is the shadow for the massive non-relativistic particle given by[22]

$$X^{+'} = t(\tau), \quad X^{-'} = \frac{\vec{r}\cdot\vec{p} - tH}{m}, \quad X^0 = \pm \left|\vec{r} - \frac{t}{m}\vec{p}\right|, \quad X^i = \vec{r}^i(\tau)$$
$$P^{+'} = m, \quad P^{-'} = H(\tau), \quad P^0 = 0 \qquad\qquad P^i = \vec{p}^i(\tau) \tag{9}$$

where $t(\tau)$ and $H(\tau)$ are shadow canonical variables just like $(\vec{r}(\tau), \vec{p}(\tau))$ as described by the action $S = \int d\tau \left(-H\partial_\tau t + \partial_\tau \vec{r}\cdot\vec{p} - \frac{1}{2}A^{22}\left(-2mH + \vec{p}^2\right)\right)$, which follows from (2) by inserting the shadow configuration above. In this case the remaining constraint $P^2 = 0$ takes the form $-2mH + \vec{p}^2 = 0$ which shows that H is the non-relativistic Hamiltonian when the constraint is solved and the final gauge choice is made $t(\tau) = \tau$. The corresponding completely gauge fixed action in $(d-1)$ *space* dimensions is $S = \int d\tau \left(\partial_\tau \vec{r}\cdot\vec{p} - \frac{\vec{p}^2}{2m}\right)$. Evidently it describes the massive non-relativitic particle. However, surprizingly (for 1T-physics) it is invariant under $SO(d,2)$, which is realized by rather complicated non-linear and τ dependent transformations generated by L^{MN}, which are the analogs[22] of Eq.(7).

Another remarkable property, for both shadows that describe the relativistic and non-relativistic particles, is the emergence of the mass parameter m as a modulus in the embedding of the shadow phase space (x, p) in the higher dimensional phase space (X, P). I still believe that mass is very likely explained by the Higgs particle as hopefully will be confirmed at the LHC. But the alternative mass generation mechanism I have just displayed must also mean something in Nature and I hope to find its meaning some day. If the Higgs gets into trouble at the LHC it might be a good idea to investigate seriously for the origin of mass in this alternative direction.

So far I displayed shadows of the free 2T particle in flat space that behave like free particles in 1T-physics. However, there are also all sorts of shadows of the same "substance" that behave like particles subject to a variety of forces, as shown with some examples in Fig.1. These include some non-relativistic potentials (Hydrogen atom, harmonic oscillator), and some curved spaces, such as the Robertson-Walker expanding universe, any conformally flat space ($AdS^{d-n} \times S^n$, maximally symmetric space), and even some singuar spaces. Furtheremore for all of these shadows new twistor formulations provide an alternative expression of the shadow phase space,[18-20] as indicated on the figure. The mathematical expressions

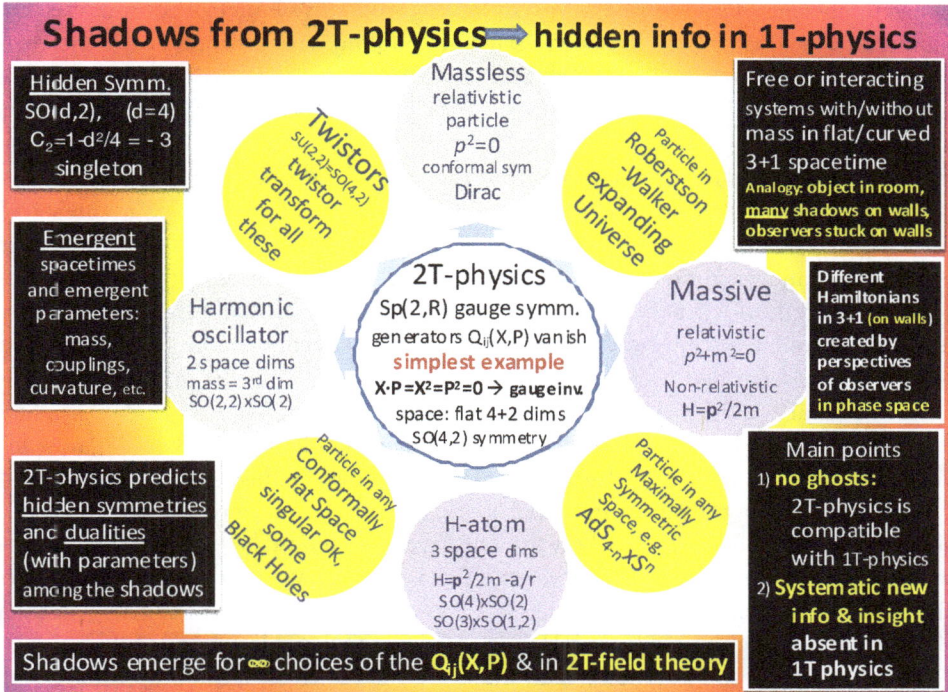

Fig. 1.   Some 1T shadows of the 2T free particle in flat spacetime.

for these shadows (similar to Eqs.(6), (8), (1)) were developed non-systematically[22] over several years and some of them are summarized in tables I, II, III in Ref. 13.

So, parameters, such as mass, coupling, curvature, emerge from the moduli for embedding (x, p) into (X, P). All of these shadows have the hidden $SO(d, 2)$ symmetry, which is realized in terms of non-linear realizations of L^{MN}. When these systems are quantized, the Casimir operators can be evaluated (they are all zero at the classical level) and show that they all give the same quantized value, such as $C_2 = \frac{1}{2} L_{MN} L^{MN} = 1 - d^2/4$. This says that this is the singleton representation of $SO(d, 2)$. All shadows are in the same representation, but each shadow is realized in unitarily equivalent bases of this symmetry.

Evidently there is a lot of information in the hidden relationships among these systems. This information resides in the guge invariant properties of the 2T "substance" in $d + 2$ dimensions which is captured holographically by each 1T shadow in d dimensions. 1T-physics treats all the shadows as different from each other and gives no clues that they may be related. By contrast 2T-physics makes the prediction that observers in d dimensions will discover the predicted relationships and hidden symmetries if they look hard enough.

The relationship between the shadows is similar to duality transformations, which in the present case amount to $Sp(2, R)$ gauge transformations from one fixed gauge to another. These transformations involve not only change of coordinates and momenta but also parameters such as mass, coupling, curvature, etc. All the relations among shadows amount to the fact that L^{MN} is gauge invariant and therefore any function $F(L)$ must have the same gauge invariant value in all the shadows as expressed in terms of the phase space for that shadow. This is the key for all the expected duality relations derivable from the free 2T particle in flat spacetime.

A complete classification of the possible shadows that emerge from the set of $Q_{ij} = (X^2, P^2, X \cdot P)$ is not known. Other forms of $Q_{ij}(X, P)$ will produce their own set of shadows. Similarly the corresponding 2T field theories[6−12] produce shadows in the form of 1T field theories.[13−14] This rich set of dualities is likely to be useful for developing coputational tools. So far this has remained largely unexplored due to lack of time and other pressing priorities.

In summary, quite generally, 2T-physics defines a "substance" that has many "shadows" in 1T-physics. Each one of them inherits holographically the gauge invariant properties of the "substance" (i.e. the theory defined by $Q_{ij}(X, P)$, and corresponding generalization in field theory, including spin, etc) in $d + 2$ dimensions, but the shadows themselves are effective systems in $(d − 1) + 1$ dimensions with only 1T. Many possible 1Ts emerge from phase space in $d + 2$ dimensions, so the 1T in a given shadow is not the same 1T in another shadow. For this reason each shadow is described by a different Hamiltonian in the usual language of 1T-physics. Automatically, these 1T dynamical systems are related to each other by their gauge invariance properties. But 1T-physics is not equipped to display those hidden relationships among the shadows, because for 1T-physics they seem like unrelated dynamical systems with separate Hamiltonians. This is how 1T-physics misses the systematic predictions of 2T-physics.

4. Gravity and Standard Model in 2T-Physics

A particle moving in arbitrary backgrounds, including electromagnetic, gravitational or other general fields in $d + 2$ dimensions is formulated in terms of more general $Q_{ij}(X, P)$.[4] This formulation treats the Maxwell-type gauge symmetries, general coordinate transformations and more general cases of gauge symmetries in a unified way, all as special forms of canonical transformations.[4] I consider here just the gravitational background given by[4,9]

$$Q_{11} = W(X), \ Q_{12} = V^M(X)P_M, \ Q_{22} = G^{MN}(X)P_MP_N. \quad (10)$$

Contrast this to the flat background in Eq.(3) to understand the significance of the background fields, noting that one specializes to the flat case with $G^{MN}_{flat}(X) = \eta^{MN}$, $W_{flat}(X) = X^2$, and $V^M_{flat} = X^M$. There is one further requirement for this to be compatible with the Sp(2, R) gauge symmetry of the action in Eq.(2), that is, these Q_{ij} must close into the Sp(2, R) Lie algebra under Poisson brackets. Consequently the background fields $W(X), V^M(X), G^{MN}(X)$ must satisfy certain equations which I have called the *"kinematic equations"*. There is no space to discuss them here, but they can be found in Ref. 9.

Next, to construct a 2T *field theory* for gravity, a dilaton field $\Omega(X)$ is also needed in addition to the fields $W(X), G_{MN}(X)$ (the field V_M can be solved as $V_M = \frac{1}{2}\partial_M W$, so it is not independent). The field theory action must be such that the "kinematic equations" mentioned in the previous paragraph must emerge as some of the equations of motion through the variational principle.[c] Furthermore, the $Sp(2,R)$ constraints $Q_{ij} \sim 0$ (gauge invariant physical states) must also be satisfied as *dynamical or kinematical field equations* when the field interactions are turned off. These requirements, combined with general coordinate invariance in $d+2$ dimensions, are so strong that they lead to a unique theory for 2T gravity as a field theory. The action for 2T gravity is[9]

$$S_{grav} = \gamma \int d^{d+2}X \sqrt{G} \left\{ \begin{array}{l} \delta(W)\left[\Omega^2 R(G) + \frac{1}{2a}\partial\Omega \cdot \partial\Omega - V(\Omega)\right] \\ +\delta'(W)\left[\Omega^2\left(4 - \nabla^2 W\right) + \partial W \cdot \partial\Omega^2\right] \end{array} \right\}. \tag{11}$$

Here $R(G)$ is the Riemann curvature scalar, a is the special constant $a \equiv \frac{d-2}{8(d-1)}$, while the potential V can only have the form $V(\Omega) = \lambda\Omega^{\frac{2d}{d-2}}$ with a dimensionless coupling λ. Other than λ there are no parameters at all. The field $W(X)$ appears in a delta function and its derivative $\delta(W)$, $\delta'(W)$, as well as in additional terms. This unusual and unique structure emerged from the underlying properties of the $Sp(2,R)$ gauge symmetry on the worldline theory as outlined above. In particular the delta functions are consistent with one of the $Sp(2,R)$ physical state requirements $Q_{11} = W(X) = 0$, while the others $Q_{12}, Q_{22} \sim 0$ emerge from the equations of motion that follow from this action. This field theoretic structure has a bunch of unusual gauge symmetries of its own, which I called 2T gauge symmetries.[9] These are just strong enough gauge symmetries to eliminate all ghosts from the 2T fields and yield shadows in two lower dimensions (analogs of Fig.1) that are ghost free physical interacting 1T field theories in d dimensions. Dualities must relate these shadow 1T field theories to each other.

This theory of gravity has no dimensionful constants, in particular there is no Newton's constant G. This emerges from the condensate of the dilaton (and other scalars, see below) *in the conformal shadow*. The action above yields a shadow 1T General Relativity in d dimensions in the form $S_{grav} = \int d^d x \sqrt{-g}\left\{\phi^2 R(g) + \frac{1}{2a}\partial\phi \cdot \partial\phi - V(\phi)\right\}$, where $\phi, g_{\mu\nu}$ are the shadows of their counterparts. In this shadow, due to the special constant a, there is an emergent local scaling (Weyl) symmetry which is a remnant of general coordinate transformations in the extra $1+1$ dimensions.[10] Since the coefficient of $R(g)$ is positive, the dilaton must have the wrong sign kinetic energy to satisfy the Weyl symmetry, so ϕ is a ghost. Using the Weyl gauge symmetry the shadow dilaton is gauge fixed to a constant ϕ_0 (thus eliminating the ghost which would also have been a Goldstone boson after condensation), yielding precisely Einstein's General Rela-

---

[c]This is analogous to string theory, where background fields are restricted by worldsheet local conformal symmetry, while the field theory must be constructed to reproduce these field equations as equations of motion derived from the field theory action.

tivity $S_{grav} = \int d^d x \sqrt{-g}[\phi_0^2 R(g) - \lambda \phi_0^{2d/d-2}]$ where the condensate ϕ_0^2 must be interpreted as Newton's constant.

Matter fields can be added, including Klein-Gordon scalars $S_i(X)$, Dirac or Weyl spinors $\Psi_\alpha(X)$ and Yang-Mills type vectors $A_M(X)$, all in $d+2$ dimensions. There are special restrictions on each one of these, on the form of their kinetic energies, and the forms of permitted interactions among themselves and with the gravitational triplet (W, Ω, G_{MN}). These restrictions emerge from the underlying $Sp(2, R)$ gauge symmetry and the corresponding physical state conditions at the worldline level.

Within these restrictions I constructed the 2T field theory for the Standard Model in 4+2 dimensions.[8] It is a perfectly consistent, ghost free 2T field theory because of the new 2T gauge symmetries[8] satisfied by these new field theoretic structures. In the conformal shadow it yields the usual Standard Model in 3+1 dimensions which is in exquisite agreement with experiment. This shadow Standard Model has some additional constraints on the scalar sector (Higgs and others) and their interaction with the dilaton. The new features are consistent with known phenomenology, but may help shed some light on Higgs physics[d] when more data becomes available, and perhaps the absence of axions[8] which awaits further clarification until the *quantum* version of 2T field theory is better understood.

In the coupling of gravity to matter there is another interesting physics prediction to be emphasized. Every scalar S_i in the complete field theory must couple to the curvature term just like the dilaton in Eq.(11), but with the opposite sign and standard normalization in the kinetic term. Then in the conformal shadow the curvature term is predicted to take the form $(\phi^2 - a \sum_i s_i^2) R(g)$ with a required relative minus sign! Hence the gravitational constant must emerge from the condensates of all the scalars, not only the dilaton's. This predicts a physical effect, that the effective gravitational constant $G \sim (\phi^2 - a \sum_i s_i^2)^{-1}$ is not really a constant, rather it must increase after every phase transition of the universe as a whole (since the dominant part of each field is the condensate after the phase transition, this quantity is approximately a constant in between the phase transitions). Thus the Newton constant we measure today cannot be the same as the analogous constant before the various transitions occured, such as inflation, grand unification, SUSY breaking, electroweak symmetry breaking. Of course the earlier ones are the dominant condensates in the sum. There is also the curious possibility that $G \sim (\phi^2 - a \sum_i s_i^2)^{-1}$ could turn negative if the other scalars dominate over the dilaton in some regions of the universe, or in the history of the universe, thus producing antigravity in those parts of spacetime. In fact, the Big Bang may be related to the vanishing of

---

[d]This is in spirit similar to the Higgs phenomenology talk we heard from J. Gunion in this conference (see also Refs. 25–28) because the Higgs in the 2T Standard Model is required to couple to at least one additional scalar,[8] which may be the dilaton, or another *electroweak neutral* scalar. Note that historically the importance of this neutral sector for phenomenlogy was pointed out based on the prediction from the 2T Standard Model[8] prior to the (less theoretically motivated) recent phenomenological studies were undertaken.

$(\phi^2 - a \sum_i s_i^2)$ at which point the effective G blows up. The effects of this idea on cosmology is curently under investigation.[24]

5. Progress in 2T-Physics

Here I will list various points with only very brief comments due to lack of space.

1) Local $Sp(2,R)$ on the worldline, as a gauge symmetry in phase space, has proven to be a physically correct general principle in both classical and quantum mechanics. The advantages of this new principle include the unification of various 1T dynamical systems under a new unifying umbrella which I called 2T-physics (as summarized in the example of Fig.1). As compared to 1T-physics, 2T-physics reveals much more *correct information on physical phenomena* which is sytematically missed in the usual formalism of physics at all scales of distance or energy.

2) The principles of 2T field theory in $d + 2$ dimensions have been established. New types of gauge symmetries eliminate all ghosts and produce a physical sector which effectively is in two lower dimensions. As in the corresponding worldline theory, the 2T field theory produces shadow 1T field theories with duality relations among them (field analogs of Fig.1). Within these principles I have constructed various physically relevant 2T field theories. These include the Standard Model, General Relativity, Grand Unified theories, in 4+2 dimensions, whose conformal shadows are basically the same as the familiar corresponding theories in $3 + 1$ dimensions, except for some additional restriction (mainly on scalar fields) which so far are consistent with phenomenology, and may even lead to measurable signals at the LHC or in cosmology.

3) Both the worldline and field theory approaches to 2T-physics have been generalized to supersymmetry. In particular, the general $N = 1, 2, 4$ SUSY 2T field theories in $4 + 2$ dimensions have been constructed.[11−12] Consequently, the SUSY generalizations of the Standard Model or GUTS in $4 + 2$ dimensions are already available. So, if SUSY phenomenolgy becomes relevant at the LHC, the constraints from 2T-physics could become interesting.

4) SUSY in higher dimensions with 2T has also been achieved. In particular the super Yang-Mills theory in 10+2 dimensions has been constructed.[29] This 12-dimensional field theory is the first one to ever go beyond 11 dimensions. It yields many interesting shadows as well as compactifications that unify various theories, from M(atrix)-theory to the hotly pursued $N = 4$ super Yang-Mills theory in 3+1 dimensions, and thus may lead to possible new insights in these 1T theories. These connections are outlined in Fig.2, and are the basis for considering the new theory as the parent of all those mentioned in the figure. Finally a connection between 2T-physics and 12D and 13D S-theory, where it all started in 1995,[1] is beginning to emerge in a clearer way.

5) Supergravity in 2T-physics is almost constructed, the path to follow is now clear. The expected maximal SUSY theory in 13 dimensions should yield 11-dimensional supergravity as the conformal shadow.

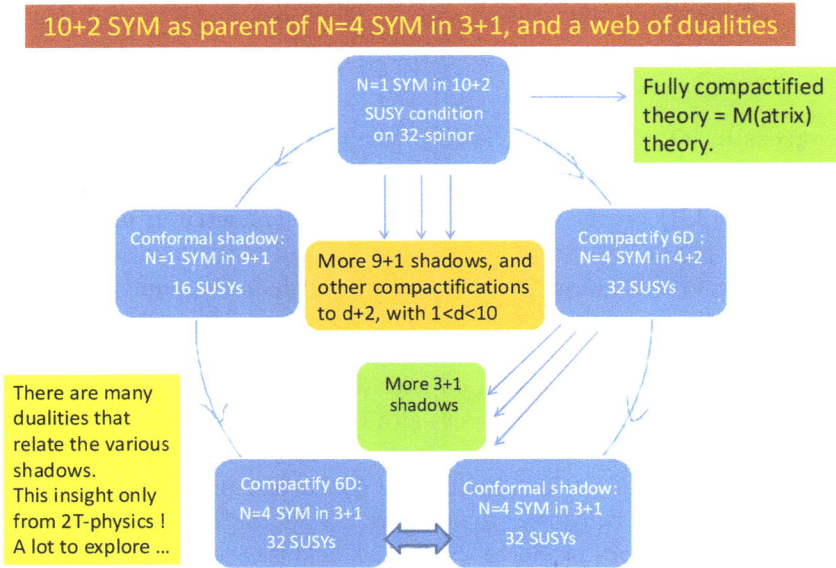

Fig. 2. SYM in 12D is the parent of 11D SYM, N=4 SYM in 3+1, N=2 SYM in 4+2, M(atrix)-theory and others.

6) For strings, branes, and more generally for M-theory in 2T-physics, there is only old partial progress for tensionless strings and branes, and more recent progress on the 2T version of the twistor string.[30] The usual tensionful string has historically resisted 2T-physics and the reason for this may have now become clear: it must be the fact that the tension is dimensionful, but as I explained above 2T gravity does not allow any dimensionful constants. The tension will have to emerge from some condensate. This new insight has not yet been implemented.

7) There is the potential of developing powerful new computational tools that take advantage of the dualities of the shadows. As in other examples of dualities, a given theory may be more easily solvable in one dual version as compared to another. Since the gauge invariant physical content is holographically captured by every shadow, it may be possible to study physical effects more easily in some shadows and transform the result to the shadow of interest (which may be the conformal shadow in the case of field theory). The shadow phenomena is much more easily handled in the worldline formalism, while in field theory so far there is limited progress because only some shadows are easy to study but others seem to be more difficult.[13–14] In any case, due to lack of time, little has been done so far on this feature of 2T-physics, but I think it is where 2T-physics may become most useful to 1T physicists as well as where most tests of 2T-physics can be developed.

8) As a final comment, I should mention that I consider all the encouraging progress in 2T field theory to be only a stepping stone toward a more comprehensive 2T theory which is based on fields in phase space, not just position space. I expect

the field theory to be non-commutative along the lines initiated in Ref. 5. When this approach can be connected to the successful 2T field theories that work at the present, I expect much more dramatic insight and progress.

6. 2T-Physics as a Unifying Framework for 1T-Physics

I would like to conclude this talk with an allegory which may be helpful to convey the basic idea of 2T-physics and its role as a completion of 1T-physics. I should warn that, the allegory is not perfect and is not a substitute for the equations. So it should not be taken too far without the corresponding equations.

As in Fig. 3, consider an object in a room. In the allegory this represents phase space X^M, P_M in $d + 2$ dimensions, with the associated $Sp(2, R)$ generators $Q_{ij}(X, P)$. Then consider the *many* shadows of this object on the surrounding walls which could be formed by shining light on it from different directions. In the allegory the shadows represent the *many* emergent physical phase spaces (x^μ, p_μ) in two lower dimensions which solve $Q_{ij} = 0$, as in Eqs. (6), (8), (9) and Fig. 1.

An essential point is that one single object in the room (in the allegory a specific set of Q_{ij}, such as Eq.(3)) has many shadows. To observers that are stuck on the wall (like we are stuck in 3+1 dimensions) the various shadows appear like different "beasts" performing different unrelated motions. However, an observer in the room immediately knows that the many shadows coming from the same object must be related. These relationships among shadows can in principle be discovered by careful observers who live on the wall, but who have no privilege of being in the room.

The gauge choices in phase space that create the shadows are the analogs of the many perspectives for observing the object in the room. The relationships among the emergent dynamical systems in Fig.1 comes from the many forms in which the same gauge invariant information in $d + 2$ dimensions is encoded *holographically* in each shadow in d dimensions. The 2T-physics formulation makes the relationships between the "shadows" evident and predicts them to the 1T physicists on the "wall" who can study them and verify them. The information in these relationships, which I called hidden "dualities" and hidden symmetries, is information about the perspectives, and therefore it is information that relates to the properties of the larger space-time in $d + 2$ dimensions. By interpreting this data correctly, and recognizing its relation to the higher dimensions, the 1T physicists on the "wall" have a probe for studying the extra dimensions, albeit indirectly.

In this way 2T-physics provides the privilege of being in the room. It gives us the ability to recognize that certain sytems are indeed related, and that the predicted relationships are interpreted as perspectives in a higher space-time.

Colleagues that have followed my work have generally been in agreement with my results. Usually I receive encouragement and never criticism. However, sometimes I am asked: "All of this is quite nice, but do we really need 2T, can't we do everything with 1T anyway?". This attitute is probably part of the reason for having doubts on whether to invest effort in 2T-physics.

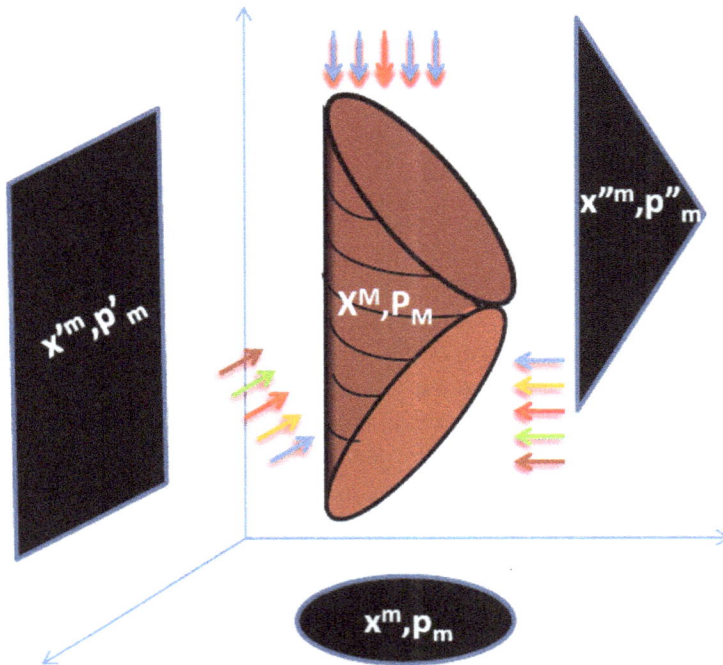

Fig. 3. An allegory. "Room"=4+2, "Walls"=3+1. Substance versus shadows.

My answer to this question is a definite, yes you do need 2T, you cannot do everything with 1T! I have already displayed examples of systems, as in Fig.1, where new information, not available in 1T-physics is obtained in 2T-physics. This shows definitely that in principle 1T-physics *systematically* misses information, while 2T-physics makes it accesible with definite predictions. 2T-physics opens new avenues and provides new information not available before. 1T-physics is clearly incomplete. Apparently this has not been fully appreciated yet by many of my colleagues.

Given that this is a fact, I would guess that, besides seeking practical applications in 1T-physics, in seeking the unified theory for everything we may find that the additional information of the extra 1+1 dimensions may lead to the "Holy Grail".

To conclude, I would like to come back to the quotation from Gell-Mann: "Anything which is not forbidden is compulsory!". Murray used this phrase in connection to the properties of the strong interactions. But I would like to adopt it to the fundamental formulation of physics.

Acknowledgment

Work partially supported by the US Department of Energy, grant number DE-FG03-84ER40168.

References

1. I Bars, "Duality and hidden dimensions", Lecture in 1995 conference, appeared in *Frontiers in quantum field theory*, Ed. H. Itoyonaka, M. Kaku. H. Kunitomo, M. Ninomiya, H. Shirokura, Singapore, World Scientific, 1996, page 52 [hep-th/9604200]; I Bars, "Supersymmetry, p-brane duality and hidden space and time dimensions," Phys. Rev. **D54** (1996) 5203, [hep-th/9604139]; I. Bars, "S-theory", Phys. Rev. **D55** (1997) 2373 [hep-th/9607112]; I. Bars, "Algebraic Structures in S-Theory", talk at conferences Strings-96 and 2nd Sakharov conference, hep-th/9608061.
2. C. Vafa, Nucl. Phys. **B469** (1996) 403 [hep-th/9602022].
3. I Bars, C. Deliduman and O. Andreev, " Gauged Duality, Conformal Symmetry and Spacetime with Two Times", Phys. Rev. **D58** (1998) 066004 [hep-th/9803188]. For reviews of subsequent work see: I. Bars, " Two-Time Physics", in the Proc. of the 22nd Intl. Colloq. on Group Theoretical Methods in Physics, Eds. S. Corney at. al., International Press, 1999, pp.2-17 [hep-th/9809034]; " Survey of two-time physics," Class. Quant. Grav. **18**, 3113 (2001) [hep-th/0008164]; " 2T-physics 2001," AIP Conf. Proc. **589** (2001), pp.18-30; AIP Conf. Proc. **607** (2001), pp.17-29 [hep-th/0106021]; "Lectures on twistors," hep-th/0601091.
4. I. Bars, "Two time physics with gravitational and gauge field backgrounds", Phys. Rev. **D62**, 085015 (2000) [hep-th/0002140]; I. Bars and C. Deliduman, " High spin gauge fields and two time physics", Phys. Rev. **D64**, 045004 (2001) [hep-th/0103042].
5. I. Bars, "$u_*(1,1)$ non-commutative gauge theory as the foundation of 2T-physics in field theory", Phys. Rev. D64 (2001) 126001 [hep-th/0106013]. I. Bars and S. Rey, "Noncommutative Sp(2,R) gauge theories: A Field theory approach to two time physics.", Phys. Rev. D64 (2001) 046005 [hep-th/0104135].
6. I. Bars, " Two-time physics in Field Theory", Phys. Rev. **D 62**, 046007 (2000), [hep-th/0003100].
7. I. Bars and Y-C. Kuo, "Interacting two-time Physics Field Theory with a BRST gauge Invariant Action", Phys. Rev. **D74** (2006) 085020, [hep-th/0605267].
8. I. Bars, "The standard model of particles and forces in the framework of 2T-physics", Phys. Rev. **D74** (2006) 085019 [hep-th/0606045]. For a summary see "The Standard Model as a 2T-physics theory", Proc. of SUSY06: 14th Int. Conference on Supersymmetry and the Unification of Fundamental Interactions, Irvine, California, 12-17 Jun 2006, hep-th/0610187.
9. I. Bars, "Gravity in 2T-Physics", Phys. Rev. **D77** (2008) 125027 [0804.1585 [hep-th]].
10. I. Bars and S-H Chen, "Geometry and Symmetry Structures in 2T Gravity", Phys. Rev. D79 (2009) 085021 [0811.2510 (hep-th)]..
11. I. Bars and Y.C. Kuo, "Field theory in 2T-physics with $N = 1$ supersymmetry" Phys. Rev. Lett. **99** (2007) 41801 [hep-th/0703002]; ibid. "Supersymmetric field theory in 2T-physics," Phys. Rev. **D76** (2007) 105028,. [hep-th/0703002].
12. I. Bars and Y.C. Kuo, " N=2,4 Supersymmetric Gauge Field Theory in 2T-physics" Phys. Rev. **D79** (2009) 025001 [0808.0537[hep-th]].
13. I. Bars, S-H. Chen and G. Quelin, "Dual Field Theories in (d-1)+1 Emergent Spacetimes from a Unifying Field Theory in d+2 Spacetime," Phys. Rev. **D76** (2007) 065016 [0705.2834 [hep-th]].
14. I. Bars, and G. Quelin, "Dualities among 1T-Field Theories with Spin, Emerging from a Unifying 2T-Field Theory", Phys. Rev. **D77** (2008) 125019 [0802.1947 [hep-th]].
15. I. Bars and S-H Chen, "Why 2T-Physics may be Special", to be published.
16. I. Bars and C. Deliduman, Phys. Rev. **D58** (1998) 106004, hep-th/9806085.

314

17. I. Bars, C. Deliduman and D. Minic, "Supersymmetric Two-Time Physics", Phys. Rev. **D59** (1999) 125004, [hep-th/9812161]; "Lifting M-theory to Two-Time Physics", Phys. Lett. **B457** (1999) 275, [hep-th/9904063].

18. I. Bars, " 2T physics formulation of superconformal dynamics relating to twistors and supertwistors," Phys. Lett. B **483**, 248 (2000) [hep-th/0004090]. "Twistors and 2T-physics," AIP Conf. Proc. **767** (2005) 3 , [hep-th/0502065].

19. I. Bars and M. Picon, "Single twistor description of massless, massive, AdS, and other interacting particles," Phys. Rev. **D73** (2006) 064002 [hep-th/0512091]; "Twistor Transform in d Dimensions and a Unifying Role for Twistors," Phys. Rev. **D73** (2006) 064033, [hep-th/0512348].

20. I. Bars, Lectures on Twistors, USC-06/HEP-B1, [hep-th/0601091].

21. I. Bars and B. Orcal, "Generalized Twistor Transform And Dualities, With A New Description of Particles With Spin, Beyond Free and Massless", Phys. Rev. **D75** (2007) 104015 [0704.0296 [hep-th]].

22. I. Bars, "Conformal symmetry and duality between free particle, H-atom and harmonic oscillator", Phys. Rev. **D58** (1998) 066006 [hep-th/9804028]; "Hidden Symmetries, $AdS_d \times S^n$, and the lifting of one-time physics to two-time physics", Phys. Rev. **D59** (1999) 045019 [hep-th/9810025].

23. I. Bars, "Twistor superstring in 2T-physics," Phys. Rev. **D70** (2004) 104022, [hep-th/0407239].

24. I. Bars and S-H Chen, "The Big Bang and inflation united with an analytic solution for the inflaton", [Archiv:].

25. M. Shaposhnikov and I. Tkachev, "The νMSM, inflation and dark matter", Phys. Lett. **B639** (2006) 414, [hep-ph/0604236].

26. [7] M. Shaposhnikov, "Is there a new physics between electroweak and Planck scales"?, 0708.3550 [hep-th].

27. W. D. Goldberger, B. Grinstein and W. Skiba, "Light scalar at LHC: The Higgs or the dilaton?", [0708.1463 [hep-ph]]

28. J. Ramón Espinosa and M. Quirós, "Novel Effects in Electroweak Breaking from a Hidden Sector", Phys.Rev.**D76** (2007) 076004 [hep-ph/0701145].

29. I. Bars and Y.C. Kuo, "Super Yang-Mills theory (SYM) in 10+2 dimensions, linking $SYM_{d=4}^{N=4}$ and M(atrix) theory to 2T-physics", in preparation.

30. I. Bars, C. Deliduman and D.Minic, "Strings, branes and two time physics", Phys. Lett. **B466** (1999) [hep-th/9906223, *ibid.* "Lifting M theory to two time physics", Phys. Lett. **B457** (1999) 275 [hep-th/9904063]; I. Bars, "Twistor superstring in 2T-physics", Phys. Rev. **D70** (2004)104022 [hep-th/0407239].

31. P.A.M Dirac, Ann. Math. **37** (1936) 429.

32. G. Mack and A. Salam, Ann. Phys. **53** (1969) 174.

33. S. Adler, Phys. Rev. **D6** (1972) 3445; ibid. **D8** (1973) 2400.

34. S. Ferrara, Nucl. Phys. **B77** (1974) 73.

35. F. Bayen, M. Flato, C. Fronsdal and A. Haidari, Phys. Rev. **D32** (1985) 2673.

36. W. Siegel, Int. J. Mod. Phys. **A3** (1988) 2713; Int. Jour. Mod. Phys. **A4** (1989) 2015.

37. C. R. Preitschopf and M. A. Vasiliev, Nucl. Phys. **B549** (1999) 450, [hep-th/9812113].

38. R. Marnelius, Phys. Rev. D20, 2091 (1979); R. Marnelius and B. Nilsson, Phys. Rev. **D22** (1980) 830; P. Arvidsson and R. Marnelius, "Conformal theories including conformal gravity as gauge theories on the hypercone" [hep-th/0612060].

COMPOSITE HIGGS PARTICLE

KOICHI YAMAWAKI

Department of Physics, Nagoya University, Japan

Higgs as formulated in the Gell-Mann–Levy (GL) linear sigma model may be a composite object as it turned out in QCD. We shall discuss a composite Higgs boson, techni-dilaton, in the walking/conformal technicolor near the Caswell–Banks–Zaks conformal fixed point which is characterized by the essential singularity scaling, breakdown of the GL effective theory, and large anomalous dimension $\gamma_m = 1$. As a remnant of (spontaneously broken) conformal symmetry, there exists a composite pseudo Nambu–Goldstone boson, techni-dilaton, whose mass arises from the scale anomaly. The techni-dilaton may be discovered at LHC.

BLACK HOLES AND ATTRACTORS IN SUPERGRAVITY

A. CERESOLE* and S. FERRARA[†]

Istituto Nazionale di Fisica Nucleare, Sezione di Torino,
Via P. Giuria 1, Torino 10125, Italy
*ceresole@to.infn.it

Physics Department, Theory Unit, CERN,
CH-1211, Geneva 23, Switzerland
[†]sergio.ferrara@cern.ch

We discuss some of the basic features of extremal black holes in four-dimensional extended supergravities. Firstly, all regular solutions display an attractor behavior for the scalar field evolution towards the black hole horizon. Secondly, they can be obtained by solving first order flow equations even when they are not supersymmetric, provided one identifies a suitable superpotential W which also gives the black hole entropy at the horizon and its ADM mass at spatial infinity. We focus on N=8 supergravity and we review the basic role played by U-duality of the underlying supergravity in determining the attractors, their entropies, their masses and in classifying both regular and singular extremal black holes.

Keywords: N=8 supergravity; black holes; attractors.

1. Extremal Black Holes

In 1976, at the dawn of the N=1 supersymmetric theory of gravity in four dimensions[1] (now called "N=1 supergravity"), when its N=2 extension had just appeared, Murray Gell-Mann was the first to remark during a seminar at Caltech that, if higher N supergravity would have indeed existed, then there would be a bound such that $N_{\max}$=8 would be the end of the story. Today, after 34 years, we are still struggling to understand this beautiful maximally extended theory,[2] its connection with superstring and M-theory,[3] its hypothetical perturbative finiteness and its non perturbative completion. This contribution focusses on the black holes that arise in N=8 supergravity in four dimensions,[4] which have been recently claimed to play a possible key role in relation to string theory and to the issue of perturbative finiteness of N=8 supergravity.[5]

Black holes, one of the most interesting outcome of General Relativity, are the typical probes of the quantum regime of any fundamental theory of gravity and as such, they are naturally investigated within the framework of superstring and M-theory. As a first approximation, they can be proficiently studied as classical solutions of the underlying extended supergravities, which arise upon compactification in the effective field theory limit. Once the background geometry of the d-dimensional spacetime and the number N of supersymmetry charges have been selected, all the important features of a given solution are encoded into the electric magnetic-duality group G acting on the vector fields A^Λ, and in the geometric properties of the moduli space G/H parametrized by the scalar fields ϕ^i.[6] When the electric and magnetic charges are quantized, the group G becomes the U-duality group which is known to dictate the string dynamics in various dimensions.[7,8]

The thermodynamical properties of black holes can be obtained from quantum mechanical attributes that are their (ADM) mass, charge, spin and scalar charges (see for instance Refs. 9–12). Unlike Schwarzschild black holes, charged (Reissner-Nordstrom) and/or spin (Kerr-Newman) black holes can be *extremal*, *i.e* with vanishing temperature for non-zero entropy, in which case their event horizon and Cauchy horizons coincide. In formulae, the extremality parameter is given by

$$c = 2ST = \frac{1}{2}(r_+ - r_-) \to 0\,, \tag{1}$$

where c measures the surface gravity and $S = log\,\mathcal{N}$ is the black hole entropy which counts the number $\mathcal{N}$ of microstates. In supergravity, it is given by the Bekenstein-Hawking area formula

$$S_{BH} = \frac{1}{2}A_H = \pi R_H^2\,, \tag{2}$$

where R_H is the effective radius of a sphere encircling the horizon. For extremal charged black holes, R_H must respect the symmetries of the theory and in particular it must depend only on the electric and magnetic charges and not on the scalar field values.[6] Therefore, also the entropy will depend only on the charges and it will take particular expressions depending on the duality symmetries of the given model.

The lack of dependence on scalar fields of the entropy can be viewed as a sort of "no hair" theorem[13] and reflects the fact that, under certain conditions, extremal black holes in N-extended supergravities enjoy a remarkable property: the scalar field trajectories $\phi^i(\tau)$, in terms of the radial evolution parameter $\tau = -1/\rho$ from asymptotic infinity ($\tau \to 0$) to the horizon ($\tau \to -\infty$), behave as dynamical systems, *i. e.* independently on their initial values $\phi_0^i = \phi^i(\tau)|_{\tau \to 0}$. The scalars evolve to a common value at the horizon, $\phi_H^i = \phi_H^i(Q)$, which they reach with zero velocity ($\dot{\phi}^i \to 0$), and where they entirely depend on the electric-magnetic charge vector Q of the asymptotic configuration. This attractive feature is called *Attractor Mechanism*,[10,11] and the attractor fixed points can be obtained as extrema of a suitable effective potential. Another important feature of extremal black holes is that their horizon geometry of spacetime is universal, and in four dimensions it is given by the $AdS_2 \times S_2$ Bertotti-Robinson metric. This is a particular case of the geometry of black p-branes in D-dimensions and it is an instance of the AdS/CFT correspondence relating the gauge theory on the boundary to bulk supergravity.[14] Thus, extremal black holes behave as solitons interpolating between maximally symmetric geometries of (super)spacetime: Minkowski for $\tau \to 0$ and the conformally flat metric for $\tau \to -\infty$.[15,16]

It is remarkable that the common radius of $AdS_2 \times S_2$ and therefore the entropy, can be actually computed using the electric-magnetic duality of the underlying supergravity theory.

For Reissner-Nordstrom black holes, with electric charge q and magnetic charge p, the entropy is $S \sim p^2 + q^2$ due to the $U(1)$ symmetry that rotates the p and q charges into each other. For more complicated objects where scalar fields are present, such as the Axion-Dilaton black hole having two $U(1)$ gauge fields and charges (q_1, p_1) and (q_2, p_2), the entropy becomes $S \sim |p_1 q_2 - p_2 q_1|$ and it is invariant under $SL(2) \times SO(2)$.[35] The appearance of a non compact symmetry group is a quite general signature of the presence of scalar fields. In $N = 4$ supergravity, one has $S \sim |p^2 q^2 - (p.q)^2|^{1/2}$, with $SL(2) \times SO(6,n)$ symmetry.[34] In the maximal $N = 8$ case, the dyonic charge vector Q^a transforms in the fundamental **56** representation of $E_{7(7)}$ (since there are 28 vector fields yielding 28 electric and 28 magnetic charges) while the scalar fields span the 70-dimensional scalar manifold $G/H = E_{7(7)}/SU(8)$. It turns out that the entropy for regular black holes is proportional to $\sqrt{|I_4|}$ where I_4 is the quartic invariant of the **56** representation of $E_{7(7)}$, $I_4 = T_{abcd} Q^a Q^b Q^c Q^d$. When $I_4 = 0$, the black hole is singular, with vanishing horizon area, and there can be 1/8, 1/4 or 1/2 supersymmetry preserved.[17,18,31]

In the last few years it has become clear that the scalar field dynamics for the extremal black holes can be entirely encoded into a real "superpotential" function $W(\phi^i, Q)$ for which ϕ^i_H is a critical point in the moduli space of the theory: $\partial W / \partial \phi^i |_{\phi^i = \phi^i_H} = 0$.

For BPS (supersymmetric) configurations, $W = |z_h| = \rho_h$ where ρ_h is the modulus of the highest among the skew eigenvalues of the central charge matrix $Z_{AB} = -Z_{BA}$ of the supersymmetry algebra,

$$\{\mathcal{Q}_{\alpha A}, \mathcal{Q}_{\beta B}\} = \epsilon_{\alpha\beta} Z_{AB}(\phi^i_0, Q) \tag{3}$$

$$\{\mathcal{Q}_{\alpha A}, \mathcal{Q}^B_{\dot{\beta}}\} = \sigma^\mu_{\alpha\dot{\beta}} P_\mu \delta^B_A . \tag{4}$$

Depending on how many of these skew eigenvalues are coincident, one has various degrees of preserved supersymmetry. For the N=8 theory, one can have four different cases ranging from 1/8, to 1/4, 1/2 and zero preserved supersymmetries. In the original setup of N=2 extremal black holes in D=4, $Z_{AB} = \epsilon_{AB} Z$ and such a superpotential function for supersymmetric configurations was to be identified with the modulus of the N=2 central charge Z, appearing in the ordinary BPS equations.[10,11] Remarkably, such a function W can be shown to exist also for non supersymmetric configurations, in which case it is called the "fake superpotential"[19] because of the similarity with the set up of "fake supergravities".[20] When the attractors are regular, the W function has a minimum for $\phi^i = \phi^i_H$, and its horizon value gives the entropy of the configuration

$$S = \frac{1}{4} A_H = \pi W^2_H(Q) = \pi W^2_{\text{crit}}(\phi^i_h(Q), Q) \tag{5}$$

according to the Bekenstein-Hawking formula. However, the if W has a runaway behavior in moduli space, $\phi_H \to \infty$ $W \to 0$ (which is not acceptable in D=4), the corresponding black hole solutions are singular. Then the scalar fields are never stabilized within the boundaries of moduli space, there are no attractors and the entropy of the extremal configuration vanishes.

In order to describe a static, spherically symmetric extremal black hole background in the extremal case, $c = 2ST = 0$, the metric ansatz reads[11]

$$ds^2 = -e^{2U}dt^2 + e^{-2U}\frac{d\tau^2}{\tau^4} + \frac{1}{\tau^2}(d\theta^2 + \sin^2\theta d\phi^2) \ . \tag{6}$$

with the field strength $F^\Lambda_{\mu\nu}$ for n_v vectors ($\Lambda = 1, \ldots, n_V$) and its dual $G_{\Lambda\mu\nu} = \frac{\delta\mathcal{L}}{\delta F^\Lambda_{\mu\nu}}$ given by

$$F = e^{2U}C\mathcal{M}(\phi^i)Qdt \wedge d\tau + Q\sin\theta d\theta \wedge d\phi \tag{7}$$

$$F = \begin{pmatrix} F^\Lambda_{\mu\nu} \\ G_{\Lambda\mu\nu} \end{pmatrix} \frac{dx^\mu dx^\nu}{2} \ . \tag{8}$$

Electric and magnetic charges are defined by

$$q_\Lambda = \frac{1}{4\pi}\int_{s_2} G_\Lambda \ , \qquad p^\Lambda = \frac{1}{4\pi}\int_{S_2} F_\Lambda \ . \tag{9}$$

$\mathcal{M}(\phi^i)$ is a $2n_v \times 2n_v$ real symmetric $Sp(2n_v, R)$ matrix, satisfying $\mathcal{M}C\mathcal{M} = C$,

$$\mathcal{M}(\phi^i) = \begin{pmatrix} I + RI^{-1}R & -RI^{-1} \\ -I^{-1}R & I^{-1} \end{pmatrix} \tag{10}$$

where $I = Im\mathcal{N}_{\Lambda\Sigma}, R = Re\mathcal{N}_{\Lambda\Sigma}$, the vector kinetic matrix $\mathcal{N}_{\Lambda\Sigma}$ depends on the scalar fields and enters the 4D lagrangian

$$\mathcal{L} = -\frac{R}{2} + \frac{1}{2}g_{ij}(\phi)\partial_\mu\phi^i\partial^\mu\phi^j + I_{\Lambda\Sigma}F^\Lambda \wedge^* F^\Sigma + R_{\Lambda\Sigma}F^\Lambda \wedge F^\Sigma \ . \tag{11}$$

The black hole effective potential[10] is given by

$$V_{BH} = -\frac{1}{2}Q^T\mathcal{M}Q = \frac{1}{2}Z_{AB}\overline{Z}^{AB} = \sum_i \rho_i^2 \ , \tag{12}$$

where A,B are SU(8) indices.[12] In the last expression, ρ_i are the moduli of the skew eigenvalues z_i of the central charge matrix Z_{AB}. It arises upon reducing the general 4D (or higher D) lagrangian to the one-dimensional almost geodesic action describing the radial evolution of the $n + 1$ scalar fields $(U(\tau), \phi^i(\tau))$:

$$S = \int \mathcal{L}d\tau = \int (\dot{U} + g_{i\bar{j}}\dot{\phi}^i\dot{\phi}^{\bar{j}} + e^{2U}V_{BH}(\phi(\tau), p, q)d\tau. \tag{13}$$

In order to have the same equations of motion of the original theory, the action must be complemented with the Hamiltonian constraint (in the extremal case)[11]

$$\dot{U}^2 + g_{ij}\dot{\phi}^i\dot{\phi}^j - e^{2U}V_{BH}(\phi(\tau), p, q) = 0 \ . \tag{14}$$

The black hole effective potential can be written in terms of the superpotential $W(\phi)$ as

$$V_{BH} = W^2 + 2g^{ij}\partial_i W\partial_j W \ . \tag{15}$$

This formula can be viewed as a differential equation defining W for a given black hole effective potential V_{BH}, and it can lead to multiple choices: only one of those will corresponds to BPS solutions, while a different one will be associated to non BPS ones. In both cases, W allows to rewrite the ordinary second order supergravity equations of motion

$$\frac{d^2 U}{d\tau^2} = e^{2U} V_{BH} \tag{16}$$

$$\frac{d^2 \phi^i}{d\tau^2} = g^{i\bar{j}} \frac{\partial V_{BH}}{\partial \phi_{\bar{j}}} e^{2U} , \tag{17}$$

as first order flow equations, defining the radial evolution of the scalar fields ϕ^i and the warp factor U from asymptotic infinity towards the black hole horizon:[19]

$$U' = -e^U W , \qquad \phi'^i = -2e^U g^{ij} \partial_j W . \tag{18}$$

The important point is that, at the prize of finding a suitable fake superpotential W, one only has to deal with these first order flow equations even for non supersymmetric solutions, where one does not have Killing spinor equations.[19,23]

Beside the horizon entropy $S_{BH} = \pi W_H^2$ and the first order flows, the value at radial infinity of the superpotential W also encodes other basic property of the extremal black hole, which are its ADM mass, given by

$$M_{ADM}(\phi_0^i, Q) = \dot{U}(\tau = 0) = W(\phi_0, Q) \tag{19}$$

and the scalar charges at infinity

$$\Sigma^i = \dot{\phi}^i(\tau = 0) = 2g^{ij}(\phi_0) \frac{\partial W}{\partial \phi^i}(\phi_0, Q) . \tag{20}$$

A finite horizon area demands that $\dot{\phi}^i(\tau = -\infty) = 0$ and thus $\frac{\partial W}{\partial \phi^i}|_{\phi_H^i} = 0$ and

$$\lim_{\tau \to -\infty} e^{-2U} = R_H^2 \tau^2 , R_H^2 = W_H^2(\phi_H, Q) \tag{21}$$

and $R_H^2(Q) = |I_4(Q)|^{1/2}$, so that indeed the effective radius is given in terms of the Cartan quartic invariant of the 56 of $E_{7(7)}$.

By using the asymptotic behaviour of the warp factor at spacial infinity $\tau \to 0$, one has $U \to -M_{ADM}\tau$ and at the horizon $\tau \to \infty$

$$e^{-2U} \to \frac{1}{4\pi} A_H \tau^2 . \tag{22}$$

Thus, one gets model independent expressions for ADM mass, scalar charges and entropy as a function of W

$$M_{ADM}^2 = W_{\tau \to 0}^2 = W_\infty^2 \tag{23}$$

$$\Sigma^i = -2g^{ij} \frac{\partial W}{\partial \phi^j}|_{\tau \to 0} = -2g^{ij} \frac{\partial W}{\partial \phi^j}|_\infty \tag{24}$$

$$S = \frac{1}{4} A_H = \pi W^2|_{\tau \to -\infty} = \pi W^2|_H . \tag{25}$$

The horizon value of the scalar fields ϕ_H^i is a critical point for W in the moduli space of the theory, $\partial W/\partial \phi^r|_{\phi^i=\phi_H^i} = 0$. It follows that for $\tau \to 0$ (radial infinity)

$$M_{ADM}^2 = V_{BH}(\phi_\infty, p, q) - \frac{1}{2}g_{ij}\Sigma^i\Sigma^j \tag{26}$$

while for $\tau \to -\infty$ (horizon)

$$S_{BH} = \frac{1}{4}A_H = \pi V_{BH}(\phi_H, p, q) = \pi V_{BH}|_{\phi_H} \tag{27}$$

showing *attractor behaviour*.[10] For BPS states

$$S_{BH} = \pi|z_h|_H^2 \qquad M_{ADM}^2 = |z_h|_\infty^2 \,, \tag{28}$$

where z_h is the skew eigenvalue of the central charge Z_{AB} with the highest modulus. Quite generally, one can prove the validity of the BPS bound along the flow

$$M_{ADM} \geq |z_i| \,. \tag{29}$$

At the horizon

$$\dot{\phi}^i|_{\tau \to -\infty} = 0 \to \partial_i W_H = 0 \quad (W_H \leftrightarrow W_{crit}) \,. \tag{30}$$

For $N \geq 2$, the fake superpotential for the non-BPS branch[21,22] has been computed for wide classes of models,[19,23–29,38] based on symmetric geometries of moduli spaces, using as a tool the U-duality symmetry of the underlying supergravity. A universal procedure for its construction in $N = 2$ special geometries has been established,[26,27] which generalizes the results obtained for the *stu* model.[37] This universal procedure can be also applied to the $N = 8$ theory using the fact that the *stu* model is both a subsector of $N = 8$ and a model in $N = 2$ where we know how to describe W in terms of invariants. The results of Ref. 27 agree with the outcome of studies of black hole evolution with the method of time-like reduction to 3 dimensions in Ref. 28. However, in both contexts, the expression for W is only given implicitly, as solution of a sixth order polynomial having for coefficients some $SU(8)$ invariant functions composed out of the $N = 8$ central charge.

2. Attractors and Duality Orbits in N=8 Supergravity

As already noticed above, U-duality of the underlying extended supergravity dictates many important features.[7] Another important point is that the absolute Cartan invariant I_4 of the 56 dimensional representation of $E_{7(7)}$ and its derivatives allow to specify the supersymmetric features of a given solution and to classify the U-duality (continuous) orbits for the charge vector Q.[30,32,33]

The area of the horizon for both regular 1/8-BPS and non-BPS attractors, is proportional to the square of the absolute Cartan quartic invariant[17]

$$I_4 = Tr(Z\overline{Z})^2 - \frac{1}{4}(Tr\, Z\overline{Z})^2 + 4(Pf\, Z + Pf\, \overline{Z}) \,. \tag{31}$$

This expression, involving the traces of the matrix $Z_{AB}\bar{Z}^{BC}$ and of its square as well as the real part of the Pfaffian of $Z_{AB}(\phi, Q)$,

$$PfZ = \frac{1}{2^4 4!} \epsilon^{abcdefgh} Z_{AB} Z_{CD} Z_{EF} Z_{GH} \tag{32}$$

actually depends only on the electric and magnetic charges $Q = (p, q)$ and not on the scalar fields. Moreover, for fixed values of I_4 in d=4 and of an analogous cubic invariant I_3 in d=5, charge vectors Q for supergravities on symmetric spaces describe related orbits and attractors.[40,41]

By an $SU(8)$ transformation, one can reach the canonical bases where the anti-symmetric central charge matrix is (skew)diagonalized and takes the form

$$Z_{AB} = \begin{pmatrix} \rho_0 & & & \\ & \rho_1 & & \\ & & \rho_2 & \\ & & & \rho_3 \end{pmatrix} \otimes \begin{pmatrix} 0 & 1 \\ -1 & 0 \end{pmatrix} e^{i\varphi/4} \tag{33}$$

in terms of its skew eigenvalues $z_i = \rho_i e^{i\varphi/4}$, $i = 0, 1, 2, 3$ which amount to 5 independent parameters. The attractor equation becomes

$$\partial_i V = 0 \rightarrow z_i z_j + z_k^* z_l^* = 0, (i \neq j \neq k \neq l) \tag{34}$$

which yields two solutions

    a) $z_h \neq 0$, $z_{i \neq h} = 0$
    b) $|z_i| = \rho$, $argPfZ = \pi$

Considering the quartic invariant in this basis,

$$I_4 = [(\rho_0 + \rho_1)^2 - (\rho_2 + \rho_3)^2][(\rho_0 - \rho_1)^2 \tag{35}$$
$$- (\rho_2 - \rho_3)^2] + 8\rho_0\rho_1\rho_2\rho_3(\cos\phi - 1) \tag{36}$$
$$\rho_0 \geq \rho_1 \geq \rho_2 \geq \rho_3 \tag{37}$$

one finds that there are two disjoint regular orbits corresponding to the a) and b) attractor solutions:
i) 1/8 BPS: with $I_4 > 0$, for $\rho_0 \neq 0$ and $\rho_1 = \rho_2 = \rho_3 = 0$. The duality orbit is $E_{7(7)}/E_{6(2)}$.
ii) non BPS: with $I_4 < 0$, for $|z_i| = \rho$ and $argPfZ = \pi$. The duality orbit is $E_{7(7)}/E_{6(6)}$.
There are three additional "small" orbits, preserving respectively 1/8, 1/4 and 1/2 supersymmetry, which are singular, having $I_4 = 0$ and thus zero horizon area.

The entropy of the two regular branches is given by

$$S_{1/8\text{BPS}} = \pi V_{BH}|_H = \pi \rho_0^2|_H, \tag{38}$$
$$S_{\text{nonBPS}} = \pi V_{BH}|_H = \pi (2\rho_0)^2|_H, \tag{39}$$

where $\rho_0|_H$ is the fixed point value of the field dependent modulus of the highest eigenvalue ρ_0. A similar classification has recently been achieved also in the $N = 2$

case, where the superpotential W was computed for large and small orbits.[44] The new feature in that case is that singular orbits for N=2 black holes can also be non-BPS and that W can always be determined by simple radicals.

An interesting outcome of the study of black holes in N=8 four dimensional supergravity is related to some recent results,[68] which point to a possible key role played by black holes with singular geometry in relation to String Theory. $N = 8$ supergravity in $D = 4$ can be obtained by dimensional reduction of D=11, N=1 supergravity (M-theory) on a 7-torus T^7, or compacifying Type-II string theory on T^6. In Ref. 67 it has been recently observed that in the process of compactification, one is left not only with the 256 massless states of the N=8, D=4 supergravity, but also with an infinite tower of stringy elementary states with arbitrarily small mass, making the decoupling impossible. On the other hand, since the degeneracies of BPS states in string and M-theory are counted by U-duality invariant formulas, it is interesting to inquire whether one can consistently decouple these extra massless states without violating the U-duality invariance of the degeneracy formula. If one can't decouple the extra states without breaking U-duality, then this would suggest that one may be able to disprove the conjecture of UV finiteness of the perturbative N=8 supergravity theory. From the 4D point of view, these additional massless states appear to correspond to classical black hole solutions carrying charges which lead to light-like orbits with $I_4 = 0$,[68] and that $N = 8$ black holes with vanishing area of the horizon can be interpreted as dynamically reduced black branes with regular geometry. For example, 1/8 BPS black holes with $I_4 = 0$ correspond to reduced (wrapped) black holes (black strings) at D=5 (with three charges q_1, q_2, q_3, $J_{MAX} = 7/2$). Conversely, 1/4 BPS black holes with $\partial I_4 = 0$ correspond to reduced and wrapped black strings at $D = 6$ (two charges q_1 and q_2, $J_{MAX} = 3$. Finally, 1/2 BPS black holes with $\partial^2 I_4 = 0$ correspond to KK states i.e. massless particles in higher dimensions $J_{MAX} = 2$. If only regular black holes, with $AdS_2 \times S_2$ horizon geometries are to be retained in a consistent theory of gravity, one may not need to open up extra dimensions:[67] weather this feature could hint to a possible finite theory of D=4, $N = 8$ supergravity[5] without the need of string theory, is a question that is presently under close investigation.

3. Moduli Spaces of Attractors and Flat Directions

It is a property of N=2 supergravity that the geometry of black holes, at least in the Einsteinian approximation, does not depend on the hypermultiplet moduli space.[10] This implies that attractive solutions exist for arbitrary v.e.v's of the hypermultiplet scalars. This phenomenon generalizes to the non-BPS N=2 regular black holes as well as to small black holes, for which there is no attractor behaviour.[42–44] In fact, it has been realized that, in terms of the fake superpotential W (or, for BPS solutions, in terms of the Z central charge), such "flat directions" are flat directions for W along the complete flow and therefore the ADM mass as well. For $N > 2$, flat directions exist even for BPS regular attractive (large) black holes and for singular solutions. In this latter case, the moduli space of flat directions is usually bigger, and in fact for the doubly critical orbits the W function just depends on one real modulus having the simple interpretation as the radius dependence of the Kaluza-Klein mass.[45,46] Let us confine again the discussion to the N=8 case. The five charge orbits are as follows[31]

orbit	susy	coset	moduli space	
$I_4 > 0$	1/8BPS	$\frac{E_{7(7)}}{E_{6(2)}}$	$\frac{E_{6(2)}}{SU(6) \times SU(2)}$	
$I_4 < 0$	nonBPS	$\frac{E_{7(7)}}{E_{6(6)}}$	$\frac{E_{6(6)}}{USp(8)}$	
$I_4 = 0, \partial I_4 \neq 0$	1/8BPS	$\frac{E_{7(7)}}{F_{4(4)} \odot T_{26}}$	$\frac{F_{4(4)}}{USp(6) \times USp(2)} \odot T_{26}$	(40)
$I_4 = 0, \partial I_4 = 0$	1/4BPS	$\frac{E_{7(7)}}{O(6,5) \odot (T_{32}+T_1)}$	$\frac{O(6,5)}{O(6) \times O(5)} \odot (T_{32} + T_1)$	
$I_4 = 0, \partial I_4 = 0, \partial^2 I_4 = 0$	1/2BPS	$\frac{E_{7(7)}}{E_{6(6)} \odot T_{27}}$	$\frac{E_{6(6)}}{USp(8)} \odot T_{27}$	

The last column refers to the moduli space of flat directions and is given by the stabilizer of the coset orbit divided by its maximal compact subgroup. One observes that small lightlike black holes, having only the condition $I_4 = 0$, apart from T_{26} translations, have the same moduli space as the 1/8 BPS large black hole in D=5.[42]

Remarkably, the 1/2 BPS black holes, apart from T_{27} additional translations, have the same moduli space as large non-BPS black holes and both are the 5D N=8 supergravity moduli space. This is due to the fact that both these configurations can be obtained without genuine 5D black hole charges but turning on the NUT charge and angular momentum , without breaking the $E_{6(6)}$ symmetry of the 5D theory.[24,40,41]

4. Conclusions and Other Recent Developments

Since the attractor mechanism was observed,[10] there has been a lively activity around the study of black holes in connection with string theory and M-theory. To begin with, an intriguing connection was conjectured between the central charge and the topological supergravity partition function at the attractor point.[48] A further important development has been the computation of the entropy through microstate counting, reproducing the Bekenstein-Hawking formula in the regime of large charges.[47]

Then, quantum corrections to the entropy formulae and to the attractor equations were computed in superstrings and supergravity with N=2 supersymmetry.[49–52] Microstate counting for black holes was considered in many papers, including the case of small black holes for which the area formula does not apply, even in the classical regime.[53–55] Several aspects concerning multicenter black holes, black hole deconstructions, split attractor flows and walls of marginal stability have been thoroughly investigated.[56–59]

The connection between 5D and 4D has been explored under different perspectives, as for instance in Refs. 40, 66 among many other contributions. Three dimensional time reduction has also been shown to be a powerful tool for black hole classification, duality orbits and possible quantization of the geodesic flows.[60–62]

As a final note, we would like to mention another more speculative aspect of black hole physics, namely the mathematical connection between black hole physics and Quantum Information Theory, in particular the relation between black hole entropy and qubits multipartite entanglement.[64]

Acknowledgments

This work is supported in part by the ERC Advanced Grant no. 226455, *"Supersymmetry, Quantum Gravity and Gauge Fields" (SUPERFIELDS)*, by MIUR-PRIN contract 20075ATT78 and by DOE Grant DE-FG03-91ER40662.

References

1. S. Ferrara, D. Z. Freedman and P. Van Nieuwenhuizen, *Progress Toward a Theory of Supergravity*, Phys. Rev. **D13** (1976) 3214; S. Deser and B. Zumino, *Consistent Supergravity*, Phys. Lett. **62B** (1976) 335.
2. E. Cremmer and B. Julia, *The SO(8) Supergravity*, Nucl. Phys. B **159** (1979) 141.
3. E. Cremmer, B. Julia and J. Scherk, *Supergravity theory in 11 dimensions* Phys. Lett. B **76** (1978) 409.
4. S. Ferrara and R. Kallosh, *On N = 8 attractors*, Phys. Rev. D **73**, 125005 (2006) [arXiv:hep-th/0603247].
5. Z. Bern, L. J. Dixon and R. Roiban, *Is N = 8 Supergravity Ultraviolet Finite?*, Phys. Lett. B **644**, 265 (2007) [arXiv:hep-th/0611086]; Z. Bern, J. J. Carrasco, L. J. Dixon, H. Johansson and R. Roiban, *The Ultraviolet Behavior of N=8 Supergravity at Four Loops*, Phys. Rev. Lett. **103**, 081301 (2009) [arXiv:0905.2326 [hep-th]].
6. M. K. Gaillard and B. Zumino, *Duality Rotations For Interacting Fields*, Nucl. Phys. B **193** (1981) 221; P. Aschieri, S. Ferrara and B. Zumino, *Duality Rotations in Nonlinear Electrodynamics and in Extended Supergravity*, Riv. Nuovo Cim. **31**, 625 (2009) [arXiv:0807.4039 [hep-th]].
7. C. M. Hull and P. K. Townsend, Nucl. Phys. B**438** 109 (1995); M. Cvetic and C. M. Hull, *Black holes and U-duality* Nucl. Phys. B **480** (1996) 296 [arXiv:hep-th/9606193].
8. E. Witten, *String theory dynamics in various dimensions*, Nucl. Phys. B **443** (1995) 85 [arXiv:hep-th/9503124].
9. S. Ferrara, K. Hayakawa and A. Marrani, *Lectures on Attractors and Black Holes*, Fortsch. Phys. **56**, 993 (2008) [arXiv:0805.2498 [hep-th]].
10. S. Ferrara, R. Kallosh and A. Strominger, *N=2 extremal black holes*, Phys. Rev. D **52** (1995) 5412; A. Strominger, *Macroscopic entropy of N=2 extremal black holes*, Phys. Lett. **B383**, 39 (1996); S. Ferrara and R. Kallosh, *Supersymmetry and attractors*, Phys. Rev. D **54**, 1514 (1996); S. Ferrara and R. Kallosh, *Universality of supersymmetric attractors*, Phys. Rev. D **54**, 1525 (1996).
11. S. Ferrara, G. W. Gibbons and R. Kallosh, *Black holes and critical points in moduli space*, Nucl. Phys. B **500**, 75 (1997).
12. L. Andrianopoli, R. D'Auria, S. Ferrara and M. Trigiante, *Extremal Black Holes in Supergravity*, Lect.Notes Phys.737:661 (2008), [arXiv:hep-th/0611345]

13. R. Kallosh, A. D. Linde, T. Ortin, A. W. Peet and A. Van Proeyen, *Supersymmetry as a cosmic censor*, Phys. Rev. D **46**, 5278 (1992) [arXiv:hep-th/9205027].

14. O. Aharony, S. S. Gubser, J. M. Maldacena, H. Ooguri and Y. Oz, *Large N field theories, string theory and gravity*, Phys. Rept. **323**, 183 (2000) [arXiv:hep-th/9905111].

15. M. J. Duff, R. R. Khuri and J. X. Lu, *String solitons*, Phys. Rept. **259**, 213 (1995) [arXiv:hep-th/9412184].

16. G. Gibbons and P. K. Townsend, *Vacuum interpolation in supergravity via super p-brane*, Phys.Rev.Lett.**71**, (1993) 3754 .

17. R. Kallosh and B. Kol, *E(7) Symmetric Area of the Black Hole Horizon*, Phys. Rev. D **53**, 5344 (1996) [arXiv:hep-th/9602014].

18. G. Arcioni, A. Ceresole, F. Cordaro, R. D'Auria, P. Fre, L. Gualtieri and M. Trigiante, *N = 8 BPS black holes with 1/2 or 1/4 supersymmetry and solvable Lie algebra decompositions*, Nucl. Phys. B **542** (1999) 273 [arXiv:hep-th/9807136].

19. A. Ceresole and G. Dall'Agata, *Flow Equations for Non-BPS Extremal Black Holes*, JHEP **0703** (2007) 110 [arXiv:hep-th/0702088].

20. D. Z. Freedman, C. Nunez, M. Schnabl and K. Skenderis, *Fake Supergravity and Domain Wall Stability*, Phys. Rev. D **69** (2004) 104027 [arXiv:hep-th/0312055]. A. Celi, A. Ceresole, G. Dall'Agata, A. Van Proeyen and M. Zagermann, *On the fakeness of fake supergravity*, Phys. Rev. D **71**, 045009 (2005) [arXiv:hep-th/0410126].

21. A. Dabholkar, A. Sen and S. P. Trivedi, *Black hole microstates and attractor without supersymmetry*, JHEP **0701**, 096 (2007) [arXiv:hep-th/0611143].

22. E. G. Gimon, F. Larsen and J. Simon, *Black Holes in Supergravity: the non-BPS Branch*, JHEP **0801** (2008) 040 [arXiv:hep-th/0710.4967].

23. L. Andrianopoli, R. D'Auria, E. Orazi and M. Trigiante, *First Order Description of Black Holes in Moduli Space*, JHEP **0711** (2007) 032 [arXiv:hep-th/0706.0712].

24. G. Lopes Cardoso, A. Ceresole, G. Dall'Agata, J. M. Oberreuter and J. Perz, *First-order flow equations for extremal black holes in very special geometry*, JHEP **0710** (2007) 063 [arXiv:hep-th/0706.3373].

25. S. Bellucci, S. Ferrara, A. Marrani and A. Yeranyan, *stu Black Holes Unveiled*, [arXiv:hep-th/0807.3503].

26. A. Ceresole, G. Dall'Agata, S. Ferrara and A. Yeranyan, *First order flows for N=2 extremal black holes and duality invariants*, Nucl.Phys.B824:239-253,2010 [arXiv:0908.1110] .

27. A. Ceresole, G. Dall'Agata, S. Ferrara and A. Yeranyan, *Universality of the superpotential for d=4 extremal black holes*, Nucl. Phys. **B** (2010), in press [arXiv:0910.2697].

28. G. Bossard, Y. Michel and B. Pioline, *Extremal black holes, nilpotent orbits and the true fake superpotential*, [arXiv:0908.1742].

29. S. Ferrara, A. Gnecchi and A. Marrani, *d=4 Attractors, Effective Horizon Radius and Fake Supergravity*, Phys. Rev. D **78** (2008) 065003 [arXiv:hep-th/0806.3196]; J. Perz, P. Smyth, T. Van Riet and B. Vercnocke, *First-order flow equations for extremal and non-extremal black holes*, JHEP **0903** (2009) 150 [arXiv:hep-th/0810.1528]; K. Goldstein and S. Katmadas, *Almost BPS black holes*, JHEP **0905** (2009) 058 [arXiv:hep-th/0812.4183]; I. Bena, G. Dall'Agata, S. Giusto, C. Ruef and N. P. Warner, *Non-BPS Black Rings and Black Holes in Taub-NUT*, JHEP **0906** (2009) 015 [arXiv:hep-th/0902.4526]; P. Galli and J. Perz, *Non-supersymmetric extremal multicenter black holes with superpotentials*, [arXiv:0909.5185].

30. S. Ferrara and J. M. Maldacena, *Branes, central charges and U-duality invariant BPS conditions*, Class. Quant. Grav. **15**, 749 (1998) [arXiv:hep-th/9706097].

31. S. Ferrara and M. Gunaydin, *Orbits of exceptional groups, duality and BPS states in string theory*, Int. J. Mod. Phys. A13 (1998) 2075;

32. S Bellucci, S. Ferrara, M. Gunaydin and A. Marrani, *Charge orbits of symmetric special geometries and attractors*, Int. J. Mod. Phys. A **21** (2006) 5043 [arXiv:hep-th/0606209].

33. L. Andrianopoli, R. D'Auria and S. Ferrara, *U-duality and central charges in various dimensions revisited*, Int. J. Mod. Phys. A **13**, 431 (1998) [arXiv:hep-th/9612105]; L. Andrianopoli, R. D'Auria and S. Ferrara, *Five dimensional U-duality, black-hole entropy and topological invariants*, Phys. Lett. B **411**, 39 (1997) [arXiv:hep-th/9705024].

34. M. Cvetic and D. Youm, *Dyonic BPS saturated black holes of heterotic string on a six torus*, Phys. Rev. **D 53** (1996) 584 [arXiv:hep-th/9507090].

35. R. Kallosh, T. Ortin and A. W. Peet, *Entropy and action of dilaton black holes*, Phys. Rev. D **47** (1993) 5400 [arXiv:hep-th/9211015].

36. P. K. Townsend, *P-brane democracy*, arXiv:hep-th/9507048.

37. M. J. Duff, J. T. Liu and J. Rahmfeld, *Four-Dimensional String-String-String Triality*, Nucl. Phys. B **459**, 125 (1996) [arXiv:hep-th/9508094]; K. Behrndt, R. Kallosh, J. Rahmfeld, M. Shmakova and W. K. Wong, *STU black holes and string triality*, Phys. Rev. D **54** (1996) 6293 [arXiv:hep-th/9608059].

38. L. Andrianopoli, R. D'Auria, E. Orazi and M. Trigiante, *First Order Description of D=4 static Black Holes and the Hamilton-Jacobi equation*, [arXiv:hep-th/0905.3938].

39. B. L. Cerchiai, S. Ferrara, A. Marrani and B. Zumino, *Duality, Entropy and ADM Mass in Supergravity*, Phys. Rev. D **79** (2009) 125010 [arXiv:0902.3973].

40. A. Ceresole, S. Ferrara and A. Gnecchi, *5d/4d U-dualities and N=8 black holes*, Phys. Rev. D **80** (2009) 125033 [arXiv:0908.1069 [hep-th]].

41. A. Ceresole, S. Ferrara, A. Gnecchi and A. Marrani, *More on N=8 Attractors*, Phys. Rev. D **80** (2009) 045020 [arXiv:0904.4506 [hep-th]].

42. S Ferrara and A. Marrani, *On the Moduli Space of non-BPS Attractors for N=2 Symmetric Manifolds*, Phys. Lett. B **652** (2007) 111 [arXiv:0706.1667 [hep-th]].

43. L. Andrianopoli, R. D'Auria, S. Ferrara and M. Trigiante, *Fake Superpotential for Large and Small Extremal Black Holes*, arXiv:1002.4340 [hep-th].

44. A. Ceresole, S. Ferrara and A. Marrani, *Small N=2 Extremal Black Holes in Special Geometry* [arXiv:1006.2007[hep-th]]

45. B. L. Cerchiai, S. Ferrara, A. Marrani and B. Zumino, *Charge Orbits of Extremal Black Holes in Five Dimensional Supergravity*, arXiv:1006.3101 [hep-th].

46. L. Borsten, D. Dahanayake, M. J. Duff, S. Ferrara, A. Marrani and W. Rubens, *Observations on Integral and Continuous U-duality Orbits in N=8 Supergravity*, arXiv:1002.4223 [hep-th].

47. A. Strominger and C. Vafa, *Microscopic Origin of the Bekenstein-Hawking Entropy*, Phys. Lett. B **379**, 99 (1996) [arXiv:hep-th/9601029].

48. H. Ooguri, A. Strominger and C. Vafa, *Black hole attractors and the topological string*, Phys. Rev. D **70** (2004) 106007 [arXiv:hep-th/0405146].

49. G. Lopes Cardoso, B. de Wit, J. Kappeli and T. Mohaupt, *Black hole partition functions and duality* JHEP **0603** (2006) 074 [arXiv:hep-th/0601108]. G. Lopes Cardoso, B. de Wit, J. Kappeli and T. Mohaupt, *Asymptotic degeneracy of dyonic N = 4 string states and black hole entropy*, JHEP **0412** (2004) 075 [arXiv:hep-th/0412287].

50. A. Dabholkar, R. Kallosh and A. Maloney, *A stringy cloak for a classical singularity*, JHEP **0412** (2004) 059 [arXiv:hep-th/0410076].

51. J M. Maldacena, A. Strominger and E. Witten, *Black hole entropy in M-theory*, JHEP **9712**, 002 (1997) [arXiv:hep-th/9711053].

52. A. Sen, *Black Hole Entropy Function, Attractors and Precision Counting of Microstates* Gen.Rel.Grav.**40** (2008) 2249-2431, e-Print: arXiv:0708.1270 [hep-th]

53. A. Dabholkar, *Black hole entropy in string theory: Going beyond Bekenstein and Hawking*, Int. J. Mod. Phys. D **15**, 1561 (2006); A. Dabholkar, F. Denef, G. W. Moore and B. Pioline, *Exact and Asymptotic Degeneracies of Small Black Holes*, JHEP **0508**, 021 (2005) [arXiv:hep-th/0502157].

54. A. Dabholkar, F. Denef, G. W. Moore and B. Pioline, *Precision counting of small black holes*, JHEP **0510**, 096 (2005) [arXiv:hep-th/0507014].

55. D. Shih, A. Strominger and X. Yin, *Counting dyons in N = 8 string theory*, JHEP **0606**, 037 (2006) [arXiv:hep-th/0506151].

56. F. Denef and G. W. Moore, *Split states, entropy enigmas, holes and halos*, arXiv:hep-th/0702146.

57. F. Denef, D. Gaiotto, A. Strominger, D. Van den Bleeken and X. Yin, *Black hole deconstruction*, arXiv:hep-th/0703252.

58. A. Sen, *Wall Crossing Formula for N=4 Dyons: A Macroscopic Derivation*, JHEP **0807**, 078 (2008) [arXiv:0803.3857 [hep-th]]; *N=8 Dyon Partition Function and Walls of Marginal Stability*, JHEP **0807**, 118 (2008) [arXiv:0803.1014 [hep-th]].

59. D. Gaiotto, G. W. Moore and A. Neitzke, *Wall-crossing, Hitchin Systems, and the WKB Approximation*, arXiv:0907.3987 [hep-th]; D. L. Jafferis and G. W. Moore, *Wall crossing in local Calabi Yau manifolds*, arXiv:0810.4909 [hep-th].

60. M. Gunaydin, A. Neitzke, B. Pioline and A. Waldron, *Quantum Attractor Flows*, JHEP **0709**, 056 (2007) [arXiv:0707.0267 [hep-th]].

61. E. Bergshoeff, W. Chemissany, A. Ploegh, M. Trigiante and T. Van Riet, *Generating Geodesic Flows and Supergravity Solutions* Nucl. Phys. B **812**, 343 (2009) [arXiv:0806.2310 [hep-th]].

62. G. Bossard, H. Nicolai and K. S. Stelle, *Universal BPS structure of stationary supergravity solutions*, JHEP **0907**, 003 (2009) [arXiv:0902.4438 [hep-th]].

63. M. Guica, T. Hartman, W. Song and A. Strominger, *The Kerr/CFT Correspondence*, Phys. Rev. D **80**, 124008 (2009) [arXiv:0809.4266 [hep-th]].

64. L. Borsten, D. Dahanayake, M. J. Duff, H. Ebrahim and W. Rubens, *Black Holes, Qubits and Octonions*, Phys. Rept. **471**, 113 (2009) [arXiv:0809.4685 [hep-th]]; M. Duff, *Black Holes And Qubits*, CERN Cour. **50N4**, 13 (2010).

65. I. Bena, G. Dall'Agata, S. Giusto, C. Ruef and N. P. Warner, *Non-BPS Black Rings and Black Holes in Taub-NUT*, JHEP **0906** (2009) 015 [arXiv:0902.4526 [hep-th]].

66. D. Gaiotto, A. Strominger and X. Yin, *5D black rings and 4D black holes*, JHEP **0602**, 023 (2006) [arXiv:hep-th/0504126]; D. Gaiotto, A. Strominger and X. Yin, *New Connections Between 4D and 5D Black Holes*, JHEP **0602**, 024 (2006) [arXiv:hep-th/0503217].

67. M. B. Green, H. Ooguri and J. H. Schwarz, *Decoupling Supergravity from the Superstring*, Phys. Rev. Lett. **99**, 041601 (2007) [arXiv:0704.0777 [hep-th]].

68. M. Bianchi, S. Ferrara and R. Kallosh, *Observations on Arithmetic Invariants and U-Duality Orbits in N =8 Supergravity*, JHEP **1003**, 081 (2010) [arXiv:0912.0057 [hep-th]]; *Perturbative and Non-perturbative N =8 Supergravity*, arXiv:0910.3674 [hep-th].

GAUGE THEORY OF GRAVITY WITH DE SITTER SYMMETRY AS A SOLUTION TO THE COSMOLOGICAL CONSTANT PROBLEM AND THE DARK ENERGY PUZZLE

PISIN CHEN[1,2]

1. Department of Physics and Graduate Institute of Astrophysics &
Leung Center for Cosmology and Particle Astrophysics,
National Taiwan University, Taipei 10617, Taiwan, R.O.C.
pisinchen@phys.ntu.edu.tw

2. Kavli Institute for Particle Astrophysics and Cosmology,
SLAC National Accelerator Laboratory,
Stanford University, Stanford, CA 94025, U.S.A.
chen@slac.stanford.edu

We propose a solution to the longstanding cosmological constant (CC) problem which is based on the fusion of two existing concepts. The first is the suggestion that the proper description of classical gravitational effects is the gauge theory of gravity in which the connection instead of the metric acts as the dynamical variable. The resulting field equation does not then contain the CC term. This removes the connection between the CC and the quantum vacuum energy, and therefore addresses the *old* CC problem of why quantum vacuum energy does not gravitate. The CC-equivalent in this approach arises from the constant of integration when reducing the field equation to the Einstein equation. The second is the assumption that the universe obeys de Sitter symmetry, with the observed accelerating expansion as its manifestation. We combine these ideas and identify the constant of integration with the inverse-square of the radius of curvature of the de Sitter space. The origin of dark energy (DE) is therefore associated with the inherent spacetime geometry, with the smallness of DE protected by symmetry. This addresses the *new* CC problem, or the DE puzzle. This approach, however, faces major challenges from quantum considerations. These are the ghost problem associated with higher order gravity theories and the quantum instability of the de Sitter spacetime. We discuss their possible remedies.

Keywords: Cosmological constant; dark energy; gauge theory of gravity; de Sitter space; ghost.

1. Introduction

It is well-known that Einstein's general relativity (GR) allows for a cosmological constant (CC) which is a priori undetermined. The quantum vacuum energy resulting from the zero point fluctuations of the quantum fields, which is a natural consequence of the uncertainty principle, satisfies every aspect as a candidate for the CC. Yet numerically the quantum vacuum energy associated with the fluctuations at the Planck scale, $\sim 10^{112} \text{eV}^4$, is about 124 orders of magnitude larger

330

than the critical density of the universe. Evidently the quantum vacuum energy should not gravitate. Otherwise the universe would not have survived until now. This conflict between GR and quantum theory is the essence of the longstanding CC problem,[1] which clearly requires a resolution. We shall refer to this as the *old* CC problem.

The dramatic discovery of the accelerating expansion of the universe in 1998[2,3] ushers in another chapter of the CC problem. The substance that is supposedly responsible for this accelerating expansion has been referred to as dark energy (DE). It has been customary to characterize DE in terms of its equation of state, $p = w\rho$, where p is the pressure and ρ the density of DE. According to GR, accelerating expansion can be accomplished if $w < -1/3$. Einstein's CC corresponds to $w = -1$. While the accuracy of the measurements to date still cannot resolve whether w varies in time and whether it is exactly equal to -1, a constant DE and thus the CC remains the simplest and most likely answer. This, however, creates a new challenge for the CC problem. After finding a way, hopefully, to cancel the CC to 124 decimal points, how do we reinstate 1 to the last digit to make it nonzero but tiny? Let us call this the *new* CC problem, or the DE puzzle.

In this Letter, we propose a solution to the problems raised above by combining two existing ideas, that is, by invoking the gauge theory of gravity as a substitute of Einstein's GR as the foundation of gravity theory and by assuming de Sitter symmetry as the underlying group property of the universe. The gauge theory of gravity, which is quadratic in the curvature tensor, results in a field equation that is second order differential equation in the connection and therefore third order in the metric. The CC therefore does not appear in this field equation. This means the quantum vacuum energy is not a source of gravity. This proposal, however, does not remove the CC entirely. In order to reduce the field equation to the Einstein equation, one must perform a integration which then induces a constant of integration that behaves like a CC. To fix this constant, we assume that the universe satisfies the de Sitter symmetry. That is, we associate the constant of integration with the radius of curvature of the de Sitter space, a new fundamental length of the universe. We suggest that this new fundamental length plays a role in the action of the gauge theory of gravity.

Neither of these two ingredients in our solution is new. Guided by the fiber bundle theory for gauge fields, Yang first formulated the gauge theory of gravity in 1974.[4] This formulation was further investigated by various authors.[5-7] In the aftermath of the discovery of accelerating expansion, recently Cook suggested the circumvention of the CC problem via the gauge theory of gravity.[8] He observed that the quantum vacuum energy cannot be a source of gravity in the third order field equation derived from the gauge theory of gravity, while the DE responsible for the accelerating expansion of the universe can be identified with the constant of integration. The history of de Sitter symmetry dates further back. The notion that the physical laws are not invariant under the Poincare group but instead the de Sitter group was first proposed by Luigi Fantappie in 1954 and reinvestigated in

1968 by Bacry and Levy-Leblond.[9] In the post dark energy era, many authors have connected this notion with the observed accelerating expansion.[10–13]

However, separately these two lines of solutions to the CC problem would be incomplete. While the gauge theory of gravity can indeed be devoid of the CC by construction, the CC reappears through the constant of integration. It may seem that one can at this point resort to the anthropic principle to settle its value. But as long as one still regards it as some form of stress energy-momentum, one would still be obliged to address the microscopic, or quantum, origin of this substance. Without a symmetry principle for the protection, this substance still has to be subject to quantum corrections, which may not help in preserving the desired smallness of the CC. The drawback of the solution that relies only on the de Sitter symmetry is more obvious. It simply does not address the old CC problem. That is, why doesn't the quantum vacuum energy gravitate? We will see that by fusing these two concepts together, the CC problem may be solved more properly.

2. Gauge Theory of Gravity and the Cosmological Constant

In Yang's formulation of gravity, the affine connection $\Gamma^\alpha_{\mu\nu}$ on the fiber bundle is the dynamical variable which determines the curvature tensor

$$R^\alpha_{\beta\mu\nu} = \partial_\mu \Gamma^\alpha_{\beta\nu} - \partial_\nu \Gamma^\alpha_{\beta\mu} + \Gamma^\alpha_{\tau\mu}\Gamma^\tau_{\beta\nu} - \Gamma^\alpha_{\tau\nu}\Gamma^\tau_{\beta\mu}. \tag{1}$$

For our purpose, we shall only consider the pure space. In close analogy with the Maxwell theory, the action of the gauge theory of gravity reads[8]

$$S_G = \kappa \int d^4x \sqrt{-g} \Big(R^{\alpha\beta\mu\nu} R_{\alpha\beta\mu\nu} + 16\pi J^{\alpha\beta}_\mu \Gamma^\mu_{\alpha\beta} \Big), \tag{2}$$

where κ is the overall coefficient which will be determined later. $J^{\alpha\beta}_\mu$ is the gravitational current defined as

$$J^{\alpha\beta}_\mu = \frac{2G}{c^4} \Big[\nabla^\alpha \bar{T}^\beta_\mu - \nabla^\beta \bar{T}^\alpha_\mu \Big], \tag{3}$$

where ∇_α is the covariant derivative and

$$\bar{T}^\alpha_\mu = T^\alpha_\mu - \frac{1}{2}\delta^\alpha_\mu T \tag{4}$$

is the Hilbert conjugate of T^α_μ and $T = T^\mu_\mu$. Varying S_G against $\Gamma^\mu_{\alpha\beta}$, we arrive at one of the field equation:

$$\nabla_\nu R^{\mu\nu}_{\alpha\beta} = -4\pi J^\mu_{\alpha\beta}. \tag{5}$$

This and the Bianchi identity,

$$\nabla_\lambda R_{\alpha\beta\mu\nu} + \nabla_\nu R_{\alpha\beta\lambda\mu} + \nabla_\mu R_{\alpha\beta\nu\lambda} = 0, \tag{6}$$

together determines the curvature tensor. The analogy of these equations to Maxwell's equations is again apparent. Note that Einstein's CC does not appear in

(6). This fact can also be appreciated intuitively if we recall that the affine connection is related to the first derivatives of the metric,

$$\Gamma_{\alpha\beta\gamma} = \frac{1}{2}[\partial_\gamma g_{\alpha\beta} + \partial_\beta g_{\alpha\gamma} - \partial_\alpha g_{\beta\gamma}], \tag{7}$$

then we see that the Bianchi identity is a third order differential equation in $g_{\mu\nu}$. This is in contrast with the Einstein equation which is second order in the metric. Since the covariant derivative of the metric is zero, a third order differential equation in $g_{\mu\nu}$ removes the CC term in principle.

A pure space that is empty of energy-momentum satisfies the condition[4]

$$\nabla_\gamma R_{\alpha\beta} - \nabla_\beta R_{\alpha\gamma} = 0, \tag{8}$$

for the Ricci tensor $R_{\alpha\beta}$. This condition then reduces the gauge-Bianchi identity to

$$\nabla_\alpha R^\alpha_{\beta\mu\nu} = 0. \tag{9}$$

Integrating this equation once over spacetime, we recover the Einstein equation with a constant of integration which has the same form as that of Einstein's CC. But in our case this constant should be associated with the boundary condition of the universe.

3. De Sitter Symmetry and the Dark Energy

We now invoke the de Sitter symmetry and assume that the universe is inherently de Sitter, where the 4-spacetime is a hyperboloid in a 5-dimensional Minkowski space under the constraint

$$-x_0^2 + x_1^2 + x_2^2 + x_3^2 + x_4^2 = l_{dS}^2, \tag{10}$$

where l_{dS} is the radius of curvature of the de Sitter space, or simply the de Sitter radius. The Hubble expansion of the universe is then viewed as a process that approaches the asymptotic limit of a pure space which is de Sitter in nature, evidenced by the current assessment that the CC-like DE substance has become dominant in the universe at late times: $\Omega_{DE} = \rho_{DE}/\rho_{cr} \simeq 0.75$, where the critical density $\rho_{cr} = 3H_0^2/8\pi G = 1.88 \times 10^{-29} h^2 \text{g/cm}^2$, with $h \equiv H_0/[100(\text{km/s})/\text{Mpc}]$ and H_0 is the Hubble constant. Based on this assumption the de Sitter radius is the asymptotic value of the Hubble distance, both in the gauge theory of gravity and in GR:

$$l_{dS} \simeq 1.33 H_0^{-1} \sim 1.5 \times 10^{28} \text{cm}. \tag{11}$$

Identifying the constant of integration as 3 times the inverse-square of the de Sitter radius, we then have

$$G_{\mu\nu} = -\frac{3}{l_{dS}^2} g_{\mu\nu}, \tag{12}$$

where $G_{\mu\nu}$ is the Einstein tensor. The only nontrivial component that satisfies this equation is a constant for the Ricci scalar,

$$R = \frac{12}{l_{dS}^2}. \tag{13}$$

The local structure is then characterized by

$$R_{\alpha\beta\mu\nu} = \frac{1}{12}[g_{\alpha\mu}g_{\beta\nu} - g_{\alpha\nu}g_{\beta\mu}]R, \tag{14}$$

which confirms that the Kretschmann scalar is a constant in the de Sitter universe,

$$R_{\alpha\beta\mu\nu}R^{\alpha\beta\mu\nu} = \frac{1}{6}R^2 = \frac{24}{l_{dS}^4}. \tag{15}$$

Based on this picture, the origin of DE is associated with the inherent spacetime geometry and not with vacuum energy or any other physical substance. Note that as an fundamental constant under de Sitter symmetry, l_{dS} is not subject to quantum corrections. The smallness of DE is therefore protected by symmetry.

In our definition of the gauge gravity action in (2), we did not specify the overall coefficient κ. In general, the overall coefficient of a classical action does not affect the physics and is therefore irrelevant. It does matter, however, if we consider the quantum fluctuations around the classical state. We note that since the curvature tensor has dimension $[L]^{-2}$, any action term that is quadratic in the curvature tensor has its dimensionality entirely cancelled upon integration over the 4-spacetime. Thus the coefficient κ must have the same dimension as the action itself. In comparison, since in the Einstein-Hilbert action the integration over the Ricci scalar has dimension $[L]^2$, its coefficient is $1/(16\pi G)$. Numerically, this factor is obtained by demanding that in the nonrelativistic limit GR reduces to the Newtonian gravity. But at the mean time since the gravity is weak, and therefore $1/G$ is large, quantum fluctuations of spacetime curvature in GR would be tiny at scales much larger than the Planck length. In the case of gauge theory of gravity, a natural choice for κ would be the Planck constant $\hbar$, which has the right dimension. This, however, would imply that the the deviation of the curvature tensor from its classical minimum would be of the order unity at all scales, which is unacceptable. A different possibility is to introduce a length parameter L so that $\kappa \propto L^2/G$. Following the same approach as in GR, we fix κ numerically by demanding that the asymptotic de Sitter limit in the gauge theory of gravity is identical to that in GR. This can be done by comparing (13) and (15) above, which are the de Sitter solutions to the Einstein-Hilbert action and the gauge theory action, respectively. By demanding that the two theories have the same de Sitter limit, we find that L is proportional to the de Sitter length l_{dS}, namely,

$$\kappa = \frac{l_{dS}^2}{96\pi G}. \tag{16}$$

4. Quantum Instability of de Sitter Spacetime

One challenge of our approach is that de Sitter space may be inherently unstable. The issue of quantum instability of de Sitter space was investigated by various authors in the 1980s in the aftermath of the introduction of the concept of inflation.

Abbott and Deser[14] have shown that de Sitter space is stable under a restricted class of classical gravitational perturbations. So any instability of de Sitter space may likely have a quantum origin. Ford[15] demonstrated through the expectation value of the energy-momentum tensor for a system with a quantum field in a de Sitter background space that in general it contains a term that is proportional to the metric tensor and grows in time. As a result the curvature of the spacetime would decrease and de Sitter space tends to decay into the flat space. Similar conclusion has also been reached by Antoniadis, Iliopoupos, and Tomaras.[16]

We note that generically the expectation value of such energy-momentum tensor has the form

$$\langle T_{\mu\nu} \rangle \propto g_{\mu\nu} H^4(Ht), \tag{17}$$

where H is the Hubble parameter in the de Sitter metric, i.e.,

$$ds^2 = dt^2 - a^2(t)dx^2, \tag{18}$$

with $a(t) = e^{Ht}$. According to (17), the decay time of this process is

$$\tau \sim H^{-1}. \tag{19}$$

In our case, this means the decay time is of the order of the de Sitter radius,

$$\tau \sim l_{dS} \simeq 1.33 H_0^{-1}. \tag{20}$$

Since the age of our universe is smaller than l_{dS}, we are still safe in observing the accelerating expansion in action.

5. Ghost Problem in Higher Order Gravity Theories

Another major challenge of our approach is the ghost problem that is generally associated with higher order gravity theories such as ours. Ghost states are quantum states having negative norms. A quantum field theory is generally considered unacceptable if it contains ghost states, because negative norm implies negative probability. There are possible ways, however, to circumvent the ghost problem.

Since our action is quadratic in $R_{\alpha\beta\mu\nu}$, which has mass dimension 2, the theory is conformally invariant. Analogous to the scalar field theory the conformal invariance can be spontaneously broken, which would induce a mass to the field. This means that our action will necessarily induce an Einstein term R, since it is proportional to $1/k^2$ in the propagator and therefore acts as a mass term.[17] This implies that our theory goes to the Einstein GR at large distance. On the other hand, with $R_{\alpha\beta\mu\nu} \propto 1/k^4$, it is renormalizable at short distance. So in general the propagator of our gravity field becomes

$$G(k) = \frac{1}{k^4 + ak^2 + b}. \tag{21}$$

It is clear that one of the poles of this propagator must be negative, and therefore there exists a ghost. However since such theory is scale invariant, we are free to choose the scale so that the ghost gets pushed to the Planck scale.[17] Although this does not completely expel the ghost, the situation should become harmless. After all QED, for example, is also not free of divergence. The well-known Landau pole would appear when the energy goes to $m_e e^{1/\alpha}$.

Another way to circumvent the ghost problem was recently proposed by Bender and Mannheim.[18] They observed that the ghost problem may be originated from the fact that in the conventional quantum theory the Hamiltonian is assumed to be Dirac Hermitian, i.e., $H = H^\dagger$. If the Hamiltonian is instead invariant under the more physical discrete symmetry of spacetime reflection, $H = H^{\mathcal{PT}}$, then the negative norm and probability would not occur to the propagator. Here parity $\mathcal{P}$ is a linear operator that performs space reflection and $\mathcal{T}$ is an antilinear operator that performs time reversal. If the energy spectrum of H is real and positive, then the $\mathcal{PT}$ symmetry is unbroken and there exissts a reflection symmetry $\mathcal{C}$ which commutes with H, and also $[\mathcal{C}, \mathcal{PT}] = 0$. Using this new inner product, it has been shown by Bender and others that the norm of a state is strictly positive.

Before leaving this issue, we should like to mention that lately there has been a revival interest[19] in the Lee-Wick theory,[20] which also involves the ghost problem, albeit not so directly related to our issue. It should be fair to say that the ghost problem in higher order field theories, though serious, may not necessarily be incurable.

6. Discussion

We see from the above discussion that the two assumptions for the solution of the CC problem, namely the gauge theory of gravity and the de Sitter symmetry, are knitted together due to the need to address both the old and the new CC problems. As often the case, a fundamental constant would manifest itself in various physical phenomena. Being an overall coefficient in the gauge gravity action, l_{dS} does not appear in the field equations (6) and (5) and therefore would not affect the gravitational dynamics other than serving as the asymptotic limit of the cosmic expansion. It does, however, reveal itself in a curious way. Unlike GR in which the matter action is detached from the Einstein-Hilbert action, in gauge theory of gravity the matter field couples with the affine connection and is therefore an integral part of the total action, similar to the case of Maxwell theory. In this approach, Newton's constant acts as the gravitational charge. But if we insist on writing this matter-gravity coupling action separately from the curvature piece, then we have to absorb κ into it and as a result we find that the actual matter-gravity coupling turns out to have the Newton's constant cancelled and is proportional to l_{dS}^2 only. What does this imply? On the one hand, the absence of Newton's constant in the matter action is the same situation as that in GR. On the other hand, unlike GR, the matter in the gauge theory of gravity couples to the geometry, i.e., the affine connection. However we now see that this coupling is mediated through the curvature of the de Sitter space.

Another implication is that in our universe the Poincare symmetry should necessarily be replaced by the de Sitter symmetry. Such modification must be minute as l_{dS} is astronomically large and therefore the deviation of spacetime from perfect flatness is small. Indeed there have been authors who look into the so-called de Sitter special relativity in recent years.[10] The migration of spacetime symmetry from the Poincare group to the de Sitter group should in principle induce additional observable effects. These aspects are beyond the scope of this brief article and should be investigated separately.

Our approach faces challenges based on quantum considerations. As we discussed, the problem of quantum instability of de Sitter space may not be fatal. Its decay time appears to be longer than the age of the universe. As for the problem of the ghost state, we argued that the negative pole may be pushed to the Planck scale and would therefore not be too harmful. These arguments are heuristic. To be sure, more in depth investigations are required before these problems can be resolved.

With regard to the gauge theory of gravity, in addition to the possibility of solving the CC problem, it may hopefully pave the way to the quantization of gravity and the unification with other gauge theories of interactions. As Yang himself commented in 1983: "In Ref. 4 I proposed that the gravitational equation should be changed to a third order differential equation. I believe today, even more than 1974, that this is a promising idea, because the third order equation is more natural than the second order one and because quantization of Einstein's theory leads to difficulties."[21]

Acknowledgments

It is a pleasure to thank R. J. Adler, H. Kleinert, I. Antoniadis, G. t'Hooft, and my students Shu-Heng Shao and Nian-An Tung for helpful discussions. This research is supported by Taiwan National Science Council under Project No. NSC 97-2112-M-002-026-MY3 and by US Department of Energy under Contract No. DE-AC03-76SF00515.

References

1. S. Weinberg, Rev. Mod. Phys. **61**, 1 (1989).
2. S. Perlmutter et al., Astrophys. J. **517**, 565 (1999) [arXiv:astro-ph/9812133].
3. A. G. Riess et al., Astron. J. **116**, 1009 (1998) [arXiv:astro-ph/9805201].
4. C. N. Yang, Phys. Rev. Lett. **33**, 445 (1974).
5. A. H. Thompson, Phys. Rev. Lett. **34**, 507 (1975).
6. V. Szczyrba, Phys. Rev. D **36**, 351 (1987).
7. F. Gronwald and F. W. Hehl, arXiv: gr-qc/9602013.
8. R. J. Cook, arXiv:0810.4495v2 [gr-qc].
9. H. Bacry, J-M. Levy-Leblond, J. Math Phys. **9**, 1605 (1968).
10. Han-Ying Guo, Chao-Guang Huang, Zhan Xu, Bin Zhou, Phys. Lett. A **331**, 1 (2004) [arXiv:hep-th/0403171].
11. R. Aldrovandi, J. G. Pereira, Found. Phys. **57**, 221 (2008) [arXiv:0711.2274].

12. S. Cacciatori, V. Gorini, A. Kamenshchik, Ann. der Phys. **17**, 728 (2008) [arXiv:08073009].

13. A. Zee, lectures given at the Asia-Pacific Winter School on Cosmology, Taipei, Jan. 18-23, 2010.

14. L. F. Abbott and S. Deser, Nucl. Phys. **B195**, 76 (1982).

15. L. H. Ford, Phys. Rev. D **31**, 710 (1985).

16. I. Antoniadis, J. Iliopoupos, and T. N. Tomaras, Phys. Rev. Lett. **56**, 1319 (1986).

17. H. Kleinert, Phys. Lett. B **196**, 355 (1987); Private communications (2010).

18. C. M. Bender and P. Mannhaim, J. Phys. A: Math. Theor. **41**, 304018 (2008); C. M. Bender and P. Mannhaim, Phys. Rev. Lett. **100**, 110402 (2008); P. Mannheim, arXiv:0912.2635 [hep-th].

19. B. Grinstein, D. O'Connell and M. B. Wise, Phys. Rev. D **77**, 025012 (2008).

20. T. D. Lee and G. C. Wick, Nucl. Phys. **B9**, 209 (1969); T. D. Lee and G. C. Wick, Phys. Rev. D **2**, 1033 (1970).

21. Chen Ning Yang, *Selected Papers, 1945-1980, With Commentary*, p.74 (W. H. Freeman, 1983).

INFLATION: THEORY AND OBSERVATIONS

VIATCHESLAV MUKHANOV

Physics Department, Ludwig-Maximilians University, Munich, Germany

I will consider the robust predictions of the accelerated stage of the universe evolution (cosmic inflation) and discuss the current status of their observational verification. In particular, I consider in details the predictions for primordial spectrum of inhomogeneities originated from quantum fluctuations.

HOLOGRAPHIC SPACE-TIME AND ITS PHENOMENOLOGICAL IMPLICATIONS

T. BANKS

NHETC, Rutgers University and SCIPP,
U.C. Santa Cruz, Santa Cruz, CA 95064, USA
banks@scipp.ucsc.edu

I briefly review the theory of Holographic Space-time and its relation to the cosmological constant problem, and the breaking of supersymmetry (SUSY). When combined with some simple phenomenological requirements, these ideas lead to a fairly unique model for Tera-scale physics, which implies direct gauge mediation of SUSY breaking and a model for dark matter as a hidden sector baryon, with nonzero magnetic dipole moment.

1. Introduction to Holographic Space-Time[1]

This paper is the written version of a talk given at the conference celebrating the 80th birthday of Murray GellMann at the Nanyang Technical University in Singapore. I'd like to thank Harald Fritzsch and the other organizers of the conference for inviting me to join in honoring one of the greatest physicists of the 20th century.

String theory models are our only rigorously established models of quantum gravity, but none of the known models apply to the real world. They do not incorporate cosmology, and they do not explain the breaking of supersymmetry (SUSY) that we observe. The theory of Holographic Space-Time is an attempt to generalize string theory in order to resolve these problems. Its basic premise is a strong form of the holographic principle, formulated by myself and W. Fischler:

- Each causal diamond in a d dimensional Lorentzian space-time has a maximal area space-like $d-2$ surface in a foliation of its boundary. The area of this *holographic screen* in Planck units is 4 times the logarithm of the dimension of the Hilbert space describing all possible measurements within the diamond.

Every pair of causal diamonds has a maximal area causal diamond in their intersection. This is identified with a common tensor factor in the Hilbert spaces of the individual diamonds

$$\mathcal{H}_1 = \mathcal{O}_{12} \otimes \mathcal{N}_1$$

$$\mathcal{H}_2 = \mathcal{O}_{12} \otimes \mathcal{N}_2\,.$$

A holographic space-time is defined by starting from a $d-1$ dimensional spatial lattice, which specifies the *topology of a particular space-like slice*. To each point $\mathbf{x}$

on this lattice, we associate a sequence of Hilbert spaces

$$\mathcal{H}(n, \mathbf{x}) = \otimes \mathcal{P}^n \,.$$

The *single pixel Hilbert space*, $\mathcal{P}$ will be specified below, and has to do with the geometry of compactified dimensions. These spaces represent the sequence of causal diamonds of a time-like observer as the proper time separation of its future tip from the point where it crosses the space-like slice increases. $N(\mathbf{x})$ is the maximal value that n attains as the proper time goes to infinity. In a future asympotically dS space time, $N(\mathbf{x})$ will be finite. In an asymptotically flat space-time or FRW universe which is matter or radiation dominated, n will go to infinity with the proper time, while in an asymptotically AdS universe n will go to infinity at finite proper time. In a Big Bang space-time the past tip of each causal diamond lies on the Big Bang hypersurface. In a time symmetric space-time we think of the diamonds as having past and future tips which are equidistant in proper time from the slice on which the lattice is placed. In either case we will refer to *the causal diamonds at a fixed time* as those carrying the same label n.

In any theory of quantum gravity, the Hilbert space formulation will refer to a particular time slicing. We have chosen slices such that the causal diamonds at any fixed time have equal area holographic screens. Such equal area slicings exist in all commonly discussed classical space-times.

The rest of the specification of holographic space-time consists of a prescription of the overlap tensor factor

$$\mathcal{O}(m, \mathbf{x}; n, \mathbf{y}) \,,$$

in $\mathcal{H}(m, \mathbf{x})$ and $\mathcal{H}(n, \mathbf{y})$. For nearest neighbor points at $m = n$ this overlap is just $\mathcal{P}$. For other pairs of points the specification of the overlap is part of the dynamical consistency condition described below. The only kinematic restriction on it is that the dimension of $\mathcal{O}$ is a non-increasing function of $d(\mathbf{x}, \mathbf{y})$, the minimum number of lattice steps between the points.

We introduce dynamics as a sequence of unitary operators $U(n, \mathbf{x})$ in $\mathcal{H}(N(\mathbf{x}), \mathbf{x})$, with the property that $U(n, \mathbf{x}) = V(n, \mathbf{x})W(n, \mathbf{x})$, where $V(n)$ is a unitary in $\mathcal{H}(n, \mathbf{x})$, while $W(n)$ is a unitary in the tensor complement of $\mathcal{H}(n, \mathbf{x})$ in $\mathcal{H}(N(\mathbf{x}), \mathbf{x})$. This requirement implements the idea that the dynamics inside a causal diamond effects only those degrees of freedom associated with the diamond. In particular, in a Big Bang space-time, it builds the concept of *particle horizon* into the dynamics of the system. Note by the way that in Big Bang space time the sequence of unitaries $U(n)$ may be thought of as a conventional time dependent Hamiltonian system with a discrete time, while for a time symmetric space-time they are instead "approximate S-matrices", $U(T, -T)$.

Starting from some initial pure state in $\mathcal{H}(N(\mathbf{x}), \mathbf{x})$, the unitaries $U(n, \mathbf{x})$ produce a sequence of density matrices $\rho(n, \mathbf{x})$ in each overlap factor involving the point $\mathbf{x}$. *The key dynamical consistency condition for a holographic space time is that*

$$\rho(n, \mathbf{y}) = U(n, \mathbf{x}; \mathbf{y})\rho(n, \mathbf{x})U^\dagger(n, \mathbf{x}; \mathbf{y}) \,,$$

for every pair of points. This staggeringly complicated set of consistency conditions is the analog in this formalism of the Dirac-Schwinger commutation relations, which guarantee the consistency of "many fingered time". The only known solution of these conditions is the dense black hole fluid (DBHF) cosmology described briefly below. In that example, the consistency conditions dictate both the choice of overlap Hilbert spaces, and the dynamics at each point in the lattice.

There are a number of very important points to understand about this formalism

- Although we have used geometrical pictures to motivate our constructions, they are entirely phrased in quantum mechanical language. The Lorentzian space-time is an emergent property of these quantum systems, useful in the limit of large causal diamonds (large dimension Hilbert spaces).
- The emergent space-time geometry is *not* a fluctuating quantum variable. Its causal structure is specified by the overlaps, and its conformal factor by the Hilbert space dimensions.
- The lattice specifies only the topology of a space-like slice in the non-compact dimensions[a] This topology does not change with time.

2. SUSY and the Holographic Screens[3]

Since space-time geometry is *not* a fluctuating quantum variable, it is natural to associate the quantum variables with the properties of the holographic screen of the causal diamond. Intuitively, the space-time orientation of an infinitesimal bit of screen is determined by the outgoing null direction, and the transverse plane in which the screen lies. That information is encoded in the Cartan-Penrose equation

$$\bar{\psi}\gamma^{\mu}\psi(\gamma_{\mu})^{\alpha}_{\beta}\psi^{\beta} = 0\,.$$

Indeed this equation implies that $\bar{\psi}\gamma^{\mu}\psi$ is a null vector, and that the hyperplanes

$$\bar{\psi}\gamma^{\mu_1\cdots\mu_k}\psi\,,$$

with $k \geq 2$ all lie in a single $d - 2$ plane. More succinctly, $\psi = (0, S_a)$: ψ is a transverse spinor in the light front frame defined by the null direction.

The Cartan-Penrose equation is conformally invariant, but our quantization procedure will violate that invariance. This is simply the statement that the Bekenstein-Hawking area formula is being used to define the conformal factor of our space-time geometry in terms of the dimension of the quantum Hilbert space. The holographic principle now implies two constraints on the quantization procedure:

We want to have independent degrees of freedom for different points on the holographic screen. This is compatible with a finite dimensional Hilbert space only if a finite area screen is "pixelated": its function algebra must be replaced by a finite dimensional algebra. If n labels a basis of the algebra, the single pixel Hilbert space

---

[a]In a holographic theory, dS space has non-compact spatial sections because one restricts attention to the causal diamond of a fixed observer.

342

$\mathcal{P}$ is the lowest dimension representation space of the algebra generated by the $S_a(n)$ variables. If we insist on transverse $SO(d-2)$ invariance, the only quantization rule having a finite dimensional representation space is

$$[S_a(n), S_b(n)]_+ = \delta_{ab}.$$

SUSY aficionadas will recognize this as the algebra of a single massless supersymmetric particle with longitudinal momentum $(1, \mathbf{\Omega})$. If $d = 11$ the smallest representation of this algebra is the SUGRA multiplet. In fewer non-compact dimensions there are non-gravitational multiplets, but, since we are trying to construct a theory of gravity, we should retain 16 real spinor generators for each pixel, in order to guarantee that there is a helicity two particle in the spectrum.

The anti-commutation relations postulated so far are invariant under $S_a(n) \rightarrow (-1)^{F(n)} S_a(n)$. This is a remnant of the rescaling symmetry of the CP equation. We treat it as a gauge symmetry. Using it, we can perform a Klein transformation so that the independent operators on different pixels anti-commute rather than commute with each other.

A convenient way to pixelate the holographic screen is to use fuzzy geometry. We replace the algebra of functions by a sequence of finite dimensional matrix algebras. The most famous example is the two sphere. The algebra of $n \times n$ matrices has a natural action of the group $SU(2)$ on it because the spin $\frac{n-1}{2}$ representation is n dimensional. The matrices carry every spin from zero up to $n - 1$ and so can be thought of as a natural cutoff of the angular momentum on the sphere. Vector bundles over the sphere are rectangular matrices. In particular $n \times n + 1$ and $n + 1 \times n$ matrices converge to the two chiral spinor bundles over the sphere. Many of the compactification spaces of string theory are Kahler manifolds, or Kahler fibrations over a one (Horava-Witten) or three dimensional (G2 manifolds which are K3 fibrations) base. These are naturally thought of as limits of finite dimensional matrix algebras. The pixel variables of such compactifications will have the quantum algebra

$$[(\psi^M)_i^A, (\psi^{\dagger N})_B^j]_+ = \delta_i^j \delta_B^A B^{MN},$$

where $i, j = 1 \cdots K$ and $A, B = 1 \cdots K + 1$, so that the fermionic matrices fill out the two spinor bundles over the fuzzy two sphere. The indices M, N also run over a set of rectangular matrices which approximate either the spinor bundle over a seven manifold, or two copies of the spinor bundle over a six manifold (and there are two possibilities, according to whether the two copies have the same or different chiralities). The B^{MN} should be interpreted as wrapped brane charges. They can be further decomposed into sums over cycles of various dimensions. That is to say, we have an algebraic way of encoding the homology of the manifold.[b] In this formalism, the problem of (kinematically) classifying four dimensional compactifications

---

[b]Though of course, we know from string duality that the interpretation as homology of a particular manifold will only be valid in certain limits. The algebra of SUSY charges (of which our algebra is an analog) is valid independently of the geometric interpretation.

reduces to classifying superalgebras such that in the limit, $K \to \infty$ they contain one copy of the $N = 1$ SUSY algebra. Equivalently, in this limit, the representation space of the algebra should contain exactly one $N = 1$ graviton supermultiplet. The super-generators are constructed by using the conformal structure of the 2-sphere, whose invariance group is SO (1, 3). Conformal Killing spinors on the sphere transform as the Dirac spinor of SO (1, 3).

If $(q_\alpha)_A^i$ are matrices that converge to the left handed conformal Killing spinor, then our kinematic condition on the algebra of pixel variables is that there exists a set of coefficients F_M such that

$$Q_\alpha \equiv F_M \mathrm{Tr} \left[\psi^M q_\alpha \right],$$

satisfies the super-Poincare algebra as $K \to \infty$. The representation space of the pixel algebra should break up into a finite number of single particle representations of the SUSY algebra, with only a single supergraviton multiplet. This is, in our formalism, the condition for a compactification with $N = 1$ SUSY. Note that, for finite K, these constructions have no continuous moduli. They are finite dimensional unitary representations of finite dimensional non-abelian super-algebras.

2.1. Particles

If we suppose that we have found such an algebra, we can now make multiple copies of our single particle Hilbert space by replacing the algebra of functions, the matrix algebra

$$\mathcal{M}_K \otimes \mathcal{A},$$

(where $\mathcal{A}$ is the algebra of matrices approximating the function algebra on the internal manifold), by a direct sum

$$\oplus_i \mathcal{M}_{Ki} \otimes \mathcal{A},$$

and take the limit $K_i \to \infty$ with $\frac{K_i}{K_j}$ fixed. As in Matrix Theory[2] the ratios are interpreted as the ratios of longitudinal momenta $P_i(1, \mathbf{\Omega_i})$ of a set of particles. Here however, each particle has its own null direction. The S_p gauge symmetry relating commuting operators to block diagonal matrices is interpreted as particle statistics. Note that a particle must have a large momentum in order to have good angular localization, but for fixed holographic screen area one can only make a finite number of particles, and the larger the momentum that each one carries the fewer particles we can make. One can argue[5] that the states with all the momentum carried by one "particle" should actually be thought of as black holes that fill the causal diamond.

2.2. Holographic cosmology

Here we give a brief description of the Dense Black Hole Fluid model of holographic cosmology.[4] In this model, one takes the overlap Hilbert spaces to be

$$\mathcal{O}(n, \mathbf{x}; n\mathbf{y}) = \mathcal{P}^{n - d(\mathbf{x}, \mathbf{y})},$$

where $d(\mathbf{x}, \mathbf{y})$ is the minimum number of lattice steps between the two points. If the exponent is negative, we interpret it as 0. The time evolution operators are identical at each lattice point, and the time dependent Hamiltonian is chosen randomly at each time n to be

$$\ln V(n, \mathbf{x}) = \sum S_a(i) S_a(i) A(n; i, j) + I(n).$$

Here the S_a satisfy fermionic commutation relations,[c] and $A(n; i, j)$ is a random $n \times n$ anti-symmetric matrix. For large n the quadratic term converges to the Hamiltonian for free massless fermions in $1 + 1$ dimensions, and $I(n)$ is chosen to be a random irrelevant perturbation of this CFT. One then argues that there is a coarse grained description of this system as a flat FRW universe, with equation of state $p = \rho$, which saturates the covariant entropy bound.

This model is used to construct more realistic cosmologies by using the Israel junction condition. We think of our own universe as a low entropy "defect" inside the DBHF. Consider first a spherical volume of $p = w\rho$ universe with $-1 < w < 1$, embedded in a $p = \rho$ universe. Consider time slices of the two geometries of equal holographic area. This means the time coordinates are proportional to each other with a fixed constant. A coordinate volume of radius L has a physical radius that grows as $t^{\frac{2}{3(1+w)}}$. Since the physical radius in the DBHF grows more slowly, we must let the coordinate L shrink with time in the $p = w\rho$ universe in order to satisfy the Israel condition that the geometry of the interface be the same in both embedding. The exception is $w = -1$. In this case, a cosmological horizon volume is bounded by a null surface of fixed holographic area. We can satisfy the Israel condition by matching to a black hole with the same area horizon, embedded in the $p = \rho$ background.

In Ref. 4 we argued that non-spherical defects could survive as $-1 < w < 1$ regions, but that the above argument about the Israel condition implies that eventually the universe must approach $w = -1$. The late time cosmological constant is determined by cosmological initial conditions, namely the number of degrees of freedom that are initially in a low entropy state. In this way of realizing dS space, it is clear that only a single horizon volume of the classical geometry is necessary to the description, and that this is described as a quantum system with a finite number of states: the representation space for the pixel algebra over the finite area holographic screen of the cosmological horizon.

3. Cosmological SUSY Breaking[1,3,5]

We have noted that in holographic cosmology, the cosmological constant Λ is a positive tunable parameter, determined by cosmological initial conditions. To discuss particle physics, we can replace the actual cosmological history with that of

---

[c] We have not yet made a cosmology compatible with the more complicated superalgebras that arise for non-trivial compactifications.

an eternal dS space. In the limit $\Lambda \to 0$ the theory of stable dS space approaches a super-Poincare invariant theory, similar to conventional string theories, but with no moduli. The theory has a discrete R symmetry, explaining the vanishing of the superpotential at the supersymmetric point. However, the way in which this limit is approached is interesting. One horizon volume of dS space approaches all of Minkowski space. The logarithm of the total number of quantum states of the dS theory is $\pi(RM_P)^2$, but only $(RM_P)^{3/2}$ of that entropy can be modeled by field theory in the horizon volume.

This entropy bound can be derived in two complementary ways.[5,6] On the one hand, we can try to maximize the particle entropy in a horizon volume, subject to the constraint that no black holes of size that scales like R are formed. The maximal entropy particle states are modeled as a cutoff CFT with cutoff μ, so that the entropy,

$$S \sim (\mu R)^3 .$$

The condition that no horizon sized black holes are formed is

$$\mu^4 R^3 < M_P^2 R ,$$

which leads to $\mu < (M_P/R)^{1/2}$, and the $(RM_P)^{3/2}$ scaling of the particle entropy. On the other hand, if we model the holographic screen of dS space as a fuzzy sphere with $K \propto RM_P$, and particle states by block diagonal fuzzy spheres of block sizes K_i with $\sum K_i = K$, then the complementary constraints of angular localization (maximizing each K_i) and maximizing the multiparticle particle entropy, lead to $K_i \sim \sqrt{K}$. If our basic unit of longitudinal momentum is $1/R$, then this gives the same scaling for entropy, momentum cutoff, and average particle number as the previous argument. These remarks also lead to a conjecture for what the other, off diagonal bands, of the matrices represent. The total entropy of dS space allows us to have $(RM_P)^{1/2}$ independent copies of the field theoretic degrees of freedom in a single horizon volume, and it is an obvious conjecture that this is the way the classical geometric result that at late global times dS space has an unbounded number of independent horizon volumes, is realized in the limit $RM_P \to \infty$. The off block diagonal bands of the $K \times K$ matrix algebra approximation to the function algebra on the 2-sphere, represent the particle degrees of freedom in different horizon volumes.

It is the field theoretic states in a single horizon volume, which approach the scattering states of the Minkowski theory. The exponentially overwhelming majority of the states of the dS theory decouple in this limit.[d] These states should be viewed as living on the cosmological horizon. However, because their number is so large, the effect of the interaction of localized particles with the horizon states may be larger than one might have imagined.

---

[d]In quantum field theory, this is the statement that the dS temperature goes to zero.

The discrete R symmetry of the $\Lambda = 0$ theory is broken by interactions with the horizon. The lightest particle in the theory carrying R charge is the gravitino. Thus, R violating interactions will be dominated by Feynman diagrams in which a gravitino propagates out to the horizon. These are suppressed by a factor $e^{-2m_{3/2}R}$. The contribution from the interaction with the horizon has the form

$$\sum_n \frac{|<\tilde{g}|V|n>|^2}{\Delta E} .$$

Note that there is no n dependence energy denominators in this formula, because the horizon states are approximately degenerate. To estimate the number of states that contributes to this formula we note that the horizon states, like degenerate Landau levels, can be localized and have a fixed entropy per unit area. The gravitino can propagate in the vicinity of the horizon, a null surface, for a proper time of order $\frac{1}{m_{3/2}}$. Quantum particles execute random walks in proper time. If we take the step size to be Planck scale, the area covered will also scale like $\frac{1}{m_{3/2}M_P}$. Thus, the contribution of this diagram is of order

$$e^{-2m_{3/2}R + \frac{bM_P}{m_{3/2}}} ,$$

where b is an unknown constant. We know that $m_{3/2} \to 0$ as $R \to \infty$. If it went to zero faster than $R^{-\frac{1}{2}}$, then the diagram would blow up exponentially. If it goes to zero more slowly than $R^{-\frac{1}{2}}$, then the diagram is exponentially small. However, it is precisely the R violating terms in the effective Lagrangian, which are supposed to be responsible for the non-zero gravitino mass. So we have a contradiction unless $m_{3/2} = K\Lambda^{1/4}$. In Ref. ? we gave an argument that the constant K is of order 10.

4. CSB and Phenomenology[10]

The relationship $m_{3/2} = K\Lambda^{1/4}$, with K of order 10, puts strong constraints on low energy phenomenological models. In low energy effective SUGRA models, SUSY breaking is parametrized by a non-vanishing F term for some chiral superfield, X. In order to obtain gaugino masses, the model must generate couplings of the form

$$c_i \frac{\alpha_i}{4\pi} \frac{X}{M} \text{tr}(W_\alpha^i)^2 .$$

Since, according to CSB, $F_X = K(\text{TeV})^2$, we cannot have M larger than a few TeV. Thus we *must* have a strongly coupled hidden sector to generate the scale M, and that sector must contain particles charged under the standard model gauge group. That is, we have a model of direct gauge mediation.

If we wish to preserve the prediction of SUSY gauge coupling unification, the new particles must be in complete multiplets of a unified group, and transform under the hidden sector group G. If the unified group contains SU(5) we get, at least R copies of the $5 + \bar{5}$, where R is a G representation. If $R \geq 5$, this leads to Landau poles below the unification scale, which implies at best a fuzzy prediction of unification. All hidden sector groups with $R < 5$ appear to predict light pseudo-Goldstone bosons that transform under the standard model and should have been seen in experiments.

The only resolution I have found to these competing exigencies is to employ *trinification*,[11] with a hidden sector group $SU_P(3)$. The resulting model has a pyramidal quiver diagram and is called The Pyramid Scheme. It has perturbative one loop unification, and no unpleasant PNGBs. The gauge group is $SU_P(3) \otimes SU_1(3) \otimes SU_2(3) \otimes SU_3(3) \rtimes Z_3$, and the matter content is

$$3 \times [(1,1,\bar{3},3) \oplus (1,3,1,\bar{3}) \oplus (1,\bar{3},3,1)],$$
$$(3,\bar{3},1,1) \oplus (3,1,\bar{3},1) \oplus (3,1,1,\bar{3}) \oplus c.c.$$

The Z_3 symmetry permutes the last 3 SU(3) subgroups. The $SU(2) \times SU(3)$ of the standard model is embedded in the indicated $SU_{2,3}(3)$ groups of the Pyramid, and the U(1) is a combination of a generator in $SU_1(3)$ and one in $SU_2(3)$. In addition, we introduce 3 singlet fields S_i. The three fields that couple both to $SU_P(3)$ and to the standard model are called *trianons* and are denoted $T_i + \tilde{T}_i$, with the index i indicating that the field is charged under $SU_i(3)$.

The underlying principle of CSB implies that the low energy Lagrangian consists of two pieces. The first, $\mathcal{L}_R$, preserves a discrete R symmetry and has a supersymmetric R symmetric minimum of its effective potential. This is the low energy Lagrangian for the supersymmetric S-matrix of the $\Lambda = 0$ limit. Experience with string theory suggests that it should satisfy the demands of field theoretic naturalness: every term consistent with hypothesized symmetries is allowed. Any term smaller or larger than would be indicated by Planck scale dimensional analysis should be explained in terms of an explicit low energy dynamical mechanism.

The second term $\delta\mathcal{L}$ arises, in a low energy effective picture, from interactions of a single gravitino with degrees of freedom on the cosmological horizon in dS space. These DOF *do not have a field theoretic description and we do not yet have a precise model of them.* We can only list some properties of these terms, which follow from general principles:

- They violate the discrete R symmetry.
- They must give us a low energy effective theory that violates SUSY, incorporating the relation $m_{3/2} = K\Lambda^{1/4}$.
- The low energy effective theory must be consistent with a model of dS space as a system with a finite number of quantum states. In particular, if the SUSY violating minimum with c.c. Λ is not the absolute minimum of the potential, then the potential must be *Above the Great Divide.*[8,10]

The last item implies that the non-gravitational low energy dynamics must have a *stable* SUSY violating ground state.[e] Results of Nelson and Seiberg,[?] when com-

---

[e]If the SUSY violating state is only meta-stable when $m_P \to \infty$, and if the difference in energy density between the meta-stable and "stable" minima is much larger than Λ, the gravitational theory is below the Great Divide.[8,9]

bined with the requirement that R symmetry is explicitly broken, then imply that the R violating part of the Lagrangian *cannot* satisfy the demands of naturalness. It must omit terms allowed by all symmetries. We have however emphasized that the origin of these terms is novel and corresponds to nothing in our experience with ordinary string theory or quantum field models that emerge from quasi-local lattice dynamics. In models of quantum gravity, the states on horizons, whether black hole horizons or the cosmological horizon in dS space, do not have a description in terms of localized bulk degrees of freedom, obeying the rules of QFT. The R violating terms in the Lagrangian for local degrees of freedom, are the residuum of interactions with a large number of horizon states, which decouple as the dS radius is taken to infinity. These terms are important, because they are the origin of supersymmetry breaking. They do not obey the constraints of naturalness.

The R preserving part of the TeV scale superpotential is the superpotential of the standard model plus

$$W_R = \sum g_i S_i \tilde{T}_i T_i + \sum y_i (T_i^3) + \tilde{y}_i (\tilde{T}_i)^3 + \sum g_{\mu i} S_i H_u H_d \,.$$

The R symmetry, which must have no gauge anomalies, is chosen so that either g_1 or g_3 vanishes, as well as one of the pairs $y_{1,3}$ and $\tilde{y}_{1,3}$. It can also be chosen such that the coefficients of all baryon and lepton number violating operators of dimensions 4 and 5, apart from neutrino seesaw terms $(H_u L)^2$, vanish. We require the vanishing of $g_{1 \text{ or } 3}$ in order to eliminate SUSY preserving minima. The vanishing of one pair of the y couplings is introduced in order to have a dark matter candidate.

The R violating superpotential, coming from interactions with the horizon, is postulated to be

$$\delta W = W_0 + \sum (m_i T_i \tilde{T}_i) + \mu_i^2 S_i + \mu H_u H_d \,.$$

I'll conclude with a brief list of the properties of the model

- It has no Supersymmetric minimum at sub-Planckian field values, and is compatible with an underlying model of dS space with a finite number of states, incorporating the CSB relation $m_{3/2} = K\Lambda^{1/4}$.
- $\mathcal{L}_R$ has a discrete R symmetry and all R preserving couplings appear with natural strength. Dimension 4 and 5 couplings that violate baryon number are absent, and the only allowed lepton number violating couplings are the neutrino seesaw terms $(H_u L)^2$. The μ term $H_u H_d$ is also forbidden by R symmetry. All CP violating angles, apart from the CKM phase, can be rotated away, and $\mathcal{L}_R$ has a dangerous axion. The non-generic terms in δW lift the axion. One can argue that if the origin of CP violation is at energies below the Planck scale, so that the thermal bath near the horizon is approximately CP invariant, the CP violating phases in δW are very small. This is a novel solution of the strong CP problem. The NMSSM couplings in $\mathcal{L}_R$ and the explicit μ term in δW give an acceptable Higgs spectrum, without tuning.
- All couplings are perturbative at the unification scale, and the model generates a dynamical scale $\Lambda_3 \sim$ a few TeV, which can explain the origin of gaugino

masses. There is freedom to separately tune different gaugino masses by using the parameters m_i in δW. The chargino decays promptly in this model, so that the Fermilab trilepton analysis bounds its mass from below by ~ 270 GeV. By making the parameter m_3 reasonably large, we insure that the gluino is not heavy enough to make dangerous modifications to the Higgs potential.

- Dark matter is the pyrma-baryon field $(T_i)_a^M (T_i)_b^N (T_i)_c^K \epsilon^{abc} \epsilon_{MNK}$, where $i = 1$ or 3. It is not a thermal relic, but can have the right relic density if an appropriate pyrma-baryon asymmetry is generated in the early universe. There will be no dark matter annihilation signals. The dark matter particle weighs tens of TeV, and has a magnetic moment. The magnetic moment leads to an interesting pattern of signals in terrestrial dark matter detectors, rather different from the signal for a convential WIMP. The details of this are being worked out.

5. Conclusions

The theory of holographic space-time seeks to generalize string theory to situations where the boundaries of space-time are not asymptotically flat or anti-de Sitter, and the quantum theory does not have a unique ground state. It builds space-time out of purely quantum data, the dimensions of Hilbert spaces and common tensor factors in a net of Hilbert spaces. The topology of a Cauchy surface is part of the specification of the formalism, and does not change with time. Space-time geometry is not a fluctuating quantum variable. Instead the quantum degrees of freedom are quantized orientations of pixels on the holographic screens of causal diamonds. Their quantum kinematics is determined by a super-algebra, whose structure incorporates the quantum remnants of the geometry of compact dimensions. Compactifications are classified in terms of possible superalgebras.

In the limit that the holographic screen area goes to infinity, and the screen approaches that of null infinity in Minkowski space, the pixel variables describe supersymmetric multiplets, including the gravity multiplet. There is, as yet, no general prescription for calculating the scattering matrix of this super-Poincare invariant theory.

The general formalism leads in particular to a completely non-singular, mathematically complete quantum description of what one might call the generic Big Bang universe. This is called the dense black hole fluid (DBHF). Heuristically, at any time, the particle horizon volume is completely filled with a single large black hole, and causally disconnected black holes merge to preserve this condition as the particle horizon expands. The coarse grained description of this situation is a flat FRW geometry with equation of state $p = \rho$. Note that flatness, homogeneity and isotropy emerge automatically, without any inflation.

An heuristic model of our own universe based on the concept of a low entropy defect in the DBHF implies that the universe *must* approach an asymptotically dS future, with c.c. determined by cosmological initial conditions. dS space is modeled as a quantum system with a finite number of states, as first envisioned by Fischler and the present author.[7]

The general formalism of holographic space-time implies that SUSY is restored as $\Lambda \to 0$. Two arguments, one of which was reviewed above, suggest that $m_{3/2} = K\Lambda^{1/4}$. The constant K has been argued to be of order 10, and is related to the ratio between the unification scale and the Planck scale. When combined with the desire to explain the apparent unification of standard model couplings, this low scale of SUSY breaking puts strong constraints on the effective Lagrangian for particle physics at TeV energy scales. So far, only a unique class of models has been found, which can satisfy these constraints. These are the Pyramid Schemes, and differ from each other only in the values of a few parameters. They all have a discrete R symmetry in the $\Lambda = 0$ limit, which is broken by interactions with states on the horizon, which decouple in this limit. The R violating terms in the effective Lagrangian do not satisfy the usual laws of naturalness.

The Pyramid Scheme resolves many of the puzzles of low energy supersymmetric particle physics, some by a novel mechanism. It has an acceptable level of flavor changing neutral currents and no dangerous B and L violating operators. It has a novel dark matter candidate, which carries an approximately conserved U(1) quantum number and can have the right relic density if an appropriate asymmetry is generated in the early universe. There are no annihilation signals. The dark matter candidate is quite heavy and has a magnetic dipole moment. Its signals in terrestrial detectors depend on the target nucleus, and are being worked out.[12] The supersymmetric and strong CP problems, the μ problem, and the little hierarchy problem are all resolved by the non-generic nature of the R violating part of the effective Lagrangian.

The theory of holographic space-time thus provides a comprehensive quantum mechanical framework for early and late universe cosmology, as well as incorporating the surprising connection between the asymptotic dS nature of the universe and low energy particle physics. The particle physics implications will be checked, at least in part, by the LHC. If the theory's predictions are verified, one would be motivated to attack the unsolved problem of formulating dynamical equations for holographic space-time.

References

1. T. Banks, *Deriving particle physics from quantum gravity: a plan*, arXiv:0909.3223; T. Banks, Holographic Space-time from the Big Bang to the de Sitter era, *J. Phys. A* **42**, 304002 (2009), arXiv:0809.3951; T. Banks, *II(infinity) factors and M-theory in asymptotically flat space-time*, arXiv:hep-th/0607007; T. Banks, *More thoughts on the quantum theory of stable de Sitter space*, arXiv:hep-th/0503066; T. Banks, Supersymmetry, the cosmological constant and a theory of quantum gravity in our universe, *Gen. Rel. Grav.* **35**, 2075 (2003), arXiv:hep-th/0305206; T. Banks, *Some Thoughts on the Quantum Theory of de Sitter Space*, arXiv:astro-ph/0305037.
2. T. Banks, W. Fischler, S. H. Shenker and L. Susskind, *Phys. Rev. D* **55**, 5112 (1997), arXiv:hep-th/9610043.
3. T. Banks, *SUSY and the holographic screens*, arXiv:hep-th/0305163; T. Banks, *Breaking SUSY on the horizon*, arXiv:hep-th/0206117; T. Banks, *The phenomenology of cosmological supersymmetry breaking*, arXiv:hep-ph/0203066.

4. T. Banks and W. Fischler, *The holographic approach to cosmology*, arXiv:hep-th/0412097; T. Banks, W. Fischler and L. Mannelli, Microscopic quantum mechanics of the $p = $ rho universe, *Phys. Rev. D* **71**, 123514 (2005), arXiv:hep-th/0408076; T. Banks and W. Fischler, *Holographic cosmology*, arXiv:hep-th/0405200; T. Banks and W. Fischler, Holographic cosmology 3.0, *Phys. Scripta* **T117**, 56 (2005), arXiv:hep-th/0310288; T. Banks and W. Fischler, *An holographic cosmology*, arXiv:hep-th/0111142; T. Banks and W. Fischler, *M-theory observables for cosmological space-times*, arXiv:hep-th/0102077.

5. T. Banks, B. Fiol and A. Morisse, *JHEP* **0612**, 004 (2006), arXiv:hep-th/0609062.

6. T. Banks, W. Fischler and S. Paban, *JHEP* **0212**, 062 (2002), arXiv:hep-th/0210160.

7. T. Banks, *Cosmological breaking of supersymmetry or little Lambda goes back to the future. II*, arXiv:hep-th/0007146; W. Fischler, *Taking de Sitter Seriously*, talk given at the symposium in honor of G. West, Los Alamos, June 2000.

8. A. Aguirre, T. Banks and M. Johnson, Regulating eternal inflation. II: The great divide, *JHEP* **0608**, 065 (2006), arXiv:hep-th/0603107; T. Banks and M. Johnson, *Regulating eternal inflation*, arXiv:hep-th/0512141.

9. R. Bousso, B. Freivogel and M. Lippert, Probabilities in the landscape: The decay of nearly flat space, *Phys. Rev. D* **74**, 046008 (2006), arXiv:hep-th/0603105.

10. T. Banks and J. F. Fortin, Tunneling Constraints on Effective Theories of Stable de Sitter Space, *Phys. Rev. D* **80**, 075002 (2009), arXiv:0906.3714; T. Banks and J. F. Fortin, A Pyramid Scheme for Particle Physics, *JHEP* **0907**, 046 (2009), arXiv:0901.3578; T. Banks, J. D. Mason and D. O'Neil, A dark matter candidate with new strong interactions, *Phys. Rev. D* **72**, 043530 (2005), arXiv:hep-ph/0506015.

11. S L. Glashow, *Trinification Of All Elementary Particle Forces*, CITATION = PRINT-84-0577-BOSTON.

12. T. Banks, J.-F. Fortin and S. Thomas, *Fermionic dark matter with electro-magnetic dipole moments*, In preparation.

MODELING THE FLYBY ANOMALIES WITH DARK MATTER SCATTERING

STEPHEN L. ADLER

School of Natural Sciences, Institute for Advanced Study,
Princeton, NJ 08540, USA
adler@ias.edu
www.sns.ias.edu/~adler/

We continue our exploration of whether the flyby anomalies can be explained by scattering of spacecraft nucleons from dark matter gravitationally bound to the earth. We formulate and analyze a simple model in which inelastic and elastic scatterers populate shells generated by the precession of circular orbits with normals tilted with respect to the earth's axis. Good fits to the data published by Anderson et al. are obtained.

Keywords: Dark matter; flyby anomaly.

1. Introduction

In this paper[a] we follow up our earlier investigation [1] of the anomalous geocentric frame orbital energy changes that are observed during earth flybys of various spacecraft, as reported by Anderson et al. [2]. Some flybys show energy decreases, and others energy increases, with the largest anomalous velocity changes of order 1 part in 10^6. While the possibility that these anomalies are artifacts of the orbital fitting method used in [2] is still being actively explored, there is also a chance that they may represent new physics. In [1] we explored the possibility that the flyby anomalies result from scattering of spacecraft nucleons from dark matter particles in orbit around the earth, with the observed velocity decreases arising from elastic scattering, and the observed velocity increases arising from exothermic inelastic scattering, which can impart an energy impulse to a spacecraft nucleon. Many constraints on this hypothesis were analyzed in [1], with the conclusion that the dark matter scenario is not currently ruled out, but requires dark matter to be non-self-annihilating, with the dark matter scattering cross section on nucleons much larger, and the dark matter mass much lighter, than usually assumed.

However, no attempt was made in [1] to construct a model for the spatial and velocity distribution functions for dark matter populations in earth orbit, to see

---

[a]See http://www.sns.ias.edu/~adler/talks/bnl/pdf for a pedagogical introduction given at the conference. See arXiv:0908.2414 for appendices, additional tables, and further details of the numerical fitting procedure, beyond those given here.

whether it can fit the flyby data reported in [2]. Formulating such a model is the aim of the present paper. Our basic assumption is to consider two populations of dark matter particles, one of which scatters on nucleons elastically, and the other of which scatters inelastically, each with a shell-like distribution of orbits generated by the precession of a tilted circular orbit around the earth's rotation axis. The formulas defining this model are developed in Sec. 2, and the results of numerical fits to the flyby data are briefly described in Sec. 3. We show that good fits to the data are possible, which leaves dark matter scattering as a viable candidate for explaining the flyby anomalies, pending further investigation of possible artifactual explanations of the flyby data, and further experiments aimed at directly detecting dark matter and determining its properties.

2. Formulas Defining the Model

2.1. *Velocity change formulas*

We recall from [1] formulas for the velocity change when a spacecraft nucleon of mass $m_1 \simeq 1\,\mathrm{GeV}$ and initial velocity $\vec{u}_1$ scatters from a primary dark matter particle of mass m_2 and initial velocity $\vec{u}_2$, into an outgoing nucleon of mass m_1 and velocity $\vec{v}_1$, and an outgoing secondary dark matter particle of mass $m_2' = m_2 - \Delta m$ and velocity $\vec{v}_2$. The inelastic case corresponds to $m_2' \neq m_2$, while in the elastic case, $m_2' = m_2$ and $\Delta m = 0$. Under the assumptions, (i) both initial particles are nonrelativistic, so that $|\vec{u}_1| << c, |\vec{u}_2| << c$, (ii) the center of mass scattering amplitude $f(\theta)$ depends only on the auxiliary polar angle θ of scattering, and (iii) in the exothermic inelastic case, $\Delta m/m_2$ and m_2'/m_2 are both of order unity, a straightforward calculation gives the outgoing nucleon velocity change, averaged over scattering angles. In the elastic scattering case, with $\Delta m = 0$, $m_2' = m_2$, we have

$$\langle \delta \vec{v}_1 \rangle = -2 \frac{m_2}{m_1 + m_2} (\vec{u}_1 - \vec{u}_2) \langle \sin^2(\theta/2) \rangle \ , \tag{1}$$

while in the inelastic case a good approximation is

$$\langle \delta \vec{v}_1 \rangle \simeq \frac{\vec{u}_1 - \vec{u}_2}{|\vec{u}_1 - \vec{u}_2|} \left(\frac{2\Delta m \, m_2'}{m_1(m_1 + m_2')} \right)^{1/2} c \langle \cos \theta \rangle \ , \tag{2}$$

with $\langle ... \rangle$ denoting the angular average over the center of mass differential scattering cross section. Since $\vec{u}_1$ and $\vec{u}_2$ are typically of order 10 km s^{-1}, the velocity change in the inelastic case is significantly larger than that in the elastic case.

2.2. *Change in outgoing spacecraft velocity*

Again as shown in [1], to get the force per unit spacecraft mass resulting from dark matter scatters, that is, the acceleration, one multiplies the velocity change in a single scatter $\langle \delta \vec{v}_1 \rangle$ by the number of scatters per unit time. This latter is given by the flux $|\vec{u}_1 - \vec{u}_2|$, times the scattering cross section σ, times the dark matter spatial and velocity distribution $\rho(\vec{x}, \vec{u}_2)$. Integrating out the dark matter velocity, one thus gets for the force acting at the point $\vec{x}(t)$ on the spacecraft trajectory with velocity $\vec{u}_1 = d\vec{x}(t)/dt$,

$$\delta \vec{F} = \int d^3 u_2 \langle \delta \vec{v}_1 \rangle |\vec{u}_1 - \vec{u}_2| \sigma \rho(\vec{x}, \vec{u}_2) \ . \tag{3}$$

Equating the work per unit spacecraft mass along a trajectory from t_i to t_f to the change in kinetic energy per unit mass (assuming that the initial and final times are in the asymptotic region where the potential energy can be neglected) we get

$$\delta \frac{1}{2}(\vec{v}_f^2 - \vec{v}_i^2) = \vec{v}_f \cdot \delta \vec{v}_f = \int_{t_i}^{t_f} dt (d\vec{x}/dt) \cdot \delta \vec{F}$$

$$= \int_{t_i}^{t_f} dt \int d^3 u_2 (d\vec{x}/dt) \cdot \langle \delta \vec{v}_1 \rangle |\vec{u}_1 - \vec{u}_2| \sigma \rho(\vec{x}, \vec{u}_2) \ . \tag{4}$$

2.3. Cross section and scattering-angle averaged kinematics

Let W be the center of mass scattering energy of the dark matter-spacecraft nucleon system. A simple calculation shows that to a good approximation we have

$$\frac{W}{(m_1 + m_2)c^2} \simeq 1 + \frac{m_1 m_2}{2(m_1 + m_2)^2} \frac{(\vec{u}_1 - \vec{u}_2)^2}{c^2} \ , \tag{5}$$

and so for $m_2 \leq m_1$ and for the nonrelativistic velocities $\vec{u}_1$, $\vec{u}_2$ of interest, the scattering is very close to threshold. Thus the cross section will be dominated by the lowest partial waves, which near threshold each have a characteristic power law dependence on the entrance channel momentum

$$k = \frac{m_1 m_2}{m_1 + m_2} |\vec{u}_1 - \vec{u}_2| \ . \tag{6}$$

For elastic scattering, the cross section is S-wave dominated, and tends to a k-independent constant $\sigma_{\rm el}$ near threshold, and the angular average $2\langle \sin^2(\theta/2) \rangle$ reduces to $1 - \langle \cos\theta \rangle = 1$. Thus when Eq. (1) is substituted into Eqs. (3) and (4), we can effectively replace $2\langle \sin^2(\theta/2) \rangle \sigma$ by the k-independent constant $\sigma_{\rm el}$.

For exothermic inelastic scattering, the leading contribution to $\langle \cos\theta \rangle$ comes from the interference term between the S- and P-waves in the cross section, which scales [3] as $k^{-2} k^{1/2} k^{3/2} \sim$ constant near threshold. Writing near threshold

$$\frac{d\sigma}{d\Omega} = \frac{A_{\rm inel}}{4\pi} k^{-1} + B_{\rm inel} \frac{3}{4\pi} \cos\theta + ..., \tag{7}$$

we have

$$\sigma \simeq A_{\rm inel} k^{-1} \ ,$$

$$\langle \cos\theta \rangle \simeq B_{\rm inel}/(A_{\rm inel} k^{-1}) \ . \tag{8}$$

So when Eq. (2) for the inelastic exothermic case is substituted into Eqs. (3) and (4), we can effectively replace $\langle \cos\theta \rangle \sigma$ by the k-independent constant $B_{\rm inel}$, remembering, however, that this is not the total cross section (which approaches $A_{\rm inel} k^{-1}$ near threshold) but is proportional to the coefficient of the S-wave P-wave interference term in the differential cross section.

2.4. The dark matter distribution function $\rho(\vec{x}, \vec{u}_2)$

We now address the task of formulating a model for the distribution function $\rho(\vec{x}, \vec{u}_2)$ that describes dark matter postulated to be in orbit around the earth. The simplest model would be a disk composed of dark matter in circular orbits in earth's equatorial plane, but attempts to fit the flyby anomaly data with such a model were unsuccessful, since for any reasonable disk inner radius, some of the flybys (such as NEAR) pass inside the disk. We thus proceed to the next simplest model, which is constructed from dark matter in a circular orbit, of radius r and tilted at an angle ψ ($0 \le \psi \le \pi$) with respect to earth's equatorial plane. If the earth were exactly spherically symmetric, its gravitational field would be strictly monopole, and such a tilted orbit would be stable. But in fact the earth's rotation produces an equatorial bulge, and so its mass distribution is only axially symmetric around its rotation axis, giving rise to quadrupole and higher moments in its gravitational field. As a result of these higher moments, the tilted orbit precesses around the earth's rotation axis, in such a way that the angular momentum component L_z along the earth's axis is conserved. Over a long period of time, this precession will smear an initial cluster of tilted orbits into a uniform shell, obtained by averaging the tilted circle over the azimuthal angle that its normal makes with respect to the earth's rotation axis.

To give this picture a mathematical description, let x, y, z be a Cartesian axis system, with positive z pointing to the earth's North pole (so that the rotation sense of the earth is from x to y). Let the normal $\hat{n}$ to the tilted orbit have polar angle ψ and azimuthal angle ϕ with respect to this system, so that $\hat{n}(\psi, \phi) = (\sin\psi\cos\phi, \sin\psi\sin\phi, \cos\psi)$, and let the angle of rotation within the plane of the dark matter orbit be θ, with increasing θ corresponding, at $\psi = 0$, to the direction of earth's rotation. Then a parametric description of the tilted circle is $\vec{P}(r, \theta, \phi) \equiv (P_x(r, \theta, \phi), P_y(r, \theta, \phi), P_z(r, \theta, \phi))$, with

$$\begin{aligned} P_x(r, \theta, \phi) &= r(\cos\theta\cos\psi\cos\phi - \sin\theta\sin\phi) \ , \\ P_y(r, \theta, \phi) &= r(\cos\theta\cos\psi\sin\phi + \sin\theta\cos\phi) \ , \\ P_z(r, \theta, \phi) &= -r\cos\theta\sin\psi \ , \quad |P_z(r, \theta, \phi)| \le r\sin\psi \ . \end{aligned} \tag{9}$$

The corresponding velocity unit vector of a dark matter particle in the tilted circular orbit is $\vec{U}(\theta, \phi) = (U_x(\theta, \phi), U_y(\theta, \phi), U_z(\theta, \phi)) = r^{-1}d\vec{P}/d\theta$, with

$$\begin{aligned} U_x &= -\sin\theta\cos\psi\cos\phi - \cos\theta\sin\phi \ , \\ U_y &= -\sin\theta\cos\psi\sin\phi + \cos\theta\cos\phi \ , \\ U_z &= \sin\theta\sin\psi \ . \end{aligned} \tag{10}$$

The velocity vector is obtained by multiplying the velocity unit vector by the velocity magnitude $(GM_\oplus/r)^{1/2}$ for a particle in a circular orbit of radius r, with G the Newton gravitational constant and $M_\oplus$ the earth mass.

Integrating the position and velocity distribution for a tilted circular orbit over the angles θ, ϕ gives the distribution for the corresponding shell, and integrating over the shell parameters r, ψ with a general weighting function $w(r, \psi)$ gives as the model for the dark matter distribution function

$$
\rho(\vec{x}, \vec{u}_2) = \int dr \int d\psi\, w(r, \psi) \int_0^{2\pi} d\theta \int_0^{2\pi} d\phi\, \delta^3(\vec{x} - \vec{P}(r, \theta, \phi))
$$
$$
\times \delta^3\left(\vec{u}_2 - (GM_\oplus/r)^{1/2}\vec{U}(\theta, \phi)\right) , \tag{11}
$$

with the corresponding total number of particles in the shell given by

$$
N \equiv \int d^3x \int d^3u_2\, \rho(\vec{x}, \vec{u}_2) = 4\pi^2 \int dr \int d\psi\, w(r, \psi) . \tag{12}
$$

Referring to Eq. (4), we have to evaluate an integral over the distribution function of the form

$$
I = \int dt \int d^3u_2 F(\vec{x}(t), d\vec{x}(t)/dt, \vec{u}_2)\rho(\vec{x}(t), \vec{u}_2) , \tag{13}
$$

with $F(\vec{x}(t), d\vec{x}(t)/dt, \vec{u}_2)$ given by

$$
F(\vec{x}(t), d\vec{x}(t)/dt, \vec{u}_2) = (d\vec{x}(t)/dt) \cdot \langle \delta\vec{v}_1 \rangle|_{\vec{u}_1 = d\vec{x}(t)/dt}\, |d\vec{x}(t)/dt - \vec{u}_2|\sigma . \tag{14}
$$

On substituting Eq. (11) and noting that the coordinate delta function constrains $r = |\vec{P}(r, \theta, \phi)| = |\vec{x}(t)| \equiv r(t)$, we obtain

$$
I = \int dt \int d\psi\, w(r(t), \psi) \int dr
$$
$$
\times \int_0^{2\pi} d\theta \int_0^{2\pi} d\phi\, F\left(\vec{x}(t), d\vec{x}(t)/dt, (GM_\oplus/r(t))^{1/2}\vec{U}(\theta, \phi)\right)\delta^3(\vec{x}(t) - \vec{P}(r, \theta, \phi)) . \tag{15}
$$

By making changes of variable one can carry out the integrations over r, ϕ and θ in Eq. (15), leaving an integral in which θ and z have been replaced, by virtue of the delta function constraints, by $\theta(\vec{x}(t))$ and $z(t) \equiv z(\vec{x}(t))$,

$$
I = \int dt \int d\psi\, w(r(t), \psi) \sum_\pm F\left(\vec{x}(t), d\vec{x}(t)/dt, (GM_\oplus/r(t))^{1/2}\vec{U}_\pm(\theta(\vec{x}(t)), \phi(\vec{x}(t)))\right)
$$
$$
\times \frac{1}{r(t)\sqrt{r(t)^2 \sin^2 \psi - z(t)^2}} . \tag{16}
$$

Note that by virtue of Eq. (9), the integration domain extends only over $|z(t)| \leq r(t)\sin\psi$, and hence the argument of the square root is nonnegative. In Eq. (16) the sum over $\pm$ is over the two roots $\theta(\vec{x}(t))$ of the equation $\cos\theta(\vec{x}(t)) = -z(t)/(r(t)\sin\psi)$, which differ in the sign of $\sin\theta$,

$$\sin\theta(\vec{x}(t)) = \pm\sqrt{1 - z(t)^2/(r(t)^2\sin^2\psi)}\,, \tag{17}$$

while the values of $\phi(\vec{x}(t))$ corresponding to these two roots $\theta(\vec{x}(t))$ are obtained by equating $\vec{x}(t)$ to $\vec{P}$ and then solving Eq. (9) for $\cos\phi$ and $\sin\phi$. The two roots correspond to the fact that a circular orbit with tilt angle ψ consists of two semicircular segments, with opposite directions of the velocity component normal to the equatorial plane. Thus the intersection of the spacecraft trajectory $\vec{x}(t)$ with the dark matter shell generated by azimuthal rotation of such a tilted circular orbit will intersect two segments of circular orbits, one up-going and one down-going relative to the equatorial plane.

It will be useful for what follows to express the unit velocities $\vec{U}_\pm(\theta(\vec{x}(t)), \phi(\vec{x}(t)))$ in terms of their components on unit vectors $\hat{n}_\parallel(t) = (\hat{z}\times\hat{x}(t))/|\hat{z}\times\hat{x}(t)|$ and $\hat{n}_\perp(t) = \hat{x}(t)\times\hat{n}_\parallel(t)$, normal to $\hat{x}(t) = \vec{x}(t)/r$, that are respectively parallel (in the sense of earth rotation) and perpendicular to the earth equatorial plane. A simple calculation shows that $\vec{U}_\pm$ are given on this basis by

$$\vec{U}_\pm(\theta(\vec{x}(t)), \phi(\vec{x}(t))) = C(t)\hat{n}_\parallel \pm D(t)\hat{n}_\perp\,, \tag{18}$$

with the coefficients $C(t)$ and $D(t)$ given by

$$C(t) = \frac{r(t)\cos\psi}{\sqrt{r(t)^2 - z(t)^2}}\,, \quad D(t) = \frac{\sqrt{r(t)^2\sin^2\psi - z(t)^2}}{\sqrt{r(t)^2 - z(t)^2}}\,, \tag{19}$$

which obey $C(t)^2 + D(t)^2 = 1$. Explicit expressions for $\hat{n}_\parallel(t)$ and $\hat{n}_\perp(t)$ in the flyby plane basis are given in the next subsection, which together with Eqs. (18) and (19) give the formulas for the unit velocities $\vec{U}_\pm(\theta(\vec{x}(t)), \phi(\vec{x}(t)))$ on the flyby plane basis needed in the numerical computations.

2.5. Flyby orbital plane kinematics

The Anderson et al. paper [2] gives the flyby orbit parameters in terms of coordinates on the celestial sphere, but it will be more convenient for our purposes to carry out all flyby orbit calculations in the flyby orbital plane. Let x_o, y_o, z_o be a Cartesian axis system, with z_o normal to the flyby orbital plane. The flyby orbit can then be written in parametric form as

$$x_o(t) = r(t) \cos \theta_o(t) \;,$$
$$y_o(t) = r(t) \sin \theta_o(t) \;,$$
$$r(t) = \frac{p}{1 + e \cos \theta_o(t)}, \qquad R_f = \frac{p}{1 + e} \;,$$
$$dx_o(t)/dt = \frac{-V_f \sin \theta_o(t)}{1 + e} = \frac{-y_o(t)}{1 + e \cos \theta_o(t)} d\theta_o(t)/dt \;,$$
$$dy_o(t)/dt = \frac{V_f(e + \cos \theta_o(t))}{1 + e} = \frac{er(t) + x_o(t)}{1 + e \cos \theta_o(t)} d\theta_o(t)/dt \;,$$
$$d\theta_o(t)/dt = \frac{R_f V_f}{r(t)^2} \;. \tag{20}$$

The scale parameter p, the eccentricity e, the velocity at closest approach to earth V_f, the radius at closest approach R_f, and the velocity at infinity can be determined from data in [2], together with the polar angle I and azimuthal angle α of the earth's north pole with respect to the x_o, y_o, z_o coordinate system. The quantities V_f and V_∞ are given directly in [2], while R_f, p, and e can be calculated from them using the formulas

$$R_f = \frac{2GM_\oplus}{V_f^2 - V_\infty^2} \;,$$

$$e = 1 + \frac{2V_\infty^2}{V_f^2 - V_\infty^2} \;,$$

$$p = \frac{4GM_\oplus}{V_\infty^2} \left[\left(\frac{V_\infty^2}{V_f^2 - V_\infty^2} \right)^2 + \frac{V_\infty^2}{V_f^2 - V_\infty^2} \right] \;. \tag{21}$$

The earth axis polar angle I is also directly given in [2], while the azimuthal angle α can be calculated from the formula

$$\cos \alpha = \frac{\sin \phi'}{\sin I} \;, \tag{22}$$

with ϕ' the geocentric latitude at closest approach (which is called ϕ in [2]; with the orbit parametrization of Eq. (20), ϕ' is the latitude of the positive x_o axis). This formula does not determine the quadrant in which α lies, but this can be fixed from the additional orbital parameters given in [2] (with some corrections supplied to me by J.K. Campbell [4]).

To carry out the computation of the flyby velocity change in the flyby plane basis x_o, y_o, z_o we will need the components of $\hat{n}_\parallel$ and $\hat{n}_\perp$ on this basis. It is easy to calculate them directly by going back to the defining cross product relations, using the components of $\vec{x}(t)$ and of the earth axis $\hat{z}$ on the flyby plane basis,

$$\vec{x}(t) = (x_o(t), y_o(t), 0) \;,$$
$$\hat{z} = (\sin I \cos \alpha, \sin I \sin \alpha, \cos I) \;. \tag{23}$$

From these we find

$$
\hat{n}_{\parallel}(t) = \frac{\hat{z} \times \hat{x}(t)}{|\hat{z} \times \hat{x}(t)|} = \frac{1}{\sqrt{r(t)^2 - z(t)^2}}
$$
$$
\times \left(-y_o(t)\cos I,\; x_o(t)\cos I,\; (y_o(t)\cos\alpha - x_o(t)\sin\alpha)\sin I \right) ,
$$
$$
\hat{n}_{\perp}(t) = \hat{x}(t) \times \hat{n}_{\parallel}(t) = \frac{1}{r(t)}\vec{x}(t) \times \hat{n}_{\parallel}(t) \tag{24}
$$

with

$$
r(t) = |\vec{x}(t)| = \sqrt{x_o(t)^2 + y_o(t)^2} ,
$$
$$
z(t) = \vec{x}(t) \cdot \hat{z} = (x_o(t)\cos\alpha + y_o(t)\sin\alpha)\sin I . \tag{25}
$$

Substituting Eq. (20) for $x_o(t)$ and $y_o(t)$ into Eq. (25) we have

$$
z(t) = r(t)\sin I \cos\left(\theta_o(t) - \alpha\right) , \tag{26}
$$

which allows one to rewrite the Jacobian factor appearing in Eq. (16) as

$$
\frac{1}{r(t)\sqrt{r(t)^2 \sin^2\psi - z(t)^2}} = \frac{1}{r(t)^2 \sin\psi \sqrt{1 - (\sin I/\sin\psi)^2 \cos^2\left(\theta_o(t) - \alpha\right)}} . \tag{27}
$$

when the argument of the square root is nonnegative.

2.6. Simplified model used for numerical work

The model as defined above involves a general weighting function $w(r, \psi)$, but for an initial survey we make the simplifying assumption of only a single tilt angle ψ_i, ψ_e for the inelastic and elastic scatterers, respectively, and Gaussian distributions in r with different centers and widths for each. Thus we take for the inelastic scatterers

$$
w_i(r, \psi) = K_i e^{-(r - R_i)^2/D_i^2} \delta(\psi - \psi_i) , \tag{28}
$$

and for the elastic scatterers

$$
w_e(r, \psi) = K_e e^{-(r - R_e)^2/D_e^2} \delta(\psi - \psi_e) . \tag{29}
$$

With this choice, the integral of Eq. (12) becomes

$$
N_\ell = 4\pi^{5/2} K_\ell D_\ell , \quad \ell = i, e . \tag{30}
$$

It is now convenient to combine the constants $K_{i,e}$ with the mass-dependent constants appearing in Eqs. (1) and (2) of Sec. 2.1, and the constants σ_{el} and B_{inel} introduced in Sec. 2.3, giving new parameters ρ_i, ρ_e characterizing the effective density times cross section for the inelastic and elastic scatterer distributions,

$$
\rho_e \equiv \frac{m_2}{m_1 + m_2} \sigma_{el} K_e ,
$$
$$
\rho_i \equiv \left(\frac{2\Delta m\, m_2'}{m_1(m_1 + m_2')} \right)^{1/2} B_{inel} K_i . \tag{31}
$$

Thus in Eq. (15) we effectively replace (see Eq. (14))

$$\int d\psi \, w(r(t), \psi) \, F(\vec{x}(t), d\vec{x}(t)/dt, \vec{u}_2) \tag{32}$$

by

$$\sum_{\ell=i,e} \left\{ |d\vec{x}(t)/dt - \vec{u}_2| (d\vec{x}(t)/dt) \cdot \vec{V}_\ell \, \rho_\ell \, e^{-(r(t)-R_\ell)^2/D_\ell^2} \right\} \Big|_{\psi=\psi_\ell} , \tag{33}$$

with $\vec{V}_\ell$ given by

$$\vec{V}_i = c \left(d\vec{x}(t)/dt - \vec{u}_2 \right) / |d\vec{x}(t)/dt - \vec{u}_2| ,$$
$$\vec{V}_e = - \left(d\vec{x}(t)/dt - \vec{u}_2 \right) , \tag{34}$$

and with $\vec{u}_2$ evaluated as $\vec{U}_\pm$ of Eqs. (16) and (18). The simplified model thus defined has eight parameters, four parameters ψ_i, ρ_i, R_i, D_i characterizing the inelastic scatterers, and four parameters ψ_e, ρ_e, R_e, D_e characterizing the elastic scatterers. Finally, we note that by combining Eqs. (30) and (31), and approximating

$$\frac{m_2}{m_1 + m_2} \sim \left(\frac{2\Delta m \, m_2'}{m_1(m_1 + m_2')} \right)^{1/2} \sim \frac{m_2}{m_1} , \tag{35}$$

we find the following estimates for the total mass in the dark matter shells,

$$M_e \equiv m_2 N_e = 4\pi^{5/2} \rho_e D_e m_1 / \sigma_{\mathrm{el}} ,$$
$$M_i \equiv m_2 N_i = 4\pi^{5/2} \rho_i D_i m_1 / B_{\mathrm{inel}} . \tag{36}$$

3. Numerical Results and Discussion

Let us turn now to numerical fitting of the eight parameter model to the flyby anomalies reported in [2]. In carrying out the needed integrals over flyby orbits, we replaced the integration over t by an integration over orbit angle θ_o, using the expression for $d\theta_o/dt$ given in Eq. (20). To utilize integration mesh points efficiently, the integrations were restricted to the parts of the orbits where the Gaussian factors $e^{-(r-R_\ell)^2/D_\ell^2}$ were larger than $e^{-9} = 0.00012$, that is, to the parts of the orbits where $|r - R_\ell| \leq 3D_\ell$.

Our numerical searches were carried out by minimizing a least squares likelihood function χ^2, defined as

$$\chi^2 = \sum_{k=1}^{6} (\delta v_{k;\mathrm{th}} - \delta v_{k;\mathrm{A}})^2 / \sigma_{k;\mathrm{A}}^2 , \tag{37}$$

where k indexes the six flybys reported by Anderson et al. [2], the $\delta v_{k;\mathrm{th}}$ are the theoretical values of the velocity discrepancies computed from our model, the $\delta v_{k;\mathrm{A}}$ are the observed values for these discrepancies reported in [2], and the $\sigma_{k;\mathrm{A}}$ are the corresponding estimated errors in these discrepancies given in [2]. Since the quoted

$\sigma_{k;A}$ values contain both systematic and statistical components, a least squares likelihood function is not a true statistical chi square function, but having a quadratic form is very convenient for the following reason. Because the theoretical values $\delta v_{k;th}$ are linear in the dark matter density times cross section parameters $\rho_{i,e}$,

$$\delta v_{k;th} = \rho_i \delta v_{k;i} + \rho_e \delta v_{k,e} \ , \tag{38}$$

with $\delta v_{k;i,e}$ the respective contributions from the inelastic and elastic scatterers computed with $\rho_{i,e} = 1$, the likelihood function is a positive semi-definite quadratic form in these two parameters. Hence for fixed values of the other six parameters $\psi_{i,e}$, $R_{i,e}$, $D_{i,e}$, the minimization of χ^2 with respect to the parameters $\rho_{i,e}$ can be accomplished algebraically by solving a pair of linear equations in the two variables $\rho_{i,e}$, with the result

$$\rho_i = \frac{C_{ee}G_i - C_{ei}G_e}{C_{ii}C_{ee} - C_{ie}C_{ei}} \ ,$$
$$\rho_e = \frac{C_{ii}G_e - C_{ie}G_i}{C_{ii}C_{ee} - C_{ie}C_{ei}} \ , \tag{39}$$

with coefficients given by

$$C_{\ell m} = \sum_{k=1}^{6} \frac{\delta v_{k;\ell} \delta v_{k;m}}{\sigma_{k;A}^2} \ , \quad \ell, m = i, e \ ,$$

$$G_{\ell} = \sum_{k=1}^{6} \frac{\delta v_{k;A} \delta v_{k;\ell}}{\sigma_{k;A}^2} \ , \quad \ell = i, e \ . \tag{40}$$

This has the effect of reducing the parameter space that must be searched numerically from an eight parameter space to a six parameter space, which results in a substantial saving of computational effort.

Our search procedure is described in detail in arXiv:0908.2414, and detailed numerical results for the fits are given there in Tables II-VII, three of which are reproduced here (and renumbered 1, 2, 3 respectively). From the products $\rho_i D_i$ and $\rho_e D_e$ for each fit, one can use Eq. (36) to estimate the total mass in the dark matter shells, in terms of the elastic and inelastic scattering parameters $\sigma_{\rm el}$ and $B_{\rm inel}$. Alternatively, given the upper bound [5] on the mass of dark matter in orbit around the earth between the LAGEOS satellite orbit and the moon's orbit, of $4 \times 10^{-9} M_\oplus \sim 1.4 \times 10^{43} {\rm GeV}/c^2$, one can turn these relations into lower bounds on $\sigma_{\rm el}$ and $B_{\rm inel}$. For example, from the values $\rho_i D_i = 0.00304\,{\rm km}^2$ and $\rho_e D_e = 19.2\,{\rm km}^2$ for fit 2d, one finds the bounds

$$\sigma_{\rm el} \geq 9.4 \times 10^{-31} {\rm cm}^2 \ ,$$
$$B_{\rm inel} \geq 1.5 \times 10^{-34} {\rm cm}^2 \ , \tag{41}$$

which are consistent with the cross section range arrived at from various constraints in [1]. The spatial constraints found in [1], which require that the dark matter should be localized well away from the earth and the moon, are also obeyed.

362

Table 1. Flyby anomaly fits.

	χ^2	GLL-I	GLL-II	NEAR	Cassini	Rosetta	Messenger
δv_A (mm/s)		3.92	-4.6	13.46	-2	1.80	0.02
σ_A (mm/s)		0.3	1.0	0.01	1	0.03	0.01
δv_{th} fits 1a–e	$< 10^{-6}$	3.92	-4.60	13.46	-2.00	1.80	0.020
δv_{th} fit 2a	2.07	3.98	-5.5	13.46	-3.1	1.79	0.021
δv_{th} fit 2b	1.68	4.15	-5.2	13.46	-2.9	1.80	0.020
δv_{th} fit 2c	1.29	4.13	-5.0	13.46	-2.8	1.80	0.020
δv_{th} fit 2d	0.51	3.90	-4.6	13.46	-2.7	1.80	0.020
δv_{th} fit 2e	0.52	3.88	-4.6	13.46	-2.7	1.80	0.020
δv_{th} fit 2f	0.70	3.84	-4.7	13.46	-2.7	1.80	0.021
δv_{th} fit 2g	7.5	3.76	-4.7	13.46	-2.8	1.73	0.028

Fits 1a–e are for the model with smoothed Jacobian and trapezoidal integration, resulting from a five-parameter fit with R_i constrained to the values shown in Table 2. Fits 2a–g are for the un-smoothed model and adaptive trapezoidal integration, resulting from a five-parameter fit with R_i constrained to the values shown in Table 3.

Table 2. Parameter values for fits 1a–e.

fit	$10^6 \times \rho_i$ (km)	$10^2 \times \rho_e$ (km)	ψ_i (rad)	ψ_e (rad)	R_i (km)	D_i (km)	R_e (km)	D_e (km)
1a	0.304	0.268	1.926	0.3939	30000	6278	28620	6303
1b	1.55	0.245	1.261	0.3945	40000	2185	27985	5890
1c	0.411	0.261	1.374	0.3952	50000	13540	28450	6299
1d	0.351	0.253	1.381	0.3946	60000	20193	28340	6334
1e	0.343	0.248	1.394	0.3942	70000	25780	28240	6367

Table 3. Parameter values for fits 2a–g.

fit	$10^6 \times \rho_i$ (km)	$10^2 \times \rho_e$ (km)	ψ_i (rad)	ψ_e (rad)	R_i (km)	D_i (km)	R_e (km)	D_e (km)
2a	0.537	0.323	1.767	0.3902	25000	3030	29370	6678
2b	0.827	0.316	1.626	0.3902	30000	3030	29370	6678
2c	0.965	0.309	1.515	0.3902	32500	3030	29370	6678
2d	1.000	0.288	1.372	0.3902	34520	3030	29370	6678
2e	0.655	0.288	1.369	0.3902	35000	4663	29370	6678
2f	0.348	0.288	1.364	0.3902	37500	9223	29370	6678
2g	0.290	0.286	1.361	0.3902	40000	11681	29370	6678

The results in the Tables show that the dark matter scattering model, with inelastic and elastic scatterers, can account for the flyby anomaly data. One could argue that the fits are too good, and are indicative of "over-fitting", since there are 8 parameters in the model, and only 6 data points. On the other hand, it was not a priori obvious that such a simple model should be able to account for data from a complicated physical process with a three-dimensional geometry, and the results shown in Table VII of arXiv:0908.2414 support the view that the success of the model is not attributable to over-fitting of the data.

Further steps in this investigation would be: (1) incorporation of further flybys into the fits, when the corresponding flyby parameters and velocity discrepancy and error values are available, or alternatively, using fit 2d (or 4a) to predict the velocity discrepancy for future flybys, given their orbital parameters; (2) incorporating constraints on residual drag coming from fitting satellite drag measurements to conventional drag sources; (3) as suggested to me by V. Toth [6], incorporating the time development of the velocity anomaly near perigree when such data becomes available from improved tracking of future flybys; (4) as suggested to me by J. Rosner [7], investigating possible constraints arising from the effect of the quadrupole moment of the dark matter shells on the precession of high-lying satellite orbits; (5) extending the model to include a general form of the weighting function $w(r, \psi)$; and (6) extending the model to include shells generated by precessing elliptical, as opposed to circular orbits, and shells generated by a precessing Schwarzschild disk [8]. The extensions (5) and (6) will require computing resources well beyond those used here to analyze the 8 parameter model. It will also be necessary to address the question of mechanisms for producing dark matter shells. According to A. Peter [9], the accumulation cascade suggested in [1] is not viable as a mechanism. Another scenario, suggested by Dr. Peter's comments and the structure of the model formulated here, would involve the gravitational capture by the earth of a dense (up to $\sim 10^{15}$ times galactic halo mean density, that is $\sim 10^{-9}$ times mean ordinary matter density) condensed ball of dark matter into an orbit tilted with respect to earth's rotation axis; breakup of this by tidal forces could then lead to population of a shell of the type we have assumed.[b] If the flyby anomalies are ultimately confirmed, detailed study of such a mechanism would be warranted.

Acknowledgments

This work was supported by the Department of Energy under grant no DE-FG02-90ER40542, and parts of this work were done during the author's stay at the Aspen Center for Physics. I wish to thank Scott Tremaine for helpful conversations about orbital dynamics, James Campbell for sending me corrections to some of the data published in [2], Michele Papucci for suggesting that I use the CERN minimization program Minuit, and Prentice Bisbal for assistance in downloading it to my computer. I also wish to thank Angelo Bassi, Annika Peter, Jonathan Rosner, and Viktor Toth for helpful comments after the initial version of this paper was posted on the arXiv.

References

1. S. L. Adler, Phys. Rev. D **79**, 023505 (2009).
2. J. D. Anderson, J. K. Campbell, J. E. Ekelund, J. Ellis, and J. F. Jordan, Phys. Rev. Lett. **100**, 091102 (2008).

---

[b]The constraints derived in [1] on the sun-bound dark matter density are not relevant for this scenario for producing dark matter shells.

364

3. S. Weinberg, "The Quantum Theory of Fields, Vol. I Foundations", Cambridge University Press (1995), pp. 156-157.
4. J. K. Campbell, private email communication (2008).
5. S. L. Adler, J. Phys. A: Math. Theor. **41**, 412002 (2008).
6. V. Toth, private email communication (2009).
7. J. Rosner, private email communication (2009).
8. J. Binney and S. Tremaine, *Galactic Dynamics*, second edition, Princeton University Press (2009), Sec. 4.4.3.
9. A. Peter, private email communication (2009).

PHYSICS OF MICRO BLACK HOLES

GEORGI DVALI

Institute for Advanced Study, New York University, USA

We shall discuss an intrinsic connection between the nature of short distance gravity and the low energy particle species (and their symmetries) and how this connection defines the underlying nature of microscopic black holes.

POSSIBLE SOLUTION OF DARK MATTER, THE SOLUTION OF DARK ENERGY AND GELL-MANN AS GREAT THEORETICIAN

PAUL HOWARD FRAMPTON

Department of Physics and Astronomy, University of North Carolina, Chapel Hill, USA
and
Institute for the Mathematics and Physics of the Universe,
University of Tokyo, Japan
frampton@physics.unc.edu, paul.frampton@ipmu.jp

This work is dedicated to Murray Gell-Mann for his 80th birthday.

This talk discusses the formation of primordial intermediate-mass black holes, in a double-inflationary theory, of sufficient abundance possibly to provide all of the cosmological dark matter. There follows my, hopefully convincing, explanation of the dark energy problem, based on the observation that the visible universe is well approximated by a black hole. Finally, I discuss that Gell-Mann is among the five greatest theoreticians of the twentieth century.

Keywords: Black hole; dark matter; holographic principle; entropy.

1. Outline of Talk

It is an honor to talk at a festschrift for Murray Gell-Mann, who dominated research in particle phenomenology for at least twenty years.

At the beginning of my talk, I shall discuss a recent paper[1] on the production of primordial intermediate-mass black holes of mass $M_{BH} = M_\odot^p$ with $-8 \le p \le +5$, providing a sufficient abundance, that the primordial IMBHs can possibly act as all the cosmological dark matter.

I then discuss my solution[2] for the difficult dark energy problem which was first identified from observations of supernovae, twelve years ago. Although I knew all the correct theoretical ingredients back then, the solution hit me only on February 6, 2010. Because this was an overwhelming human experience, I self-indulgently discuss it.

Finally, I discuss why Gell-Mann, who must himself have experienced a similar personally fulfilling moment, for the Ω^- particle,[3,4] is to be correctly, regarded as among the five greatest theoreticians, of the twentieth century.

2. Possible Solution for Dark Matter

If the dark matter (DM) is made of a weakly interacting massive particle (WIMP), we may be able to observe collider, direct and indirect DM signatures; the DM particles may be produced at LHC, and the next-generation direct search experiments will probe a significant portion of parameter space predicted by various theoretical DM models. In spite of thorough DM searches using widely different techniques, the results are negative so far. If no DM signature is found in the future experiments, it may suggest that the basic assumption that the DM is made of unknown particles is simply wrong.

There actually *is* a DM candidate in the framework of SM, namely, a primordial black hole (PBH). In the early Universe PBHs can form when the density perturbation becomes large, and it has been known that a PBH of mass greater than 10^{15} g survives evaporation, and therefore contributes to the DM density.

In consideration of the entropy of the universe it was pointed out in Ref. 5 that if all DM were in the form of $10^5 M_\odot$ black holes it would contribute a thousand times more entropy than the supermassive black holes at galactic centers and hence be a statistically favored configuration. Here we consider primordial black holes (PBHs) with masses from $10^5 M_\odot$ to $10^{-8} M_\odot$ and, subject to observational constraints, any of these masses can comprise all DM although the entropy argument favors the heaviest $10^5 M_\odot$ mass.

There are several ways to realize large density fluctuations leading to PBH formation. One possibility is the production of PBHs from density fluctuations generated during inflation. Since the blue spectrum with a spectral index $n_s > 1$ is disfavored by the WMAP data, a single inflation may not be able to produce large density fluctuations at small scales unless some dynamics is introduced during inflation. On the other hand, the density fluctuations can be easily enhanced at small scales in a double inflation model.

In Ref. 1, we discuss a double inflation model that consists of a smooth-hybrid inflation and a new inflation. In this set-up PBHs with a narrow mass distribution are formed as a result of an explosive particle production between the two inflations. We show that the PBH mass can take a wide range of values from $10^{-8} M_\odot$ up to $10^5 M_\odot$. Also, the resultant PBH mass has a correlation with running of spectral index. We numerically calculated the correlation, which can be tested by future observations.

The black hole mass, and the formation epoch, are related to each other, due to the causality. In the early Universe, the mass contained in the Hubble horizon sets an upper bound on the PBH mass formed at that time. Assuming that the whole mass in the horizon is absorbed into one black hole, we obtain

$$M_{\text{BH}} = \frac{4\pi\sqrt{3}M_P^3}{\sqrt{\rho_f}} \simeq 0.05\, M_\odot \frac{g_*}{100}^{-\frac{1}{2}} \frac{T_f}{\text{GeV}}^{-2},$$

$$\simeq 1.4 \times 10^{13}\, M_\odot \frac{g_*}{100}^{-\frac{1}{6}} \frac{k_f}{\text{Mpc}^{-1}}^{-2}, \tag{1}$$

where M_{BH} is the black hole mass, $M_P \simeq 2.4 \times 10^{18}$ GeV is the reduced Planck mass, $M_\odot \simeq 2 \times 10^{33}$ g is the solar mass, g_* counts the light degrees of freedom in thermal equilibrium, ρ_f, T_f and k_f are the energy density, the plasma temperature and the comoving wavenumber corresponding to the Hubble horizon at the formation, respectively. The radiation domination was assumed in the second equality.

As is well known, any black holes have a temperature inversely proportional to its mass and evaporates in a finite time τ_{BH},

$$\tau_{\mathrm{BH}} \simeq 10^{64} \frac{M_{\mathrm{BH}}}{M_\odot} 3 \mathrm{yr}. \tag{2}$$

Thus the black holes with mass less than 10^{15} g must have evaporated by now. PBHs which remain as (a part of) DM must therefore be created at a temperature below 10^9 GeV. In the following we assume that PBHs account for all DM in our Universe.

The cosmological effects of PBHs have been extensively studied so far. While PBHs with masses below 10^{15} g are significantly constrained, it is very difficult to detect PBHs heavier than 10^{15} g because of negligible amount of the radiation. The MACHO and EROS collaborations monitored millions of stars in the Magellanic Clouds to search for microlensing events caused by MAssive Compact Objects (MACHOs) passing near the line of sight. The MACHO collaboration excluded the objects in the mass range $0.3 M_\odot$ to $30 M_\odot$, and the latest result of the EROS-1 and EROS-2 excluded the mass range $0.6 \times 10^{-7} M_\odot < M < 15 M_\odot$, as the bulk component of the galactic DM. On the other hand, if we assume that the PBH formation occurs before the big bang nucleosynthesis (BBN) epoch, the PBH mass should be lighter than $10^5 M_\odot$. Therefore we consider PBHs with masses (i) $M_{\mathrm{BH}} < 10^{-7} M_\odot$ and (ii) $30 M_\odot < M_{\mathrm{BH}} < 10^5 M_\odot$.

The above observational constraints provide us with information on the PBH formation. If PBHs are produced at different times, the mass function tends to be broad, thereby making it difficult to be consistent with observations. In order to realize the PBH mass function with a sharp peak, most of the PBHs should be produced at the same time. Thus the production mechanism must involve such a dynamics that only the density fluctuation of a certain wavelength rapidly grows.

What kind of dynamics can create PBHs? First of all, density perturbation must become large for PBHs to be formed. There are several ways to realize large density fluctuations leading to the PBH formation. One possibility is the production of PBHs from density fluctuations generated during inflation. In the standard picture of inflation, the inflation driven by a slow-rolling scalar field lasts for more than about 60 e-foldings to solve theoretical problems of the big bang cosmology. Then no dynamics for producing a sharp peak in the density perturbation is expected. However, there is no a priori reason to believe that our Universe experienced only one inflationary expansion. Indeed, the cosmological gravitino or modulus problem can be relaxed if the energy scale of the last inflation is rather low, and it is then quite likely that there was another inflation before the last one. If the multiple inflation is a common phenomenon, we expect that explosive particle production between the successive inflation periods may produce a sharp peak in the density perturbation at the desired scales, which leads to the PBH formation at a later time. In the next section, we show that this is actually feasible using a concrete double inflation model.

We provide a double inflation model, producing PBHs with a sharp mass function, as an existence proof. The first inflation is realized by smooth hybrid inflation. The smooth hybrid inflation model is built in framework of supergravity and the superpotential and Kähler potential are given by

$$W_H = S \left(\mu^2 + \frac{(\bar{\Psi}\Psi)^m}{M^{2(m-1)}} \right) \qquad (m = 2, 3, \ldots), \tag{3}$$

$$K_H = |S|^2 + |\Psi|^2 + |\bar{\Psi}|^2, \tag{4}$$

where S is the inflaton superfield, Ψ and $\bar{\Psi}$ are waterfall superfields, μ is the inflation scale and M is the cut-off scale which controls the nonrenormalizable term. From the above superpotential and Kähler potential together with phase redefinition and the D-flat condition, we obtain the scalar potential as

$$V_H(\sigma, \psi) \simeq \left(1 + \frac{\sigma^4}{8} + \frac{\psi^2}{2} \right) \left(-\mu^2 + \frac{\psi^4}{4M^2} \right)^2 + \frac{\sigma^2 \psi^6}{16M^4}, \tag{5}$$

where $\sigma \equiv \sqrt{2} ReS$ and $\psi \equiv 2Re\Psi = 2Re\bar{\Psi}$. Here and in what follows we use the Planck unit $M_P = 1$ and take $m = 2$ for simplicity. Although the scalar potential (5) is derived in the framework of supergravity, one may start with (5) without assuming supersymmetry. The potential (5) has a true vacuum at $\sigma = 0$ and $\psi = 2\sqrt{\mu M}$. For $\sigma \gtrsim \sqrt{\mu M}/2$, however, the potential for ψ has a σ-dependent minimum at

$$\psi_{\min} \simeq \frac{2}{\sqrt{3}} \frac{\mu M}{\sigma}. \tag{6}$$

Note that ψ quickly settles down at the minimum during inflation since its mass is larger than the Hubble parameter. Then we can integrated out ψ and obtain the effective potential for σ as

$$V(\sigma) = \mu^4 \left(1 + \frac{\sigma^4}{8} - \frac{2}{27} \frac{\mu^2 M^2}{\sigma^4} \right) = \mu^4 + \frac{\mu^4}{8} \left(\sigma^4 - \sigma_d^4 \left(\frac{\sigma_d}{\sigma} \right)^4 \right), \tag{7}$$

where $\sigma_d \equiv \sqrt{2}/3^{3/8} (\mu M)^{1/4}$. If the scalar potential is dominated by the first term, the inflaton σ slow rolls and therefore inflation occurs.

According to the WMAP 5yr data, the curvature perturbation $\mathcal{R}$, the spectral index n_s and its running $dn_s/d\ln k$ at the pivot scale $k_* = 0.002 \mathrm{Mpc}^{-1}$ are

$$\mathcal{R} = 4.9 \times 10^{-5}, \tag{8}$$

$$n_s = 1.031 \pm 0.055, \tag{9}$$

$$\frac{dn_s}{d\ln k} = -0.037 \pm 0.028. \tag{10}$$

From the effective potential, we obtain

$$\mathcal{R} = \frac{V^{3/2}}{\sqrt{3}\pi V'} = \frac{\mu^2}{\sqrt{3}\pi} \left[\sigma_*^3 + \sigma_d^3 \left(\frac{\sigma_d}{\sigma_*} \right)^5 \right]^{-1}, \tag{11}$$

$$n_s - 1 \simeq 2\frac{V''}{V} = \left[3\sigma_*^2 - 5\sigma_d^2 \left(\frac{\sigma_d}{\sigma_*} \right)^6 \right], \tag{12}$$

$$\frac{dn_s}{d\ln k} \simeq -2\frac{V'''V'}{V^2} = -3 \left[\sigma_*^3 + \sigma_d^3 \left(\frac{\sigma_d}{\sigma_*} \right)^5 \right] \left[\sigma_* + 5\sigma_d \left(\frac{\sigma_d}{\sigma_*} \right)^7 \right], \tag{13}$$

where σ_* is the field value of the inflaton when the fluctuation corresponding to the pivot scale exits the Hubble horizon.

The fluctuation corresponding to the pivot scale k_* exits the horizon at $t = t_*$ when $k_*/a(t_*) = H_H = \mu^2/\sqrt{3}$ (H_H: hubble during the smooth hybrid inflation). Thus the scale factor $a_* = a(t_*)$ is given by

$$\ln a_* = -2\ln\mu - 136. \tag{14}$$

The e-folding number between the horizon exit of the pivot scale and the end of the smooth hybrid inflation is estimated as

$$N_*(\sigma) = \int_{\sigma_e}^{\sigma_*} d\sigma \frac{V}{V'}$$

$$\simeq \frac{4}{3\sigma_d^2} - \frac{1}{\sigma_*^2} \quad (\sigma_* > \sigma_d)$$

$$\simeq \frac{\sigma_*^6}{3\sigma_d^8} \quad (\sigma_* < \sigma_d) \tag{15}$$

where $\sigma_e (\ll \sigma_d)$ denotes the field value when the smooth hybrid inflation ends.

After the smooth hybrid inflation, σ and ψ oscillate about their minima and decay into the σ and ψ quanta via self-couplings and mutual coupling of the two fields. Since their effective masses depend on the field amplitudes and therefore time-dependent, specific modes of the σ and ψ quanta are strongly amplified by parametric resonance. To see this, let us write down the evolution equation for the Fourier modes of fluctuations σ_k from (5) as

$$\sigma_k'' + 3H\sigma_k' + \left[\frac{k^2}{a^2} + m_\sigma^2 + 3m_\sigma^2 \frac{\tilde{\psi}}{\sqrt{\mu M}} \cos(m_\sigma t) \right] \sigma_k \simeq 0, \tag{16}$$

where $m_\sigma = \sqrt{8\mu^3/M}$ and $\tilde{\psi}$ is the amplitude of the ψ oscillations. ($\tilde{\psi} \sim \sqrt{\mu M}$ at the beginning of the oscillations.) Neglecting the cosmic expansion, Eq. (16) has a form similar to the Mathieu equation which is known to have a exponentially growing solution. The detailed numerical simulation showed that the wave number for the fastest growing mode is given by

$$\frac{k_p}{a_{\text{osc}}} \simeq 0.3\,m_\sigma. \tag{17}$$

The fluctuations amplified by the parametric resonance eventually produce PBHs when they reenter the horizon after inflation. The mass of the PBH is approximately given by the horizon mass when the fluctuations reenter the horizon. Thus the PBH mass is estimated as

$$M_{\mathrm{BH}} \simeq 1.4 \times 10^{13} \, M_{\odot} \left(\frac{k_p}{\mathrm{Mpc}^{-1}} \right)^{-2}. \tag{18}$$

From Eqs. (17) and (18) the scale factor at the beginning of the oscillation phase is estimated as

$$\ln a_{\mathrm{osc}} = -114 - \ln m_{\sigma} - 0.5 \ln(M_{\mathrm{BH}}/M_{\odot}). \tag{19}$$

Because the e-folding number N_* is equal to $\ln a_{\mathrm{osc}} - \ln a_*$, we obtain

$$N_* = 21 + 0.5 \ln(\mu M) - 0.5 \ln(M_{\mathrm{BH}}/M_{\odot}). \tag{20}$$

For a fixed black hole mass M_{BH}, there are two parameters in the model, i.e., μ and M, one of which can be removed by using the WMAP normalization (8). Therefore observable quantities can be expressed in terms of one free parameter, leading to a non-trivial relation between n_s and $dn_s/d\ln k$. In practice, we adopt μM as the free parameter, and solve Eqs. (15) and (20) for σ_* in terms of μM. Then μ and M are determined with use of Eqs. (11) and (8) for a fixed μM. Thus, varying μM, we obtain sets of model parameters which are consistent with the observed curvature perturbations.

After σ and ψ decay, the second inflation (= new inflation) starts. As mentioned before, the role of the new inflation is to stretch the fluctuations produced during the smooth hybrid inflation and subsequent preheating phase to appropriate cosmological scales. The effective potential for the new inflation is given by

$$V_{\mathrm{new}} = v^4 \left(1 - \frac{c}{2}\phi^2 \right) - \frac{g}{2} v^2 \phi^4 + \frac{g^2}{16}\phi^8, \tag{21}$$

where ϕ is the inflaton of the new inflation, v is the scale of the new inflation and g and c are constants. The scale factor a_f at the end of the new inflation is estimated as

$$\ln a_f = -68 + \frac{1}{3}\ln\left(\frac{T_R}{10^9 GeV} \right) - \frac{4}{3}\left(\frac{v}{10^{15} GeV} \right), \tag{22}$$

where T_R is the reheating temperature after the new inflation. Therefore, the new inflation should provide the total e-fold number $\simeq (\ln a_f - \ln a_{\mathrm{osc}})$.

What makes the PBH particularly attractive as a DM candidate is that it is naturally long-lived due to the gravitationally suppressed evaporation rate. No discrete symmetries need to be introduced in an ad hoc manner. Also the PBH DM may be motivated from the arguments based on entropy of the Universe.[5]

3. The Solution for Dark Energy

At the beginning of the twenty-first century, there existed a problem in mathematics which was considered so difficult that it was expected that the century might end without solution. The problem was the Poincaré Conjecture in topology.

In fundamental theoretical physics, there was, at the beginning of the twenty-first century, an equally impossible seeming problem which likewise might not be solved for a hundred years. The problem was the Dark Energy in cosmology.

The creativity of *homo sapiens* had been underestimated. The Poincaré Conjecture was proved by Perelman, in less than three years. The Dark Energy problem was solved, by myself, in less than ten years.

In my Festschrift from 2003, there is a photograph[6] of a four-year-old boy with three special properties - a talent for mathematics, infinite chutzpah and he said he is cleverer than Newton. The talent meant that if the young boy were given a three-digit number, say, 506, he could, within seconds, answer 22x23; at most, one per cent of four-year-olds could do, similarly. At that time, in 1948, Newton was better known, even than any of the monarchs, except possibly the then monarch, King George VI. Surely, Newton was among the top one percent of human intelligence, so to be cleverer would require further reality checks. One would be forthcoming in 1965.

On the road from 1948 to 2010, I will make mercifully brief rest stops at 1957, 1965 and 2006. The first of these, 1957, is when I learned, at King Charles I School, about the universal law of gravitation. This was a key stage, because I clearly recall looking up at the Moon and feeling my own weight, and being so impressed by the idea that I decided, then and there, that I would, one day, have a grander idea, than Newton's. At about the same time, in 1957, my French teacher recommended, to my parents, a career, as a university professor, in linguistics. I might have done that, were it not for the call of Newton. Finally, in 1957, it was a memorable year because I met, for twelve seconds in Kidderminster Town Hall, the monarch, Queen Elizabeth II. Having bowed, I was ready to answer absolutely any question but all she said was that it was very nice to meet me. I should have worn a sign, soliciting a royal question.

In 1965, it was my turn for the opportunity of the Oxford Final Honors Schools (OFHS) with its six three-hour examinations, two each on Wednesday, Thursday and Friday June 6 - 8, 1965. The three morning exams were conventional while the afternoon OFHS exams were open-ended essay questions, with no instruction, even on how many questions to answer.

For the four months February - May, 1965 I did nothing, except study and make extensive notes, and memory cards. I was sequestered, in Frewin Hall, and talked to nobody, except college servants who could bring me food, or physics books from Blackwell's. What is pertinent to the sequel, in 2007 and 2010, is that of the hundred physics books I accumulated in Frewin Hall, my personal favorite was always Tolman's *Relativity, Thermodynamics and Cosmology*, a clear and endearingly modest discussion, of the role of entropy in cyclic cosmology. I do recall spending hours then intrigued by the apparent contradiction, between the attractive idea of cyclic cosmology, and the second law of themodynamics; the contents of Tolman's book, however, did not appear on my examinations.

For the OFHS paper on Thursday afternoon (June 7, 1965) my strategy was to answer only one essay question. I had retained extensive material on a dozen topics, with a good probability at least one of them would appear on the question paper. There it was: X-ray diffraction. In three hours, I produced a meticulously-detailed 100-page monograph on X-ray diffraction, later described by an experienced examiner, as the most detailed answer, he had ever seen. This required some of Gell-Mann's attributes: clear thinking, profound understanding and extensive retention. Incidentally, it also needed fast handwriting. My OFHS grades on my six papers were $\alpha, \alpha, \alpha, \alpha^+, \alpha, \alpha$. This is called straight alphas. Two alphas were necessary for First Class Honours. The unprecedented α^+ led to some discussion, in the Brasenose College (BNC) senior common room, and the BNC Fellows decided to allow me dining rights, on High Table, for as long as I would remain at BNC, as a doctoral student. The α^+ did support my being in the top one percent of human intelligence, just like Newton. At High Table dinners, I befriended a philologist ,who had collected numerous honorary doctorates, and could understand a hundred languages. He once mentioned that he had met, dining in BNC, just the previous evening, Gell-Mann who had explained his ideas, about the origin, of the Basque language. Therefore, I could have first met Gell-Mann in 1965, in BNC, had I attended that dinner. Instead I first met Gell-Mann the following year, 1966, as discussed in the next section.

More than fourty years after my OFHS experience, and after the accelerated cosmic expansion had been discovered, in 1998, I took on a new PhD student at UNC-Chapel Hill, Lauris Baum, in 2006 and suggested that he study, assiduously, existing papers on cyclic cosmology. This he did, and we discussed, at length, the issue of the Tolman conundrum, which had first piqued my intellectual curiosity, in 1965. The result was the first, and still only, solution to the 75-year-old conundrum.[7,8] In 2010, at Tokyo, on Thursday, February 4, Hirosi Ooguri who is a distinguished professor at the California Institute of Technology and, like me, a professor at the University of Tokyo (I am also a distinguished professor in Chapel Hill) wrote, to inform me,[a] that, on Saturday, a Todai visitor, Professor Dam Son, would give three lectures on the holographic principle at Hongo campus, starting at 1:30 PM. Son's lectures exceeded expectations. During the lectures (February 6, 2010), I realized, writing in my notebook, that the visible universe is approximated by a black hole, and that this leads to a resolution of the dark energy problem.[2]

Consider the Schwarzschild radius (r_s), and the physical radius (R), of the Sun $(\odot)$. They are $(r_s)_\odot = 3km$ and $R_\odot = 800,000km$. Their ratio is $(\rho)_\odot \equiv (R/r_s)_\odot = 2.7 \times 10^5$. One can readily check that, for the Earth or for the Milky Way, that the ratio $\rho = (R/r_s)$ is likewise much larger than one: $\rho >> 1$. Such objects are

---

[a]A useful communication, from Ooguri san, at IPMU, is acknowledged.

nowhere close to being black hole. Now consider the visible universe (VU), with mass $M_{VU} = 10^{23} M_\odot$. It has $(r_s)_{VU} = 30 Gly$, and $(R)_{VU} = 48 Gly$, hence $(\rho)_{VU} = 1.6$. The visible universe, within which we all live, is close to being a black hole. The solution to the dark energy problem follows, providing I so approximate the visible universe. At the horizon, there is a PBH temperature,[9–11] T_β, which I can estimate as

$$T_\beta = \frac{\hbar}{k_B} \frac{H}{2\pi} \sim 3 \times 10^{-30} K. \tag{23}$$

This temperature of the horizon information screen leads to a concomitant FDU acceleration[12–14] $a_{Horizon}$, outward, of the horizon given by the relationship

$$a_{Horizon} = \left(\frac{2\pi c k_B T_\beta}{\hbar} \right) = cH \sim 10^{-9} m/s^2. \tag{24}$$

When T_β is used in Eq. (24), I arrive at a cosmic acceleration which is essentially in agreement with the observations.[15,16]

It would be a wonderful to have lunch, may be at L'Atelier de Joël Robuchon in Roppongi Hills, with Murray Gell-Mann, Isaac Newton, and Grigori Perelman to compare notes on personal fulfillment. What does Grigori Perelman mean, when he tells journalist, in turning down a million dollars, *I have all I want. I'm not interested in money or fame*? This seems to baffle some americans, whose idea of happiness, as an inalienable right, is a three-comma net worth. Yet, a two-comma net worth suffices, for all practical purposes. Fame can hardly exceed that of the singer and entertainer, Elvis Presley (1935-1977), whose name, from my non-scientific studies in public transportation, is still recognizable by one billion people. He died, when he was only forty-two, so his fame was not very useful.

After Son's lectures on February 6, 2010, I went to the nearby Yushima Shrine around 6:00 PM and, impossibly, hoped that one of the many Japanese strolling around the shrine was Nambu sensei, to tell him. One ramification was that most of the work on quantum gravity, since the discovery of quantum mechanics, was called into question. There was an indescribable feeling of personal fulfillment, that the 66 years and 98 days, so far, of my life, had a significance. This was/is a totally individual experience which, unlike money or fame, involves no other person, and is therefore different. Because the visible universe is much bigger than the Solar System,[b] I had vindicated my claim, as a four-year-old, to be cleverer than Newton. Because, in my opinion, time travel into the past will forever be impossible, I cannot return to Isaac Newton in 1686 and forewarn him that a cleverer person will be born on October 31, 1943; nor can I return to 1948 and tell the four-year-old on a tricycle that he is right to say he is cleverer than Newton. The first reaction is to want to achieve the personal fulfillment again, and again. I am certain that Perelman is presently pursuing the six other Clay prolems, in alphabetical order: Birch and Swinnerton-Dyer Conjecture, Hodge Conjecture, Navier-Stokes Equations, P vs NP,

---

[b]A useful discussion, with Gerard 't Hooft, at the Gell-Mann Festschrift, is acknowledged.

Riemann Hypothesis and Yang-Mills Theory. More likely, Perelman is considering a more profound direction, in mathematics.

Newton finished Book I of *Principia*, entitled *De Motu Corporum I*, in 1686; then Book II (*De Motu Corporum II*) and Book III (*De Systemata Mundi*) in 1687. Book III adds more empiricism. I now explain why PHF would write, even then in 1685, a better *Principia* than Newton. PHF would start, in 1685, with Book II (Newton's grade, B-), knowing that Book I (Newton's grade, A+) was easier.

In order to explain why his sound speed formula $v_s = \sqrt{p/\rho}$ gives $v_s = 290m/s$ whereas the experimental value for v_s at one atmospheric pressure and $T = 20^0C$ is $v_s = 343m/s$, Newton needed a large correction. About a half of this correction arises, according to *Principia* Book II, from Newton's crassitude, where the sound propagates instantaneously through particles in the air. The remaining discrepancy leads to, surely, the most confused passage, in all of the *Principia*. Although I know that Gell-Mann[c] reads Latin as well as I do, other non-British-educated theoreticians may not, so I quote, instead of Latin, an English translation of a Scholium:

Moreover, the vapors floating in the air, being of another spring, and a different tone, will hardly, if at all, partake of the motion of the true air in which the sounds are propagated. Now if these vapors remain unmoved, that motion will be propagated the swifter through the true air alone, and that in the subduplicate ratio of the defect of the matter. So if the atmosphere consists of ten parts of true air and one part of vapors, the motion of sounds will be swifter in the subduplicate ratio of 11 to 10, or very nearly in the entire ratio of 21 to 20, than if it were propagated through eleven parts of true air: and therefore the motion of sounds above discovered must be increased in that ratio.

Newton was not only clever in mathematics, he was also a brilliant experimentalist. He himself measured the speed of sound in Nevile's Court[d] at Trinity College in Cambridge by hitting the paving stone with a hammer at such a frequency that the echo coincided with the next hit. Other experimentalists, such as Sauveur, cited by Newton, had determined v_s, so there was no doubt the theory was wrong.

The cleverer PHF would have thought more deeply, than Newton, about the ratio $r = (v_s)_{expt}/(v_s)_{theory}$. It is not too difficult to see, that this requires the isothermal Boyle's equation of state for an ideal gas, $PV = constant$ to become the adiabatic $PV^{r^2} = constant$. From theory and experiment, one knows $r^2 = 7/5$ and then, via diatomic molecules and statistical mechanics, I arrive smoothly at the entropy defined by Clausius, whose birthname[e] was Gottlieb, in 1865. What emerges is Boltzmann's equation $S = k \ln W$ as more profound than the equations of Newton, like $F = Gm_1m_2/R^2$, or those of the, still future, Einstein, like $E = mc^2$. Here the emphasis is not on exactitude, but on profundity.[f]

---

[c]A useful discussion, with Murray Gell-Mann, at the Gell-Mann Festschrift, is acknowledged.
[d]A useful discussion, with Bernard Carr, at IPMU, is acknowledged.
[e]A useful discussion, with Finn Ravndal, at the Gell-Mann Festschrift, is acknowledged.
[f]A useful discussion, with Murayama san, at IPMU, is acknowledged.

376

The aforementioned solution, of the dark energy problem, not only solves a cosmological problem, it casts a completely new light, on the nature of the gravitational force. Since the expansion of the universe, including the acceleration thereof, can only be a gravitational phenomenon, I arrive at the viewpoint, that gravity is a classical result, of the second law of thermodynamics. This means that gravity cannot be regarded as, on a footing with, the electroweak and strong interactions. Although this can be the most radical change, in gravity theory, for over three centuries, it is worth emphasizing, that general relativity remains unscathed.

My result calls into question, almost all of the work done on quantum gravity, since the discovery of quantum mechanics. For gravity, there is no longer necessity for a graviton. In the case of string theory, the principal motivation[17,18] for the profound, and historical, suggestion, by Scherk and Schwarz, that string theory be reinterpreted, not as a theory of the strong interaction, but instead as a theory of the gravitational interaction, came from the natural appearance, of a massless graviton, in the closed string sector. I am not saying that string theory is dead. What I am saying is, that string theory cannot be a theory of the fundamental gravitational interaction, since there is no fundamental gravitational interaction.

The way this new insight emerged, and the solution of the dark energy problem itself, was as a natural line of thought, following the discovery of a cyclic model in Ref. 8, and the subsequent investigations[5,20–23] of the entropy of the universe, including a possible candidate for dark matter.[1,5]

Another ramification, of my solution of the dark energy, problem is the status, fundamental versus emergent, of the three spatial dimensions, that we all observe every day. Because the solution assumes the holographic principle,[24] at least one spatial dimension appears as emergent.[g,h] Regarding the visible universe as a sphere, with radius of about 48 Gly, the emergent space dimension is then, in spherical polar coordinates, the radial coordinate, while the other two coordinates, the polar and azimuthal angles, remain fundamental. Physical intuition, related to the isotropy of space, may suggest that, if one space dimension is emergent, then so must be all three. This merits further investigation, and may require a generalization of the holographic principle in Ref. 24. On the other hand, a fundamental time coordinate is useful in dynamics. This present discussion is merely one step towards the goal of a cyclic model, in which time never begins or ends.

4. Gell-Mann in Twentieth Century Physics

Whereas I have published research in particle phenomenology for fourty years, and whereas I will not include my own name in the list, these are sufficient credentials to assess the greatest theoreticians of the twentieth century. I shall arrive, at a top-ten list, which includes: four very distinguished Europeans, four truly brilliant

[g]A useful discussion, with John Schwarz, at the Gell-Mann Festschrift, is acknowledged.
[h]A useful discussion, with Sugimoto san, at IPMU, is acknowledged.

Americans all born, by coincidence, in the great state of New York and two living-legend Asians, of whom, only Yang has been significantly influenced, in adult life, by the Confucian analects.[i]

In alphabetical order, the top five, with two chosen accomplishments:

Paul Dirac (`antimatter` _`and`_ `g = 2`)
Albert Einstein (`relativity` _`and`_ `photoelectric effect`)
Murray Gell-Mann (Ω^- _`and`_ `quarks`)
Gerard 't Hooft (`holographic principle` _`and`_ `renormalizability`)
Yang Zhenning (`gauge theories` _`and`_ `parity violation`)

The next five are, again in alphabetical order: Richard Feynman, Sheldon Glashow, Werner Heisenberg, Nambu sensei and Julian Schwinger. Below these, the ordering becomes more subjective, but my top ten choices, I believe, are close to the general opinion.

It should be noted that, in 1948, Nambu sensei, independently of the late Julian Schwinger, derived the one-loop quantum electrodynamics correction to $(g-2)$. That would give Nambu sensei (`symmetry breaking` _`and`_ `(g-2)`) which is very strong, and could displace one of the top five. However, Nambu sensei did not[j] publish, possibly because he did not want to overshadow Tomonaga sensei, fourteen years senior, chronological age being all-important in Japanese society.

I first met Murray (if I may) in 1966 when I was starting research in particle phenomenology and my Oxford doctoral adviser, J.C. Taylor, considered it worth driving ten miles to the Rutherford Laboratory. It was indeed worthwhile. Murray spoke with infinite self-confidence, and, in answering questions, provided information, like a computer download, reflecting encyclopaedic knowledge. In those times, Murray's prescient paper[25] *Symmetries of Baryons and Mesons* was a standard reference for Oxford students.

Murray has many first-rate accomplishments. Equally impressive, is the sheer number of new results, sometimes several in the same year[k] of which I can mention, in the time available, just a hint of Murray's gigantic contributions, with the renormalization group,[26] the sigma model[27] and the invention[28] of the theory of strong interactions. As one speaker at this Festschrift put it, everything in particle phenomenology was either by Murray, or named by Murray, who enriched the field, with erudite names, like strangeness, and quark.[29] In his monumental quark paper, perhaps the best two pages ever printed in Physics Letters B, Murray's infinite self confidence wobbled, when he discussed *non*-existence of real quarks.

Murray, I wish you many more years of creativity. You are, forever, a giant in particle phenomenology.

---

[i]A useful discussion, with Yang Zhenning, at the Gell-Mann Festschrift, is acknowledged.
[j]A useful discussion, with Nambu sensei, at Osaka University, is acknowledged.
[k]A useful discussion, with Kenneth Wilson, at the Gell-Mann Festschrift, is acknowledged.

Acknowledgments

This work was supported in part by the World Premier International Research Center Initiative (WPI initiative), MEXT, Japan and by U.S. Department of Energy Grant No. DE-FG02-05ER41418.

References

1. P.H. Frampton, M. Kawasaki, F. Takahashi and T. Yanagida. IPMU 09-0157. JCAP **04**:023 (2010). `arXiv:1001.2308 [hep-ph]`.
2. P. H. Frampton, IPMU 10-0055. `arXiv:1004.1285[astro-ph.CO]`.
3. M. Gell-Mann, *Proceedings of the International Conference on High-Energy Nuclear Physics*. CERN Scientific Information Service, Geneva, Switzerland (1962). Page 805.
4. V. Barnes, *et al.*, Phys. Rev. Lett. **12**, 204 (1964).
5. P.H. Frampton, JCAP **10**:016 (2009). `arXiv:0905.3632[hep-th]`.
6. http://www.physics.unc.edu/ frampton/Frampton9.jpg
7. R. C. Tolman, Phys. Rev. 38, 1758 (1931).
8. L. Baum and P.H. Frampton, Phys. Rev. Lett. **98**, 071301 (2007). `hep-th/0610213`.
9. L. Parker, Phys. Rev. **183**, 1057 (1969).
10. J.D. Bekenstein, Phys. Rev. **D7**, 2333 (1973).
11. S.W. Hawking, Commun. Math. Phys. **43**, 199 (1975).
12. S.A. Fulling, Phys. Rev. **D7**, 2850 (1973).
13. P.C.W. Davies, J. Phys. **A8**, 609 (1975);
14. W.G. Unruh, Phys. Rev. **D14**, 870 (1976).
15. S. Perlmutter et al., Supernova Cosmology Project, Astrophys. J. **517**, 565 (1998). `astro-ph/9812133`.
16. A.G. Reiss et al., Supernova Search Team, Astrn. J. **116**, 1009 (1998). `astro-ph/9805201`.
17. T.Yoneya, Prog.Theor.Phys. **51**, 1907 (1974).
18. J. Scherk and J.H. Schwarz, Nucl. Phys. **B81**, 118 (1974).
19. C.A. Egan and C.H. Lineweaver, Astrophys. J. **710**, 1825 (2010). `arXiv:0909.3983[astro-ph.CO]`.
20. P.H. Frampton and T.W. Kephart, JCAP, **06**:008 (2008). `arXiv:0711.0193[gr-qc]`.
21. P.H. Frampton, S.D.H. Hsu, T.W. Kephart and D. Reeb, Class. Quant. Grav. **26**, 145005 (2009). `arXiv:0801.1847 [hep-th]`
22. D.A. Easson, P.H. Frampton and G.F. Smoot. `arXiv:1002.4278[hep-th]`
23. D.A. Easson, P.H. Frampton and G.F. Smoot. `arXiv:1003.1528[hep-th]`
24. G. 't Hooft, in Salamfestschrift, editors: A. Ali, J. Ellis and S. Randjbar-Daemi, World Scientic (1994), pg. 284 `arXiv:gr-qc/9310026`.
25. M. Gell-Mann, Phys. Rev. **125** 1067 (1962).
26. M. Gell-Mann and F.E. Low, Phys. Rev. **95**, 1300 (1954).
27. M. Gell-Mann and M. Levy, Nuovo Cimento **16**, 705 (1960).
28. H. Fritzsch and M. Gell-Mann, Proceedings of the 16th International Conference on High Energy Physics, Fermilab. Editors J.D. Jackson and A. Roberts (1972).
29. M. Gell-Mann, Phys. Lett. **8**, 214 (1964)

REASONING BY ANALOGY: ATTEMPTS TO SOLVE THE COSMOLOGICAL CONSTANT PARADOX

RAFAEL A. PORTO and A. ZEE

Kavli Institute for Theoretical Physics,
University of California, Santa Barbara, CA 93106, USA
and
Department of Physics, University of California,
Santa Barbara, CA 93106, USA

Talk given by one of us (A. Zee) at Murray Gell-Mann's 80th Birthday Celebration held in Singapore, February 2010. Based on R. Porto and A. Zee, *Class. Quant. Grav.* **27**, 065006 (2010) [arXiv:0910.3716 [hep-th]].

The cosmological constant paradox could be summarized as follows. Expected value: $\Lambda \sim m^4 = m/(1/m)^3$, an enormous energy density even if we take m to be the electron mass, let alone Planck mass. Decreed value: mathematically 0, but an exact symmetry was never found. Observed value: tiny $\sim (10^{-3} \text{ eV})^4$ but not 0. (For the purpose of this talk, we will assume that the dark energy represents the cosmological constant.)

Can we learn something arguing by analogy? The history of physics is full of examples of analogies providing a guiding light. The story of proton decay may provide such an analogy, as one of us proposed[1] some 27 years ago. Suppose that long ago, before Murray Gell-Mann thought of quarks, perhaps in another civilization in another galaxy, a young theorist decided to calculate the rate for proton decay into $e^+ + \pi^0$. It would have been natural to write down the effective Lagrangian $\mathcal{L} \sim f \pi \bar{e} p$ and to compare this with $\mathcal{L} \sim g \pi \bar{n} p$, thus concluding that $f \sim \alpha g$ with a factor of α to account for isospin breaking.

The story of the proton decay rate would then be similar to the story of the cosmological constant as follows. Expected value: enormous since a priori Nature gave us no reason to suspect that f should be tiny. Decreed value: mathematically 0, with a proof by authority (let's say Wigner) dressed up with words like "baryon number conservation." Observed value: We could easily imagine that the particle physicists in the other galaxy were not as unlucky as we were, and the proton was soon observed to decay, with an extremely tiny but non-zero rate.

As is often the case in physics, the solution to the proton decay paradox did not come from thinking about the mechanism for proton decay, but from hadron spectroscopy: Quarks! (Gell-Mann, Zweig, Greenberg, Han and Nambu). With quarks, the proton decays via a dimension 6 rather than dimension 4 operator in the effective Lagrangian: $\mathcal{L} \sim \frac{1}{M^2}\bar{e}qqq$ instead of $\pi\bar{e}p$. Remarkably, promotion from dimension 4 to 6 is enough to solve the paradox. The reason is of course that the numbers 4 and 6 appear in the exponential! This simple argument, basically dimensional analysis, could be made more respectable using modern notions of renormalization group flow and scaling (developed by Gell-Mann and Low, Wilson, and others.)

Could we try the same trick and promote[2] the dimension of the cosmological constant term to make it less relevant at large distances compared with the Einstein–Hilbert curvature term? One problem is that both terms, $\int d^4x \sqrt{g} R$ and $\int d^4x \sqrt{g}$, are made of the same kind of stuff. In the proton decay story, the recognition that hadrons and leptons are distinct provided the first step. The only difference (within our present understanding of gravity) is that curvature involves derivatives while volume doesn't. (This suggests that perhaps a foam-like structure could distinguish between the two.) In the proton decay story we somehow managed to promote the proton decay term $\mathcal{L} \sim f\pi\bar{e}p$ without promoting the $\mathcal{L} \sim g\pi\bar{n}p$ term. The "secret" of course is that the the proton decay term metamorphosed into a term involving a Yang–Mills gauge field, with dimension staying at 4.

An interesting step was taken by Polyakov who considered (in a paper titled "Beyond spacetime"[3]) a conformally flat universe (in Euclidean signature) $g_{\mu\nu} = \phi(x)^2 \delta_{\mu\nu}$. Plugging this into the Einstein–Hilbert action, one gets (in units with $G_N = 1$) the action $S \int d^4x[(\partial\phi)^2 + \Lambda\phi^4]$ after the unwanted minus sign is removed via analytic continuation $\phi \to i\phi$ (Lorentz group is non-compact).

Interestingly, the cosmological term has been promoted to dimension 4 and so becomes marginal. It would be screened in the IR, but only logarithmically. Logarithmic running is not enough to bring the huge cosmological constant down to its observed value. Another serious objection is that gravity doesn't have a scalar mode! Thus the effect that Polyakov looked at appears to be purely a gauge artifact.

Einstein said "Physics should be as simple as possible, but not any simpler." To this we say "The solution to the cosmological constant paradox should be as crazy as possible, but not any crazier."

Let us think out of the box, but not stray too far away from it. In a recent paper,[4] we speculated that gravity departs from general relativity at cosmological distance scales (in what we called the "extreme ultra infra-red" EuIR regime). Note that work in quantum gravity and string theory has focussed almost exclusively on short distance thus far, while our proposal is in the opposite regime.

In condensed matter physics we are used to systems scaling differently in space and time: $t \to b^z t$, $x \to bx$. Lorentz invariant would tell us that $z = 1$. Here we speculate that in the EuIR gravity breaks Lorentz Invariance and scales with a dynamical critical exponent z not equal to one in Fourier space: $\omega \to b^{1/z}\omega$, $k \to \frac{1}{b}k$. We exploit a possible dynamical critical behavior of gravity in the EuIR regime to scale the cosmological constant. Our proposal is highly speculative but not outrageous.

After some simple manipulations, we find that the cosmological constant scales like $\Lambda(\text{EuIR}) \sim \Lambda(\text{IR})(\frac{l_{\text{IR}}}{l_{\text{EuIR}}})^{z-1}$. (Notice that for $z = 1$ we recover Polyakov's logarithmic scaling.) With $l_{\text{EuIR}} \sim 10^4$ Mpc the EuIR scale of the visible universe, and $l_{\text{IR}} \sim 1\text{--}10^3$ kpc the galactic or cluster scale, we have $l_{\text{EuIR}}/l_{\text{IR}} \sim 10^4\text{--}10^7$. To screen the cosmological constant to the desirable value, we need $z_{\text{EuIR}} \sim 20\text{--}30$, within an order magnitude of unity and at least not outrageously large. By splitting spacetime into space and time we have managed to make the cosmological constant irrelevant!

Many difficulties (see our paper for a long and tortured discussion[4]) remain. For example, we spoke of each local region of the universe trying to expand and pressing against each other in "rebellious symphony" perhaps something like a cluster of soap bubbles. An intriguing feature is that our action is non-local in time at cosmological distances. Perhaps a die-hard optimist could think that this could provide a hint about the nature of time. Of course, we do not have a theory, only a speculative proposal. By splitting spacetime into space and time at (almost inconceivably) vast distances and durations, in a regime with which physics has little direct experience, we see a glimmer of a hope of understanding the cosmological constant problem.

We end with another possibly relevant historical analogy regarding the inverse light speed $\zeta \equiv c^{-1}$. Consider the expected value before it was measured, again in some civilization in a galaxy far far away. The expected value is enormous in "natural units", if propagation in the ether is assumed to be similar to say sound waves in ordinary materials, let alone ocean waves (in other words, by the "naturalness" dogma, we might have expected ζ to be comparable to ζ_{sound} and we would have been off by some 6 orders of magnitude). On the contrary, the decree (proof by authority) is that ζ is mathematically 0. Finally, it was observed by the extragalactic version of Romer: the observed value turned out to be tiny but not 0 (as both Galileo and Newton thought).

How was this ζ paradox resolved? It was resolved by making c part of the kinematics. We went from the Galilean to the Lorentz group, and c became a "conversion factor" between space and time. The unification of spacetime allows us to chose units in which $c = 1$, a value protected by Lorentz invariance. In other words, it does not get renormalized! (In contrast, in non-relativistic theories c would get renormalized). Quantum fluctuations do not affect $\zeta \equiv c^{-1}$ thanks to its being part of an algebra.

Does this analogy tell us anything? To solve the ζ paradox, we had to go from the Galilean group to the Lorentz group. Perhaps we need to go one step farther and extend the Lorentz group to the de Sitter group! The cosmological constant Λ, like c before it, would then become a fundamental constant of nature. Just as c is a fixed constant in the Lorentz algebra, Λ then becomes a fixed constant in the de Sitter algebra. In this sense, the question why the cosmological constant is so small compared to what the "naturalness dogma" would lead us to expect might eventually turn out to be the wrong question to ask, or at least the wrong way of phrasing the question. (We then have to "explain" why the Planck mass is so large, but at least "vacuum fluctuations" would be absorbed into an overall wave-function renormalization).

We have told this story happily over three birthday parties: Dirac's 80th in Coral Gables,[1] Yang's 85th in Singapore,[2] and now Gell-Mann's 80th in Singapore.[5]

Happy Birthday, Murray!

Acknowledgments

During the course of our speculations we have benefited from enlightening conversations with Nima Arkani-Hamed, David Berenstein, Raphael Flauger, Sean Hartnoll, João Penedones and Joe Polchinski. We are supported in part by NSF under Grant No. 04-56556.

References

1. A. Zee, "Remarks on the cosmological constant paradox," in *High Energy Physics in Honor of P. A. M. Dirac in his Eightieth Year*, eds. S. L. Mintz and A. Perlmutter, Proceedings of the 20th Orbis Scientiae (Plenum Press, New York, 1983), p. 211.
2. A. Zee, "Gravity and its mysteries," in *Proc. Conference in Honor of C. N. Yang's 85th Birthday*, eds. M.-L. Ge, C. H. Oh and K. K. Phua, page 131 (World Scientific, 2008); A. Zee, *Int. J. Mod. Phys. A* **23**, 1295 (2008).
3. A. M. Polyakov, "Beyond space-time," arXiv:hep-th/0602011.
4. R. A. Porto and A. Zee, "Relaxing the cosmological constant in the extreme ultra-infrared," *Class. Quant. Grav.* **27**, 065006 (2010), arXiv:0910.3716.
5. The present volume; Proceedings of the Conference in Honor of Murray Gell-Mann's 80th Birthday, World Scientific.

MANIFESTATIONS OF SPACE-TIME VARIATION OF COUPLING CONSTANTS AND FUNDAMENTAL MASSES

V. V. FLAMBAUM and J. C. BERENGUT

School of Physics, University of New South Wales,
Sydney, NSW 2052, Australia

We present an overview of recent works that discuss observable manifestations of space-time variation of fundamental "constants" in a variety of physical systems. Studies in this direction are motivated by theories unifying gravity with other interactions; these often suggest the possibility of temporal and spatial variation of the fundamental constants in an expanding Universe. The effects of such variations could be observable in phenomena covering the lifespan of the Universe, from Big Bang nucleosynthesis to modern atomic clocks.

1. Introduction

It is widely believed that the Standard Model of elementary particles is a low-energy manifestation of a more complete theory that unifies gravity with the other interactions (electromagnetic, weak nuclear, and strong nuclear forces). Many well-motivated extensions to the Standard Model include variation of fundamental constants as a possibility, or even as a necessity in an expanding Universe (see e.g. review 1). For example, extra spatial dimensions are a feature of many of the "grand unified theories" that seek to unite gravity with the electromagnetic, weak nuclear, and strong nuclear interactions of the Standard Model. Any change in the size of these dimensions would manifest itself as variation of fundamental constants in our observable Universe. Another possibility is the existence of scalar fields that vary in space and time and can lead to, for example, a variable vacuum dielectric constant. These may be related to the accelerated expansion of the Universe (dark energy). Additionally, the fundamental constants may be slightly different near massive bodies (see e.g. review 2).

Some hints that variation of the fundamental constants occur have been seen in quasar absorption spectra[3-5] and Big Bang nucleosynthesis[6,7] data, however the majority of publications report limits on the variations (see e.g. recent reviews 8, 9).

We can only detect the variation of dimensionless fundamental constants. In this review we will discuss variation of the fine-structure constant $\alpha = e^2/\hbar c$, the proton g-factor g_p, and the dimensionless mass ratios $X_e = m_e/\Lambda_{QCD}$ and $X_q = m_q/\Lambda_{QCD}$ where m_e and m_q are the electron and quark masses and Λ_{QCD} is the quantum chromodynamics (QCD) scale, defined as the position of the Landau pole in the logarithm of the running strong coupling constant, $\alpha_s(r) \sim 1/\ln(\Lambda_{QCD}r/\hbar c)$. The proton mass m_p is proportional to Λ_{QCD}, therefore the relative variation of $\mu = m_e/m_p$ is equal to the relative variation of $X_e = m_e/\Lambda_{QCD}$ (if we neglect a small contribution of quark masses, $m_q \sim 5$ MeV, to the proton mass, $m_p = 938$ MeV). In the Standard Model electron and quark masses are proportional to the vacuum expectation value of the Higgs field.

Grand Unification implies that variations in the different fundamental constants may be related.[10–14] We can obtain a simple estimate of the relations in the following way. The strong (i=3), and electroweak (i=1,2) inverse coupling constants have the following dependence on the scale ν and normalization point ν_0:

$$\alpha_i^{-1}(\nu) = \alpha_i^{-1}(\nu_0) + b_i ln(\nu/\nu_0) \tag{1}$$

In the Standard Model, $2\pi b_i = 41/10, -19/6, -7$; the electromagnetic $\alpha^{-1} = (5/3)\alpha_1^{-1} + \alpha_2^{-1}$ and the strong $\alpha_s = \alpha_3$. In the Grand Unification Theories (GUT) all coupling constants are equal at the unification scale, $\alpha_i(\nu_0) \equiv \alpha_{GUT}$. If we assume that α_{GUT} varies, then Eq. (1) gives us the same shifts for all inverse couplings:

$$\delta\alpha_1^{-1} = \delta\alpha_2^{-1} = \delta\alpha_3^{-1} = \delta\alpha_{GUT}^{-1} . \tag{2}$$

We see that the variation of the strong interaction constant $\alpha_3(\nu)$ at low energy ν is much larger than the variation of the elecromagnetic constant α, since $\delta\alpha_3/\alpha_3 = (\alpha_3/\alpha_{1,2})\delta\alpha_{1,2}/\alpha_{1,2}$ and $\alpha_3 \gg \alpha_{1,2}$. The predicted relationship between variations in constants is

$$\frac{\delta X_q}{X_q} \sim 35\frac{\delta\alpha}{\alpha} . \tag{3}$$

The coefficient here is model-dependent, but again we note that large values are generic for models in which variations come from high energy scales. If these ideas are correct, the variation in $X_{e,q}$ or μ may be easier to detect than α-variation.

One may also consider variation of physical constants near massive bodies (violation of local position invariance).[2] The reason gravity is so important at large scales is that its effect is additive. The same should be true for massless (or very light) scalars: their effect near a large body is proportional to the number of particles in it. For not-too-relativistic objects, like regular stars or planets, both the total mass M and the total scalar charge Q are simply proportional to the number of nucleons in the object. Therefore the scalar field near such an object is simply proportional to the gravitational potential, and we expect that the fundamental constants would also depend on the position via the gravitational potential at the measurement point. The gravitational potential on Earth changes due to the ellipticity of its orbit. Thus we may interpret measurements of half-year variation in the ratio of different clocks in terms of variation in fundamental constants with respect to gravitational potentials. For limits, see Ref. 2.

In Sec. 2 we discuss variation of fundamental constants in Big Bang nucleosynthesis. The factor of three disagreement between the calculations and measurements of the BBN abundance of ^{7}Li may, in principle, be explained by the variation of

X_q at the level of $\sim 10^{-2}$; there have also been several publications studying variation of α and gravitational constant in BBN which we will not consider here, see review 1. The claim of detection of variation of fundamental constants based on the Oklo data in Ref. 15 is not confirmed by recent studies[16–18] which give a stringent limit on the possible variation of the resonance in ^{150}Sm during the last two billion years (Sec. 3). Atomic and molecular spectra can be used to probe variation of constants over cosmological timescales by comparison with astrophysical data (Sec. 4) and in laboratory-based clocks (Sec. 5).

2. Big Bang Nucleosynthesis

Big Bang Nucleosynthesis (BBN) provides one of the few quantitatively testable probes of the early Universe. Furthermore, with the independent WMAP measurement of the baryon-to-photon mass ratio and known number of neutrino flavours, it has become an essentially parameter-free theory. In 2004 we suggested that a reduced deuteron binding energy of $\Delta Q/Q = -0.019 \pm 0.005$ would yield a better fit to observational data, namely the WMAP value of barion-to-photon ratio η and measured primordial deuterium (d), ^{4}He, and ^{7}Li abundances.[6] Using our calculations[19] we interpreted this as a variation in strange quark mass at the 10^{-3} level.

More recently the sensitivity of primordial d, ^{4}He and ^{7}Li abundances to the variation of nuclear binding energies of d, t, ^{3,4}He, ^{6,7}Li and ^{7}Be were calculated in a linear approximation.[20] We have calculated the dependence of these binding energies on the light quark mass.[21–24] With the observational data and BBN binding-energy sensitivities of Ref. 20 we obtain the following equations for d, ^{4}He and ^{7}Li:[21]

$$1 + 7.7x = 1.07 \pm 0.15 , \tag{4}$$
$$1 - 0.95x = 1.005 \pm 0.036 , \tag{5}$$
$$1 - 50x = 0.33 \pm 0.11 , \tag{6}$$

where $x = \delta X_q/X_q$. These equations yield 3 consistent values of x: 0.009 ± 0.019, -0.005 ± 0.038 and 0.013 ± 0.002. The statistically weighted average of $\delta X_q/X_q = 0.013 \pm 0.002$ is dominated by the ^{7}Li data.

New measurements of the important ^{3}He$(\alpha, \gamma)^7$Be reaction[25] increase the discrepancy between observation and theory for ^{7}Li abundance. With the WMAP value of η, the discrepancy is now a factor of $2.4 - 4.3$ at the $4 - 5\sigma$ level.[26] We have performed an accurate calculation that takes into account the effect of the ^{8}Be binding energy variation (which is not included in Ref. 20), the variation of the virtual ^{1}S$_0(np)$ level in deuterium, and non-linear corrections in x which are important for ^{7}Li.[7] Our results suggest that to obtain the observed primordial abundances of ^{7}Li/H $= (1.5 \pm 0.5) \times 10^{-10}$ requires a variation of $\delta X_q/X_q = 0.016 \pm 0.005$. Allowing for the theoretical uncertanties, we should understand this BBN result as $\delta X_q/X_q = K \cdot (0.016 \pm 0.005)$ where $K \sim 1$ and the expected accuracy in K is about a factor of 2. Note that here we neglected effects of the strange quark mass variation. A rough estimate of these effects on BBN due to the deuteron binding energy variation was made in Refs. 19, 27.

In addition, we showed in Ref. 7 that including changes in the resonance positions of nuclear reactions might affect our conclusion. The predicted ^{7}Li abundance has a strong dependence on the cross-section of the resonant reactions ^{3}He $(d,\ p)\ ^4$He and $t\ (d,\ n)\ ^4$He. We show that changes in X_q at the time of BBN could shift the position of these resonances, leading to an increased production of ^{7}Li and exacerbating the lithium problem.

3. Oklo Natural Nuclear Reactor

Strong limits on the temporal variation of fundamental constants are obtained from consideration of the Oklo natural nuclear reactor, which ran approximately 1.8 billion years ago. These results are based on the measurement of the position of a very low energy resonance ($E_r = 0.1$ eV) in neutron capture by ^{149}Sm nucleus. The shift of this resonance induced by variation of α was estimated a long time ago in Refs. 28, 29. We derived a simple formula to estimate the effect of the variation of X_q on resonance or weakly-bound energy levels in Refs. 19, 27, 30. More recently we obtained a more accurate result by extrapolating from light nuclei to the resonance in ^{150}Sm.[31] The final result is

$$\delta E_r = 10 \left(\frac{\delta X_q}{X_q} - 0.1 \frac{\delta \alpha}{\alpha} \right) \ \text{MeV} \ . \tag{7}$$

Refs. 16–18 have all found that the limit of the variation in the Oklo reactor is $|\delta E_r| < 0.1$ eV. This gives us a limit on fundamental constants of[31]

$$\left| \frac{\delta X_q}{X_q} - 0.1 \frac{\delta \alpha}{\alpha} \right| \lesssim 4 \times 10^{-9} \tag{8}$$

The contribution of α variation to this equation is very small and should be neglected since the accuracy of the calculation of the main term is low. Thus, the Oklo data can not give any limit on the variation of α. Assuming linear time dependence during last 1.8 billion years we obtain the best terrestrial limit on the variation of fundamental constants

$$\left| \frac{\dot{X}_q}{X_q} \right| < 2.2 \times 10^{-18} \ \text{yr}^{-1} \ . \tag{9}$$

One may obtain even stronger limits by considering variation of the strange quark mass.[19]

4. Comparison of Quasar Absorption Spectra with Laboratory Spectra

4.1. *Optical atomic spectra*

It is natural to analyse fine-structure intervals in the search of variation of α. Measurements of α-variation by comparison of cosmic and laboratory optical spectra were first performed by Savedoff.[32] There were numerous works successfully implementing this "alkali-doublet" method (see review 1).

Later we developed a different approach: the many-multiplet method.[33,34] The relative value of any relativistic corrections to atomic transition frequencies is proportional to α^2. These corrections can exceed the fine-structure interval between the excited levels by an order of magnitude (for example, an s-wave electron does not have the spin-orbit splitting but it has the maximal relativistic correction to energy). We can express the α-dependence as

$$\omega = \omega_0 + q \left(\frac{\alpha^2}{\alpha_0^2} - 1 \right) , \tag{10}$$

where α_0 is the laboratory value of α and ω_0 is the laboratory frequency of a particular transition. The coefficients q can vary strongly from atom to atom and can have opposite signs in different transitions (for example, in s–p versus d–p transitions). Thus, any variation of α may be revealed by comparing different transitions in different atoms in cosmic and laboratory spectra. A statistical gain is also realised because many more spectral lines in different elements can be used. This method improves the sensitivity to any variation of α by more than an order of magnitude compared to the alkali-doublet method.

Relativistic many-body calculations are used to reveal the dependence of atomic frequencies on α^2 (the q coefficients). We have performed accurate many-body calculations of the q coefficient for all transitions of astrophysical interest.[33–39] These are strong E1 transtions from the ground state in Mg I, Mg II, Fe I, Fe II, Cr II, Ni II, Al II, Al III, Si II, Zn II, Mn II, as well as many other atoms and ions which are seen in quasar absorption spectra, but have not yet been used in the quasar measurements because of the absence of accurate UV transition laboratory wavelengths. For a "shopping list" of needed measurements, see Ref. 40.

Hints of α-variation were reported from the first analyses of quasar data from the Keck telescope that utilised the new method.[3,4] The largest analysis, with three independent samples of data containing 143 absorption systems spread over redshift range $0.2 < z < 4.2$, gave[5] $\delta\alpha/\alpha = (-0.543 \pm 0.116) \times 10^{-5}$. If one assumes a linear time-dependence of α over the time interval of about 12 billion years, the fit of the data gives $\dot{\alpha}/\alpha = (6.40 \pm 1.35) \times 10^{-16}$ yr^{-1}. A very extensive search for possible systematic errors has shown that known systematic effects cannot explain the result.[41]

Our method and calculations[33–38] were used by two other groups[42–44] to analyse data obtained on the VLT. However, they have not detected any variation of α. Recently, the results of Ref. 42 were questioned in a reanalysis of the same spectral data.[45,46] The re-analysis revealed flawed parameter estimation methods; a more accurate fit gives $\delta\alpha/\alpha = (-0.64 \pm 0.36) \times 10^{-5}$ (instead of the $\delta\alpha/\alpha = (-0.06 \pm 0.06) \times 10^{-5}$ reported in Ref. 42). However, even this revised result may require further revision.

One systematic effect that is still not completely excluded is that the effect of α variation may be imitated by a large change in relative isotope abundance during last 10 billion years. Spurious observation of variation in α due to a change in the relative isotope abundance of any one element has been ruled out. Nevertheless, an improbable "conspiracy" of several elements could mimic the observed effect. We have performed very complicated calculations of these isotopic shifts.[47-52] However, as shown in Ref. 52, calculations in atoms and ions with an open d-shell (like Fe II, Ni II, Cr II, Mn II, Ti II) are difficult, and our accuracy may be very low. Therefore measurements for at least a few lines are needed in order to benchmark calculations. Additionally, these measurements are needed to study the evolution of isotope abundances in the Universe, and to test models of nuclear reactions in stars and supernovae.

4.2. *Comparison of hydrogen hyperfine and optical transitions*

A comparison of the hyperfine transition in atomic hydrogen with optical transitions in ions was done in Refs. 53, 54. This method allows one to study time-variation of the parameter $x = \alpha^2 \mu g_p$. Analysis of 9 quasar spectra with redshifts $0.23 \leq z \leq 2.35$ gave

$$\delta x/x = (6.3 \pm 9.9) \times 10^{-6}, \tag{11}$$

$$\dot{x}/x = (-6 \pm 12) \times 10^{-16} \ \mathrm{yr}^{-1}. \tag{12}$$

4.3. *Molecular rotational quasar spectra*

Taking advantage of newly available H_2 wavelengths, Reinhold *et al.*[55] reported a non-zero cosmological variation of μ in quasar spectra using molecular hydrogen transitions in the Ly-α forest. The authors obtained $\Delta\mu/\mu = (2.4 \pm 0.6) \times 10^{-5}$ at redshifts $z \approx 2.6 - 3.0$, i.e. a decrease of μ in the past 12 Gyr at the 3.5σ confidence level. More recently, however, this result has been challenged. By carefully controlling systematics and accounting for known calibration errors in the VLT, King *et al.* obtained[56] $\Delta\mu/\mu = (0.26 \pm 0.30) \times 10^{-5}$, a much more stringent null constraint.

4.4. *Comparison of hydrogen hyperfine and molecular rotational transitions*

The frequency of the hydrogenic hyperfine line is proportional to $\alpha^2 \mu g_p$; molecular rotational frequencies are proportional to μ. Comparison places limits on the variation of the parameter $F = \alpha^2 g_p$.[57] Recently a similar analysis was repeated by Murphy *et al.*[58] using more accurate data for the same object at $z = 0.247$ and for a more distant object at $z = 0.6847$, and the following limits for the relative variation of F were obtained:

$$\delta F/F = (-2.0 \pm 4.4) \times 10^{-6} \tag{13}$$

$$\delta F/F = (-1.6 \pm 5.4) \times 10^{-6} \tag{14}$$

The object at $z = 0.6847$ is associated with the gravitational lens toward quasar B0218+357 and corresponds to lookback time ~ 6.5 Gyr.

4.5. *Enhancement of variation of μ in the inversion spectrum of ammonia*

A few years ago van Veldhoven *et al.* suggested using a decelerated molecular beam of ND$_3$ to search for the variation of μ in laboratory experiments.[59] The ammonia molecule has a pyramidal shape and the inversion frequency depends on the exponentially small tunneling of three hydrogens (or deuteriums) through the potential barrier. Because of that, it is very sensitive to any changes of the parameters of the system, particularly to the reduced mass for this vibrational mode.

High precision data on the redshifts of NH$_3$ inversion lines exist for the previously mentioned object B0218+357 at $z \approx 0.6847$.[60] Comparing them with the redshifts of rotational lines of CO, HCO$^+$, and HCN molecules from Ref. 61 one can get the following limit:

$$\frac{\delta\mu}{\mu} = \frac{\delta X_e}{X_e} = (-0.6 \pm 1.9) \times 10^{-6}. \tag{15}$$

Assuming a linear time dependence over the 6.5 Gyr (corresponding to $z \approx 0.68$), we obtain the most stringent present limit for the variation of μ and X_e:[62]

$$\dot{\mu}/\mu = \dot{X}_e/X_e = (1 \pm 3) \times 10^{-16} \text{ yr}^{-1}. \tag{16}$$

This result is combined with atomic clock results (Sec. 5.3) to give a strong limit on the variation of α (Equation 18). A more accurate measurement[63] utilising new rotational spectra gives a limit $|\delta\mu/\mu| < 1.8 \times 10^{-6}$ (95% confidence level).

5. Clocks

5.1. *Optical atomic clocks*

Just like in quasar spectra (Sec. 4.1), optical atomic clocks include transitions which have positive, negative or small constributions of the relativistic corrections to frequencies. We used the same methods of relativistic many-body calculations used in the quasar absorption studies to calculate the dependence on α of different clocks.[34,35,64–67] A summary of the results for the coefficients q is presented in Ref. 68. The relativistic effects are proportional to $(Z\alpha)^2$, therefore the q coefficients for optical clock transitions may be substantially larger than in cosmic transitions since the clock transitions are often in heavy atoms (Hg II, Yb II, Yb III, etc.) while cosmic spectra contain mostly light atoms ($Z < 33$).

The frequency ratio of Hg II and Al II clocks was precisely measured several times over the course of a year.[69] With our calculations, the best current constraint on time variation of α was achieved: $\dot{\alpha}/\alpha = (-1.6 \pm 2.3) \times 10^{-17} \text{ yr}^{-1}$.

5.2. *Enhanced effect of α-variation in Dy atom*

Enhancement of the relative effect of α-variation can be obtained in transitions between the almost degenerate levels in Dy atom.[35,67] These levels move in opposite directions if α varies. The relative variation may be presented as $\delta\omega/\omega = K\delta\alpha/\alpha$ where the coefficient K exceeds 10^8 ($q = 30,000$ cm^{-1}, $\omega \sim 10^{-4}$ cm^{-1}). Specific values of $K = 2q/\omega$ are different for different hyperfine components and isotopes which have different ω. An experiment is currently underway to place limits on α variation using this transition.[70,71] The current limit is $\dot\alpha/\alpha = (-2.7 \pm 2.6) \times 10^{-15}$ yr^{-1}. Unfortunately, one of the levels has quite a large linewidth and this limits the accuracy. Several other enhanced effects of α variation in atoms have been calculated.[72,73]

5.3. *Atomic microwave clocks*

Hyperfine microwave transitions may be used to search for α-variation.[74] Karshenboim[75] has pointed out that measurements of ratios of hyperfine-structure intervals in different atoms are also sensitive to variations in nuclear magnetic moments. However, the magnetic moments are not the fundamental parameters and cannot be directly compared with any theory of the variations. Atomic and nuclear calculations are needed for the interpretation of the measurements. We have performed both atomic calculations of the α-dependence[34,35,64–67] and nuclear calculations of the X_q-dependence[76] (see also Ref. 24) for all microwave transitions of current experimental interest including hyperfine transitions in ^{133}Cs, ^{87}Rb, ^{171}Yb$^+$, ^{199}Hg$^+$, ^{111}Cd, ^{129}Xe, ^{139}La, ^{1}H, ^{2}H and ^{3}He. The results for the dependence of the transition frequencies on variation of α, X_e and X_q are presented in Ref. 76. Also, one can find there experimental limits on these variations which follow from the recent measurements. The accuracy is approaching 10^{-15} per year. This may be compared to the sensitivity $\sim 10^{-5}-10^{-6}$ per 10^{10} years obtained using the quasar absorption spectra.

According to Ref. 76 the frequency ratio Y of the 282 nm ^{199}Hg$^+$ optical clock transition to the ground state hyperfine transition in ^{133}Cs has the following dependence on the fundamental constants:

$$\dot Y/Y = -6\dot\alpha/\alpha - \dot\mu/\mu - 0.01\dot X_q/X_q \tag{17}$$

This ratio has been measured[77] as $\dot Y/Y = (0.37 \pm 0.39) \times 10^{-15}$ yr^{-1}. Assuming a linear time dependence we obtained from the quasar result[62] (see Sec. 4.5) $\dot\mu/\mu = \dot X_e/X_e = (1 \pm 3) \times 10^{-16}$ yr^{-1}. Combining this result and the atomic clock result for Y (neglecting the small contribution of X_q) gives a strong limit on the variation of α:

$$\dot\alpha/\alpha = (-0.8 \pm 0.8) \times 10^{-16} \text{ yr}^{-1} . \tag{18}$$

5.4. *Proposals for enhanced effects in diatomic molecules*

Diatomic molecules provide a system where relative effects of the variation may be enhanced by several orders of magnitude in transitions between very close narrow levels Such levels may occur due to cancellation between the hyperfine and rotational structures,[78] or between the fine and vibrational structures of the electronic ground state.[62] The intervals between the levels are conveniently located in microwave frequency range and the level widths are very small, typically $\sim 10^{-2}$ Hz. Enhancements also exist in transitions between close levels in Cs_2[79] and HfF^{+}[80] and similar ions due to cancellation between electronic and vibrational excitations.

5.5. *Enhanced effects of α-variation in highly-charged ions*

Sensitivity to α-variation increases with ion charge as $(Z_i + 1)^2$.[81] The most sensitive atomic systems will maximize the contributions from three factors: high nuclear charge Z, high ionization degree, and significant differences in the configuration composition of the states involved. Unfortunately, the interval between different energy levels in an ion also increases as $\sim (Z_i + 1)^2$, which can quickly take the transition frequency out of the range of lasers as Z_i increases. The phenomena of Coulomb degeneracy and configuration crossing can be used to combat this tendency. In a neutral atom, an electron orbital with a larger angular momentum is significantly higher than one with smaller angular momentum but with the same principal quantum number n. On the other hand, in the hydrogen-like limit orbitals with different angular momentum but the same principal quantum number are nearly degenerate. Therefore somewhere in between there can be a crossing point where two levels with different angular momentum and principal quantum number can come close together: in such cases the excitation energy may be within laser range.

In Ref. 81 we showed, using the Ag isoelectronic sequence as an example, why high q-values can occur in highly-charged ions, and how the tendency of such systems towards large transition frequencies could be overcome. A two-valence-electron ion, Sm^{14+}, was identified, which has optical transitions that are the most sensitive to potential variation of α ever found. While atomic spectroscopy in electron beam ion traps is currently not competitive with optical frequency standards (see, e.g., 82, 83 and review 84) the technology continues to improve, and with the enhancements in sensitivity, highly-charged ions may prove to be a good system for detecting variation of α.

5.6. *Enhanced effect of variation in UV transition of ^{229}Th nucleus*

The ^{229}Th nucleus has the lowest known excited state, lying just 7.6 ± 0.5 eV above the ground state.[85] The position of this level was determined from the energy differences of many high-energy γ-transitions to the ground and first-excited states.

The subtraction produces the large uncertainty in the position of the 7.6 eV excited state. The width of this level is estimated to be about 10^{-4} Hz,[86] which explains why it is so hard to find the direct radiation in this very weak transition. Nevertheless, the search for the direct radiation continues.

Because the ^{229}Th transition is very narrow and can be investigated with laser spectroscopy, it is a possible reference for an optical clock of very high accuracy.[87] The near degeneracy of these isomers is a result of cancellation between very large energy contributions (order of MeV). Since these contributions would have different dependences on fundamental constants, this transition would be a very sensitive probe of possible variation of fundamental constants.[88] A rough estimate for the relative variation of the ^{229}Th transition frequency is

$$\frac{\delta\omega}{\omega} \approx 10^5 \left(0.1\frac{\delta\alpha}{\alpha} + \frac{\delta X_q}{X_q} \right) \qquad (19)$$

Therefore, the experiment would have the potential of improving the sensitivity to temporal variation of the fundamental constants by many orders of magnitude.

More accurate nuclear calculations give different values for the sensitivity of this transition to α. Refs. 89, 90 claim that both isomers have identical deformations and therefore there is no enhancement of α-variation. Other calculations give enhancement factors in the range $10^2 - 10^5$, depending on particulars of the model used.[91–94] To resolve this, we have proposed a method of extracting sensitivity to α-variation using direct laboratory measurements of the nuclear mean-square radius and electric quadrupole moment.[95]

From Eq. (19), we obtain the following energy shift in the 7.6 eV ^{229}Th transition:

$$\delta\omega \approx \frac{\delta X_q}{X_q} \text{ MeV} \qquad (20)$$

This corresponds to the frequency shift $\delta\nu \approx 3 \cdot 10^{20}\, \delta X_q/X_q$ Hz. The width of this transition is 10^{-4} Hz so one may hope to get the sensitivity to the variation of X_q about 10^{-24} per year. This is 10^{10} times better than the current atomic clock limit on the variation of X_q.

Note that there are other narrow low-energy levels in nuclei, for example the 76 eV level in ^{235}U with lifetime 26.6 minutes is the second-lowest known. One may expect a similar enhancement there. Unfortunately, this level cannot be reached with usual lasers. In principle, it may be investigated using a free-electron laser or synchrotron radiation. However, the accuracy of the frequency measurements is much lower in this case.

6. Enhancement of Variation of Fundamental Constants in Ultracold Atom and Molecule Systems near Feshbach Resonances

The scattering length A, which can be measured in Bose-Einstein condensate and Feshbach molecule experiments, is extremely sensitive to the variation of the electron-to-proton mass ratio $\mu = m_e/m_p$ or $X_e = m_e/\Lambda_{QCD}$:[96]

$$\frac{\delta A}{A} = K\frac{\delta\mu}{\mu} = K\frac{\delta X_e}{X_e}, \tag{21}$$

where K is the enhancement factor. For example, for Cs-Cs collisions we obtained $K \sim 400$. With the Feshbach resonance, however, one is given the flexibility to adjust position of the resonance using external fields. Near a narrow magnetic or an optical Feshbach resonance the enhancement factor K may be increased by many orders of magnitude.

References

1. J.-P. Uzan, *Rev. Mod. Phys.* **75**, 403 (2003).
2. V. V. Flambaum and E. V. Shuryak, *AIP Conf. Proc.* **995**, 1 (2008), arXiv:physics/0701220.
3. J. K. Webb, V. V. Flambaum, C. W. Churchill, M. J. Drinkwater and J. D. Barrow, *Phys. Rev. Lett.* **82**, 884 (1999).
4. J. K. Webb, M. T. Murphy, V. V. Flambaum, V. A. Dzuba, J. D. Barrow, C. W. Churchill, J. X. Prochaska and A. M. Wolfe, *Phys. Rev. Lett.* **87**, 091301 (2001).
5. M. T. Murphy, J. K. Webb and V. V. Flambaum, *Mon. Not. R. Astron. Soc.* **345**, 609 (2003).
6. V. F. Dmitriev, V. V. Flambaum and J. K. Webb, *Phys. Rev. D* **69**, 063506 (2004).
7. J. C. Berengut, V. V. Flambaum and V. F. Dmitriev, *Phys. Lett. B* **683**, p. 114 (2010).
8. V. V. Flambaum and J. C. Berengut, *Int. J. Mod. Phys. A* **24**, 3342 (2009).
9. S. N. Lea, *Rep. Prog. Phys.* **70**, 1473 (2007).
10. W. J. Marciano, *Phys. Rev. Lett.* **52**, 489 (1984).
11. X. Calmet and H. Fritzsch, *Eur. Phys. J. C* **24**, 639 (2002).
12. P. Langacker, G. Segrè and M. J. Strassler, *Phys. Lett. B* **528**, 121 (2002).
13. C. Wetterich, *J. Cosmol. Astropart. Phys.* **JCAP10**, 002 (2003).
14. T. Dent and M. Fairbairn, *Nucl. Phys. B* **653**, 256 (2003).
15. S. K. Lamoreaux and J. R. Torgerson, *Phys. Rev. D* **69**, 121701 (2004).
16. C. R. Gould, E. I. Sharapov and S. K. Lamoreaux, *Phys. Rev. C* **74**, 024607 (2006).
17. Y. V. Petrov, A. I. Nazarov, M. S. Onegin, V. Y. Petrov and E. G. Sakhnovsky, *Phys. Rev. C* **74**, 064610 (2006).
18. Y. Fujii, A. Iwamoto, T. Fukahori, T. Ohnuki, M. Nakagawa, H. Hidaka, Y. Oura and P. Möller, *Nucl. Phys. B* **573**, 377 (2000).
19. V. V. Flambaum and E. V. Shuryak, *Phys. Rev. D* **67**, 083507 (2003).
20. T. Dent, S. Stern and C. Wetterich, *Phys. Rev. D* **76**, 063513 (2007).
21. V. V. Flambaum and R. B. Wiringa, *Phys. Rev. C* **76**, 054002 (2007).
22. V. V. Flambaum, A. Höll, P. Jaikumar, C. D. Roberts and S. V. Wright, *Few-Body Sys.* **38**, 31 (2006).
23. A. Höll, P. Maris, C. D. Roberts and S. V. Wright, *Nucl. Phys. B Proc. Suppl.* **161**, 87 (2006).
24. V. V. Flambaum, D. B. Leinweber, A. W. Thomas and R. D. Young, *Phys. Rev. D* **69**, 115006 (2004).
25. R. H. Cyburt and B. Davids, *Phys. Rev. C* **78**, 064614 (2008).
26. R. H. Cyburt, B. D. Fields and K. A. Olive, *J. Cosmol. Astropart. Phys.* **11**, p. 012 (2008).

27. V. F. Dmitriev and V. V. Flambaum, *Phys. Rev. D* **67**, 063513 (2003).
28. A. I. Shlyakhter, *Nature* **264**, 340 (1976).
29. T. Damour and F. Dyson, *Nucl. Phys. B* **480**, 37 (1996).
30. V. V. Flambaum and E. V. Shuryak, *Phys. Rev. D* **65**, 103503 (2002).
31. V. V. Flambaum and R. B. Wiringa, *Phys. Rev. C* **79**, p. 034302 (2009).
32. M. P. Savedoff, *Nature* **178**, 689 (1956).
33. V. A. Dzuba, V. V. Flambaum and J. K. Webb, *Phys. Rev. Lett.* **82**, 888 (1999).
34. V. A. Dzuba, V. V. Flambaum and J. K. Webb, *Phys. Rev. A* **59**, 230 (1999).
35. V. A. Dzuba, V. V. Flambaum and M. V. Marchenko, *Phys. Rev. A* **68**, 022506 (2003).
36. V. A. Dzuba and V. V. Flambaum, *Phys. Rev. A* **71**, 052509 (2005).
37. V. A. Dzuba, V. V. Flambaum, M. G. Kozlov and M. Marchenko, *Phys. Rev. A* **66**, 022501 (2002).
38. J. C. Berengut, V. A. Dzuba, V. V. Flambaum and M. V. Marchenko, *Phys. Rev. A* **70**, 064101 (2004).
39. A. Dzuba and V. V. Flambaum, *Phys. Rev. A* **77**, 012514 (2008).
40. J. C. Berengut, V. A. Dzuba, V. V. Flambaum, J. A. King, M. G. Kozlov, M. T. Murphy and J. K. Webb, *Mem. Soc. Astron. It.* **80**, p. 795 (2009).
41. M. T. Murphy, J. K. Webb, V. V. Flambaum, C. W. Churchill and J. X. Prochaska, *Mon. Not. R. Astron. Soc.* **327**, 1223 (2001).
42. R. Srianand, H. Chand, P. Petitjean and B. Aracil, *Phys. Rev. Lett.* **92**, 121302 (2004).
43. S. A. Levshakov, M. Centurión, P. Molaro and S. D'Odorico, *Astron. Astrophys.* **434**, 827 (2005).
44. S. A. Levshakov, M. Centurión, P. Molaro, S. D'Odorico, D. Reimers, R. Quast and M. Pollmann, *Astron. Astrophys.* **449**, 879 (2006).
45. M. T. Murphy, J. K. Webb and V. V. Flambaum, *Phys. Rev. Lett.* **99**, 239001 (2007).
46. M. T. Murphy, J. K. Webb and V. V. Flambaum, *Mon. Not. R. Astron. Soc.* **384**, 1053 (2008).
47. M. G. Kozlov, V. A. Korol, J. C. Berengut, V. A. Dzuba and V. V. Flambaum, *Phys. Rev. A* **70**, 062108 (2004).
48. J. C. Berengut, V. A. Dzuba, V. V. Flambaum and M. G. Kozlov, *Phys. Rev. A* **69**, 044102 (2004).
49. J. C. Berengut, V. A. Dzuba and V. V. Flambaum, *Phys. Rev. A* **68**, 022502 (2003).
50. J. C. Berengut, V. V. Flambaum and M. G. Kozlov, *Phys. Rev. A* **72**, 044501 (2005).
51. J. C. Berengut, V. V. Flambaum and M. G. Kozlov, *Phys. Rev. A* **73**, 012504 (2006).
52. J. C. Berengut, V. V. Flambaum and M. G. Kozlov, *J. Phys. B* **41**, 235702 (2008).
53. P. Tzanavaris, J. K. Webb, M. T. Murphy, V. V. Flambaum and S. J. Curran, *Phys. Rev. Lett.* **95**, 041301 (2005).
54. P. Tzanavaris, J. K. Webb, M. T. Murphy, V. V. Flambaum and S. J. Curran, *Mon. Not. R. Astron. Soc.* **374**, 634 (2007).
55. E. Reinhold, R. Buning, U. Hollenstein, A. Ivanchik, P. Petitjean and W. Ubachs, *Phys. Rev. Lett.* **96**, 151101 (2006).
56. J. A. King, J. K. Webb, M. T. Murphy and R. F. Carswell, *Phys. Rev. Lett.* **101**, 251304 (2008).
57. M. J. Drinkwater, J. K. Webb, J. D. Barrow and V. V. Flambaum, *Mon. Not. R. Astron. Soc.* **295**, 457 (1998).
58. M. T. Murphy, J. K. Webb, V. V. Flambaum, M. J. Drinkwater, F. Combes and T. Wiklind, *Mon. Not. R. Astron. Soc.* **327**, 1244 (2001).
59. J. van Veldhoven, J. Küpper, H. L. Bethlem, B. Sartakov, A. J. A. van Roij and G. Meijer, *Eur. Phys. J. D* **31**, 337 (2004).

60. C. Henkel, N. Jethava, A. Kraus, K. M. Menten, C. L. Carilli, M. Grasshoff, D. Lubowich and M. J. Reid, *Astron. Astrophys.* **440**, 893 (2005).

61. F. Combes and T. Wiklind, *Astrophys. J.* **486**, L79 (1997).

62. V. V. Flambaum and M. G. Kozlov, *Phys. Rev. Lett.* **98**, 240801 (2007).

63. M. T. Murphy, V. V. Flambaum, S. Muller and C. Henkel, *Science* **320**, 1611 (2008).

64. V. A. Dzuba and V. V. Flambaum, *Phys. Rev. A* **61**, 034502 (2000).

65. E. J. Angstmann, V. V. Flambaum and S. G. Karshenboim, *Phys. Rev. A* **70**, 044104 (2004).

66. E. J. Angstmann, V. A. Dzuba and V. V. Flambaum, *Phys. Rev. A* **70**, 014102 (2004).

67. V. A. Dzuba and V. V. Flambaum, *Phys. Rev. A* **77**, 012515 (2008).

68. V. V. Flambaum and V. A. Dzuba, *Can. J. Phys.* **87**, 25 (2009).

69. T. Rosenband, D. B. Hume, P. O. Schmidt, C. W. Chou, A. Brusch, L. Lorini, W. H. Oskay, R. E. Drullinger, T. M. Fortier, J. E. Stalnaker, S. A. Diddams, W. C. Swann, N. R. Newbury, W. M. Itano, D. J. Wineland and J. C. Bergquist, *Science* **319**, 1808 (2008).

70. A.-T. Nguyen, D. Budker, S. K. Lamoreaux and J. R. Torgerson, *Phys. Rev. A* **69**, 022105 (2004).

71. A. Cingöz, A. Lapierre, A.-T. Nguyen, N. Leefer, D. Budker, S. K. Lamoreaux and J. R. Torgerson, *Phys. Rev. Lett.* **98**, 040801 (2007).

72. V. A. Dzuba and V. V. Flambaum, *Phys. Rev. A* **72**, 052514 (2005).

73. E. J. Angstmann, V. A. Dzuba, V. V. Flambaum, S. G. Karshenboim and A. Yu. Nevsky, *J. Phys. B* **39**, 1937 (2006).

74. J. D. Prestage, R. L. Tjoelker and L. Maleki, *Phys. Rev. Lett.* **74**, 3511 (1995).

75. S. G. Karshenboim, *Can. J. Phys.* **78**, 639 (2000).

76. V. V. Flambaum and A. F. Tedesco, *Phys. Rev. C* **73**, 055501 (2006).

77. T. M. Fortier, N. Ashby, J. C. Bergquist, M. J. Delaney, S. A. Diddams, T. P. Heavner, L. Hollberg, W. M. Itano, S. R. Jefferts, K. Kim, F. Levi, L. Lorini, W. H. Oskay, T. E. Parker, J. Shirley and J. E. Stalnaker, *Phys. Rev. Lett.* **98**, 070801 (2007).

78. V. V. Flambaum, *Phys. Rev. A* **73**, 034101 (2006).

79. D. DeMille, S. Sainis, J. Sage, T. Bergeman, S. Kotochigova and E. Tiesinga, *Phys. Rev. Lett.* **100**, 043202 (2008).

80. A. N. Petrov, N. S. Mosyagin, T. A. Isaev and A. V. Titov, *Phys. Rev. A* **76**, 030501 (2007).

81. J. C. Berengut, V. A. Dzuba and V. V. Flambaum, Enhanced laboratory sensitivity to variation of the fine-structure constant using highly-charged ions, arXiv:1007.1068, (2010).

82. I. Draganić, J. R. Crespo López-Urrutia, R. DuBois, S. Fritzsche, V. M. Shabaev, R. S. Orts, I. I. Tupitsyn, Y. Zou and J. Ullrich, *Phys. Rev. Lett.* **91**, p. 183001 (2003).

83. J. R. Crespo López-Urrutia, *Can. J. Phys.* **86**, p. 111 (2008).

84. P. Beiersdorfer, *Phys. Scr.* **T134**, p. 014010 (2009).

85. B. R. Beck, J. A. Becker, P. Beiersdorfer, G. V. Brown, K. J. Moody, J. B. Wilhelmy, F. S. Porter, C. A. Kilbourne and R. L. Kelley, *Phys. Rev. Lett.* **98**, 142501 (2007).

86. E. V. Tkalya, A. N. Zherikhin and V. I. Zhudov, *Phys. Rev. C* **61**, 064308 (2000).

87. E. Peik and Chr. Tamm, *Europhys. Lett.* **61**, 181 (2003).

88. V. V. Flambaum, *Phys. Rev. Lett.* **97**, 092502 (2006).

89. A. C. Hayes and J. L. Friar, *Phys. Lett. B* **650**, 229 (2007).

90. A. C. Hayes, J. L. Friar and P. Möller, *Phys. Rev. C* **78**, 024311 (2008).

91. X.-t. He and Z.-z. Ren, *J. Phys. G* **34**, 1611 (2007).

92. X.-t. He and Z.-z. Ren, *J. Phys. G* **35**, 035106 (2008).

93. V. V. Flambaum, N. Auerbach and V. F. Dmitriev, *Europhys. Lett.* **85**, 50005 (2009).

94. E. Litvinova, H. Feldmeier, J. Dobaczewski and V. V. Flambaum, *Phys. Rev. C* **79**, p. 064303 (2009).
95. J. C. Berengut, V. A. Dzuba, V. V. Flambaum and S. G. Porsev, *Phys. Rev. Lett.* **102**, p. 210801 (2009).
96. C. Chin and V. V. Flambaum, *Phys. Rev. Lett.* **96**, 230801 (2006).

CLASSICAL CELLULAR AUTOMATA AND QUANTUM FIELD THEORY

GERARD 'T HOOFT

Institute for Theoretical Physics, Utrecht University, the Netherlands
and
Spinoza Institute, Post Box 80.195, 3508 TD Utrecht, the Netherlands
g.thooft@uu.nl http://www.phys.uu.nl/~thooft/

It is pointed out that a mathematical relation exists between cellular automata and quantum field theories. Although the proofs are far from perfect, they do suggest a new look at the origin of quantum mechanics, and an essential role for the gravitational force in these considerations is suspected.

1. Introduction. Primitive Quantization

Systems that exhibit "complexity" at small distance scales can only be treated statistically at large distance scales. The theory considered in this paper is that Quantum Mechanics might result from just such a situation: a complex system, such as a cellular automaton, might develop quantum statistical features. From a mathematical point of view there are reasons for such a suspicion, as we will attempt to demonstrate.

We start from "primordial" or "primitive" quantization, a procedure proposed by the author some time ago.[1–7] Imagine some set of classical dynamical equations of motion. One could think of the motion of the hands of a clock, but we can also imagine other completely classical objects varying from particles, strings or branes following classical equations of motion, to planets obeying Newton's laws. One then starts from a description of the entire space of all classical states, and subsequently promotes each and every one of these states to an element of the basis of a Hilbert space. The time evolution of these states from time t_1 to t_2 is then generated by an operator that we will call the evolution operator $U(t_1, t_2)$. If the original world that we wanted to describe is strictly continuous, the number of states in Hilbert space is non denumerable, and this sometimes causes complications in pursuing this program. Often a fundamentally discrete model can be used, which then makes our work easier. The evolution operator can then be constructed systematically.

If the system considered is *time reversible*, which means that the past can be reconstructed from the future as easily as the future can be derived from the past, the evolution operator is also *unitary*, and this enables us to construct an other operator H, called "Hamiltonian," such that

$$U(t_1, t_2) = e^{-i(t_2 - t_1)H} \,, \tag{1.1}$$

where H is Hermitian, and its positivity and its symmetry structures can be studied. If the Hamiltonian that emerges from such calculations resembles the one used in familiar quantum theories, we have a mathematical system that on the one hand is based on an entirely classical model while on the other hand it allows for a description entirely in line with Quantum Mechanics.

Therefore, quantum theories of this sort could be interpreted as "hidden variable theories." The question then arises how to deal with Bell's inequalities.[8,9] These are inequalities that describe boundaries for measurements of physical features such as spin of quantum entangled objects. If the system is a classical one, the boundaries cannot be surpassed, whereas they are surpassed in a quantum theory. For many investigators, this is a sufficient reason to categorically reject all hidden variable theories. However, the procedure just described appears to give us a number of interesting models that could perfectly well serve as good examples for "hidden variables." What is wrong with them? This is the subject of this work. The situation is quite delicate and interesting. Our bottom line will be that it may well be possible to construct classical theories underlying Quantum Mechanics along these lines, but, quite remarkably, the gravitational force and General Relativity might be essential for a deeper understanding of the underlying structures.

For future reference, we will consider those degrees of freedom that describe the original classical states of the system as *beables*.[10] Beables $B(t)$ are operators that, at all times t, commute with all other beables:

$$[B(t_1), B(t_2)] = 0 \,, \quad \forall \, t_1, t_2 \,. \tag{1.2}$$

A *changeable* $C(t)$ is an operator that maps a beable onto another single beable:

$$C(t)B_1(t) = B_2(t)C(t) \,. \tag{1.3}$$

Finally, a *superimposable* S is an operator that can be any quantum superposition of any set of beables and/or changeables.

The eigenstates of the Hamiltonian defined in Eq. (1.1) will always be quantum superpositions. If we limit ourselves to low energy states only, this means that, from the start, we only talk of quantum entangled states. This is what distinguishes our models from other "hidden variable" theories: we never attempt to reform a set of states such as the ones used in the "Standard Model," into totally classical states. Only the suspected dynamical laws are classical, but the states considered are always quantum states, and usually highly entangled.

We do not completely resolve the issue with the Bell inequalities, but we suspect that they will be a lot harder to use as a "no go" theorem for hidden variables than usually assumed.

2. The Hamiltonian of a Deterministic System

According to the theory just described, we have discrete sets of physical data that evolve according to nonquantum mechanical, deterministic laws. We consider the case that also the time evolution is fundamentally discrete. The theory then can be defined in terms of an evolution operator U_0 that describes one step in time. Hence, we would like to write

$$(U_0)^k = e^{-iHk}, \tag{2.1}$$

where U_0 acts on the states $|n\rangle$ that are treated as basis vectors of a Hilbert space, so that H plays the role of a *quantum* Hamiltonian. We get conventional quantum mechanics if we can assure that H is bounded from below,

$$\langle \psi | H | \psi \rangle \geq \langle \psi_0 | H | \psi_0 \rangle, \tag{2.2}$$

for one state $|\psi_0\rangle$ and all states $|\psi\rangle$, where $|\psi\rangle$ and $|\psi_0\rangle$ may be any superposition of states $|n\rangle$, in particular any eigenstate of H. In principle, an operator obeying this demand is not difficult to construct. Using the Fourier transform

$$x = \pi - \sum_{k=1}^{\infty} \frac{2}{k} \sin kx, \quad 0 < x < 2\pi, \tag{2.3}$$

we derive:

$$H = \pi + \sum_{k=1}^{\infty} \frac{i}{k} (U_0^k - U_0^{-k}). \tag{2.4}$$

This Hamiltonian has only eigenvalues between 0 and 2π, and it reproduces Eq. (2.1). The importance of having a lower bound of the Hamiltonian eigenvalues is that the lowest state can be identified as the "vacuum state," and the first excited states can be interpreted as states containing particles. Thermodynamics gives us mixed states with probabilities

$$\varrho = Ce^{-E/kT}, \tag{2.5}$$

However, in *extensive* systems, such as Fock space for a quantum field theory, this Hamiltonian is not good enough, for two (related) reasons. One is that the very high k contributions in Eq. (2.4) refer to large times, and this implies that these contributions are nonlocal. If interactions spread with the speed of light, the Hamiltonian will generate direct interactions over spatial distances proportional to k. This necessitates a cutoff: if the time steps are assumed to be one Planck time, then also k time steps may perhaps be considered to be short enough to reproduce local physics at the Standard Model scale. Using a cutoff in Eq. (2.4) gives an energy spectrum as sketched in Fig. 1. It is derived from the fact that, if a system has periodicity N, the eigenvalues of U_0 are $e^{2\pi in/N}$.

In this figure, a smooth cutoff has been applied (cutting off the large k values with a Gaussian exponential). We see that, as a consequence, the lowest energy states are severely affected; their energy eigenvalues are now quadratic in the

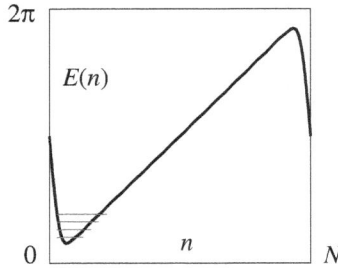

Fig. 1.   The energy spectrum $E(n)$ after a cutoff at large k. The lowest energy states (small lines) are severely affected by the cutoff.

momenta k, so that, here, the Hamiltonian does not reproduce the correct evolution operator (2.1). This region, however, is important when applying the laws of thermodynamics, since these states dominate in the Boltzmann expression (2.5).

The second problem is that we expect a Hamiltonian to decouple when states are considered that are spatially separated: $H = H_1 + H_2$. In that case, one cannot maintain that the eigenvalues of the entire Hamiltonian stay within the bounds $(0, 2\pi)$. We return to this point in Sec. 5.

3. A Cellular Automaton

The construction of an extensive Hamiltonian was suggested in Ref. 7. Consider a cellular automaton. Space and time[11] are both discrete here: we have a D-dimensional space, where positions are indicated by integers: $\mathbf{x} = (x^1, x^2, \ldots, x^D)$, where $x^i \in \mathbb{Z}$. Also time t will be indicated by integers, and time evolution takes place stepwise. The physical variables $F(\mathbf{x}, t)$ in the model could be assumed to take a variety of forms, but the most convenient choice is to take these to be integers modulo some integer $\mathbb{N}$. We now write down an explicit model, where these physical degrees of freedom are defined to be attached only to the even lattice sites:

$$\sum_{i=1}^{D} x^i + t = \text{even} \,. \tag{3.1}$$

Furthermore, the data can be chosen freely at two consecutive times, so for instance at $t = 0$, we can choose the initial data to be $\{F(\mathbf{x}, t = 0), \ F(\mathbf{x}, t = 1)\}$.

The dynamical equations of the model can be chosen in several ways, provided that they are time reversible. To be explicit, we choose them to be as follows:

$$F(\mathbf{x}, t + 1) = F(\mathbf{x}, t - 1) + Q\big(F(x^1 \pm 1, x^2, \ldots, x^D, t), \ldots,$$

$$F(x^1, \ldots, x^D \pm 1, t)\big) \ \text{Mod} \ \mathbb{N} \ \text{when} \ \sum_i x^i + t \ \text{is odd} \,, \tag{3.2}$$

where the integer Q is some arbitrary given function of all variables indicated: all nearest neighbors of the site $\mathbf{x}$ at time t. This is time reversible because we can find $F(\mathbf{x}, t - 1)$ back from $F(\mathbf{x}, t + 1)$ and the neighbors at time t. Assuming Q to be a sufficiently irregular function, one generally obtains quite nontrivial cellular automata this way. Indeed, this category of models have been shown to contain examples that are computationally universal.[12] Models of this sort are often considered in computer animations.

We now discuss the mathematics of this model using Hilbert space notation. We switch from the Heisenberg picture, where states are fixed, but operators such as the beables $F(\mathbf{x}, t)$ are time dependent, to the Schrödinger picture. Here, we call the operators F on the even sites $X(\mathbf{x})$, and the ones on the odd sites $Y(\mathbf{x})$. As a function of time t, we alternatingly update $X(\mathbf{x})$ and $Y(\mathbf{x})$, so that we construct the evolution operator over two time steps. Keeping the time parameter t even:

$$U(t, t-2) = A \cdot B, \tag{3.3}$$

where A updates the data $X(\mathbf{x})$ and B updates the data $Y(\mathbf{x})$.

Updating the even sites only, is an operation that consists of many parts, each defined on an even space coordinate $\mathbf{x}$, and all commuting with one another:

$$A = \prod_{\mathbf{x}\,\text{even}} A(\mathbf{x}), \quad [A(\mathbf{x}), A(\mathbf{x}')] = 0, \tag{3.4}$$

whereas the B operator refers only to the odd sites,

$$B = \prod_{\mathbf{x}\,\text{odd}} B(\mathbf{x}), \quad [B(\mathbf{x}), B(\mathbf{x}')] = 0. \tag{3.5}$$

Note now, that the operators $A(\mathbf{x})$ and $B(\mathbf{x}')$ do not all commute. If $\mathbf{x}$ and $\mathbf{x}'$ are neighbors, then

$$\mathbf{x} - \mathbf{x}' = \mathbf{e}, \quad |\mathbf{e}| = 1 \rightarrow [A(\mathbf{x}), B(\mathbf{x}')] \neq 0. \tag{3.6}$$

It is important to observe here that both the operators $A(\mathbf{x})$ and $B(\mathbf{x})$ only act in finite subspaces of Hilbert space, and they are all unitary, so we can easily write them as follows:

$$A(\mathbf{x}) = e^{-i\mathfrak{a}(\mathbf{x})}, \qquad B(\mathbf{x}) = e^{-i\mathfrak{b}(\mathbf{x})}. \tag{3.7}$$

Note that $A(\mathbf{x})$ and $B(\mathbf{x})$ are changeables, while $\mathfrak{a}(\mathbf{x})$ and $\mathfrak{b}(\mathbf{x})$ will be superimposables, in general, and they are Hermitian. We can write

$$\mathfrak{a}(\mathbf{x}) = \mathcal{P}_x(\mathbf{x})Q(\{Y\}), \quad \mathfrak{b}(\mathbf{x}) = \mathcal{P}_y(\mathbf{x})Q(\{X\}), \tag{3.8}$$

where $\mathcal{P}_x(\mathbf{x})$ is the generator for a one-step displacement of $X(\mathbf{x})$:

$$e^{i\mathcal{P}_x(\mathbf{x})}|X(\mathbf{x})\rangle \stackrel{\text{def}}{=} |X(\mathbf{x}) - 1 \bmod \mathbb{N}\rangle, \tag{3.9}$$

and, similarly, $\mathcal{P}_y(\mathbf{x})$ generates one step displacement of the function $Y(\mathbf{x})$.

As an example, we give the matrix P for the case $\mathbb{N} = 5$. Defining the numerical coefficients $\alpha = 2\sin(\pi/5) + \sin(2\pi/5)$ and $\beta = 2\sin(2\pi/5) - \sin(\pi/5)$, we have

$$
P = \frac{4\pi i}{25}
\begin{pmatrix}
0 & -\alpha & \beta & -\beta & \alpha \\
\alpha & 0 & -\alpha & \beta & -\beta \\
-\beta & \alpha & 0 & -\alpha & \beta \\
\beta & -\beta & \alpha & 0 & -\alpha \\
-\alpha & \beta & -\beta & \alpha & 0
\end{pmatrix} ;
$$

$$
e^{iP} =
\begin{pmatrix}
0 & 1 & 0 & 0 & 0 \\
0 & 0 & 1 & 0 & 0 \\
0 & 0 & 0 & 1 & 0 \\
0 & 0 & 0 & 0 & 1 \\
1 & 0 & 0 & 0 & 0
\end{pmatrix} .
$$

$$(3.10)$$

We see that

$$
[\mathfrak{a}(\mathbf{x}), \mathfrak{a}(\mathbf{x}')] = 0 , \quad [\mathfrak{b}(\mathbf{x}), \mathfrak{b}(\mathbf{x}')] = 0 , \quad \forall\, (\mathbf{x}, \mathbf{x}') , \tag{3.11}
$$

$$
[\mathfrak{a}(\mathbf{x}), \mathfrak{b}(\mathbf{x}')] = 0 \quad \text{only if} \quad |\mathbf{x} - \mathbf{x}'| > 1 . \tag{3.12}
$$

A consequence of Eqs. (3.11) is that also the products A in Eq. (3.4) and B in Eq. (3.5) can be written as

$$
A = e^{-i\sum_{\mathbf{x}\,\text{even}}\mathfrak{a}(\mathbf{x})} , \quad B = e^{-i\sum_{\mathbf{x}\,\text{odd}}\mathfrak{b}(\mathbf{x})} . \tag{3.13}
$$

However, now A and B do not commute. Nevertheless, we wish to compute the total evolution operator U for two consecutive time steps, writing it as

$$
U = A \cdot B = e^{-i\mathfrak{a}}e^{-i\mathfrak{b}} = e^{-2iH} . \tag{3.14}
$$

For this calculation, we could use the power expansion given by the Baker–Campbell–Hausdorff formula,[13]

$$
e^{P}e^{Q} = e^{R} ,
$$

$$
R = P + Q + \frac{1}{2}[P, Q] + \frac{1}{12}[P, [P, Q]] \tag{3.15}
$$

$$
+ \frac{1}{12}[[P, Q], Q] + \frac{1}{24}[[P, [P, Q]], Q] + \cdots ,
$$

a series that continues exclusively with commutators.[13] Replacing P by $-i\mathfrak{a}$, Q by $-i\mathfrak{b}$ and R by $-2iH$, we find a series for the "Hamiltonian" H in the form of an infinite sequence of commutators. Now note that the commutators between the local operators $\mathfrak{a}(\mathbf{x})$ and $\mathfrak{b}(\mathbf{x}')$ are nonvanishing only if $\mathbf{x}$ and $\mathbf{x}'$ are neighbors, $|\mathbf{x} - \mathbf{x}'| = 1$. Consequently, if we insert the sums (3.13) into Eq. (3.15), we obtain again a sum:

$$
H = \sum_{\mathbf{x}} \mathcal{H}(\mathbf{x}) ,
$$

$$
\mathcal{H}(\mathbf{x}) = \frac{1}{2}\mathfrak{a}(\mathbf{x}) + \frac{1}{2}\mathfrak{b}(\mathbf{x}) + \mathcal{H}_2(\mathbf{x}) + \mathcal{H}_3(\mathbf{x}) + \cdots , \tag{3.16}
$$

where

$$\mathcal{H}_2(\mathbf{x}) = -\frac{1}{4} i \sum_{\mathbf{y}} \left[\mathfrak{a}(\mathbf{x}), \mathfrak{b}(\mathbf{y}) \right],$$

$$\mathcal{H}_3(\mathbf{x}) = -\frac{1}{24} \sum_{\mathbf{y}_1, \mathbf{y}_2} \left[\mathfrak{a}(\mathbf{x}) - \mathfrak{b}(\mathbf{x}), [\mathfrak{a}(\mathbf{y}_1), \mathfrak{b}(\mathbf{y}_2)] \right], \quad \text{etc.}$$

(3.17)

All these commutators are only nonvanishing if the coordinates $\mathbf{y}$, $\mathbf{y}_1$, $\mathbf{y}_2$, etc., are all neighbors of the coordinate $\mathbf{x}$. It is true that, in the higher order terms, next-to-nearest neighbors may enter, but still, one may observe that these operators are all local functions of "field operators" $\Phi(\mathbf{x}, t)$, and thus we arrive at a Hamiltonian H that can be regarded as the sum over D-dimensional space of a Hamilton density $\mathcal{H}(\mathbf{x})$, which has the property that

$$[\mathcal{H}(\mathbf{x}), \mathcal{H}(\mathbf{x}')] = 0, \quad \text{if} \quad |\mathbf{x} - \mathbf{x}'| \gg 1. \tag{3.18}$$

The $\gg$ symbol here means that at the nth order in the BCH series, $\mathbf{x}$ and $\mathbf{x}'$ must be further than n steps away from one another.

At every finite order of the series, the Hamilton density $\mathcal{H}(\mathbf{x})$ is a finite-dimensional Hermitian matrix, and therefore, it will have a lowest eigenvalue h. In a large but finite volume V, the total Hamiltonian H will therefore also have a lowest eigenvalue, obeying

$$E_0 > hV. \tag{3.19}$$

The associated eigenstate $|0\rangle$ might be identified with the "vacuum." This vacuum is stationary, even if the automaton itself may have no stationary solution. The next-to-lowest eigenstate may be a one-particle state. In a Heisenberg picture, the fields $F(\mathbf{x}, t)$ may create a one-particle state out of the vacuum. Thus, we arrive at something that resembles a genuine quantum field theory. The states are quantum states in complete accordance with a Copenhagen interpretation. The fields $\mathfrak{a}(\mathbf{x}, t)$ and $\mathfrak{b}(\mathbf{x}, t)$ should obey the Wightman axioms.

There are three ways, however, in which this theory differs from conventional quantum field theories. One is, of course, that space and time are discrete. Well, maybe there is an interesting "continuum limit," in which the particle mass(es) is(are) considerably smaller than the inverse of the time quantum.

Secondly, no attempt has been made to arrive at Lorentz invariance, or even Gallilei invariance. Thus, the dispersion relations for these particles, if they obey any at all, may be nothing resembling conventional physical particles. Do note, however, that no physical information can travel faster than velocity one in lattice units. This is an important constraint that the model still has in common with special relativity.

But the third difference is more profound. It was tacitly assumed that the Baker–Campbell–Hausdorff formula converges. This is often not the case. In Sec. 4, we argue that the series will converge well only if sandwiched between two eigenstates

$|E_1\rangle$ and $|E_2\rangle$ of H, where E_1 and E_2 are the eigenvalues, that obey

$$2|E_1 - E_2| < 2\pi\hbar/\Delta t, \qquad (3.20)$$

where Δt is the time unit of our clock, and the first factor 2 is the one in Eq. (3.14). ("Planck's constant," $\hbar$, has been inserted merely to give time and energy the usual physical dimensions.)

This may seem to be a severe restriction, but, first, one can argue that $2\pi\hbar/\Delta t$ here is the Planck energy, and in practice, when we do quantum mechanics, we only look at energies, or rather energy differences, that indeed are much smaller than the Planck energy. Does this mean that transitions with larger energy differences do not occur? We must realize that energy is perhaps not exactly conserved in this model. Since time is discrete, energy at first sight seems to be only conserved modulo π, and this could indicate that our "vacuum state" is not stable after all. The energy might jump towards other states by integer multiples of π. In Section 4 however, we argue that such violations of energy conservation will *not* occur, and the existence of an Hamiltonian density is a more profound property of all cellular automata that allow time reversal (so that the evolution is obviously unitary).

The conclusion we are able to draw now, is that procedures borrowed from genuine quantum mechanics can be considered, and they may lead to a rearrangement of the states in such a way that beables, changeables and superimposables naturally mix, leaving an effective description of a system at large time and distance scales for which only quantum mechanical language applies. This, we think, is all we really need to understand why it is quantum mechanics that seems to dominate the world of atoms and other tiny particles, which, though small compared to humans, are still very large compared to the Planck scale.

At this point we like to remark that there is a precedent. Quantum field theory in fact can be used conveniently to solve a completely classical problem: the two-dimensional Ising model.[14,15]

4. Convergence of the BCH Expansion

When the operators A and B are bounded below and above, the BCH series expansion is expected to have a finite radius of convergence, but it certainly does not converge in general. To understand the situation, let us consider a quick derivation of the expansion. Given the operators A and B, consider the definition of an operator $C(\sigma)$ as a continuous function of σ, obeying

$$e^{iC(\sigma)} = e^{i\sigma A}e^{iB}; \qquad C(0) = B. \qquad (4.1)$$

Differentiating with respect to σ gives

$$\int_0^1 dx\, e^{ixC(\sigma)} \frac{d}{d\sigma} C(\sigma) e^{-ixC(\sigma)} = A. \qquad (4.2)$$

Diagonalizing $C(\sigma)$ at a given point $\sigma = \sigma_0$, we define

$$C(\sigma_0)|E\rangle \stackrel{\text{def}}{=} E|E\rangle\,, \qquad \langle E_1| \frac{\mathrm{d}}{\mathrm{d}\sigma} F(\sigma)|E_2\rangle\Big|_{t=t_0} \stackrel{\text{def}}{=} F'_{12}\,. \tag{4.3}$$

From Eq. (4.2) one then derives

$$C'_{12} = \frac{i(E_1 - E_2)}{e^{i(E_1-E_2)} - 1} A_{12}$$

$$= \sum_{n=0}^{\infty} \frac{i^n B_n}{n!} (E_1 - E_2)^n A_{12}$$

$$= \sum_{n=0}^{\infty} \frac{i^n B_n}{n!} [C, [C, \ldots, [C, A]], \ldots]_{12}\,, \tag{4.4}$$

where B_n are the Bernouilli numbers. Recursively, this defines $C(\sigma)$. It is clear that this series (4.4) converges precisely when

$$|E_1 - E_2| < 2\pi\,. \tag{4.5}$$

Note that this does not imply that the BCH series itself converges when sandwiched between two states $|E_1\rangle$ and $|E_2\rangle$ that are less than 2π apart, because the condition (4.5) must hold at all σ, while the states $|E_i\rangle$ are σ-dependent. The derivation merely suggests that, if during the entire calculation, only states are considered whose energies are much less separated than 2π in natural units, *at all stages*, the series may be expected to converge. This condition may nevertheless be considered to be a weak one, likely to be fulfilled in many cases, because the natural unit here is the Planck unit, $E_{\text{Planck}}/c^2 \approx 21\mu g$, which is always much larger than any experiment done in quantum mechanics.

An interesting aside is the observation that the BCH expansion (3.15) can be rewritten in terms of a series that contains much fewer terms. Write

$$-i\mathfrak{a} = P + Q\,, \qquad -i\mathfrak{b} = P - Q\,. \tag{4.6}$$

and note that we are really only interested in the conjugacy classes of H, not H itself:

$$e^{P+Q}e^{P-Q} = e^F e^R e^{-F}\,, \tag{4.7}$$

where F can be chosen with certain amounts of freedom. Noting that interchanging $\mathfrak{a}$ and $\mathfrak{b}$ should give us a Hamiltonian that is just as good as H, and certainly in the same conjugacy class, we search for an F such that

$$R(P, Q) = R(P, -Q)\,, \qquad R(-P, Q) = -R(P, Q)\,. \tag{4.8}$$

Using the short hand notation $QP^3Q = [Q,[P,[P,[P,Q]]]]$, etc., one finds

$$R = 2P - \frac{1}{12}QPQ + \frac{1}{960}Q(8P^2 - Q^2)PQ$$
$$+ \frac{1}{60480}Q\left(-51P^4 - 76QPQP + 33Q^2P^2 + 44PQ^2P - \frac{3}{8}Q^4\right)PQ + \mathcal{O}(P,Q)^9 .$$

(4.9)

Already the third term in this expression goes beyond the expansion (3.15), but (4.9) does not converge much faster.

5. Quantum Gravity

To get quantum particles that have a dispersion law:

$$E(p) \rightarrow \frac{p^2}{2m} ,$$

(5.1)

one needs at least some form of Galilean invariance — so that particles can have a velocity. Since cellular automata have a limiting speed of the transfer of information, this would have to be replaced by Lorentz invariance from the start, so, special relativity is hard to avoid even at the early stages of constructing models. However, both the Galilei group and the Lorentz group are *noncompact*, and this makes it very difficult to turn these symmetries into even approximately reasonable symmetries for cellular automata.[a]

Not only special relativity, but also *General Relativity* may perhaps not be left out before obtaining a true understanding of Quantum Mechanics as a description of the statistical features of a classical system. We saw in our calculations that, first, the construction (3.16) for our Hamiltonian would fail to produce an extensive expression for the energy. The total energy of the Universe would always be less than $2\pi/T_{\text{Planck}}$ (For simplicity, we take our fundamental time unit to be the Planck time). Yet, we wish to describe widely separated regions $(1, 2, 3, \ldots)$ of the Universe in terms of a total Hamiltonian H_{tot} that is approximately written as

$$H_{\text{tot}} = H_1 + H_2 + H_3 + \cdots ,$$

(5.2)

so that, inevitably, the total energy can become much larger than the limit $2\pi/T_{\text{Planck}}$.

Our treatment of the cellular automaton conveniently produced for us a Hamilton density, so that the global energy is extensive, but divergence of the BCH series forced us to limit ourselves to quantum states whose energies do stay closer together than the Planck energy.

It would be much better if a procedure could be found where such limitations do not have to be imposed. This would be the case if we had another way to define

---

[a]One could have hoped that an approximate symmetry turns into an exact symmetry in the continuum limit.

Hamilton density. This is where General Relativity comes in. When the gravitational force is taken into account, we may have a space-dependent gravitational potential, which leads to space-dependent dynamical operators $\tau(\mathbf{x})$ that determine the local clock speed. Thus, the evolution operator becomes

$$U(\tau(\mathbf{x})) = e^{-i \int \tau(\mathbf{x})\mathcal{H}(\mathbf{x})\mathrm{d}^3\mathbf{x}} . \tag{5.3}$$

Thus, such a theory allows for a direct definition of Hamilton densities. By comparing the evolution operators at different gravitational potentials, one derives $\mathcal{H}(\mathbf{x})$. In General Relativity,

$$\tau(\mathbf{x}) = t\sqrt{-g^{00}(\mathbf{x})} . \tag{5.4}$$

6. Conclusions

The importance of these calculations is that each of the commutators in the BCH series are non vanishing only if they consist of neighboring operators; therefore, the resulting Hamiltonian can be written as the sum of Hamilton densities. Therefore, it seems as if the Hamiltonian is constructed just as in a quantized field theory. Distant parts of this "universe" evolve independently. The total Hamiltonian has eigenvalues much greater than 2π, and are all bounded from below. If now we concentrate on the lowest lying states, we see that these consist of localized particles resembling what we see in the real world. They have positive energies, so that thermodynamics applies to them.

But, as we saw, there is a caveat: the BCH series (3.15) does not converge. To be precise, the operator H is ambiguous when two of its eigenvalues get further separated than $2\pi/T_{\mathrm{Planck}}$, as can be seen directly from its definition (2.1). That is where the series' radius of convergence ends.

One must deduce that the series (3.15) cannot be truncated easily; it will have to be resummed carefully, and whether or not a resummation exists remains questionable. We can argue that these difficulties only refer to matrix elements between states whose energy eigenvalues are more than the Planck energy apart, which is a domain of quantum physics that has never been addressed experimentally anyway, so maybe they can safely be ignored. Yet this situation is unsatisfactory, so more work is needed.

We do conclude with an important conjecture: perhaps our world is not quantum mechanical, but only our perception of it.

References

1. G. 't Hooft, *Class. Quantum Grav.* **16**, 3263 (1999), arXiv:gr-qc/9903084.
2. G. 't Hooft, *Int. J. Theor. Phys.* **42**, 355 (2003), arXiv:hep-th/0104080.
3. H. Th. Elze, J. Phys. Conf. Ser. **33**, 399 (2006), arXiv:gr-qc/0512016v1.
4. M. Blasone, P. Jizba and H. Kleinert, *Ann. Phys.* **320**, 468 (2005), arXiv:quant-ph/0504200.
5. M. Blasone, P. Jizba and H. Kleinert, *Braz. J. Phys.* **35**, 497 (2005), arXiv:quant-ph/0504047.

408

6. G. 't Hooft, Emergent quantum mechanics and emergent symmetries, presented at *PASCOS 13*, Imperial College, London, July 6, 2007; ITP-UU-07/39, SPIN-07/27, arXiv:hep-th/0707.4568.

7. G. 't Hooft, Entangled quantum states in a local deterministic theory, *2nd Vienna Symposium on the Foundations of Modern Physics*, June 2009, ITP-UU-09/77, SPIN-09/30, arXiv:0908.3408v1.

8. A. Einstein, B. Podolsky and N. Rosen, *Phys. Rev.* **47**, 777 (1935).

9. J. S. Bell, *Speakable and Unspeakable in Quantum Mechanics* (Cambridge University Press, Cambridge, 1987).

10. G. 't Hooft, The mathematical basis for deterministic quantum mechanics, in *Beyond the Quantum*, ed. Th. M. Nieuwenhuizen *et al.* (World Scientific, 2007), pp. 3–19, arXiv:quant-ph/0604008.

11. A. P. Balachandran and L. Chandar, *Nucl. Phys. B* **428**, 435 (1994).

12. D. B. Miller and E. Fredkin, Two-state, reversible, universal cellular automata in three dimensions, in *Proc. 2nd Conf. on Computing Frontiers*, Ischia, Italy (ACM, 20???), p. 45, arXiv:nlin/0501022.

13. A. A. Sagle and R. E. Walde, *Introduction to Lie Groups and Lie Algebras* (Academic Press, New York, 1973).

14. B. Kaufman, *Phys. Rev.* **76**, 1232 (1949).

15. B. Kaufman and L. Onsager, *Phys. Rev.* **76**, 1244 (1949).

COULD TESTING OF THE LAWS OF PHYSICS EVER BE COMPLETE?

KENNETH G. WILSON

Ohio State University, USA

Collaborators:
GEORGE E. SMITH, CONSTANCE K. BARSKY, and STANISLAW D. GLAZEK

I need, first of all, to explain my choice of title. Physicists attending this conference live and work in two worlds. There is the world of the theorist, an imaginary world in which there are a set of "laws of physics" that are presumed to be exact, except for very small distances or in the cosmic domain.

But there is also the real world of experimental or observational physics and astronomy. In this real world, all claimed results are marred by sources of uncertainty that can be far more difficult to estimate than many theorists realize. An international organization called CODATA exists to help clarify issues of uncertainty plaguing real data. I also call your attention to my father's 1952 book *An Introduction to Scientific Research*, which has helped many experimentalists become exceptionally cautious in dealing with uncertainties of measurement. In this real world, no law about continuum quantities (such as time or distance or energy) can be established, through actual experimental tests, to be *exact*. This cannot be done today. I presume it cannot be done in the foreseeable future either, although *uncertainty estimates* can certainly be lowered in the future as compared to what they are today.

The greatest challenge for elementary particle physicists, theorists and experimentalists alike, is to discover new phenomena. But exciting new phenomena do not have to be confined to either very small distances or the cosmic domain. My talk will provide a humorous illustration of the assertion.

Now turn to the talk itself.

My talk will have four parts:

1. Introductory remarks.
2. A review of initially bizzare but now experimentally confirmed concepts, one example being confined quarks.
3. An anecdote to set the mood for the rest of my talk.
4. A humorous science fiction tale: an account of Murray's exceptional career as contrasted with my own more plodding one.

In part 2, I will raise a question: could at least *one* even more bizzare development than confined quarks be both proposed and later confirmed beyond question over, say, the next *century* of physics research?

The possibility I will raise is that ordinary life could be invaded by unexpected and highly transient phenomena emerging from a world of higher dimension than four, but a world that is currently completely *unknown and unexpected*.

In part 4, comprising the bulk of my talk: I will provide a humorous form of science fiction inspired by the question raised in part 2.

My science fiction will include encounters with three of the scientists I admire most: Murray mostly, but also *invented* versions of Isaac Newton and Aristotle. All three are (were) noted for their eagerness to know everything.

I will offer an imaginary interpretation of a number of episodes of my own more pedestrian career as intertwined with the more notable career of Murray, with Newton and Aristotle appearing as needed for my fable.

PARTON DISTRIBUTION FUNCTIONS IN DIFFRACTIVE DEEP INELASTIC SCATTERING

S. TAHERI MONFARED,* A. KHORRAMIAN and S. ATASHBAR TEHRANI

Physics Department, Semnan University, Semnan, Iran
* sara.taherimonfared@gmail.com

School of Particles and Accelerators,
Institute for Research in Fundamental Sciences (IPM),
P.O. Box 19395-5531, Tehran, Iran

In this paper The current understanding of the structure of the proton is reviewed, along with a framework within which diffractive deep inelastic processes may be studied. More over, the presence of diffraction in deep inelastic scattering is suggested, confirmed and quantified. At the end, our conventional DGLAP analysis which results in a good description of diffractive structure functions at higher values of Q^2 is detailed and the present level of understanding of the partonic structure of the diffractive exchange is discussed.

Keywords: Diffractive deep inelastic scattering; parton distribution functions; DGLAP equation.

1. Introduction

High energy diffraction, and more specifically diffractive DIS, constitutes a very extended topic, and only the most relevant aspects of the theory necessary to follow of this study are discussed in this paper. About 10% of the DIS events measured at HERA are characterized by large region devoid of activity or 'rapidity gap' between the hadronic final state observed in the main detector and the proton or its dissociation products forming a low mass hadronic state which passes unobserved into the forward beam pipe. This class of DIS events is called 'diffractive' since the collisions are elastic or quasi-elastic. The exchange of a colorless object between the proton and the photon was introduced to explain the presence of the rapidity gap. The existence of such an exchange with vacuum quantum numbers (C=1, P=1), called the pomeron, had originally been postulated to describe the rise with the center of mass energy of the total cross sections observed in hadron collisions.[1]

2. Kinematic Variables

The discussion of diffractive DIS requires additional kinematic variables to describe the three new degrees of freedom introduced by the presence of the rapidity gap between the outgoing proton and diffracted photon. A schematic representation of

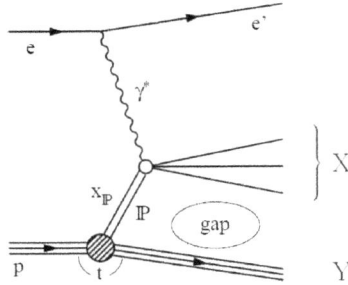

Fig. 1.   Diagram of diffractive scattering in DIS.

diffractive DIS is shown in Figure 1. In what follows, X is used to label the hadronic state into which the photon dissociates, and Y refers to the scattered proton or the proton remnant.[2] The invariant mass of the two systems are called M_X and M_Y, moreover p_X and p_Y designate their 4-vectors. The 4-momentum transfer squared across the proton vertex is defined by:

$$t = (p - p_Y)^2. \tag{1}$$

Two further diffractive variables are introduced:

$$x_{I\!P} = \frac{q.(p - p_Y)}{q.p} \approx \frac{Q^2 + M_X^2}{Q^2 + W^2}, \tag{2}$$

and

$$\beta = \frac{x}{x_{I\!P}} = \frac{-q^2}{2q.(p - p_Y)} \approx \frac{Q^2}{Q^2 + M_X^2}. \tag{3}$$

The mass of the proton and the squared four momentum transfer t have been neglected in the approximation of equations 2 and 3 since $m_p^2 \ll Q^2, W^2$ and $|t| \ll Q^2, M_X^2$ for the limits appropriate to this analysis. $x_{I\!P}$ is the fraction of the 4-momentum of the proton transferred to the pomeron, and β is interpreted as the fraction of the 4-momentum of the pomeron carried by the struck quark when a partonic structure is ascribed to the colourless exchange. β is therefore the analogue of x in inclusive DIS. Thus, the five independent kinematic variables typically used to describe diffractive scattering are Q^2, t, $x_{I\!P}$, β and M_Y.

3. Diffractive Cross Section

The cross section for $ep \rightarrow eXp$ in one-photon exchange approximation can be written in terms of the diffractive structure functions $F_2^{D(4)}$ and $F_L^{D(4)}$ as

$$\frac{d\sigma^{ep \rightarrow eXp}}{d\beta \, dQ^2 \, dx_{I\!P} \, dt} = \frac{4\pi\alpha_{em}^2}{\beta Q^4}[(1 - y + \frac{y^2}{2})F_2^{D(4)}(\beta, Q^2, x_{I\!P}, t) - \frac{y^2}{2}F_L^{D(4)}(\beta, Q^2, x_{I\!P}, t)]. \tag{4}$$

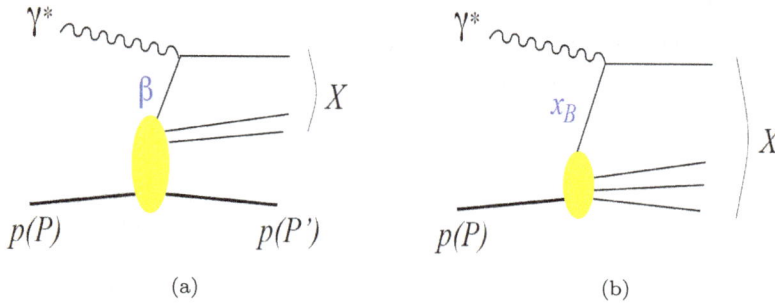

Fig. 2. Parton model diagrams for deep inelastic diffractive (a) and inclusive (b) scattering. The variable β is the momentum fraction of struck quark with respect to $P-P'$, and x_{Bj} its momentum fraction with respect to P.

In analogy with the way $d\sigma^{ep\rightarrow eX}/(dx_{Bj}dQ^2)$ is related to the structure function F_2 and F_L for inclusive DIS, $ep \rightarrow eX$. Here y is the fraction of energy lost by the incident lepton in the proton rest frame. The structure function $F_L^{D(4)}$ corresponds to longitudinal polarization of the virtual photon, its contribution to the cross section is small in a wide range of the experimentally accessible kinematic region (in particular at low y). The structure function $F_2^{D(3)}$ is obtained from $F_2^{D(4)}$ by integrating over t.

In a parton model picture, inclusive diffraction $\gamma^*p \rightarrow Xp$ proceeds by the virtual photon scattering on a quark, in analogy to inclusive scattering (see Fig. 2). In this picture, β is the momentum fraction of the struck quark with respect to the exchanged momentum $P-P'$ (indeed the allowed kinematical range of β is between 0 and 1). The diffractive structure function describes the proton structure in these specific processes with a fast proton in the final state. F_2^D may also be viewed as describing the structure of whatever is exchanged in $t-$channel in diffraction, i.e. of the Pomeron. It is however important to bear in mind that the Pomeron in QCD cannot be interpreted as a particle on which the virtual photon scatters.[3] The data on $F_2^{D(3)}$ have two remarkable features:

- F_2^D is largely flat in the measured β range. Keeping in mind the analogy between β in diffractive DIS and x_{Bj} in inclusive DIS, this is very different from the behavior of the "usual" structure function F_2, which strongly decreases for $x_{Bj} \geq 0.2$.
- The dependence on Q^2 (see Fig. 3 and 4) is logarithmic. The structure function F_2^D increases with Q^2 for all β values except for the highest. This is reminiscent of the scaling violations of F_2, except that F_2 rises with Q^2 only for $x_{Bj} < 0.2$ and that the scaling violations become negative at higher x_{Bj}. In the proton, negative scaling violations reflect the presence of the valence quarks radiating gluons, while positive scaling violations are due to the increase of the sea quarks and gluon densities as the proton is probed with higher resolution. The F_2^D data thus suggest that the partons resolved in diffractive events are predominantly gluons.

4. Diffractive Parton Distributions

The conclusion just reached can be made quantitative by using the QCD factorization theorem for inclusive diffraction, $\gamma^* p \to Xp$. According to this theorem, the diffractive structure function, in the limit of large Q^2 at fixed β, $x_{I\!P}$ and t, can be written as[4–6]

$$F_2^{D(4)}(\beta, Q^2, x_{I\!P}, t) = \Sigma_i \int_\beta^1 \frac{dz}{z} C_i(\frac{\beta}{z}) f_i^D(z, x_{I\!P}, t; Q^2), \tag{5}$$

where the sum is over partons of type i. The coefficient functions C_i describe the scattering of the virtual photon on the parton and are exactly the same as in inclusive DIS. In analogy to the usual parton distribution functions (PDFs), the diffractive PDFs $f_i^2(z, x_{I\!P}, t; Q^2)$ can be interpreted as conditional probabilities to find a parton i with fractional momentum $z x_{I\!P}$ in a proton, probed with resolution Q^2 in a process with a fast proton in the final state. Several fits of the available F_2^D data are available which are based on the factorization formula 5 at next-to-leading order (NLO) in α_s.[7]

5. The Procedure of QCD Analysis

The DPDFs $f_i^D(z, Q^2, x_{I\!P}, t)$ are parameterised following the fit procedure of the inclusive analysis.[8] They are factorized into a pomeron flux $f_{I\!P/p}(x_{I\!P}, t)$ and parton densities of the pomeron $f_i(z, Q^2)$ using the proton vertex factorisation ansatz

$$f_i^D(z, Q^2, x_{I\!P}, t) = f_{I\!P/p}(x_{I\!P}, t) \cdot f_i(z, Q^2). \tag{6}$$

The parton densities f_i are modeled as a singlet distribution $\Sigma(z, Q^2)$ consisting of the three light quark and corresponding antiquark distributions, which are all assumed to be of equal magnitude, and a gluon distribution $g(z, Q^2)$. Here z is the longitudinal momentum fraction of the parton entering the hard subprocess with respect to the diffractive exchange, such that $z = \beta = x/x_{I\!P}$ for the lowest order quark parton model process in inclusive diffraction.[9–11] The parton densities $f_i(z, Q^2)$ are parameterized at a starting scale and are evolved to higher factorization scales using a numerical solution of the NLO DGLAP evolution equations.[12,13] The singlet and gluon distributions are parameterized at the starting scale as

$$f_i(z, Q_0^2) \equiv A_i \cdot z^{B_i} \cdot (1 - z)^{C_i}. \tag{7}$$

The DPDFs as defined in equation 7 are multiplied by a term $e^{-\frac{0.01}{1-z}}$ in order to ensure that they vanish at $z = 1$, as required for the evolution equations to be solvable. The parameters C_q and C_g thus have the freedom to take negative as well as positive values. Modifying the argument of the exponential term within reasonable limits has no visible influence on the fit quality or the extracted DPDFs in the range of the measurement.

The pomeron flux is parameterized as in Ref. 8 using a form motivated by Regge theory:

$$f_{I\!P/p}(x_{I\!P}, t) = A_{I\!P} \left(\frac{1}{x_{I\!P}} \right)^{2\alpha_{I\!P}(t)-1} e^{B_{I\!P}t}. \tag{8}$$

The normalization parameter $A_{I\!P}$ is defined as in Ref. 8. The pomeron trajectory $\alpha_{I\!P}(t)$ is assumed to be linear:

$$\alpha_{I\!P}(t) = \alpha_{I\!P}(0) + \alpha'_{I\!P} \cdot t. \tag{9}$$

For comparison with the data, all DPDFs are integrated over the measured range $|t| < 1$ GeV^2. To properly describe the data, especially at high $x_{I\!P}$, it is necessary to include a sub-leading exchange (the so called Reggeon, $I\!R$, for details see Ref. 8). As in Refs. 2, 14 this contribution is assumed to factorize similarly to the pomeron, so that the definition of the diffractive parton densities is modified to

$$f_i^D(z, Q^2, x_{I\!P}, t) = f_{I\!P/p}(x_{I\!P}, t) \cdot f_i(z, Q^2) + n_{I\!R} \cdot f_{I\!R/p}(x_{I\!P}, t) \cdot f_i^{I\!R}(z, Q^2). \tag{10}$$

The reggeon flux $f_{I\!R/p}(x_{I\!P}, t)$ is parameterized in the same way as the pomeron flux.[15] The parton densities $f_i^{I\!R}(z, Q^2)$ are taken from a parametrization of pion structure function data.[16] Choosing a different parametrization does not affect the fit results significantly.[17]

The free parameters of the fit are the A, B and C parameters which determine the quark singlet and gluon distributions (equation 7), together with $\alpha_{I\!P}(0)$, which controls the $x_{I\!P}$ dependence and $n_{I\!R}$, which controls the normalization of the sub-leading exchange contribution. In order to constrain these parameters, a χ^2 function as defined in Ref. 18 is minimized. All other parameters are fixed using the same values and uncertainties as in Ref. 8.

For the quark singlet distribution, the data require the inclusion of all three parameters A_q, B_q and C_q in equation 7. By comparison, the gluon density is weakly constrained by the data, which are found to be insensitive to the B_g parameter. The gluon density is thus parameterized at Q_0^2 using only the A_g and C_g parameters. This fit is referred to as the First Scenario Fit in the figures.

As discussed in Ref 8, the Q^2 dependence of the data at fixed β and $x_{I\!P}$ determines the gluon density well at low β. However, At the highest β values, where the Q^2 evolution is driven by quarks, the Q^2 dependence of $\sigma_r^{D(3)}$ becomes insensitive to the gluon density. The results for the gluon density at large z are thus determined principally by the data at lower z coupled with the parametrization choice. This lack of sensitivity is confirmed by repeating the fit with the parameter C_g, which determines the high z behavior, set to zero. Apart from the exponential term, the gluon density is then a simple constant at the starting scale for evolution. This fit is referred to as the Second Scenario Fit in the figures.

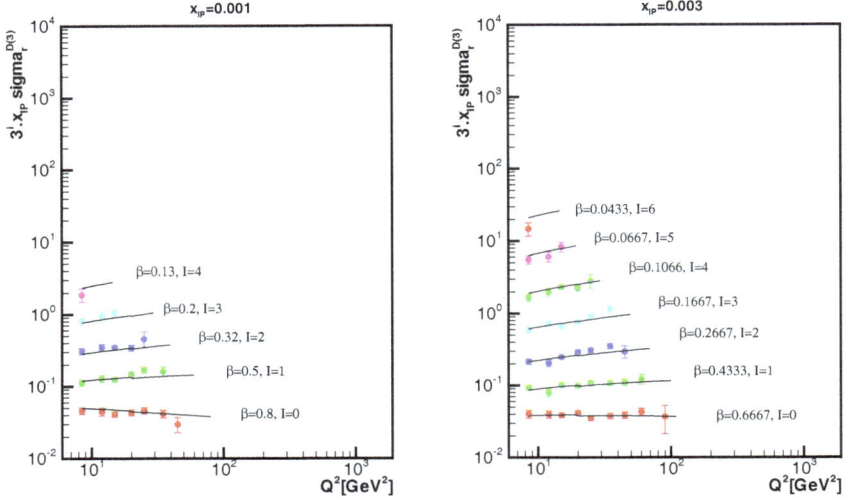

Fig. 3. The Q^2 dependence of the diffractive reduced cross section $\sigma_r^{D(3)}$ multiplied by $x_{I\!P} = 0.001$ (left) and $x_{I\!P} = 0.003$ (right) at various values of β. The cross sections are multiplied by powers of 3 for better visibility. The data points are taken from the publication.[8] Only data points included in the DPDF fits are shown. The data are compared to NLO QCD predictions based on the First scenario DPDF, which are shown as solid lines.

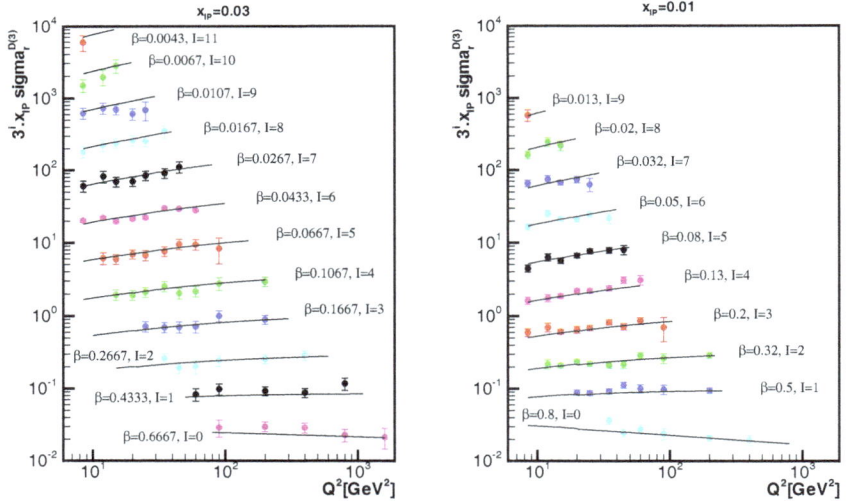

Fig. 4. The Q^2 dependence of the diffractive reduced cross section $\sigma_r^{D(3)}$ multiplied by $x_{I\!P} = 0.01$(left) and $x_{I\!P} = 0.03$ (right) at various values of β. See caption of Fig. 3 for further details.

Table 1.   The central values of the parameters extracted in the first and second scenarios, and the corresponding uncertainties.

Fit Parameter	First scenario	Second scenario
$\alpha_{I\!P}(0)$	1.10±0.005	1.111±0.005
$n_{I\!R}$	$(1.03\pm0.1)\times10^{-3}$	$(1.97\pm0.19)\times10^{-3}$
A_q	1.11±0.20	0.51±0.14
B_q	2.21±0.20	1.26±0.12
C_q	0.64±0.08	0.41±0.11
A_g	0.13±0.005	0.60±0.003
C_g	−0.91±0.02	0 (fixed)
χ^2/ndf	0.77	0.86

6. Fit Ansatz

A good description of the data is obtained by performing a conventional Next-to-Leading order (NLO) DGLAP analysis of data at small x and $Q^2 > Q_0^2$ throughout the fitted range $Q^2 \geq 8.5$ GeV^2, $\beta \leq 0.8$ and $M_x > 2$ GeV by both First and Second Scenario Fit. Figures 3 and 4 show the inclusive data points in the form of the product $x_{I\!P}.\sigma_r^{D(3)}(x_{I\!P}, \beta, Q^2)$. The results for the fit parameters are given in Table 1.

Even with this very simple parametrization of the gluon density in Second Scenario Fit, the χ^2/ndf variable increases only slightly to 0.86 from its value in First Scenario Fit, 0.77. The overall value of χ^2/ndf indicates that both scenarios describe the data well. The diffractive quark singlet and gluon distributions from first and second scenario fit are shown on a linear z scale in Figure 5.

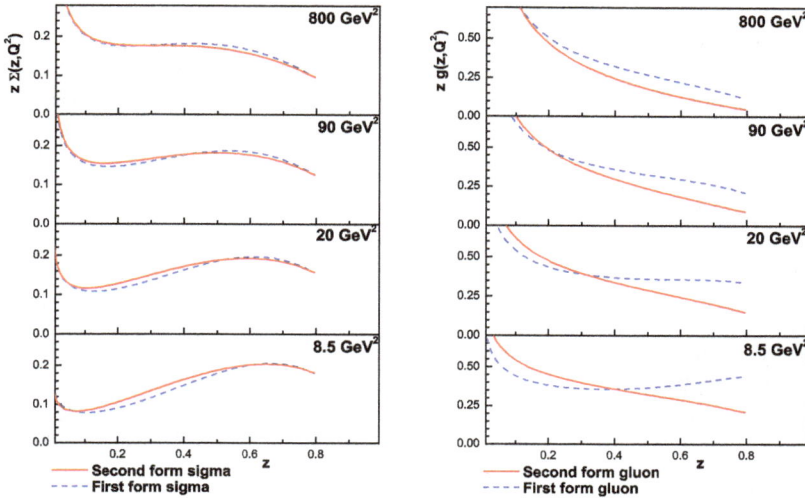

Fig. 5.   Comparison of a linear z scale between the total quark singlet and gluon distributions obtained from the First scenario fit and the Second scenario fit. These two fits differ in the parametrization chosen for the gluon density at the starting scale for QCD evolution.

418

7. Conclusion

Finally, the important qualitative feature of the calculation which should be emphasized is that the diffractive gluon distribution is much larger than the diffractive quark distribution. This result reflected in the pattern of scaling violations of $F_2^{D(3)}$. Figures 3 and 4 show the measurements of $\sigma_r^{D(3)}$ as a function of Q^2 for different values of β and $x_{I\!P}$, together with the NLO predictions based on the First scenario fit. The results of fit to the inclusive data alone are also shown. A very good description is obtained with our fit.

Acknowledgments

We are especially grateful to A. De Roeck for guidance and constructive comments.

References

1. A. Donnachie and P. Landshoff, Phys. Lett. B **296** (1992) 227 [hep-ph/9209205].
2. C. Adloff *et al.* [H1 Collaboration], Z. Phys. C **76** (1997) 613 [hep-ex/9708016].
3. M. Arneodo and M. Diehl, [hep-ph/0511047].
4. L. Trentadue and G. Veneziano, Phys. Lett. B **323** (1994) 201.
5. A. Berera and D. Soper, Phys. Rev. D **53** (1996) 6162 [hep-ph/9509239].
6. J. Collins, Phys. Rev. D **57** (1998) 3051 [Erratum-ibid. D **61** (2000) 019902] [hep-ph/9709499].
7. S. Alekhin *et al.*, arXiv:hep-ph/0601013.
8. A. Aktas *et al.* [H1 Collaboration], Eur. Phys. J. C **48** (2006) 715 [hep-ex/0606004].
9. A. Martin, W. Stirling and R. Thorne, Phys. Lett. B **636** (2006) 259 [hep-ph/0603143].
10. A. Buras and K. Gaemers, Nucl. Phys. B **132** (1978) 249.
11. A. Martin, R. Roberts, W. J. Stirling and R. Thorne, Eur. Phys. J. C **23** (2002) 73 [hep-ph/0110215].
12. V. Gribov and L. Lipatov, Sov. J. Nucl. Phys. **15** (1972) 438 [Yad. Fiz. **15** (1972) 781].
    V. Gribov and L. Lipatov, Sov. J. Nucl. Phys. **15** (1972) 675 [Yad. Fiz. **15** (1972) 1218].
    Y. Dokshitzer, Sov. Phys. JETP **46** (1977) 641 [Zh. Eksp. Teor. Fiz. **73** (1977) 1216].
    G. Altarelli and G. Parisi, Nucl. Phys. B **126** (1977) 298.
13. W. Furmanski and R. Petronzio, Z. Phys. C **11** (1982) 293.
14. H1 Collaboration, *"Diffractive Deep-Inelastic Scattering with a Leading Proton at HERA"*, DESY 06-048, submitted to Eur. Phys. J. C.
15. A. D. Martin, R. G. Roberts, W. J. Stirling and R. S. Thorne, Eur. Phys. J. C **28** (2003) 455 [hep-ph/0211080].
16. J. Owens, Phys. Rev. D **30** (1984) 943.
17. M. Glück, E. Reya and A. Vogt, Z. Phys. C **53** (1992) 651.
18. C. Adloff *et al.* [H1 Collaboration], Eur. Phys. J. C **21** (2001) 33 [hep-ex/0012053].
    J. Cudell, K. Kang and S. Kim, Phys. Lett. B **395** (1997) 311 [hep-ph/9601336].

FLUID QCD APPROACH FOR QUARK-GLUON PLASMA IN STELLAR STRUCTURE

T. P. DJUN[a] and L. T. HANDOKO[a,b]

[a] *Group for Theoretical and Computational Physics,*
Research Center for Physics, Indonesian Institute of Sciences,
Kompleks Puspiptek Serpong, Tangerang 15310, Indonesia

[b] *Department of Physics, University of Indonesia,*
Kampus UI Depok, Depok 16424, Indonesia

The quark-gluon plasma in stellar structure is investigated using the fluid-like QCD approach. The classical energy momentum tensor relevant for high energy and hot plasma having the nature of fluid bulk of gluon sea is calculated within the model. The transition of gluon field from point particle field inside stable hadrons to relativistic fluid field in hot plasma and vice versa is briefly discussed. The results are applied to construct the equation of state using the Tolman–Oppenheimer–Volkoff equation to describe the hot plasma dominated stellar structure.

Keywords: Quark-gluon-plasma; fluid QCD; hydrodynamics model; relativistic nuclear collision.

1. Introduction

Recent experiments in the last decades on relativistic nuclear collisions shed light on the phenomena of hot plasma formed by dense quarks and gluons. Those experiments suggest that the quark gluon matter behaves more like a deconfined quark-gluon plasma (QGP) liquid.[1,2] A comprehensive review on this matter is given by E. Shuryak.[3]

This fact immediately encourages some models based on the (relativistic) hydrodynamic approaches. In particular, dissipative ideal hydrodynamics has been used to fit some experimental data at high energy heavy ion program at the Relativistic Heavy Ion Collider (RHIC).[4] The successful fit requires the models to take into account very small value of the ratio shear viscosity over entropy.[5–10] However the puzzle must still be confirmed by the next coming experiments at the Large Hadron Collider (LHC).[11]

Since QGP is containing many quark-anti-quarks and gluons, it is considerable to treat it using the well-established quantum chromodynamics (QCD). In pure QCD, QGP is described as a quark soup before hadronization which is a phase of QCD, and exists at extremely high temperature and/or density. It is argued that this phase consists of almost free quarks and gluons. Therefore, the phase transi-

420

tion from the deconfined QGP to the hadronic matters or vice versa gets particular interest in this approach. It unfortunately turns out to the many body problems with large color charge which cannot be calculated analytically using perturbation. As a result, the main theoretical tools to explore the QGP within QCD is lattice gauge theory. The lattice calculation predicts that the phase transition occurs at approximately 175 MeV.[12,13]

On the other hand concerning that the QGP is a strongly interacting elementary particle system which should be governed by strong interaction, while it also dissolves into an almost perfect dense fluid of quarks and gluons,[14] it is plausible to describe it as a fluid system. In this sense, there are approaches based on unifying or hybridizing the charge field with flow field.[15–22] Recently, some works have constructed the models in a lagrangian with certain non-Abelian gauge symmetry to the matter inside the fluid.[23,24]

In this paper, the energy momentum tensor of QGP within the recent fluid QCD model[24] is investigated. Further, the equation of state relevant for hot plasma dominated stellar structure is constructed using the so-called Tolman–Oppenheimer–Volkoff (TOV) equation.

The paper is organized as follows. First we briefly introduce the underlying model of gauge invariant fluid lagrangian and discuss the relevant physical scale and region within the model. Then, the energy momentum tensor in the model is derived and investigated. Subsequently it is followed with relevant equation of state in a particular geometry using TOV equation.[25,26] Finally, the paper is ended with a summary.

2. The Model

Let us adopt the model developed by Sulaiman et al.[24] It describes the QGP as a strongly interacting gluon sea with the matters of quarks and anti-quarks inside. The model deploys the conventional QCD lagrangian with SU(3) color gauge symmetry, that is,

$$\mathcal{L} = i\bar{Q}\gamma^\mu\partial_\mu Q - m_Q\bar{Q}Q - \frac{1}{4}S^a_{\mu\nu}S^{a\mu\nu} + g_s J^a_\mu U^{a\mu}. \tag{1}$$

Here Q and U_μ represent the quark (color) triplet and gauge vector field. g_s is the strong coupling constant, $J^a_\mu = \bar{Q}T^a\gamma_\mu Q$ and T^a's belong to the SU(3) Gell-Mann matrices. The strength tensor is $S^a_{\mu\nu} = \partial_\mu U^a_\nu - \partial_\nu U^a_\mu + g_s f^{abc}U^b_\mu U^c_\nu$ with f^{abc} is the structure constant of SU(3) group respectively. It should be noted that the quarks and anti-quarks feel the electromagnetic force due to the U(1) field A_μ, but the size is suppressed by a factor of $e/g = \sqrt{\alpha/\alpha_s} \sim O(10^{-1})$.

Following the original model,[24] the gluon fluid is put to have a particular form in term of relativistic velocity as,[a]

$$U^a_\mu = (U^a_0, \mathbf{U}^a) \equiv u^a_\mu \phi, \tag{2}$$

---

[a]This form was first proposed in the early work of Sulaiman et al.in 2005 available in arXiv:physics/0508219. The form was then adopted in the work of Bambah et al.in arXiv:hep-th/0605204, but the citation to the original work disappeared in their published version.[22]

with $u_\mu^a \equiv \gamma_{\mathbf{v}^a}(1, \mathbf{v}^a)$ and $\gamma_{\mathbf{v}^a} = (1 - |\mathbf{v}^a|^2)^{-1/2}$. ϕ is a dimension one scalar field to keep correct dimension and should represent the field distribution. It is argued that taking this form leads to the equation of motion (EOM) for a single gluon field as follow,[24]

$$\frac{\partial}{\partial t} \left(\gamma_{\mathbf{v}^a} \mathbf{v}^a \phi\right) + \nabla \left(\gamma_{\mathbf{v}^a} \phi\right) = -g_s \oint d\mathbf{x} \left(\mathcal{J}_0^a + F_0^a\right), \tag{3}$$

where $\mathcal{J}_\mu^a$ is the covariant current of gluon field, and F_μ^a is an auxiliary function which can be found in the original paper.[24] It has been concluded that Eq. (3) should be a general relativistic fluid equation, since at the non-relativistic limit Eq. (3) coincides to the classical Euler equation.

More precisely, Eq. (3) provides a clue that a single gluonic field U_μ^a may behave as a fluid at certain scale, beside its point particle properties with a polarization vector ϵ_μ in the form of $U_\mu^a = \epsilon_\mu^a \phi$. One can consider that there is a kind of "phase transition",

$$\underbrace{\text{hadronic state}}_{\epsilon_\mu^a} \longleftrightarrow \underbrace{\text{QGP state}}_{u_\mu^a} . \tag{4}$$

As the gluon field behaves as a point particle, it is in a stable hadronic state and is characterized by its polarization vector. On the other hand in the pre-hadronic state (before hadronization) like hot QGP, the gluon field behaves as a highly energized flow particle and the properties are dominated by its relativistic velocity.

One should also recall that the wave function U_μ for a free particle satisfies $\left[g^{\nu\mu}(\partial^2 + m_U^2) - \partial^\nu \partial^\mu\right] U_\mu = 0$ with a solution $U_\mu \sim \epsilon_\mu \exp(-ip_\nu x^\nu)$ where p_ν is the 4-momentum. For a massive vector particle, $i.e. m_U \neq 0$, we have no choice but to take $\partial^\mu U_\mu = 0$. It is not a gauge condition like the case of massless particle. This then demands $p^\mu \epsilon_\mu = 0$. Therefore the number of independent polarization vectors is reduced from four to three in a covariant fashion. In contrast with this, in the case of massless bosonic particles like gluon there are only two degrees of freedom remain. Therefore, one should keep in mind that in the present model the spatial velocity has only 2 degrees of freedom, that means one component must be described by another two vector components. Fortunately, in real applications in cosmology or compact star, this requirement is satisfied by the assumption that the system under consideration is isotropic.

From now, throughout the paper let us focus only on the gluon sea of plasma. This means one should consider only the related gluonic terms in Eq. (1),

$$\mathcal{L}_g = -\frac{1}{4} S_{\mu\nu}^a S^{a\mu\nu} + g_s J_\mu^a U^{a\mu} . \tag{5}$$

3. Energy Momentum Tensor

Now we are ready to proceed with deriving the energy momentum tensor within the model. It should be pointed out that once the hot (high energy) QGP state is achieved, the system is assumed to be predominated by the classical motion rather than the quantum effects.

Therefore the total action of matter for non-gravitational fields in a general geometry of space-time $\mathcal{R}$ is $S_g = \int_{\mathcal{R}} \mathrm{d}^4 x \sqrt{-g}\, \mathcal{L}_g$, where g is the determinant of metric $g_{\mu\nu}$. It is well-known that the variation of S_g in the metric is given by $\delta S_g = -\frac{1}{2}\int_{\mathcal{R}} \mathrm{d}^4 x \sqrt{-g}\, \mathcal{T}_{\mu\nu}\, \delta g^{\mu\nu}$. Since the energy momentum tensor density is,

$$\mathcal{T}_{\mu\nu} = \frac{2}{\sqrt{-g}}\frac{\delta \mathcal{L}_g}{\delta g^{\mu\nu}}, \tag{6}$$

one obtains,

$$\mathcal{T}_{\mu\nu} = S^a_{\mu\rho} S^{a\,\rho}_{\ \ \nu} - g_{\mu\nu}\mathcal{L}_g + 2g_s J^a_\mu U^a_{\ \nu}. \tag{7}$$

It is clear that Eq. (7) is symmetric as expected to fulfill the Einstein gravitational EOM. The total energy momentum tensor $T_{\mu\nu}$ is given by integrating out Eq. (7) in term of total volume in the space-time under consideration. $T_{\mu\nu}$ is a result of bulk of gluons flow in the system.

Furthermore, in a general space-time coordinates, the components of energy momentum tensor determine the total energy density (T_{00}), the heat conduction ($T_{0i,i0}$), the isotropic pressure (T_{ii}) and the viscous stresses (T_{ij} with $i \neq j$) of the gluonic plasma. Of course, in this case the derivative ∂_μ inside the strength tensor $S_{\mu\nu}$ should be replaced by the covariant one, ∇_μ. Also, the energy momentum tensor satisfies the conservation condition, $\nabla_\mu T^{\mu\nu} = 0$. Nevertheless, one can trivially conclude that the model induces non-zero viscosity since generally $T_{ij} \neq 0$ for $i \neq j$. From the experimental clues, however the size should be small such that it is always treated perturbatively in most hydrodynamics models.[5–10]

Before going further to apply these results, one should determine the quark current J^a_μ in Eq. (7). This can be simply calculated by considering the EOM (Dirac equation) of a single colored quark (q) or anti-quark ($\bar{q}$) with 4-momentum p_μ. Since the solution of the EOM is $q(p,x) = u(p)\exp(-ip\cdot x)$, one immediately gets $\bar{u}\gamma_\mu u = 4p_\mu$. Assuming that all colored quarks / anti-quarks have the same momenta and the velocity of gluons are homogeneity, approximately $J^a_\mu U^{a\mu} \propto 4p_\mu U^\mu = 4m_Q\phi$ since $u_\mu u^\mu = u^2 = 1$.

4. Equation of State for Stellar Structure

This section is dedicated to provide an example on the applications of the present model to describe the stellar interior, in particular the compact stars which are still dominated by hot plasma before transforming itself into neutron star.

The stellar structure is commonly described as a static spherically symmetric space-time represented by Schwarzschild geometry. This means one deals with the relativistic gravitational equations for the interior of spherically symmetric plasma distribution. In the region under consideration the presence of the flow gluonic fields induces non-zero energy momentum tensor which is making up the star. This is the phase before the neutron star is getting mature. Starting from the stellar nebula made of hot plasma which is gradually getting colder as the hadronization occurs from the colder surface, while the inner core is still in pure hot QGP state.

As a consequence of the diagonal metric of Schwarzschild space-time, the model falls back to the perfect fluid without viscosity and heat conduction, $i.e. T_{0i} = T_{ij} = 0$ for $i \neq j$. Also, since the plasma distribution should be spherically isotropic, it is considerable to put $v_1 = v_2 = v_3 = v$ as constant for all colored gluons. This assumption is consistent with the degree of freedom counting discussed in the preceding section. Moreover, the vanishing off-diagonal components of the Ricci tensor, R_{i0}, actually forces the spatial 3-velocity of the fluid must vanish everywhere. Hence particular assumption for v_i is indeed not necessary. However, the gluon distribution still depends on the radius length, $\phi = \phi(r)$.

For the sake of simplicity one can put homogeneous gluon fields for all color states, $i.e. U_\mu^a = U_\mu$ for all $a = 1, \cdots, 8$. This yields,

$$
\mathcal{T}_{\mu\nu} = \left[8\, g_s\, f_Q\, m_Q\, \phi(r) + g_s^2\, f_g^2\, \phi(r)^4 \right] u_\mu u_\nu
$$
$$
- \left[4\, g_s\, f_Q\, m_Q\, \phi(r) - \frac{1}{4} g_s^2\, f_g^2\, \phi(r)^4 \right] g_{\mu\nu} , \tag{8}
$$

where f_g is the factor of summed colored gluon states from the structure constant f^{abc}, while f_Q is the factor of summed colored quark states from $J_\mu^a U^{a\mu}$. Remind that the energy momentum tensor for perfect fluid takes the form,

$$
\mathcal{T}_{\mu\nu} = (\mathcal{E} + \mathcal{P})\, u_\mu u_\nu - \mathcal{P}\, g_{\mu\nu} . \tag{9}
$$

Here $\mathcal{E}$ and $\mathcal{P}$ denote the density and isotropic pressure for single fluid field, each is related to the total density and pressure of the system through $\rho = \oint d^4 x\, \mathcal{E}$ and $P = \oint d^4 x\, \mathcal{P}$ respectively. Obviously, from Eqs. (8) and (9) one can obtain the density and pressure in the model as follows,

$$
P(r) = \int_0^\beta dt \int dV \left[4\, g_s\, f_Q\, m_Q\, \phi(r) - \frac{1}{4} g_s^2\, f_g^2\, \phi(r)^4 \right]
$$
$$
= \frac{4\, g_s\, f_Q\, m_Q}{T} \int dV \left[1 - \frac{g_s\, f_g^2}{16\, f_Q\, m_Q} \phi(r)^3 \right] \phi(r) , \tag{10}
$$

$$
\rho(r) = \int_0^\beta dt \int dV \left[4\, g_s\, f_Q\, m_Q\, \phi(r) + \frac{5}{4} g_s^2\, f_g^2\, \phi(r)^4 \right]
$$
$$
= \frac{4\, g_s\, f_Q\, m_Q}{T} \int dV \left[1 + \frac{5\, g_s\, f_g^2}{16\, f_Q\, m_Q} \phi(r)^3 \right] \phi(r) , \tag{11}
$$

at a finite temperature $\beta = 1/T$ in a 3-dimensional spatial volume V.

The proper spatial volume element for Schwarzschild geometry is $dV = \sqrt{B(r)} r^2 \sin\theta\, dr\, d\theta\, d\varphi$ with radius r and two angles θ and φ in spherical coordinates. The solution for $B(r)$ is given by,

$$
B(r) = \left[1 - \frac{2Gm(r)}{r} \right]^{-1} , \tag{12}
$$

and $m(r) = 4\pi \int_0^r \mathrm{d}\bar{r} \rho(\bar{r}) \, \bar{r}^2$ is the 'bare mass'. This generates the proper integrated mass $\tilde{m}(r)$ contained within a coordinate radius r inside the star. On the other hand, the stellar structure with Schwarzschild geometry is well known as the TOV equation which relates density and pressure in a unique way,[25,26]

$$\frac{\mathrm{d}P(r)}{\mathrm{d}r} = -\frac{1}{r^2} \left[\rho(r) + P(r)\right] \left[4\pi G \, P(r) \, r^3 + G \, m(r)\right] \left[1 - \frac{2Gm(r)}{r}\right]^{-1}. \quad (13)$$

Substituting Eqs. (11) and (10) into Eq. (13) provides a direct relationship between density and pressure. In another word, one can obtain a contour of the equation of state of gluonic plasma in term of distribution function $\phi(r)$ and temperature T.

5. Summary

The phase transition between stable hadronic state and highly energized QGP state is briefly discussed using recently developed fluid QCD lagrangian. It is argued that inside the hadronic state the gluons behave as point particles and the properties are determined by its polarization vectors. However, in the hot QGP state, the gluon should behave as fluid particles and characterized by its relativistic velocities.

Then, the energy momentum tensor for the fluid particle is investigated. In particular, a detailed derivation has been worked out in the case of Schwarzschild space-time that is relevant for stellar structure of unmatured stars composed of hot QGP. In principle one can obtain a kind of equation of state of gluonic plasma inside the star through the TOV equation. This approach would enable us to describe the stellar structure without assuming a particular relation between pressure and density.

Further works, especially on the numerical analysis, are still in progress and will be reported elsewhere.

Acknowledgments

The authors thank A. O. Latief, C. S. Nugroho and M. K. Nurdin for fruitful discussion during the final stage of this paper. This work is funded by Riset Kompetitif LIPI in fiscal year 2010 under Contract no. 11.04/SK/KPPI/II/2010.

References

1. C. Adler *et.al.* (STAR Collaboration), *Phys. Rev. C* **66**, p. 034904 (2002).
2. K. Adcox *et.al.* (PHENIX Collaboration), *Nucl. Phys. A* **757**, p. 184 (2005).
3. E. Shuryak, *Nucl.Phys. A* **774**, p. 387 (2006).
4. M. Harrison, T. Ludlam and S. Ozaki, *Nucl. Inst. Meth. Phys. Res. A* **499**, 235 (2003).
5. D. Teaney, J. Lauret and E. V. Shuryak, *Phys. Rev. Lett.* **86**, p. 4783 (2001).
6. P. Huovinen, P. F. Kolb, U. W. Heinz, P. V. Ruuskanen and S. A. Voloshin, *Phys. Lett. B* **503**, 58 (2001).
7. P. F. Kolb, U. W. Heinz, P. Huovinen, K. J. Eskola and K. Tuominen, *Nucl. Phys. A* **696**, 197 (2001).

8. P. F. Kolb and R. Rapp, *Phys. Rev. C* **67**, p. 044903 (2003).

9. T. Hirano and K. Tsuda, *Phys. Rev. C* **66**, p. 054905 (2002).

10. R. Baier and P. Romatschke, *Eur. Phys. J. C* **51**, 677 (2007).

11. J. Jowett, *LHC Lead Ion Beam Commissioning in LHC Design Report*, tech. rep., CERN (February 2009).

12. S. Gottlieb, *J. Phys. Conf. Ser.* **78**, p. 012023 (2007).

13. P. Petreczky, *Europ. Phys. J. Special Topics* **155**, 1951 (2008).

14. W. A. Zajc, *Nucl. Phys. A* **805**, 283 (2008).

15. U. Heinz, *Phys. Rev. Lett.* **51**, p. 351 (1983).

16. D. D. Holm and B. A. Kupershmidt, *Phys. Rev. D* **30**, p. 2557 (1984).

17. Y. Choquet-Bruhat, *J. Math. Phys.* **33**, p. 1782 (1992).

18. J. P. Blaizot and E. Iancu, *Nucl. Phys. B* **421**, 565 (1994).

19. B. Bistrovic, R. Jackiw, H. Li, V. P. Nair and S. Y. Pi, *Phys. Rev. D* **67**, p. 025013 (2003).

20. S. M. Mahajan, *Phys. Rev. Lett.* **90**, p. 035001 (2003).

21. C. Manuel and S. Mrowczynski, *Phys. Rev. D* **74**, p. 105003 (2006).

22. B. A. Bambah, S. M. Mahajan and C. Mukku, *Phys. Rev. Lett.* **97**, p. 072301 (2006).

23. Marmanis, *Phys. of Fluid* **10**, 1428 (1998).

24. A. Sulaiman, A. Fajarudin, T. P. Djun and L. T. Handoko, *Int. J. Mod. Phys. A* **24**, 3630 (2009).

25. R. C. Tolman, *Proc. Nat. Acad. Sci.* **20**, 169 (1934).

26. J. R. Oppenheimer and G. M. Volkoff, *Phys. Rev.* **55**, 374 (1939).

HEAVY QUARK CONTRIBUTIONS TO THE PROTON STRUCTURE FUNCTION

H. KHANPOUR* and ALI. N. KHORRAMIAN†

Physics Department, Semnan University, P.O. Box 35195-363, Semnan, Iran
School of Particles and Accelerators,
Institute for Research in Fundamental Sciences (IPM),
P.O. Box 19395-5531, Tehran, Iran
*Hamzeh_Khanpour@nit.ac.ir
†Khorramiana@theory.ipm.ac.ir
http://particles.ipm.ir/

S. ATASHBAR TEHRANI

School of Particles and Accelerators,
Institute for Research in Fundamental Sciences (IPM),
P.O. Box 19395-5531, Tehran, Iran
Atashbar@ipm.ir

Our aim in this article is to perform the standard next-to-leading order (NLO) parton distribution functions QCD analysis of the inclusive neutral-current deep-inelastic-scattering (NC DIS) world data. Parton distribution functions (PDFs) are important for the theoretical description of hard QCD processes at hadron colliders such as forthcoming LHC. In the present paper we restrict the analysis to the NLO heavy flavor corrections and extract heavy quark flavors distributions. These analysis are undertake within the framework of the so called "variable flavor number scheme" (VFNS) parton model predictions at high energy colliders. We generate VFNS parton distributions where the heavy quark flavors h = c,b,t are considered as massless partons within the nucleon. By studying the role of these distributions in the production of heavy particles at high energy ep and pp colliders, we show that our results for parton distribution functions and proton structure function are in good agreement with the DIS data and other theoretical models.

Keywords: Parton distribution functions; heavy quark flavors; variable flavor number scheme.

1. Introduction

Deeply inelastic electron-nucleon scattering at large momentum transfer allows to measure the parton distribution functions (PDFs) of the nucleons together with the $\alpha_s(M_Z^2)$ in order to provide one of the cleanest possibilities to test the predictions of the quantum chromodynamics (QCD). In this paper we study proton structure function in neutral-current deep-inelastic scattering (NC DIS) of leptons to the nucleons. We focus on the contributions of heavy quarks, like charm, which proceeds within perturbative QCD. We specifically include QCD corrections to DIS heavy-quark production and we wish to determine their impact on our information about the parton distribution functions (PDFs) extracted from global fits including the most recent neutral-current DIS data. The presently available DIS data allows for high precision extractions of PDFs in global fits. The treatment of the charm contribution in these fits is an important issue as it can induce potentially large effects also in the PDFs of light quarks and the gluon obtained from these global fits. Precision predictions of these PDFs are very essential for all measurements at hadron colliders.[1] Determination of PDFs both theoretically and experimentally appears to be one of the most important concern for the high energy particle physicists. Therefore it is a hot topic to be looked for in high energy colliders such as LHC.

This paper is organized as follows. In Section 2, we outline the theoretical formalism which describes the DIS structure functions. The heavy quark contributions to proton structure function and the formulation of VFN scheme is performed in Section 3. In Section 4, we present the results of our NLO PDF fit to the DIS world data using correlated errors to determine the PDF parameters and $\alpha_s(M_Z^2)$. Summary and conclusions are given in Section 5.

2. Theoretical Framework

The NLO analysis presented here is performed in the common modified minimal subtraction ($\overline{\text{MS}}$) factorization and renormalization scheme. Heavy quarks (c,b,t) are considered as massless partons within the nucleon. This defines the so-called 'variable flavor number scheme' (VFNS), which is fully predictive in the heavy quark sector. A VFN scheme has to be used in global fits of hadron collider data if the cross sections of the corresponding processes are not available in the 3-flavor scheme. VFN schemes are only applicable at asymptotically large momentum transfers.[1] In the common $\overline{\text{MS}}$ factorization scheme the relevant structure function F_2^p as extracted from the DIS ep process can be, up to NLO, written as[1-7]

$$F_2^p(x, Q^2) = F_{2,\text{NS}}^+(x, Q^2) + F_{2,S}(x, Q^2) + F_2^{(c,b)}(x, Q^2, m_{c,b}^2),\qquad (1)$$

with the non–singlet contribution for three active (light) flavors being given by

$$\frac{1}{x} F_{2,\text{NS}}^+(x, Q^2) = \left[C_{2,q}^{(0)} + a C_{2,\text{NS}}^{(1)} \right] \otimes \left[\frac{1}{18} q_8^+ + \frac{1}{6} q_3^+ \right] (x, Q^2),\qquad (2)$$

where $a = a(Q^2) \equiv \alpha_s(Q^2)/4\pi$, $C_{2,q}^{(0)}(z) = \delta(1 - z)$, $C_{2,\text{NS}}^{(1)}$ is the common NLO coefficient function.[8-10] The NLO Q^2-evolution of the flavor non-singlet combinations $q_3^+ = u + \bar{u} - (d + \bar{d}) = u_v - d_v$ and $q_8^+ = u + \bar{u} + d + \bar{d} - 2(s + \bar{s}) = u_v + d_v + 2\bar{u} + 2\bar{d} - 4\bar{s}$, where $\bar{u} \neq \bar{d}$ and $s = \bar{s}$, is related to the LO (1-loop) and NLO (2-loop) splitting functions $P_{\text{NS}}^{(0)}$ and $P_{\text{NS}}^{(1)+}$.[8,9] Notice that we consider sea breaking effects ($\bar{u} \neq \bar{d}$) since the HERA data used, and thus our analysis, are sensitive to such corrections. Because our analysis is based on the DIS structure function data rather than the *reduced cross sections*, we aren't able to include fixed-target Drell-Yan data,[11,12]

428

which allows the flavor separation of the sea-quark distributions. Since sea quarks cannot be neglected for small x, we supposed the following $\bar{d} - \bar{u}$ distribution[3,7]

$$x(\bar{d} - \bar{u})(x, Q_0^2) = 7.28 x^{1.27}(1 - x)^{18.75}(1 - 6.31 x + 18.30 x^2), \qquad (3)$$

at $Q_0^2 = 2\,\mathrm{GeV}^2$ which gives a good description of the Drell-Yan dimuon production data. In our analysis we used the above distribution for considering the symmetry breaking of sea quarks. Since these data sets we are using are also insensitive to the specific choice of the strange quark distributions, we consider a symmetric strange sea. In the "standard" approach we choose as usual the strange quark distribution in the symmetric form $s(x, Q_0^2) = \bar{s}(x, Q_0^2) = \frac{1}{4}(\bar{u}(x, Q_0^2) + \bar{d}(x, Q_0^2))$.[1-4] The flavor singlet contribution in Eq. (1) reads

$$\frac{1}{x} F_{2,S}(x, Q^2) = \frac{2}{9} \left\{ \left[C_{2,q}^{(0)} + a C_{2,q}^{(1)} \right] \otimes \Sigma + a C_{2,g}^{(1)} \otimes g \right\}(x, Q^2), \qquad (4)$$

with $\Sigma(x, Q^2) \equiv \Sigma_{q=u,d,s}(q + \bar{q}) = u_v + d_v + 2\bar{u} + 2\bar{d} + 2\bar{s}$, $C_{2,q}^{(1)} = C_{2,\mathrm{NS}}^{(1)}$ and the additional common NLO gluonic coefficient function $C_{2,g}^{(1)}$ can be again found in Ref. 8, for example. We have performed all Q^2-evolutions in Mellin n-moment space and used the QCD-PEGASUS program[13] for the NLO evolutions. Our another method of QCD analysis during last years was based on the expansion of the structure function in terms of orthogonal polynomials. This method was developed and applied for non-singlet QCD analysis up to N³LO.[14-20] The same method has also been applied in polarized case.[21-23]

3. Heavy Quark Contributions

For many hard processes at high energies, heavy flavor production forms a significant part of the scattering cross section. In all precision measurement, a detailed treatment of the heavy flavor contributions is required. Generally speaking, heavy quark production in deep inelastic scattering is calculable in QCD and provides information on the gluonic content of the proton. Nowadays, heavy quarks are produced in several experiments of high energy physics and the production and decay properties of the heavy quarks are extremely interesting. Heavy quark contributions and testing the different behavior of them are the main way in order to get information about the reliability of the perturbative QCD calculations, on threshold effects and on higher order corrections.

The heavy flavor contributions $F_2^{c,b}$ have been added in the present analysis are taken as in Refs. 24–28. We also considered the small bottom contribution, however it turn out to be negligible. For the quark masses we have chosen:

$$m_c = 1.3\,\mathrm{GeV}, \qquad m_b = 4.2\,\mathrm{GeV}, \qquad m_t = 175\,\mathrm{GeV}, \qquad (5)$$

which turn out to be the optimal choices for all our subsequent analysis, in particular for heavy quark production. The LO expression for the h-contribution to F_2 simplifies to

$$\frac{1}{x} F_2^{h,LO} = e_h^2 [h(x, Q^2) + \bar{h}(x, Q^2)], \qquad (6)$$

where e_h is the charge of the heavy quark, with h=c,b. The higher order extension of above equation for the heavy quark contributions are well known and taken as Refs. 1, 4. It can be shown that the resulting predictions for $F_2^{c,b}$ are in perfect agreement with all available recent DIS data on heavy quark production[11,29–31] and are furthermore perturbatively stable.

4. Quantitative Results

For the present analysis, the proton PDFs are parametrized at the input $Q_0^2 = 2$ GeV^2 in the VFN scheme. At the starting scale, the following functions are used for the valence quark, gluon and sea-quark distributions:

$$
\begin{aligned}
xu_v(x, Q_0^2) &= N_u\, x^{\alpha_u}(1-x)^{\beta_u}(1+\gamma_u\,\sqrt{x}+\eta_u\,x) \\
xd_v(x, Q_0^2) &= N_d\, x^{\alpha_d}(1-x)^{\beta_d}(1+\gamma_d\,\sqrt{x}+\eta_d\,x) \\
x\Sigma(x, Q_0^2) &= N_\Sigma\, x^{\alpha_\Sigma}(1-x)^{\beta_\Sigma}(1+\gamma_\Sigma\,\sqrt{x}+\eta_\Sigma\,x) \\
xg(x, Q_0^2) &= N_g\, x^{\alpha_g}(1-x)^{\beta_g}
\end{aligned}
\tag{7}
$$

where $u_v \equiv u-\bar{u}$, $d_v \equiv d-\bar{d}$ and $\Sigma \equiv \bar{u}+\bar{d}$. The distributions are further constrained by quark number and momentum conservation sum rules:

$$
\int_0^1 u_v\, dx = 2, \qquad \int_0^1 d_v\, dx = 1, \qquad \int_0^1 x\big(u_v + d_v + 2(\bar{u}+\bar{d}+\bar{s}) + g\big)dx = 1.
$$

which, as usual, we use to determine $N_{u,d,g}$, which therefore are not free parameters in our fits. We have tried different forms for these parametrizations, including $(1 + c\,x^d)$ and $(1 + c\sqrt{x} + d\,x + e\,x^{1.5})$ for the polynomials, without finding any improvement because the polynomials used in above distributions provide sufficient flexibility of the PDF-parametrization with respect to the analyzed data and no additional terms are required to improve the fit quality. In particular all of our fits did not require the polynomial for the gluon distribution. The values obtained for the parameters of the input distributions in our NLO QCD fit are given in Table 1. Our results for valence and gluon distributions are shown in Figures 1 and 2. They have been parametrized according to Eq. (7) with the parameters given in Table 1.

Table 1. Parameters of our input distributions as parametrized in Eq. (7) referring to an input scale of $Q_0^2 = 2$ GeV^2 in the VFN scheme at NLO.

	u_v	d_v	$\bar{u}+\bar{d}$	g
N	0.1747	0.2603	0.1997±0.0033	6.1983
α	0.4190±0.0029	0.2841±0.0039	-0.2667±0.0041	0.2446±0.0061
β	3.5839±0.0085	6.8998±0.066	10.104±0.3469	6.9498±0.0420
γ	29.9418	1.9978	4.8059	-
η	25.4905	25.5864	13.0965	-
		$\chi^2/\mathrm{dof} = 1.15$		
		$\alpha_s(M_Z^2) = 0.1154 \pm 0.0026$		

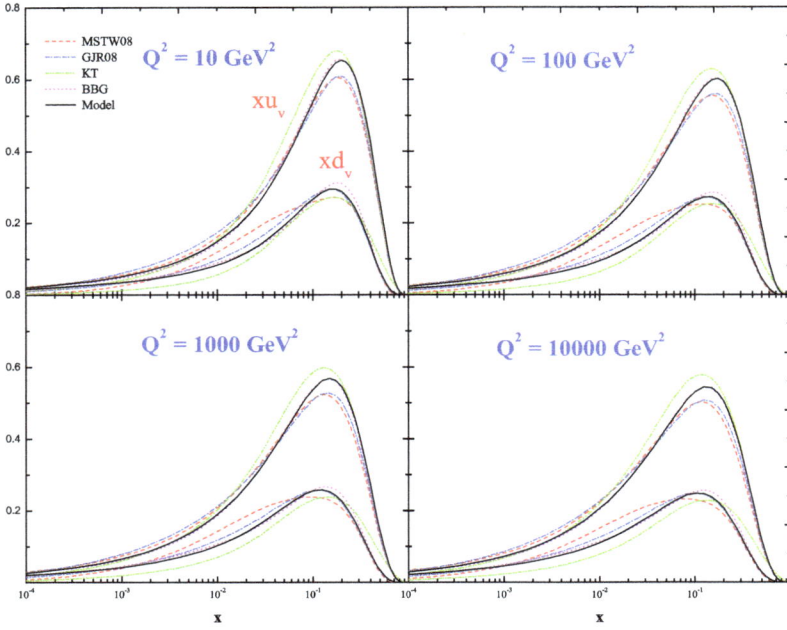

Fig. 1. Comparison of our standard NLO ($\overline{\text{MS}}$) results for $xu_v(x, Q^2)$ and $xd_v(x, Q^2)$ evolved up to $Q^2 = 10000 GeV^2$ with the results obtained by MSTW08,[2] GJR08,[7] KT[16] and BBG.[32]

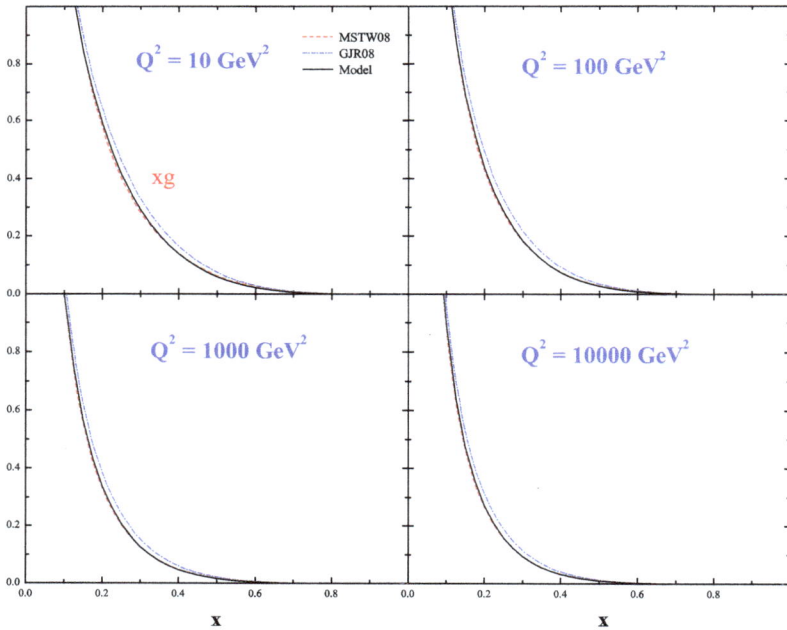

Fig. 2. As in Figure 1 but for gluon distribution $xg(x, Q^2)$.

Table 2.   Data sets fitted in our NLO
QCD analysis.

Data sets	NLO
H1 MB 97 e^+p NC	59 [33]
H1 low Q^2 96–97 e^+p NC	71 [33]
H1 high Q^2 99–00 e^+p NC	132 [34]
H1 high Q^2 94–97 e^+p NC	130 [35]
H1 high Q^2 98–99 e^-p NC	126 [36]
ZEUS SVX 95 e^+p NC	30 [37]
ZEUS 96–97 e^+p NC	242 [38]
SLAC ep F_2	37 [39]
BCDMS μp F_2	164 [40]
E665 μp F_2	53 [41]
NMC μp F_2	123 [42]
All data sets	**1167**

For comparison the results obtained by MSTW08,[2] GJR08,[7] KT[16] and BBG[32] are displayed as well.

The statistically most significant data that we use are the HERA (H1 and ZEUS) measurements, the small-x and large-x H1 and ZEUS F_2^p data for $Q^2 \geq 2$ GeV^2.[33–38] In addition, we have used the fixed target F_2^p data of SLAC,[39] BCDMS,[40] E665[41] and NMC[42] all subject to the standard cuts $Q^2 \geq 4$ GeV^2 and $W^2 = Q^2(\frac{1}{x} - 1) + m_p \geq 10$ GeV^2. The data sets included in the analysis are listed in Table 2 and are ordered according to the type of process. This amounts to the total of 1167 data points.

The minimization procedure follows the usual chi-squared method with χ^2 defined as:

$$\chi^2 = \sum_{i=1}^{N^{data}} \left(\frac{F_{2,i}^{data} - F_{2,i,p}^{theor}}{\Delta F_{2,i}^{data}}\right)^2 , \qquad (9)$$

where p denotes the set of 16 independent parameters in the fit, including $\alpha_s(M_Z^2)$, and N^{data} is the number of data points included; $N = 1167$ for the NLO fits. The errors include systematic and statistical uncertainties, being the total experimental error evaluated in quadrature. It is worth to mention that we include $\alpha_s(M_Z^2)$ as a free parameter in our fits, and determine its value together with the parton distributions. The minimization of the above χ^2 value to determine the best parametrization of the parton distribution functions is done using the program MINUIT.[43] It should be emphasized that the our perturbatively stable QCD predictions are in perfect agreement with other theoretical models and with all recent high statistics measurements of the Q^2-dependance of $F_2(x, Q^2)$ in wide range of x and Q^2.

5. Summary and Conclusion

To summarize, recent deep inelastic data for the structure function F_2^p have been analyzed in the standard parton model approach at next-to-leading order of perturbative QCD. The precision of these DIS world data has reached a level which requires precise analysis to determine the PDFs and to measure the strong coupling constant $\alpha_s(M_Z^2)$. In the analysis, we have taken into account correlated errors whenever available. In total, 16 parameters including $\alpha_s(M_Z^2)$ have been extracted from fitting to deep inelastic proton structure function F_2^p data. Perturbative NLO QCD evolutions of parton distribution functions and proton structure function are fully compatible with all recent high-statistics measurements of the Q^2-dependence of F_2^p in that region. We compared our results with other theoretical model, such as MSTW08[2] and GJR08,[7] and we found that our results are in good agreements with these models. The strong coupling constant obtained from our standard NLO analysis is $\alpha_s(M_Z^2) = 0.1154 \pm 0.0026$ to be compared with $\alpha_s(M_Z^2) = 0.1202^{+0.0012}_{-0.0015}$ [2] and $\alpha_s(M_Z^2) = 0.1178 \pm 0.0021$ [7]. During the last years, our understanding of PDFs has steadily improved at the NNLO level, and upcoming high-precision data from hadron colliders will continue in this direction. So in order to perform a consistent QCD analysis of the DIS world data and other hard scattering data, a next-to-next-to-leading order (NNLO) analysis is required and this is our motivation to do the NNLO QCD analysis in early future.

Acknowledgments

We would like to thanks Johannes Bluemlein for fruitful discussions. We thanks Semnan University for partial financial support of this project. We acknowledge the Institute for Research in Fundamental Sciences (IPM) for financially supporting this project.

References

1. S. Alekhin, J. Blumlein, S. Klein and S. Moch, Phys. Rev. D **81**, 014032 (2010) [arXiv:0908.2766 [hep-ph]].
2. A. D. Martin, W. J. Stirling, R. S. Thorne and G. Watt, Eur. Phys. J. C **63**, 189 (2009) [arXiv:0901.0002 [hep-ph]].
3. P. Jimenez-Delgado and E. Reya, Phys. Rev. D **79**, 074023 (2009) [arXiv:0810.4274 [hep-ph]].
4. M. Gluck, C. Pisano and E. Reya, Eur. Phys. J. C **50**, 29 (2007) [arXiv:hep-ph/0610060].
5. M. Gluck, E. Reya and A. Vogt, Eur. Phys. J. C **5**, 461 (1998) [arXiv:hep-ph/9806404].
6. C. Pisano, Nucl. Phys. Proc. Suppl. **191**, 35 (2009) [arXiv:0812.3250 [hep-ph]].
7. M. Gluck, P. Jimenez-Delgado and E. Reya, Eur. Phys. J. C **53**, 355 (2008) [arXiv:0709.0614 [hep-ph]].
8. W. L. van Neerven and A. Vogt, Nucl. Phys. B **588**, 345 (2000) [arXiv:hep-ph/0006154].
9. W. L. van Neerven and A. Vogt, Nucl. Phys. B **568**, 263 (2000) [arXiv:hep-ph/9907472].

433

10. J. A. M. Vermaseren, A. Vogt and S. Moch, Nucl. Phys. B **724** (2005) 3 [arXiv:hep-ph/0504242].
11. J. C. Webb *et al.* (NuSea Collaboration), arXiv: hep-ex/0302019.
12. R. S. Towell *et al.* (FNAL E866/NuSea Collaboration), *Phys. Rev.* **D64** (2001) 052002.
13. A. Vogt, Comput. Phys. Commun. **170**, 65 (2005) [arXiv:hep-ph/0408244].
14. A. N. Khorramian, H. Khanpour and S. A. Tehrani, Phys. Rev. D **81**, 014013 (2010) [arXiv:0909.2665 [hep-ph]].
15. H. Khanpour, A. N. Khorramian, S. Atashbar Tehrani and A. Mirjalili, Acta Phys. Polon. B **40**, 2971 (2009).
16. A. N. Khorramian and S. A. Tehrani, Phys. Rev. D **78**, 074019 (2008) [arXiv:0805.3063 [hep-ph]].
17. A. N. Khorramian, S. Atashbar Tehrani and M. Ghominejad, Acta Phys. Polon. B **38**, 3551 (2007).
18. A. N. Khorramian and S. A. Tehrani, J. Phys. Conf. Ser. **110**, 022022 (2008).
19. A. N. Khorramian and S. A. Tehrani, AIP Conf. Proc. **1006** (2008) 118.
20. S. Atashbar Tehrani and A. N. Khorramian, Nucl. Phys. Proc. Suppl. **186**, 58 (2009).
21. S. Atashbar Tehrani and A. N. Khorramian, JHEP **0707**, 048 (2007) [arXiv:0705.2647 [hep-ph]].
22. A. Mirjalili, A. N. Khorramian and S. Atashbar-Tehrani, Nucl. Phys. Proc. Suppl. **164**, 38 (2007).
23. A. Mirjalili, S. Atashbar Tehrani and A. N. Khorramian, Int. J. Mod. Phys. A **21**, 4599 (2006) [arXiv:hep-ph/0608224].
24. M. Gluck, P. Jimenez-Delgado, E. Reya and C. Schuck, Phys. Lett. B **664**, 133 (2008) [arXiv:0801.3618 [hep-ph]].
25. M. Gluck, E. Reya and M. Stratmann, Nucl. Phys. B **422**, 37 (1994).
26. E. Laenen, S. Riemersma, J. Smith and W. L. van Neerven, Phys. Lett. B **291**, 325 (1992).
27. S. Riemersma, J. Smith and W. L. van Neerven, Phys. Lett. B **347**, 143 (1995) [arXiv:hep-ph/9411431].
28. E. Laenen, S. Riemersma, J. Smith and W. L. van Neerven, Nucl. Phys. B **392**, 162 (1993).
29. C. Adloff *et al.* [H1 Collaboration], Phys. Lett. B **528**, 199 (2002) [arXiv:hep-ex/0108039].
30. A. Aktas *et al.* [H1 Collaboration], Eur. Phys. J. C **40**, 349 (2005) [arXiv:hep-ex/0411046].
31. A. Aktas *et al.* [H1 Collaboration], Eur. Phys. J. C **45**, 23 (2006) [arXiv:hep-ex/0507081].
32. J. Blumlein, H. Bottcher and A. Guffanti, Nucl. Phys. B **774**, 182 (2007) [arXiv:hep-ph/0607200].
33. C. Adloff *et al.* (H1 Collaboration), *Eur. Phys. J.* **C21** (2001) 33.
34. C. Adloff *et al.* (H1 Collaboration), *Eur. Phys. J.* **C30** (2003) 1.
35. C. Adloff *et al.* (H1 Collaboration), *Eur. Phys. J.* **C13** (2000) 609.
36. C. Adloff *et al.* (H1 Collaboration), *Eur. Phys. J.* **C19** (2001) 269.
37. J. Breitweg *et al.* (ZEUS Collaboration), *Eur. Phys. J.* **C7** (1999) 609.
38. S. Chekanov *et al.* (ZEUS Collaboration), *Eur. Phys. J.* **C21** (2001) 443.
39. L. W. Whitlow *et al.*, *Phys. Lett.* **B282** (1992) 475.
40. A. C. Benvenuti *et al.* (BCDMS Collaboration), *Phys. Lett.* **B223** (1989) 485.
41. M. R. Adams *et al.* (E665 Collaboration), *Phys. Rev.* **D54** (1996) 3006.
42. M. Arneodo *et al.* (New Muon Collaboration), *Nucl. Phys.* **B483** (1997) 3.
43. F. James, CERN Program Library, Long Writeup D506 (MINUIT).

NUCLEUS-NUCLEUS CHOU–YANG CORRELATIONS WITH GENERALIZED MULTIPLICITY DISTRIBUTION

J. K. JASVANTLAL[*], A. DEWANTO[*], A. H. CHAN[*,†] and C. H. OH[*,†]

[*]*Department of Physics,*
National University of Singapore Faculty of Science,
2 Science Drive 3 Singapore 117542
[†]*Institute of Advanced Studies,*
Nanyang Technological University, Singapore 639798

The Chou–Yang model has been successful in describing the forward-backward multiplicity distributions for Hadron–Hadron collisions. The model is extended to the case of Nucleus–Nucleus collisions where geometry is incorporated into the Generalized Multiplicity Distribution component. This contribution is investigated for forward-backward multiplicity measured at 130 GeV Au–Au collisions by calculating the correlation coefficient obtained with the model. Results are also produced for various oxygen-nucleus collisions. Finally the model is applied to the case of Pb–Pb collisions which can be extended to predict correlation coefficients in the TeV range.

Keywords: Chou–Yang correlations; multiplicity distribution.

1. Introduction

The Chou–Yang Model (CYM) was conceived by T. T. Chou and C. N. Yang to examine inelastic scattering in high energy collisions and their distributions. This model was in very good agreement with experiment, following which the model has also been successful at explaining other phenomena like diffraction dissociation, limiting fragmentation and multiparticle production.

T. T. Chou and C. N. Yang[1] studied the data from UA5 collaboration in terms of n_F and n_B, the number of forward and backward particles respectively. It was found that for collisions at 540 GeV, the distribution with respect to the charge asymmetry parameter $z = n_F - n_B$ is binomial. The sum of $n_F + n_B$ is given as n the total multiplicity. They evaluated the value of $\langle z^2 \rangle$ for fixed n using the preliminary data of UA5 for the central region of rapidity $|\Delta\eta| < 4$ and obtained a simple relationship

$$[\langle z^2 \rangle \text{ at fixed } n] = 2n. \tag{1}$$

This allows one to write the two-dimensional forward-backward multiplicity as

$$P(n, z) = \psi(n/\bar{n})C^{n/2}_{(n+z)/2}[B(n)]^{-1} \tag{2}$$

where $C^{n/2}_{(n+z)/2}$ is the binomial coefficient and $\psi(n/\bar{n})$ is the KNO scaling function and $[B(n)]^{-1}$ is the normalization constant. It was later shown that the KNO scaling function is not generally valid whereas the Negative Binomial Distribution (NBD) is a better fit to experimental data. It was shown by S. L. Lim et al.[2] by examining various forward-backward correlation data at CM energies of 24 GeV to 900 GeV that replacing the KNO scaling function with the NBD gave better fits to experimental data.

$$P^{\text{NBD}}(n, z) = p(n)C^{n/2}_{(n+2)/2}[B(n)]^{-1} \tag{3}$$

where the distributionis $p(n)$ given by the NBD

$$p(n) = \frac{\Gamma(n+k)}{\Gamma(n+1)\Gamma(k)}\left(\frac{k}{k+\bar{n}}\right)^k \left(\frac{\bar{n}}{k+\bar{n}}\right)^{\bar{n}}. \tag{4}$$

Here k is a free parameter, $\bar{n}$ is the mean charge-particle multiplicity and Γ is the usual gamma function.

2. Generalized Multiplicity Distribution

The Negative Binomial Distribution provides good fits to the limited pseudo-rapidity range and full phase space at 200, 546 and 900 GeV at small rapidity intervals. At high energies, the NBD is known not to describe the 900 GeV non-single-diffractive data well for large pseudo-rapidity intervals in the peak region of the distribution. The Generalized Multiplicity Distribution (GMD) was used by W. C. Lai et al.[3] as a replacement to the alternatives to NBD such as the double NBD[4] which are pure empirical fits where the parameters have no clear physical interpretations. The GMD used is as follows

$$P^{\text{GMD}}(\bar{n}, k', k) = \frac{\Gamma(n+k)}{\Gamma(n-k'+1)\Gamma(k'+k)}\left(\frac{\bar{n}-k'}{k+\bar{n}}\right)^{n-k'} \left(\frac{k'+k}{k+\bar{n}}\right)^{k'+k}. \tag{5}$$

Here k is interpreted as the quark contribution, k' is the gluon contribution.

The GMD can provide a better parameterization of charged particle multiplicity distribution at higher energies since it provides better fit to SPS energies than NBD does. The physical interpretations of the parameters k and k' also reduced to the case of e^+e^- experimental data where k, the quark contribution dominates and k' the gluon contribution is zero, which the NBD is known to describe well without considering the gluon term. W. C. Lai et al. produced good results for various ISR and SPS energies and also yielded a prediction of the correlation at 14 TeV for $p\bar{p}$ data. For studies at ISR and CERN, the produced charged particle multiplicities yielded a linear correlation at each energy

$$\langle n_B \rangle_{n_F} = a + bn_F. \tag{6}$$

The mean backward multiplicity at a particular n_F depends linearly on n_F. Here a and b are constants with a being interpreted as a measure of uncorrelated particles and b is the correlation coefficient. This distribution is extended in this paper to the case of Nucleus–Nucleus collisions.

3. Nucleus–Nucleus Collisions

Nucleus–Nucleus collisions are qualitatively different from Hadron–Hadron collisions in that geometry plays an important role in this type of collision, we cannot simply use the distribution given in Eq. (5) as the distribution with respect to n. This is because a small change in impact parameter greatly affects the number of participants in an instance of collision, as argued by W. Q. Chao et al.[5] Drawing from the works of A. H. Chan and C. K. Chew,[6] by considering how an incoming nucleon would carve out a tube of nucleons in a target nucleus, we are able to write down the contribution of the interacting region at particular impact parameter β. Using Cartesian coordinates with the z direction being the colliding axis, the total number of nucleon contribution is given by the volume of the cylinders carved on each nucleus by the other and can be obtained from the interaction volume depending on the region that the impact parameter falls in.

$$8 \int_0^\zeta dx \int_{\beta-\sqrt{R_B^2-x^2}}^{\sqrt{R_B^2-x^2}} dx \left[\int_0^{z_A} dz_A \times \int_0^{z_B} dz_B \right]. \tag{7}$$

Equation (7) describes the interaction region $R_A > \beta > R_B$, where R_A is the radius of the target and R_B is the radius of the projectile. Here $\zeta = \sqrt{R_A^2 - (\frac{R_A^2+\beta^2-R_B^2}{2\beta})^2}$.

$$8 \int_0^\zeta dx \int_{\beta-\sqrt{R_B^2-x^2}}^{\sqrt{R_A^2-x^2}} dy \left[\int_0^{Z_A} dz_A \times \int_0^{Z_B} dz_B \right] + 8 \int_\zeta^{R_B} dx \int_{\beta-\sqrt{R_B^2-x^2}}^{\beta+\sqrt{R_B^2-x^2}} dy \left[\int_0^{Z_A} dz_A \times \int_0^{Z_B} dz_B \right]. \tag{8}$$

Equation (8) describes the interaction region $R_A > \beta > R_B$.

$$8 \int_0^{R_B} dx \int_{\beta-\sqrt{R_B^2-x^2}}^{\beta+\sqrt{R_B^2-x^2}} dy \left[\int_0^{Z_A} dz_A \times \int_0^{z_B} dz_B \right]. \tag{9}$$

Equation (9) describes the interaction region $\beta < R_B < R_A$.

To obtain the expression for the total number of collisions as a function of impact parameter, the interacting volume is multiplied by σ_{nn}, the nucleon–nucleon cross-section.

$$N_T(\beta) = \sigma_{nn} \int dx \int dy \left[\rho_A dz_A \times \int \rho_B dz_B \right]. \tag{10}$$

This model involves the distribution arising from the collision to be convoluted over the number of collisions, meaning the final cross-sections obtained for given impact parameter are calculated by folding of the distributions for each pair of projectile-target nucleons. This distinguishes the model from others that only consider the number of participants rather than the number of nucleon–nucleon collisions. Finally, after impact parameter averaging, the scheme used by A. H. Chan

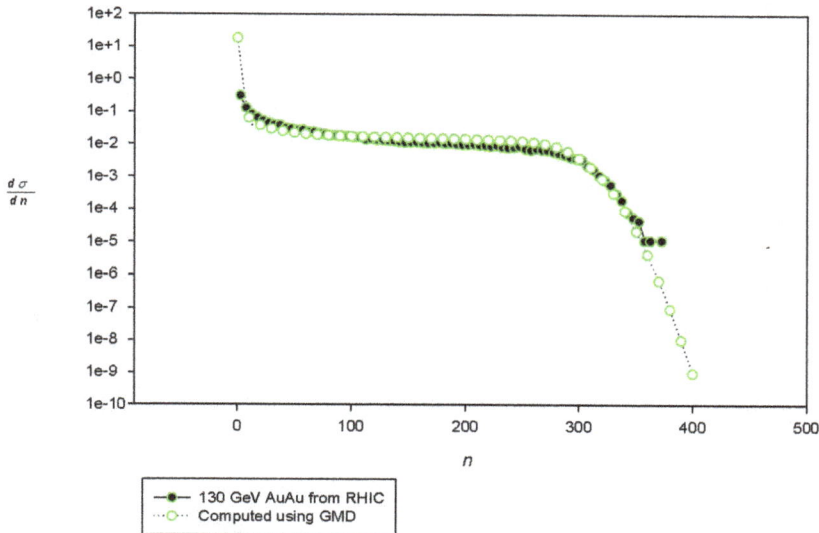

Fig. 1.　Au–Au H^- spectra compared with data at 130 GeV.[9]

and C. K. Chew to calculate the differential cross-section yields

$$
\frac{d\sigma}{dn} = \frac{2}{(R_A + R_B)^2} \int\limits_{0}^{R_A+R_B} \frac{\Gamma(n + k\frac{N_T}{\varepsilon})}{\Gamma(n - \frac{N_T}{\varepsilon}k' + 1)\Gamma(\frac{N_T}{\varepsilon}(k + k'))}
$$

$$
\times \left(\frac{\bar{n} - k'}{\bar{n} + k}\right)^{n - \frac{N_T}{\varepsilon}k'} \left(\frac{k' + k}{\bar{n} + k}\right)^{\frac{N_T}{\varepsilon}(k'+k)} \beta d\beta \tag{11}
$$

where ε is the sole adjustable parameter, which can be interpreted as the energy density according to G. Baym $et\ al.$[7] The parameter that gives the initial number of gluons is $k' = 0.5$ while the parameter related to the number of quarks is given by $1/k = -0.2256 + 0.1435 \ln(\bar{n})$ which is a parameterization of ISR energies.[8] This scheme has successfully described various Oxygen–Nuclei data at 60 and 200 GeV by A. H. Chan and C. K. Chew.[6]

3.1.　Results for multiplicity distributions using GMD

The above scheme is used to describe Au–Au negative Hadron spectra at 130 GeV data.[9]

All the features of the multiplicity data including the narrow neck, plateau and steep tail decline are reproduced well in this description using the GMD as the distribution. The scheme is also used for Pb–Pb collision data at 158 GeV.[10]

438

4. Chou–Yang Distribution with GMD

W. C. Lai *et al.* had already shown that the GMD was a good alternative for the NBD in describing multiplicity distributions for Hadron–Hadron collisions at various energies.

We modify the distribution that they used, to account for geometric considerations in Nucleus–Nucleus collisions yielding Eq. (12).

$$P^{\mathrm{GMD}}(n,z) = p_\sigma(n)C_{(n+z)/r}^{n/r}[B(n/r)]^{-1}. \tag{12}$$

Here the parameter r is introduced according the arguments given by S. L. Lim *et al.*[2] that the parameter in the binomial term might not be fixed at 2 as used by T. T. Chou and C. N. Yang in their 1984 paper. This parameter r, will be varied and interpreted as the size of intermediate clusters formed before hadronization. The distribution with respect to n contains the GMD

$$p_\sigma = \frac{\sigma(n)}{\sigma_T}. \tag{13}$$

And σ_T is the total cross-section as given by the Glauber theory. $\sigma(n)$ is obtained by integrating the differential cross-section in Eq. (11) over relevant values of n, after impact parameter averaging, hence:

$$\sigma(n) = \int_0^n \frac{d\sigma}{dn}dn. \tag{14}$$

4.1. *Correlation coefficients and cluster size r*

Using T. T. Chou and C. N. Yang's procedure to calculate the correlation coefficient using the distribution in Eq. (12), we are able to write down the average backward multiplicity at a fixed forward multiplicity by taking the summation over relevant n_B values as

$$\langle n_b \rangle_{n_f} = \frac{\sum_{nb} n_b P(r; n_f, n_b)}{\sum_{n_b} P(r; n_f, n_b)} \tag{15}$$

where the distribution is summed over even n_B for even n_F and odd n_B for odd n_F.

4.2. *Au–Au results and discussions*

The distributions obtained at various values of the parameters r yielded different correlations b. Linear results that agreed well with experimental linear relationship given by Eq. (6) were produced. Some of these plots are referenced in Fig. 2. Plotting the correlation coefficient b against r, yields Fig. 3.

It is noted that as cluster size parameter r is increased, the obtained correlation coefficient decreased. The increase in cluster size also increase the value of a, which can be interpreted as the number of uncorrelated particles. An estimate of the cluster size involved can be obtained by assuming that the correlation coefficient

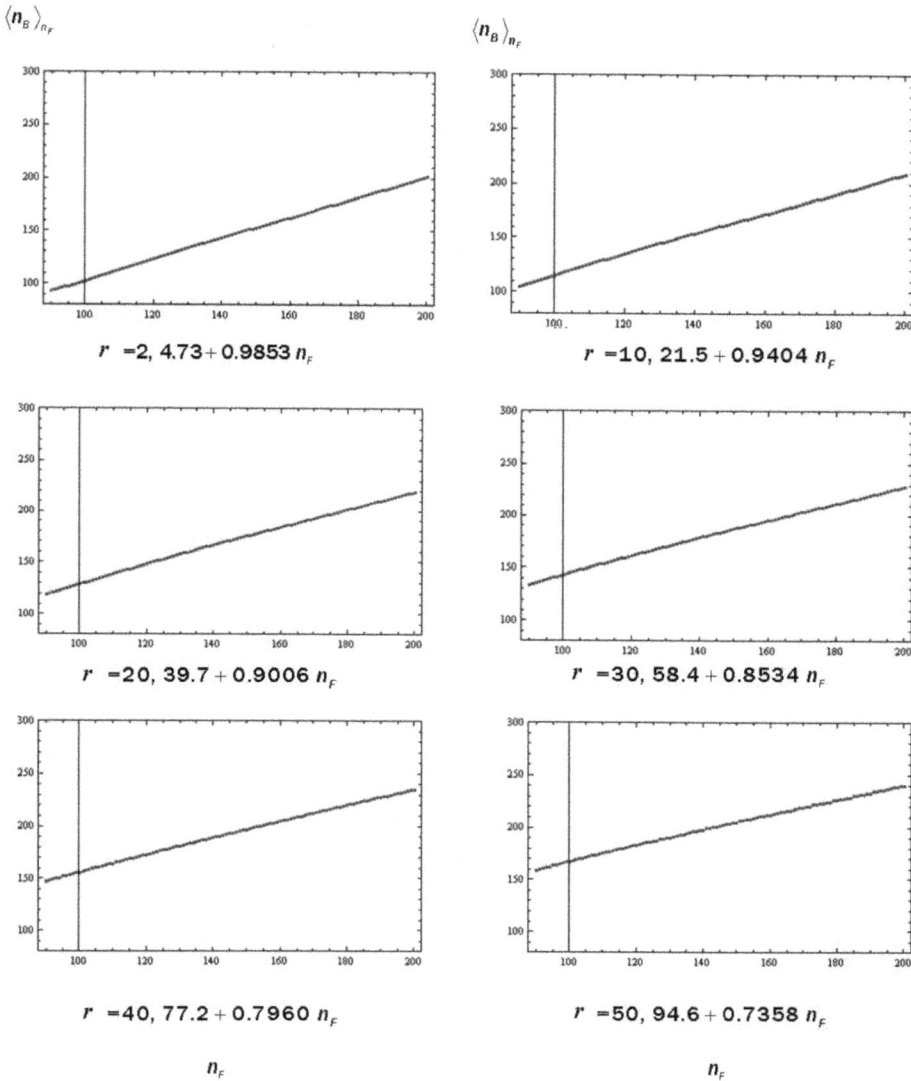

Fig. 2   Mathematica Plots of $\langle n_B \rangle_{n_F}$ vs n_F for various values of r in 130 GeV Au–Au collisions.

for this collision does not differ drastically from 200 GeV collisions obtained by STAR[11] which is used as a reference to the calculated values. Here it is estimated that the value of r that would yield the same correlation coefficient as those from 200 GeV data is around 80. An exact value would be able to be gleaned if the correlation coefficients were experimentally available for 130 GeV collisions, which are unfortunately unavailable. If the r could be parameterized, we would be able to extend this model to predictions for correlations coefficients for TeV energies.

440

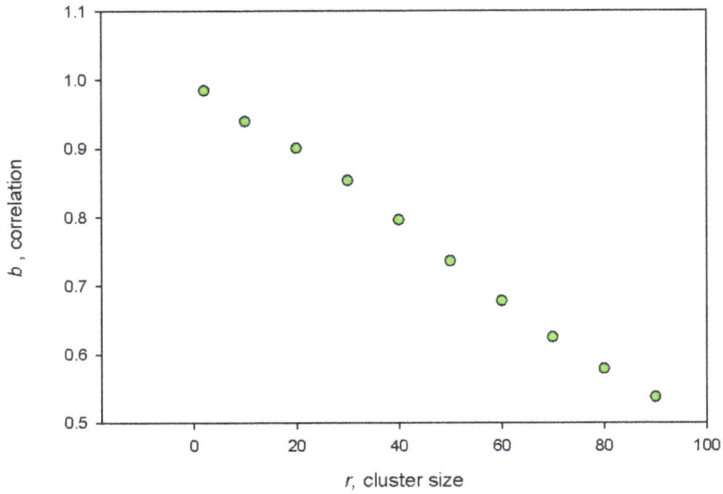

Fig. 3. Plots of b against r input for 130 GeV Au–Au collisions.

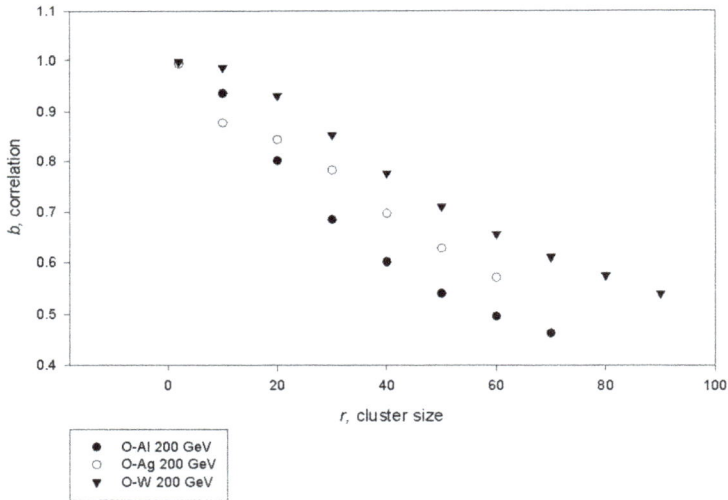

O-Al 200 GeV
O-Ag 200 GeV
O-W 200 GeV

Fig. 4. Correlation coefficients for various Oxygen–Nucleus collisions at 200 GeV.

4.3. O-Al/Ag/W results

Parameters taken from A.H. Chan and C. K. Chew's[12] treatment of O-Nuclei collisions at 200 GeV using GMD were also applied to this case of the CYM. Similar to the Au–Au case, linear results were obtained from the Chou–Yang distribution. Varying the parameter r and calculating the value of b at each r also yielded similar curves. Refer to Fig. 4 for the resulting distributions.

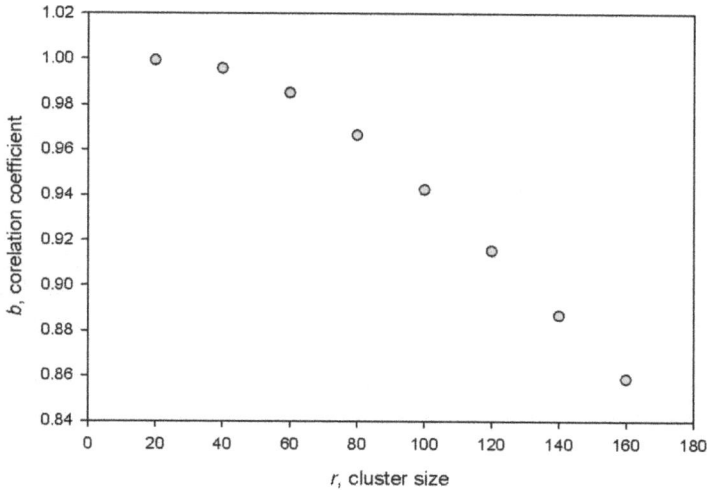

Fig. 5. Correlation coefficients for Pb–Pb collisions at 158 GeV.

As in the case of Au–Au collisions, it is noted that as cluster-size parameter increased, the obtained correlation coefficient decreased. It is also noted that for larger number of involved nucleons, the larger the parameter r required to obtain the same correlation b, which is an indication that r is dependent on the total energy of collision as the total energy depends on the average number collisions, since the energy of each collision is fixed (on average) per nucleon.

4.4. Pb–Pb results

Finally, the b–r distribution was produced for Pb–Pb collision at 158 GeV by fitting the GMD to the differential cross-section produced from fitting to NA 57 data. This can be found in Fig. 5.

It is noted that the distribution with respect to b is flatter than in other cases and is evaluated to be related to the spread in the multiplicity distribution with respect to n, which is also wider in this case than the other cases.

5. Cluster Size r

It was discussed in W. C. Lai *et al.* that the correlation b can be calculated analytically using

$$b = \frac{D^2(n) - \langle d_n^2(z) \rangle}{D^2(n) + \langle d_n^2(z) \rangle}.$$ (16)

Here $D^2(n)$ is the variance of $n = n_F + n_B$ and $\langle d_n^2(z) \rangle$ is the variance of $z = n_F - n_B$ at fixed values of n. It was also discussed that b is defined as the variance of the charge asymmetry variable z. Thus for the Hadron–Hadron case, it is possible to

442

directly obtain the correlation b on the assumption that the multiplicity obeys the GMD. Alternatively, r can be calculated for values of b according to

$$r = \frac{(\bar{n} + k)(\bar{n} - k')}{k + k'} \left(\frac{1 - b}{\bar{n}(1 + b)} \right) . \tag{17}$$

However, for Nucleus–Nucleus case, the distribution does not simply follow the GMD in n, but is built into the derivative of σ_n which is the differential cross-section. Hence Eq. (17) will not work in this case. In fact it will underestimate the cluster size at each correlation b, because $\bar{n}$ is the mean multiplicity which in the case of Hadron–Hadron collisions is just one collision, but varies according to impact parameter for Nucleus–Nucleus collisions. Hence r must be experimentally parameterized in the latter case to further the ability of this model to describe correlation coefficients. Compared to Hadron–Hadron cases, cluster sizes in Nucleus–Nucleus cases are much larger. The interpretation offered here is that during large scale collisions at higher energies, clusters formed are larger in size and hence have a larger value at each calculation for b. Further work could be explored in this area to glean a more complete understanding of this parameter in order to yield predictions for more massive Nucleus–Nucleus collisions or those at higher energies.

6. Summary

The GMD has been successfully used to describe multiplicity distributions of Au–Au H^- spectra at 130 GeV and Pb–Pb collisions at 158 GeV. The differential cross-sections obtained agreed well with experimental data from RHIC and NA 57 respectively. The distributions with the GMD are then integrated to obtain the total cross-section which is input as the distribution with respect to n in the Chou–Yang model. The average value of n_B was evaluated at fixed n_F to yield straight lines that agree well with experimental linear relationships. By varying the extra parameter r, interpreted as the cluster size, various values of b, the correlation were obtained for Au–Au, Pb–Pb and O–Nuclei collisions at 130, 158 and 200 GeV respectively. It was found that the cluster sizes obtained were of larger magnitudes to those in Hadron–Hadron collisions, due to the differences in spectra yielded by each type of collisions.

The GMD has a major advantage over other alternatives of the NBD because there are physical interpretations of the parameters k and k' which are the number of quarks and gluons respectively, whereas the alternatives have no physical interpretations of the extra parameters used. This is critical as gluon activities are expected to increase at higher energies.

The GMD appears to be successful in this attempt to describe Nucleus–Nucleus correlations and further works to parameterize the cluster size parameter can yield useful predictions at higher energies. However, like the NBD, the GMD is purely statistical in nature when used to describe multi-particle production. It is unable to account for the mechanics of hadronization which for example requires first principle derivation from QCD which is still unachievable at the moment.

Acknowledgment

This work is supported by NUS research grant, WBS: R-144-000-178-112.

References

1. T. T. Chou and C. N. Yang, *Phys. Lett.* **135**, 175 (1984).
2. S. L. Lim, Y. K. Lim, C. H. Oh and K. K. Phua, *Z. Phys. C. Part. Fields* **43**, 621 (1989).
3. W. C. Lai, A. H. Chan and C. H. Oh, *Int. J. Mod. Phys. A* **24**, 3552 (2009).
4. L. K. Chen, D. Kiang and C. K. Chew, *Phys. C* **76**, 263 (1997).
5. W. Q. Chao and L. Bo, *Nucl. Phys. A* **470**, 669 (1987).
6. A. H. Chan and C. K. Chew, *Il Nuovo Cimento* **104A**, 81 (1991).
7. G. Baym, P. Braun-Munzinger and V. Ruuskanen, *Phys. Rev. Lett. B* **190**, 29 (1987).
8. W. Thome *et al.*, *Nucl. Phys. B* **129**, 365 (1977).
9. C. Alder *et al.*, *Phys. Rev. Lett.* **87**, 112303 (2001).
10. F. Antinori *et al.*, *J. Phys. G. Nucl. Part. Phys.* **31**, 321 (2005).
11. B. J. Srivasta *et al.*, *Int. J. Mod. Phys. E* **16**, 3371 (2007).
12. A. H. Chan and C. K. Chew, *Il Nuovo Cimento* **105**, 941 (1992).
13. A. D. Jackson and H. Boggild, *Nucl. Phys. A* **470**, 669 (1987).

PADÉ APPROXIMATIONS AND NSQD'S UP TO 4-LOOP USING F_2 DEEP INELASTIC WORLD DATA

A. N. KHORRAMIAN* and H. KHANPOUR[†]

Physics Department, Semnan University, Semnan, Iran
School of Particles and Accelerators,
Institute for Research in Fundamental Science (IPM),
P.O. Box 19395-5531, Tehran, Iran
*Khorramiana@theory.ipm.ac.ir
[†]hamzeh_khanpour@nit.ac.ir

S. ATASHBAR TEHRANI

School of Particles and Accelerators,
Institute for Research in Fundamental Science (IPM),
P.O. Box 19395-5531, Tehran, Iran
Atashbar@ipm.ir

In this paper the nonsinglet quark distributions (NSQD's) from the precise next-to-next-to-next-to leading order QCD fit is presented. In this regards 4-loop anomalous dimension can be obtained from the Padé approximations. The analysis is based on the Jacobi polynomials expansion of the structure function. Our results for NSQD's up to N^3LO are in good agreement with available theoretical models.

Keywords: Perturbative QCD; non-singlet structure function; QCD fits.

1. Introduction

In our quest for the understanding of the structure of matter, scattering experiments have always played an outstanding role. DIS processes have played and still play a very important role for our understanding of QCD and nucleon structure. For quantitatively reliable predictions of DIS and hard hadronic scattering processes, perturbative QCD corrections at the N^2LO and the next-to-next-to-next-to-leading order (N^3LO) need to be taken into account. Based on our experience obtained in a series of LO, NLO and N^2LO analysis[1] of NSQD's, here we extend our work to N^3LO accuracy in perturbative QCD. In this QCD analysis, we apply the method of the structure function reconstruction over their Mellin moments, which is based on the expansion of the structure function in terms of Jacobi polynomials. This method was developed and applied for different QCD analyses in Refs. [2–18]. The same method has also been applied in the polarized case in Refs. [19–24]. In the present paper we perform a QCD analysis of the flavor nonsinglet unpolarized deep–inelastic charged $e(\mu)p$ and $e(\mu)d$ world data[26–30] at N^3LO and derived parameterizations of valence quark distributions $x u_v(x, Q^2)$ and $x d_v(x, Q^2)$ at a starting scale Q_0^2 together with the QCD–scale Λ_{QCD} by using the Jacobi polynomial expansions. We have therefore used the 3-loop splitting functions and Padé approximations[31–35] for the evolution of NSQD's of hadrons.

The organization of this paper is as follows. Section 2 introduces theoretical formalism in the context of deep inelastic scattering to higher order for nonsinglet sector of proton structure function F_2. A description of the Jacobi polynomials is illustrated in section 3. Our parametrization for NSQD's, procedure of the QCD fit of F_2 data and numerical results are illustrated in section 4.

2. Theoretical Formalism

The combinations of parton densities in the nonsinglet regime and the valence region $x \geq 0.3$ for F_2^p in LO is

$$F_2^p(x, Q^2) = \frac{4}{9} x\, u_v(x, Q^2) + \frac{1}{9} x\, d_v(x, Q^2) \ . \tag{1}$$

In the region $x \geq 0.3$, sea quarks can be neglected. Also in this region, the combinations of parton densities for F_2^d are also given by

$$F_2^d(x, Q^2) = \frac{5}{18} x(u_v + d_v)(x, Q^2) \ , \tag{2}$$

where $F_2^d = (F_2^p + F_2^n)/2$ if we ignore the nuclear effects here.

In the region $x \leq 0.3$ for the difference of the proton and deuteron data we use

$$
\begin{aligned}
F_2^{NS}(x, Q^2) &\equiv 2(F_2^p - F_2^d)(x, Q^2) \\
&= \frac{1}{3} x(u_v - d_v)(x, Q^2) + \frac{2}{3} x(\bar{u} - \bar{d})(x, Q^2) \ ,
\end{aligned}
\tag{3}
$$

since sea quarks can not be neglected for x smaller than about 0.3.

The first clear evidence for the flavor asymmetry combination of light parton distributions $x(\bar{d} - \bar{u})$ in nature came from the analysis of NMC at CERN to study of the Gottfried sum rule.[36] In our calculation we supposed the $\bar{d} - \bar{u}$ distribution[37–40] at $Q_0^2 = 4$ GeV^2 which gives a good description of the Drell-Yan dimuon production data.[41]

By using the solution of the nonsinglet evolution equation for the parton densities up to 4– loop order, the nonsinglet structure functions are given by[38]

$$
F_2^k(N, Q^2) = \left(1 + a_s\, C_{2,\mathrm{NS}}^{(1)}(N) + a_s^2\, C_{2,\mathrm{NS}}^{(2)}(N) + a_s^3\, C_{2,\mathrm{NS}}^{(3)}(N)\right) F_2^k(N, Q_0^2)
$$

$$
\times \left(\frac{a_s}{a_0}\right)^{-\hat{P}_0(N)/\beta_0} \left\{1 - \frac{1}{\beta_0}(a_s - a_0)\left[\hat{P}_1^+(N) - \frac{\beta_1}{\beta_0}\hat{P}_0(N)\right]\right.
$$

446

$$-\frac{1}{2\beta_0}\left(a_s^2 - a_0^2\right)\left[\hat{P}_2^+(N) - \frac{\beta_1}{\beta_0}\hat{P}_1^+(N) + \left(\frac{\beta_1^2}{\beta_0^2} - \frac{\beta_2}{\beta_0}\right)\hat{P}_0(N)\right]$$

$$+\frac{1}{2\beta_0^2}\left(a_s - a_0\right)^2\left(\hat{P}_1^+(N) - \frac{\beta_1}{\beta_0}\hat{P}_0(N)\right)^2$$

$$-\frac{1}{3\beta_0}\left(a_s^3 - a_0^3\right)\left[\hat{P}_3^+(N) - \frac{\beta_1}{\beta_0}\hat{P}_2^+(N) + \left(\frac{\beta_1^2}{\beta_0^2} - \frac{\beta_2}{\beta_0}\right)\hat{P}_1^+(N)\right.$$

$$+\left(\frac{\beta_1^3}{\beta_0^3} - 2\frac{\beta_1\beta_2}{\beta_0^2} + \frac{\beta_3}{\beta_0}\right)\hat{P}_0(N)\Bigg]$$

$$+\frac{1}{2\beta_0^2}\left(a_s - a_0\right)\left(a_0^2 - a_s^2\right)\left(\hat{P}_1^+(N) - \frac{\beta_1}{\beta_0}\hat{P}_0(N)\right)$$

$$\times\left[\hat{P}_2(N) - \frac{\beta_1}{\beta_0}\hat{P}_1(N) - \left(\frac{\beta_1^2}{\beta_0^2} - \frac{\beta_2}{\beta_0}\right)\hat{P}_0(N)\right]$$

$$-\frac{1}{6\beta_0^3}\left(a_s - a_0\right)^3\left(\hat{P}_1^+(N) - \frac{\beta_1}{\beta_0}\hat{P}_0(N)\right)^3\Bigg\}. \tag{4}$$

Here $a_s(= \alpha_s/4\pi)$ and a_0 denotes the strong coupling constant in the scale of Q^2 and Q_0^2 respectively. $k = p, d$ and NS also denotes the three above cases, i.e. proton, deuteron and nonsinglet structure function. $C_{2,NS}^{(m)}(N)$ are the nonsinglet Wilson coefficients in $O(a_s^m)$ which can be found in Refs. [42–44] and $\hat{P}_m$ denote also the Mellin transforms of the $(m+1)-$ loop splitting functions.

In spite of the unknown 4-loop anomalous dimensions, one can obtain the nonsinglet parton distributions and Λ_{QCD} by estimating uncalculated fourth-order corrections to the nonsinglet anomalous dimension. On the other hand the 3–loop Wilson coefficients are known[44] and now it is possible to know, which effect has the 4-loop anomalous dimension if compared to the Wilson coefficient. In this case the 4-loop anomalous dimension may be obtained from Padé approximations. According to Padé approximations, it is easy to obtain the results for $\hat{P}_m^{[1/1]}(N)$ or $\hat{P}_m^{[0/2]}(N)$ for the $(m+1)$-th order quantities $\hat{P}_{(m+1)}$.[25]

3. The Method

According to Jacobi polynomials expansion method, one can relate the F_2 structure function with its Mellin moments

$$F_2^{k,N_{max}}(x, Q^2) = x^\beta(1-x)^\alpha\sum_{n=0}^{N_{max}}\Theta_n^{\alpha,\beta}(x)\sum_{j=0}^{n}c_j^{(n)}(\alpha,\beta)F_2^k(j+2, Q^2), \tag{5}$$

where N_{max} is the number of polynomials, k denotes the three cases, i.e. $k = p, d, NS$. In the above, $c_j^{(n)}(\alpha,\beta)$ are the coefficients expressed through Γ-functions and satisfying the orthogonality relation and $F_2(j+2, Q^2)$ are the moments determined in the previous section. N_{max}, α and β have to be chosen so as to achieve the fastest convergence of the series on the right-hand side of Eq. (5) and to reconstruct F_2 with the required accuracy. In our analysis we use $N_{max} = 9$, $\alpha = 3.0$ and $\beta = 0.5$. The same method has been applied to calculate the nonsinglet structure function xF_3 from their moments[7-10] and for polarized structure function xg_1.[19-21] Obviously the Q^2-dependence of the polarized structure function is defined by the Q^2-dependence of the moments.

4. Parametrization

In the present QCD analysis, we choose the following parametrization

$$xq_v(x, Q_0^2) = \mathcal{N}_q \, x^{a_q} (1 - x)^{b_q} (1 + c_q \sqrt{x} + d_q \, x) , \tag{6}$$

for the valence quark densities in the input scale of $Q_0^2 = 4$ GeV^2. Here $q = u, d$ and the normalization factors $\mathcal{N}_u$ and $\mathcal{N}_d$ are fixed by $\int_0^1 u_v dx = 2$ and $\int_0^1 d_v dx = 1$, respectively.

By QCD fits of the world data for $F_2^{p,d}$, we can extract valence quark densities using the Jacobi polynomials method. For the nonsinglet QCD analysis presented in this paper we use the structure function data measured in charged lepton-proton and deuteron deep-inelastic scattering. The experiments contributing to the statistics are BCDMS,[26] SLAC,[27] NMC,[28] H1,[29] and ZEUS.[30] In our QCD analysis we use three data samples : $F_2^p(x, Q^2)$, $F_2^d(x, Q^2)$ in the nonsinglet regime and the valence quark region $x \geq 0.3$ and $F_2^{NS} = 2(F_2^p - F_2^d)$ in the region $x < 0.3$.

The valence quark region may be parameterized by the nonsinglet combinations of parton distributions, which are expressed through the parton distributions of valence quarks. Only data with $Q^2 > 4$ GeV^2 were included in the analysis and a cut in the hadronic mass of $W^2 \equiv \left(\frac{1}{x} - 1\right) Q^2 + m_N^2 > 12.5$ GeV^2 was applied in order to widely eliminate higher twist (HT) effects from the data samples. After these cuts we are left with 762 data points, 322 for F_2^p, 232 for F_2^d, and 208 for F_2^{NS}. By considering the additional cuts on the BCDMS ($y > 0.35$) and on the NMC data($Q^2 > 8$ GeV^2) the total number of data points available for the analysis reduce from 762 to 551, because we have 227 data points for F_2^p, 159 for F_2^d, and 165 for F_2^{NS}.

Now the sums in χ^2_{global} run over all data sets and in each data set over all data points. The minimization of the χ^2 value to determine the best parametrization of the unpolarized parton distributions is done using the program MINUIT.[45]

In Table 1 we summarize the N^{2}LO,[1] N^{3}LO with using Padé [1/1] fit results for the parameters of the parton densities $xu_v(x, Q_0^2)$, $xd_v(x, Q_0^2)$ and $\Lambda_{\text{QCD}}^{N_f=4}$. The values without error have been fixed after a first minimization since the data do not constrain these parameters well enough.

In Fig. 1 we show the evolution of the valence quark distributions $xu_v(x, Q^2)$ and $xd_v(x, Q^2)$ from $Q^2 = 1$ GeV^2 to $Q^2 = 10^4$ GeV^2 in the region $x \in [10^{-4}, 1]$ at N^{3}LO. In this figure we also compared our results with the nonsinglet QCD analysis from Ref. [38].

Table 1. Parameter values of the N^2LO from Ref. [1] and N^3LO nonsinglet QCD fit at $Q_0^2 = 4$ GeV^2 for Padé [1/1].

q_v	N^2LO	N^3LO
a_u	0.7434 ± 0.009	0.7772 ± 0.009
b_u	4.0034 ± 0.033	4.02637 ± 0.0402
c_u	0.1000	0.0940
d_u	1.1400	1.1100
a_d	0.7858 ± 0.043	0.80927 ± 0.0621
b_d	3.6336 ± 0.244	3.76847 ± 0.3499
c_d	0.3899	0.1838
d_d	-1.2152	-1.1200
$\Lambda_{QCD}^{N_f=4}$, MeV	239.9 ± 27	241.44 ± 29
χ^2/ndf	$506/546 = 0.9267$	$491.07/546 = 0.8994$

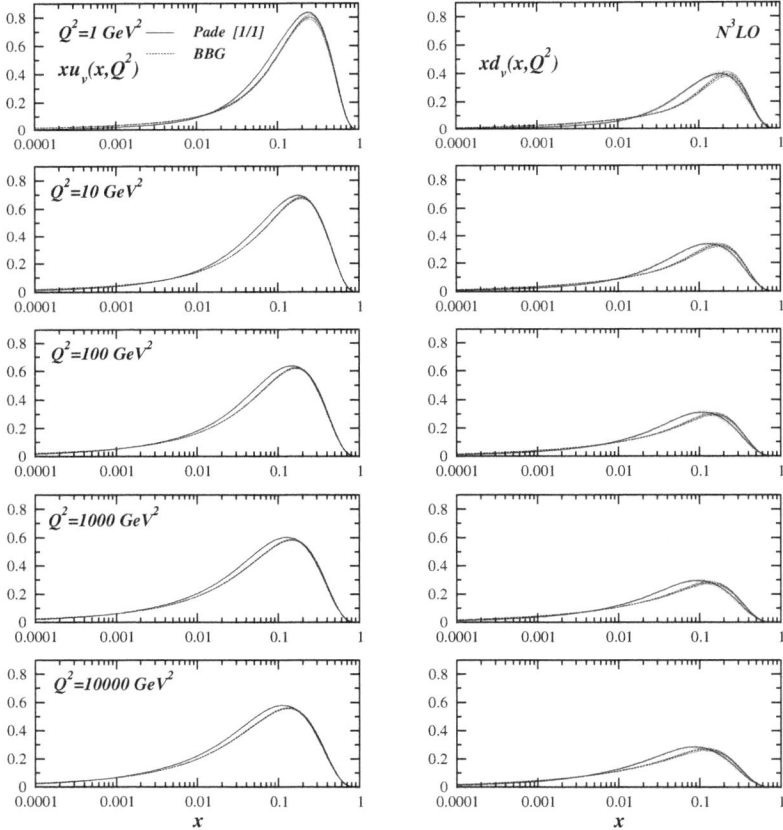

Fig. 1. The parton densities xu_v and xd_v at N^3LO evolved up to $Q^2 = 10000$ GeV^2 (solid lines) compared with results obtained by BBG[38] (dashed line).

References

1. A. N. Khorramian and S. A. Tehrani, Phys. Rev. D **78**, 074019 (2008) [arXiv:0805.3063 [hep-ph]].
2. G. Parisi and N. Sourlas, *Nucl. Phys.* **B151** (1979) 421;
   I. S. Barker, C. B. Langensiepen and G. Shaw, *Nucl. Phys.* **B186** (1981) 61.
3. V. G. Krivokhizhin, S. P. Kurlovich, V. V. Sanadze, I. A. Savin, A. V. Sidorov and N. B. Skachkov, Z. Phys. C **36** (1987) 51.
4. V. G. Krivokhizhin *et al.*, Z. Phys. C **48**, 347 (1990).
5. J. Chyla and J. Rames, Z. Phys. C **31** (1986) 151.
6. I. S. Barker, C. S. Langensiepen and G. Shaw, Nucl. Phys. B **186** (1981) 61.
7. A. L. Kataev, A. V. Kotikov, G. Parente and A. V. Sidorov, Phys. Lett. B **417**, (1998) 374 [arXiv:hep-ph/9706534].
8. A. L. Kataev, G. Parente and A. V. Sidorov, arXiv:hep-ph/9809500.
9. A. L. Kataev, G. Parente and A. V. Sidorov, Nucl. Phys. B **573**, (2000) 405 [arXiv:hep-ph/9905310].
10. A. L. Kataev, G. Parente and A. V. Sidorov, Phys. Part. Nucl. **34**, (2003) 20 [arXiv:hep-ph/0106221];
    A. L. Kataev, G. Parente and A. V. Sidorov, Nucl. Phys. Proc. Suppl. **116** (2003) 105 [arXiv:hep-ph/0211151].
11. A. N. Khorramian, S. Atashbar Tehrani and M. Ghominejad, Acta Phys. Polon. B **38**, 3551 (2007).
12. A. N. Khorramian and S. A. Tehrani, J. Phys. Conf. Ser. **110**, 022022 (2008).
13. A. N. Khorramian and S. A. Tehrani, AIP Conf. Proc. **1006** (2008) 118.
14. S. Atashbar Tehrani and A. N. Khorramian, Nucl. Phys. Proc. Suppl. **186**, 58 (2009).
15. A. N. Khorramian, S. Atashbar Tehrani, H. Khanpour and S. Taheri Monfared, Hyperfine Interactions **194**, 337 (2009).
16. A. N. Khorramian, S. Atashbar Tehrani, M. Soleymaninia. and S. Batebi, Hyperfine Interactions **194**, 341 (2009).
17. S. Atashbar Tehrani and A. N. Khorramian, Hyperfine Interactions **194**, 331 (2009).
18. S. Atashbar Tehrani and A. N. Khorramian, Applied Mathematics & Information Sciences (2009), 367-373.
19. E. Leader, A. V. Sidorov and D. B. Stamenov, Int. J. Mod. Phys. A **13**, 5573 (1998) [arXiv:hep-ph/9708335].
20. S. Atashbar Tehrani and A. N. Khorramian, JHEP **0707**, 048 (2007) [arXiv:0705.2647 [hep-ph]].
21. A. N. Khorramian and S. Atashbar Tehrani, arXiv:0712.2373 [hep-ph].
22. A. N. Khorramian and S. Atashbar Tehrani, AIP Conf. Proc. **915**, 420 (2007).
23. A. Mirjalili, A. N. Khorramian and S. Atashbar-Tehrani, Nucl. Phys. Proc. Suppl. **164**, 38 (2007).
24. A. Mirjalili, S. Atashbar Tehrani and A. N. Khorramian, Int. J. Mod. Phys. A **21**, 4599 (2006) [arXiv:hep-ph/0608224].
25. A. N. Khorramian, H. Khanpour and S. A. Tehrani, Phys. Rev. D **81**, 014013 (2010) [arXiv:0909.2665 [hep-ph]].
26. A.C. Benvenuti *et al.* [BCDMS Collaboration], Phys. Lett. B **237** (1990) 592;
    A.C. Benvenuti *et al.* [BCDMS Collaboration], Phys. Lett. **B223** (1989) 485; Phys. Lett. **B237** (1990) 592.
    A.C. Benvenuti *et al.* [BCDMS Collaboration], Phys. Lett. B **237** (1990) 599.
27. L. W. Whitlow, E. M. Riordan, S. Dasu, S. Rock and A. Bodek, Phys. Lett. B **282** (1992).

28. M. Arneodo *et al.* [New Muon Collaboration], Nucl. Phys. B **483** (1997) 3 [arXiv:hep-ph/9610231].

29. C. Adloff *et al.* [H1 Collaboration], Eur. Phys. J. C **21** (2001) 33 [arXiv:hep-ex/0012053];
    C. Adloff *et al.* [H1 Collaboration], Eur. Phys. J. C **30** (2003) 1 [arXiv:hep-ex/0304003].

30. J. Breitweg *et al.* [ZEUS Collaboration], Eur. Phys. J. C **7** (1999) 609 [arXiv:hep-ex/9809005];
    S. Chekanov *et al.* [ZEUS Collaboration], Eur. Phys. J. C **21** (2001) 443 [arXiv:hep-ex/0105090].

31. M.A. Samuel, J. Ellis and M. Karliner, Phys. Rev. Lett. 74 (1995) 4380

32. J. Ellis, E. Gardi, M. Karliner and M.A. Samuel, Phys. Lett. B366 (1996) 268

33. J. Ellis, E. Gardi, M. Karliner and M.A. Samuel, Phys. Rev. D54 (1996) 6986

34. G.A. Baker, Jr. *Essentials of Padé Approximants*, Academic Press, 1975.

35. C.M. Bender and S.A. Orszag, *Advanced Mathematical Methods for Scientists and Engineers*, McGraw-Hill, 1978.

36. P. Amaudruz *et al.* [New Muon Collaboration], Phys. Rev. Lett. **66**, 2712 (1991); M. Arneodo *et al.* [New Muon Collaboration], Phys. Rev. D **50**, 1 (1994); M. Arneodo *et al.* [New Muon Collaboration], Nucl. Phys. B **487**, 3 (1997) [arXiv:hep-ex/9611022].

37. M. Glück, E. Reya and C. Schuck, arXiv:hep-ph/0604116.

38. J. Blumlein, H. Bottcher and A. Guffanti, Nucl. Phys. B **774**, 182 (2007) [arXiv:hep-ph/0607200].

39. A.D. Martin et al., *Eur. Phys. J.* **C23** (2002) 73.

40. J. Blümlein, H, Böttcher, and A. Guffanti, *Nucl. Phys. B* (Proc. Suppl.) **135** (2004) 152.

41. R.S. Towell et al., E866 Collab., *Phys. Rev.* **D64** (2001) 052002.

42. W. Furmanski and R. Petronzio, Z. Phys. C **11** (1982) 293.

43. W. L. van Neerven and E. B. Zijlstra, Phys. Lett. B **272** (1991) 127;
    E. B. Zijlstra and W. L. van Neerven, Nucl. Phys. B **383** (1992) 525.

44. J. A. M. Vermaseren, A. Vogt and S. Moch, Nucl. Phys. B **724** (2005) 3 [arXiv:hep-ph/0504242].

45. F. James, CERN Program Library, Long Writeup D506 (MINUIT).

A PHENOMENOLOGICAL ANALYSIS OF THE LONGITUDINAL PROTON STRUCTURE FUNCTION F_L

M. SOLEYMANINIA

Physics Department, Semnan University, Semnan, Iran
Maryam.Soleimaninia@gmail.com

A. N. KHORRAMIAN* and S. ATASHBAR TEHRANI†

Physics Department, Semnan University, Semnan, Iran
School of Particles and Accelerators,
Institute for Research in Fundamental Science (IPM),
P.O. Box 19395-5531, Tehran, Iran
* Khorramiana@theory.ipm.ac.ir
† Atashbar@ipm.ir

The study about the $O(\alpha_s^2)$ corrections to the massless quarks and heavy flavor contributions to the longitudinal structure function $F_L(x, Q^2)$ of proton is presented. We use the heavy flavor coefficient functions and massive operator matrix elements in unpolarized deeply inelastic scattering in the region $Q^2 \gg m^2$. This representation is carried in Mellin space QCD evolution programs. Numerical illustrations for the longitudinal structure function are compared with available experimental data.

Keywords: Perturbative QCD; longitudinal structure function; coefficient function; heavy quark; distribution function.

1. Introduction

The ep collider at HERA has played a crucial role in understanding of the proton structure. It operated from 1992 until 2007. Also the process of deep inelastic scattering has been the key for basic information on the internal structure of nucleons. Measurement of the cross section of DIS were performed at HERA by both H1 and ZEUS collaborations. The results revealed that the proton structure is described by perturbative QCD. H1 and ZEUS have made the first direct measurements of the longitudinal structure function F_L.[1-3] The inclusive e^-p scattering cross section, σ^{e^-p}, can be expressed in terms of the two structure functions, F_2 and F_L

$$\frac{d^2\sigma}{dxdQ^2} = \frac{2\pi\alpha^2}{xQ^4}\left(Y_+ F_2(x, Q^2) - y^2 F_L(x, Q^2)\right), \qquad Y_+ = 1 + (1-y)^2. \quad (1)$$

Quark parton model (QPM) predicts $\sigma_L = 0$, which leads to the so-called Callan-Gross relation $F_L = 0$. The naive QPM has to be modified in QCD as quarks interact through gluons. Thus, in QCD the longitudinal structure function F_L is non-zero.[4]

The large gluon density, determined from scaling violation of the F_2 data using QCD fits, implies that the structure function F_L must be significant at low x. Due to its origin, F_L is directly dependent on the gluon distribution in the proton and therefore the measurement of F_L provides a sensitive test of perturbative QCD.

We need to use corresponding massless Wilson coefficients in leading (LO) and next-to-leading order (NLO)[5-7] but it was calculated up to next-to-next-to leading order (NNLO).[8-10] Since the longitudinal structure function $F_L(x, Q^2)$ contains rather large heavy flavor contributions in the small x region, we calculate heavy contributions of longitudinal structure function in leading order and next-to-leading order by using massive Wilson coefficients in the asymptotic region $Q^2 \gg m^2$. All logarithmic terms and the constant term of the heavy flavor Wilson coefficients are obtained due to a factorization of this quantity into the massive operator matrix elements.

The outline of this article is as follows. In Section 2 we briefly recall the formalism, based on perturbative QCD. In Section 3 we study heavy quark contributions for $F_L^{Q\bar{Q}}(x, Q^2)$ to $O(\alpha_s^2)$ in the rejoin $Q^2 \gg m^2$. Numerical results and conclusion also are summarized in final Sections.

2. General QCD Formalism

In QCD corrections the longitudinal structure function, to all orders in perturbation theory due to the factorization theorem, is defined as Mellin convolution between distribution functions $f_j(x, \mu^2)$ and Wilson coefficient functions $C_L^j(x, \frac{Q^2}{\mu^2})$[11,12]

$$F_L(x, Q^2) = \sum_j C_L^j\left(x, \frac{Q^2}{\mu^2}\right) \otimes f_j(x, \mu^2) , \qquad (2)$$

Here $j = NS, S, g$ and μ^2 denotes the factorization scale and the Mellin convolution is given by the integral

$$[A \otimes B](x) = \int_0^1 dx_1 \int_0^1 dx_2 \, \delta(x - x_1 x_2) \, A(x_1) B(x_2) . \qquad (3)$$

For determination of the parton density functions, they are parameterized by smooth analytical functions at a low starting scale Q_0^2 as a function of x with a certain number of free parameters. In the present paper, we use two different kinds of initial distribution functions. The first scenario is included the two six-parameter standard forms[13]

$$x f_i(x, \mu_0^2) = N_i \, p_{i,1} \, x^{p_{i,2}} (1-x)^{p_{i,3}} \left[1 + p_{i,5} \, x^{p_{i,4}} + p_{i,6} \, x\right] , \qquad (4)$$

and

$$x f_i(x, \mu_0^2) = N_i \, p_{i,1} \, x^{p_{i,2}} (1-x)^{p_{i,3}} \left[1 + p_{i,4} \, x^{0.5} + p_{i,5} \, x + p_{i,6} \, x^{1.5}\right] . \qquad (5)$$

Then the parametrization for valence quarks, gluon and sea quarks is

$$xu_v(x, \mu_0^2) = 5.107200\, x^{0.8}\, (1-x)^3\ ,$$
$$xd_v(x, \mu_0^2) = 3.064320\, x^{0.8}\, (1-x)^4\ ,$$
$$xg\,(x, \mu_0^2) = 1.700000\, x^{-0.1}(1-x)^5\ ,$$
$$x\bar{d}\,(x, \mu_0^2) = 0.1939875\, x^{-0.1}(1-x)^6\ ,$$
$$x\bar{u}\,(x, \mu_0^2) = (1-x)\, x\bar{d}\,(x, \mu_0^2)\ ,$$
$$xs\,(x, \mu_0^2) = x\bar{s}\,(x, \mu_0^2)\ =\ 0.2\,x(\bar{u}+\bar{d})(x, \mu_0^2)\ , \tag{6}$$

with

$$\alpha_s(Q^2 = 2\ \text{GeV}^2)\ =\ 0.35\ . \tag{7}$$

Having very few amounts of experimental data for longitudinal structure function, we cannot perform a fit to the existing data. According to the fact that parton distribution functions are universal, we can use $F_2(x, Q^2)$ experimental data and form initial distribution functions.[14–17] Base on that we predict longitudinal structure function F_L. In second scenario, longitudinal structure function is predicted with parametrization form that are gotten from F_2 experimental data[18]

$$xu_v(x, Q_0^2) = N_u x^{\alpha_u} (1-x)^{\beta_u} (1 + \gamma_u \sqrt{x} + \eta_u x)\ ,$$
$$xd_v(x, Q_0^2) = N_d x^{\alpha_d} (1-x)^{\beta_d} (1 + \gamma_d \sqrt{x} + \eta_d x)\ ,$$
$$x\Delta(x, Q_0^2) = N_\Delta x^{\alpha_\Delta} (1-x)^{\beta_\Delta} (1 + \gamma_\Delta \sqrt{x} + \eta_\Delta x)\ ,$$
$$x\Sigma(x, Q_0^2) = N_\Sigma x^{\alpha_\Sigma} (1-x)^{\beta_\Sigma} (1 + \gamma_\Sigma \sqrt{x} + \eta_\Sigma x)\ ,$$
$$xg(x, Q_0^2) = N_g x^{\alpha_g} (1-x)^{\beta_g}\ . \tag{8}$$

Then the parton distributions at higher scales are obtained by evolution.

3. Heavy Quark Contributions

One of the important areas of research at accelerators is the study of heavy flavor production. Heavy flavors can be produced in electron-positron, hadron-hadron, photon-hadron and lepton-hadron interactions. We concentrate on the latter and in particular on electron-proton collisions which investigate experimentally at HERA and recently at LHC. The data for heavy quark (c, b) production, being theoretically described in the fixed–flavor number factorization scheme by the fully predictive fixed–order perturbation theory, that it was calculated for F_2^c and F_2^b.[19–21] The study heavy flavor production in deep inelastic electron-proton scattering is given by the reaction

$$e^-(l_1) + P(p) \rightarrow e^-(l_2) + Q(p_1)\,(\bar{Q}(p_2)) + X\ , \tag{9}$$

X stands for any final hadronic state.

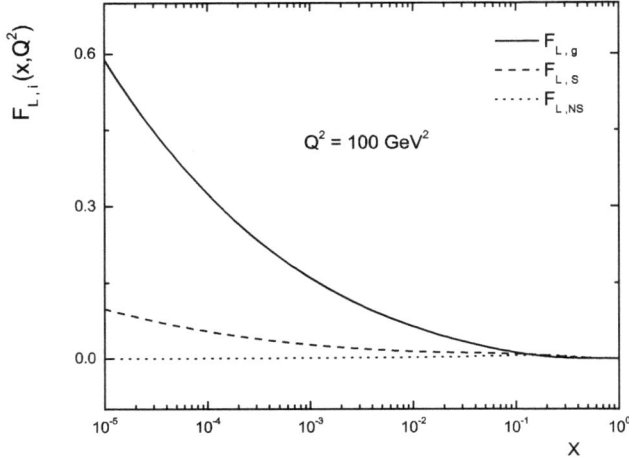

Fig. 1.   The gluon, singlet and non-singlet contributions due to scenario 2 to $F_L(x, Q^2)$ at $O(\alpha_s^2)$.

The longitudinal structure function $F_L(x, Q^2)$ consists of three parts that it is shown in Fig. 1

$$F_L(x, Q^2) = C_L^{\rm NS}\left(x, a_s, \frac{Q^2}{\mu^2}\right) \otimes q_{\rm NS}(x, \mu^2) + C_L^{\rm S}\left(x, a_s, \frac{Q^2}{\mu^2}\right) \otimes q_{\rm S}(x, \mu^2)$$
$$+ C_L^g\left(x, a_s, \frac{Q^2}{\mu^2}\right) \otimes g(x, \mu^2) \ . \tag{10}$$

Heavy quarks (c, b, t) will not be considered as partons and the heavy flavor effects are contained in the Wilson coefficients only

$$C_L^{(i)}\left(x, a_s, \frac{Q^2}{\mu^2}\right) = C_L^{(i),{\rm light}}(x, a_s) + H_L^{(i),{\rm heavy}}\left(x, a_s, \frac{Q^2}{m^2}\right), \quad i = {\rm S, NS}, g \ . \tag{11}$$

This defines the so–called 'fixed–flavor number scheme' (FFNS). We are interested in the massive contributions in the region $Q^2 \gg m^2$.

As already noted, the LO and NLO heavy quark contributions $F_L^{c,b}$ are calculated in the FFNS and contribute to the total structure function as

$$F_L(x, Q^2) = F_L^{\rm light} + F_L^{\rm heavy} \ , \tag{12}$$

where 'light' refers to the common u, d, s (anti)quarks and gluon initiated contributions,[22] and $F_L^{\rm heavy} = F_L^c + F_L^b$. The light and heavy flavor contributions to $F_L(x, Q^2)$ is shown in Fig. 2. The perturbative predictions for $F_L(x, Q^2)$ can be written as[23,24]

$$x^{-1}F_L = C_{L,ns} \otimes q_{ns} + \frac{2}{9}\left(C_{L,q} \otimes q_s + C_{L,g} \otimes g\right) + x^{-1}F_L^c \ , \tag{13}$$

where $\otimes$ in the $n_f = 3$ light quark flavor sector denotes the common convolution, q_{ns} stands for the usual flavor non–singlet combination and $q_s = \sum_{q=u,d,s}(q + \bar{q})$ is the

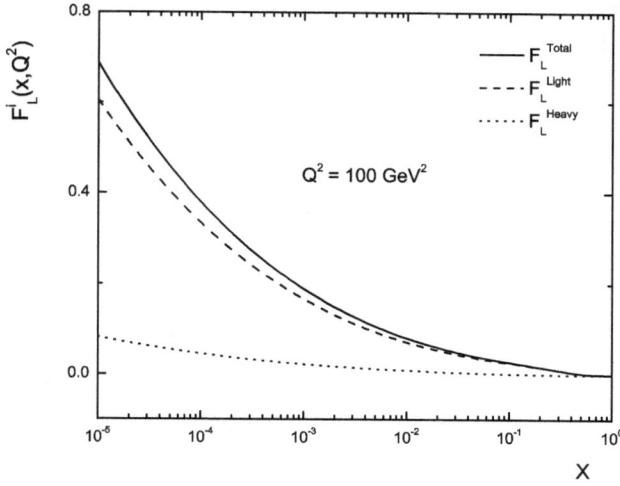

Fig. 2. The light flavor and heavy flavor contributions due to scenario 2 to $F_L(x, Q^2)$ at $O(\alpha_s^2)$.

corresponding flavor–singlet quark distribution. In the limit $Q^2 \gg m^2$ the massive Wilson coefficients $H_{2,L,i}^{S,NS}(Q^2/m^2, m^2/\mu^2, x)$, describe base on Wilson coefficients $C_{L,k}^{S,NS}(Q^2/\mu^2, x)$ accounting for light flavors only and massive operator matrix elements $A_{k,i}^{S,NS}(m^2/\mu^2, x)$

$$H_{2,L,i}^{S,NS}\left(\frac{Q^2}{m^2}, \frac{m^2}{\mu^2}, x\right) = C_{2,L,k}^{S,NS}\left(\frac{Q^2}{\mu^2}, x\right) \otimes A_{k,i}^{S,NS}\left(\frac{m^2}{\mu^2}, x\right) . \tag{14}$$

The massive operator matrix elements to $O(\alpha_s^2)$ allow to calculate the heavy quark Wilson coefficients in the asymptotic region for $F_L(x, Q^2)$ to $O(\alpha_s^2)$. The general structure of the Wilson coefficients is

$$H_{L,g}^{S}\left(\frac{Q^2}{m^2}, \frac{m^2}{\mu^2}\right) = a_s \widehat{C}_{L,g}^{(1)}\left(\frac{Q^2}{\mu^2}\right)$$
$$+ a_s^2 \left[A_{Q,g}^{(1)}\left(\frac{\mu^2}{m^2}\right) \otimes C_{L,q}^{(1)}\left(\frac{Q^2}{\mu^2}\right) + \widehat{C}_{L,g}^{(2)}\left(\frac{Q^2}{\mu^2}\right) \right] , \tag{15}$$

$$H_{L,q}^{PS}\left(\frac{Q^2}{m^2}, \frac{m^2}{\mu^2}\right) = a_s^2 \widehat{C}_{L,q}^{PS,(2)}\left(\frac{Q^2}{\mu^2}\right) , \tag{16}$$

$$H_{L,q}^{NS}\left(\frac{Q^2}{m^2}, \frac{m^2}{\mu^2}\right) = a_s^2 \widehat{C}_{L,q}^{NS,(2)}\left(\frac{Q^2}{\mu^2}\right) . \tag{17}$$

Singlet part of Wilson coefficients is denoted by

$$C_{L,q}^{S} = C_{L,q}^{NS} + C_{L,q}^{PS} , \tag{18}$$

and

$$H_{L,q}^{S} = H_{L,q}^{NS} + H_{L,q}^{PS} . \tag{19}$$

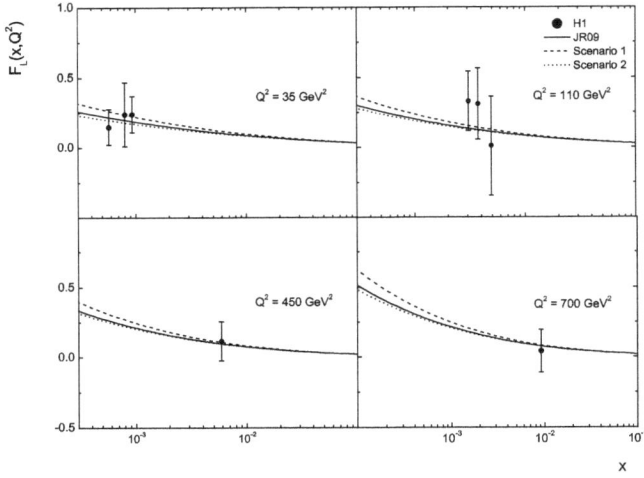

Fig. 3. F_L as a function of x at various Q^2, compared with QCD predictions[25] and H1 data.

Also heavy flavor effects in Wilson coefficients appear in $C_{L,k}\left(\frac{Q^2}{\mu^2}\right)$ where N_H, N_L are the number of heavy and light flavors,

$$\widehat{C}_{L,k}\left(\frac{Q^2}{\mu^2}\right) = C_{L,k}\left(\frac{Q^2}{\mu^2}, N_L + N_H\right) - C_{L,k}\left(\frac{Q^2}{\mu^2}, N_L\right) . \qquad (20)$$

All our calculations are done in Mellin-space

$$a(N) = \int_0^1 dx\, x^{N-1}\, a(x) . \qquad (21)$$

The advantage of this transformation is that it turns the Mellin convolutions into simple products,

$$[a \otimes b](N) = a(N)\, b(N) , \qquad (22)$$

which greatly simplifies all further manipulations.

4. Numerical Results

According to the last section we calculated longitudinal structure function up to NLO just for charm quark ($N_H = 1$). The largest contribution is related to gluon that it is shown in Fig. 1. It grows towards the small x region. Also non-singlet contribution is very small and the asymptotic heavy flavor singlet and non-singlet contributions together with the light flavor terms are shown in Fig. 1. The result of two scenarios are compared in Fig. 3 and they are in good agreement with JR09 model[25] and experimental data.

5. Conclusion

We have calculated the heavy flavor contributions to the structure function $F_L(x, Q^2)$ in the region $Q^2 \gg m^2$ at $O(\alpha_s^2)$. The heavy quark effects are just appeared in Wilson coefficient functions and distribution functions refer to massless partons.[26] The heavy flavor Wilson coefficients are determined by a convolution of the light flavor Wilson coefficients and operator matrix elements, which contain the information on the heavy quarks and mass effects. Also we used the F_2 experimental data to make parametrization form. The gluon contribution grows towards the small x region and it has the largest contribution in this structure function. On the other hand non-singlet contribution is very small because heavy flavor contribute is little.[27] Our result is in good agreement with experimental data and other models.[25]

References

1. T. Namsoo, arXiv:0905.4658 [hep-ex].
2. S. Chekanov *et al.* [ZEUS Collaboration], Phys. Lett. B **682**, 8 (2009) [arXiv:0904.1092 [hep-ex]].
3. R. S. Thorne, arXiv:0808.1845 [hep-ph].
4. M. Buza, Y. Matiounine, J. Smith, and W.L. van Neerven, Nucl. Phys. **B485** (1997) 420.
5. A. Zee, F. Wilczek, and S.B. Treiman, Phys. Rev. **D10** (1974) 2881.
6. W.L. van Neerven and E.B. Zijlstra, Phys. Lett. **B272** (1991) 127;
   E.B. Zijlstra and W.L. van Neerven, Phys. Lett. **B273** (1991) 476; Nucl. Phys. **B383** (1992) 525.
7. S. Moch and J.A.M. Vermaseren, Nucl. Phys. **B573** (2000) 853.
8. S.A. Larin, T. van Ritbergen, and J.A.M. Vermaseren, Nucl. Phys. **B427** (1994) 41;
   S.A. Larin, P. Nogueira, T. van Ritbergen, and J.A.M. Vermaseren, Nucl. Phys. **B492** (1997) 338;
   A. Retey and J.A.M. Vermaseren, Nucl. Phys. **B604** (2001) 281;
   J. Blümlein and J.A.M. Vermaseren, Phys. Lett. **B606** (2005) 130.
9. S.-O. Moch, J.A.M. Vermaseren, and A. Vogt, Phys. Lett. **B606** (2005) 123.
10. J.A.M. Vermaseren, A. Vogt, and S.-O. Moch, Nucl. Phys. **B724** (2005) 3.
11. J. Blumlein, A. De Freitas, W. L. van Neerven and S. Klein, Nucl. Phys. B **755**, 272 (2006) [arXiv:hep-ph/0608024].
12. I. Bierenbaum, J. Blumlein and S. Klein, Nucl. Phys. B **780**, 40 (2007) [arXiv:hep-ph/0703285].
13. A. Vogt, Comput. Phys. Commun. **170**, 65 (2005) [arXiv:hep-ph/0408244].
14. A. N. Khorramian and S. A. Tehrani, Phys. Rev. D **78**, 074019 (2008) [arXiv:0805.3063 [hep-ph]].
15. A. N. Khorramian and S. A. Tehrani, AIP Conf. Proc. **1006** (2008) 118.
16. A. N. Khorramian and S. A. Tehrani and M. Ghominejad, Acta Phys. Polon. B **38**, 3551 (2007).
17. A. N. Khorramian, H. Khanpour and S. A. Tehrani, Phys. Rev. D **81**, 014013 (2010) [arXiv:0909.2665 [hep-ph]].
18. H. Khanpour, Ali N. Khorramian, S. Atashbar, Will be appear in Int. J. Mod. Phys. A, (2010).
19. S. Chekanov et al, ZEUS Collab, Phys. Rev. **D69**, 012004 (2004).
20. C. Adloff et al, H1 Collab, Phys. Lett. **B528**, 199 (2002).

458

21. A. Atkas et al, H1 Collab, Eur. Phys. J. **C40**, 349 (2005); **C45**, 23 (2006).

22. W. Furmanski, R. Petronzio, Z. Phys. **C11**, 293 (1982).

23. M. Gluck, P. Jimenez-Delgado and E. Reya, Eur. Phys. J. C **53**, 355 (2008) [arXiv:0709.0614 [hep-ph]].

24. M. Gluck, C. Pisano and E. Reya, Phys. Rev. D **77**, 074002 (2008) [Erratum-ibid. D **78**, 019902 (2008)] [arXiv:0711.1248 [hep-ph]].

25. P. Jimenez-Delgado and E. Reya, Phys. Rev. D **79**, 074023 (2009) [arXiv:0810.4274 [hep-ph]].

26. A. N. Khorramian, S. A. Tehrani, M. Soleymaninia, S. Batebi, Hyperfine Interactions DOI 10.1007/S 10751-009-0091-9.

27. A. N. Khorramian, S. Atashbar Tehrani and A. Mirjalili, Nucl. Phys. Proc. Suppl. **186**, 379 (2009).

HELICITY CONTRIBUTIONS OF W^+-BOSON IN ENERGY DISTRIBUTION OF B-HADRON IN TOP QUARK DECAY

S. M. MOOSAVI NEJAD

Physics Faculty, Yazd University, Yazd, Iran
School of Particles and Accelerators,
Institute for Research in Fundamental Science (IPM),
P.O. Box 19395-5531, Tehran, Iran
mmoosavi@yazduni.ac.ir

B. A. KNIEHL and G. KRAMER

II. Institut für Theoretische Physik, Universität Hamburg,
Luruper Chaussee 149, Hamburg 22761, Germany

We calculate the energy spectrum of the inclusive b-flavored hadrons in the decay of an unpolarized top quark into a bottom quark and a W^+-boson at next-to-leading order (NLO). The helicity contributions of the W^+-boson in the energy distribution of the B-hadron are determined. The helicities of the W^+-boson are specified as longitudinal, transverse-plus and transverse-minus. In our calculation we apply the zero-mass variable-flavor-number scheme (ZM-VFNS) using realistic non-perturbative fragmentation functions obtained through a global fit to e^+e^- data from CERN LEP1 and SLAC SLC.

Keywords: Perturbative QCD; fragmentation function; helicity; ZM-VFNS.

1. Introduction

Clearly, to investigate the full structure of partonic interactions one needs to do polarization measurements. When the particle decays, its polarization measurements are particularly simple. The angular decay distribution of the decay products reveals information on the state of polarization of the decaying particle. This information is maximal when the particle decay is weak. The fact that the angular decay distribution reveals information on the polarization of the decaying particle is sometimes referred to such that the particle decay is self-analyzing.

According to the CKM matrix elements, the decay width of the top quark is expected to be dominated by the two-body channel $t \to b + W^+$. It is suggested that the final state with leptons, coming from the W^+ decay, and the b-flavored hadron, coming from fragmentation of the b-quark, will be a nice channel to study top decay and reconstruct the top mass. Therefore, to obtain some information on the polarization states of the W^+-boson it is considered the helicity contributions of the W^+-boson in the energy distribution of the b-flavored hadron. The helicities of the W^+-boson are specified as longitudinal, transverse-plus and transverse-minus.

Since the b-quark fragments into the b-flavored hadron and this process is not calculable in perturbative QCD, thus the b-quark fragmentation is the largest source of uncertainty in our calculation. In order to describe the hadronization of the b-quark into the b-flavored hadron, several phenomenological models have been proposed, see Ref. 1. Non-perturbative fragmentation functions (non-pFFs), describing hadronization of a parton into a hadron, contain parameters which are required to be fitted to the experimental data. Since the hadronization mechanism is universal and independent of the perturbative process, one can exploit the experimental data on special event process to fit such models and apply these models in the other processes. In this work we apply the non-pFFs obtained through a global fit to e^-e^+ annihilation at the Z-boson resonance ($e^-e^+ \rightarrow \gamma, Z \rightarrow B + X$) from ALEPH, OPAL and SLD data.

2. Theoretical Formalism

According to the factorization theorem,[2,3] the energy distribution of the B-hadron in the decay of an on-shell top quark at NLO in α_s

$$t \rightarrow b + W^+(g) \rightarrow B + X, \quad (1)$$

can be expressed as the convolution of the parton-level spectrum with the non-pFF $D_b^B(x_B/x_b, \mu_F)$:

$$\frac{d\Gamma}{dx_B}(x_B, m_t, m_W, m_b) = \int_{x_B}^1 \frac{dx_b}{x_b} \frac{d\Gamma}{dx_b}(x_b, m_t, m_W, m_b, \mu, \mu_F) D_b^B(\frac{x_B}{x_b}, \mu_F), \quad (2)$$

where $d\Gamma/dx_b$ is the parton-level differential width, μ and μ_F are the renormalization and factorization scales, respectively. It is also defined the normalized energy fractions of the B-hadron and the b-quark in top quark rest frame as,

$$x_B = \frac{2E_B}{m_t(1-\omega)} \quad , \quad x_b = \frac{2E_b}{m_t(1-\omega)}, \quad (3)$$

where we have $x_{B,min} = 2m_B/(m_t(1-\omega))$ and $x_{B,max} = 1$. We defined $\omega = m_W^2/m_t^2$ and m_B stands for the mass of B-hadron.

To calculate the energy distribution of the B-hadron, we need a theoretical framework. The QCD-improved parton model implemented in the $\overline{MS}$ renormalization and factorization scheme is a suitable framework. In this framework, we apply zero-mass variable-flavor-number scheme for NLO calculations in perturbative QCD where all parton masses are neglected. In this scheme, the non-zero value of b-quark mass only enter through the initial condition of the non-pFFs.

In Eq.(2), the functional form of the non-pFF is not yet calculable in each scale but having the non-pFF at some initial fragmentation scale μ_0, the numerical values of the non-pFF at any other scale μ_F can be obtained by solving the DGLAP evolution equations, See Refs. 4, 5. Several models have been proposed to describe the non-perturbative transition from a parton into a hadron state, see Ref. 1. These models are suited to determine initial condition of non-pFFs at the DGLAP evolution. In this work we apply the Power model which consists of a simple power functional form as:

$$D(x; \mu_0, \alpha, \beta) = Nx^\alpha(1-x)^\beta, \qquad (4)$$

where the coefficients (N, α, β) should be specified experimentally. In this article we use the results extracted in Ref. 6, where a combined fit is done to the data sets obtained by the ALEPH, OPAL, and SLD collaborations for the inclusive B-meson production in e^-e^+ annihilation on the Z-boson resonance using $\mu_0 = m_b = 4.5$ GeV. The values reported for the parameters in Eq.(4) are: $N = 4684.1$, $\alpha = 16.87$ and $\beta = 2.628$.

To obtain our theoretical predictions for the helicity contributions of the W^+-boson in the energy distribution of the B-hadron in top decay, via Eq.(2), we need to know the parton level differential width. In the next section we present our results for W^+- helicity fractions in top decay.

2.1. W^+-Helicity Fractions and Parton Level Results

We are now in the situation to calculate the angular distribution of the differential decay width to produce a bottom quark in the cascade decay of top quark at NLO($t \to b + W^+ (\to e^+ + \nu_e)$). For unpolarized top decay, the angular distribution of the differential width is determined by the transverse-plus, transverse-minus and the longitudinal helicity components of the W^+-boson. Using fixed x_b, we have:

$$\frac{1}{\Gamma_0} \frac{d^2\hat{\Gamma}}{dx_b \, d\cos\theta} = \hat{H}_{++} \cdot \frac{3}{8}(1 + \cos\theta)^2 + \hat{H}_{--} \cdot \frac{3}{8}(1 - \cos\theta)^2 + \hat{H}_{00} \cdot \frac{3}{4}\sin^2\theta, \quad (5)$$

where $\hat{H}_{++}, \hat{H}_{--}$ and $\hat{H}_{00}$ stand for the transverse-plus, transverse-minus and the longitudinal helicity components of the top quark differential decay rate. The polar angle θ, which is measured in the W^+ rest frame, denotes the angle between the W^+ momentum direction and the outgoing positron momentum, see Fig. 1.

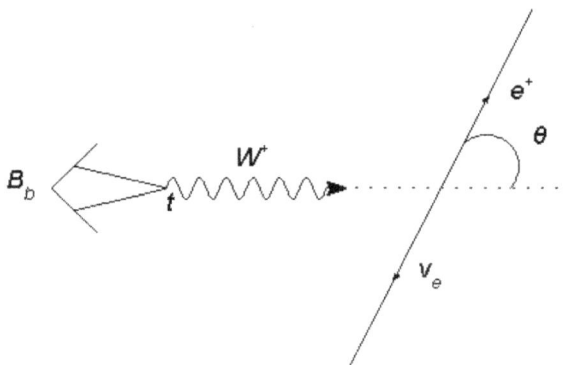

Fig. 1.   Definition of the polar angle θ in the W^+ rest frame.

According to the approach proposed in Ref. 7, using covariant projectors one can work out the various helicity components of the W^+ boson in differential decay width. According to this approach, instead of the completeness relation:

$$\sum_{\lambda=0,\pm} \epsilon^\mu(\lambda)\epsilon^{\nu\star}(\lambda) = -g^{\mu\nu} + \frac{p_W^\mu p_W^\nu}{m_W^2}, \tag{6}$$

which is used to calculate the differential decay rate when the W^+-boson is unpolarized, we use the following relations to extract the longitudinal, transverse-plus and the transverse-minus helicities:

$$\epsilon^\mu(0)\epsilon^{\nu\star}(0) = \frac{\omega}{|\overrightarrow{P_W}|^2}\left(p_t^\mu - \frac{p_t \cdot p_W}{m_W^2}p_W^\mu\right)\left(p_t^\nu - \frac{p_t \cdot p_W}{m_W^2}p_W^\nu\right),$$

$$\epsilon^\mu(\pm)\epsilon^{\nu\star}(\pm) = \frac{1}{2}\Bigg(-g^{\mu\nu} + \frac{p_W^\mu p_W^\nu}{m_W^2}$$

$$-\frac{\omega}{|\overrightarrow{P_W}|^2}\left(p_t^\mu - \frac{p_t \cdot p_W}{m_W^2}p_W^\mu\right)\left(p_t^\nu - \frac{p_t \cdot p_W}{m_W^2}p_W^\nu\right) \mp \frac{i\epsilon^{\mu\nu\alpha\beta}}{m_t|\overrightarrow{P_W}|}(p_t)_\alpha(p_W)_\beta\Bigg), \tag{7}$$

where $\epsilon^{0123} = 1$ and $|\overrightarrow{P_W}|^2 = (m_t - E_b - E_g)^2 - m_W^2$ and $\omega = m_W^2/m_t^2$.

Considering a massless b-quark, for the contribution of the differential decay width into a longitudinal W^+-boson in $\overline{MS}$ scheme, we have:

$$\hat{H}_{00} = \frac{1}{1+2\omega}\Big\{\delta(1-x_b) + \frac{\alpha_S C_F}{2\pi}A(x_b)\Big\}, \tag{8}$$

where,

$$A(x_b) = \delta(1-x_b)$$

$$\left(-\frac{3}{2}\log\frac{\mu^2}{m_t^2} + 2\log\omega\log(1-\omega) + \frac{2\omega}{\omega-1}\log\omega + 4Li_2(1-\omega) - \frac{2\pi^2}{3} - 3\frac{2+5\omega}{1+2\omega}\right)$$

$$+\frac{1}{(1-x_b)_+}\bigg((1+x_b^2)\Big[2\log x_b - \log\frac{\mu^2}{m_t^2} + 2\log(1-\omega)\Big] + 2(1-\omega)x_b^3 +$$

$$\frac{8\omega^2-4\omega-3}{1+2\omega}x_b^2 - 2(1+\omega)x_b + \frac{1}{1+2\omega}\bigg) + 2(1+x_b^2)\left(\frac{\log(1-x_b)}{1-x_b}\right)_+$$

$$-\frac{2x_b(1-x_b)(2-x_b(1-\omega))^2}{(1-\omega)x_b^2-4x_b+4} + \frac{2\sqrt{\omega}}{(\omega-1)((1-\omega)x_b^2-4x_b+4)^2}\bigg($$

$$(1+\sqrt{\omega})^2(x_b(1-\sqrt{\omega})^2 + 2\sqrt{\omega})(x_b^2(1-\omega) + x_b(\sqrt{\omega}-3) + 2)^2\log(1-x_b(1-\sqrt{\omega}))-$$

$$(1-\sqrt{\omega})^2(x_b(1+\sqrt{\omega})^2 - 2\sqrt{\omega})(x_b^2(1-\omega) - x_b(\sqrt{\omega}+3) + 2)^2\log|1-x_b(1+\sqrt{\omega})|\bigg). \tag{9}$$

For the transverse-minus helicity component of differential decay rate, defining $R = \log(1 + (S-1)x_b + \sqrt{S(Sx_b^2 - 2x_b + 2)})$, $T = \log(-2S^2x_b^3 + 4Sx_b^2 - (1+3S)x_b + 1 + |2Sx_b^2 - 2x_b + 1|\sqrt{S(Sx_b^2 - 2x_b + 2)})$, $D = \log((1-S)x_b^2 - x_b + 1/2 + |2Sx_b^2 - 2x_b + 1|/2)$ and $M = \log(2S^2x_b^2 - S(1+2x_b) + 1 - S|2Sx_b^2 - 2x_b + 1|)$, we find:

$$\hat{H}_{--} = \frac{2\omega}{1+2\omega}\Big\{\delta(1-x_b) + \frac{\alpha_S C_F}{4\pi}B(x_b)\Big\}, \tag{10}$$

where,

$$
B(x_b) = \delta(1 - x_b)\left(-\frac{3(7 + 18\omega)}{2(1 + 2\omega)} - \frac{4\omega}{1 - \omega}\log\omega - 3\log\frac{\mu^2}{m_t^2} \right.
$$

$$
\left. -\frac{2(1 - \omega)}{\omega}\log(1 - \omega) + 4\log\omega\log(1 - \omega) + 8Li_2(1 - \omega) - \frac{4\pi^2}{3} \right) +
$$

$$
+\frac{2}{(1 - x_b)_+}\left((1 + x_b^2)\left[2\log x_b - \log\frac{\mu^2}{m_t^2} + 2\log(1 - \omega)\right] - Dx_b^3 +
$$

$$
\frac{x_b^3\sqrt{1 - \omega}}{\sqrt{(1 - \omega)x_b^2 - 4x_b + 4}}(T - R)\right) + 2(1 + x_b^2 + 2x_b^3)\left(\frac{\log(1 - x_b)}{1 - x_b}\right)_+ +
$$

$$
(-D + 2\log(1 - x_b))\left(1 + x_b + 2x_b^2\right) + 2\frac{(x_b(1 + \omega) - 2)|(1 - \omega)x_b^2 - 2x_b + 1|}{(1 - x_b)(1 - \omega)((\omega - 1)x_b^2 + 4x_b - 4)}
$$

$$
+2(\frac{1 + \omega}{1 - \omega} - x_b)\left(\log(1 - x_b(1 - \omega)) - M + \log\omega \right) +
$$

$$
+\frac{1}{(1 - x_b)(1 + 2\omega)((1 - \omega)x_b^2 - 4x_b + 4)(1 - x_b(1 - \omega))}\left(- (1 - 2\omega)(1 - \omega)^2 x_b^5 \right.
$$

$$
+(\omega - 1)(4\omega^2 + 12\omega - 5)x_b^4 - (10\omega^3 - 21\omega^2 - 2\omega + 21)x_b^3 +
$$

$$
(-66\omega^2 + 75\omega + 47)x_b^2 + 2\frac{20\omega^3 - 80\omega^2 + 31\omega + 23}{\omega - 1}x_b + 4\frac{14\omega^2 - 7\omega - 4}{\omega - 1}\right) +
$$

$$
\frac{1}{\sqrt{\omega}(\omega - 1)((1 - \omega)x_b^2 - 4x_b + 4)^2}\left(
$$

$$
(1 - \sqrt{\omega})^2(x_b(1 + \sqrt{\omega})^2 - 2\sqrt{\omega})(x_b^2(1 - \omega) - x_b(\sqrt{\omega} + 3) + 2)^2\log|1 - x_b(1 + \sqrt{\omega})| -
$$

$$
(1 + \sqrt{\omega})^2(x_b(1 - \sqrt{\omega})^2 + 2\sqrt{\omega})(x_b^2(1 - \omega) + x_b(\sqrt{\omega} - 3) + 2)^2\log(1 - x_b(1 - \sqrt{\omega}))\right)
$$

$$
+\frac{R - T}{(((1 - \omega)x_b^2 - 4x_b + 4)(1 - \omega))^{\frac{3}{2}}} \times
$$

$$
\left((\omega - 1)^3 x_b^4 + (\omega - 3)(\omega - 1)^2 x_b^3 + 2(\omega - 13)(\omega - 1)x_b^2 - 4(\omega^2 - 8\omega + 9)x_b + 8\right).
$$

$$
(11)
$$

It is simple to show that in the massless b-quark case, there is no transverse-plus helicity contribution for the W^+ decay at leading order (LO). However in the higher order QCD corrections this contribution vanishes no longer. At NLO, this contribution reads:

$$\hat{H}_{++} = \frac{\alpha_S \omega}{2\pi(1+2\omega)} C_F \left\{ \frac{3}{2}\delta(1-x_b) + 2(1+x_b^2-2x_b^3)\left(\frac{\log(1-x_b)}{1-x_b}\right)_+ + \right.$$

$$\frac{2}{(1-x_b)_+}\left(Dx_b^3 + \frac{(R-T)\sqrt{1-\omega}x_b^3}{\sqrt{(1-\omega)x_b^2-4x_b+4}}\right) + (1+x_b+2x_b^2)(D-2\log(1-x_b))$$

$$+2(x_b - \frac{1+\omega}{1-\omega})\left(-M + \log(1-x_b(1-\omega)) + \log\omega\right)$$

$$+\frac{1}{(1-x_b)((1-\omega)x_b^2-4x_b+4)(1-x_b(1-\omega))}\left((1-\omega)^2 x_b^5 + \right.$$

$$(1-\omega)(2\omega-9)x_b^4 + 3(\omega-7)(\omega-1)x_b^3 - \frac{27\omega^2-46\omega+19}{1-\omega}x_b^2$$

$$+2\frac{6\omega^2-11\omega+3}{1-\omega}x_b + 4\frac{\omega}{1-\omega}\right) - 2\frac{(x_b(\omega+1)-2)|(1-\omega)x_b^2-2x_b+1|}{(1-x_b)(1-\omega)((\omega-1)x_b^2+4x_b-4)}$$

$$+\frac{1}{\sqrt{\omega}(\omega-1)((1-\omega)x_b^2-4x_b+4)^2}\left(\right.$$

$$(1-\sqrt{\omega})^2(x_b(1+\sqrt{\omega})^2-2\sqrt{\omega})(x_b^2(1-\omega)-x_b(\sqrt{\omega}+3)+2)^2 \log|1-x_b(1+\sqrt{\omega})| -$$

$$(1+\sqrt{\omega})^2(x_b(1-\sqrt{\omega})^2+2\sqrt{\omega})(x_b^2(1-\omega)+x_b(\sqrt{\omega}-3)+2)^2 \log(1-x_b(1-\sqrt{\omega}))\right)$$

$$+\frac{T-R}{(((1-\omega)x_b^2-4x_b+4)(1-\omega))^{\frac{3}{2}}} \times$$

$$\left((\omega-1)^3 x_b^4 + (\omega-3)(\omega-1)^2 x_b^3 + 2(\omega-13)(\omega-1)x_b^2 - 4(\omega^2-8\omega+9)x_b + 8\right)\right\}.$$

$$(12)$$

The results presented in Eqs.(8,10,12) are in complete agreement with Ref. 7 if one integrates over $x_b(0 \le x_b \le 1)$. It can also be shown that integrating over $\cos\theta(-1 \le \cos\theta \le 1)$ in Eq.(5), our results lead to the unpolarized differential decay rate given in Ref. 8.

3. Theoretical Predictions

We are now in a position to discuss our predictions. According to the approach explained in the previous section, to make our predictions we use the ZM-VFN scheme considering the Power model for the hadronization of the b-quark. In order to be consistent at NLO, the strong coupling constant α_s has to be taken at NLO using the two-loop formula with $n_f = 5$ quark flavors. In our calculation, for the typical QCD scale we adopt the NLO value $\Lambda_{\overline{MS}}^{(5)} = 227$ MeV appropriate for $n_f = 5$ which corresponds to $\alpha_s^{(5)}(m_t) = 0.1071$, i.e this corresponds to $\alpha_s^{(5)}(m_Z) = 0.1181$.[9]

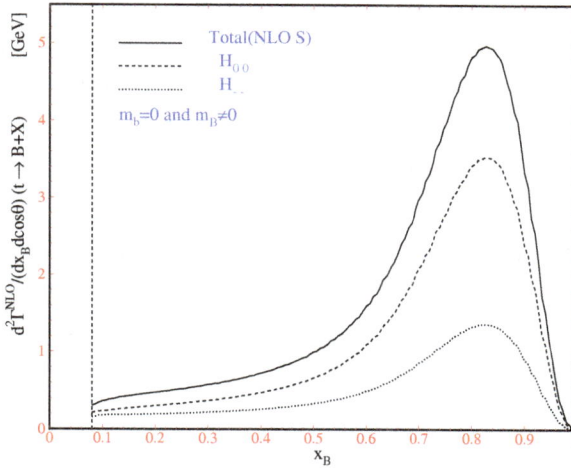

Fig. 2. Comparison of the NLO contributions of the longitudinal (dashes) and the transverse-minus (dots) helicity components of the W^+-boson in the B-hadron energy distribution using the Power model. The solid line shows the summation of all helicity contributions. The initial factorization scale is $\mu_0 = 4.5$ GeV and the b-quark is considered to be massless.

Fig. 2, considering the helicity components of the W^+-boson, shows the differential decay rate of inclusive B-hadron production in top decay at $\sqrt{s} = m_t = 174$GeV. We set $m_W = 80$ GeV, $m_B = 5.28$ GeV and $\mu_0 = 4.5$ GeV . Since the contribution of the transverse-plus helicity of W^+ is tiny therefore we did not plot it, specially this contribution vanishes at the LO. In Fig. 2, the solid line shows the energy distribution of the B-hadron when the W^+-boson is unpolarized from the beginning.

4. Conclusion

To perform accurate studies of the top-quark properties and a precise measurement of its mass at the Tevatron accelerator and at the LHC, a reliable description of the b-quark fragmentations in top quark decay will be necessary. In this work we presented results on the $\mathcal{O}(\alpha_s)$ radiative corrections to the three helicity rates in unpolarized top quark decay. The results can be determined from an analysis of the shape of the lepton spectrum or from doing an angular analysis on the decay products. The radiative corrections to the unpolarized transverse-minus and longitudinal rates are sizable and the radiative correction to the transverse-plus rate is tiny, i.e. most produced charged leptons in the W^+-boson decay are due to a W^+-boson with a longitudinal or transverse-minus helicity. The presented results show that decay width has an angular distribution which peaks at $\theta = \pi/2$. It is also observed that the charged leptons which are scattered in angles $\theta = 0$ or $\theta = \pi$ are due to the decay of a transverse W^+-boson. If one can reconstruct the $\cos\theta$ distributions of top quark decay, these unique shapes can be used for the measurement of W^+-boson polarizations.

References

1. G. Colangelo and P. Nason, *Phys. Lett. B* **285** (1992) 167; M. Cacciari and M. Greco, *Phys. Rev. D* **55** (1997) 7134; M. Cacciari, M. Greco, S. Rolli and A. Tanzini, *Phys. Rev. D* **55** (1997) 2736; C. Peterson, D. Schlatter, I. Schmitt and P. M. Zerwas, *Phys. Rev. D* **27** (1983) 105.
2. J. C. Collins, D. E. Soper and G. Sterman, *in Perturbative QCD*, edited by A. H. Mueller (World Scientific, Singapore, 1989), P.1.
3. J. C. Collins, *Phys. Rev. D* **66** (1998) 094002.
4. G. Altarelli and G. Parisi, *Nucl. Phys. B* **126** (1977) 298.
5. V. N. Gribov and L. N. Lipatov, *Sov. J. Nucl. Phys* **15** (1972) 438; L. N. Lipatov, *Sov. J. Nucl. Phys* **20** (1975) 94; Yu. L. Dokshitzer, *Sov. Phys. JETP* **46** (1977) 641.
6. B. A. Kniehl, G. Kramer, I. Schienbein and H. Spiesberger, *Phys. Rev. D* **77** (2008) 014011 [arXiv:0705.4392 [hep-ph]].
7. M. Fischer, S. Groote, J. G. Körner and M. C. Mauser, *Phys. Rev. D* **63** (2001) 031501 [arXiv:hep-ph/0011075].
8. G. Corcella and A. D. Mitov, *Nucl. Phys. B* **623** (2002) 247.
9. K. Hagiwara et al, *Phys. Rev. D* **66** (2002) 010001.

THE KOLMOGOROV REVERSE EQUATION AND HIGH ENERGY MULTIPLICITY RELATIONS

A. DEWANTO,[1,*] A. H. CHAN[1,2] and C. H. OH[1,2]

[1] *Department of Physics, National University of Singapore,*
2 Science Drive 3, Singapore 117542
[2] *Institute of Advanced Studies, Nanyang Technological University,*
Nanyang Executive Centre, 60 Nanyang View #02-18, Singapore 639798
* phyda@nus.edu.sg

Another solution to the Kolmogorov backward partial differential equation for stationary branching is proposed leading to a new multiplicity relation. This relation is applied to CERN's ISR and SPS energies, in which excellent fits are obtained and compared to Chaudhuri,[1] Matinyan and Prokhorenko,[2] and Golyak.[3] Further prediction to the highly-anticipated 14 TeV data for pp collision is also made.

Keywords: Kolmogorov equation; multiparticle production; multiplicity distribution.

1. Introduction

It is now common knowledge that the generating function G, for the most elementary scenario for branching process with one species of particles (or clans) produced at high energies satisfies the Kolmogorv reverse equation,[4] namely

$$\frac{\partial G}{\partial t} = -f(G, t) \tag{1}$$

where t being some relevant evolution parameter.

Chliapnikov and Tchikilev[5] were the first to use the popular Negative Binomial Distribution (NBD) law for G and hence derive the following equation

$$\frac{dG}{dt} = G \ln G \left(\frac{1}{k} \frac{dk}{dt} \right) + G(1 - G^{\frac{1}{k}}) \left(\frac{k}{m} \frac{dm}{dt} \right). \tag{2}$$

By arguing that a pure birth branching process satisfies the boundary conditions $f(G = 1, t) = 0$ and $f(G, t) \to 0$ at $G \to 0$, they obtained

$$\frac{1}{k} = a + b \ln m \tag{3}$$

where $m = \frac{\bar{n}}{k}$, $\bar{n}$ is mean multiplicity and a,b are unknown parameters, if it is a stationary branching process and the function $f(G, t)$ partitions into G and t.

Chaudhuri,[1] Matinyan and Prokhorenko[2] have studied the $f(G, t)$ factorization in detail. These authors eliminate the $G^{\frac{1}{k}}$ by considering $G \sim 1$ so that $f(G, t)$

factorizes exactly. However, we will take the course of Golyak.[3] This paper is to point out that Hwa and Lam's proposal[6] of using Furry-Yule Distribution (FYD) to study high energy scattering also satisfies Eq. (2) and we further derived another relation to describe ISR and SPS data and also extend it to future LHC data.

2. Discussion

One can write down the FYD generating function $(G(x,t) = \sum_{n=0}^{\infty} x^n P_n(t))$

$$G_{FYD}(x) = \left[1 + m\left(\frac{1-x}{x}\right)\right]^{-k'} \tag{4}$$

where $m = \frac{\bar{n}}{k'}$, while

$$G_{NBD}(x) = [1 + m(1-x)]^{-k} . \tag{5}$$

One strong justification to using Eq. (4) is that at high energies, especially SPS energies, the Generalized Multiplicity Distribution (GMD)[7,8] will approach FYD due to more significant gluon contribution. k' is interpreted as initial cluster size[6] and the normalized dispersion relation for FYD is $\left(\frac{D}{\bar{n}}\right)^2 = \frac{1}{k'} - \frac{1}{\bar{n}}$ as contrasted to $\left(\frac{D}{\bar{n}}\right)^2 = \frac{1}{k'} + \frac{1}{\bar{n}}$ for NBD where $D = \left(\overline{n^2} - \bar{n}^2\right)^{\frac{1}{2}}$.

It is straight forward to show that Eq. (4) satisfies Eq. (2) by simply replacing k with k' and together with boundary condition as mentioned earlier, one immediately gets 2 independent solutions

$$\frac{1}{k'}\frac{dk'}{dt} = b\frac{k'}{m}\frac{dm}{dt} \tag{6}$$

and

$$c \ln G = \left(1 - G^{\frac{1}{k}}\right) \tag{7}$$

where b and c are arbitrary constants. Note $b \neq c$.[3]

Solving Eq. (6), we have

$$\frac{1}{k'} = a + b \ln m \tag{8}$$

where a is the constant of inetgration. Once can easily find a and b to give a better fit to the data ($\sqrt{s} \sim 10$ GeV to 900 GeV)[5] than the generally accepted UA5 Collaboration's empirical formula

$$\frac{1}{k} = a + b \ln s \tag{9}$$

but this k^{-1} parameter from NBD which increases phenomenologically with c.m. $\sqrt{s}$ energy still has no clear theoretical understanding. We remark that authors[5] have neglected Eq. (7) in their derivation that results in a similar equation in Eq. (8) (replace k' by k for NBD). This discrepancy was also pointed out by Golyak.[3]

Substituting Eq. (4) into Eq. (7) and setting $x \to 0$ and after some algebra, we obtain

$$\frac{1}{k'} = c\left(\frac{1+m}{m}\right)\ln(1+m).\qquad(10)$$

Finally the existence of independent solution Eq. (8) and Eq. (10) implies that the complete solution is

$$\frac{1}{k'} = a + b\ln m + c\left(\frac{1+m}{m}\right)\ln(1+m).\qquad(11)$$

This 3-parameter equation is also obtained by Golyak[3] for the NBD case (replace k' in Eq. (11) with k, $a = $ -0.040, $b = 0.030$, $c = 0.125$). Equation (11) indeed describes a wide spectrum of energies much better than UA5's Eq. (9). The author also used their 2 mechanism model to show that $c = 1/8$ (notice that c from their fit is also $\sim 1/8$). With more degrees of freedom it should not be at all surprising to find it a better fir then $\frac{1}{k} = a + \ln s$, i.e. Eq. (9).

As for us, having also derived Eq. (11) from FYD, we will adopt Hwa and Lam's conjecture that k' is independent of energy and therefore quite different from k which behaves as in Eq. (9). From previous works,[8,9] we reckon that $k' \sim 3$ to 4 (for the range; 30.4 GeV $< \sqrt{s} < 900$ GeV; ISR and SPS data). In fact, Hwa and Lam[6] also conjectured that in the cluster picture $k' = 3$. So from Eq. (10), we can calculate c by observing from Table 1 that $m \gg 1$ at high energies, hence

$$c = \frac{1}{3\ln(1+m)}.\qquad(12)$$

In short, we have reduced the complete solution in Eq. (11) to a 2-parameter relation

$$\frac{1}{k'} = a + b\ln m + \frac{1+m}{3m}.\qquad(13)$$

The result of this parametrization is given in Table 1 and Fig. 1 with $a = -0.2888$ and $b = 0.1068$. One perceives that it describes both the ISR and SPS data well, especially SPS (200 GeV to 900 GeV). By utilizing FYD generating function (4), one also gets equations $\frac{1}{k'} = am^{-b}$ and $\frac{1}{k'} = m(1-a) - b$ using the factorizing scheme discussed by Chaudhuri,[1] Matinyan and Prokhrorenko[2] mentioned earlier. We compare these equations with Eq. (11) obtained by Golyak (using the NBD k for thier 2 component model). Our observation reveals equally good description by the 4 equations.

Note that, in the case of 1.8 TeV and 14 TeV, in the absence of experimental data, we use

$$\bar{n}(\sqrt{s}) = 3.01 - 0.474\ln(\sqrt{s}) + 0.754\ln^2(\sqrt{s})\qquad(14)$$

to extrapolate the mean-value $\bar{n}$.[11] The results are shown in the last two lines of Table 1 and the last two points in Fig. 1. In anticipation of the upcoming 14 TeV pp collision at LHC, these data provide an excellent benchmark.

Table 1.   The parantheses show the number of multiplicity points considered at each energy. We compare our model (Eq. (13)) with works done by Golyak,[3] Matinyan and Prokhorenko[2] and Chaudhuri[1]

$\sqrt{s}$ (GEV)	$\bar{n}$	k'	m	$1/k'$ exp-fit	$1/k'$ Eq. (13)	Golyak	Matinyan	Chaudhuri
30.4 (17)	10.54	3.70	2.8486	0.2702	0.2733	0.2684	0.2703	0.2662
44.5 (19)	12.08	3.50	3.4514	0.2857	0.2734	0.2743	0.2747	0.2745
52.6 (21)	12.76	3.60	3.5440	0.2777	0.2737	0.2752	0.2754	0.2756
62.6 (20)	13.63	3.90	3.4948	0.2564	0.2735	0.2747	0.2750	0.2750
200 (27)	21.40	3.28	6.5243	0.3048	0.2959	0.3005	0.2970	0.3026
546 (40)	29.40	3.13	9.3929	0.3194	0.3192	0.3196	0.3178	0.3199
900 (49)	35.60	3.02	11.788	0.3311	0.3363	0.3328	0.3351	0.3312
1800 (Tevatron)	41.82	3.00	13.94	-	0.3498	0.3432	0.3501	0.3398
14000 (LHC)	67.21	3.00	22.40	-	0.3915	0.3748	0.4110	0.3653

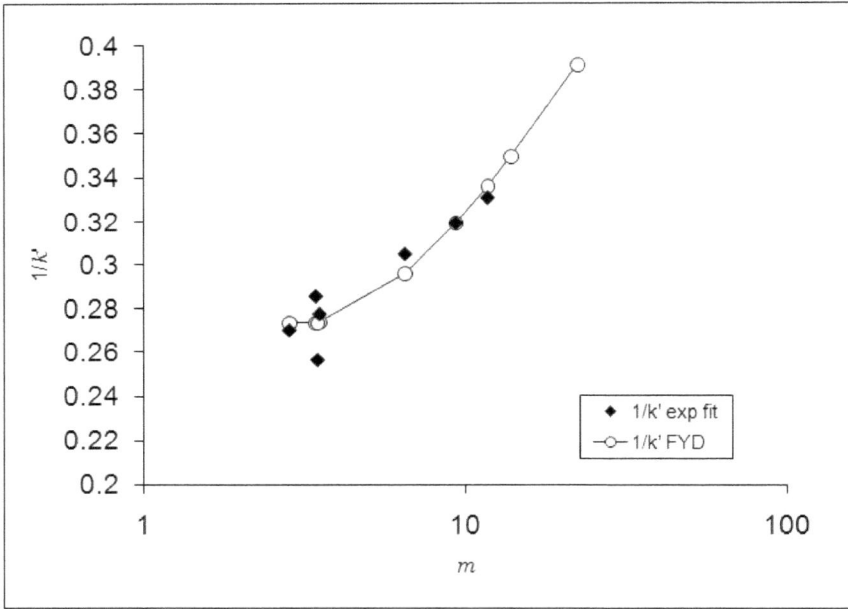

Fig. 1.   A comparison between experimentally (ISR and SPS data) fitted $\frac{1}{k'}(m)$ with $\frac{1}{k'}(m)$ as given by Eq. (13). The last 2 points refer to prediction made at 1.8 TeV and 14 TeV respectively by Eq. (13).

3. Conclusion

In conclusion, we have pointed out that the Furry-Yule Distribution is another solution to the Kolmogorov backward partial differential equation leading to a new multiplicity relation in $k'(m)$, k' being the initial number of clusters and by conjecturing that it is 3, we get an excellent fit to the ISR and SPS data. Although FYD law provides a less precise description at lower energies but it has more predictive power because of the energy independent nature of k'.[6]

We would like to caution that a stochastic evolution normally involves real time that has no counterparts in particle physics, as pointed out by Hwa.[10] Here it is sufficient to mention that it is plausible that the evolution parameter t is related to energy in an indirect way. Traditionally, it refers to the QCD evolution parameter.

Acknowledgments

This work is supported by NUS academic research grant No. WBS: R-144-000-178-112.

References

1. A.K. Chaudhuri. *Phys. Rev.* **D45** (1992) 4057
2. S.G. Matinyan and and E.B. Prokhorenko. *Phys. Rev.* **D48** (1993) 5127
3. I. Golyak. *Nucl. Part. Phys., J. Phys.* **G20** (1994) 565
4. N.A. Dmitriev and A.N. Kolmogorov. *Dok. Akad. Nauk* **SSSR 56** (1947) 7
5. P.V. Chliapnikov and O.G. Tchikilev. *Phys. Lett.* **B222** (1989) 152
6. R.C. Hwa and C.S. Lam. *Phys. Lett.* **B173** (1986) 346
7. C.K. Chew, D. Kiang and H. Zhou. *Phys. Lett.* **B186(3)** (1987) 411
8. A.H. Chan and C.K. Chew. *Nuovo Cimento* **101A** (1989) 409
9. A.H. Chan and C.K. Chew. *Physs. Rev.* **D41** (1990) 851
10. R.C. Hwa. in *Hadronic Multiparticle Production, edited by P. Carruthers (World Scientific 1998)*
11. A. Giovannini and R. Ugoccioni. *Phys. Rev.* **D59** (1999) p094020

CONFORMATION CHANGES AND PROTEIN FOLDING INDUCED BY ϕ^4 INTERACTION

M. JANUAR

Department of Physics, University of Indonesia,
Kampus UI Depok, Depok 16424, Indonesia

A. SULAIMAN

Badan Pengkajian dan Penerapan Teknologi, BPPT Bld. II (19th floor),
Jl. M.H. Thamrin 8, Jakarta 10340, Indonesia
asulaiman@webmail.bppt.go.id
sulaiman@teori.fisika.lipi.go.id

L. T. HANDOKO

Group for Theoretical and Computational Physics,
Research Center for Physics, Indonesian Institute of Sciences,
Kompleks Puspiptek Serpong, Tangerang, Indonesia

Department of Physics, University of Indonesia,
Kampus UI Depok, Depok 16424, Indonesia
handoko@teori.fisika.lipi.go.id
handoko@fisika.ui.ac.id
laksana.tri.handoko@lipi.go.id

A model to describe the mechanism of conformational dynamics in protein based on matter interactions using lagrangian approach and imposing certain symmetry breaking is proposed. Both conformation changes of proteins and the injected non-linear sources are represented by the bosonic lagrangian with an additional ϕ^4 interaction for the sources. In the model the spring tension of protein representing the internal hydrogen bonds is realized as the interactions between individual amino acids and nonlinear sources. The folding pathway is determined by the strength of nonlinear sources that propagate through the protein backbone. It is also shown that the model reproduces the results in some previous works.

Keywords: Protein dynamics; protein folding; Lagrangian; ϕ^4 interaction.

1. Introduction

The pathway of proteins are determined by the sequences of its amino acid constituents. The time ordered of protein folding sequence leads from the primary to the secondary and subsequent structures. The secondary structure consists of the shape representing each segment of a polypeptide tied by hydrogen bonds, van der Walls forces, electrostatic interaction and hydrophobic effects. It is moreover formed around a group of amino acids considered as the ground state. Then it is extended to include adjacent amino acids till the blocking amino acids are reached, and the whole protein chain along the polypeptide adopted its preferred secondary structure.

Our understanding on the underlying above-mentioned mechanism has unfortunately not been at the satisfactory level. For instance, the studies based on statistical analysis of identifying the probabilities of locating amino acids in each secondary structure are still at the level of less than 75% accuracy. Moreover, the main mechanism responsible for a structured folding pathway have not yet been identified at all. On the other hand, it is known that the protein misfolding has been identified as the main cause of several diseases like cancers and so on.[1]

Recently, Mingaleev et.al. have shown that the nonlinear excitations play an important role in conformational dynamics by decreasing the effective bending rigidity of a biopolymer chain leading to a buckling instability of the chain.[2] Following this understanding, a model to explain the transition of a protein from a metastable to its ground conformation induced by solitons has been proposed.[3] In the model the mediator of protein transition is the Davydov solitons propagating through the protein backbone.

At present, the most reliable theoretical explanation for this kind of the conformational dynamics of biomolecules is the so-called ab initio quantum chemistry approach. This however requires astronomical computational power to deal with realistic biological systems.[4,5] In contrary, there are some phenomenological model describing the folding pathway as a result of the interplay between the energy transfer from a solitary solution that travels along the protein backbone and string tension.[6] There are also some attempts to describe the dynamics in term of elementary biomatter using field theory approach[7] and open quantum system.[8,9]

This paper follows the later approach, but starting from the first principle using the lagrangian method to derive the responsible interactions and to clarify its origins. The paper is organized as follows. First, the model and the underlying assumptions are explained in detail. It is then followed by the derivation of relevant equation of motions (EOMs). Summary and conclusion based on the numerical analysis are given at the end of the paper.

2. The Models

The model is an extension of the toy model proposed in Ref. 10. More than considering a self-interaction mechanism as proposed in Ref. 10 and subsequently developed in Refs. 3, 6, more realistic model is introduced. In the model, the dynamics of amino acids forming proteins is initially considered as a free and linear system of bosonic matters. Further, external nonlinear sources, like laser or light bunch, are introduced. The sources which propagate through the protein backbone interact each other with the amino acids to induce conformation changes.

The model describes the conformation changes as the dynamics of amino acids using a free and massive (relativistic) bosonic lagrangian as below,

$$\mathcal{L}_c = (\partial_\mu \phi_c)^\dagger (\partial^\mu \phi_c) + \frac{1}{2} m_{\phi_c}^2 \phi_c^\dagger \phi_c \,, \tag{1}$$

where ϕ_c represents the conformation field. The hermite conjugate is $\phi^\dagger \equiv (\phi^*)^T$ for a general complex field ϕ. On the other hand, the nonlinear sources represented by the field ϕ_s are also governed by a massless bosonic lagrangian,

$$\mathcal{L}_s = (\partial_\mu \phi_s)^\dagger (\partial^\mu \phi_s) + V(\phi_s) \,, \tag{2}$$

with an additional potential $V(\phi_s)$ taking the typical ϕ^4- self-interaction,

$$V(\phi_s) = \frac{1}{4} \lambda (\phi_s^\dagger \phi_s)^2 \,, \tag{3}$$

where λ is the coupling constant. It should be noted that both scalar fields, $\phi_c = \phi_c(t,x)$ denotes the local curvature of the conformation at position x with $\phi_c(x) = 1$ or 0 for α or $\beta-$helix.

The choice of interactions in Eqs. (1) and (2) are justified by the following considerations,

- The conformation changes are assumed to be linear. It is actually not necessarily massive. Although one can put by hand the mass term $m_{\phi_c}^2 \phi_c^\dagger \phi_c$ in the lagrangian as written above, the massive conformational field could also be generated dynamically through certain symmetry breaking as shown later.
- The source is assumed to be massless concerning the laser or light source injected to the protein chains to induce the foldings.
- Its non-linearity is realized by introducing the ϕ_s self-interaction which leads to the non-linear EOM.
- For the sake of simplicity, the lagrangian is imposed to be symmetry under certain transformations, for instance in the present case is time and parity symmetry, i.e. $\phi(t,x) \to -\phi(-t,-x)$ for one-dimensional space.

We should remark here that the model is although written in a relativistic form, after deriving relevant EOMs one can take its non-relativistic limits to obtain final EOMs describing the desired dynamics. Secondly, instead of using the vector electromagnetic field A_μ to represent the nonlinear sources, like laser for instance, it is more convenient to consider the nonlinear source as a bunch of light or laser such that one might represent it in a 'macrosocopic' scalar field ϕ_s.

Considering the dimensional counting and the invariance on time-parity symmetry, the most general interaction between the conformation field and nonlinear sources is,

$$\mathcal{L}_{\text{int}} = -\Lambda \, (\phi_c^\dagger \phi_c)(\phi_s^\dagger \phi_s) \,, \tag{4}$$

with Λ denotes the strength of the interaction. Eqs. (3) and (4) lead to the total potential in the model,

$$V_{\text{tot}} = \frac{1}{4}\lambda\,(\phi_s^\dagger\phi_s)^2 - \Lambda\,(\phi_c^\dagger\phi_c)(\phi_s^\dagger\phi_s)\,. \tag{5}$$

Eqs. (1), (2) and (5) provide the underlying interactions in the model.

Concerning the minima of the total potential in term of source field, that is

$$\left.\frac{\partial V_{\text{tot}}}{\partial \phi_s}\right|_{\langle\phi_s\rangle,\langle\phi_c\rangle} = 0\,, \tag{6}$$

at the vacuum expectation values (VEV) of the fields yields the non-trivial solution,

$$\langle\phi_s\rangle = \sqrt{\frac{2\Lambda}{\lambda}}\langle\phi_c\rangle\,. \tag{7}$$

Imposing certain local symmetry, namely the phase or U(1) symmetry to the above total lagrangian, the VEV in Eq. (7) obviously breaks the symmetry. The symmetry breaking at the same time shifts the mass term for ϕ_c as follow,

$$m_{\phi_c}^2 \to \overline{m}_{\phi_c}^2 \equiv m_{\phi_c}^2 - \frac{2\Lambda^2}{\lambda}\langle\phi_c\rangle^2\,, \tag{8}$$

from Eq. (4).

On the other hand, Eq. (7) induces the 'tension force' which plays an important role to enable folded pathway in the present model. This will be discussed in the following section.

3. EOMs and Its Behaviours

Having the total lagrangian at hand, one can derive the EOM's using the Euler-Lagrange equation,

$$\frac{\partial \mathcal{L}_{\text{tot}}}{\partial \phi} - \partial_\mu \frac{\partial \mathcal{L}_{\text{tot}}}{\partial(\partial_\mu \phi)} = 0\,, \tag{9}$$

where $\mathcal{L}_{\text{tot}} = \mathcal{L}_c + \mathcal{L}_s + \mathcal{L}_{\text{int}}$.

Substituting Eqs. (1), (2) and (4) into Eq. (9) in term of ϕ_c and ϕ_s, one immediately obtains a set of EOMs,

$$\left(\frac{\partial^2}{\partial x^2} - \frac{1}{c^2}\frac{\partial^2}{\partial t^2} - \frac{1}{\hbar^2}m_{\phi_c}^2 c^2 + 2\Lambda\,\phi_s^2\right)\phi_c = 0\,, \tag{10}$$

$$\left(\frac{\partial^2}{\partial x^2} - \frac{1}{c^2}\frac{\partial^2}{\partial t^2} + 2\Lambda\,\phi_c^2 - 3\lambda\,\phi_s^2\right)\phi_s = 0\,. \tag{11}$$

Here the natural unit is restored to make the light velocity c and $\hbar$ reappear in the equation.

The last term in Eq. (11) determines the non-linearity of the EOM of source. One should also put an attention in the last term of Eq. (10), *i.e.* $\sim k\,\phi_c$ with $k \sim 2\Lambda\langle\phi_s\rangle^2$. This actually induces the tension force in the dynamics of conformational field enabling the folded pathway as expected.

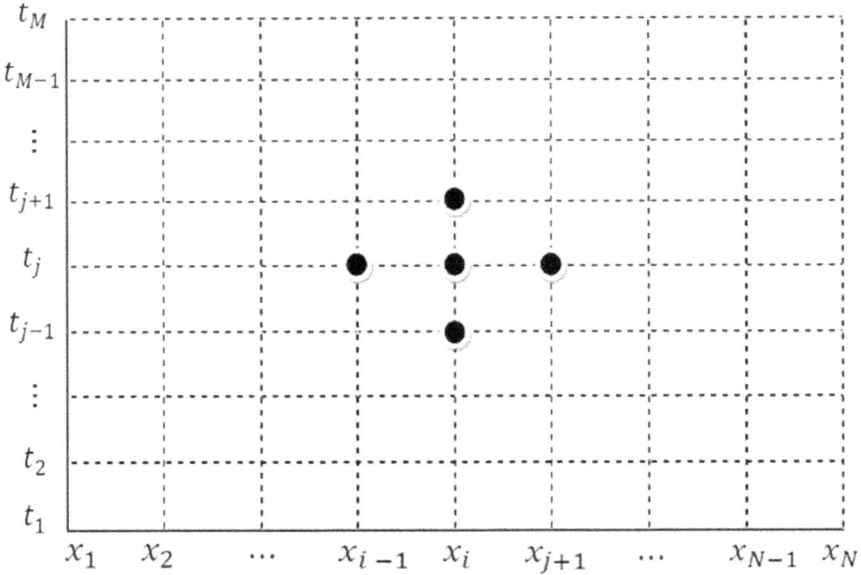

Fig. 1.   The discretized grid for solving the EOMs over the coordinate space R.

Hence, solving both EOMs in Eqs. (10) and (11) simultaneously would provide the contour of conformational changes in term of time and one-dimensional space components.

4. Numerical Analysis

Since the EOMs under consideration involves non-linear term, one should solve them numerically. The numerical analysis and simulation in the present paper are done using the finite difference method.[11] Throughout numerical works, non-relativistic limit $v = \partial x/\partial t \ll c$ and the following boundary conditions for both fields are deployed,

$$\phi_s(0,t) = \phi_s(L,t) = 0 \text{ and } \phi_c(0,t) = \phi_c(L,t) = 0 \text{ for } 0 \leq t \leq b ,$$

$$\phi_s(x,0) = f(x) \text{ and } \phi_c(x,0) = p(x) \qquad \text{for } 0 \leq x \leq L , \qquad (12)$$

$$\frac{\partial \phi_s(x,0)}{\partial t} = g(x) \text{ and } \frac{\partial \phi_c(x,0)}{\partial t} = q(x) \qquad \text{for } 0 < x < L ,$$

with $f(x)$, $p(x)$, $g(x)$ and $q(x)$ are newly introduced auxiliary functions. In finite difference scheme, it is more convenient to replace ϕ_s and ϕ_c with u and w respectively, and rewrite them in discrete forms. Then, let us consider the coordinate space $R = \{(x,t) : 0 \leq x \leq L, 0 \leq t \leq b\}$ discretized on a grid consisting of $(N-1) \times (M-1)$ rectangles with side length $\Delta x = \delta$ and $\Delta t = \epsilon$ shown in Fig. 1. Solving the equations over the grid gives us the desired numerical solutions.

Both coupled EOMs in Eqs. (10) and (11) are rewritten in explicit discrete forms as follows,

$$u_{i,j+1} = 2u_{i,i} - u_{i,j-1} + c^2\epsilon^2 \left(\frac{u_{i+1,j} - 2u_{i,j} + u_{i-1,j}}{\delta^2} + 2\Lambda w_{i,j}^2 u_{i,j} - 3\lambda u_{i,j}^3 \right), \quad (13)$$

$$w_{i,j+1} = 2w_{i,i} - w_{i,j-1} + c^2\epsilon^2 \left(\frac{w_{i+1,j} - 2w_{i,j} + w_{i-1,j}}{\delta^2} + 2\Lambda u_{i,j}^2 w_{i,j} - \frac{c^2}{\hbar^2} m_{\phi_c}^2 w_{i,j} \right),$$
$$(14)$$

for $i = 2, 3, \cdots, N - 1$ and $j = 2, 3, \cdots, M - 1$. In order to calculate all values of Eqs. (13) and (14), the initial values for two lowest rows in Fig. 1 must be given. On the other hand, the value at t_1 is fixed by the boundary conditions in Eq. (12). The second order of Taylor expansion can also be used to determine the values in the second row. Therefore, the values at t_2 are determined by,

$$u_{i,2} = f_i - \epsilon g_i + \frac{c^2\epsilon^2}{2} \left(\frac{f_{i+1} - 2f_i + f_{i-1}}{\delta^2} + 2\Lambda p_i^2 f_i - 3\lambda f_i^3 \right), \quad (15)$$

$$w_{i,2} = p_i - \epsilon q_i + \frac{c^2\epsilon^2}{2} \left(\frac{p_{i+1} - 2p_i + p_{i-1}}{\delta^2} + 2\Lambda f_i^2 p_i - \frac{c^2}{\hbar^2} m_{\phi_c}^2 p_i \right), \quad (16)$$

for $i = 2, 3, \cdots, N - 1$.

For the initial stage, suppose the nonlinear sources has a particular form $f(x) = 2\mathrm{sech}(2x)\,e^{i2x}$ and $g(x) = 1$ to generate the α-helix, while $g(x) = q(x) = 0$ for the sake of simplicity. Then, one can obtain the initial values in this case using Eqs. (15) and (16). The subsequent values are generated by substituting the preceeding values into Eqs. (13) and (14). The higher order values can be obtained using iterative procedure.

The result is given in Fig. 2. The left figure in each box describes the propagation of nonlinear sources in protein backbone, while the right one shows how the protein is folded. As can be seen in the figure, the protein backbone is initially linear before the nonlinear source injection. As the soliton started propagating over the backbone, the conformational changes appear. It should be remarked that the result is obtained up to the second order accuracy in Taylor expansion. In order to guarantee that the numerical solutions do not contain large amount of truncation errors, the step sizes δ and ϵ are kept small enough. Nevertheless, this should be good approximation to describe visually the mechanism of protein folding.

5. Conclusion

An extension of phenomenological model describing the conformational dynamics of proteins is proposed. The model based on the matter interactions among the relevant constituents, namely the conformational field and the nonlinear sources represented as the bosonic fields ϕ_c and ϕ_s. It has been shown that from the relativistic bosonic lagrangian with ϕ_s^4 self-interaction, the nonlinear and tension force terms appear naturally as expected in some previous works.[6]

478

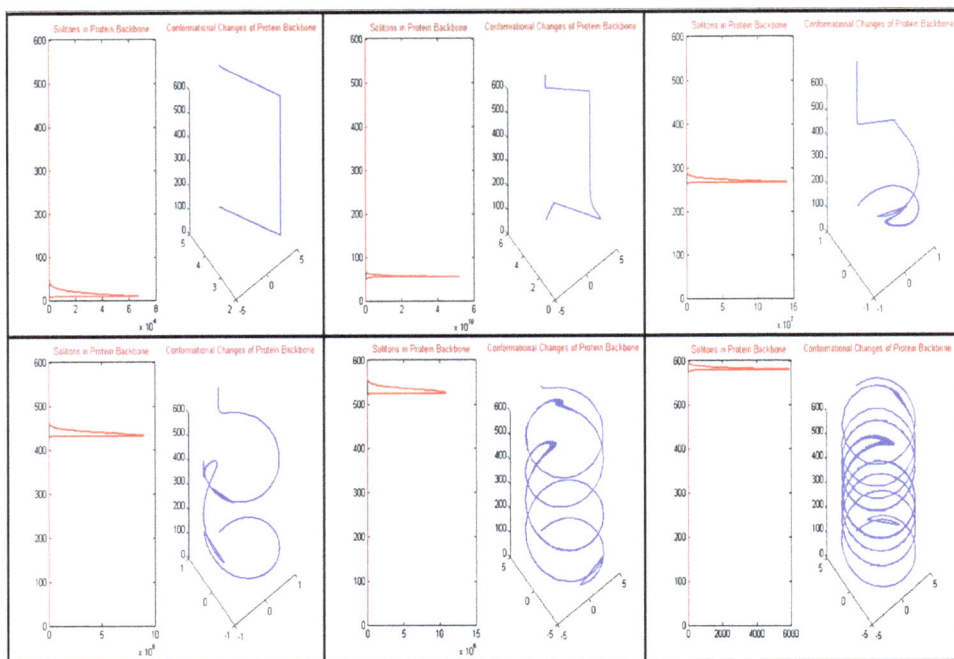

Fig. 2.   The soliton propagations and conformational changes on the protein backbone inducing protein folding. The vertical axis in soliton evolution denotes time in second, while the horizontal axis denotes its amplitude. The conformational changes are on the (x, y, z) plane.

However, the present model has different contour since the EOMs governing the whole dynamics are the linear and nonlinear Klein-Gordon equations. Note that the original model by Berloff deployed the linear Klein-Gordon and nonlinear Schrodinger equations.

Moreover, the present model has inhomogenous tension force, in contrast with the homogeneous tension force in the Berloff's model, due to simultaneous solutions of Eqs. (10) and (11). These lead to wrigling folded pathways as shown in Fig. 2 which should be more natural than the homogeneous one.

Acknowledgments

The authors greatly appreciate fruitful discussion with T.P. Djun throughout the work. AS thanks the Group for Theoretical and Computational Physics LIPI for warm hospitality during the work. This work is partially funded by the Indonesia Ministry of Research and Technology and the Riset Kompetitif LIPI in fiscal year 2010 under Contract no. 11.04/SK/KPPI/II/2010.

References

1. C. M. Dobson, *Nature* **426**, p. 884 (2003).

2. S. F. Mingaleev, Y. B. Gaididei, P. L. Christiansen and Y. Kivshar, *Europhys. Lett.* **59**, p. 403 (2002).
3. S. Caspi and E. Ben-Jacob, *Phys. Lett. A* **272**, p. 124 (2000).
4. A. Garcia and J. Onuchi, *Proc. Natl. Acad. Sci. USA* **100**, p. 13898 (2003).
5. J. N. Onuchi and P. G. Wolynes, *Curr. Opin. Struct. Biol.* **14**, p. 70 (2004).
6. N. G. Berloff, *Phys. Lett. A* **337**, p. 391 (2005).
7. A. Sulaiman and L. T. Handoko, *J. Compt. Theor. Nanoscience* **in press** (2010).
8. A. Sulaiman, F. P. Zen, H. Alatas and L. T. Handoko, *Phys. Rev. E* **81**, p. 061907 (2010).
9. A. Sulaiman, F. P. Zen, H. Alatas and L. T. Handoko, *Int. J. Mod. Phys. A* **in press** (2010).
10. S. Caspi and E. Ben-Jacob, *Europhys. Lett.* **47**, p. 522 (1999).
11. J. H. Mathews and K. D. Fink, *Numerical Methods using Matlab 4th Ed.* (Prentice-Hall, 2004).

NON-RELATIVISTIC NEUTRINO OSCILLATION IN DENSE MEDIUM

H. T. HE,[*,‡] Z. Y. LAW,[*] A. H. CHAN[*,†] and C. H. OH[*,†]

[*]*Physics Department, National University of Singapore,*
2 Science Drive, Singapore 115742
[†]*Institute of Advanced Studies, Nanyang Technological University,*
Singapore 639798
[‡]*haitao.he@live.com*

We first review the study of neutrino oscillation under non-relativistic energy and dense medium assumption. We then develop a computational model to study neutrino oscillation under this general scenario. In particular, we look at both the uniform medium and the linear medium cases, where various oscillation features are exhibited. The transition probability is found to depend on helicity for such cases. The neutrino mass dependence of the solution is also investigated.

Keywords: Neutrino oscillation; non-relativistic; dense medium.

1. Introduction

In a 1986 paper, Halprin[1] examined neutrino oscillation by studying charged current interactions of neutrino with electrons in ordinary matter. The derivation started from first-principle Dirac equation, and it was assumed that the neutrino is of relativistic energy and the medium is of ordinary density. More recently, a paper by Law *et al.*[2] removed some of Halprin's assumptions and examined the phenomena in a more general scenario. This result is applicable to non-relativistic neutrino propagating in dense medium. However, as the final result is a second order non-homogenous differential system, only some very simple cases are analytically solvable.

In this paper, we try to develop a computational model to validate and implement Law's idea. In Sec. 2 we first briefly review results from Ref. 2, paying special attention to the underlying assumptions he made. In Sec. 3 we then proceed to demonstrate how to implement this model computationally. In Sec. 4 we will be studying a uniform matter density case. We compare our results with the analytical results, and hence validate our model. In Sec. 5 we will apply this model to linear matter density case, where an analytical solution is not available.

2. Neutrino Oscillation in the General Scenario

Following the treatment from Ref. 2, we start with the total effective Lagrangian

$$L = -\sqrt{2}G\bar{v}[\gamma^\mu J_\mu^{CC} N + J_\mu^{CC} N^{NC}]v \tag{1}$$

where we have a charged current term

$$J_\mu^{CC} = 2 \sum \overline{f_L} \gamma_\mu f_L \quad \text{charged current}$$

$$N^{CC} = \begin{pmatrix} 1 & 0 \\ 0 & 0 \end{pmatrix} \tag{2}$$

and a neutral current term

$$J_\mu^{NC} = 2 \sum \overline{f_L} \gamma_\mu f_L (T_{3f} - 2Q_f \sin^2 \theta_w) \quad \text{neutral current}$$

$$N^{NC} = \begin{pmatrix} 1 & 0 \\ 0 & 1 \end{pmatrix}. \tag{3}$$

In the above equations, f are the fermionic fields in matter, Q and T are their corresponding charge and isospin respectively, and θ_w is the Weinberg angle.

Note that a general background may contain various kinds of baryons and leptons. However, in the following discussion we assume that the background only consists of ordinary matter, which includes electrons, protons and neutrons.

Eliminating the right handed field, we arrive at the left handed field equation

$$-\partial^2 v_L - M^+ M v_L - i\sqrt{2} G_F \gamma^v \partial_v \{\gamma^\mu (J_\mu^{CC} N^{CC} + J_\mu^{NC} N^{NC}) v_L\} = 0. \tag{4}$$

If we assume the background particles are static and unpolarized, the fermion fields could be represented by the fermion density. Further simplifying the above field equation by combining the terms into a density matrix A, we get

$$\{\partial_t^2 - \nabla^2 + M^2 + i[(\partial_t + \vec{\sigma} \cdot \vec{\nabla}) A + A(\partial_t + \vec{\sigma} \cdot \vec{\nabla})]\} v_L = 0$$

$$A = \begin{pmatrix} 2(\alpha - \beta) & 0 \\ 0 & -2\beta \end{pmatrix}$$

$$\alpha = \frac{G_F}{\sqrt{2}} \rho_e \quad \beta = \frac{G_F}{\sqrt{2}} \sum_f \rho_f (T_{3f_L} - 2Q_f \sin^2 \theta_w). \tag{5}$$

We study a one dimensional case and assume the neutrino is in a spin eigenstate. Moreover, we further assume the solution to be an energy eigenstate.

$$v_L(x, t) = v_L(x) e^{-iEt}. \tag{6}$$

Substitute (6) into Eq. (5), and note that the density profile has no time dependence, we get the time independent differential equation

$$\{(-E^2 + M^2) + (is\partial_x A + AE) + iAs\partial_x - \partial_x^2\} v_L = 0 \quad s = \pm 1. \tag{7}$$

Before we proceed to solve this differential system, we should take note that unlike Halprin's case, where the final differential equation is of first order in x, this differential equation is second order in x. Hence, we should not expect the usual probability expression of the wave solution to hold for our case, and must look for another conserved quantity.

This is done in Ref. 2, where we have the conservation equation

$$\partial_t(v_L^{+a}(i\overleftrightarrow{\partial}_t - A)v_L^b) + \vec{\nabla} \cdot (v_L^{+a}(-i\overleftrightarrow{\nabla} - \vec{\sigma}A)v_L^b) = 0 \,. \tag{8}$$

Note that in our case, we are dealing with a time independent solution as in Eq. (7), therefore we have

$$\partial_t(v_L^+(i\overleftrightarrow{\partial}_t - A)v_L) = 0 \,.$$

Consequently,

$$v_L^+(-i\overleftrightarrow{\partial}_x - \sigma_x A)v_L = v_e^+(-i\overleftrightarrow{\partial}_x - \sigma_x A)v_e + v_\mu^+(-i\overleftrightarrow{\partial}_x - \sigma_x A)v_\mu = \text{constant} \,.$$

We then interpret the two terms as the probability for the neutrino to be in electron state and muon state respectively (Assuming 2-flavor oscillation).

$$v_j^+(-i\overleftrightarrow{\partial}_x - \sigma_x A)v_j = \text{probability} \quad (j = e, \mu) \,. \tag{9}$$

For varying matter distribution, the above holds if and only if

$$(\partial_t + \vec{\sigma} \cdot \vec{\nabla})A \approx 0 \,. \tag{10}$$

In the derivation of the conservation equation, this term is assumed to be negligible (A reasonable assumption, if neutrino wavelength is much smaller that scale of density variation[1,2]).

With the above probability measure, we could now proceed to solve Eq. (7). Imposing condition (10), the equation is further simplified into

$$\{(-E^2 + M^2) + AE + iAs\partial_x - \partial_x^2\}v_L = 0 \quad s = \pm 1 \,. \tag{11}$$

This is the equation we are trying to solve in Sec. 3. Note that although we have discarded the first order derivative of the density profile, all the information about the density distribution is encoded in matrix A.

3. The Computational Approach

Equation (11) is a second order differential system with non-constant coefficients. In general, we cannot expect to solve it analytically. Although in Ref. 2 gave an analytical expression for the case of uniform matter distribution, even for this simple case, the solution is very much complicated. We will make use of their result in the subsequent section for verification purpose. However, here, we take up a computational approach to study the behavior of the solution, which could be applied to more general scenarios.

We first combine the terms of Eq. (11) and adopt a new notation convention

$$\{\Omega + \Delta(x) + \Sigma(x)\partial_x - \partial_x^2\}\phi(x) = 0 \,,$$

$$\begin{aligned}
\Omega &= -E^2 + M^2 && \text{is the constant non-homogenous term}\,, \\
\Delta(x) &= A(x)E && \text{is the } x\text{-dependent non-homogenous term}\,, \\
\Sigma(x) &= isA(x) && \text{is the } x\text{-dependent first order coefficient}\,.
\end{aligned}$$

Note that in flavor basis our solution takes the form

$$\phi = \begin{pmatrix} \phi_e \\ \phi_\mu \end{pmatrix}.$$

With the above notations, the equation to be solved could be written out explicitly as

$$\begin{pmatrix} \Omega_{ee}\phi_e + \Omega_{e\mu}\phi_\mu \\ \Omega_{\mu e}\phi_e + \Omega_{\mu\mu}\phi_\mu \end{pmatrix} = \begin{pmatrix} \partial_x^2\phi_e - \Sigma_{ee}(x)\partial_x\phi_e - \Delta_{ee}(x)\phi_e \\ \partial_x^2\phi_\mu - \Sigma_{\mu\mu}(x)\partial_x\phi_\mu - \Delta_{\mu\mu}(x)\phi_\mu \end{pmatrix}. \tag{12}$$

The programming and numerical calculation of the solution to the above system is done in MATLAB.

4. Propagation in Uniform Medium

We first study the simplest case, where the density profile is constant,

$$\alpha(x) = \alpha_0, \quad \beta(x) = \beta_0, \quad 0 \le x \le 100.$$

A typical probability plot of neutrino with $s = +1$ is as shown in Fig. 1. The plot is a perfect sinusoidal curve with certain oscillation amplitude and oscillation length. This agrees with the result from the analytical solutions both in Halprin's case and Law's case, as it should be.

Changing the energy of the neutrino changes both the oscillation length and oscillation amplitude. Larger neutrino energy results in longer oscillation length. This is exactly as expected from the analytical result. Changing the density changes both the oscillation length and oscillation amplitude, although the change is smaller compared to changing the energy. A denser medium results in a shorter oscillation length. Increasing the energy of neutrino has the same effect as reducing the density of the profile.

A more dramatic change in the solution takes place if we change spin number s to -1, as shown in Fig. 2. This dramatic change is due to the solution to the dispersion equation. With spin number $+1$, the two wave numbers are very close to each other, and therefore the oscillation length is relatively long. However, with

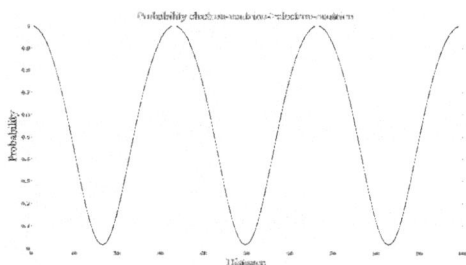

Fig. 1. $(s = 1, m_0, E_0, \alpha_0, \beta_0)$.

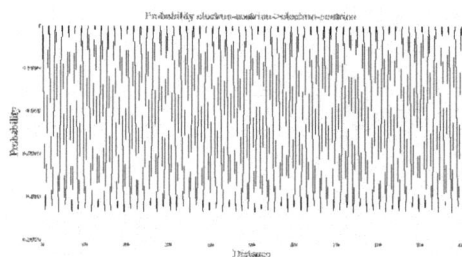

Fig. 2. $(s = -1, m_0, E_0, \alpha_0, \beta_0)$.

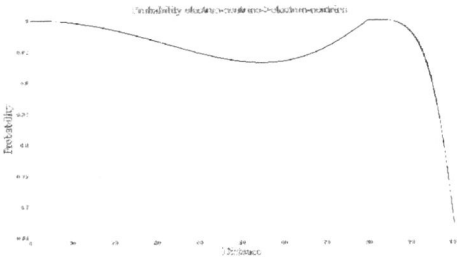

Fig. 3. $(s = 1, m_0, E_0\ \alpha_0, \beta_0)$.

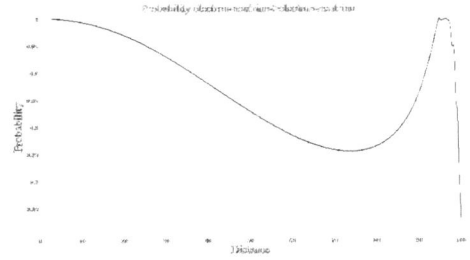

Fig. 4. $(s = 1, m_0, 0.25m_0, \alpha_0, \beta_0)$.

$s = -1$, the two wave number are generally very different, which results in a much shorter oscillation length.

Lastly, changing the absolute input neutrino mass has no effect on the solution. This is expected because all the analytical solutions depend only on the neutrino mass square difference (See Ref. 3 for standard treatment).

To summarize, the uniform matter density case is as expected from the analytical results in the limiting cases. Therefore, our model is validated.

5. Propagation in Linear Medium

We then study the case of linear medium, where the density profile is given by,

$$\alpha(x) = \alpha_0(100 - x), \quad \beta(x) = \beta_0(100 - x), \quad 0 \le x \le 100.$$

A typical probability plot of neutrino with $s = +1$ is as shown is Fig. 3. We see that the solution exhibits much more complex behavior compared to the uniform density case. The probability plot could generally be classified into three phases. In phase one, the probability first decreases gradually and then increases back to the maximum value 1. The second phase of the probability plot is a plateau, where the probability remains at the maximum value 1 for a distance. In phase three, the probability decreases with many small oscillations.

Figures 4 to 6 shows the probability plot in the case of neutrino with energy $0.25E_0, 2E_0, 4E_0$ respectively. Comparing to the reference case, we see that the first

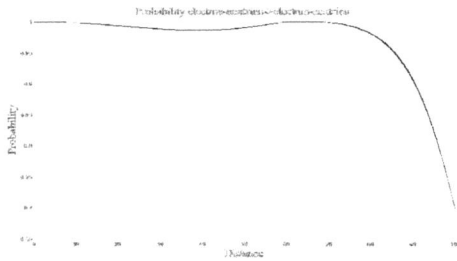

Fig. 5. $(s = 1, m_0, 2E_0, \alpha_0, \beta_0)$.

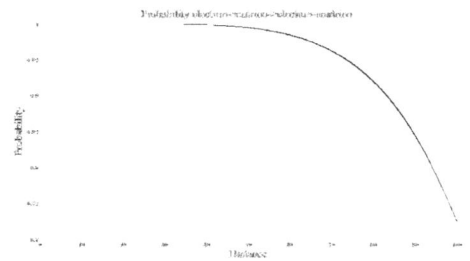

Fig. 6. $(s = 1, m_0, 4E_0, \alpha_0, \beta_0)$.

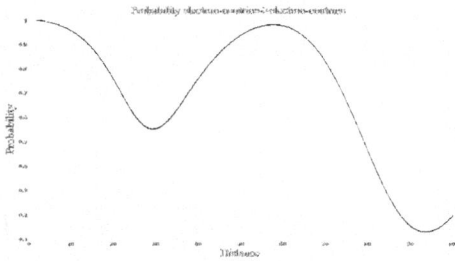

Fig. 7. $(s = 1, m_0, E_0, 0.1\alpha_0, 0.1\beta_0)$.

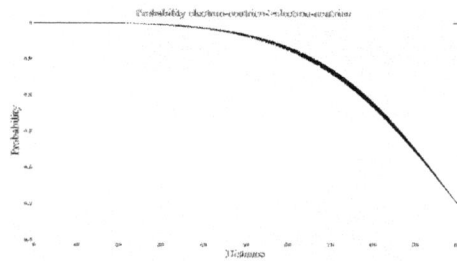

Fig. 8. $(s = 1, m_0, E_0, 0.2\alpha_0, 0.2\beta_0)$.

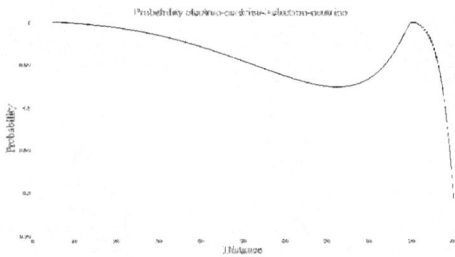

Fig. 9. $(s = 1, m_0, E_0, 2\alpha_0, 2\beta_0)$.

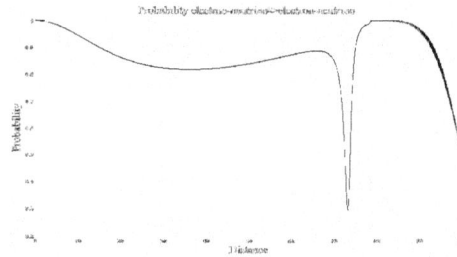

Fig. 10. $(s = 1, 2m_0, E_0, \alpha_0, \beta_0)$.

phase is more significant both in its amplitude and its length with lower energy. In fact, for the case with $0.25E_0$, the plateau in phase two and the oscillation features in phase three become negligible, while for the case with $4E_0$, the first phase disappears completely and the plot only includes an initial plateau and an oscillating decreasing phase.

Figures 7 to 9 shows the probability plot in the case of density $(\alpha = 0.1\alpha_0, \beta = 0.1\beta_0)$, $(\alpha = 0.2\alpha_0, \beta = 0.2\beta_0)$, $(\alpha = 2\alpha_0, \beta = 2\beta_0)$ respectively. From the high similarity between the case $(\alpha = 0.2\alpha_0, \beta = 0.2\beta_0)$ and the case $4E_0$, the case $(\alpha = 2\alpha_0, \beta = 2\beta_0)$ and the case $0.5E_0$, we see that reducing the density has essentially the same effect as increasing the energy of the neutrino. However, it should also be noted that further reducing the density, the decreasing phase three will become the starting point of an oscillating curve, as in Fig. 7.

More complicated changes occur if we change the neutrino mass, which in this case affects the solution. Figures 10 and 11 shows the probability plot in the case of mass $m = 2m_0$ and $m = 4m_0$. Comparing to the base case, the first obvious feature is that for m greater than m_0, some additional sudden and deep dips appear in phase one of the curve. Also, we see that as we increases the neutrino mass, the oscillation in phase three is of greater amplitude. The physical origin of the features is worth further investigation.

Lastly, if we change the spin number s to -1, the overall shape of the curve is changed, as is shown in Fig. 12. This curve could also be classified into three phases.

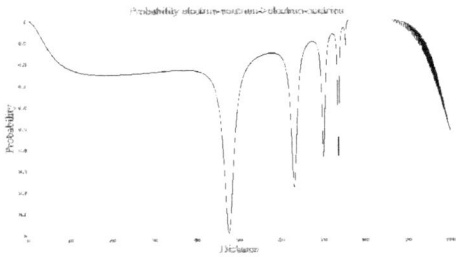

Fig. 11. $(s = 1, em_0, E_0, \alpha_0, \beta_0)$.

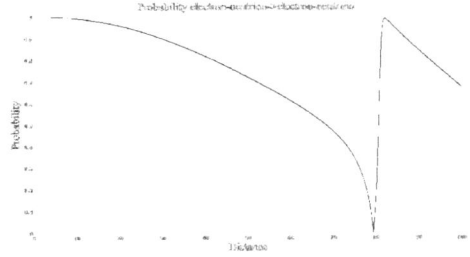

Fig. 12. $(s = -1, m_0, E_0, \alpha_0, \beta_0)$.

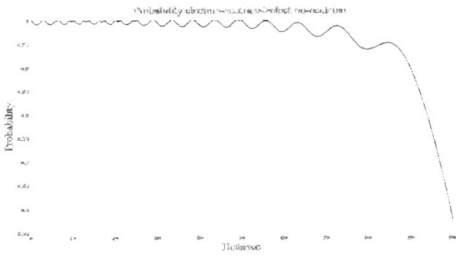

Fig. 13. $(s = -1, m_0, E_0, 0.01\alpha_0, 0.01\beta_0)$.

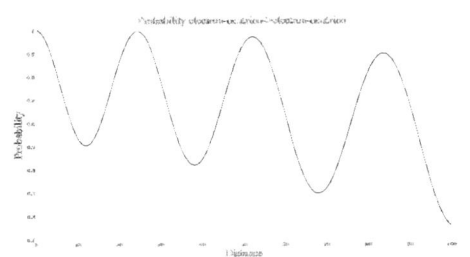

Fig. 14. $(s = -1, m_0, E_0, 0.001\alpha_0, 0.001\beta_0)$.

The first is a decreasing phase, from the maximum value 1 to the minimum value 0. The second phase is an increasing edge, where the probability increases from 0 to a certain maximum value over a very short distance. The third phase is again a decreasing phase, where the probability decreases monotonically until the edge of the density profile.

Note that in the relativistic energy and low density medium cases studied by Halprin,[1] the transition probability does not depend on spin number s. Therefore, for our computational model, we expect the behavior of the solution for the $s = -1$ case to be more similar to the $s = 1$ case as we decrease the density of the profile (or increase the neutrino energy). This is indeed the case. Figures 13 and 14 show the case for $(\alpha = 0.01\alpha_0, \beta = 0.0.1\beta_0)$ and $(\alpha = 0.001\alpha_0, \beta = 0.001\beta_0)$ respectively. We notice Fig. 13 has similar general features as Fig. 3. This shows that the behavior of the solution for $s = 1$ and $s = -1$ are similar for the high energy or low medium density case. If we further decrease the density profile, we see that Fig. 14 is similar to Fig. 7. This shows that at ultra-relativistic energy or very low density limits, the features of both cases approaches that of the vacuum case.

6. Conclusion

We computationally studied neutrino oscillation in a more general scenario and validated our model by comparing our results for the uniform density case to the analytical results. In the linear density case, we observed some new features in the oscillation curve, in particular the helicity and mass dependence. It should be noted that this computational model could be applied to more general matter distributions.

Acknowledgments

This work is supported by the National University of Singapore academic research grant No. WBS: R-144-000-178-112.

References

1. A. Halprin, *Phys. Rev. D* **34**, 3462 (1986).
2. Z. Y. Law, A. H. Chan and C. H. Oh, *Int. J. M. Phys. A* **24**, 3483 (2009).
3. S. M. Bilenky, C. Giunti and W. Grimus, http://www.arXiv.hep-ph/9812360v4 (1999).

QUARK SPIN IN THE PROTON

STEVEN D. BASS

Institute for Theoretical Physics, University of Innsbruck,
Technikerstrasse 25, A6020 Innsbruck, Austria
Steven.Bass@uibk.ac.at

The proton spin puzzle has challenged our understanding of QCD for the last 20 years. We survey new developments in theory and experiment. The proton spin puzzle seems to be telling us about the interplay of valence quarks with chiral dynamics and the complex vacuum structure of QCD.

Keywords: Proton spin puzzle.

1. Introduction

Protons behave like spinning tops. Unlike classical tops, however, the spin of these particles is an intrinsic quantum mechanical phenomenon. This spin is responsible for many fundamental properties of matter, including the proton's magnetic moment, the different phases of matter in low-temperature physics, the properties of neutron stars, and the stability of the known universe. How is the proton's spin built up from the spin and orbital angular momentum of the quarks and gluons inside?

Polarized deep inelastic scattering experiments have revealed a small value for the nucleon's flavour-singlet axial-charge $g_A^{(0)}|_{\rm pDIS} \sim 0.3$ suggesting that the quarks' intrinsic spin contributes little of the proton's spin. The challenge to understand the spin structure of the proton[1,2] has inspired a vast programme of theoretical activity and new experiments. Why is the quark spin content $g_A^{(0)}|_{\rm pDIS}$ so small ?

We start by recalling the g_1 spin sum-rules, which are derived starting from the dispersion relation for polarized photon-nucleon scattering and, for deep inelastic scattering, the light-cone operator product expansion. One finds that the first moment of the g_1 spin structure function is related to the scale-invariant axial-charges of the target nucleon by

$$\int_0^1 dx\, g_1^p(x, Q^2) = \left(\frac{1}{12} g_A^{(3)} + \frac{1}{36} g_A^{(8)} \right) \left\{ 1 + \sum_{\ell \geq 1} c_{{\rm NS}\ell}\, \alpha_s^\ell(Q) \right\}$$

$$+ \frac{1}{9} g_A^{(0)}|_{\rm inv} \left\{ 1 + \sum_{\ell \geq 1} c_{{\rm S}\ell}\, \alpha_s^\ell(Q) \right\} + \mathcal{O}(\frac{1}{Q^2}) + \beta_\infty. \quad (1)$$

Here $g_A^{(3)}$, $g_A^{(8)}$ and $g_A^{(0)}|_{\rm inv}$ are the isovector, SU(3) octet and scale-invariant flavour-singlet axial-charges respectively. The flavour non-singlet $c_{\rm NS\ell}$ and singlet $c_{\rm S\ell}$ Wilson coefficients are calculable in ℓ-loop perturbative QCD.[3] The term β_∞ represents a possible leading-twist subtraction constant from the circle at infinity when one closes the contour in the complex plane in the dispersion relation.[1] If finite, the subtraction constant affects just the first moment. The first moment of g_1 plus the subtraction constant, if finite, is equal to the axial-charge contribution.

In terms of the flavour dependent axial-charges

$$2Ms_\mu \Delta q = \langle p, s|\bar{q}\gamma_\mu\gamma_5 q|p, s\rangle \tag{2}$$

the isovector, octet and singlet axial-charges are:

$$g_A^{(3)} = \Delta u - \Delta d$$
$$g_A^{(8)} = \Delta u + \Delta d - 2\Delta s$$
$$g_A^{(0)}|_{\rm inv}/E(\alpha_s) \equiv g_A^{(0)} = \Delta u + \Delta d + \Delta s. \tag{3}$$

Here $E(\alpha_s) = \exp \int_0^{\alpha_s} d\tilde{\alpha}_s\, \gamma(\tilde{\alpha}_s)/\beta(\tilde{\alpha}_s)$ is a renormalization group factor which corrects for the (two loop) non-zero anomalous dimension $\gamma(\alpha_s)$ of the singlet axial-vector current,[4] $J_{\mu 5} = \bar{u}\gamma_\mu\gamma_5 u + \bar{d}\gamma_\mu\gamma_5 d + \bar{s}\gamma_\mu\gamma_5 s$, and which goes to one in the limit $Q^2 \to \infty$; $\beta(\alpha_s)$ is the QCD beta function. The singlet axial-charge, $g_A^{(0)}|_{\rm inv}$, is independent of the renormalization scale μ and corresponds to $g_A^{(0)}(Q^2)$ evaluated in the limit $Q^2 \to \infty$. The axial-charges $g_A^{(3)}$ and $g_A^{(8)}$ are renormalization group invariants.

If one assumes no twist-two subtraction constant ($\beta_\infty = O(1/Q^2)$) the axial-charge contributions saturate the first moment at leading twist. The isovector axial-charge is measured independently in neutron β-decays ($g_A^{(3)} = 1.270 \pm 0.003$[5]) and the octet axial-charge is commonly taken to be the value extracted from hyperon β-decays assuming a 2-parameter SU(3) fit ($g_A^{(8)} = 0.58 \pm 0.03$[6]). Using the sum-rule for the first moment of g_1, given in Eq. (1), polarized deep inelastic scattering experiments have been interpreted in terms of a small value for the flavour-singlet axial-charge. If we take $g_A^{(8)} = 0.58 \pm 0.03$, then inclusive g_1 data with $Q^2 > 1$ GeV^2 give[7]

$$g_A^{(0)}|_{\rm pDIS,Q^2\to\infty} = 0.33 \pm 0.03({\rm stat.}) \pm 0.05({\rm syst.}) \tag{4}$$

— considerably smaller than the value of $g_A^{(8)}$ quoted above.

In the naive parton model $g_A^{(0)}|_{\rm pDIS}$ is interpreted as the fraction of the proton's spin which is carried by the intrinsic spin of its quark and antiquark constituents. When combined with $g_A^{(8)} = 0.58 \pm 0.03$ the value of $g_A^{(0)}|_{\rm pDIS}$ in Eq.(4) corresponds to a negative strange-quark polarization

$$\Delta s_{Q^2\to\infty} = \frac{1}{3}(g_A^{(0)}|_{\rm pDIS,Q^2\to\infty} - g_A^{(8)}) = -0.08 \pm 0.01({\rm stat.}) \pm 0.02({\rm syst.}) \tag{5}$$

— that is, polarized in the opposite direction to the spin of the proton.

What physics separates the values of the octet and singlet axial-charges ?

2. Spin and the Singlet Axial-Charge $g_A^{(0)}$

There are two key issues: the physics interpretation of the flavour-singlet axial-charge $g_A^{(0)}$ and possible SU(3) breaking in the extraction of $g_A^{(8)}$ from hyperon β-decays.

First consider $g_A^{(0)}$. Gluonic information feeds into $g_A^{(0)}$ through the QCD axial anomaly. QCD theoretical analysis leads to the formula[1,8–11]

$$g_A^{(0)} = \left(\sum_q \Delta q - 3\frac{\alpha_s}{2\pi}\Delta g \right)_{\mathrm{partons}} + \mathcal{C}_\infty. \tag{6}$$

Here $\Delta g_{\mathrm{partons}}$ is the amount of spin carried by polarized gluons in the polarized proton ($\alpha_s \Delta g \sim$ constant as $Q^2 \to \infty$[8,9]) and $\Delta q_{\mathrm{partons}}$ measures the spin carried by quarks and antiquarks carrying "soft" transverse momentum $k_t^2 \sim P^2, m^2$ where P is a typical gluon virtuality and m is the light quark mass. The polarized gluon term is associated with events in polarized deep inelastic scattering where the hard photon strikes a quark or antiquark generated from photon-gluon fusion and carrying $k_t^2 \sim Q^2$.[10,11] $\mathcal{C}_\infty$ denotes a potential non-perturbative gluon topological contribution which is associated with the possible subtraction constant in the dispersion relation for g_1 and Bjorken $x = 0$:[1] $g_A^{(0)}|_{\mathrm{pDIS}} = g_A^{(0)} - \mathcal{C}_\infty$.

The subtraction constant, if finite, is a non-perturbative effect and vanishes in perturbative QCD. It is sensitive to the mechanism of axial U(1) symmetry breaking and the realization of axial U(1) symmetry breaking by instantons: spontaneous U(1) symmetry breaking by instantons naturally generates a subtraction constant whereas explicit symmetry breaking does not.[12] The QCD vacuum is a Bloch superposition of states characterised by non-vanishing topological winding number and non-trivial chiral properties. When we put a valence quark into this vacuum it can act as a source which polarizes the QCD vacuum with net result that the spin "dissolves" and some fraction of the spin of the constituent quark is associated with non-local gluon topology with support only at Bjorken $x = 0$.

Possible explanations for the small value of $g_A^{(0)}|_{\mathrm{pDIS}}$ extracted from the polarized deep inelastic experiments include screening from positive gluon polarization, negative strangeness polarization in the nucleon, a subtraction at infinity in the dispersion relation for g_1 associated with non-perturbative gluon topology and connections to axial U(1) dynamics,[13–16] as well as possible SU(3) breaking in $g_A^{(8)}$ – possibly as large as 20%.[17,18] The QCD axial anomaly decouples from the non-singlets $g_A^{(3)}$ and $g_A^{(8)}$.

One would like to understand the dynamics which appears to suppress the singlet axial-charge extracted from polarized deep inelastic scattering relative to the OZI prediction $g_A^{(0)} = g_A^{(8)}$ and also the sum-rule for the longitudinal spin structure of the nucleon

$$\frac{1}{2} = \frac{1}{2}\sum_q \Delta q + \Delta g + L_q + L_g \tag{7}$$

where L_q and L_g denote the orbital angular momentum contributions. There is presently a vigorous programme to disentangle the different contributions. Key experiments include semi-inclusive polarized deep inelastic scattering (COMPASS and HERMES) and polarized proton-proton collisions (PHENIX and STAR at RHIC), as well as deeply virtual Compton scattering to learn about total angular momentum.

3. The Shape of g_1

To understand the proton spin puzzle, it is interesting to look at the x dependence of the measured g_1 spin structure function. Deep inelastic measurements of g_1 have been performed in experiments at CERN, DESY, JLab and SLAC. There is a general consistency among all data sets. COMPASS are yielding precise new data at small x, down to $x \sim 0.004$. JLab are focussed on the large x region.

Precise measurements of the deuteron spin structure function g_1^d show the remarkable feature that g_1^d is consistent with zero in the small x region between 0.004 and 0.02.[7] In contrast, the isovector part of g_1 is observed to rise at small x as $\sim x^{-0.22\pm0.07}$ and is much bigger than the isoscalar part of g_1.[19] This is in sharp contrast to the situation in the unpolarized structure function F_2 where the small x region is dominated by isoscalar pomeron exchange. The g_1^{p-n} data are consistent with quark model and perturbative QCD predictions in the valence region $x > 0.2$.[20] The size of $g_A^{(3)}$ forces us to accept a large contribution from small x and the observed rise in g_1^{p-n} is in excellent agreement with the prediction $g_1^{p-n} \sim x^{-0.22}$ of hard Regge exchange - in particular a possible a_1 hard-pomeron cut[21] involving the hard-pomeron which seems to play an important role in unpolarized deep inelastic scattering.[22]

The "missing spin" is associated with a "collapse" in the isosinglet part of g_1 to something close to zero instead of a valence-like rise for x less than about 0.02. This isosinglet part is the sum of SU(3)-flavour singlet and octet contributions. If there were a large positive polarized gluon contribution to the proton's spin, this would act to drive the small x part of the singlet part of g_1 negative[23] – that is, acting in the opposite direction to any valence-like rise at small x. However, gluon polarization measurements at COMPASS, HERMES and RHIC constrain this spin contribution to be small in measured kinematics meaning that the sum of valence and sea quark contributions is suppressed at small x. (Soft Regge theory predicts that the singlet term should behave as $\sim N \ln x$ in the small x limit, with the coefficient N to be determined from experiment.[24,25])

There is presently a vigorous programme to disentangle the different contributions involving experiments in semi-inclusive polarized deep inelastic scattering and polarized proton-proton collisions.[26–29] These direct measurements show no evidence for negative polarized strangeness in the region $x > 0.006$ (in apparent contrast to the extraction of negative strangeness polarization extracted from inclusive measurements of g_1). For gluon polarization, present measurements suggest

$|-3\frac{\alpha_s}{2\pi}\Delta g| < 0.06$ corresponding to $|\Delta g| < 0.4$ with $\alpha_s \sim 0.3$. That is, they are not able to account for the difference $(g_A^{(0)}|_{\text{pDIS}} - g_A^{(8)}) \sim -0.25$ obtained via Eq.(4). An independent measurement of the strange-quark axial-charge could be made through neutrino-proton elastic scattering.[30] The axial-charge measured in νp elastic scattering is independent of any assumptions about the presence or absence of a subtraction at infinity in the dispersion relation for g_1 and the $x \sim 0$ behaviour of g_1. Further measurements to push the small x frontier in polarized deep inelastic scattering would be possible with a polarized ep collider.[31]

4. SU(3) Breaking and $g_A^{(8)}$

Given that the contributions to $g_A^{(0)}$ from the measured distribution Δs and from $-3\frac{\alpha_s}{2\pi}\Delta g$ are small, it is worthwhile to ask about the value of $g_A^{(8)}$. The value 0.58 is extracted from a 2 parameter fit to hyperon β-decays in terms of the SU(3) constants $F = 0.46$ and $D = 0.80$[6] – see Table 1. The fit is good to $\sim 20\%$ accuracy.[18,33] The uncertainty quoted for $g_A^{(8)}$ has been a matter of some debate. There is considerable evidence that SU(3) symmetry may be badly broken and some have suggested that the error on $g_A^{(8)}$ should be as large as 25%.[18] More sophisticated fits will also include chiral corrections. Calculations of non-singlet axial-charges in relativistic constituent quark models are sensitive to the confinement potential, effective colour-hyperfine interaction,[34–37] pion and kaon clouds plus additional wavefunction corrections[38] chosen to reproduce the physical value of $g_A^{(3)}$.

This physics has recently been investigated by Bass and Thomas[17] within the Cloudy Bag model (CBM)[38,39] which has the attractive feature that when pion cloud and quark mass effects are turned off the model reproduces the SU(3) analysis. One finds that chiral corrections significantly reduce the value of $g_A^{(8)}$. This, in turn, has the effect of increasing the value of $g_A^{(0)}|_{\text{pDIS}}$ and consequently reducing the absolute value of the "polarized strangeness" extracted from inclusive polarized deep inelastic scattering.

The Cloudy Bag[40] was designed to model confinement and spontaneous chiral symmetry breaking, taking into account pion physics and the manifest breakdown of chiral symmetry at the bag surface in the MIT bag. If we wish to describe proton spin data including matrix elements of $J_{\mu5}^3$, $J_{\mu5}^8$ and $J_{\mu5}$, then we would like to know that the model versions of these currents satisfy the relevant Ward identities.

Table 1. g_A/g_V from β-decays with $F = 0.46$ and $D = 0.80$, together with the mathematical form predicted in the MIT Bag with effective colour-hyperfine interaction (see text and Ref. 32).

Process	measurement	SU(3) combination	Fit value	MIT + OGE
$n \to p$	1.270 ± 0.003	$F + D$	1.26	$\frac{5}{3}B' + G$
$\Lambda^0 \to p$	0.718 ± 0.015	$F + \frac{1}{3}D$	0.73	B'
$\Sigma^- \to n$	-0.340 ± 0.017	$F - D$	-0.34	$-\frac{1}{3}B' - 2G$
$\Xi^- \to \Lambda^0$	0.25 ± 0.05	$F - \frac{1}{3}D$	0.19	$\frac{1}{2}B' - G$
$\Xi^0 \to \Sigma^+$	1.21 ± 0.05	$F + D$	1.26	$\frac{5}{3}B' + G$

For the scale-invariant non-singlet axial-charges $g_A^{(3)}$ and $g_A^{(8)}$, corresponding to the matrix elements of partially conserved currents, the model is well designed to make a solid prediction.

The effective colour-hyperfine interaction has the quantum numbers of one-gluon exchange (OGE). In models of hadron spectroscopy this interaction plays an important role in the nucleon-Δ and $\Sigma - \Lambda$ mass differences, as well as the nucleon magnetic moments[37] and the spin and flavor dependence of parton distribution functions.[41] It shifts total angular-momentum between spin and orbital contributions and, therefore, also contributes to model calculations of the octet axial-charges.[34–36] In Bag model calculations one also needs to include wavefunction corrections associated with the well known issue that, for the MIT and Cloudy Bag models, the nucleon wavefunction is not translationally invariant and the centre of mass is not fixed. To compare the model results with experiment we take the view[38] that, in principle, the model - with corrections - should give the experimental value of $g_A^{(3)}$. We therefore choose the centre-of-mass factor phenomenologically to give the experimental value of $g_A^{(3)}$. This then fixes the parameters of the model and allows us to use it to make a model prediction for $g_A^{(8)}$.

Without pion cloud corrections the MIT Bag with centre of mass corrections reproduces the SU(3) analysis of the axial-charges extracted from β-decays. This is illustrated in Table 1. Without additional physics input, e.g. pion chiral corrections, there is a simple algebraic relation between the SU(3) parameters F and D, the bag parameter B' and the OGE correction G: $F = \frac{2}{3}B' - \frac{1}{2}G$ and $D = B' + \frac{3}{2}G$. The numerical agreement is very good.[17]

The pion cloud of the nucleon also renormalizes the nucleon's axial-charges by shifting intrinsic spin into orbital angular momentum.[34,38] In the Cloudy Bag Model (CBM),[40] the nucleon wavefunction is written as a Fock expansion in terms of a bare MIT nucleon, $|N\rangle$, and baryon-pion, $|N\pi\rangle$ and $|\Delta\pi\rangle$, Fock states. The probabilities to find the nucleon in each Fock component are determined phenomenologically by fitting to a wealth of nucleon observables.[42] The expansion converges rapidly and we may safely truncate the Fock expansion at the one pion level. When we calculate the pion and kaon cloud chiral corrections to $g_A^{(8)}$ we also have to choose the chiral representation, in particular whether to use the original surface coupling or the later volume coupling version of the Cloudy Bag model.

The extent of the reduction in $g_A^{(8)}$ depends upon the version of the CBM used, lying in the range 0.49 ± 0.02 for the original CBM and 0.42 ± 0.02 for the volume coupling version.[17] These changes alone raise the value of $g_A^{(0)}|_{\text{pDIS},Q^2 \to \infty}$ derived from the experimental data from 0.33 ± 0.03(stat.)±0.05(syst.) to 0.35 ± 0.03(stat.)± 0.05(syst.) and 0.37 ± 0.03(stat.) ± 0.05(syst.), respectively. Both of these values have the effect of reducing the level of OZI violation associated with the difference $g_A^{(0)}|_{\text{pDIS}} - g_A^{(8)}$ from -0.25 ± 0.07 to just -0.14 ± 0.06 and -0.05 ± 0.06, respectively. It is this OZI violation which eventually needs to be explained in terms of singlet degrees of freedom: effects associated with polarized glue and/or a topological effect associated with $x = 0$.

The uncertainty in this model calculation lies in the small ambiguity between the two chiral representations that one can choose. In order to quote an overall value that properly encompasses these possibilities we follow the Particle Data Group procedure[43] for combining data that *may* not be compatible to estimate the overall error, finding a combined value of $g_A^{(8)} = 0.46 \pm 0.05$ (with the corresponding semi-classical singlet axial-charge or spin fraction being 0.42 ± 0.07 before inclusion of gluonic effects).[17] Note that the error ± 0.02 on $g_A^{(8)}$ quoted for each version of the model follows from varying over the phenomenological range of possible pion parameters within first order perturbation theory. In terms of analogy to experimental errors, the ± 0.02 is like a statistical error and the final ± 0.05 error includes systematic effects. With this final value for $g_A^{(8)}$ the corresponding experimental value of $g_A^{(0)}|_{\text{pDIS}}$ would increase to $g_A^{(0)}|_{\text{pDIS}} = 0.36 \pm 0.03 \pm 0.05$.

5. Towards Possible Understanding

Where are we in our understanding of the spin structure of the proton and the small value of $g_A^{(0)}|_{\text{pDIS}}$? Measurements of valence, gluon and sea polarization suggest that the polarized glue term $-3\frac{\alpha_s}{2\pi}\Delta g_{\text{partons}}$ and strange quark contribution $\Delta s_{\text{partons}}$ in Eq.(6) are unable to resolve the small value of $g_A^{(0)}|_{\text{pDIS}}$. Two explanations are suggested within the theoretical and experimental uncertainties depending upon the magnitude of SU(3) breaking in the nucleon and hyperon axial-charges. One is a value of $g_A^{(8)} \sim 0.5$ (as suggested by the surface coupling model) plus an axial U(1) topological effect at $x = 0$ associated with a finite subtraction constant in the g_1 dispersion relation. The second is a much larger pion cloud reduction of $g_A^{(8)}$ to a value ~ 0.4 (as suggested by the volume coupling model in first order pion cloud perturbation theory). Combining the theoretical error on the pion cloud chiral corrections embraces both possibilities. The proton spin puzzle seems to be telling us about the interplay of valence quarks with chiral dynamics and the complex vacuum structure of QCD.

Acknowledgments

The research of SDB is supported by the Austrian Science Fund (FWF grant P20436). I thank A. W. Thomas for collaboration on physics reported here and H. Fritzsch for the invitation to this stimulating meeting in honour of the 80th birthday of Prof. Murray Gell-Mann.

References

1. S. D. Bass, Rev. Mod. Phys. **77** (2005) 1257.
2. S. D. Bass, *The Spin structure of the proton* (World Scientific, 2008).
3. S. A. Larin, T. van Ritbergen and J. A. M. Vermaseren Phys. Lett. B **404** (1997) 153.
4. J. Kodaira, Nucl. Phys. B **165** (1980) 129.
5. Particle Data Group: C. Amsler *et al.*, Phys. Lett. B **667** (2008) 1.
6. F. E. Close and R. G. Roberts, Phys. Lett. B **316** (1993) 165.

7. COMPASS Collab. (V. Yu. Alexakhin *et al.*), Phys. Lett. B **647** (2007) 8.
8. G. Altarelli and G. G. Ross, Phys. Lett. B **212** (1988) 391.
9. A. V. Efremov and O. Teryaev, JINR Report No. E2-88-287.
10. R. D. Carlitz, J. C. Collins and A. Mueller, Phys. Lett. B **214** (1988) 229.
11. S D. Bass, B. L. Ioffe, N. N. Nikolaev and A. W. Thomas, J. Moscow Phys. Soc. **1** (1991) 317.
12. S D. Bass, Mod. Phys. Lett. A **13** (1998) 791.
13. S Narison, G. M. Shore and G. Veneziano, Nucl. Phys. B **433** (1995) 209.
14. G. M. Shore, hep-ph/0701171.
15. H. Fritzsch, Phys. Lett. B **229** (1989) 122.
16. S. D. Bass, Phys. Lett. B **463** (1999) 286.
17. S. D. Bass and A. W. Thomas, Phys. Lett. B **684** (2010) 216.
18. R. L. Jaffe and A. Manohar, Nucl. Phys. B **337** (1990) 509.
19. COMPASS Collab. (M. G. Alekseev *et al.*), arXiv:1001.4654 [hep-ex]
20. S. D. Bass, Eur. Phys. J. A **5** (1999) 17.
21. S. D. Bass, Mod. Phys. Lett. A **22** (2007) 1005.
22. J R. Cudell, A. Donnachie and P. V. Landshoff, Phys. Lett. B **448** (1999) 281.
23. S. D. Bass and A. W. Thomas, J. Phys. G **19** (1993) 925.
24. S. D. Bass and P. V. Landshoff, Phys. Lett. B **336** (1994) 537.
25. F. E. Close and R. G. Roberts, Phys. Lett. B **336** (1994) 257.
26. S. D. Bass, Mod. Phys. Lett. A **24** (2009) 1087.
27. G. K. Mallot, hep-ex/0612055.
28. HERMES Collab. (A. Airapetian *et al.*), Phys. Rev. D **71** (2005) 012003.
29. COMPASS Collab. (M. Alekseev *et al.*), Phys. Lett. B **680** (2009) 217.
30. S. D. Bass, R. J. Crewther, F. M. Steffens and A. W. Thomas, Phys. Rev. D **66** (2002) 031901 (R).
31. S. D. Bass and A. De Roeck, Nucl. Phys. B (Proc. Suppl.) **105** (2002) 1.
32. E. Hogaasen and F. Myhrer, Z Phys. C **48** (1990) 295.
33. E. Leader and D. B. Stamenov, Phys. Rev. D **67** (2003) 037503.
34. F. Myhrer and A. W. Thomas, Phys. Rev. D **38** (1988) 1633.
35. F. Myhrer and A. W. Thomas, Phys. Lett. B **663** (2008) 302.
36. A. W. Thomas, these proceedings.
37. F. E. Close, *An Introduction to Quarks and Partons* (Academic, N.Y., 1978).
38. A. W. Schreiber and A. W. Thomas, Phys. Lett. B **215** (1988) 141.
39. T. Yamaguchi, K. Tsushima, Y. Kohyama and K. Kubodera, Nucl. Phys. A **500** (1989) 429.
40. A. W. Thomas, Adv. Nucl. Phys. **13** (1984) 1.
41. F. E. Close and A. W. Thomas, Phys. Lett. B **212** (1988) 227.
42. A. W. Thomas, Prog. Theor. Phys. **168** (2007) 614.
43. C. Amsler *et al.*, (Particle data Group), Phys. Lett. B **667** (2008) 1; see sect. 5.2.2.

IS THERE UNIFICATION IN THE 21ST CENTURY?

YUAN K. HA

Department of Physics, Temple University,
Philadelphia, Pennsylvania 19122, USA
yuanha@temple.edu

In the last 100 years, the most important equations in physics are Maxwell's equations for electrodynamics, Einstein's equation for gravity, Dirac's equation for the electron and Yang–Mills equation for elementary particles. Do these equations follow a common principle and come from a single theory? Despite intensive efforts to unify gravity and the particle interactions in the last 30 years, the goal is still to be achieved. Recent theories have not answered any question in physics. We examine the issues involved in this long quest to understand the ultimate nature of spacetime and matter.

Keywords: Unification; gravity; spacetime; matter.

1. Introduction

At the end of the 19th century, physicists were very confident that they had the laws of nature at hand. Classical mechanics had been firmly established for 200 years. Celestial mechanics was highly developed. Electrodynamics was discovered. Thermodynamics was understood. The euphoria was so evident that in a speech given by Albert Michelson in 1894 the following remarks was said:[1]

> "*The more important fundamental laws and facts of physical science have all been discovered, and these are so firmly established that the possibility of their ever being supplanted in consequence of new discoveries is exceedingly remote ... Our future discoveries must be looked for in the sixth place of decimals.*"

Since then, physicists have identified four fundamental interactions together with their interaction strengths:

(1) Electromagnetic interactions — 10^{-2}.
(2) Weak interactions — 10^{-5}.
(3) Strong interactions — 10^0.
(4) Gravitational interactions — 10^{-38}.

They have also discovered a set of fundamental particles: the quarks (u, d, s, c, b, t); the leptons $(e, \mu, \tau, \nu_e, \nu_\mu, \nu_\tau)$; the gauge bosons $(\gamma, W^+, W^-, Z^0, G^{\alpha\beta}$, and the still anticipated Higgs boson H^0. These interactions and particles are governed by four fundamental equations:

(1) Maxwell's equations (1864).
(2) Einstein's equation (1915).
(3) Dirac's equation (1928).
(4) Yang–Mills equation (1954).

The theories constructed for the fundamental interactions are gauge theories based on various Lie groups:

(1) Quantum Electrodynamics — U(1).
(2) Quantum Electroweak Theory — SU(2)×U(1).
(3) Quantum Chromodynamics — SU(3).
(4) Classical General Relativity — SO(3, 1).

The goal of unification is not simply to combine the various fundamental interactions in a consistent mathematical framework. It should entail a unifying principle and produce an interlocking structure. It should answer long-standing questions and make new predictions. The search for unification would force physicists to confront fundamental issues, to abandon old dogmas and to recognize new realities. The quantum theory of elementary particles has been quite successful in that it can explain accurately a number of phenomena in the strong, weak and electromagnetic interactions. On the other hand, it is gravity which is the most challenging and the least understood of the four interactions. We shall therefore focus our attention in this article to the problems in gravity. Without a deeper understanding of the nature of gravity and the theories which claim to explain it, unification is pointless as recent attempts to unify the interactions have not answered any question in physics.

2. Supersymmetry

The goal of supersymmetry is to unify spacetime and internal symmetries of elementary particles, thereby evading the Coleman–Mandula theorem which states that all possible symmetries of the S-matrix under general assumptions can only be a direct product of the Poincare algebra and an internal symmetry algebra. In supersymmetry, there exists a symmetry between fermions and bosons and the prediction is for every boson there exists a corresponding fermion of the same mass and quantum numbers. The role of supersymmetry is to cancel divergences in the perturbative calculations of quantum field theory since fermion and boson have opposite signs in loop corrections. In the standard model, quadratic corrections to the Higgs mass due to Yukawa interactions appear that cause its mass to diverge. Supersymmetry does away with the corrections by supplying terms with a minus sign. This scheme works through some physical cutoff mechanism, and there is a scale associated with it. Within the dimensional regularization approach, however, quadratic divergences do not exist and it is not clear what purpose would be served by a supersymmetric theory. From another point of view, the goal of supersymmetry is not to double the number of fundamental particles. The doubling of particles has already been achieved by the existence of antiparticles. Antiparticles are crucial in virtual particle pair creations and annihilations in quantum field theory. So far there is no irrefutable evidence that supersymmtry is a symmetry of nature after 40 years. According to Veltman:[2]

> "*The concept of naturalness is usually cited as the underlying motivation for supersymmetry. We will challenge that concept, and in any case need to point out that there is nothing natural about the development of the theory itself. Its main success is its agility in dodging the facts. The dubious explanation of the convergence of the three scale coupling constants into a single point can not be taken seriously. It is just another fit, using some of the many free parameters.*"

It should be pointed out that coupling constant unification does not prove unification of the strong, weak and electromagnetic interactions. There are other particles that can produce coupling constant unification.

3. Higher Dimensions

Many unification theories involve higher spacetime dimensions. There is nothing compelling about higher dimensions themselves. They may simply be a book-keeping device to account for the number of observed gauge fields. Gauge transformations are coordinate transformations in higher-dimensional space. In general, higher-dimensional theories suffer from instability and causality problems. There are negative energy solutions of the field equation. In Kaluza–Klein type theories of pure gravity in higher dimensions, the difficulty is noticeable at the classical level. Analysis of the perihelion shift of planets in the solar system shows that the shift depends on the total number of spatial dimensions in these theories.[3] The decomposition of metric assumes only a compact internal space with the geometry of tori. The result is independent of the size of the extra dimension, even if it is of sub-millimeter scale. Starting from the multidimensional Einstein equation, a nonrelativistic limit of the metric in four dimensions can be obtained. The metric coefficients are found to depend explicitly on the total number of spatial dimensions D and they affect the equation of motion in general relativity. In the perihelion shift calculation of the planet Mercury, the resulting formula is given by

$$\frac{D}{(D-2)}\frac{\pi}{2}\frac{m^2c^2R_s^2}{M^2}, \tag{1}$$

where m is the mass of the planet; M, the mass of the Sun; R_s, the Schwarzschild radius of the Sun and c is the speed of light. The observed discrepancy for Mercury is 43.11 ± 0.21 arcsec per century. Only the ordinary three-dimensional case $D = 3$ gives a satisfactory result $42.94''$ which is within the measurement accuracy. For $D = 4$, the result is $28.63''$ and for $D = 9$ it is $18.40''$. Thus all multidimensional case $D > 3$ contradict observations.

In the deflection of light by the Sun, a corresponding analysis provides the formula[4]

$$\frac{D-1}{D-2}\frac{R_s}{R},\tag{2}$$

where R is the radius of the Sun. The observed deflection of a light ray that grazes the Sun surface has a historical value of 1.75 arcsec. For the three-dimensional case $D = 3$ the above formula reproduces this value accurately. For $D = 4$, the result is $1.31''$ and for $D = 9$ it is $1.00''$. Again, the multidimensional case shows a severe problem with the classical tests of general relativity. The implication of incorporating Kaluza-Klein type theories in unification is rather obvious.

4. Higher Derivative Gravity Theories

A number of theories known as higher derivative gravity theories have the goal of constructing a renormalizable theory of gravity explicitly in four dimensions. The Lagrangians contain higher order curvature invariants in Riemannian geometry such as those of scalar curvature R^2, Ricci curvature $R_{\alpha\beta}R^{\alpha\beta}$, Riemann curvature $R_{\alpha\beta\mu\nu}R^{\alpha\beta\mu\nu}$ and other combinations of these terms, including Weyl curvature invariant $C_{\alpha\beta\mu\nu}C^{\alpha\beta\mu\nu}$, in order that the equations be invariant under general coordinate transformations. These theories generally have problems with stability, unitarity, ghosts and nonlocality.[5] None of them is yet successful as a quantum theory of gravity. A further problem of higher derivative theories at the classical level is that none admits Birkhoff's theorem,[6] which states that spherically symmetric solution is unique and time-independent. The failure of Birkhoff's theorem in higher derivative gravity theories means that spherically symmetric solution is time-dependent and dynamical. A similar failure of Birkhoff's theorem in a generalization of Einstein's gravity called $f(R)$ theory, in which the action is a nonlinear function of the scalar curvature R, also shows that spherically symmetric solutions are time-dependent.[7] As a result, black holes in these theories are dynamical. Their horizons disappear and a naked singularity will emerge.[8] In some $f(R)$ models, relativistic stars cannot exist due to the dynamics of the effective scalar degree of freedom and there are doubts about the viability of these models.[9]

5. Alternative Gravity Theories

There are still other efforts to modify Einstein's gravity theory in order to achieve a finite and consistent theory of quantum gravity. These are generally known as modified gravity theories.[10] The modification can take place both at the microscopic scale and at the macroscopic scale.[11] At very small distances near the Planck length, modifications have included discrete spacetime, breaking of discrete symmetries, Lorentz symmetry violation, nonlocal interaction, extra dimensions, and non-commutative coordinates. However, these modification effects are extremely tiny to be noticed at their current observational levels. Decoupling at the Planck scale prevents these effects from being observed at low energies. At large distances, modifications have included varying speed of light, varying gravitational constant, modifying Newton's Second Law of motion, non-symmetric metric and incorporating scalar, vector, and tensor particles into Einstein's gravity. The difficulty at this end is to obtain agreement with all astrophysical and solar system observations. So far none of these alternative theories of gravity has succeeded in replacing general relativity as the best theory of gravity.

A more fruitful approach to understand gravity is to develop quantum field theory of particles in curved spacetime.[12] This is done by treating spacetime classically and matter fields quantum mechanically. It is possible to study particle creation in strong gravitational fields. This has led to the prediction of Hawking radiation in which particles are emitted from a black hole with a thermal spectrum;[13] the Unruh effect in which an observer under acceleration in vacuum sees a thermal collection of particles;[14] and interpreting Einstein's equation as a thermodynamic equation of state of spacetime and matter,[15] thereby realizing toward an emergent theory of gravity.[16]

6. Is Spacetime Quantum?

In special relativity, the Lorentz transformation is a pseudo-rotation in four-dimensional Minkowski spacetime. It is not possible to include the Planck constant or any other parameter into the transformation. It is a purely mathematical transformation. Therefore there is no such theory to be called quantum theory of special relativity. This term has a completely different meaning from relativistic quantum mechanics which is a description of matter. Similarly, it is not possible to include Planck's constant in general coordinate transformations, or to have a quantum theory of Riemannian spacetime. Since geometry is gravity in general relativity, this calls into the question whether gravity really needs to be quantized.[17,18] Spacetime originally is a macroscopic concept. Is it possible that Einstein's equation is similar in nature to Navier-Stokes equation in fluid mechanics as a macroscopic theory?[19]

The investigation of quantum black holes[20] shows that they are extremely microscopic objects with a macroscopic mass. Their Schwarzschild radius is equal to their Compton wavelength. They exist at the boundary between classical and quantum regions. They obey the Laws of Thermodynamics and they decay into elementary particles. A quantum black hole of the size of the Planck length 1.6×10^{-33} cm has a mass of 2.2×10^{-5} gm. Like the nucleus of a heavy atom, quantum black holes may require the use of quantum mechanics but not necessarily quantum field theory for their description. The difference between quantum mechanics and quantum field theory is tremendous — it is the creation and annihilation of particles. There are no anti-black holes in general relativity. Therefore there are no virtual pair creations and annihilations of black holes as in ordinary particles. Two black holes combine to form another black hole according to the area non-decrease theorem. The resulting black hole evaporates according to Hawking's description with a temperature. Quantum black holes are intrinsically semi-classical objects.

7. Quantum Gravity in Crisis

An important result in cosmology was obtained recently which can elucidate the nature of spacetime down to the smallest scale. This is the observation of the highest energy gamma rays from a gamma ray burst GRB 090510 by the Fermi Gamma-Ray Space Telescope.[21] A single 31-GeV photon was detected from a source at a redshift of $z = 0.903$ which corresponds to a distance of 7.3 billion light years from Earth. It was the last of the seven pulses in a short burst that lasted for 0.829 s. One of the two postulates of Einstein's special relativity is Lorentz invariance in that all observers measure exactly the same speed of light in vacuum, independent of the motion of the source and of the photon energy. In certain quantum theories of gravity, there is great interest in the possibility that Lorentz invariance might be broken near the Planck scale due to quantum fluctuation of spacetime and the notion of spacetime foam. A variation of photon speed is an indication that Lorentz invariance is violated. This may be revealed by observing the sharp features in the gamma ray burst light-curves. If the spread in travel time of less than 0.9 s between the highest and lowest-energy gamma rays in the burst GRB 090510 is all attributed to quantum effects, then a thorough analysis shows that any quantum effects in which the speed is linearly proportional to energy do not show up until the distance is down to about $0.8L_{Pl}$, which is below the Planck length. This result therefore rules out a number of quantum gravity models that predict such linear variation with energy.

The gamma ray burst reported above is significant in that it allows for the exploration of spacetime near Planck length by using effects accumulated over cosmological distances since direct access to Planck energy in experiments is not possible. The result indicates that there is no evidence so far of any quantum nature of spacetime above the Planck length. Spacetime there is smooth and continuous. The speed of light is constant and special relativity is right. At the Planck length, quantum black holes would appear in observation and they effectively provide a natural cutoff to spacetime. For observable purpose, it is not necessary to consider theories below the Planck length. Further detections using gamma ray bursts with even higher energy photons will settle the question of quantum spacetime definitively. It would be amazing that in effect spacetime is classical and there is no need for a quantum theory of gravity. There would be an underlying theory for gravity which is not gravity, just as statistical mechanics is the underlying theory of thermodynamics. Unification would have a very different meaning from the current understanding involving quantum gravity as a fundamental premise.

References

1. A. Michelson — Opening of Ryerson Laboratory at University of Chicago.
2. M. Veltman, *Acta Phys. Pol. B* **25**, 1399 (1994).
3. M. Eingorn and A. Zhuk, arXiv: 0912.2698.
4. M. Eingorn and A. Zhuk, arXiv: 1003.5690.
5. T. Chiba, *JCAP* **03**, 008 (2005).

6. P. Havas, *Gen. Rel. Grav.* **8**, 631 (1977).
7. V. Faraoni, *Phys. Rev. D* **81**, 044002 (2010).
8. V. Faraoni, arXiv: 1005.5398.
9. T. Kobayashi and K. Maeda, *Phys. Rev. D* **78**, 064019 (2008).
10. J. W. Moffat, *Reinventing Gravity* (HaperCollins, New York, 2008).
11. C. P. Burgess, arXiv: 0912.4295.
12. R. M. Wald, *Quantum Field Theory in Curved Spacetime and Black Hole Thermodynamics* (The University of Chicago Press, Chicago, 1994).
13. S. W. Hawking, *Commun. Math. Phys.* **43**, 199 (1975).
14. W. H. Unruh, *Phys. Rev. D* **14**, 870 (1976).
15. T. Jacobson, *Phys. Rev. Lett.* **75**, 1260 (1995).
16. T. Padmanabhan, *AIP Conf. Proc.* **939**, 114 (2007).
17. S. Carlip, *Class. Quant. Grav.* **25**, 154010 (2008).
18. S. Boughn, *Found. Phys.* **39**, 331 (2009)
19. Y. K. Ha, *Int. J. Mod. Phys. A* **24**, 3577 (2009).
20. Y. K. Ha, arXiv: 0812.5012.
21. A. Abdo *et al.*, *Nature* **462**, 331 (2009).

NUCLEON POLARIZED STRUCTURE FUNCTION

F. ARBABIFAR

Physics Department, Semnan University, Semnan, Iran
Arbabifar_f@yahoo.com

A. N. KHORRAMIAN

Physics Department, Semnan University, Semnan, Iran
School of Particles and Accelerators,
Institute for Research in Fundamental Sciences (IPM),
P.O. Box 19395-5531,Tehran, Iran
Khorramiana@theory.ipm.ac.ir

S. ATASHBAR TEHRANI

School of Particles and Accelerators,
Institute for Research in Fundamental Sciences (IPM),
P.O. Box 19395-5531, Tehran, Iran
Atashbar@ipm.ir

A. NAJAFGHOLI

Physics Department, Semnan University, Semnan, Iran

The alternative approach to QCD analysis for quark distributions and structure functions for polarized deep inelastic scattering is presented. We use very recently experimental data to parameterize our model for quark and gluon distributions up to NLO approximation. Our calculation is based on Jacobi polynomials expansion of the polarized structure functions. The final results are in good agreement with the other theoritical models and experimental data.

Keywords: Jacobi polynomials; NLO approximation.

1. Introduction

The concept of the structure of the proton has appeared more than 30 years ago and is still very important phenomenological tool for analyzing processes involving nucleons. A major goal in the study of Quantum Chromo Dynamics (QCD) in recent years has been the detailed investigation of the spin structure of the nucleon and determination of the partonic composition of its spin projection. One of the cleanest processes to access Polarized Parton Distribution Functions (PPDF's) and also Polarized Structure Functions (PSF's) is deep inelastic scattering in which one lepton scatters off a polarized nucleon. The spin structure function of the proton has been measured for the first time almost 15 years ago by the SMC experiment at

CERN. Recently some theoretical and experimental studies on the spin structure of the nucleon has been discussed in great detail in several recent reviews[1-6] and comprehensive analysis of the polarized deep inelastic scattering data, based on next-to-leading-order quantum chromodynamics, have appeared too.[7-9] In these analysis the polarized parton distribution functions are written in terms of the well-known parameterizations of the unpolarized parton distribution functions or parameterized independently, the unknown parameters are determined by fitting the polarized DIS data.

In this paper we study the impact of the recent very precise HERMES, EMC, SMC, E142, E143, E144, E155, COMPASS[10-15,17-20] inclusive polarized DIS data on the determination of polarized parton distributions in the nucleon. These experiments give important information about the nucleon structure in quite different kinematic regions.

Since these new experimental data are just for different values of Q^2 and not fixed Q^2, one can not use the Bernstein polynomial expansion method. In this method it is needed to use experimental data for each bin of Q^2 separately, so it has seemed suitable to use the Jacobi polynomial expansion method. In our last calculations[21,22] we applied the Jacobi polynomials to determine the polarized valon distributions.

The following sections describe in more detail the general layout: Section 2 contains the method of the QCD analysis of polarized structure function, based on Jacobi polynomials. Parametrization of parton densities are written down in Section 3. In Sections 4 and 5 we present the used experimental data and a brief review of QCD fit procedure. Conclusions and comparison with other models and experimental data are shown in Section 6.

2. The Method

In this section we will discuss how to calculate the Q^2-dependence of the PPD's provided at a certain reference point Q_0^2.

Our method is based on Jacobi polynomial to determine PPD's up to NLO approximation. Then we will obtain some unknown parameters to parameterize PPD's at Q_0^2. By using Jacobi polynomials we are able to apply for all data points and not just a series of data points and this is our motivation to use Jacobi polynomials and not Bernstein polynomial to study spin dependent of parton distribution functions. Jacobi polynomials also allow us to factor out an essential part of the x-dependence of the structure function into the weight function.[23] Therefore polarized structure function xg_1 can be reformed in a new form of the series[24,25]

$$xg_1(x, Q^2) = x^\beta (1-x)^\alpha \sum_{n=0}^{N_{max}} a_n(Q^2)\Theta_n^{\alpha,\beta}(x) , \qquad (1)$$

where we have

$$\Theta_n^{\alpha,\beta}(x) = \sum_{j=0}^{n} c_j^{(n)}(\alpha, \beta)x^j , \qquad (2)$$

which is the Jacobi polynomials of order n and N_{max} is the number of polynomials, $c_j^{(n)}(\alpha, \beta)$ are the coeficients that expressed through Γ-functions and satisfy the orthogonality relation as following

$$\int_0^1 dx\, x^\beta (1-x)^\alpha \Theta_k^{\alpha,\beta}(x) \Theta_l^{\alpha,\beta}(x) = \delta_{k,l} \ . \tag{3}$$

By inserting Eq. (2) into Eq. (1) and using Eq. (3), we obtain Q^2 dependence of the Jacobi moments by inverting the final form of Eq. (1)

$$a_n(Q^2) = \int_0^1 dx\, x g_1(x, Q^2) \Theta_k^{\alpha,\beta}(x)$$

$$= \sum_{j=0}^n c_j^{(n)}(\alpha, \beta) \mathbf{M}[x g_1, j+2] \ . \tag{4}$$

Now we can relate polarized strucure function with its moments using above equations

$$x g_1^{N_{max}}(x, Q^2) = x^\beta (1-x)^\alpha \sum_{n=0}^{N_{max}} \Theta_n^{\alpha,\beta}(x) \sum_{j=0}^n c_j^{(n)}(\alpha, \beta) \mathbf{M}[x g_1, j+2] \ , \tag{5}$$

where $\mathbf{M}[x g_1, j+2]$ is the moments of PSF in Mellin-N space. Here we choose $N_{max} = 9$, $\alpha = 3.0$ and $\beta = 0.5$ to acheive the fastest convergence of the series on the R.H.S of equation above and also to get the most accurate $x g_1$.

3. Parametrization of the Polarized Parton Distributions

In the chosen small input scale $Q_0^2 = 4.0$ GeV^2, the input hadronic distribution functions are parameterized as follows

$$x \delta u_v = A_{u_v} \eta_{u_v} x^{a_{u_v}} (1-x)^{b_{u_v}} (1 + c_{u_v} x)$$

$$x \delta d_v = A_{d_v} \eta_{d_v} x^{a_{d_v}} (1-x)^{b_{d_v}} (1 + c_{d_v} x)$$

$$x \delta \bar{q} = A_s \eta_s x^{a_s} (1-x)^{b_s}$$

$$x \delta g = A_g \eta_g x^{a_g} (1-x)^{b_g} \ , \tag{6}$$

where the term x^{a_i} stands for the low-x behavior of the parton densities and $(1-x)^{b_i}$ is for large value of x. The reminder polynomial factor accounts for the additional medium-x degree of freedom. We can express the normalization factors A_i as functions of a_i, b_i and c_i

$$A_i^{-1} = \left(1 + c_i \frac{a_i}{a_i + b_i + 1}\right) B\left(a_i, b_i + 1\right) \tag{7}$$

to make η_i the first moments of $\Delta q_i(x, Q_0^2)$, $\eta_i = \int_0^1 dx \Delta q_i(x, Q_0^2)$. The function $B(a, b)$ is Euler Beta function expressed by

$$B(a, b) = \Gamma(a)\Gamma(b)/\Gamma(a+b) \ . \tag{8}$$

Notice that in above parametrization we considered $SU(3)$ flavor symmetry.

In the present approach the QCD-evolution equations are solved in Mellin space and the Mellin transform of the parton densities is performed and Mellin n moment are calculate for complex argument N:

$$\mathbf{M}[\delta f_i(x, Q_0^2)](N) = \int_0^1 x^{N-1} \delta f_i(x, Q_0^2) dx$$
$$= \eta_i A_i \left(1 + c_i \frac{N - 1 + a_i}{N + a_i + b_i}\right) B(N - 1 + a_i, b_i + 1) , \quad (9)$$

here $f_i = q_v, d_v, \bar{q}, g$. At the LO approximation the Mellin moments for the polarized structure function $g_1^p(N, Q^2)$ is related to $g_1^p(x, Q^2)$, contained the polarized distribution functions, as follows

$$\mathbf{M}[g_1(x, Q^2)](N) = g_1^p(N, Q^2) = \int_0^1 x^{N-1} g_1^p(x, Q^2) dx , \quad (10)$$

whereas at the NLO approximation $g_1(N, Q^2)$ can be represented in terms of the polarized parton densities, δq and $\delta \bar{q}$, and the coefficient functions δC_i^N in the Mellin -N space

$$g_1^p(N, Q^2) = \frac{1}{2} \sum_q e_q^2 \{(1 + \frac{\alpha_s}{2\pi} \Delta C_q^N)[\Delta q(N, Q^2) + \Delta \bar{q}(N, Q^2)] + \frac{\alpha_s}{2\pi} 2\Delta C_g^N \Delta g(N, Q^2)\} ,$$

$$(11)$$

where N_f denotes the number of active flavors and α_s stands for NLO running coupling constant.[26] The Wilson coefficients ΔC_q^N and ΔC_g^N have been calculated in perturbative QCD in Mellin-N space:

$$\Delta C_q^N = \frac{4}{3} \left[-S_2(N) + (S_1(N))^2 + \left(\frac{3}{2} - \frac{1}{N(N+1)}\right) S_1(N) + \frac{1}{N^2} + \frac{1}{2N} + \frac{1}{N+1} - \frac{9}{2} \right]$$

$$(12)$$

and

$$\Delta C_g^N = \frac{1}{2} [-\frac{N-1}{N(N+1)} (S_1(N) + 1) - \frac{1}{N^2} + \frac{2}{N(N+1)}] \quad (13)$$

with definite $S_k(N)$.[2]

In LO, Eq. (11) reduces to

$$g_1^p(N, Q^2) = \frac{1}{2} \sum_q e_q^2 [\Delta q(N, Q^2) + \Delta \bar{q}(N, Q^2)] , \quad (14)$$

since then ΔC_q^N and ΔC_g^N vanish.

The singlet and non-singlet parton and glueon densities evolution which accur in Eq. (11) are expressed by

$$\delta q_{NS\pm}^n(Q^2) = \left\{ 1 + \frac{\alpha_s(Q^2) - \alpha_s(Q_0^2)}{2\pi} \left(-\frac{2}{\beta_0}\right) \left(\delta P_{NS\pm}^{(1)n} - \frac{\beta_1}{2\beta_0} \delta P_{qq}^{(0)n}\right) \right\}$$
$$\times L^{-(2/\beta_0)\delta P_{qq}^{(0)n}} \delta q_{NS\pm}^n(Q_0) + \mathcal{O}(\alpha_s^2) \quad (15)$$

and

$$\begin{pmatrix} \delta\Sigma^n(Q^2) \\ \delta g^n(Q^2) \end{pmatrix} = \left\{ L^{-(2/\beta_0)\delta\widehat{P}^{(0)n}} + \frac{\alpha_s(Q^2)}{2\pi}\widehat{U}L^{-(2/\beta_0)\delta\widehat{P}^{(0)n}} - \frac{\alpha_s(Q_0^2)}{2\pi}L^{-(2/\beta_0)\delta\widehat{P}^{(0)n}}\widehat{U} \right\}$$
$$\times \begin{pmatrix} \delta\Sigma^n(Q_0^2) \\ \delta g^n(Q_0^2) \end{pmatrix} + O(\alpha_s^2) . \tag{16}$$

The definitions of $\widehat{U}$ and $\delta\widehat{P}^{(i)n}$ are determined.[2] Now we are able to do QCD fits using the available data to extract 15 unknown parameters involved in reconstructed form of $g_1^p(N, Q^2)$.

Table 1. Published data points above $Q^2 = 1.0$ GeV^2.

Experiment	x-range	Q^2-range [GeV^2]	number of data points	Ref.
E143(p)	0.031 −0.749	1.27 − 9.52	28	[17]
HERMES(p)	0.023 − 0.660	1.01 − 7.36	19	[10]
HERMES(p)	0.023 − 0.660	2.5 (Fixed)	20	[10]
SMC(p)	0.005 − 0.480	1.30 − 58.0	12	[13]
EMC(p)	0.015 − 0.466	3.50 − 29.5	10	[14]
E155(p)	0.015 − 0.750	1.22 − 34.72	24	[18]
HERMES(p)	0.026 − 0.731	1.12 −14.29	51	[11]
COMPASS(p)	0.005 − 0.479	1.3 − 54.8	12	[19]
proton			176	
E143(d)	0.031 − 0.749	1.27 − 9.52	28	[17]
E155(d)	0.015 − 0.750	1.22 − 34.79	24	[18]
E142(d)	0.035 − 0.466	1.10 − 5.50	8	[16]
HERMES(d)	0.026− 0.731	1.12 − 14.29	51	[12]
deuteron			111	
HERMES(n)	0.033 − 0.464	1.22 − 5.25	9	[11]
E154(n)	0.017 − 0.564	1.20 − 15.0	17	[16]
HERMES(n)	0.026 − 0.731	1.12 − 14.29	51	[12]
COMPASS(n)	0.004 − 0.566	1.10 − 62.1	15	[19]
neutron			92	
total			379	

4. Data

The remarkable growth of experimental data on inclusive polarized deep-inelastic scattering of leptons off nucleons over the last years allows to perform refined QCD fit of polarized structure functions in order to reveal the spin-dependent partonic structure of the nucleon. For the QCD analysis presented in the present paper the following data sets are used: the HERMES proton,neutron and deuteron data,[10–12] the SMC proton and deuteron data,[13] the EMC proton data,[14,15] the E142 neutron data,[16] the E143 proton and deuteron data,[17] the E154 neutron data,[16] the E155 proton and deuteron data,[18] the COMPASS proton data.[19] The number of the published data points for the different data sets are summarized in Table 1 for data on g_1 together with the $x-$ and Q^2-ranges for different experiments.

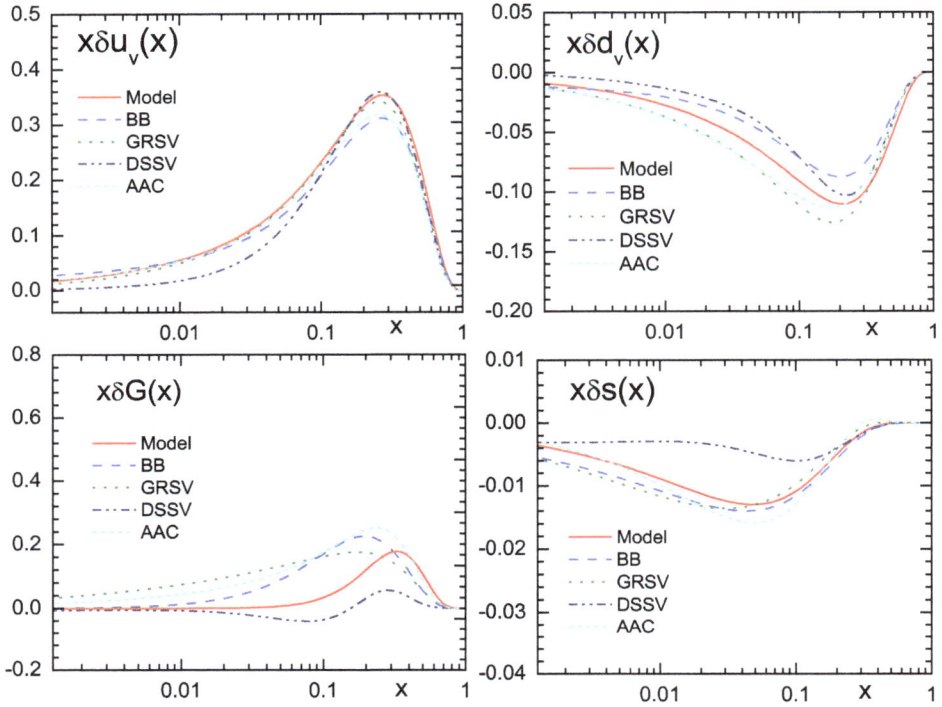

Fig. 1. NLO polarized parton distributions at the input scale Q_0^2=4.0 GeV^2 (solid line) compared to results obtained by BB[27] (dashed line), GRSV[29] (dotted line), DSSV[30] (dashed-dotted-dotted line) and AAC[31] (short dashed line).

5. QCD Fit Procedure

In this section we briefly present the results of the QCD analysis of the polarized structure function g_1^p at the NLO approximation. In the fitting procedure we started with the 15 unknown parameters and $\alpha_s(Q_0^2)$ to be determined. It is essential to reduce the number of parameters to meet the quality of present data and the reliability of the fitting program by some constraints given in the last studies.[27] We have performed a global χ^2 fit for all 379 set of experimental data points using the CERN subroutine MINUIT[28]. Systematic and statistical errors were also added in quadrature. We allow for a relative normalization shift between the different data sets within the normalization uncertainties ΔN_n coming from the measurement of the luminosity and the beam target polarization. Thus the normalization shift N_i enters an additional term in the χ^2 expression for each data set

$$\chi^2_{global} = \sum_n \omega_n \chi^2_n , \quad \text{(n labels the different experiments)}$$

$$\chi^2_n = (\frac{1 - N_n}{\Delta N_n})^2 + \sum_i (\frac{N_n g_{1,i}^{data} - g_{1,i}^{theor}}{N_n \Delta g_{1,i}^{data}})^2 , \tag{17}$$

where $g_{1,i}^{data}$, $\Delta g_{1,i}^{data}$ and $g_{1,i}^{theor}$ stand for data value, measurement uncertainty and theoritical value for the ith data point respectively, the default value of weighting factor ω_n is 1. Using program MINUIT we found an acceptable fit with minimum $\chi^2/\text{n.d.f.} = 0.73$ in NLO approximation.

6. Conclusions

We have investigated the QCD analysis of the inclusive deep-inelastic charged lepton-nucleon scattering data up to next to leading order by using Jacobi polynomials. We also derived parameterizations of polarized parton distributions at a starting scale Q_0^2 together with $\alpha_s(Q_0^2)$, in deriving polarized distributions some unknown parameters were introduced which should be determined by fitting to experimental data. As shown in Fig. 1, the results of the present analysis are in good agreement with other models.

References

1. M. Anselmino, A. Efremov and E. Leader, Phys. Rept. **261** (1995) 1.
2. B. Lampe and E. Reya, Phys. Rept. **332** (2000) 1 [arXiv:hep-ph/9810270].
3. E. W. Hughes and R. Voss, Ann. Rev. Nucl. Part. Sci. **49** (1999) 303.
4. B. W. Filippone and X. D. Ji, Adv. Nucl. Phys. **26** (2001) 1 [arXiv:hep-ph/0101224].
5. A. Mirjalili, A. N. Khorramian and S. Atashbar-Tehrani, Nucl. Phys. Proc. Suppl. **164** (2007) 38.
6. A. Mirjalili, S. Atashbar Tehrani and A. N. Khorramian, Int. J. Mod. Phys. A **21** (2006) 4599 [arXiv:hep-ph/0608224].
7. J. Blumlein and H. Bottcher, Nucl. Phys. B **636** (2002) 225 [arXiv:hep-ph/0203155].
8. A. N. Khorramian, A. Mirjalili and S. A. Tehrani, JHEP **0410** (2004) 062.
9. A. Mirjalili, A. N. Khorramian, S. Atashbar Tehrani and H. Mahdizadeh Saffar, Acta Phys. Polon. B **40** (2009) 2965.
10. A. Airapetian *et al.* [HERMES Collaboration], Phys. Lett. B **442** (1998) 484.
11. [HERMES Collaboration], [arXiv:0609039 [hep-ex]].
12. [HERMES Collaboration], Phys. Rev. D **75** (2007) 012007.
13. B. Adeva *et al.* [Spin Muon Collaboration], Phys. Rev. D **58** (1998) 112001.
14. J. Ashman *et al.* [European Muon Collaboration], Phys. Lett. B **206** (1988) 364.
15. J. Ashman *et al.* [European Muon Collaboration], Nucl. Phys. B **328** (1989) 1.
16. P. L. Anthony *et al.* [SLAC E142 Collaboration], Phys. Rev. D **54** (1996) 6620; K. Abe *et al.* [SLAC/E154 Collaboration], Phys. Rev. Lett. **79** (1997) 26.
17. K. Abe *et al.* [E143 collaboration], Phys. Rev. D **58**, 112003 (1998).
18. P. L. Anthony *et al.* [SLAC E155 Collaboration], Phys. Lett. B **463** (1999) 339.
19. and M. G. Alekseev [The COMPASS Collaboration], [arXiv:1001.4654 [hep-ex]].
20. D. Adams *et al.* [Spin Muon Collaboration (SMC)], Phys. Rev. D **56** (1997) 5330.
21. S. Atashbar Tehrani and A. N. Khorramian, JHEP **0707**, 048 (2007).
22. A. N. Khorramian and S. Atashbar Tehrani, AIP Conf. Proc. **915** (2007) 420.
23. G. Parisi and N. Sourlas, *Nucl. Phys.* **B151** (1979) 421.
24. I. S. Barker and B. R. Martin, *Z. Phys.* **C24** (1984) 255; S. P. Kurlovich, A. V. Sidorov and N. B. Skachkov, JINR Report E2-89-655, Dubna, 1989.
25. I. S. Barker, C. S. Langensiepen and G. Shaw, Nucl. Phys. B **186** (1981) 61.
26. J. Hejbal, (2008) [arXiv:0811.2382v1 [hep-ph]].

27. J. Blumlein and H. Bottcher, [arXiv:1005.3113 [hep-ph]].
28. F. James, CERN Program Library Long Writeup D506; F. James and M. Roos, Comput. Phys. Commun. **10** (1975) 343.
29. M. Gluck, E. Reya, M. Stratmann and W. Vogelsang, Phys. Rev. D **63** (2001) 094005.
30. D. de Florian, R. Sassot, M. Stratmann and W. Vogelsang, Phys. Rev. Lett. **101** (2008) 072001; Phys. Rev. D **80** (2009) 034030.
31. Y. Goto et al. [Asymmetry Analysis collaboration], Phys. Rev. D **62** (2000) 034017; M. Hirai, S. Kumano and N. Saito [Asymmetry Analysis Collaboration], Phys. Rev. D **69** (2004); M. Hirai, S. Kumano and N. Saito, Phys. Rev. D **74** (2006) 014015; M. Hirai and S. Kumano [Asymmetry Analysis Collaboration], Nucl. Phys. B **813** (2009) 106.

GELL-MANN'S ANGULAR MOMENTUM AND CURRENTS

KESHAV N. SHRIVASTAVA

Department of Physics, Faculty of Science, University of Malaya,
Kuala Lumpur 50603, Malaysia
keshav1001@yahoo.com

In 1965, Gell-Mann has shown that the magnetic moments of the neutron and the proton are related to the linear combinations of vectors and the axial vector, $\mathcal{F}_{i\alpha}$ and $\mathcal{F}_{i\alpha}{}^5$ which obey the same commutation relations as $\mathcal{G}_{i\alpha}$ and $\mathcal{G}_{i\alpha}{}^5$. These currents are also expressed as products of Dirac $\gamma_\alpha \lambda_i$ and $\gamma_\alpha \gamma_5 \lambda_i$ matrices. The index i varies from 0 to 8 with $\lambda_0 = (2/3)^{1/2}1$. This explains that $\mu_n/\mu_p = -2/3$. It is possible to express the magnetic moments in terms of g values, g_n and g_p. The ratio of the magnetic moments can be treated as the ratio of g values. We find that $g = 2j+1/2l+1$ gives the effective charge of a particle in a given state. Although the charge of a particle in high energy physics depends on isospin, baryon number, etc, it does not depend on spin. We show that the charge of the electron in different states is derivable from the spin. For spin $1/2$, the effective charge of the electron in a magnetic field becomes $(1/2)ge = [(l + (1/2) \pm s)/(2l + 1)]e$ which for positive sign gives $(l + 1)/(2l + 1)$ which is $2/3$ for $l = 1$ and for the negative sign, $l/2l+1$ gives 0 for $l = 0$ and $1/3$ for $l = 1$. Using the helicity, which changes the sign of s, we can obtain the charges of 0, ± 1 from the spin and further values can be developed by changing the orbital angular momentum quantum number. Hence, fractional charge of the electron can be explained by the angular momentum.

Keywords: Fractional charge; angular momentum; spin-charge relation.

1. Introduction

Gell-Mann developed the SU_2, SU_3 and SU_6 to describe the fundamental particles by angular momentum algebra. First, the proton and the neutron are described by SU_2 and later 8 baryons, all of spin $1/2$ and positive parity are obtained. The set of 8 mesons, all of spin zero and negative parity are obtained. Further development of this model leads to the correct prediction of particles. A set of three particles of charges 0, ± 1 is obtained by means of isospin which can be extended to particles of other masses by inventing the hypercharge. Hence, the charge, Q, of a particle is determined by the isospin, I, and the hypercharge, Y, by the expression, $Q = (1/2)Y + I_3$ which can be further corrected by introducing more quantum numbers. In this theory the set of 8 particles have a constant spin so that the charge does not depend on the spin. Gell-Mann obtained the correct ratio for the magnetic moments of the neutron and the proton by using the matrix representation. The ratio of the g-values of the neutron and the proton is correctly predicted by Gell-Mann.[1,2] We extend this theory to the effective charge of the electron in various states. The predicted values agree with the experimental values of the effective charge of the electron in various states. We describe a theory of charge of the electron in various states in a magnetic field which does use the true spin. Since the charge occurs in the Bohr magneton, the g values multiply the charge and hence determine the effective charge of the electron. The calculated values of the charge of the electron are in good agreement with those measured in the Hall effect experiments.

2. Gell-Mann's Theory

In 1965, Gell-Mann explained the magnetic moment ratio of neutron to proton, $\mu_n/\mu_p = -2/3$ by using A_{ej} matrices. The matrices $A_{io}(i = 1, \ldots, 8)$ are just $(3/2)^{1/2} F_i$ which are components of F connected by SU(3). The F_i matrices are given below:

$$F_1 = \frac{1}{2} \begin{pmatrix} 0 & 1 & 0 \\ 1 & 0 & 0 \\ 0 & 0 & 0 \end{pmatrix}, \quad F_2 = \frac{1}{2} \begin{pmatrix} 0 & -i & 0 \\ i & 0 & 0 \\ 0 & 0 & 0 \end{pmatrix}, \quad F_3 = \frac{1}{2} \begin{pmatrix} 1 & 0 & 0 \\ 0 & -1 & 0 \\ 0 & 0 & 0 \end{pmatrix}$$

$$F_4 = \frac{1}{2} \begin{pmatrix} 0 & 0 & 1 \\ 0 & 0 & 0 \\ 1 & 0 & 0 \end{pmatrix}, \quad F_5 = \frac{1}{2} \begin{pmatrix} 0 & 0 & -i \\ 0 & 0 & 0 \\ i & 0 & 0 \end{pmatrix}, \quad F_6 = \frac{1}{2} \begin{pmatrix} 0 & 0 & 0 \\ 0 & 0 & 1 \\ 0 & 1 & 0 \end{pmatrix} \quad (1)$$

$$F_7 = \frac{1}{2} \begin{pmatrix} 0 & 0 & 0 \\ 0 & 0 & -i \\ 0 & i & 0 \end{pmatrix}, \quad F_8 = \frac{1}{2\sqrt{3}} \begin{pmatrix} 1 & 0 & 0 \\ 0 & 1 & 0 \\ 0 & 0 & -2 \end{pmatrix}$$

The space integrals of all components of the vector current octet $\mathcal{F}_{i\alpha}$ and the axial vector current octet $F_{i\alpha}{}^5$ obey the same commutation rules as the currents, $\mathcal{G}_{i\alpha}$ and $\mathcal{G}_{i\alpha}{}^5$, respectively. In terms of quark field, q,

$$\mathcal{G}_{i\alpha} = (1/2)i\bar{q}\gamma_\alpha\lambda_i q \quad (2a)$$

$$\mathcal{G}_{i\alpha}{}^5 = (1/2)i\bar{q}\gamma_\alpha\gamma_5\lambda_i q. \quad (2b)$$

For the algebra to be closed, a ninth pair of currents is added by introducing, $\lambda_o = (2/3)^{1/2}$ (1) with $i = 0, 1, \ldots, 8$. For $\alpha = 1, 2, 3$ and 4 and nine values of I, $\alpha \times I$ gives $9 \times 4 = 36$ values. Similarly, $\gamma_\alpha\gamma_5\lambda_I$ also has 36 values so that there are a total of 72 values. We rearrange $\int \mathcal{G}_{i\alpha} d^3x$ and $\int \mathcal{G}_{i\alpha}{}^5 d^3x$ to obtain 72 Hermitian operators,

$$\mathcal{G}_{ij}^{\pm} = (3/32)^{1/2} \int q^+ \lambda_i \sigma_j (1 \pm \gamma_5) q d^3x \quad (3)$$

$i = 0, \ldots, 8$, $j = 0, \ldots, 3$. Here $\sigma_o = 1$ and one set of λ matrices is given by Eq. (4). The linear combinations of $\int \mathcal{F}_{i\alpha} d^3x$ and $\int \mathcal{F}_{i\alpha}{}^5 d3^x$ are called $A_{ij}{}^{\pm}$ which generate the algebra of U(6)⊗U(6). $G_{ij}{}^{\pm}$ also generate the U(6)⊗U(6).

$$\lambda_1 = \begin{pmatrix} 0 & 1 & 0 \\ 1 & 0 & 0 \\ 0 & 0 & 0 \end{pmatrix}, \quad \lambda_2 = \begin{pmatrix} 0 & -i & 0 \\ i & 0 & 0 \\ 0 & 0 & 0 \end{pmatrix}, \quad \lambda_3 = \begin{pmatrix} 1 & 0 & 0 \\ 0 & -1 & 0 \\ 0 & 0 & 0 \end{pmatrix}$$

$$\lambda_4 = \begin{pmatrix} 0 & 0 & 1 \\ 0 & 0 & 0 \\ 1 & 0 & 0 \end{pmatrix}, \quad \lambda_5 = \begin{pmatrix} 0 & 0 & -i \\ 0 & 0 & 0 \\ i & 0 & 0 \end{pmatrix}, \quad \lambda_6 = \begin{pmatrix} 0 & 0 & 0 \\ 0 & 0 & 1 \\ 0 & 1 & 0 \end{pmatrix} \tag{4}$$

$$\lambda_7 = \begin{pmatrix} 0 & 0 & 0 \\ 0 & 0 & -i \\ 0 & i & 0 \end{pmatrix}, \quad \lambda_8 = \begin{pmatrix} 1/\sqrt{3} & 0 & 0 \\ 0 & 1/\sqrt{3} & 0 \\ 0 & 0 & -2/\sqrt{3} \end{pmatrix} \lambda_o = (2/3)^{1/2} \begin{pmatrix} 1 & 0 & 0 \\ 0 & 1 & 0 \\ 0 & 0 & 1 \end{pmatrix}.$$

3. Angular Momentum

$A_{ij}{}^{\pm}$ form a system of approximate symmetries of hadrons. There is a term in the energy density, $\mathcal{H} = -\theta_{44}$ which is

$$(1/2i)(q^+\alpha \cdot \nabla q - \nabla q^+ \cdot \alpha q). \tag{5}$$

It breaks SU(6) down to SU(3). Some other interaction, $\mathcal{H}''$ can break SU(6) to SU(3) and split a multiplet even with $L = 0$. When $L = 2$ is added vectorially to $S = 1/2$ for the octet and $S = 3/2$ for the decimet, the resonances are obtained. Amongst the generators A_{ij}, A_{oo} are proportional to the baryon number and it commutes with other generators which give SU(6). The generators A_{io} ($i = 1, 2, \dots, 8$) are just $(3/2)^{1/2} F_i$ where F_i are the components of the F spin connected by SU(3) symmetry. The generators, $S_j = A_{oj}(j = 1, 2, 3)$ act exactly like a spin angular momentum since they obey the rules,

$$[S_i, S_j] = ie_{ijk}S_k \tag{6a}$$

$$[J_i, S_i] = ie_{ijk}S_k. \tag{6b}$$

The total angular momentum is $\mathbf{J}$ so that the difference is, $\mathbf{L} = \mathbf{J} - \mathbf{S}$ which obeys the rules. $\mathbf{L}$ need not be a purely orbital angular momentum. It can include the intrinsic spin of the extra particles. The operator μ_j transforms like $e_{jkl} \int x_k \mathcal{F}_{el} d^3 x$, where the index e refers to the charge direction in SU(3) space. To first order in $\mathcal{H}''$, we can get an effective μ_j operator that has $\mathbf{L} = 0$ and transforms under U(6). The effective μ_j has the form,

$$\mu_j = a_1 A_{ej} + a_2 e_{jkl}\{A_{lk}, A_{ol}\}. \tag{7}$$

The second term vanishes under time reversal invariance. Thus, the nucleon magnetic moments obey the rule,

$$\mu_n/\mu_p = -2/3 = -0.666. \tag{8}$$

characteristic of the first term alone. The experimental values of the magnetic moments of the neutron and the proton are almost equal to this value. Let us write the proton and neutron magnetic moments in terms of g values. For neutrons,

$g_s = -3.826$ and for protons, $g_l = 1$ and $g_s = 5.585$. The ratio, $-3.826/5.585 = -0.685$ is very near the value of $-2/3$ predicted by Gell-Mann. Hence, we find that g values play an important role in determining the magnetic moments.

4. The Electrons: Shrivastava's Theory

Gell-Mann introduced the idea of using the g values to obtain the magnetic moments but avoided the use of spin in determining the masses and charges of particles. We use the idea of g values to get the effective charge of the electron in a state but we use the real spin. We also use the idea of two-particle states and resonances to obtain the charge of more than 101 electron states. We take the g values of the electron as,

$$g = \frac{2j+1}{2l+1} \tag{9}$$

and $j = l \pm s$. Hence, the g value can be written as,

$$g = \frac{2(l \pm s)+1}{2l+1} . \tag{10}$$

For $l = 0$, $s = 1/2$ for the positive sign, $g = 2$. Hence it is convenient to define $(1/2)g$ as,

$$\frac{1}{2}g = \frac{l + \frac{1}{2} \pm s}{2l+1} . \tag{11}$$

The definition of the Bohr magneton is $\mu_B = e\hbar/2mc$ where m is the mass of the electron. The Hamiltonian of the electron in a magnetic field is $\mathcal{H}_o = g\mu_B$ H.S where H is the magnetic field. We observe that the g factor multiplies the Bohr magneton which has the charge of the electron. Hence, we define the effective charge of the electron as,

$$e_{\text{eff}} = \frac{1}{2}ge . \tag{12}$$

For $l = 0$, $s = 1/2$ and positive sign, $e_{\text{eff}} = e$. Hence, the spin-only value of the effective charge is the same as the charge of the electron. For $s = 1/2$, $l = 0$, there are two quasiparticles. For positive sign, the effective charge is 1 and for the negative sign, it is zero. Hence, we predict the quasiparticles of charges 0 and 1. By changing the helicity, we predict a quasiparticle of charge -1. Hence, the effective charge expression predicts three quasiparticles of charges 0, $+1$ and -1. Now for $l = 0$, we obtain the effective charge of,

$$e_{\text{eff}}/e = (1/2) \pm s \tag{13}$$

which shows the real spin-charge relationship. For $s = 1/2$, there are two series of charges,

$$\frac{1}{2}g_+ = \frac{l+1}{2l+1} \quad \text{and} \quad \frac{1}{2}g_- = \frac{l}{2l+1} . \tag{14}$$

We tabulate these charges in the Table 1.

Table 1. For $s = 1/2$, and the various values of l, the predicted charges.

S. No.	l	$(1/2)g_-$	$(1/2)g_+$
1	0	0	1
2	1	1/3	2/3
3	2	2/5	3/5
4	3	3/7	4/7
5	4	4/9	5/9

In this way, we generate a series of effectively charged quasiparticles for electrons in a magnetic field.[3-6] Thus we have first obtained the quasiparticles of charges 0, ±1 and then a series of charged quasiparticles. These quasiparticles assume that mass of the electron in the Bohr magneton is kept constant. We introduce the notation, $\nu_\pm = (1/2)g_+$. We take two, quasiparticles of effective charges ν_1 and ν_2. Then we predict a two-particle state of charge $\nu_1 + \nu_2$. For example, we take $\nu_1 = 1/3$ and $\nu_2 = 4/9$. Then a two particle state with charge $1/3 + 4/9$ is predicted. Since the charge is proportional to the energy, we define the charge of a resonance state by $\nu_1 - \nu_2 = (1/3) - (4/9) = -1/9$ or $(4/9) - (1/3) = 1/9$. Hence, the two-particle states as well as resonances are predicted. We have tabulated about 101 effectively charged quasiparticles. So far we have not made use of Landau levels which introduces one more quantum number through the harmonic oscillator type energy levels for electrons in a magnetic field, $\hbar\omega(n + 1/2)$. The quantization of the magnetic field in the Hall effect produces plateaus in the resistivity as a function of magnetic field. Replacing the charge of the electron by the effective charge and quantizing the field produces the "quantum Hall effect" correctly. More than 101 predicted effective charges are the same as those found in the experimental data.[7,8] In Gell-Mann's theory, the angular momentum using SU_{2-6} produces all of the hadrons with the correct charges and masses without the need of Coulomb correlations.

5. Laughlin's Theory

Laughlin[9] has the idea that $\langle \psi_m | \mathcal{H} | \psi_m \rangle$ will form an energy lower than that of the charge-density waves with m leading to fractional charge when flux-quantization is used along with incompressibility. Here $\mathcal{H}$ contains the kinetic energy of the electrons and the repulsive Coulomb interactions along with the positive potential. Laughlin's theory also does not use the idea of isospin or real spin but it uses the idea of incompressibility, $a_o = 1$. Here, $a_o{}^2 = 1$ cm^2 puts the magnetic field at 4.135667×10^{-7} Gauss which is too small compared with the magnetic field used in the experiments. As a case of theory of compressible states, the field at which incompressibility occurs is only 0.413 micro Gauss. The algebraic equations without incompressibility can not be solved because then the area and the charge are two variables whereas there is only one equation of flux quantization, $Ba_o{}^2 = hc/e$, where B is the magnetic induction field, $a_o{}^2$ is the area in which flux is quantized, h is the Planck's constant and c is the velocity of light. Thus Laughlin's theory can not resolve whether the area is to be fractionalized or the charge will become fractional. The value of m in Laughlin's wave function ψ_m can not be calculated whether it is 3 or 103. The ψ_m also does not have the particle-hole symmetry. It does not provide an expression of the form, $\nu_1 + \nu_2 = 1$ where $\nu_1 = 1/m_1$ and $\nu_2 = 1/m_2$ so the fractions do not add up to 1. Laughlin's ground state energy is about 91 percent of the Wigner value. It is very clear that Laughlin's wave function does not represent the ground state of the Coulomb Hamiltonian. It was found that the Laughlin's wave function has a very small range so that it is the ground state of a potential, $V_2(|r|) = \Sigma_{j=0} c_j b^{2j} \nabla^{2j} \delta^2(r)$ which is positive (repulsive). As the range tends to zero, $b \to 0$, only the leading term contributes to energy. The average $\langle c_o \delta^2(r) \rangle$ vanishes for antisymmetric wave function so that the pure δ-function never contributes. The Laughlin's wave function gives the zero-energy ground state exactly with the interaction given by, $V(r) = \lambda a_o^2 \nabla^2 \delta(r)$.[10,11] That is not a serious problem when we are receptive to a new Hamiltonian. Hence the more important question is whether it fractionalizes the charge? The flux quantization is $Ba_o^2 = n'hc/e$. The density is given by, $\sigma_m = (m2\pi a_o^2)^{-1}$. If the charge is fractionalized, e becomes e/m so that the flux quantization condition becomes, $Ba_o^2 = n'hc/(e/m)$ which can also be written as $B = n'hc/\{e(a_o^2/m)\}$. Hence, charge need not fractionalize. In order to obtain fractional charge Laughlin used an additional equation of constraint, $a_o^2 = 1$, which is not justified and makes the states "incompressible". So that the charge will change to e/m for incompressible states at a field of 0.4 micro Gauss. For compressible states the charge remains a constant and only the area a_o^2 will change. If m changes from 1 to 3, the value of a_o^2/m will change and charge need not change. Laughlin's wave function is independent of spin and there is no Boltzmann's factor between the two spin states. Usually, the spin-orbit interaction gives a small contribution to the ground state energy. However, if the spin occurred in a more complicated way, there may be a serious effect of spin on the unperturbed Hamiltonian. Hence, neglect of spin in the Laughlin's paper is not justified. We thus find that the Laughlin's wave function need not give the ground state of the Coulomb Hamiltonian. It need not fractionalize the area in which flux is quantized. Apparently, the Laughlin's wave function has a large overlap with that of a system with small, such as 2 or 3 or 10, number of electrons. Hence the applicability to a many-body system is missing. Laughlin's wave function is the ground state of a δ function Hamiltonian but there is no way to transform these δ interactions to a Coulomb interaction. What Laughlin got is the (i) wave function of small range, (ii) a new δ function Hamiltonian, (iii) overlaps with small systems, (iv) spinless particles but (v) not the fractional charge.

6. Conclusions

Gell-Mann has correctly predicted the ratio of magnetic moments of neutron and proton from a theory based on SU_2. The ratio of the magnetogyric factors is found to be proportional to the ratio of magnetic moments. Gell-Mann's theory describes the charges and masses emerge from the hypercharge and other quantum numbers. In the original Gell-Mann theory, the true spin is independent of the charge of the particles. We have also studied the charge in various states of the electron and we find that particles of charge 0, ± 1 can be predicted by using true spin. We increase the number of quasiparticles from three of charges, 0, ± 1, by changing the value of the orbital angular momentum. The two-particle states as well as the resonances are found. The calculated electron states agree with the states found for the electron in a magnetic field.[12–14] Flux quantization and the spin, predict quasiparticle states which agree with those found in the experimental measurements of the quantum Hall effect in solids.

Acknowledgments

This work is continued from the University of Hyderabad, India. It is supported by the Malaysian Academy of Sciences through the Scientific Advancement Grants Allocation (SAGA). We acknowledge support from the Fundamental Research Grants Scheme (FRGS) of the Ministry of Higher Education, Malaysia.

References

1. M. Gell-Mann, *Phys. Rev. Lett.* **14**, 77 (1965).
2. M. Gell-Mann, *Phys. Rev.* **125**, 1067 (1962).
3. K. N. Shrivastava, *Phys. Lett. A* **113**, 435 (1986); *115*, 495 (1986) (E).
4. K. N. Shrivastava, *Phys. Lett. A* **326**, 469 (2004).
5. K. N. Shrivastava, *AIP Conf. Proc.* **1017**, 422 (2008).
6. K. N. Shrivastava, *AIP Conf. Proc.* **1150**, 59 (2009).
7. H. L. Stormer, *Rev. Mod. Phys.* **71**, 875(1999).
8. W. Pan *et al.*, *Phys. Rev. B* **77**, 075307 (2008).
9. R. B. Laughlin, *Phys. Rev. Lett.* **50**, 1395 (1983).
10. S. A. Trugman and S. Kivelson, *Phys. Rev. B* **31**, 5280 (1985).
11. E. H. Rezayi and A. H. MacDonald, *Phys. Rev. B* **44**, 8395 (1991).
12. K. N. Shrivastava, *AIP Conf. Proc.* **1250**, 261 (2010).
13. K. N. Shrivastava, *AIP Conf. Proc.* **1250**, 27 (2010).
14. K. N. Shrivastava, *AIP Conf. Proc.* **1169**, 48 (2009).

SPIN AND FLAVOR CONTENT OF NUCLEON IN THE CHIRAL QUARK MODEL

MANMOHAN GUPTA,[1,*] HARLEEN DAHIYA,[2] GULSHEEN AHUJA[1] and J. M. S. RANA[3]

[1] *Department of Physics, Centre of Advanced Study, P.U., Chandigarh, India*

[2] *Department of Physics, Dr. B.R. Ambedkar N.I.T., Jalandhar, India*

[3] *Department of Physics, H.N.B. Garhwal University, India*

mmgupta@pu.ac.in

The chiral constituent quark model with configuration mixing (χCQM$_{\text{config}}$) is known to be successful in describing the spin and flavor content of the nucleon. A brief description of the model along with its application for the quark distribution functions, quark polarizations as well as the strangeness quark content of the nucleon is carried out.

Keywords: Chiral constituent quark model; spin and flavor content of the nucleon.

The chiral constituent quark model with configuration mixing (χCQM$_{\text{config}}$)[1–3] is able to provide a satisfactory description of spin and flavor structure of nucleon as well as several features such as $\bar{u} - \bar{d}$ asymmetry,[4–6] existence of significant strange quark content $\bar{s}$ in the nucleon, various quark flavor contributions to the proton spin, baryon magnetic moments and hyperon β–decay parameters, etc. Similarly, the χCQM$_{\text{config}}$, coupled with the quark sea polarization and orbital angular momentum, is also able to describe some of the subtle features of baryon magnetic moments.

Recently, several experiments[7–10] have indicated that the strangeness contribution to magnetic moment of the proton ($\mu(p)^s$) may be non zero, in contrast to its zero value predicted by the naive quark model. In fact, some recent calculations by Dong et al.[11] as well as some lattice based calculations[12] also indicate it to be non zero. This immediately brings up the issue of the presence of strange quarks in the nucleon.

The purpose of the present communication is to briefly describe the basic formalism of χCQM$_{\text{config}}$ and to discuss its prediction regarding the contribution of strange quarks to polarization functions as well as to quark distribution functions. To this end, we have not carried out a detailed and extensive analysis, rather we have emphasized only the key features.

To begin with, the key to understand the "proton spin problem", in the chiral constituent quark model,[1] is the fluctuation process

$$q^{\pm} \rightarrow \text{GB} + q^{'\mp} \rightarrow (q\bar{q}^{'}) + q^{'\mp}, \tag{1}$$

where GB represents the Goldstone boson and $q\bar{q}' + q'$ constitute the "quark sea". The effective Lagrangian describing interaction between quarks and a nonet of GBs, can be expressed as

$$\mathcal{L} = g_8 \bar{\mathbf{q}} \Phi \mathbf{q} + g_1 \bar{\mathbf{q}} \frac{\eta'}{\sqrt{3}} \mathbf{q} = g_8 \bar{\mathbf{q}} \left(\Phi + \zeta \frac{\eta'}{\sqrt{3}} I \right) \mathbf{q} = g_8 \bar{\mathbf{q}} \left(\Phi' \right) \mathbf{q}, \qquad (2)$$

where $\zeta = g_1/g_8$, g_1 and g_8 are the coupling constants for the singlet and octet GBs, respectively, I is the 3×3 identity matrix. The GB field which includes the octet and the singlet GBs is written as

$$\Phi' = \begin{pmatrix} \frac{\pi^0}{\sqrt{2}} + \beta \frac{\eta}{\sqrt{6}} + \zeta \frac{\eta'}{\sqrt{3}} & \pi^+ & \alpha K^+ \\ \pi^- & -\frac{\pi^0}{\sqrt{2}} + \beta \frac{\eta}{\sqrt{6}} + \zeta \frac{\eta'}{\sqrt{3}} & \alpha K^0 \\ \alpha K^- & \alpha \bar{K}^0 & -\beta \frac{2\eta}{\sqrt{6}} + \zeta \frac{\eta'}{\sqrt{3}} \end{pmatrix} \text{ and } q = \begin{pmatrix} u \\ d \\ s \end{pmatrix}. \quad (3)$$

SU(3) symmetry breaking is introduced by considering $M_s > M_{u,d}$ as well as by considering the masses of GBs to be non degenerate ($M_{K,\eta} > M_\pi$), whereas the axial U(1) breaking is introduced by $M_{\eta'} > M_{K,\eta}$. The parameter $a(= |g_8|^2)$ denotes the probability of chiral fluctuation $u(d) \to d(u) + \pi^{+(-)}$, whereas $\alpha^2 a$, $\beta^2 a$ and $\zeta^2 a$ respectively denote the probabilities of fluctuations $u(d) \to s + K^{-(0)}$, $u(d,s) \to u(d,s) + \eta$, and $u(d,s) \to u(d,s) + \eta'$.

The details of χCQM$_{config}$ have already been discussed in Ref. 3, however for the sake of readability of the manuscript some essential details of configuration mixing as well as chiral constituent quark model have been presented in the sequel. The most general configuration mixing in the case of octet baryons can be expressed as[13-15]

$$|B\rangle = \left(|56, 0^+\rangle_{N=0} \cos\theta + |56, 0^+\rangle_{N=2} \sin\theta \right) \cos\phi +$$

$$\left(|70, 0^+\rangle_{N=2} \cos\theta' + |70, 2^+\rangle_{N=2} \sin\theta' \right) \sin\phi, \qquad (4)$$

where ϕ represents the $|56\rangle - |70\rangle$ mixing, θ and θ' respectively correspond to the mixing among $|56, 0^+\rangle_{N=0} - |56, 0^+\rangle_{N=2}$ states and $|70, 0^+\rangle_{N=2} - |70, 2^+\rangle_{N=2}$ states. For the present purpose, it is adequate[3,14,16,17] to consider the mixing only between $|56, 0^+\rangle_{N=0}$ and the $|70, 0^+\rangle_{N=2}$ states and the corresponding "mixed" octet of baryons is expressed as

$$|B\rangle \equiv \left| 8, \frac{1}{2}^+ \right\rangle = \cos\phi |56, 0^+\rangle_{N=0} + \sin\phi |70, 0^+\rangle_{N=2}, \qquad (5)$$

for details of the spin, isospin and spatial parts of the wavefunction, we refer the reader to Ref. 18.

To understand the quark distribution functions and the spin polarization functions in the χCQM$_{config}$, we present some of the details which are affected by the presence of configuration mixing given in equation (5). In this context, the spin structure of a nucleon is defined as[1,3,19,20]

$$\hat{B} \equiv \langle B|N|B\rangle, \qquad (6)$$

where $|B\rangle$ is the nucleon wavefunction defined in Eq. (5) and N is the number operator given by

$$N = n_{u+}u^+ + n_{u-}u^- + n_{d+}d^+ + n_{d-}d^- + n_{s+}s^+ + n_{s-}s^-\,, \qquad (7)$$

where $n_{q\pm}$ are the number of $q^\pm$ quarks. The spin structure of the "mixed" nucleon, defined through the Eq. (5), is given by

$$\left\langle 8, \frac{1}{2}^+ \Big| N \Big| 8, \frac{1}{2}^+ \right\rangle = \cos^2\phi \langle 56, 0^+|N|56, 0^+\rangle + \sin^2\phi \langle 70, 0^+|N|70, 0^+\rangle. \qquad (8)$$

The contribution to the proton spin in $\chi\mathrm{CQM}_{\mathrm{config}}$, given by the spin polarizations defined as $\Delta q = q^+ - q^-$, can be written as

$$\Delta u = \cos^2\phi \left[\frac{4}{3} - \frac{a}{3}(7 + 4\alpha^2 + \frac{4}{3}\beta^2 + \frac{8}{3}\zeta^2) \right] + \sin^2\phi \left[\frac{2}{3} - \frac{a}{3}(5 + 2\alpha^2 + \frac{2}{3}\beta^2 + \frac{4}{3}\zeta^2) \right], \qquad (9)$$

$$\Delta d = \cos^2\phi \left[-\frac{1}{3} - \frac{a}{3}(2 - \alpha^2 - \frac{1}{3}\beta^2 - \frac{2}{3}\zeta^2) \right] + \sin^2\phi \left[\frac{1}{3} - \frac{a}{3}(4 + \alpha^2 + \frac{1}{3}\beta^2 + \frac{2}{3}\zeta^2) \right], \qquad (10)$$

$$\Delta s = -a\alpha^2\,. \qquad (11)$$

After having formulated the spin polarizations of various quarks, we consider several measured quantities which are expressed in terms of the above mentioned spin polarization functions. The flavor non-singlet components usually calculated in the chiral constituent quark model are the Δ_3 and Δ_8, obtained from the neutron β−decay and the weak decays of hyperons respectively.

$$\Delta_3 = \Delta u - \Delta d = -\frac{1}{9}(5\cos^2\phi + \sin^2\phi)(-3 + a(3 + 3\alpha^2 + \beta^2 + 2\zeta^2))\,, \qquad (12)$$

$$\Delta_8 = \Delta u + \Delta d - 2\Delta s = -\frac{1}{3}(-3 + a(9 - 3\alpha^2 + \beta^2 + 2\zeta^2))\,. \qquad (13)$$

The flavor non-singlet component Δ_3 is related to the well known Bjorken sum rule.[21] Another quantity which is usually evaluated is the flavor singlet component of the total quark spin content and is defined as

$$\Delta\Sigma = \frac{1}{2}(\Delta u + \Delta d + \Delta s) = -\frac{1}{6}(-3 + a(9 + 6\alpha^2 + \beta^2 + 2\zeta^2))\,, \qquad (14)$$

in the $\Delta s = 0$ limit, this reduces to the Ellis-Jaffe sum rule.[22]

Apart from the above mentioned spin polarization we have also considered the quark distribution functions which have implications for ζ as well as for other chiral constituent quark model parameters. For example, the antiquark flavor contents of the "quark sea" can be expressed as[1,3,19,20]

$$\bar{u} = \frac{1}{12}[(2\zeta + \beta + 1)^2 + 20]a\,, \quad \bar{d} = \frac{1}{12}[(2\zeta + \beta - 1)^2 + 32]a\,, \quad \bar{s} = \frac{1}{3}[(\zeta - \beta)^2 + 9\alpha^2]a\,, \qquad (15)$$

and

$$u - \bar{u} = 2\,, \qquad d - \bar{d} = 1\,, \qquad s - \bar{s} = 0\,. \tag{16}$$

The deviation of Gottfried sum rule,[6] related to the $\bar{u}(x)$ and $\bar{d}(x)$ quark distributions, is expressed as

$$I_G = \frac{1}{3} + \frac{2}{3} \int_0^1 [\bar{u}(x) - \bar{d}(x)]dx = 0.254 \pm 0.005\,. \tag{17}$$

In terms of the symmetry breaking parameters a, β and ζ, this deviation is given as

$$\left[I_G - \frac{1}{3} \right] = \frac{2}{3} \left[\frac{a}{3}(2\zeta + \beta - 3) \right]\,. \tag{18}$$

Similarly, $\bar{u}/\bar{d}^{23}$ measured through the ratio of muon pair production cross sections σ_{pp} and σ_{pn}, is expressed in the present case as follows

$$\bar{u}/\bar{d} = \frac{(2\zeta + \beta + 1)^2 + 20}{(2\zeta + \beta - 1)^2 + 32}\,. \tag{19}$$

Similarly, the other important quantities having implications for the strangeness contribution to the nucleon are the quark flavor fractions $f_q = \frac{q+\bar{q}}{\sum_q (q+\bar{q})}$ which are expressed in terms of the chiral constituent quark model parameters as

$$f_u = \frac{12 + a(21 + \beta^2 + 4\zeta + 4\zeta^2 + \beta(2 + 4\zeta))}{3(6 + a(9 + \beta^2 + 6\alpha^2 + 2\zeta^2))}\,,$$

$$f_d = \frac{6 + a(33 + \beta^2 - 4\zeta + 4\zeta^2 + \beta(-2 + 4\zeta))}{3(6 + a(9 + \beta^2 + 6\alpha^2 + 2\zeta^2))}\,,$$

$$f_s = \frac{4a(\beta^2 + +9\alpha^2 - 2\beta\zeta + \zeta^2)}{3(6 + a(9 + \beta^2 + 6\alpha^2 + 2\zeta^2))}\,. \tag{20}$$

It is clear from the above expressions that the non zero value of the parameters a, α, β and ζ implies $f_s \neq 0$ as well as modify f_u and f_d due to the strangeness contributions coming from the "quark sea". Further, the ratio of the functions

$$f_3 = f_u - f_d\,, \qquad f_8 = f_u + f_d - 2f_s\,, \tag{21}$$

and the ratios

$$\frac{2\bar{s}}{(u+d)} = \frac{4a(9\alpha^2 + \beta^2 - 2\beta\zeta + \zeta^2)}{18 + a(27 + \beta^2 + 4\beta\zeta + 4\zeta^2)}\,, \qquad \frac{2\bar{s}}{(\bar{u}+\bar{d})} = \frac{4(9\alpha^2 + \beta^2 - 2\beta\zeta + \zeta^2)}{27 + \beta^2 + 4\beta\zeta + 4\zeta^2}\,, \tag{22}$$

have also been measured, therefore providing an opportunity to check the strange quark content of the nucleon.

For checking the strangeness contribution to the magnetic moment, we first consider different contributions to the magnetic moment of the proton $\mu(p)^s$. Since there are no 'strange' valence quarks, therefore $\mu(p)^s$ receives contributions only from the "quark sea" and is expressed as

$$\mu(p)^s = \mu(p)^s_{\text{spin}} + \mu(p)^s_{\text{orbit}}\,, \tag{23}$$

the details pertaining to $\mu(p)^s_{\text{spin}}$ and $\mu(p)^s_{\text{orbit}}$ can be found in Ref. 3.

The $\chi\text{CQM}_{\text{config}}$ involves five parameters, four of these a, $a\alpha^2$, $a\beta^2$, $a\zeta^2$ representing respectively the probabilities of fluctuations to pions, K, η, η', following the hierarchy $a > \alpha > \beta > \zeta$, while the fifth representing the mixing angle. The mixing angle ϕ is fixed from the consideration of neutron charge radius[14,24] and several other low energy properties, whereas for the other parameters we would like to update our analysis using the latest data.[5,25–27] In this context, we find it convenient to use Δu, Δ_3, $\bar{u} - \bar{d}$ and $\bar{u}/\bar{d}$ as inputs with their latest values given in Tables 1 and 2.

Before carrying out the fit to the above mentioned parameters, it is convenient to find the ranges of the parameters. To this end, the range of the symmetry breaking parameter a can be easily found by considering the spin polarization function Δu, by giving the full variation to the parameters α, β and ζ, for example, one finds $0.10 \lesssim a \lesssim 0.14$. The range of the parameter ζ can be found from $\bar{u}/\bar{d}$ using the latest experimental measurement[5] and it comes out to be $-0.70 \lesssim \zeta \lesssim -0.10$. Using the above found ranges of a and ζ as well as the latest measurement of $\bar{u} - \bar{d}$ asymmetry,[5] β comes out to be in the range $0.2 \lesssim \beta \lesssim 0.7$. Similarly, the range of α can be found by considering the flavor non-singlet component Δ_3 and it comes out to be $0.2 \lesssim \alpha \lesssim 0.5$.

After finding the ranges of the symmetry breaking parameters, we have carried out a fine grained analysis using the above ranges as well as considering $\alpha \approx \beta$ by fitting Δu, Δ_3[26] as well as $\bar{u} - \bar{d}$, $\bar{u}/\bar{d}$[5] leading to $a = 0.13$, $\zeta = -0.10$, $\alpha = \beta = 0.45$ as the best fit values. The parameters so obtained have been used to calculate the spin polarization functions and the quark distribution functions. The calculated quantities pertaining to spin polarization functions have been corrected by including the gluon polarization effects[1,4] and symmetry breaking effects.[1] Similarly, the quark distribution functions have been corrected by including the symmetry breaking effects. The orbital angular momentum contributions to magnetic moment are characterized by the parameters of χCQM as well as the masses of the GBs. For the u and d quarks, we have used their most widely accepted values in hadron spectroscopy,[1,16] for example, $M_u = M_d = 330$ MeV. For evaluating the contribution of GBs, we have used its on mass shell value in accordance with several other similar calculations.[4]

In Tables 1 and 2, we have presented the results of our calculations pertaining to the strangeness dependent parameters in $\chi\text{CQM}_{\text{config}}$. For comparison sake, we have also given the corresponding quantities in Constituent Quark Model (CQM). To begin with, we first discuss the quality of fit pertaining to the spin polarization functions. In Table 1, we have presented the strangeness incorporating spin polarization functions and the weak axial vector couplings. Using Δu, Δ_3 along with $\bar{u} - \bar{d}$, $\bar{u}/\bar{d}$ from Table 2 as inputs, we find that we are able to achieve an excellent fit in the case of spin polarization functions and the weak axial vector couplings. In particular, the agreement in terms of the magnitude as well as the sign in the case of Δs is in good agreement with the latest data. An excellent agreement in the case of Δ_8 and Δ_0, which receives contribution from Δs also, not only justify the

Table 1. The calculated values of the strangeness dependent spin polarization functions and weak axial vector couplings in the CQM and χCQM$_{\text{config}}$.

Parameter	Data	CQM	χCQM$_{\text{config}}$
Δu^*	0.85 ± 0.05[25]	1.333	0.867
Δd	-0.41 ± 0.05[25]	-0.333	-0.392
Δs	-0.10 ± 0.04[25]	0	-0.08
Δ_3^*	1.267 ± 0.0035[26]	1.666	1.267
Δ_8	0.58 ± 0.025[26]	1	0.59
Δ_0	0.19 ± 0.025[26]	0.50	0.19
F/D	0.575 ± 0.016[26]	0.673	0.589

* Input

Table 2. The calculated values of the strangeness dependent quark flavor distribution functions and related parameters in the CQM and χCQM$_{\text{config}}$.

Parameter	Data	CQM	χCQM$_{\text{config}}$
$\bar{s}$	—	0	0.10
$\bar{u} - \bar{d}^*$	-0.118 ± 0.015[5]	0	-0.117
$\bar{u}/\bar{d}^*$	0.67 ± 0.06[5]	0	0.67
$\frac{2\bar{s}}{u+d}$	$0.099^{+0.009}_{0.006}$[27]	0	0.09
$\frac{2\bar{s}}{\bar{u}+\bar{d}}$	$0.477^{+0.063}_{0.053}$[27]	0	0.44
f_s	0.10 ± 0.06[27]	—	0.08
f_3	—	—	0.21
f_8	—	—	1.03
f_3/f_8	0.21 ± 0.05[27]	0.33	0.20

* Input

success of χCQM$_{\text{config}}$ but also strengthen our conclusion regarding Δs. Similarly, the excellent agreement obtained in the case of the ratio F/D again reinforces our conclusion that χCQM$_{\text{config}}$ is able to generate qualitatively as well as quantitatively the requisite amount of strangeness in the nucleon.

After finding that the χCQM$_{\text{config}}$ is able to give an excellent account of the spin dependent polarization functions, in Table 2, we have presented the results of quark distribution functions having implications for strangeness in the nucleon. In line with the success of χCQM$_{\text{config}}$ in describing the spin dependent polarization functions, in this case also we are able to give an excellent account of most of the measured values. The agreement in the case of $\frac{2s}{u+d}$ and $\frac{f_3}{f_8}$ indicates that, in the χCQM$_{\text{config}}$, we are able to generate the right amount of strange quarks through chiral fluctuation. A refinement in the case of the strangeness dependent quark ratio $\frac{2s}{\bar{u}+\bar{d}}$ would have important implications for the basic tenets of chiral constituent quark model. The observed result for the case of f_s in the present case also indicates that the strange sea quarks play a significant role in the nucleon.

Coming to the strangeness spin and orbital contributions pertaining to the magnetic moment of the nucleon, using equation (23) one finds that the strangeness contribution to the magnetic moment is coming from spin and orbital angular momentum of the "quark sea" with opposite signs. These contributions are fairly significant and they cancel in the right direction to give the right sign and magnitude to $\mu(p)^s$, For example, the spin contribution in this case is $-0.09\mu_N$ and the contribution coming from the orbital angular momentum is $0.05\mu_N$. These contributions cancel to give $-0.04\mu_N$ which is very close to the observed HAPPEX results $(-0.038 \pm 0.042\mu_N)$.[10] Interestingly, in the case of $\mu(n)^s$ the magnetic moment is dominated by the orbital part. Therefore, an observation of this would not only justify the Cheng-Li mechanism[1] but would also suggest that the chiral fluctuations is able to generate the appropriate amount of strangeness in the nucleon.

To summarize, the $\chi\text{CQM}_{\text{config}}$ is able to provide an excellent description of the spin dependent polarization functions and quark distribution functions. Further, the model is also able to predict the strangeness dependent polarization and quark distribution parameters. In particular, the model's prediction pertaining to the strangeness contribution to the magnetic moment of proton is in agreement with latest lattice based calculations as well as with the data. In conclusion, we would like to state that at the leading order constituent quarks and the weakly interacting Goldstone bosons constitute the appropriate degrees of freedom in the nonperturbative regime of QCD and the "quark sea" generation in the $\chi\text{CQM}_{\text{config}}$ through the chiral fluctuation is the key in understanding the strangeness content of the nucleon.

Acknowledgments

HD and GA would like to thank Department of Science and Technology, Government of India for financial support. GA would also like to acknowledge the Chairman, Department of Physics for providing facilities to work.

References

1. T.P. Cheng and Ling Fong Li, Phys. Rev. Lett. **74**, 2872 (1995); hep-ph/9709293; Phys. Rev. **D 57**, 344 (1998).
2. T.P. Cheng and Ling Fong Li, Phys. Rev. Lett. **80**, 2789 (1998).
3. H. Dahiya and M. Gupta, Phys. Rev. **D 64**, 014013 (2001); *ibid.* **D 66**, 051501(R) (2002); *ibid.* **D 67**, 074001 (2003); *ibid.* **67**, 114015 (2003).
4. New Muon Collaboration, P. Amaudruz *et al.*, Phys. Rev. Lett. **66**, 2712 (1991); M. Arneodo *et al.*, Phys. Rev. **D 50**, R1 (1994).
5. E866/NuSea Collaboration, E.A. Hawker *et al.*, Phys. Rev. Lett. **80**, 3715 (1998); J.C. Peng *et al.*, Phys. Rev. **D 58**, 092004 (1998); R. S. Towell *et al.*, *ibid.* **64**, 052002 (2001).
6. K. Gottfried, Phys. Rev. Lett. **18**, 1174 (1967).
7. SAMPLE Collaboration, D.T. Spayde *et al.*, Phys. Lett. **B 583**, 79 (2004).
8. G0 Collaboration, D. Armstrong *et al.*, Phys. Rev. Lett. **95**, 092001 (2005).
9. A4 Collaboration, F. E. Maas *et al.*, Phys. Rev. Lett. **94**, 152001 (2005).

10. HAPPEX Collaboration, K.A. Aniol *et al.*, Phys. Rev. Lett. **98**, 032301 (2007); *ibid.* Eur. Phys. J. **A 31**, 597 (2007).

11. S.J. Dong, K.F. Liu, A. G. Williams, Phys. Rev. **D 58**, 074504 (1998); N. Mathur, S.J. Dong, Nucl. Phys. B, Proc. Suppl. **94**, 311 (2001); R. Lewis, W. Wilcox, R. M. Woloshyn, Phys. Rev. **D 67**, 013003 (2003).

12. D.B. Leinweber, S. Boinepalli, I. C. Cloet, A.W. Thomas, A.G. Williams, R.D. Young, J.B. Zhang, J.M. Zanotti, Eur. Phys. J. **A2452**, 79 (2005); *ibid.* Phys. Rev. Lett. **94**, 212001 (2005).

13. N. Isgur, G. Karl and R. Koniuk, Phys. Rev. Lett. **41**, 1269 (1978); N. Isgur and G. Karl, Phys. Rev. **D 21**, 3175 (1980); N. Isgur *et al.*, Phys. Rev. **D 35**, 1665 (1987); P. Geiger and N. Isgur, Phys. Rev. **D 55**, 299 (1997).

14. A. Le Yaouanc, L. Oliver, O. Pene and J.C. Raynal, Phys. Rev. **D 12**, 2137 (1975); *ibid.* **15**, 844 (1977).

15. P.N. Pandit, M.P. Khanna and M. Gupta, J. Phys. G **11**, 683 (1985).

16. M. Gupta and N. Kaur, Phys. Rev. **D 28**, 534 (1983); M. Gupta, J. Phys. G: Nucl. Phys. **16**, L213 (1990).

17. M. Gupta, S.K. Sood, A. N. Mitra, Phys. Rev. **D 16**, 216 (1977); *ibid.* **19**, 104 (1979).

18. A. Le Yaouanc *et al.*, *Hadron Transitions in the Quark Model*, Gordon and Breach, 1988.

19. X. Song, J.S. McCarthy and H.J. Weber, Phys. Rev. **D 55**, 2624 (1997); X. Song, Phys. Rev. **D 57**, 4114 (1998).

20. J. Linde, T. Ohlsson and Hakan Snellman, Phys. Rev. **D 57**, 452 (1998); T. Ohlsson and H. Snellman, Eur. Phys. J., **C 7**, 501 (1999).

21. J.D. Bjorken, Phys. Rev. **148**, 1467 (1966); Phys. Rev. **D 1**, 1376 (1970).

22. J. Ellis and R.L. Jaffe, Phys. Rev. **D 9**, 1444 (1974); *ibid.* **10**, 1669 (1974).

23. NA51 Collaboration, A. Baldit *et al.*, Phys. Lett. **253B**, 252 (1994).

24. M. Gupta and A.N. Mitra, Phys. Rev. **D 18**, 1585 (1978); N. Isgur, G. Karl and D.W.L. Sprung, *ibid* **23**, 163 (1981).

25. SMC Collaboration, B. Adeva *et al.*, Phys. Rev. D **58**, 112001 (1998).

26. W.-M.Yao *et al.*, J. Phys. G **33**, 1 (2006).

27. CCFR Collaboration and NuTeV Collaboration, A.O. Bazarko *et al.*, Z. Phys **C 65**, 189 (1995);

PSEUDOSCALAR-PHOTON INTERACTIONS, AXIONS, NON-MINIMAL EXTENSIONS, AND THEIR EMPIRICAL CONSTRAINTS FROM OBSERVATIONS

WEI-TOU NI,[*,‡] A. B. BALAKIN[†,§] and HSIEN-HAO MEI[*,¶]

*Center for Gravitation and Cosmology, Department of Physics,
National Tsing Hua University, Hsinchu, Taiwan, 30013, Republic of China
†Kazan State University, Kremlevskaya Street 18, 420008, Kazan, Russia
‡weitou@gmail.com
§Alexander.Balakin@ksu.ru
¶mei@phys.nthu.edu.tw

Dedicated to Murray Gell-Mann on his 80th birthday.

Pseudoscalar-photon interactions were proposed in the study of the relations among equivalence principles. The interaction of pseudoscalar axion with gluons was proposed as a way to solve the strong CP problem. Subsequent proposal of axion as a dark matter candidate has been a focus of search. Motivation from superstring theories add to its importance. After a brief introduction and historical review, we present (i) the current status of our optical experiment using high-finesse Fabry-Perot resonant cavity — Q & A experiment — to detect pseudoscalar-photon interactions, (ii) the constraints on pseudoscalar-photon interactions from astrophysical and cosmological observations on cosmic polarization rotation, and (iii) theoretical models of non-minimal interactions of gravitational, electromagnetic and pseudoscalar (axion) fields, and their relevance to cosmology.

Keywords: Pseudoscalar-photon interactions; axions; non-minimal extensions; cosmic polarization rotation.

1. Introduction

One (WTN) of us was a student in Gell-Mann's class of "Topics on Particle Theories" in the late 1960's, and learned ups, downs and the spirit of constructing modern particle theories. Working on thesis in Thorne's group in 1969–1972, he learned relativistic astrophysics and the spirit of phenomenological study of gravitation. With this background, he started to work on the theoretical study and phenomenological study of equivalence principles in the late 1972 in the quite environment of Bozeman, Montana in Nordtvedt's group. The theoretical work reached *two milestones*, one in 1973 for finding a non-metric theory with pseudoscalar-photon interaction[1] (axion interaction) which observes the Galileo Equivalence Principle and the other in 1974 for proving this is the only non-metric theory in the general χ-g framework of charged particle-electromagnetism system in gravity.[2] The complete paper was written and published[3] after he moved to National Tsing Hua University where he continued to work on both theoretical and phenomenological aspects, and started to work on experimental aspects.[4] In Sec. 2, we review different motivations to reach pseudoscalar interactions and axions together with related developments. In Sec. 3, we present the current status of our precision ellipsometry experiments using high-finesse Fabry-Perot cavity. In the first theoretical *milestone* addressed above, we found that the non-metric pseudoscalar-photon interactions could induce polarization rotation in electromagnetic wave propagation and proposed to use long-distance astrophysical propagation to test the theory.[1] It is fortunate that, due to technological and observational developments, this test has been improved in great precision recently.[4,5] We discuss and compile the recent results in Sec. 4. In the early universe (inflationary or equivalent period, earlier period, etc.), the physics was not yet determined and studies in non-minimal coupling of photons and axions would be warranted. In Sec. 5, we review our recent work on this.[6]

2. Pseudoscalar-Photon Interaction and Axions

In the 5th Patras Workshop on Axions, WIMPs and WISPs held at the University of Durham on 13–17 July 2009, three motivations were presented. In the bottom-up approach,[7] axion is considered as a Goldstone boson associated with spontaneously broken PQ symmetry[8] to fix the strong CP problem. The name of axion was proposed by Wilczek as detergent AXION (AXION is a commercial brand of detergent) to clean up the strong CP problem. As the original axions[9,10] had not been found, invisible axions[11–14] were proposed.

Top-down motivation[15] comes from superstring theory. In supersymmetry/ supergravity, an appropriate action connects gauge and axionic couplings through a single holomorphic function. In type IIA/B superstring theory, axion comes from a Ramond–Ramond antisymmetrical field reduced on the cycle (compactified space). Axions also arise for heterotic string and M-theory. In superstring theory, "the model-independent axion appears in every perturbative string theory, and is closely related to the graviton and dilaton."[16]

The gravity-related motivation[17,18] (historically the first approach as described in the introduction) comes from a phenomenological study of equivalence principles. In 1973, we studied the relationship of Galilio Equivalence Principle (WEP I) and Einstein Equivalence Principle in a framework (the χ-g framework) of electromagnetism and charged particles, and found the following interaction Lagrangian density

$$\mathcal{L}_I = -(1/16\pi)g^{ik}g^{jl}F_{ij}F_{kl} - (1/16\pi)\varphi F_{ij}F_{kl}e^{ijkl} - A_{kj}{}^k(-g)^{(1/2)}$$
$$- \Sigma_I m_I(ds_I)/(dt)\delta(\boldsymbol{x} - \boldsymbol{x}_I)\,, \tag{1}$$

as an example which obeys WEP I, but not EEP.[1-3] (e^{ijkl} is the completely anti-symmetric symbol.) The nonmetric part of this theory is

$$\mathcal{L}^{(NM)}{}_{\text{int}} = -(1/16\pi)(-g)^{1/2}\varphi\varepsilon^{ijkl}F_{ij}F_{kl}$$
$$= -(1/4\pi)(-g)^{1/2}\varphi_{,i}\varepsilon^{ijkl}A_jA_{k,l}\,(\text{mod div})\,, \qquad (2)$$

where 'mod div' means that the two Lagrangian densities are related by partial integration in the action integral. ($\varepsilon^{ijkl} \equiv (-g)^{-1/2}e^{ijkl}$.) The modified Maxwell equations[1-3] are

$$F^{ik}{}_{|k} + \varepsilon^{ikml}F_{km}\varphi_{,l} = -4\pi j^i\,, \qquad (3)$$

where the derivation | is with respect to the Christoffel connection of g^{ik}. The Lorentz force law is the same as in metric theories of gravity or general relativity. Gauge invariance and charge conservation are guaranteed. The Maxwell Eq. (3) are also conformally invariant. Axial-photon interaction induces energy level shift in atoms and molecules, and polarization rotations in electromagnetic wave propagation. Empirical tests of the pseudoscalar-photon interaction (2) were analyzed in 1973; at that time it was only loosely constrained.[1] Now we have effective constraints on polarization rotation in the electromagnetic wave propagation from astrophysical polarization observations and CMB polarization observations for massless or nearly massless case.[4,5,18] Axion with mass is a viable candidate for dark matter search. Recently laboratory experiments have started to give constraints on them.[17]

The rightest term in Eq. (2) is reminiscent of Chern–Simons[19] term $e^{\alpha\beta\gamma}A_\alpha F_{\beta\gamma}$. There are two differences:

(i) Chern–Simons term is in 3 dimensional space;
(ii) Chern–Simons term in the integral is a total divergence.

A term similar to the one in Eq. (2) (axion-gluon interaction) occurs in QCD in an effort to solve the strong CP problem with the electromagnetic field replaced by gluon field.[8-10] Carroll, Field and Jackiw[20] proposed a modification of electrodynamics with an additional $e^{ijkl}V_iA_jF_{kl}$ term with V_i a constant vector. This term is a special case of the term $e^{ijkl}\varphi F_{ij}F_{kl}$ (mod div) with $\varphi_{,i} = -1/2V_i$. *Various terms in the Lagrangians discussed here are listed in Table 1 for comparison.*

In the Peccei–Quinn–Weinberg–Wilczek models, axion-photon interaction may or may not be induced. In terms of Feynman diagram, the interaction (2) gives a two-photon-pseudo-scalar vertex and vacuum becomes birefringent and dichroic.[21-25] These effects are quantum in origin, while the cosmic polarization rotation effect discussed following (3) is classical.

Dichroic materials have the property that their absorption constant varies with polarization. For axion models with (2), photon interacts with magnetic field has a cross section to be converted into axion and leaks away. The vacuum with magnetic field becomes absorptive. Since the cross section depends on polarization, so is the absorption. When polarized light goes through vacuum with magnetic field, its

Table 1. Various terms in the Lagrangian and their meaning.

Term	Dimension	Reference	Meaninge
$e^{\alpha\beta\gamma}A_\alpha F_{\beta\gamma}$	3	Chern–Simons[19] (1974)	Intergrand for topological invariant
$e^{ijkl}\varphi F_{ij}F_{kl}$	4	Ni[1-3] (1973, 1974, 1977)	Pseudoscalar-photon coupling
$e^{ijkl}\varphi F^{QCD}{}_{ij}F^{QCD}{}_{kl}$	4	Peccei–Quinn[8] (1977) Weinberg[9] (1978) Wilczek[10] (1978)	Pseudoscalar-gluon coupling
$e^{ijkl}V_i A_j F_{kl}$	4	Carroll–Field–Jackiw[19] (1990)	External constant vector coupling

polarization is rotated due to difference in absorption in two principal directions of the vacuum for the two polarization components. The polarization rotation ε of the photon beam for light entering the magnetic-field region polarized at an angle of θ to the magnetic field is

$$\varepsilon = (B^2\omega^2 M^{-2}m_\varphi{}^{-4})\sin^2(m_\varphi{}^2 L/4\omega)\sin(2\theta) \approx B^2 L^2/(16 M^{-2})\sin(2\theta)\,, \qquad (4)$$

where m_φ is mass of the axion, M the axion-photon interaction energy scale, ω photon circular frequency and L the magnetic-region length. The approximation is valid in the limit

$$m_\varphi{}^2 L/4\omega \ll 1\,. \qquad (5)$$

Since axions do not reflect at the mirrors, for multi-passes, the rotation effect increases by number N of passes. Therefore for the case condition (5) is satisfied, the polarization rotation effect is proportional to NB^2L^2. Axion leaking away has a cross section to interact with another magnetic field to regenerate photon. Current optical experiments to measure the dichroism and to detect photon regeneration are listed in Table 2 of Ref. 26. In the following section we discuss the method of using high finesse cavity ellipsometry to measure the dichroism and report on the current status of our Q & A experiment.

3. High Finesse Cavity Ellipsometry and Q & A Experiment

The standard technique to measure birefringence and dichroism is ellipsometry. To measure minute birefringence and minute dichroism, a high finesse cavity is used for enhancing the physical effects. In 1994, two groups — PVLAS (Polarizzazione del Vuoto con LASer) and Q & A (QED [Quantum Electrodynamics] and Axion experiment) — started to build apparatuses using high finesse Fabry-Perot cavity to measure QED birefringence and search for pseudoscalar-photon interactions. PVLAS adapted the earlier scheme proposed in 1979,[27] and had experiences from the participation of some of their members in the BFRT (Brookhaven-Fermilab-Rochester-Trieste) experiment[28] which had used multipass to enhance the physical effects.

Table 2. Constraints on cosmic polarization rotation from CMB (cosmic microwave background).

Reference	Constraint [mrad]	Source data
Ni[39,40]	±100	WMAP1[41]
Feng, Li, Xia, Chen, and Zhang[42]	-105 ± 70	WMAP3[43] & BOOMERANG (B03)[44]
Liu, Lee, Ng[45]	±24	BOOMERANG (B03)[44]
Kostelecky and Mews[46]	209 ± 122	BOOMERANG (B03)[44]
Cabella, Natoli and Silk[47]	-43 ± 52	WMAP3[43]
Xia, Li, Wang, and Zhang[48]	-108 ± 67	WMAP3[43] & BOOMERANG (B03)[44]
Komatsu, et al.[49]	-30 ± 37	WMAP5[49]
Xia, Li, Zhao, and Zhang[50]	-45 ± 33	WMAP5[49] & BOOMERANG (B03)[44]
Kostelecky and Mews[51]	40 ± 94	WMAP5[49]
Kahniashvili, Durrer, and Maravin[52]	±44	WMAP5[49]
Wu[53]	$9.6 \pm 14.3 \pm 8.7$	QuaD (2nd and 3rd sessions)[54]
Brown et al.[55]	$11.2 \pm 8.7 \pm 8.7$	QuaD[55]
Komatsu et al.[56]	$-19 \pm 22 \pm 26$	WMAP7[56]

PVLAS group started to build a vertical Fabry-Perot cavity to accommodate a rotating cryogenic superconducting dipole magnet. We started to build a 3.5 m/7 m prototype Fabry-Perot cavity with a horizontally rotating permanent dipole magnet for measuring vacuum birefringence and improving the sensitivity of axion search as part of our continuing effort in precision interferometry. In close contact with ground gravitational-wave detection community, we use a number of techniques developed by the community.[29] In June 1994, we met the PVLAS people in the Marcel Grossmann Meeting at Stanford, exchanged a few ideas and encouraged each other. BMV (Biréfringence Magnétique du Vide) group started to construct their experiment[30] using high magnetic field pulses in 2000. Both PVLAS group[31] and Q & A group[32] have used their apparatuses to measure the Cotton-Mouton birefringence of various dilute gases as applications to chemical physics and as calibrations of their apparatuses. The results of two groups in the common cases agree with each other within 1.2σ. BMV group has also done this recently.[33]

The 2006 report of PVLAS group on the positive detection of dichroism[34] stirred up a lot of experimental activities on optical detection of axions, minicharged particles and related interactions. Among groups working on LSW (Light Shining through the Wall [photon regeneration]) experiments, OSQAR (Optical Search of QED vacuum magnetic birefringence, Axions and photon Regeneration) collaboration also started to build high-finesse ellipsometry for birefringence and dichroism measurement.[35] The original PVLAS results were soon found disfavored by the results of Q & A experiment,[36] and ruled out by further and more careful measurements of PVLAS.[37] Now all groups are working on measuring QED vacuum birefringence as their immediate goal. After this is achieved, the sensitivity for searching axions and other relevant particles would be improved by several orders of magnitude.

We now report on our Q & A experiment. The schematic of the setup of our second phase (2002–2008) is shown in Fig. 1 of Ref. 26. The 3.5 m prototype interferometer is formed using a high-finesse Fabry–Perot interferometer together with a high-precision ellipsometer. The two high-reflectivity mirrors of the 3.5 m prototype interferometer are suspended separately from two X-pendulum–double pendulum suspensions mounted on two isolated tables fixed to ground using bellows inside two vacuum chambers. The sub-systems are described in Ref. 36. Our results in this phase give $(-0.2 \pm 2.8) \times 10^{-13}$ rad/pass with 18,700 passes through a 2.3 T 0.6 m long magnet for vacuum dichroism measurement, and limit pseudo-scalar-photon interaction and millicharged fermions meaningfully.[36]

We are currently upgrading our interferometer from 3.5 m armlength to 7 m armlength in the 3rd phase. We have installed a new 1.8 m 2.3 T permanent magnet capable of rotation up to 13 cycles per second to enhance the physical effects. We are working with 532 nm Nd:YAG laser as light source with cavity finesse around 100,000, and aim at 10 nrad/Hz$^{1/2}$ optical sensitivity. *With all these achieved and the upgrading of vacuum, vacuum dichroism measurement would be improved in precision by 3–4 orders of magnitude, and QED birefringence would be measured to 28% in about 50 days.* To enhance the physical effects further, another 1.8 m magnet will be added in the future.

4. Constraints on Cosmic Polarization Rotation from Observations

For the modified Maxwell Eq. (3), electromagnetic propagation induces a polarization rotation of angle $\alpha = \Delta\varphi = \varphi_2 - \varphi_1$ where φ_1 and φ_2 are the values of the scalar field at the beginning and end of the wave.[1] When the propagation distance is over a large part of our observed universe, we call this phenomenon cosmic polarization rotation.[4,5]

In the CMB (Cosmic Microwave Background) observations, there are variations and fluctuations. The variations and fluctuations due to scalar-modified propagation can be expressed as $\delta\varphi(2) - \delta\varphi(1)$, where 1 denotes a point at the last scattering surface in the decoupling epoch and 2 observation point. $\delta\varphi(2)$ is the variation/fluctuation at the last scattering surface. $\delta\varphi(1)$ at the present observation point is zero or fixed. Therefore the covariance of fluctuation $\langle [\delta\varphi(2) - \delta\varphi(1)]^2 \rangle$ gives the covariance of $\delta\varphi^2(2)$ at the last scattering surface. Since our Universe is isotropic to $\sim 10^{-5}$, this covariance is $\sim (\xi \times 10^{-5})^2$ where the parameter ξ depends on various cosmological models.[5,38] Electromagnetic propagation from different directions may acquire different polarization rotations depending on the cosmological structure of the gradient of φ. If we assume the constant part of the gradient of φ dominates, the constraints[4] of CMB observations on the cosmic polarization rotation α from various analyses are updated in the following Table 2.

Constraints from observations on individual polarization sources are discussed in Refs. 4 and 5. The most recent results are the ultraviolet polarization observations of distant radio galaxies.[57] No polarization rotation is detected within a few degrees

Table 3. Ten non-minimal (NM) coupling parameters are divided into four subgroups: the first q_1, q_2, q_3; second Q_1, Q_2, Q_3; third η_1, η_2, η_3; and fourth $\eta_{(A)}$. In the second column the terms in the Lagrangian are given in front of which the corresponding coupling parameters are introduced; the parameters of the first subgroup introduce the terms without the pseudoscalar field φ; the parameters of the second subgroup relate to the terms linear in φ; the terms indicated by η_1, η_2, η_3, are quadratic in the four-gradient of φ; and finally, $\eta_{(A)}$ introduces the term quadratic in φ. In the last column, we point out the physical meaning of these non-minimal terms, based on decompositions of the constitutive tensors for the electromagnetic (EM) and pseudoscalar fields.

	Term in the Lagrangian	Physical meaning
q_1	$(1/2)RF^{mn}F_{mn}$	NMEM susceptibility linear in the Ricci scalar
q_2	$R^{mn}F_{mk}F_n{}^k$	NMEM susceptibility linear in the Ricci tensor
q_3	$(1/2)R^{ikmn}F_{ik}F_{mn}$	NMEM susceptibility linear in the Riemann tensor
Q_1	$(1/2)\varphi R F^{mn} F^*{}_{mn}$	NMEM susceptibility induced by the axion with the Ricci scalar
Q_2	$\varphi R^{mn} F_{mk} F^{*k}{}_n$	NMEM susceptibility induced by the axion with the Ricci tensor
Q_3	$(1/2)\varphi R^{ikmn} F_{ik} F^*{}_{mn}$	NMEM susceptibility induced by the axion with the Riemann tensor
η_1	$-R^n{}_l F^{ml} \nabla_m \varphi \nabla_n \varphi$	NMEM current induced by the axion field gradient
η_2	$-R \nabla^m \varphi \nabla_m \varphi$	NM axion-graviton derivative coupling with the Ricci scalar
η_3	$-R^{mn} \nabla_m \varphi \nabla_n \varphi$	NM axion-graviton derivative coupling with the Ricci tensor
$\eta_{(A)}$	$R\varphi^2$	NM correction to the mass square of the axion

for each galaxy and overall fitting for a constant scalar gradient/constant vector direction is comparable to the best CMB constraint. More works in this direction are important as they could detect/constrain directional dependence and distinguish cosmological models.

In our original pseudoscalar model, the natural coupling strength φ is of order 1. However, the isotropy of our observable universe to 10^{-5} may leads to a change of $\Delta\varphi$ over cosmological distance scale 10^{-5} smaller. Hence, observations to test and measure $\Delta\varphi$ to 10^{-6} is very significant. A positive result may indicate that our patch of inflationary universe has a "spontaneous polarization" in the fundamental law of electromagnetic propagation influenced by neighboring patches and by a determination of this fundamental physical law we could 'observe' our neighboring patches.

The Planck Surveyor was launched in May, 2009.[58] Better sensitivity to $\Delta\varphi$ of 10^{-2}–10^{-3} (1–10 mrad) is expected. A dedicated future experiment on cosmic microwave background polarization[59–61] may reach 10^{-5}–10^{-6} $\Delta\varphi$-sensitivity.[39] As-trophysical observ-ations of cosmologically distant objects in various directions will give $\Delta\varphi$ in various directions and will compliment the CMB polarization measure-ment. Future observations to test and measure $\Delta\varphi$ to 10^{-6} and to give $\Delta\varphi$ in various directions are promising.

5. Non-minimal Coupling of Gravitational, Electromagnetic and Axion Fields

To complete the axion interaction theory (1) as a gravitational theory, we have to add a gravitational Lagrangian. This is illustrated by Eqs. (28)–(30) in Ref. 4. In the early universe, although inflationary scenario gives the right structure formation, the inflationary physics was not clear and non-minimal extensions of the coupling of photons and axions is worth study. We have formulated a ten-parameter non-minimal extension.[6] The ten non-minimal terms in the Lagrangian together with their physical meaning are compiled in Table 3. R^{ikmn}, R^{mn}, and R are Riemann tensor, Ricci tensor and Ricci scalar of g^{mn} respectively. $F^*{}_{mn}$ is the dual tensor of F^{kl}. References 6 gives a complete account of these terms together with derivations and some exact solutions. Empirical constraints on coupling parameters from astrophysical birefringence and polarization rotation observations have also been studied. We are currently pursuing on this line further.

Acknowledgments

We are grateful to the National Science Council (NSC 98-2112-M-007-009) for support.

References

1. W.-T. Ni, A Nonmetric Theory of Gravity, preprint, Montana State University, Bozeman, USA (1973), http://astrod.wikispaces.com/file/view/A+Non-metric+Theory+of+Gravity.pdf.
2. W.-T. Ni, *Bull. Am. Phys. Soc.* **19**, 655 (1974).
3. W.-T. Ni, *Phys. Rev. Lett.* **38**, 301 (1977).
4. W.-T. Ni, *Rept. Prog. Phys.* **73**, 056901 (2010).
5. W.-T. Ni, *Prog. Theor. Phys. Suppl.* **172**, 49 (2008), arXiv:0712.4082.
6. A. B. Balakin and W.-T. Ni, *Class. Quantum Grav.* **27**, 055003 (2010).
7. M. Ahlers, Axions, WIMPs and WISPs: Bottum-Up Motivation, *5th Patras Workshop on Axions, WIMPs and WISPs held at the University of Durham on 13-17 July 2009*, http://axion-wimp.desy.de/e30/e52240/e54383/MarkusAhlers.pdf; arXiv:0910.2211.
8. R. D. Peccei and H. R. Quinn, *Phys. Rev. Lett.* **38**, 1440 (1977).
9. S. Weinberg, *Phys. Rev. Lett.* **40**, 233 (1978).
10. F. Wilczek, *Phys. Rev. Lett.* **40**, 279 (1978).
11. J. Kim, *Phys. Rev. Lett.* **43**, 103 (1979).
12. M. Dine *et al.*, *Phys. Lett. B* **104**, 199 (1981).
13. M. Shifman *et al.*, *Nucl. Phys. B* **166**, 493 (1980).
14. M. Yu. Khlopov, *Cosmoparticle Physics* (World Scientific, 1999); and ref's therein.
15. J. Conlon, Axions, WIMPs and WISPs: Top-Down Motivation, *5th Patras Workshop on Axions, WIMPs and WISPs, 2009*, http://axion-wimp.desy.de/e30/e52240/e54379/JoeConlon.pdf.
16. See, e.g., J. Polchinski, *String Theory*, volume 1 & 2 (Cambridge 1998).
17. H.-H. Mei, *et al.*, The status and prospects of the Q & A exp. with some applications, *Proc. of the 5th Patras Workshop on Axions, WIMPs and WISPs, Durham, UK, 13–17 July, 2009*, eds. J. Jaeckel, A. Lindner and J. Redondo (Verlag DESY 2010), p. 108; and references therein.

534

18. W.-T. Ni, Constraints on Pseudoscalar-Photon Interaction from CMB Polarization Observations, *Proc. of the 5th Patras Workshop on Axions, WIMPs and WISPs, Durham, UK, 13–17 July, 2009*, eds. J. Jaeckel, A. Lindner and J. Redondo (Verlag DESY 2010), p. 175.

19. S.-S. Chern and J. Simons, *The Annals of Mathematics*, 2nd Ser. **99**, 48 (1974).

20. S. M. Carroll, G. B. Field and R. Jackiw, *Phys. Rev. D* **41**, 1231 (1990).

21. P. Sikivie, *Phys. Rev. Lett.* **51**, 1415 (1983).

22. A. A. Anselm, *Yad. Fiz.* **42**, 1480 (1985).

23. M. Gasperini, *Phys. Rev. Lett.* **59**, 396 (1987).

24. L. Maiani, R. Petronzio and E. Zavattini, *Phys. Lett. B* **175**, 359 (1986).

25. G. Raffelt and L. Stodolsky, *Phys. Rev. D* **37**, 1237 (1988).

26. H.-H. Mei, *et al.* (Q & A Collaboration), *Mod. Phys. Lett. A* **25**, 983 (2010).

27. E. Iacopini and E. Zavattini, *Phys. Lett. B* **85**, 151 (1971).

28. R. Cameron *et al.*, *Phys. Rev. D* **47**, 3707 (1993).

29. W.-T. Ni *et al.*, *Mod. Phys. Lett. A* **6**, 3671 (1991).

30. S. Askenazy *et al.*, *QED and Physics of Vacuum*, ed. G. Cantatore (AIP, 2001), p. 115.

31. M. Bregant *et al.*, *Chem. Phys. Lett.* **477**, 415 (2009); and references therein.

32. H.-H. Mei, *et al.* (Q & A Collaboration), *Chem. Phys. Lett.* **471**, 216 (2009).

33. M. Fouché, Paul Berceau, R. Battesti and C. Rizzo (BMV group), The QED Vacuum Magnetic Birefringence (BMV) Experiment, presented in *2010 Conference on Precision Electromagnetic Measurements, June 13-18, 2010, Daejeon, Korea*.

34. E. Zavattini *et al.* (PVLAS Collaboration), *Phys. Rev. Lett.* **96**, 110406 (2006).

35. P. Pugnat *et al.* (OSQAR Collaboration), *Phys. Rev. D* **78**, 092003 (2008).

36. S.-J. Chen, H.-H. Mei and W.-T. Ni (Q & A Collaboration), *Mod. Phys. Lett. A* **22**, 2815 (2007).

37. E. Zavattini *et al.* (PVLAS Collaboration), *Phys. Rev. D* **77**, 032006 (2008).

38. W.-T. Ni, *Int. J. Mod. Phys. A* **24**, 3493 (2009), arXiv:0903.0756.

39. W.-T. Ni, *Chin. Phys. Lett.* **22**, 33 (2005), arXiv:gr-qc/0407113.

40. W.-T. Ni, *Int. J. Mod. Phys. D* **14**, 901 (2005), arXiv:gr-qc/0504116.

41. C. L. Bennett *et al.*, *Astrophys. J. Suppl.* **148**, 1 (2003); and references therein.

42. B. Feng, M. Li, J.-Q. Xia, X. Chen and X. Zhang, *Phys. Rev. Lett.* **96**, 221302 (2006).

43. D. N. Spergel *et al.*, *Astrophys. J. Suppl.* **170**, 377 (2007); and references therein.

44. T. E. Montroy *et al.*, *Astrophys. J.* **647**, 813 (2006); and references therein.

45. G. C. Liu, S. Lee and K. W. Ng, *Phys. Rev. Lett.* **97**, 161303 (2006); private communication.

46. A. Kostelecky and M. Mewes, *Phys. Rev. Lett.* **99**, 011601 (2007).

47. P. Cabella, P. Natoli and J. Silk, *Phys. Rev. D* **76**, 123014 (2007).

48. J.-Q. Xia, H. Li, X. Wang and X. Zhang, *Astron. Astrophys.* **483**, 715 (2008).

49. E. Komatsu *et al.*, *Astrophys. J. Suppl.* **180**, 330 (2009).

50. J.-Q. Xia, H. Li, G.-B. Zhao and X. Zhang, *Astrophys. J.* **679**, L61 (2008).

51. V. A. Kostelecky and M. Mewes, *Astrophys. J.* **689**, L1 (2008).

52. T. Kahniashvili, R. Durrer and Y. Maravin, *Phys. Rev. D* **78**, 123009 (2008).

53. E. Y. S. Wu *et al.*, *Phys. Rev. Lett.* **102**, 161302 (2009).

54. C. Pryke *et al.*, *Astrophys. J.* **692**, 1247 (2009).

55. M. L. Brown *et al.*, *Astrophys. J.* **705**, 978 (2009).

56. E. Komatsu *et al.*, Seven-Year Wilkinson Microwave Anisotropy Probe (WMAP) Observations: Cosmological Interpretation, astro-ph.CO/1001.4538.

57. S. di Serego Alighieri, F. Finelli and M. Galaverni, *Astrophys. J.* **715**, 33 (2010).
58. http://www.rssd.esa.int/index.php?project=planck
59. J. Bock *et al.*, The Experimental Probe of Inflationary Cosmology (EPIC): A Mission Concept Study for NASA's Einstein Inflation Probe, arXiv:0805.4207v1 (2008).
60. KEK, Lite (light) Satellite for the Studies of B-mode Polarization and Inflation from Cosmic Background Radiation Detection (2008).
61. P. de Bernardis *et al.*, *Exp. Astron.* **23**, 5 (2009).

GAUGE FIELD THEORY OF HORIZONTAL SU(2) × U(1) SYMMETRY — DOUBLET PLUS SINGLET SCHEME

I. S. SOGAMI

Maskawa Institute, Kyoto Sangyo University,
Kyoto, 603-8555, Japan
sogami@cc.kyoto-su.ac.jp

Gauge field theory of a horizontal symmetry of the group $G_H = SU(2)_H \times U(1)$ is developed so as to generalize the standard model of particle physics. All fermion and scalar fields are assumed to belong to doublets and singlets of the group in high energy regime. Mass matrices with four texture zeros of Dirac and Majorana types are systematically derived. In addition to seven scalar particles, the theory predicts existence of one peculiar vector particle which seems to play important roles in astrophysics and particle physics.

Keywords: Quark and lepton; mass matrix; gauge field theory; horizontal symmetry.

1. Introduction

To generalize the standard model of particle physics, we develop a gauge field theory of a horizontal (H) symmetry of the group $G_H = SU(2)_H \times U(1)$. Above a high energy scale $\bar{\Lambda}$ which is much higher than the electroweak (EW) scale Λ, fundamental fermions, quarks and leptons, are postulated to form doublets and singlets of the group G_H. Classification of the fundamental fermions into chiral sectors consisting of EW doublets ($f = q, \ell$) and singlets ($f = u, d; \nu, e$) is assumed to hold also in the high energy regime.

Breakdown of the symmetry at the scale $\bar{\Lambda}$ (Λ) necessitates Higgs fields of doublet and singlet of the H symmetry which belong to the singlets (doublets) of the EW symmetry. The doublet and singlet composition of the EW and H symmetries for both of the fermion and scalar fields simplifies the formalism and enables us to reduce the number of Yukawa coupling constants. In this theory, mass matrices with four texture zeros of Dirac and Majorana types are systematically derived, and unphysical modes of bosonic fields are excluded by properly adjusting values of parameters in the Higgs potentials.

For the sake of distinction, we use the symbols $\{\tau_1, \tau_2, \tau_3\}$ and Y for the isospin and hypercharge of the EW symmetry, and the symbols $\{\bar{\tau}_1, \bar{\tau}_2, \bar{\tau}_3\}$ and $\bar{Y}$ for the "isospin" and "hypercharge" of the H symmetry. The color degrees of freedom are not specified for simplicity. We introduce a symbol $\bar{t}$ to indicate the operation of transposition for degrees of the H symmetry.

2. Doublet and Singlet Composition

The gauge fields of the EW symmetry, $A_\mu^{(2)j}$ ($j = 1, 2, 3$) and $A_\mu^{(1)}$, interact to the currents of EW-isospin τ_j and hypercharge Y with coupling constants g_2 and g_1. In contrast, we introduce gauge fields of the H symmetry, $\bar{A}_\mu^{(2)j}$ ($j = 1, 2, 3$) and $\bar{A}_\mu^{(1)}$, which interact vectorially to the currents generated by H-isospin $\bar{\tau}_j$ and H-hypercharge $\bar{Y}$ with coupling constants $\bar{g}_2$ and $\bar{g}_1$.

In high-energy region ($> \bar{\Lambda}$), fundamental fermions in sector f ($= q, u, d; \ell, \nu, e$) are postulated to belong to the doublet and singlet of the group G_H as follows:

$$\psi_d^f = {}^t\left(\psi_1^f, \psi_2^f \right), \quad \psi_s^f = \left(\psi_3^f \right), \tag{1}$$

whose components are either the EW chiral doublets as

$$\psi_i^q = \begin{pmatrix} \psi_i^u \\ \psi_i^d \end{pmatrix}_L ; \psi_i^\ell = \begin{pmatrix} \psi_i^\nu \\ \psi_i^e \end{pmatrix}_L \tag{2}$$

or the EW chiral singlets as

$$(\psi_i^u)_R, (\psi_i^d)_R; (\psi_i^\nu)_R, (\psi_i^e)_R. \tag{3}$$

In the low-energy region ($\leq \Lambda$), the doublet ψ_d^f and the singlet ψ_s^f turn out to constitute, respectively, main components of the first and second generations and the third generation of fundamental fermions.

To properly break the H and EW symmetries, two types of H multiplets of Higgs fields are presumed to exist. For the H symmetry breaking around the high-energy scale $\bar{\Lambda}$, a set of doublet and singlet of Higgs fields are introduced as

$$\phi_d = {}^t\left(\phi_1, \phi_2 \right), \quad \phi_s = \left(\phi_3 \right), \tag{4}$$

where complex fields ϕ_1 and ϕ_2 and a real field ϕ_3 belong to EW singlets. These scalar fields do not couple with the fundamental fermion fields in (1) except for the right-handed neutrino fields $\psi_{d,s}^\nu$. It is this character of ϕ_a that protects the fundamental fermion fields from acquiring Dirac masses of the scale $\bar{\Lambda}$. A dual doublet of ϕ_d is defined by $\tilde{\phi}_d = i\bar{\tau}_2\phi_d^*$.

To form the Yukawa interaction and break its symmetry at the scale Λ, a set of H doublet and singlet consisting of three EW doublets must exist as

$$\varphi_d = {}^t\left(\varphi_1, \varphi_2 \right) = {}^t\left(\begin{pmatrix} \varphi_1^+ \\ \varphi_1^0 \end{pmatrix}, \begin{pmatrix} \varphi_2^+ \\ \varphi_2^0 \end{pmatrix} \right), \quad \varphi_s = \begin{pmatrix} \varphi_3^+ \\ \varphi_3^0 \end{pmatrix}, \tag{5}$$

which, respectively, have dual multiplets $\tilde{\varphi}_d = (i\bar{\tau}_2)(i\tau_2)\varphi_d^*$ and $\tilde{\varphi}_s = i\tau_2\varphi_s^*$.

3. Lagrangian Density

The Lagrangian density for the fermion and scalar interactions consists of the Yukawa and Majorana parts. The density of the Yukawa interaction, $\mathcal{L}_Y^f$, consists of the EW×H invariants of the multiplets ψ_a and φ_a ($a = d, s$) as follows:

$$\mathcal{L}_Y^f = \mathcal{Y}_1^f\, \bar{\psi}_d^{f'}\, \tilde{\varphi}_s\psi_d^f + \mathcal{Y}_2^f\, \bar{\psi}_d^{f'}\, \tilde{\varphi}_d\psi_s^f + \mathcal{Y}_3^f\, \bar{\psi}_s^{f'}{}^t\tilde{\varphi}_d i\bar{\tau}_2\psi_d^f + \mathcal{Y}_4^f\, \bar{\psi}_s^{f'}\, \tilde{\varphi}_s\psi_s^f + \text{h.c.} \tag{6}$$

for the EW up- sectors ($f' = q$, $f = u$) and ($f' = \ell$, $f = \nu$), and

$$\mathcal{L}_Y^f = \mathcal{Y}_1^f \, \bar{\psi}_d^{f'} \varphi_s \psi_d^f + \mathcal{Y}_2^f \, \bar{\psi}_d^{f'} \varphi_d \psi_s^f + \mathcal{Y}_3^f \, \bar{\psi}_s^{f'\,\bar{t}} \varphi_d i \bar{\tau}_2 \psi_d^f + \mathcal{Y}_4^f \, \bar{\psi}_s^{f'} \varphi_s \psi_s^f + \text{h.c.} \quad (7)$$

for the EW down-sectors ($f' = q$, $f = d$) and ($f' = \ell$, $f = e$). Four unknown complex coupling constants $\mathcal{Y}_{fi}$ ($i = 1, \cdots, 4$) exist in each sector. The Lagrangian density for the Majorana interaction, $\mathcal{L}_M$, is given by

$$\mathcal{L}_M = \bar{g} \left(\overline{\psi_d^{\nu c}} \, \bar{\tau}_2 \phi_d \psi_s^\nu + \overline{\psi_s^{\nu c}} \, {}^{\bar{t}} \phi_d \bar{\tau}_2 \psi_d^\nu \right) + \bar{M}_d \overline{\psi_d^{\nu c}} \, \bar{\tau}_2 \, \psi_d^\nu + \bar{M}_s \overline{\psi_s^{\nu c}} \, \psi_s^\nu, \quad (8)$$

where $\psi_a^{\nu c}$ are the charge conjugate fields, and $\bar{g}$ and $\bar{M}_a$ ($a = d, s$) are the Majorana coupling constant and masses.

The Lagrangian density for the scalar fields, $\mathcal{L}_{\text{scalar}}$, takes the form

$$\mathcal{L}_{\text{scalar}} = \sum_{a=d,s} (\mathcal{D}^\mu \phi_a)^\dagger (\mathcal{D}_\mu \phi_a) + \sum_{a=d,s} (\mathcal{D}^\mu \varphi_a)^\dagger (\mathcal{D}_\mu \varphi_a) - V_T(\phi, \varphi), \quad (9)$$

where $V_T(\phi, \varphi)$ is the total Higgs potential including all Higgs fields. The covariant derivatives $\mathcal{D}_\mu$ for the scalar multiplets ϕ_a and φ_a are given, respectively, by[a]

$$\mathcal{D}_\mu \phi_d = \left(\partial_\mu - i\bar{g}_2 \, \bar{A}_\mu^{(2)j} \frac{1}{2} \bar{\tau}_j - i\bar{g}_1 \, \bar{A}_\mu^{(1)} \frac{1}{2} \right) \phi_d \,, \quad (10)$$

$$\mathcal{D}_\mu \phi_s = \partial_\mu \phi_s \,, \quad (11)$$

$$\mathcal{D}_\mu \varphi_d = \left(\partial_\mu - ig_2 A_\mu^{(2)j} \frac{1}{2} \tau_j - ig_1 A_\mu^{(1)} \frac{1}{2} - i\bar{g}_2 \, \bar{A}_\mu^{(2)j} \frac{1}{2} \bar{\tau}_j - i\bar{g}_1 \, \bar{A}_\mu^{(1)} \frac{1}{2} \bar{Y}_{\varphi_d} \right) \varphi_d, \quad (12)$$

and

$$\mathcal{D}_\mu \varphi_s = \left(\partial_\mu - ig_2 A_\mu^{(2)j} \frac{1}{2} \tau_j - ig_1 A_\mu^{(1)} \frac{1}{2} - i\bar{g}_1 \, \bar{A}_\mu^{(1)} \frac{1}{2} \bar{Y}_{\varphi_s} \right) \varphi_s \,. \quad (13)$$

The total Higgs potential of the multiplets ϕ_a and φ_a, $V_T(\phi, \varphi)$, can be separated into the sum of three parts as follows:

$$V_T(\phi, \varphi) = V_1(\phi) + V_2(\varphi) + V_3(\varphi; \phi). \quad (14)$$

The potential $V_1(\phi)$ of the self-interactions of the multiplets ϕ_a is given by

$$V_1(\phi) = -\bar{m}_d^2 \, \phi_d^\dagger \phi_d - \bar{m}_s^2 \, \phi_s^2 + \frac{1}{2} \bar{\lambda}_d \left(\phi_d^\dagger \phi_d \right)^2 + \frac{1}{2} \bar{\lambda}_s \, \phi_s^4 + \bar{\lambda}_{ds} \left(\phi_d^\dagger \phi_d \right) \phi_s^2, \quad (15)$$

where $\bar{\lambda}_s, \bar{\lambda}_d$ and $\bar{\lambda}_{ds}$ are positive coupling constants satisfying $\bar{\lambda}_d \bar{\lambda}_s > \bar{\lambda}_{ds}^2$. Using this density, we analyze the breakdown of the H symmetry around the scale $\bar{\Lambda}$. The potential $V_2(\varphi)$ of the self-interactions of the multiplets φ_a is expressed as

$$V_2(\varphi) = -m_d^2 \varphi_d^\dagger \varphi_d - m_s^2 \varphi_s^\dagger \varphi_s + \frac{1}{2} \lambda_d (\varphi_d^\dagger \varphi_d)^2 + \frac{1}{2} \lambda_{d1} |\varphi_d^\dagger \tilde{\varphi}_d|^2$$
$$+ \frac{1}{2} \lambda_{d2} \overline{\text{Tr}} \left({}^{\bar{t}} \varphi_d^\dagger {}^{\bar{t}} \varphi_d \right)^2 + \frac{1}{2} \lambda_{d3} \overline{\text{Tr}} \left({}^{\bar{t}} \varphi_d^\dagger \varphi_d \, {}^{\bar{t}} \tilde{\varphi}_d^\dagger \tilde{\varphi}_d \right) + \frac{1}{2} \lambda_s (\varphi_s^\dagger \varphi_s)^2$$
$$+ \lambda_{ds} \left(\varphi_d^\dagger \varphi_d \right) \left(\varphi_s^\dagger \varphi_s \right) + \lambda_{ds1} |\varphi_d^\dagger \varphi_s|^2 + \lambda_{ds2} |\varphi_d^\dagger \tilde{\varphi}_s|^2, \quad (16)$$

---

[a]The H-hypercharge of ϕ_d is chosen to be 1 by adjusting the value of the coupling constant $\bar{g}_2$.

where $\overline{\mathrm{Tr}}$ means to take the trace operation with respect to the H-degrees of freedom. For the potential of mutual interactions between the multiplets φ_a and ϕ_a, we obtain

$$V_3(\varphi; \phi) = \dot{\lambda}_1(\phi_d^\dagger\phi_d)(\varphi_d^\dagger\varphi_d) + \dot{\lambda}_2\,\phi_s^2(\varphi_d^\dagger\varphi_d) + \dot{\lambda}_3(\phi_d^\dagger\phi_d)(\varphi_s^\dagger\varphi_s)$$

$$+\dot{\lambda}_4\,\phi_s^2(\varphi_s^\dagger\varphi_s) + \dot{\lambda}_5|\tilde{\phi}_d^\dagger\varphi_d|^2 + \dot{\lambda}_6|\phi_d^\dagger\varphi_d|^2. \tag{17}$$

4. Symmetry Breakdown at High-Energy Scale $\bar{\Lambda}$

In the broken phase of the H symmetry around and below the scale $\bar{\Lambda}$, the doublet and singlet, ϕ_d and ϕ_s, are decomposed into the following forms:

$$\phi_d = {}^{\bar{t}}\left(0,\ \bar{v}_d + \tfrac{1}{\sqrt{2}}\xi_d\right), \quad \phi_s = \bar{v}_s + \frac{1}{\sqrt{2}}\xi_s, \tag{18}$$

where $\bar{v}_d$ and $\bar{v}_s$ are vacuum expectation values (VEVs), and ξ_d and ξ_s are real component scalar fields. Up to the second order, the potential $V_1(\phi)$ takes the form

$$V_1(\phi) = V_1(\bar{v}) + \bar{\lambda}_d\,\bar{v}_d^2\xi_d^2 + \bar{\lambda}_s\bar{v}_s^2\,\xi_s^2 + 2\bar{\lambda}_{ds}\bar{v}_d\bar{v}_s\,\xi_d\xi_s = V_1(\bar{v}) + \frac{1}{2}m_{\xi_1}^2\xi_1^2 + \frac{1}{2}m_{\xi_2}^2\xi_2^2, \tag{19}$$

where new real fields ξ_i ($i = 1, 2$) are introduced by

$$\xi_d = \cos\bar{\theta}\,\xi_1 - \sin\bar{\theta}\,\xi_2, \quad \xi_s = \sin\bar{\theta}\,\xi_1 + \cos\bar{\theta}\,\xi_2. \tag{20}$$

The mixing angle $\bar{\theta}$ is subject to

$$\tan 2\bar{\theta} = \frac{2\bar{\lambda}_{ds}\bar{v}_d\bar{v}_s}{\bar{\lambda}_d\bar{v}_d^2 - \bar{\lambda}_s\bar{v}_s^2} \tag{21}$$

and the mass of the field ξ_i is obtained by

$$m_{\xi_i}^2 = \bar{\lambda}_d\bar{v}_d^2 + \bar{\lambda}_s\bar{v}_s^2 + (-1)^i\sqrt{\left(\bar{\lambda}_d\bar{v}_d^2 - \bar{\lambda}_s\bar{v}_s^2\right)^2 + 4\left(\bar{\lambda}_{ds}\bar{v}_d\bar{v}_s\right)^2}. \tag{22}$$

The symmetry breaking at the scale $\bar{\Lambda}$ metamorphoses the gauge fields $\bar{A}_\mu^{(2)j}$ and $\bar{A}_\mu^{(1)}$ into new fields. Estimation of the action of the covariant derivative on the scalar doublet at the stationary state with the VEV $\bar{v}_d$ results in

$$(\mathcal{D}_\mu\langle\phi_d\rangle)^\dagger(\mathcal{D}^\mu\langle\phi_d\rangle) = M_{\bar{W}}^2\bar{W}_\mu\bar{W}^\mu + \frac{1}{2}M_{\bar{Z}}^2\bar{Z}_\mu\bar{Z}^\mu, \tag{23}$$

where $\bar{W}_\mu$ is the complex vector field

$$\bar{W}_\mu = \frac{\bar{A}_\mu^{(2)1} - i\bar{A}_\mu^{(2)2}}{\sqrt{2}} \tag{24}$$

with the mass $M_{\bar{W}}^2 = \frac{1}{2}\bar{g}^2\bar{v}^2$, and $\bar{Z}_\mu$ is the neutral vector field

$$\bar{Z}_\mu = \frac{\bar{g}_2\bar{A}_\mu^{(2)3} - \bar{g}_1\bar{A}_\mu^{(1)}}{\sqrt{\bar{g}_2^2 + \bar{g}_1^2}} = \bar{A}_\mu^{(2)3}\cos\vartheta - \bar{A}_\mu^{(1)}\sin\vartheta \tag{25}$$

carrying the mass

$$M_{\bar{Z}}^2 = \frac{1}{2}(\bar{g}_2 + \bar{g}_1^2)\bar{v}_d^2 = \frac{M_{\bar{W}}^2}{\cos^2\vartheta}. \tag{26}$$

There exists another vector field X_μ, being orthogonal to $\bar{Z}_\mu$, with the configuration

$$X_\mu = \frac{\bar{g}_1 \bar{A}_\mu^{(2)3} + \bar{g}_2 \bar{A}_\mu^{(1)}}{\sqrt{\bar{g}_2^2 + \bar{g}_1^2}} = \bar{A}_\mu^{(2)3} \sin\vartheta + \bar{A}_\mu^{(1)} \cos\vartheta, \tag{27}$$

which remains massless down to the scale Λ and makes gauge interaction to the vector currents of a new charge $\bar{Q} = \frac{1}{2}\bar\tau_3 + \frac{1}{2}\bar{Y}$ of the H symmetry with the unit of strength, $\bar{e}$, defined by

$$\bar{e} = \bar{g}_2 \sin\vartheta = \bar{g}_1 \cos\vartheta . \tag{28}$$

Substitution of the decompositions of ϕ_a in (18) into (8) leads to the effective Lagrangian density of neutrino species as

$$\mathcal{L}_{\mathrm{M}} \to \mathcal{L}_{\mathcal{M}}^{\mathrm{M}} = \overline{\Psi_L^{\nu c}} \bar{\mathcal{M}}_\nu \Psi_R^\nu + \cdots , \tag{29}$$

where $\Psi_{L,R}^\nu$ are chiral neutrino triplets in the interaction mode, $\bar{\mathcal{M}}_\nu$ is the Majorana mass matrix

$$\bar{\mathcal{M}}_\nu = \begin{pmatrix} 0 & -i\bar{M}_d & -i\bar{g}\bar{v}_d \\ i\bar{M}_d & 0 & 0 \\ i\bar{g}\bar{v}_d & 0 & \bar{M}_s \end{pmatrix} , \tag{30}$$

and the ellipsis means interactions between the neutrinos and scalar fields ξ_i.

5. Symmetry Breakdown at Low-Energy Scale Λ

To go down to the low-energy region around the scale Λ, the effects of the renormalization group must properly be taken into account for all of the physical quantities. In particular, all coupling constants run down to the values at the scale Λ. For the sake of simplicity, the same symbols are used here for the quantities including all these effects.

In the broken phase of EW symmetry around the scale Λ, the multiplets φ_a are postulated to take the forms

$$\varphi_d = \bar{\iota}\left(\begin{pmatrix} \zeta_1^+ \\ \zeta_1^0 \end{pmatrix}, \begin{pmatrix} \zeta_2^+ \\ v_d + \frac{1}{\sqrt{2}}\eta_d \end{pmatrix} \right), \quad \varphi_s = \begin{pmatrix} 0 \\ v_s + \frac{1}{\sqrt{2}}\eta_s \end{pmatrix} \tag{31}$$

with VEVs v_d and v_s, where ζ_1^+, ζ_2^+ and ζ_1^0 are complex component fields, and η_d and η_s are real component fields. To examine the dynamics around the scale Λ, it is necessary to examine the sum of the potential $V_2(\varphi)$ and also the potential $V_3(\varphi, \bar{v})$ which reflects the influence of the H symmetry breakdown at the high-energy scale $\bar{\Lambda}$. We obtain the stationary conditions as follows:

$$\begin{aligned} (\lambda_d + \lambda_{d2})v_d^2 + (\lambda_{ds} + \lambda_{ds1})v_s^2 &= m_d'^2 \equiv m_d^2 - \dot\lambda_1 \bar{v}_d^2 - \dot\lambda_2 \bar{v}_s^2 - \dot\lambda_6 \bar{v}_d^2, \\ (\lambda_{ds} + \lambda_{ds1})v_d^2 + \lambda_s v_s^2 &= m_s'^2 \equiv m_s^2 - \dot\lambda_3 \bar{v}_d^2 - \dot\lambda_4 \bar{v}_s^2. \end{aligned} \tag{32}$$

Accordingly, for the phase transition to take place ($v_d, v_s \neq 0$), reduced quantities $m_d'^2$ and $m_s'^2$ must be positive.

Around the stationary point, the sum of the Higgs potential is decomposed with respect to the component scalar fields, up to the second order, as

$$V_2(\varphi) + V_3(\varphi; \bar{v}_d) = V_2(v) + V_3(v; \bar{v}_d)$$
$$+ m_{\zeta_1^+}^2 |\zeta_1^+|^2 + m_{\zeta_2^+}^2 |\zeta_2^+|^2 + m_{\zeta_1^0}^2 |\zeta_1^0|^2 + \tfrac{1}{2} m_1^2 \eta_1^2 + \tfrac{1}{2} m_2^2 \eta_2^2 \cdots , \tag{33}$$

where η_i $(i = 1, 2)$ are new real fields introduced by

$$\eta_d = \cos\theta\,\eta_1 - \sin\theta\,\eta_2, \quad \eta_s = \sin\theta\,\eta_1 + \cos\theta\,\eta_2 . \tag{34}$$

The masses of three complex fields ζ_1^+, ζ_2^+ and ζ_1^0 are calculated to be

$$m_{\zeta_1^+}^2 = (2\lambda_{d1} - \lambda_{d2} + \lambda_{d3})v_d^2 + m_{\zeta_2^+}^2 + m_{\zeta_1^0}^2,$$
$$m_{\zeta_2^+}^2 = (\lambda_{ds2} - \lambda_{ds1})v_s^2, \tag{35}$$
$$m_{\zeta_1^0}^2 = (\dot{\lambda}_5 - \dot{\lambda}_6)\bar{v}_d^2.$$

The two real fields η_i $(i = 1, 2)$ have the masses as

$$m_{\eta_i}^2 = D + S + (-1)^i \sqrt{(D - S)^2 + 4(\lambda_{ds} + \lambda_{ds1})^2 v_d^2 v_s^2}, \tag{36}$$

and the mixing angle $\bar{\theta}$ is subjects to

$$\tan 2\theta = \frac{2(\lambda_{ds} + \lambda_{ds1})v_d v_s}{D - S}, \tag{37}$$

where the abbreviations

$$D = (\lambda_d + \lambda_{d2})v_d^2 - \frac{1}{2}(\lambda_{ds} + \lambda_{ds1})v_s^2, \quad S = \lambda_s v_s^2 \tag{38}$$

are used.

Results in (35), (36) and (38) demonstrate that the Higgs coupling constants must satisfy inequality relations so that all complex and real scalar fields are in physical modes. For example, the inequality relations $\lambda_{ds2} > \lambda_{ds1}$ and $\dot{\lambda}_5 > \dot{\lambda}_6$ must hold to make the masses of the fields ζ_2^+ and ζ_1^0 positive-definite. From (36), it is shown that the real field η_i with lighter mass must be identified with the so-called Higgs particle. Note that stringent conditions on the flavor changing neutral (charged) currents give strong restrictions on the Higgs coupling constants.

The symmetry breaking at the scale Λ changes the gauge fields $A_\mu^{(2)j}$, $A_\mu^{(1)}$ and X_μ into massive vector fields by transferring the four degrees of freedom of the scalar multiplets φ_a. To determine configurations of the vector fields, we calculate the action of the covariant derivative on the scalar multiplets φ_d and φ_s at their stationary state with the VEVs of v_d and v_s obtaining

$$\sum_{a=d,s} (\mathcal{D}_\mu \langle \phi_a \rangle)^\dagger (\mathcal{D}^\mu \langle \phi_a \rangle) = \frac{1}{2} g_2^2 (v_d^2 + v_s^2) W_\mu W^\mu$$
$$+ \frac{1}{4} v_d^2 \left[\frac{g_2}{\cos\theta_W} Z_\mu + \bar{e}(1 - \bar{Y}_{\varphi_d}) X_\mu \right] \left[\frac{g_2}{\cos\theta_W} Z^\mu + \bar{e}(1 - \bar{Y}_{\varphi_d}) X^\mu \right] \tag{39}$$
$$+ \frac{1}{4} v_s^2 \left[\frac{g_2}{\cos\theta_W} Z_\mu - \bar{e}\bar{Y}_{\varphi_s} X_\mu \right] \left[\frac{g_2}{\cos\theta_W} Z^\mu - \bar{e}\bar{Y}_{\varphi_s} X^\mu \right] + \cdots ,$$

where the charged vector field

$$W_\mu = \frac{A_\mu^{(2)1} - iA_\mu^{(2)2}}{\sqrt{2}}, \tag{40}$$

and the neutral vector field

$$Z_\mu = \frac{g_2 A_\mu^{(2)3} - g_1 A_\mu^{(1)}}{\sqrt{g_2^2 + g_1^2}} = A_\mu^{(2)3} \cos\theta_W - A_\mu^{(1)} \sin\theta_W \tag{41}$$

and

$$A_\mu = \frac{g_1 A_\mu^{(2)3} + g_2 A_\mu^{(1)}}{\sqrt{g_2^2 + g_1^2}} = A_\mu^{(2)3} \sin\theta_W + A_\mu^{(1)} \cos\theta_W \tag{42}$$

are introduced, in exactly the same way with the Weinberg-Salam theory by using the Weinberg angle θ_W related to the unit e of electromagnetic interaction as

$$g_2 \sin\theta_W = g_1 \cos\theta_W = e. \tag{43}$$

The ellipsis in (39) implies mass corrections to the super-massive vector fields $\bar{W}_\mu$ and $\bar{Z}_\mu$, and mixing interactions of the field $\bar{Z}_\mu$ with the fields Z_μ and X_μ.

The charged vector field W_μ possesses the mass

$$M_W^2 = \frac{1}{2} g_2^2 (v_d^2 + v_s^2). \tag{44}$$

The quadratic part of the neutral fields Z_μ and X_μ in (39) is readily diagonalized provided that

$$(1 - \bar{Y}_{\varphi_d}) v_d^2 = \bar{Y}_{\varphi_s} v_s^2. \tag{45}$$

Under this condition, the masses of the fields Z_μ and X_μ are determined to be

$$M_Z^2 = \frac{1}{2} \frac{g_2^2}{\cos^2\theta_W} (v_d^2 + v_s^2) = \frac{M_W^2}{\cos^2\theta_W} \tag{46}$$

and

$$M_X^2 = \frac{1}{2} \frac{\bar{e}^2 v_s^2}{v_d^2} (v_d^2 + v_s^2) \bar{Y}_{\varphi_s}^2 = \frac{\bar{e}^2 v_s^2}{g_2^2 v_d^2} \bar{Y}_{\varphi_s}^2 M_W^2. \tag{47}$$

The mass relation in (46) proves that the firmly-established experimental criterion for the standard model, $\rho = M_W^2 / M_Z^2 \cos^2\theta_W = 1$, holds also in the present theory.

Through the breakdown of EW symmetry at the scale Λ, the fermion fields acquire Dirac type masses. Substitution of the decomposition of φ_a in (31) into (6) and (7) leads to the effective Lagrangian density for the fermion fields in the low-energy region as

$$\mathcal{L}_Y \rightarrow \mathcal{L}_\mathcal{M}^Y = \sum_{f=u,d,\nu,e} \bar{\Psi}_L^f \mathcal{M}_f \Psi_R^f + \text{h.c.} + \cdots, \tag{48}$$

where $\Psi_{L,R}^f$ are the chiral triplets of f-sector in the interaction mode, and $\mathcal{M}_f$ are the Dirac mass matrices. The ellipsis stands for the interactions of fermion and scalar fields. For the up-sectors $(f = u, \nu)$ of EW symmetry, we deduce the Dirac mass matrices as follows:

$$\mathcal{M}_f = \begin{pmatrix} \mathcal{Y}_1^u v_s & 0 & \mathcal{Y}_2^u v_d \\ 0 & \mathcal{Y}_1^u v_s & 0 \\ 0 & \mathcal{Y}_3^u v_d & \mathcal{Y}_4^u v_s \end{pmatrix}. \tag{49}$$

Likewise, for the down-sectors $(f = d, e)$ of EW symmetry, we obtain

$$\mathcal{M}_f = \begin{pmatrix} \mathcal{Y}_1^d v_s & 0 & 0 \\ 0 & \mathcal{Y}_1^d v_s & \mathcal{Y}_2^d v_d \\ -\mathcal{Y}_3^d v_d & 0 & \mathcal{Y}_4^d v_s \end{pmatrix}. \tag{50}$$

Both matrices which have four texture zeros are characterized by four independent parameters. It should be recognized that all the parameters except for two can be set to be real by adjusting phases of the doublets and singlets of the fermion and scalar fields, ψ_a^f and φ_a, in the Yukawa interactions in (6) and (7).

6. Discussion

Thanks to the unique construction of the present theory in which all the fermion and scalar fields are presumed to belong to the doublet and singlet representations of the H and EW symmetries, we have succeeded systematically to obtain simple mass matrices with four texture zeros as in (30), (49) and (50). Accordingly, it is tempting to inquire why there exists such a kind of duality that the symmetry SU(2)×U(1) holds both in the H and EW degrees of freedom.

For the quark sector, the eigenvalue problems for $\mathcal{M}_f \mathcal{M}_f^\dagger$ $(f = u, d)$ which have ten adjustable parameters provide sufficient information on the mass spectra and flavor mixing matrix. Smallness of neutrino masses is usually explained by the seesaw mechanism in which the inverse of the Majorana mass matrix with large components works to suppress the Dirac matrix. Remark that the determinant of the Majorana mass matrix in (30) is calculated to be $|\bar{\mathcal{M}}_\nu| = -\bar{M}_d^2 \bar{M}_s$. Therefore, the seesaw suppression occurs exclusively by the Majorana masses $\bar{M}_a$ independently of the VEV $\bar{v}_d$. This observation reveals that the present scheme might be reinterpreted to have three energy scales, $\bar{M}_a \gg \bar{\Lambda} \gg \Lambda$, rather than two scales, $\bar{\Lambda} \gg \Lambda$.

In this theory, values of the coupling constants in the Higgs potential must be tuned so that the symmetry breakdowns properly take place and all bosonic fields acquire positive masses. Furthermore, those coupling constants must obey much stronger conditions so that the stringent experimental criteria of the flavor changing neutral (charged) currents are fulfilled.

In addition to the seven scalar particles related to the fields ξ_1, ξ_2; ζ_1^+, ζ_2^+, ζ_1^0, η_1 and η_2, the theory predicts existence of one peculiar particle described by the vector field X_μ interacting with current of the charge $\bar{Q} = \frac{1}{2}\bar{\tau}_3 + \bar{Y}$ of the horizontal symmetry. Search for the signals of these particles are expected as possible targets for the LHC experiment. Through massless and massive stages, the exotic field X_μ seems to play important roles in astrophysics and particle physics.

PROTON DECAY IN 5D–SU(6) GUT WITH ORBIFOLD S^1/Z_2 BREAKING IN SCHERK-SCHWARZ MECHANISM

A. HARTANTO,[a] F. P. ZEN,[a,b] J. S. KOSASIH[a,b] and L. T. HANDOKO[c,d]

[a] *Theoretical Physics Laboratory, THEPI, Faculty of Mathematics and Natural Sciences, Institut Teknologi Bandung, Jl. Ganesha 10 Bandung 40135, Indonesia*

[b] *Indonesia Center for Theoretical and Mathematical Physics, Jl. Ganesha 10, Bandung 40132, Indonesia*

[c] *Group for Theoretical and Computational Physics, Research Center for Physics, Indonesian Institute of Sciences, Kompleks Puspiptek Serpong, Tangerang 15310, Indonesia*

[d] *Department of Physics, University of Indonesia, Kampus UI Depok, Depok 16424, Indonesia*

Proton decay within 5-dimensional SU(6) GUT with orbifold S^1/Z_2 breaking is investigated using Scherk-Schwarz mechanism. It is shown that in the model neither leptoquark like heavy gauge bosons nor violation of baryon number conservation are allowed due to the orbifold breaking parity splitting. These results prevent too short proton lifetime within the model.

Keywords: Proton decay; SU(6); GUT; symmetry breaking; extra dimension.

1. Introduction

Despite the Standard Model (SM) of particle physics containing the electroweak $SU(2)_L \otimes U(1)_Y$ and the strong $SU(3)_C$ theories are in impressive agreements with most of experimental observables,[1] both theories are not yet unified in a single symmetry. The SM with symmetry $SU(3)_C \otimes SU(2)_L \otimes U(1)_Y$, is lacking of explaining the unification of three gauge couplings at a particular scale assuming that our nature should be explained by a single unified theory, the so-called grand unified theory (GUT).

There are many types of GUT models, but most of them deploys larger symmetry than $SU(3)_C \otimes SU(2)_L \otimes U(1)_Y$ to accommodate at least all three interactions at the particle elementer scale. One of models containing $SU(3)_C \otimes SU(2)_L \otimes U(1)_Y$ as a part of its subgroups at electroweak scale is the SU(6) GUT.[2] Unfortunately, the model is suffered from realizing the desired breaking patterns through Higgs mechanism. It has been concluded that the allowed Higgs multiplets in the model are not able to reproduce all particle spectrums within the experimental bounds.

Nevertheless, it has been found that the symmetry breaking for SU(6) GUT can be realized by introducing an extra dimension at the SU(6) scale.[3] This non-Higgs mechanism is adopting the Scherk-Schwarz mechanism to dynamically break the symmetry induced by the orbifold of extra dimension.[4-6] The effect on compactifying the extra dimension is considered to induce the Higgs bosons itself, and is known as the Higgs – gauge boson unification.[7] Later, the generated Higgs bosons are utilized to break the subsequent symmetry to the electroweak scale.[3] This paper is focused on investigating the proton decay within the model with the above-mentioned breaking mechanism. Because proton decay is the most severe constraint for any models beyond the SM.

The paper is organized as follows. First the symmetry breaking to $SU(3)_C \otimes SU(3)_H \otimes U(1)_C$ due to dimensional reduction with orbifold S^1/Z_2 is briefly reviewed. Subsequently it is shown that the adopted mechanisms in this case naturally suppress the proton decay since the leptoquark like interactions are not allowed at tree level as happened in the experimentally excluded conventional SU(5) GUT. Finally the paper is concluded with a short summary.

2. 5-Dimensional SU(6)

The 4-dimensional (4D) SU(6) model is already given in our previous work,[2] but in the present paper let us consider the extended model laying on 5D space-time. The 5D bulk in the model is divided into two branes due to the properties of orbifold breaking. One brane contains the 5D gauge bosons, A_M, while another particle contents live in the remaining 4D one. Each brane corresponds to the orbifold fixed point $y = 0$ and πR, and the parity transformation operator Z_2 is taken as,

$$Z_2^{(0)} = Z_2^{(1)} = \begin{pmatrix} 1 & 0 & & & & \\ 0 & 1 & & & & \\ & & 1 & 0 & & \\ & & 0 & -1 & & \\ & & & & -1 & 0 \\ & & & & 0 & -1 \end{pmatrix}. \tag{1}$$

This yields $U = Z_2^{(1)} Z_2^{(0)} = I_6$ is a unitary matrix.

Following the gauge-Higgs unification principle, one can define scalar boson sextet Φ in such a way that $A_M \equiv \Phi$ and $A_M^{\pm}(x, y) = \tilde{\Phi}_{\pm}(x, y)$. Here one may also define two vacuum expectation values (VEV's), v and v' at one fixed point $(y = 0)$ as below,

$$v = \begin{pmatrix} 0 \\ 0 \\ f_1 \\ 0 \\ 0 \\ 0 \end{pmatrix}, \quad \text{and} \quad v' = \begin{pmatrix} 0 \\ 0 \\ 0 \\ 0 \\ 0 \\ f_2 \end{pmatrix}. \tag{2}$$

Then, the SU(6) Little Higgs in 4D world is found to be,[8-10]

$$
\tilde{\Phi}_+^{(1)'} = \frac{1}{\sqrt{\pi R}} \exp\left[\frac{if_2}{ff_1}\begin{pmatrix} (0)_{3\times3} & \begin{pmatrix} 0 & 0 & h \\ 0 & 0 & \\ h'^\dagger & 0 \end{pmatrix} \\ \begin{pmatrix} 0 & 0 & h' \\ 0 & 0 & \\ h^\dagger & 0 \end{pmatrix} & (0)_{3\times3} \end{pmatrix}\right]\begin{pmatrix} 0 \\ 0 \\ f_1 \\ 0 \\ 0 \\ 0 \end{pmatrix}, \tag{3}
$$

$$
\tilde{\Phi}_+^{(2)'} = \frac{1}{\sqrt{\pi R}} \exp\left[-\frac{if_1}{ff_2}\begin{pmatrix} (0)_{3\times3} & \begin{pmatrix} 0 & 0 & h \\ 0 & 0 & \\ h'^\dagger & 0 \end{pmatrix} \\ \begin{pmatrix} 0 & 0 & h' \\ 0 & 0 & \\ h^\dagger & 0 \end{pmatrix} & (0)_{3\times3} \end{pmatrix}\right]\begin{pmatrix} 0 \\ 0 \\ 0 \\ 0 \\ 0 \\ f_2 \end{pmatrix}, \tag{4}
$$

Now we are ready to proceed with the first symmetry breaking of the 5D SU(6) GUT.

3. Symmetry Breaking: SU(6) → SU(3) ⊗ SU(3) ⊗ U(1)

According to the Scherk-Schwarz mechanism, the first breaking is performed by using the commutative and anti-commutative relations,[6]

$$
[Q', Z_2^{(0)}] = [\lambda^a, Z_2^{(0)}] = 0, \tag{5}
$$
$$
\{Q, Z_2^{(0)}\} = \{\lambda^{\hat{a}}, Z_2^{(0)}\} = 0. \tag{6}
$$

Here $Q' = \lambda^a$ is the unbroken SU(6) generators, while $Q' = \lambda^{\hat{a}}$ is the broken ones with $a = 1, 2, \cdots, 8, 27, 28, \cdots, 34, 35$ and $\hat{a} = 9, 10, \cdots, 26$. These induce the symmetry breaking,

$$
\underbrace{\text{SU}(6)}_{\text{5D}} \rightarrow \underbrace{\text{SU}(3) \otimes \text{SU}(3) \otimes \text{U}(1)}_{\text{4D}}, \tag{7}
$$

and the parity is splitted into: even parity for A_μ^a and $A_y^{\hat{a}}$; and odd parity for $A_\mu^{\hat{a}}$ and A_y^a. In this case the fermions are also even. Because the fields are transformed as follows,

$$
A_\mu(x, -y) = Z_2^{(0)} A_\mu(x, y) Z_2^{(0)}, \tag{8}
$$
$$
A_y(x, -y) = Z_2^{(0)} A_y(x, y) Z_2^{(0)}, \tag{9}
$$
$$
\psi^f(x, -y) = \gamma_5 Z_2^{(0)} \psi^f(x, y), \tag{10}
$$

where $Z_2^{(0)}$ fulfills the consistency condition, $U Z_2^{(0)} U = Z_2^{(0)}$.[5] Therefore, the remaining zero-modes at the low energy 4D effective theory are $A_\mu^{a(0)}$, $A_y^{a(0)}$, $\psi_{jR}^{(0)}$ $(j = 1, 2, 3)$ and $\psi_{jL}^{(0)}$ $(j = 4, 5, 6)$.

Applying the gauge-Higgs unification one obtains the scalar triplet bosons $\phi^{(i)}$ $(i = 1, 2)$ which can be derived from the upper and lower triplets of SU(6) Little Higgs in Eqs. (3) and (4),[10]

$$\phi^{(1)} = \frac{1}{\sqrt{\pi R}} e^{\frac{if_2}{ff_1} \left(\begin{smallmatrix} 0 & 0 & h' \\ 0 & 0 & \\ h^\dagger & & 0 \end{smallmatrix} \right)} \begin{pmatrix} 0 \\ 0 \\ f_1 \end{pmatrix}, \tag{11}$$

$$\phi^{(2)} = \frac{1}{\sqrt{\pi R}} e^{-\frac{if_1}{ff_2} \left(\begin{smallmatrix} 0 & 0 & h \\ 0 & 0 & \\ h'^\dagger & & 0 \end{smallmatrix} \right)} \begin{pmatrix} 0 \\ 0 \\ f_2 \end{pmatrix}. \tag{12}$$

Rescaling the VEV by a factor of $1/\sqrt{\pi R}$, and defining $(f_1')^2 + (f_2')^2 = (f')^2$ and $f' = f/\sqrt{\pi R}$, Eqs. (11) and (12) become,

$$A_y^{\hat{a}(1)} = \phi^{(1)} = e^{\frac{if_2'}{f_1'f'} \left(\begin{smallmatrix} 0 & 0 & H' \\ 0 & 0 & \\ H^\dagger & & 0 \end{smallmatrix} \right)} \begin{pmatrix} 0 \\ 0 \\ f_1' \end{pmatrix}, \tag{13}$$

$$A_y^{\hat{a}(2)} = \phi^{(2)} = e^{-\frac{if_1'}{f_2'f'} \left(\begin{smallmatrix} 0 & 0 & H \\ 0 & 0 & \\ H'^\dagger & & 0 \end{smallmatrix} \right)} \begin{pmatrix} 0 \\ 0 \\ f_2' \end{pmatrix}, \tag{14}$$

where $H = h/\sqrt{\pi R}$ and $H' = h'/\sqrt{\pi R}$. Eqs. (13) and (14) provide the scalar triplets required in 4D SU(3) $\otimes$ SU(3). The simplest little Higgs scenario is exactly reproduced by the special case with $H = H'$.[8,11,12]

Before moving forward, let us consider the particle assignment in the branes. The IR-brane at fixed point $y = 0$ contains all even particles: unbroken SU(3) gauge bosons, scalar triplet bosons and chiral fermions. Therefore the SM gauge bosons and fermions with even parity reside in IR-brane and the symmetry can accordingly labeled as usual: SU(3)$_C$ $\otimes$ SU(3)$_H$ $\otimes$ U(1)$_{C_1/C_2}$. Consequently the massive SU(3) gauge bosons corresponding to the broken generators $A_\mu^{\hat{a}}$ must reside in another UV-brane at fixed point $y = \pi R$ with off-diagonal 3×3 matrices. The lower-left is basically the hermitian conjugate of the upper-right which justifies the symmetry: SU(3)$_L$ $\otimes$ SU(3)$_R$ $\otimes$ U(1)$_{C_2}$. Remind that C_1 and C_2 bosons appear from SU(6) C−boson with generators,[2]

$$\lambda_{C_1} = \frac{1}{3}\sqrt{6} \begin{pmatrix} 1 & & \\ & 1 & \\ & & 1 \end{pmatrix}, \quad \lambda_{C_2} = \frac{\sqrt{6}}{3} \begin{pmatrix} -1 & & \\ & -1 & \\ & & -1 \end{pmatrix}. \tag{15}$$

Same with the odd particle $A_\mu^{\hat{a}}$, there is also another particle with odd-parity, that is A_y^a. This is a massless fifth-dimensional gauge boson, and under Z_2 symmetry fulfills the condition $\oint dy A_y^a = 0$. Moreover, the fifth-dimensional gauge invariance leads to $A_y^a = 0$ which also satisfies the condition, and therefore it cannot stay in UV-brane. Since fermions and scalar bosons live in IR-brane, it is obvious that UV-brane should contain only odd heavy gauge bosons from the broken SU(6) and its subsequent SU(3) symmetries.

4. Symmetry Breaking: $SU(3)_H \to SU(2)_L \otimes U(1)_{B_1/B_2}$ and $SU(3)_L \otimes SU(3)_R \to SU(3)_V$

The first breaking involves B_1 / B_2 boson that is generated from $SU(6)$ generators. Both are combined with $C_{1,2}$ bosons to form hypercharges for quarks and leptons through $U(1)_{B_1/B_2} \otimes U(1)_{C_1/C_2} \to U(1)_{Y_q/Y_l}$. This can be achieved by putting $B_1 = 0$ and $B_2 = \lambda_{34}$.

The second breaking can be realized by bringing the little Higgs triplet,

$$\phi^{(1)'} = f_1' e^{\frac{i}{f_1'} \left(\begin{smallmatrix} 0 & 0 & H' \\ 0 & 0 & \\ H^\dagger & & 0 \end{smallmatrix} \right)} \begin{pmatrix} 1 \\ & 1 \\ & & 1 \end{pmatrix}, \tag{16}$$

where the VEV is rewritten as $f_1' \mathbf{1}_{3\times 3}$ from C_1 boson in Eq. (15) to achieve non-vanishing VEV and $H \sim H'$. From the results, there are two important properties: 1) the heavy gauge bosons have zero / integer charge and not fractional one like quark charges; 2) no tree-level interaction among heavy gauge bosons and SM particles due to opposite parity and the separated branes.

5. Proton Decay

The main concern as developing the alternative GUT is the severe constraint from proton decay. In contrast to the present case, the original 4D $SU(5)$ GUT contains leptoquarks. The exotic bosons interacts with the SM particles at tree level. This immediately provide too large contribution to the proton decay which later on exceeds the experimental bounds.

The sextet and decapentuplet in $SU(6)$ consist of the same quark (anti-quark) – anti-lepton (lepton) for $\bar{6}$ and 6. On the other hand, for $\overline{15}$ and 15 both consist of quark (anti-quark) – anti-quark (quark) – anti-lepton (lepton). The structures are the same as $SU(5)$ except that there are additional heavy neutrinos N (N^c).[2] Therefore, without any manipulation $SU(6)$ must be suffered from the same problem with $SU(5)$.

Fortunately, as already discussed in the preceding section, the unique properties of fifth-dimensional orbifold might totally improve the situation. The symmetry breaking via Scherk-Schwarz mechanisms with orbifold breaking induces parity splitting such that the unbroken gauge bosons A_μ^a and the broken fifth-dimensional component gauge bosons $A_y^{\hat{a}}$ with even parity reside in IR-brane ($y = 0$). While the broken gauge bosons $A_\mu^{\hat{a}}$ and the unbroken fifth-dimensional component gauge bosons A_y^a with odd parity are projected out of IR-brane and reside in UV-brane ($y = \pi R$). This property prevents the broken heavy gauge bosons to be leptoquark like bosons. Subsequently the tree level interactions among particles reside in both branes are not allowed. This guarantees that all corresponding processes in $SU(5)$ contribute to the proton decay can not take place. Of course this fact also conserves the baryon number.

6. Summary

The symmetry breaking of SU(6) GUT model with one extra dimension through Scherk-Schwarz mechanism with orbifold S^1/Z_2 breaking has been briefly introduced. The mechanism has succeeded in realizing two steps of breakings up to the SM scale, where the second breaking is realized by Higgs mechanism using little Higgs scalar bosons generated dynamically from the first breaking.

It has been shown that the above mentioned breaking mechanism avoids the emerging leptoquark bosons as happened in SU(5) GUT, and also prevents tree level interactions among particles reside in different branes.

Acknowledgments

AH thanks the Group for Theoretical and Computational Physics, Research Center for Physics, Indonesian Institute of Sciences for warm hospitality during this work. The work of LTH is supported by the Riset Kompetitif LIPI 2010 under contract no 11.04/SK/KPPI/II/2010, while the work of FPZ is supported by Research KK-ITB 2010 and Hibah Kompetensi DIKTI 2010.

References

1. C. Amsler et al. (Particle Data Group), *Phys. Lett. B* **667**, p. 1 (2008).
2. A. Hartanto and L. T. Handoko, *Phys. Rev. D* **71**, p. 095013 (2005).
3. A. Hartanto, F. P. Zen, J. S. Kosasih and L. T. Handoko, *AIP Conf. Proc.* **1244**, p. 272 (2010).
4. G. Burdman and Y. Nomura, *Phys. Rev. D* **69**, p. 115013 (2004).
5. L. J. Hall, H. Murayama and Y. Nomura, *Nucl. Phys. B* **645**, p. 85 (2002).
6. M. Quiros, TASI Lecture: New ideas in symmetry breaking arXiv:hep-ph/0302189, (2003).
7. G. Burdman and Y. Nomura, *Nucl. Phys. B* **656**, p. 3 (2003).
8. M. Schmaltz, *JHEP* **08**, p. 056 (2004).
9. D. Kaplan and M. Schmaltz, *JHEP* **0310**, p. 039 (2003).
10. A. Hartanto, F. P. Zen, J. S. Kosasih and L. T. Handoko, *Phys. Rev. D* , submitted (2010).
11. M. Schmaltz, *Nucl. Phys. Proc. Suppl.* **117**, p. 40 (2003).
12. M. Schmaltz and D. Tucker-Smith, *Ann. Rev. Nucl. Part. Science* **55**, p. 229 (2005).

MICRO BLACK HOLES PHYSICS FROM WORLD-CRYSTAL UNCERTAINTY PRINCIPLE

PETR JIZBA

FNSPE, Czech Technical University in Prague,
Břehová 7, 115 19 Praha 1, Czech Republic
p.jizba@fjfi.cvut.cz

HAGEN KLEINERT

ITP, Freie Universität Berlin, Arnimallee 14 D-14195 Berlin, Germany
kleinert@physik.fu-berlin.de

FABIO SCARDIGLI

LeCosPA, Department of Physics, National Taiwan University, Taipei 106, Taiwan
fabio@phys.ntu.edu.tw

We formulate generalized uncertainty relations in a crystal-like universe whose lattice spacing is of order of Planck length — a "world crystal". For energies near the border of the Brillouin zone, i.e., for Planckian energies, the uncertainty relation for position and momentum does not pose any lower bound. We apply these results to micro black holes physics, where we derive a new mass-temperature relation for Schwarzschild micro black holes. In contrast to standard results based on Heisenberg and stringy uncertainty relations, our mass-temperature formula predicts both a finite Hawking's temperature and a zero rest-mass remnant at the end of the black hole evaporation. We also briefly mention some connections of the world crystal paradigm with 't Hooft's quantization and double special relativity.

PACS: 04.70.Dy, 03.65.-w

1. Introduction

The idea of a discrete structure of the space-time can be traced back to the mid 50's when John Wheeler presented his *space-time foam* concept.[1] Recent advances in physics indicate that the notion of space-time as a continuum is, indeed, likely to be superseded at the Planck scale, where gravitation and strong-electro-weak interactions become comparable in strength. In fact, discrete structures of space-time, as a rule, appear in many quantum-gravity models. Loop quantum gravity, non-commutative geometry or cosmic cellular automata may serve as examples. Perhaps a simplest toy-model for planckian physics is a *discrete lattice*.

Recently one of us (HK) proposed a model of a discrete, crystal-like universe "world crystal". There, the geometry of Einstein and Einstein-Cartan spaces can be considered as being a manifestation of the defect structure of a crystal whose lattice spacing is of the order of Planck length.[2] Curvature is due to rotational defects, torsion due to translational defects. There the memory of the crystalline structure is lost over large distances leading thus to ordinary Einstein's and Einstein-Cartan's theory.

In this paper we study the generalized uncertainty principle[3] (GUP) associated with quantum mechanics formulated on a world crystal and derive physical consequences for the corresponding world-crystal micro black holes. As micro black holes could be created at LHC (in TeV range), our results could shed some light on the fabric of planckian space-time.

2. Differential Calculus on a Lattice

On a 1D lattice the sites are at $x_n = n\epsilon$, with $n \in \mathcal{Z}$. With a 1D lattice one can affiliate two fundamental derivatives, namely

$$(\nabla f)(x) = \frac{1}{\epsilon}[f(x+\epsilon) - f(x)] \quad \text{and} \quad (\bar{\nabla} f)(x) = \frac{1}{\epsilon}[f(x) - f(x-\epsilon)] \, . \quad (1)$$

Corresponding integration is performed as a summation:

$$\int dx\, f(x) \equiv \epsilon \sum_x f(x) \, , \quad (2)$$

where x runs over all x_n. For periodic functions on the lattice or for functions vanishing at the boundary of the world crystal, the lattice derivatives can be subjected to the lattice version of integration by parts:

$$\sum_x f(x)\nabla g(x) = -\sum_x g(x)\bar{\nabla} f(x) \, , \quad (3)$$

$$\sum_x f(x)\bar{\nabla} g(x) = -\sum_x g(x)\nabla f(x) \, . \quad (4)$$

One can also define the lattice Laplacian as

$$\nabla\bar{\nabla} f(x) = \bar{\nabla}\nabla f(x) = \frac{1}{\epsilon^2}[f(x+\epsilon) - 2f(x) + f(x-\epsilon)] \, , \quad (5)$$

which reduces in the continuum limit to an ordinary Laplace operator ∂_x^2. The above calculus can be naturally extended to any number D of dimensions.

3. Position and Momentum Operator on a 1D Lattice

To build Quantum Mechanics on a 1D lattice we define a *scalar product* as

$$\langle \psi_1 | \psi_2 \rangle = \epsilon \sum_x \psi_1^*(x)\psi_2(x) \, . \quad (6)$$

It follows from Eq. (3) that

$$\langle f | \nabla g \rangle = -\langle \bar{\nabla} f | g \rangle \, , \quad (7)$$

so that $(i\nabla)^\dagger = i\bar{\nabla}$, and neither $i\nabla$ nor $i\bar{\nabla}$ are hermitian operators. The lattice Laplacian (5), however, is hermitian under the scalar product (6).

The position operator $\hat{X}_\epsilon$ acting on wave functions of x is defined by a multiplication with x

$$(\hat{X}_\epsilon f)(x) = xf(x). \tag{8}$$

To ensure hermiticity of the lattice momentum operator $\hat{P}_\epsilon$, we relate it to the symmetric lattice derivative. Using (1) we have

$$(\hat{P}_\epsilon f)(x) = \frac{\hbar}{2i}[(\nabla f)(x) + (\bar{\nabla} f)(x)] = \frac{\hbar}{2i\epsilon}[f(x+\epsilon) - f(x-\epsilon)]. \tag{9}$$

For small ϵ, this reduces to the ordinary momentum operator $\hat{p} \equiv -i\hbar\partial_x$:

$$\hat{P}_\epsilon = \hat{p} + \mathcal{O}(\epsilon^2). \tag{10}$$

The "canonical" commutator between $\hat{X}_\epsilon$ and $\hat{P}_\epsilon$ on the lattice thus reads

$$\left([\hat{X}_\epsilon, \hat{P}_\epsilon]f\right)(x) = \frac{i\hbar}{2}[f(x+\epsilon) + f(x-\epsilon)] \equiv i\hbar(\hat{I}_\epsilon f)(x). \tag{11}$$

The operators $\hat{X}_\epsilon$, $\hat{P}_\epsilon$, and $\hat{I}_\epsilon$ are hermitian under the scalar product (6). They form a $E(2)$ Lie algebra, which reduces to the standard Weyl-Heisenberg algebra in the limit $\epsilon \to 0$: $\hat{X}_\epsilon \to \hat{x}$, $\hat{P}_\epsilon \to \hat{p}$, $\hat{I}_\epsilon \to \hat{1}$. Ordinary QM is thus obtained from lattice QM by a contraction of the $E(2)$ algebra via the limit $\epsilon \to 0$.

Reminding that $\exp(\epsilon\partial_x) f(x) = f(x+\epsilon)$, we can rewrite the commutation relation (11) as

$$\left([\hat{X}_\epsilon, \hat{P}_\epsilon]f\right)(x) = i\hbar\cos\left(\epsilon\hat{p}/\hbar\right) f(x). \tag{12}$$

The latter allows to identify the lattice unit operator $\hat{I}_\epsilon$ with $\cos\left(\epsilon\hat{p}/\hbar\right)$.

4. Uncertainty Relations on Lattice

Let us define the uncertainty of an observable A in a state ψ as the standard deviation, i.e., $(\Delta A)_\psi \equiv [\langle\psi|(\hat{A} - \langle\psi|\hat{A}|\psi\rangle)^2|\psi\rangle]^{1/2}$. Via the Schwarz inequality we get the Uncertainty Relation on a lattice in the form

$$\Delta X_\epsilon \Delta P_\epsilon \geq \frac{1}{2}\left|\langle\psi|[\hat{X}_\epsilon, \hat{P}_\epsilon]|\psi\rangle\right| = \frac{\hbar}{2}\left|\langle\psi|\hat{I}_\epsilon|\psi\rangle\right| = \frac{\hbar}{2}\left|\langle\psi|\cos\left(\epsilon\hat{p}/\hbar\right)|\psi\rangle\right|. \tag{13}$$

We denote $\langle\psi|\cdots|\psi\rangle$ as $\langle\cdots\rangle_\psi$ and study two critical energy regimes of (13):

• I) The long-wavelengths regime, where $\langle\hat{p}\rangle_\psi \to 0$. In this regime the GUP (13) can be written for Planckian lattices as (cf. (10))

$$\Delta X_\epsilon \Delta P_\epsilon \geq \frac{\hbar}{2}\left(1 - \frac{\epsilon^2}{2\hbar^2}(\Delta P_\epsilon)^2\right). \tag{14}$$

• II) The regime near the boundary of the first Brillouin zone, where the averaged momentum takes its maximum value $\langle\hat{p}\rangle_\psi \to \pi\hbar/2\epsilon$. Here, using again (10), we can write

$$\Delta X_\epsilon \Delta P_\epsilon \geq \frac{\hbar}{2}\left(\frac{\pi}{2} - \frac{\epsilon}{\hbar}\langle\hat{P}_\epsilon\rangle_\psi\right). \tag{15}$$

We see that, as the momentum reaches the boundary of the Brillouin zone, the right-hand side of (15) vanishes, so that lattice quantum mechanics at short wavelengths *can* exhibit classical behavior — there is no irreducible lower bound for uncertainties of two complementary observables! This result is analogous to the one found, on a different ground, by Magueijo and Smolin in *deformed special relativity*.[6] So we see that the world crystal physics can become "classical" for energies close to the Brillouin zone, i.e., for Planckian energies. This is also a recurrent theme in 't Hooft's "deterministic" quantum mechanics,[7] where a deterministic theory at the Planck scale, supplemented with a dissipation mechanism, can give rise to a genuine quantum mechanical behavior at larger scales.

5. World-Crystal Micro Black Hole Physics

Following a strategy outlined in Ref. 4 we can now use (14) and (15) to obtain the mass-temperature relation for the world crystal micro black holes. Heisenberg's microscope argument implies that the smallest resolvable spatial detail δx goes roughly as the wavelength of the probing photon. If E is the (average) energy of photons used in the microscope, then standard Heisenberg principle dictates

$$\delta x = \frac{\hbar c}{2E} . \tag{16}$$

From (14), the lattice version of the Heisenberg formula (16) reads

$$\delta X_\epsilon \simeq \frac{\hbar c}{2E_\epsilon} \left[1 - \frac{\epsilon^2}{2\hbar^2 c^2} (E_\epsilon)^2 \right]. \tag{17}$$

This links the (average) wavelength of a photon to E_ϵ. The position uncertainty δX_ϵ of unpolarized photons of Hawking radiation just outside the event horizon is of the order of the Schwarzschild radius R_S. This follows directly from the geometry, reminding that the average wavelength of the Hawking radiation is of the order of the geometrical size of the hole. So, we can write $\delta X_\epsilon \simeq 2\mu R_S = 2\mu \ell_p m$ with $R_S = \ell_p m$, where $m = M/M_p$ is the black hole mass in Planck units ($M_p = \mathcal{E}_p/c^2$, $\mathcal{E}_p = \hbar c/2\ell_p$), and μ is a constant to be specified. According to the equipartition law, the energy E_ϵ of unpolarized photons is $E_\epsilon \simeq k_B T$. For a lattice spacing $\epsilon = a\ell_p$ the constant μ can be fixed by the condition that in the continuum limit $\epsilon \to 0$ the formula (17) should predict the standard semiclassical Hawking temperature $T = \hbar c/(4\pi k_B R_S)$. This gives $\mu = \pi$. Again using Planck units for the temperature $\mathcal{E}_p = k_B T_p/2$, $\Theta = T/T_p$, we may finally rewrite (17) as

$$2m = \frac{1}{2\pi\Theta} - \zeta^2 \, 2\pi\Theta \tag{18}$$

where we define the deformation parameter $\zeta = a/2\sqrt{2}\pi$.

The result coming from ordinary Heisenberg uncertainty principle, to which, in the continuum limit $\epsilon \to 0$, Eq.(17) reduces, is

$$m = \frac{1}{4\pi\Theta} , \tag{19}$$

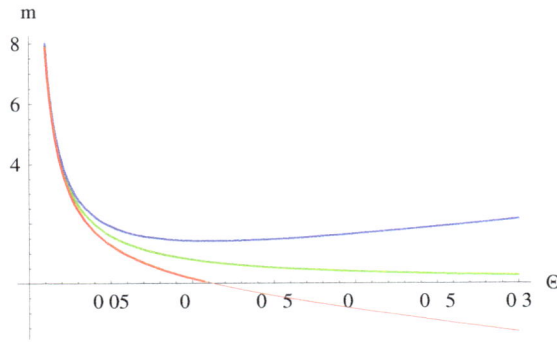

Fig. 1.   Diagrams for the three *mass-temperature* relations, ours (red), Hawking's (green), and stringy GUP result (blue), with $\zeta = \sqrt{2}$, as an example. As a consequence of the lattice uncertainty principle the evaporation ends at a *finite* temperature with a zero rest-mass remnant.

which is the dimensionless version of Hawking's formula for macroscopic black holes. The result implied by the stringy uncertainty relation[5] is

$$2m \;=\; \frac{1}{2\pi\Theta} + \zeta^2\, 2\pi\Theta\,. \tag{20}$$

The phenomenological consequences of these three relations are quite different. In Fig. 1 we compare the three results, also including the curve for the ordinary Hawking relation (19). Considering m and Θ as functions of time, we can follow the evolution of a micro black hole from the curves plotted there.

For the stringy GUP, we have a maximum temperature $\Theta_{\max} = 1/(2\pi\zeta)$, and a minimum rest mass $m_{\min} = \zeta$. The end of the evaporation process is reached after a finite time, the final temperature is finite, and there is a remnant of a finite rest mass. Should we have used the standard Heisenberg uncertainty principle we would have find the usual Hawking formula. There it is known that the evaporation process ends, after a finite time, with a zero mass and a worrisome infinite temperature. In contrast to these results, our lattice GUP predicts a finite final temperature $\Theta_{\max} = 1/(2\pi\zeta)$, with a *zero-mass remnant*. Our result (18) coming from the lattice GUP formula (14) avoids at once the difficulties of an infinite final temperature and the existence of finite-mass black hole remnants in the universe. The analysis with the GUP short wave limit (15) fully confirms the results previously obtained.

6. Entropy

From the first law of black hole thermodynamics[8] we know that the differential of the thermodynamical entropy of a Schwarzschild black hole reads

$$dS \;=\; \frac{dE}{T_H}\,, \tag{21}$$

where dE is the amount of energy swallowed by a black hole with Hawking temperature T_H. In Eq. (21) the increase in the internal energy is equal to the added heat because a black hole makes no mechanical work when its entropy/surface changes (expanding surface does not exert any pressure). Rewriting Eq. (21) with the dimensionless variables m and Θ we get $dS = k_B \, dm/(2\Theta)$. Inserting here formula (18) we find

$$dS = \frac{k_B}{2}\frac{dm}{\Theta} = -\frac{k_B}{4}\left(\frac{1}{2\pi\Theta^3} + \frac{2\pi\zeta^2}{\Theta}\right) d\Theta. \tag{22}$$

By integrating dS we obtain $S = S(\Theta)$. Just as formula (18), the relation (22) can be trusted only for $\Theta \ll \Theta_{\max} = 1/2\pi\zeta$. Thus, when integrating (22), we should do this only up to a cutoff $\tilde{\Theta}_{\max} \ll \Theta_{\max}$. The additive constant in S can then be fixed by requiring that $S = 0$ when $\Theta \to \tilde{\Theta}_{\max}$. This is equivalent to what is usually done when calculating the Hawking temperature for a Schwarzschild black hole. There, one fixes the additive constant in the entropy integral to be zero for $m = 0$, so that $S(m = 0) = S(\Theta \to \infty) = 0$ (the minimum mass attainable in the standard Hawking effect is $m = 0$). The integral yields

$$S(\Theta) = \frac{k_B}{16\,\pi}\left(\frac{1}{\Theta^2} - \frac{1}{\tilde{\Theta}_{\max}^2} + 8\pi^2\zeta^2 \log\frac{\tilde{\Theta}_{\max}}{\Theta}\right). \tag{23}$$

The entropy is always positive, and $S \to 0$ for $\Theta \to \tilde{\Theta}_{\max}$.

7. Heat Capacity

With entropy formulae (22) and (23) at hand we can now compute the heat capacity of a (micro) black hole in the world-crystal. This will give us important insights on the final stage of the evaporation process. Again, we shall obtain formulae valid only for $\Theta \ll \Theta_{\max} = 1/(2\pi\zeta)$. The heat capacity C of a black hole is defined via the relation $dQ = dE = CdT$. The pressure exerted on the environment by the expanding black hole surface is zero. Hence we do not need to specify which C is meant. Therefore we have $C = T\left(\frac{dS}{dT}\right) = \Theta\left(\frac{dS}{d\Theta}\right)$, which yields

$$C = -\frac{\pi k_B}{2}\left[\zeta^2 + \frac{1}{(2\pi\Theta)^2}\right]. \tag{24}$$

From this clearly follows that C is always negative. Most condensed-matter systems have $C > 0$. However, because of instabilities induced by gravity this is generally not the case in astrophysics,[9] especially in black hole physics. A Schwarzschild black hole has $C < 0$ which indicates that the black hole becomes hotter by radiating. The result (24) implies that this scenario holds also for micro black holes in the world-crystal. In case of stringy GUP, we have to use Eq. (20) as the mass-temperature formula. The expression for the heat capacity then reads

$$C = \frac{\pi k_B}{2}\left[\zeta^2 - \frac{1}{(2\pi\Theta)^2}\right]. \tag{25}$$

Since also here $0 < \Theta < \Theta_{\max} = 1/2\pi\zeta$, black holes have negative specific heat also according to the stringy GUP. However, stringy GUP displays a striking difference with respect to lattice GUP. In fact, since in principle we can trust Eq. (20) also when $\Theta \simeq \Theta_{\max} = 1/2\pi\zeta$, then from (25) we have, in such limit, $C = 0$. This means that for the stringy GUP the specific heat vanishes at the end point of the evaporation process in a finite time, so that the black hole at the end of its evolution cannot exchange energy with the surrounding space. In other words, the black hole stops to interact thermodynamically with the environment. The final stage of the Hawking evaporation, according to the stringy GUP scenario, contains a Planck-size remnant with a maximal temperature $\Theta = \Theta_{\max}$, but thermodynamically inert. The remnant behaves like an elementary particle — there are no internal degrees of freedom to excite in order to produce a heat absorption or emission.

8. Connection with Double Special Relativity

A different application relates to the idea of double (or doubly or deformed) special relativity (DSR) (see, e.g. Ref. 6). The general idea is that if the Planck length is a truly universal quantity, then it should look the same to any inertial observer. This demands a modification (deformation) of the Lorenz transformations, to accommodate an invariant length scale. In Ref. 6 the nonlinearity of the deformed Lorenz transformations lead the authors to novel commutators between spacetime coordinates and momenta, depending on the energy

$$\left[\hat{x}^i, \hat{p}_j\right] = i\hbar \left(1 - \frac{E}{\mathcal{E}_p}\right) \delta^i_j, \tag{26}$$

where E is the energy scale of the particle to which the deformed Lorenz boost is to be applied, while $\mathcal{E}_p$ is the Planck energy. This suggests that they have an energy-dependent Planck "constant" $\hbar(E) = \hbar(1 - E/\mathcal{E}_p)$. Their model also implies that $\hbar(E) \to 0$ for $E \to \mathcal{E}_p$. For energies much below that Planck regime, the usual Heisenberg commutators are recovered, but when $E \simeq \mathcal{E}_p$ one has $\hbar(\mathcal{E}_p) \simeq 0$. So the Planck energy is not only an invariant in this model, but the world looks also apparently classical at the Planck scale, similarly as in 't Hooft's proposal.

The connections of the DSR model with our proposal are at this point self evident. Our GUP (13) implies that, at the boundary of the Brillouin zone, when $\langle \hat{p} \rangle_\psi \to \hbar\pi/2\epsilon$, i.e. for Planck energies $E_\epsilon \simeq (2\sqrt{2}/a)\mathcal{E}_p$, the fundamental commutator vanishes $[\hat{X}_\epsilon, \hat{P}_\epsilon] \simeq 0$, and since $\Delta X_\epsilon \Delta P_\epsilon \gtrsim 0$, lattice quantum mechanics at short wavelengths allows for classical behavior, that is uncertainties of two complementary observables *can* be *simultaneously* zero. However, if we express the fundamental commutator (12) of our model in terms of energy, using the exact dispersion relation inferred from the lattice Laplacian (5), we find (for $\epsilon = a\ell_p$)

$$\left[\hat{X}_\epsilon, \hat{P}_\epsilon\right] f(x) = i\hbar \left(1 - \frac{a^2}{8} \frac{\hat{E}^2}{\mathcal{E}_p^2}\right) f(x). \tag{27}$$

This means that the deforming term in our model is quadratic in the energy, instead of being linear in the energy as the DSR model (26).

9. Discussion and Summary

It should be noted that the present lattice generalization of the uncertainty principle is not an approximate description, but it is an exact formula necessarily implied by our model of lattice space time. The great majority of the GUP research has always borrowed the deformed commutator $[\hat{x}, \hat{p}] = i\hbar(1 + \kappa\hat{p}^2)$ either from string theory, or from heuristic arguments about black holes.[5] To be precise, even in string theory[5] the formula expressing the GUP is not derived from the basic features of the model, but instead it is deduced from high-energy gedanken experiments of string scatterings. In contrast to this, we have *derived* all the results from a simple lattice model of spacetime, and from the analytic structure of the basic commutator (12).

We have calculated the uncertainties on a crystal-like universe whose lattice spacing is of the order of Planck length — the so-called world crystal. When the energies lie near the border of the Brillouin zone, i.e., for Planckian energies, the uncertainty relations for position and momenta do not pose any lower bound. Hence the world crystal universe can become "deterministic" at Planckian energies. In this high-energy regime, our lattice uncertainty relations resemble the double special relativity result of Magueijo and Smolin. The scenario in which the universe at Planckian energies is deterministic rather than being dominated by quantum fluctuations is a starting point in 't Hooft's "deterministic" quantum mechanics.

Using this new generalized uncertainty relation we derived a new mass-temperature relation for Schwarzschild micro black holes. In contrast to standard results based on Heisenberg or stringy uncertainty relations, our mass-temperature formula predicts both finite Hawking's temperature and a zero rest-mass remnant at the end of the evaporation process. Especially the absence of remnants is a welcome bonus which allows to avoid such conceptual difficulties as entropy/information problem or why we do not experimentally observe the remnants that must have been prodigiously produced in the early universe.

Finally, connections between double special relativity and our generalized uncertainty relation have been illustrated. The fundamental commutator in DSR as well as in our lattice GUP goes to zero at Planck energy. The world should therefore be manifestly "classical" in the Planck regime, a feature very different from the common believe. Moreover, this aspect presents also a strong resemblance with the results obtained in the research line of "deterministic" quantum mechanics.

References

1. see e.g., J.A. Wheeler and K.W. Ford, *Geons, Black Holes, and Quantum Foam*, (W W Norton & Co Inc, New York, 2000).
2. H. Kleinert, *Multivalued Fields*, (World Scientific, Singapore, 2008).
3. C N. Yang, *Phys. Rev.* **72**, (1947) 874. F. Scardigli, *Phys. Lett. B* **452**, (1999) 39.

4. R.J.Adler, P.Chen, D.I.Santiago, *Gen. Rel. Grav.* **33**, (2001) 2101.
5. G. Veneziano, *Europhys Lett.* **2** (1986) 199; D.J. Gross, P.F. Mende, *Phys. Lett. B* **197**, (1987) 129; K. Konishi, G. Paffuti, P. Provero, *Phys. Lett. B* **234**, (1990) 276.
6. J. Magueijo and L. Smolin, *Phys. Rev. D* **67**, (2003) 044017.
7. G. 't Hooft, *Class. Quant. Grav.* **16**, (1999) 3263. M. Blasone, P. Jizba, G. Vitiello, *Phys. Lett. A* **287**, (2001) 205.
8. J.M. Bardeen, B. Carter, S.W. Hawking, *Commun. Math. Phys.* **31** (1973) 161.
9. W. Thirring, *Z. Physik* **239** (1970) 339.

CURVED D-BRANEWORLD ACTION IN 4D AND BLACK HOLES

SUPRIYA KAR,* K. PRIYABRAT PANDEY, SUNITA SINGH and ABHISHEK K. SINGH

Department of Physics & Astrophysics,
University of Delhi, New Delhi 110 007, India
* skkar@physics.du.ac.in

In the talk, we explore a plausible scenario for an AdS_5 black hole, in an emergent gravity, underlying a two form $U(1)$ gauge theory within the realm of type IIB superstring theory. Using AdS_5/CFT correspondence, a dual D_3-brane action governed by a nonlinear $U(1)$ gauge field strength is proposed. The possibility of a gauge non-linearity in the open string boundary due to a torsion is discussed. Interestingly, we obtain a Reissner-Nordstrom geometry underlying "non-Riemannian" curvatures, which may alternately be viewed as a nonlinear gauge curvatures on a D_3-braneworld.

1. Introduction

The bulk/boundary duality becomes evident in presence of a fifth dimension in an open manifold. Importantly, the duality takes a precise form in Anti de Sitter (AdS) gravity and is established as a familiar AdS/CFT correspondence[1,2] in string theory. The very idea was motivated by the consideration of AdS black holes in five dimensions. Though the gravity theory in five dimensions is obtained from the $SL(2, Z)$ invariant curvatures in type IIB superstring on $AdS_5 \times S^5$, it may schematically be described by a Kaluza-Klein gravity in presence of a cosmological constant. For a review on AdS/CFT, see Ref. 3.

On the other hand, the D-branes under the AdS gravity/gauge theory duality may enlighten us on a plausible aspect of duality symmetry, between the geometry and matter sectors, inherent in GTR. In fact, the non-linearity in the $U(1)$ gauge field strength is essentially due to the zero mode of a two form on a D_3-brane.[4] The nonlinear gauge sector on a D_3-brane may seen to be described by a nontrivial geometry underlying the non-commutative space-time. Interestingly, the AdS_5/D_3-brane gauge theory correspondence in presence of a constant two form has been obtained in Ref. 5. Nonlinear electric charge in the $U(1)$ gauge theory has been explored for black hole geometries in $4D$ with broken spherical symmetry in the past years.[6–19]

The talk is motivated from the recent interesting idea of emergent gravity by Verlinde.[20] The presentation is primarily based on our work.[21,22] Here, we attempt to construct an emergent theory of gravity from a partially gauge fixed two form, underlying an $U(1)$ gauge symmetry in five dimensions. The fact that a three form gauge curvature is Hodge dual to the electromagnetic field in five dimensions together with AdS_5/D_3-brane $U(1)$ gauge theory, possibly allows one to go to a gauge choice for the metric field in Kaluza-Klein gravity. We explore the AdS_5/D_3-brane gauge theory correspondence in our formalism to obtain a plausible curved D_3-brane action and the Reissner-Nordstrom black hole solution underlying a non-Riemannian geomety. The nontrivial curvature on a curved D_3-brane is essentially due to the nonlinear $U(1)$ gauge field strength. However, the field strength may be seen to arise out of a combination of electromagnetic field strength and a torsion due to the two form in the formalism. A priori, the non-linear field strength does not seem to be gauge invariant. We revisit the open string world-sheet boundary dyanmics on a D_3-brane to show the $U(1)$ gauge invariance of the nonlinear gauge field strength in presence of a dynamical space-time torsion.

2. AdS_5 Underlying Torsion Geometrodynamics

2.1. Type IIB superstring on $AdS_5 \times S^5$

In this section, we explore some of the effective theoretical framework in five dimensions underlying an AdS_5 geometry. The boundary of AdS_5 gravity theory is believed to describe an $U(1)$ gauge theory on a D_3-brane. Schematically, an AdS_5 geometry may be obtained in a Kaluza-Klein gravity with a non-vanishing cosmological constant. We may ignore the Gibbons-Hawking-York surface term and write the action in the usual form:

$$S_{\text{KK}} = \frac{1}{16\pi G_N} \int d^5x \sqrt{-g} \, (R - 2\Lambda) \,. \tag{1}$$

Λ possesses its origin in the self-dual five form field strength in type IIB superstring theory. Primarily, an emergent gravity leading to AdS_5 vacua may seen to be governed by the NS-NS sector, though Λ is obtained from the RR-sector. The fact becomes obvious in a two form gauge theory. A consistent truncation of the type IIB superstring on $AdS_5 \times S^5$ may be performed to yield

$$S_{\text{NS-NS}} = \frac{1}{16\pi G_N} \int d^5x \sqrt{-g} \, e^{-2\Phi} \left(R + 4(\partial\Phi)^2 - \frac{1}{12}H^2 \right) \,. \tag{2}$$

Where R is Ricci scalar, Φ is dilaton and the field strength for a two form is:

$$H_{\mu\nu\lambda} = (\nabla_\mu B_{\nu\lambda} + \nabla_\nu B_{\lambda\mu} + \nabla_\lambda B_{\mu\nu}) \,. \tag{3}$$

It is needless to mention that the bulk string action underlies GTR. However, the action possesses a striking analogy to the Einstein-Cartan theory (ECT) of gravity in five dimensions. For some recent developments in ECT, see Refs. 23, 24. Interestingly, torsion underlying geometrodynamics is discussed in nonlinear sigma models in the literatures.[25,26] By revisiting the string worldsheet dynamics for the background two form, it may be plausible to treat the two form as a gauge connection, which in-turn shall describe an underlying ECT. A dynamical two form would incorporate a nontrivial space-time torsion and hence lead to geometrodyanmics. In other words, the new geometry may be seen to arise at the expense of three form gauge curvature in the formalism.

2.2. *Emergent metric: A gauge choice in two form theory*

At this point, our goal is two fold. Firstly, we would like to establish an equivalence between the dynamical gauge theories independently described by $B_{\mu\nu}$ and A_μ gauge fields in five dimensions. Secondly, we would like to incorporate an emergent metric description in the $B_{\mu\nu}$ gauge theory. As a result, the $B_{\mu\nu}$ gauge theory in five dimensions may provide a interconnection between the AdS_5 gravity and the gauge theory at its boundary. On the one hand, the equivalence in local gauge degrees of freedom between the two gauge theories would, a priori, imply a reduction in five redundant degrees of freedom in $B_{\mu\nu}$ gauge theory. On the other, an emergent gravity from a two form gauge theory would imply a drastic change in the notion of field and its underlying dynamics. In other words, a geometric notion needs to be incorporated through a dynamical two form gauge field. The transition from gauge theory to geometric theory (Einstein's gravity) would possibly require two essential conditions on $B_{\mu\nu}$ field. They are: (i) vanishing field strength for $B_{\mu\nu}$ and (ii) new connections due to the $B_{\mu\nu}$. Interestingly, both the goals may be achieved by incorporating an emergent metric field in the two form gauge theory. It is given by

$$G_{\mu\nu} = \left(g_{\mu\nu} - B_{\mu\lambda}g^{\lambda\rho}B_{\rho\nu}\right) , \tag{4}$$

where $g_{\mu\nu}$ is a constant metric. The emergent metric dynamics is at the expense of $B_{\mu\nu}$ gauge dynamics. Five Bianchi identities on $G_{\mu\nu}$, would further constrain the $B_{\mu\nu}$ field, reduce its gauge redundancy. The local degrees of freedom of the $B_{\mu\nu}$ field becomes equal to the A_μ field. In other words, a covariantly constant dynamical two form gauge theory may possibly generate an emergent dynamical theory of gravity.

The emergent metric is uniquely fixed by

$$\mathcal{D}_\lambda G_{\mu\nu} = 0 . \tag{5}$$

$\mathcal{D}_\lambda$ is a gauge covariant derivative in presence of a torsion. Explicitly,

$$\mathcal{D}_\lambda B_{\mu\nu} = \partial_\lambda B_{\mu\nu} - \left(\Gamma^\rho_{\lambda\mu} + \mathbf{\Gamma}_{\lambda\mu}{}^\rho\right) B_{\rho\nu} - (\Gamma^\rho_{\lambda\nu} - \mathbf{\Gamma}_{\lambda\nu}{}^\rho) B_{\rho\mu} , \tag{6}$$

where $\mathbf{\Gamma}_{\mu\nu}{}^\rho = g^{\rho\lambda}\mathbf{\Gamma}_{\mu\nu\lambda}$ are the generic connections precisely due to the two form. The field strength $H_{\mu\nu\lambda}$ is modified in presence of the new connections in the gauge fixed two form gauge theory. It may be given by

$$\begin{aligned}\mathcal{H}_{\mu\nu\lambda} &= (\nabla_\mu B_{\nu\lambda} + 2\mathbf{\Gamma}_{\mu\nu}{}^\rho B_{\rho\lambda}) + \text{cyclic in } (\mu, \nu, \lambda) \\ &= (\ \partial_\mu B_{\nu\lambda} - T_{\mu\nu}{}^\rho B_{\rho\lambda}) + \text{cyclic} , \end{aligned} \tag{7}$$

where $T_{\mu\nu}{}^\rho$ is a torsion. It is completely antisymmetric in its space-time indices and hence its trace vanishes in the theory. The gauge conditions $\mathcal{H}_{\mu\nu\lambda} = 0$ may be re-expressed as

$$H_{\mu\nu\lambda} = (T_{\mu\nu}{}^\rho B_{\rho\lambda} + \text{cyclic}) = -2\mathbf{\Gamma}_{\mu\nu\lambda} . \tag{8}$$

It is important to note that the role of nontrivial gauge curvature, before gauge fixing, is precisely reproduced by the space-time torsion in a gauge fixed $B_{\mu\nu}$ gauge theory. It implies that $H_{\mu\nu\lambda}$ in an effective closed string theory (2) may equivalently be replaced by a torsion. The geometric transition from a gauge theory is thought provoking and possibly enlighten one to incorporate ECT underlying string theories.

2.3. Non-Riemannian curvatures

In this section, we work out for the generic space-time curvatures by considering the commutator of appropriate gauge covariant derivatives in the theory. In presence of a gauge field, the appropriate covariant derivative becomes $D_\mu = (\mathcal{D}_\mu - A_\mu)$. The nontrivial curvatures may be worked out to yield:

$$\left[D_\mu \, , \, D_\nu\right]A_\lambda = \mathcal{F}_{\nu\mu}A_\lambda + K_{\mu\nu\lambda}{}^\rho A_\rho + T^\rho{}_{\mu\nu}\, \mathcal{D}_\rho A_\lambda$$
$$+\frac{1}{2}\left(\Gamma^\rho_{\mu\sigma}T^\sigma{}_{\nu\lambda} + \Gamma^\sigma_{\nu\lambda}T^\rho{}_{\mu\sigma} - \Gamma^\sigma_{\mu\lambda}T^\rho{}_{\nu\sigma} - \Gamma^\rho_{\nu\sigma}T^\sigma{}_{\mu\lambda}\right)A_\rho \, , \qquad (9)$$

where $\mathcal{F}_{\mu\nu} = (F_{\mu\nu} + T_{\mu\nu}{}^\rho A_\rho)$.

There are two nontrivial curvatures and they are $F_{\mu\nu}$ and $K_{\mu\nu\lambda}{}^\rho$ or $T_{\mu\nu}{}^\rho$. Naively, the $U(1)$ field strength $F_{\mu\nu}$ may be combined with a torsion to yield a nonlinear $U(1)$ gauge curvature $\mathcal{F}_{\mu\nu}$ in five dimensional gauge theory. However, $\mathcal{F}_{\mu\nu}$ is not gauge invariant in presence of a torsion in five dimensions. A careful analysis reveals that the $U(1)$ gauge theory is either described by the A_μ field or by the $B_{\mu\nu}$ field, not both fields together. So the correct dynamics is either governed by the gauge curvature $F_{\mu\nu}$ or by the geometric curvature $K_{\mu\nu\lambda}{}^\rho$ independently.

Though, the new curvature tensor takes an identical form to the Riemannian-Christoffel curvature tensor $R_{\mu\nu\lambda\rho}$, it is a non-Riemannian tensor. Explicitly, it is given by

$$K_{\mu\nu\lambda}{}^\rho = \partial_\nu\Gamma^\rho_{\mu\lambda} - \partial_\mu\Gamma^\rho_{\nu\lambda} + \Gamma^\sigma_{\mu\lambda}\Gamma^\rho_{\nu\sigma} - \Gamma^\sigma_{\nu\lambda}\Gamma^\rho_{\mu\sigma} \, . \qquad (10)$$

$K_{\mu\nu\lambda\rho}$ is antisymmetric under the exchange of indices within a pair. It is a 4th order non-Riemannian tensor. In the case of a non-dynamical torsion, $K_{\mu\nu\lambda\rho} \to R_{\mu\nu\lambda\rho}$.

2.4. Emergent "Kaluza-Klein" gravity

In principle, a dynamical theory of emergent gravity obtained in a two form gauge theory, may be described by a typical Kaluza-Klein theory with $\Lambda = 0$ in eq.(1). However the emergent metric would be accompanied by five degrees of freedom, rather than the natural ten degrees of freedom. As a result the emergent gravity black holes, if any, would be without any angular momentum.

On the other hand, one may attempt to describe the emergent gravity by simply working within the realm of $B_{\mu\nu}$ gauge theory in five dimensions. As we have seen, the field strength would incorporate a completely antisymmetric torsion, which in turn would generate an irreducible scalar curvature tensor K. Since the inherent gauge curvature, in disguise, acts as a torsion and governs K, the cosmological constant becomes insignificant in the gauge theory. Then, the Kaluza-Klein gravity in presence of $\Lambda \neq 0$, in a two form gauge theory, may be given by

$$S = \frac{1}{16\pi G_N} \int d^5x \sqrt{-g} \, K \; . \tag{11}$$

We recall that both the gauge theories, underlying the $U(1)$ gauge symmetry, possess five local degrees of freedom. The relevant $B_{\mu\nu}$ field equations of motion are worked out to yield

$$\partial_\lambda \Gamma^{\lambda\mu\nu} - \frac{1}{2}\Big(g^{\alpha\beta}\partial_\lambda \, g_{\alpha\beta}\Big)\Gamma^{\lambda\mu\nu} = 0 \; . \tag{12}$$

2.5. Deformed AdS₅ black hole

We consider an S^3-symmetric AdS_5 vacua for the two form ansatz to obtain an S^3 deformed AdS_5 geometry in the formalism. The two form ansatz may be given by

$$B_{t\psi} = B_{r\psi} = r_0 \; ,$$
$$B_{\theta\psi} = q(\cot\theta \sin^2\psi)$$
$$\text{and } \; B_{\psi\phi} = B_{\theta\psi}\sin\theta \; , \tag{13}$$

where (r_0 and q) are constants. The nontrivial torsion connections are:

$$\boldsymbol{\Gamma}_{\theta\psi\phi} = \frac{q}{2}(\sin\theta\sin^2\psi)$$
$$\text{and } \; \boldsymbol{\Gamma}_{t\theta\phi} = \boldsymbol{\Gamma}_{r\theta\phi} = -\frac{(qr_0)}{2r^2}(\sin\theta\sin^2\psi) \; . \tag{14}$$

The relevant AdS_5 geometry is given by

$$ds^2 = -\left(1 + \frac{r^2}{b^2} - \frac{r_0^2}{r^2}\right)dt^2 + \left(1 + \frac{r^2}{b^2} - \frac{r_0^2}{r^2}\right)^{-1}dr^2 + r^2\left(1 + \frac{2f^2}{r^4}\right)d\psi^2$$
$$+ r^2\sin^2\psi\left(1 + \frac{f^2}{r^4}\right)d\Omega^2 \; , \tag{15}$$

where $f = q(\cot\theta\sin\psi)$ and $(0 < \theta \leq \pi)$, $(0 < \psi \leq \pi)$, $(0 \leq \phi \leq 2\pi)$. The AdS_5 black hole is associated with an S^3 deformed geometry. The deformation to the typical Riemannian AdS_5 black hole geometry (15) is an essential feature of nontrivial $\Gamma_{\mu\nu\lambda}$ and possibly signifies the presence of an underlying ECT. For a constant $\Gamma_{\theta\psi\phi} \neq 0$, the conformal factor and the S^3 deformation geometry becomes trivial in the AdS_5 black hole. In the limit, the Reimannian notion of curvature becomes manifest.

3. AdS₅ Boundary Dynamics and Geometry

3.1. Dual D₃-brane action

We recall the underlying symmetry $SO(2,4)$ in AdS_5. The fifteen parameter group also acts as a conformal group on the AdS_5 boundary leading to a D_3-brane. Interestingly, the boundary dynamics may be worked out in a two form gauge theory by considering the torsion geometrodynamics (11) on S^1. Since the source for the emergent metric field is antisymmetric, it yields a trivial dilaton in $4D$. This in-turn implies that the underlying effective closed string theory is at the self dual coupling g_s =1. Then the relevant dynamics may be incorporated into a dual D_3-brane action. It is given by

$$S = -\frac{1}{4\lambda^2} \int d^4x \sqrt{-g} \left(F_{\mu\nu} F^{\mu\nu} - \frac{\lambda^2}{4\pi G_N} K^{(4)} \right)$$
$$= -\frac{1}{4\lambda^2} \int \sqrt{-g}\, \mathcal{F}_{\mu\nu} \mathcal{F}^{\mu\nu} \,, \tag{16}$$

where $K^{(4)}$ is the scalar "non-Riemannian" curvature on a dual braneworld. When the five dimensional A_μ theory is in radiation gauge, its dual, the two form gauge theory leading to an emergent gravity would be governed by three local degrees of freedom. A priori, the local degrees are shared among the electromagnetic potential and the axion (the torsion) in $4D$. Then, the nonlinear $U(1)$ gauge field strength in eq.(9) may be seen to be governed by three local degrees. As a result, the axion may be absorbed by the gauge field to yield the notion of nonlinear $U(1)$ gauge theory on a D_3-brane. Thus, the nonlinear dynamics on the brane may be viewed as a consequence of spontaneous symmetry breaking of the global $U(1)$. On the other hand, incorporating the boundary perspective of an open string, $\mathcal{F}_{\mu\nu}$ may seen to be a gauge invariant on a D_3-brane. Under an $U(1)$ gauge symmetry:

$$A_\mu \rightarrow A_\mu + \partial_\mu \epsilon \,. \tag{17}$$

The gauge field variation of the open string world-sheet, boundary part in the sigma model, action for an adiabatic $\mathcal{F}_{\mu\nu}$ becomes

$$\delta S = \int d\tau\, \delta A_\mu \partial_\tau X^\mu$$
$$= -\frac{1}{2} \int d\tau\, (\delta \mathcal{F}_{\mu\nu} X^\nu) \, \partial_\tau X^\mu$$
$$= -\frac{1}{2} \int d\tau\, T_{\mu\nu}{}^\lambda \delta A_\lambda\, X^\nu \partial_\tau X^\mu$$
$$= -\frac{1}{2} \int d\tau\, T_\nu X^\nu\, \partial_\tau \epsilon = 0 \,. \tag{18}$$

The $U(1)$ gauge invariance of $\mathcal{F}_{\mu\nu}$ on the D_3-braneworld is restored due to the vanishing trace of the torsion.

3.2. Reissner-Nordstrom D_3-braneworld black hole

In this section, we obtain a Reissner-Nordstrom black hole underlying the "non-Riemannian" geometry in an emergent gravity scenario (16). The gauge fields ansatz may be given by

$$B_{t\phi} = -M \cos\theta \,, \quad A_t = \frac{Q_e}{r} \,, \tag{19}$$

where M and Q_e are constants. The nontrivial connection becomes

$$\mathbf{\Gamma}_{t\theta\phi} = \frac{1}{2}(M\sin\theta) \ . \tag{20}$$

We use the dual field, *i.e.* an axion χ, which may be treated as a gravitational potential in a non-relativistic limit. The axion potential turns out to yield $\chi = -(M/r)$. We consider $G_{tt} = G^{rr}$, around an S^2-symmetric isolated vacua, in a non-relativistic limit. The G_{tt} component may be obtained from a geodesic expansion represented by the derivative corrections of gauge potentials of increasing rank in each term. It is given by

$$G_{tt} = \left(g_{tt} + h_{tt}^{(0)} + h_{tt}^{(1)} + h_{tt}^{(2)} + \dots\right) \ . \tag{21}$$

The sources in 2nd, 3rd and 4th terms are distinct scalar contributions, respectively, out of the tensor potentials of increasing rank ($= 0, 1, 2, \dots$). In the case, the nontrivial sources are due to the the axion field and the Lorentz force. Explicitly, they are:

$$h_{tt}^{(0)} = 2\chi \quad \text{and} \quad h_{tt}^{(1)} = |Q_e F_e| = \frac{Q_e^2}{r^2} \ . \tag{22}$$

As a result, the dual D_3-braneworld may describe a typical Reissner-Nordstrom black hole underlying a "non-Riemannian" geometry. The nontrivial vacua common to two distinct tensor formulations of gravity in string theory is remarkable.

4. Concluding Remarks

To conclude, we have explored a plausible scenario leading to an emergent gravity underlying a torsion geometrodynamics in the realm of AdS_5/D_3-brane $U(1)$ gauge theory in type IIB string theory. The NS-NS two-form is exploited for a geometric construction to obtain a generalized curvature, which in turn is shown to govern an AdS_5 black hole with broken S^3-symmetry. The AdS_5 boundary dynamics was revisted to obtain a nonlinear $U(1)$ gauge theory on a D_3-brane. It was shown that the non-linearity may arise from a space-time torsion and an open string boundary dynamics enforces the gauge invariance. Interestingly, Riessner-Nordstrom black hole was constructed in the emergent "non-Riemannian" geometry. It may suggest a need for a deeper understanding of the $SL(2, Z)$ symmetry interchanging the two form in NS-NS and RR-sectors. Our analysis may provokes thought to speculate a possibility of Einstein-Cartan geometry underlying string theories with a string charge.

Acknowledgments

S.K. would like to thank the organizers for warm hospitalities at IAS, Nanayang Technological University, Singapore during the conference. He would like to acknowledge the financial grant extended by the University of Delhi, India to participate in the conference.

References

1. J.M. Maldacena, Adv.Theor.Math.Phys.**2**, 231 (1998).
2. E. Witten, Adv.Theor.Math.Phys.**2**, 253 (1998).
3. E. D'Hoker and D. Freedman, TASI 2001 Lectures, hep-th/0201253 (2002).
4. N. Seiberg and E. Witten, Jour.High Ener.Phys.**09**, 032 (1999).
5. M. Li, Y-S. Wu, Phys.Rev.Lett.**84**, 2084 (2000).
6. G.W. Gibbons and K. Hashimoto, Jour.High Ener.Phys.**09**, 013 (2000).
7. M. Mars, J.M.M. Senovilla and R. Vera, Phys.Rev.Lett.**86**, 4219 (2001).
8. G.W. Gibbons and A. Ishibashi, Class. & Quan.Grav.21, 2919 (2004).
9. S. Kar and S. Majumdar, Phys.Rev. **D74**, 066003 (2006).
10. S. Kar and S. Majumdar, Int.J.Mod.Phys. **A21**, 6087 (2006).
11. S. Kar and S. Majumdar, Int.J.Mod.Phys. **A21**, 2391 (2006).
12. S. Kar, Phys.Rev. **D74**, 126002 (2006).
13. S. Kar, Jour.High Ener.Phys.**0610**, 052 (2006).
14. L-H. Liu, B. Wang and G-H Yang, Phys.Rev. **D76**, 064014 (2007).
15. S. Chen, B. Wang and R-K Su, Phys.Rev. **D77**, 024039 (2008).
16. J. Zhang, Phys.Lett. **B668**, 353 (2008).
17. R. Casadio, P. Nicolini, Jour.High Ener.Phys.**0811**, 072 (2008).
18. S. Kar, Int.J.Mod.Phys. **A24**, 3571 (2009).
19. E. Spallucci, A. Smailagic and P. Nicolini, Phys.Lett. **B670**, 449 (2009).
20. E. Verlinde, arXiv:1001.0785 [hep-th] (2010).
21. S. Kar, K.P. Pandey, S. Singh and A.K. Singh, arXiv:1002.1906 [hep-th] (2010).
22. S. Kar, K.P. Pandey, S. Singh and A.K. Singh, arXiv:1002.3976 [hep-th] (2010).
23. A. Saa, Class. & Quan.Grav., L85 (1995).
24. A. Saa, Gen.Rel.Grav.**29**, 205 (1997).
25. D. Freedman, P.K. Townsend, Nucl.Phys. **B177**, 282 (1981).
26. E. Braaten, T. Curtright and C. Zachos, Nucl.Phys. **B260**, 630 (1985).

D-BRANEWORLD BLACK HOLES

ABHISHEK K. SINGH,* K. PRIYABRAT PANDEY, SUNITA SINGH and SUPRIYA KAR

Department of Physics & Astrophysics, University of Delhi, New Delhi 110 007, India
* aksingh@physics.du.ac.in

We investigate an emergent AdS_5 geometry, underlying torsion geometrodynamics in a two form $U(1)$ gauge theory. We obtain a D_3-braneworld black hole using the AdS_5/CFT from the emergent AdS_5 black hole. Perspective of a dual D_3-brane $U(1)$ gauge dynamics, with a torsion, leading to an amazing braneworld black hole is worked out. A complementarity is argued between the magnetic field, on a D_3-braneworld, and the torsion on its dual braneworld. Our analysis may provide an urge to go beyond Riemannian curvatures underlying string theories with a string charge.

1. Introduction

Intrinsic notion of geometry, governing space-time curvature, is manifested beautifully in the success of the General Theory of Relativity (GTR). Importantly, the Einstein's field equations confirm the presence of non-linear matter and establishes a precise relation between the matter and geometry in GTR. With the help of an extra dimension to Einstein's theory, the matter/geometry relation may be explored to view as a gauge/gravity duality or in particular AdS_5-gravity/gauge theory duality established in literatures.[1–3]

Very recently, it has been conjectured that gravity emerges at a macroscopic level due to the entropic force described by the matter field, which in turn results in a deformed geometry.[4] Similar to AdS/CFT, the emergent gravity is believed to be projected along a preferred holographic direction in space, which in turn may describe an accelerated frame and establishes the notion of temperature.[5]

In the talk, we focus on the notion of emergent gravity due to a plausible torsion geometrodynamics[6–9] underlying a D_3-braneworld.[10,11] A plausible interplay between the irreducible gauge and gravity curvatures is exploited to obtain a notion of dual D_3-braneworld. The deformation geometries leading to black hole on a D_3-braneworld and its dual are obtained. In the context, the non-linearity in the $U(1)$ gauge field strength on a D-brane has extensively been addressed in the last decade.[12–26]

2. Emergent AdS_5

Let us recall a plausible scenario leading to an emergent AdS_5 geometry presented in the previous talk by Kar. It was shown that a dynamical two form may alternately be

568

described by a torsion geometrodyanmics. As a result, presumably, string theories with background two-form may be seen to incorporate an underlying "Einstein-Cartan Theory" (ECT).[8,9] It is speculative that under such a circumstance, one may possibly decouple Einstein's curvature in presence of Cartan's and vice-versa. Interestingly, the scenario is identical to the gravity decoupling limit leading to a D-brane in presence of a noncommutative geometry.

We consider a D=10 type IIB superstring theory on $AdS_5 \times S^5$ and primarily focus on the NS-NS sector. The reason being, the emergent AdS_5 is precisely governed by the two form gauge connections and hence the cosmological constant becomes insignificant in the formalism. On the one hand, the theory is Hodge dual to a typical A_μ gauge theory and on the other, the $B_{\mu\nu}$ may be gauge fixed partially to yield an emergent metric in five dimensions. In other words, the emergent gravity scenario leading to AdS_5 geometry and its dual gauge theory on a D_3-brane, may alternately be governed by single two form gauge theory in five dimensions.

2.1. U(1) gauge theories

We begin with a typical $U(1)$ gauge theory described by $A_\mu(x)$ in five dimensions. The gauge curvature becomes $F_{\mu\nu} = \nabla_\mu A_\nu - \nabla_\nu A_\mu$. The gauge field dynamics may be given by

$$S_A = -\frac{1}{4\lambda^2} \int d^5x \, \sqrt{-g} \, F_{\mu\nu} F^{\mu\nu} \,. \tag{1}$$

Naively, A_μ has five local gauge degrees of freedom and $F_{\mu\nu}$ possesses ten independent components. In radiation gauge, there are three local degrees of freedom. On the other hand, the $F_{\mu\nu}$ can be re-expressed in term of its Hodge dual three form gauge curvature leading to an alternate two form gauge theory. Then, the dynamical $B_{\mu\nu}$ is describe by

$$S_B = -\frac{1}{12\lambda^2} \int d^5x \, \sqrt{-g} \, H_{\mu\nu\lambda} H^{\mu\nu\lambda} \,, \tag{2}$$

$$\text{where} \quad H_{\mu\nu\lambda} = \nabla_\mu B_{\nu\lambda} + \nabla_\nu B_{\lambda\mu} + \nabla_\lambda B_{\mu\nu} \,. \tag{3}$$

The two-form gauge theory action is invariant under an underlying $U(1)$ gauge symmetry. The gauge transofrmations are given by

$$\delta B_{\mu\nu} = \partial_\mu \Lambda_\nu - \partial_\nu \Lambda_\mu \,. \tag{4}$$

2.2. Geometric transition from gauge theory

Though the $H_{\mu\nu\lambda}$ has equal number of components to that of its Hodge dual, the $B_{\mu\nu}$ has five more local degrees of freedom than A_μ. The apparent problem may seen to be resolved with a gauge choice for $B_{\mu\nu}$. This in turn, leads to a notion of emergent metric in the gauge theory. It is given by

$$G_{\mu\nu} = g_{\mu\nu} - \left(Bg^{-1}B\right)_{\mu\nu} \,, \tag{5}$$

where $g_{\mu\nu}$ is a constant metric. Interestingly even in presence of completely anti-symmetric torsion connections $T_{\mu\nu}{}^{\lambda}$, the emergent metric satisfies

$$\mathcal{D}_{\lambda}G_{\mu\nu} = \left(\nabla_{\lambda}G_{\mu\nu} + \frac{1}{2}T_{\lambda\mu}{}^{\rho}G_{\rho\nu} + \frac{1}{2}T_{\lambda\nu}{}^{\rho}G_{\rho\mu}\right) = 0 \, , \tag{6}$$

where $\mathcal{D}_{\lambda}$ is the covariant derivative. The gauge choice leads to a covariantly constant but dynamical $B_{\mu\nu}$ in the theory. As a result, the modified field strength vanishes, $i.e.$ $H_{\mu\nu\lambda} \to \mathcal{H}_{\mu\nu\lambda} = 0$. The gauge choice may be re-interpreted as a transition point of the gauge curvature into the torsion connections. It is worked out to yield:

$$H_{\mu\nu\lambda} = T_{\mu\nu}{}^{\rho}B_{\rho\lambda} + T_{\nu\lambda}{}^{\rho}B_{\rho\mu} + T_{\lambda\mu}{}^{\rho}B_{\rho\nu} \, . \tag{7}$$

Then, the irreducible curvatures due to the torsion geometrodynamics are worked out with the gauge choice to obtain a "Kaluza-Klien" theory of gravity. The action takes a form:

$$S = \frac{1}{16\pi G_{N}} \int d^{5}x\sqrt{-g}\, K \, . \tag{8}$$

The scalar curvature K may be expressed in terms of the torsion connections and is given by

$$K = -\boldsymbol{\Gamma}_{\mu\nu\lambda}\boldsymbol{\Gamma}^{\mu\nu\lambda} \, ,$$
$$\text{where } \boldsymbol{\Gamma}_{\mu\nu\lambda} = -\frac{1}{2}T_{\mu\nu}{}^{\rho}B_{\rho\lambda} + \text{cyclic} \, . \tag{9}$$

In the first order formalism, the field equations reduce to one of the vacuum equations in ECT, $i.e.$ $\boldsymbol{\Gamma}_{\mu\nu\lambda} = 0$. On the other hand, the $B_{\mu\nu}$ equations of motion may be worked out for the dynamical scalar curvature K. It is given by

$$\left(\partial_{\sigma}g^{\rho\nu}\boldsymbol{\Gamma}^{\sigma\mu\lambda} - \partial_{\sigma}g^{\rho\mu}\boldsymbol{\Gamma}^{\sigma\nu\lambda} + \partial_{\sigma}g^{\rho\sigma}\,\boldsymbol{\Gamma}^{\mu\nu\lambda}\right)B_{\lambda\rho}$$
$$-\frac{1}{2}g_{\alpha\beta}\partial_{\sigma}g^{\alpha\beta}\left(g^{\rho\nu}\boldsymbol{\Gamma}^{\sigma\mu\lambda} - g^{\rho\mu}\boldsymbol{\Gamma}^{\sigma\nu\lambda} + g^{\rho\sigma}\boldsymbol{\Gamma}^{\mu\nu\lambda}\right)B_{\lambda\rho}$$
$$+g^{\mu\rho}\partial_{\lambda}\left(B_{\sigma\rho}\boldsymbol{\Gamma}^{\lambda\sigma\nu}\right) - g^{\nu\rho}\partial_{\lambda}\left(B_{\sigma\rho}\boldsymbol{\Gamma}^{\lambda\sigma\mu}\right) + g^{\rho\lambda}\partial_{\lambda}\left(B_{\sigma\rho}\boldsymbol{\Gamma}^{\mu\nu\sigma}\right)$$
$$+g^{\mu\rho}\left(\partial_{\lambda}B_{\sigma\rho}\right)\boldsymbol{\Gamma}^{\lambda\sigma\nu} - g^{\nu\rho}\left(\partial_{\lambda}B_{\sigma\rho}\right)\boldsymbol{\Gamma}^{\lambda\sigma\mu}$$
$$+\frac{1}{2}g^{\mu\lambda}\left(\partial_{\lambda}B_{\sigma\rho}\right)\boldsymbol{\Gamma}^{\sigma\rho\nu} - \frac{1}{2}g^{\nu\lambda}\left(\partial_{\lambda}B_{\sigma\rho}\right)\boldsymbol{\Gamma}^{\sigma\rho\mu} = 0 \, . \tag{10}$$

The field equations are nonlinear. Naively, the torsion connections in eq.(9) take an identical form to the Riemann-Christoffel connections. However a careful analysis reveals that $\boldsymbol{\Gamma}_{\mu\nu\lambda}$, iteratively, incorporate all higher orders staring from first order in $B_{\mu\nu}$ field. Formally, we may refer the first order connections to a gauge (or matter) dominated scenario and all the connections higher order in $B_{\mu\nu}$ to a gravity dominated one.[10,11] In fact, they may be seen to govern two dual descriptions, which is evident from the gauge conditions (7).

At this point, we recall that both the gauge theories, underlying an $U(1)$ gauge symmetry, possess five local degrees of freedom. In radiation gauge, the A_{μ} field describes a massless gauge particle. It can have three gauge degrees of freedom in five dimensions. In the gauge, $B_{\mu\nu}$ may also be described up to three of its nontrivial components. It can be checked that while the matter dominated scenario is described by a single $B_{\mu\nu}$ component, the gravity dominated one requires two or more $B_{\mu\nu}$ components. Nevertheless, both the descriptions are dual to each other and hence they describe the same physics. Thus a theory with a geometric notion of curvature described by the $B_{\mu\nu}$ field equations (10) may equivalently be described by a simpler looking field equations in its dual scenario. The relevant $B_{\mu\nu}$ field equations of motion are worked out to yield:

$$\partial_\lambda \Gamma^{\lambda\mu\nu} - \frac{1}{2}\left(g^{\alpha\beta}\partial_\lambda \, g_{\alpha\beta}\right)\Gamma^{\lambda\mu\nu} = 0 \; . \tag{11}$$

3. D_3-Brane Geometry

3.1. Dual braneworld black hole

Let me begin with the dual D_3-brane dynamics presented in the previous talk by Kar. A priori, the irreducible gauge and gravity curvatures on the dual braneworld may be given by

$$S = -\frac{1}{4\lambda^2} \int d^4x \sqrt{-g} \left(F^2 - \frac{\lambda^2}{4\pi G_N}K^{(4)}\right)$$
$$= -\frac{1}{4\lambda^2} \int d^4x \sqrt{-g} \, \mathcal{F}_{\mu\nu}\mathcal{F}^{\mu\nu} \; , \tag{12}$$

where $K^{(4)}$ is the scalar curvature in presence of a space-time torsion. $\mathcal{F}_{\mu\nu}$ is a non-linear $U(1)$ field strength on the dual D_3-braneworld. It takes a form:

$$\mathcal{F}_{\mu\nu} = \mathcal{D}_\mu A_\nu - \mathcal{D}_\nu A_\mu \; ,$$
$$= F_{\mu\nu} + T_{\mu\nu}^\lambda A_\lambda \; , \tag{13}$$

where $\mathcal{D}_\mu = (D_\mu - A_\mu)$. The dynamical curvatures on the dual braneworld (12) are not the decoupled ones even though they appear so. In fact, they may seen to be coupled intrinsically due to the gauge Chern-Simons coupling. The gauge field equations of motion are modified in presence of torsion due to the Chern-Simons term in the theory. The ansatz for the gauge fields are:

$$B_{t\phi} = -M\cos\theta \; , \quad A_t = \frac{Q_e}{r} \; , \tag{14}$$

where M and Q_e are mass and electric charge. The nontrivial torsion becomes

$$T_{\theta\phi}{}^t = M\sin\theta \; . \tag{15}$$

Then, an emergent geometry in the non-relativistic limit becomes

$$ds^2 = -\left(1 - \frac{M^2}{r^2} - \frac{Q_e^2}{r^4}\right)dt^2 + \left(1 - \frac{Q_e^2}{r^4}\right)dr^2 + r^2\left(1 - \frac{M^2}{r^2}\right)d\Omega^2 + M^2 d\theta^2 \; , \tag{16}$$

where $d\Omega^2$ is defined by (θ, ϕ)-coordinates on an S^2. The dual braneworld describes a four diemnsional charged black hole with a curvature singulairity at $r=0$. It is characterized, a priori, by two parameters M and Q_e together. However, the black hole possesses a single horizon at

$$r = r_h = \frac{M}{\sqrt{2}}\left(1 + \sqrt{1 + \frac{4Q_e^2}{M^4}}\right)^{1/2} . \tag{17}$$

Since both nontrivial M and Q_e are vital to the deformation geometry, leading to a black hole, the role of torsion and electric field may seen to be inseparable for a braneworld black hole geometry. In fact, two gauge degrees of freedom are shared between the $B_{\mu\nu}$ and A_μ gauge fields. These two local degrees, described by two independent gauge fields on the dual D_3-braneworld may be viewed as that of an $U(1)$ gauge field, in radiation gauge, on a D_3-brane. Apparently, the magnetic field on a D_3-braneworld may seen to behave as a torsion on its dual braneworld. Formally,

$$B_{t\phi} \longleftrightarrow A_\phi . \tag{18}$$

Analysis re-assures the role played by a two form and a one form gauge field on a braneworld. A two form gauge field in disguise of a magentic field incorporates deformation in spherical symmetry in particular, while an electric field deforms the longitunidal (rt)-space.

3.2. Braneworld black hole from AdS_5

An ansatz for the two form, leading to a generalized AdS_5 geometry, may be given by

$$B_{t\psi} = B_{r\psi} = r_0 ,$$
$$B_{\theta\psi} = q\left(\cot\theta\sin^2\psi\right)$$
$$\text{and } B_{\psi\phi} = q\left(\cos\theta\sin^2\psi\right) , \tag{19}$$

where $(r_0$ and $q)$ are two arbitrary constants. Interestingly, the ansatz is also governed by two nontrivial gauge degrees of freedom in $B_{\theta\psi}$ and in $B_{\psi\phi}$. The emergent AdS_5 black hole turns out to break the S^3-symmetry. It is given by

$$ds^2 = -\left(1 + \frac{r^2}{b^2} - \frac{r_0^2}{r^2}\right)dt^2 + \left(1 + \frac{r^2}{b^2} - \frac{r_0^2}{r^2}\right)^{-1}dr^2 + \left(1 + \frac{f^2}{r^4}\right)r^2\sin^2\psi\, d\Omega^2$$
$$+ \left(1 + \frac{2f^2}{r^4}\right)r^2 d\psi^2 , \tag{20}$$

where $f = q(\cot\theta\sin\psi)$. The angular coordinates describe an S^3 and they are defined as:

$$0 < \theta \leq \pi , \quad 0 < \psi \leq \pi \quad \text{and} \quad 0 \leq \phi \leq 2\pi . \tag{21}$$

The AdS_5 boundary, leading to a gauge theory on a D_3-brane, may be obtained through Kaluza-Klein compactification, as the dilaton turns out to be trivial in the formalism.

A priori, if AdS_5 is compactified along its ψ-coordinate, the gauge field ansatz on the braneworld may be generated from the two form ansatz (19) in the case. The gauge field anstaz may be worked out to yield:

$$A_t = A_r = r_0 \ ,$$
$$A_\theta = q \cot \theta \ ,$$
$$A_\phi = -q \cos \theta$$
$$\text{and } B_{\mu\nu} = 0 \ . \tag{22}$$

The nontrivial local degrees of freedom in the two form (19) appear to be manifested in the A_θ and A_ϕ and the gauge theory on D_3-brane describes an electromagnetic field. However, the electromagnetic field $F_{\mu\nu}$ possesses only one nontrivial component, i.e. a radial component of the magnetic field B_r. Explicitly,

$$F_{\theta\phi} = q \sin \theta$$
$$\text{and } B_r = \frac{q}{r^2} \ . \tag{23}$$

Surprisingly, one dynamical degree of freedom seems to be missing from the gauge theory on the braneworld. Further analysis reveals nonzero torsion due to the gauge Chern-Simons term. They are given by

$$T_{\theta\phi}{}^r = -T_{\theta\phi}{}^t = (qr_0) \sin \theta \ . \tag{24}$$

The problem of missing gauge degree does not resolve as the topological nature of Chern-Simons coupling do not incorporate any local degrees of freedom, though at times topological coupling may yield nontriviality in a gravity theory. Interestingly, the $U(1)$ field strength (13) in presence of torsion (24) may be simplified, in the case, to yield:

$$\mathcal{F}_{\mu\nu} = F_{\mu\nu} \ . \tag{25}$$

The analysis confirms that the constant q in four dimensions can be interpreted as a nonlinear magnetic charge.[27] In the case, a D_3-braneworld black hole derived from an AdS_5 geometry (20), a priori, takes a form:

$$ds^2 = -\left(1 + \frac{r^2}{b^2} - \frac{r_0^2}{r^2}\right) dt^2 + \left(1 + \frac{r^2}{b^2} - \frac{r_0^2}{r^2}\right)^{-1} dr^2 + \left(1 + \frac{A_\theta^2}{r^4}\right) r^2 d\Omega^2 \ . \tag{26}$$

The emergent geometry describes a $4D$ black hole though the spherical symmetry is broken. Naively, the deformation geometry from S^2-symmetry may be argued due to the nontrivial gauge potential A_θ accounting for the missing local degree. As a result, a large number of deformed geometries may be seen to arise underlying the $U(1)$ gauge symmetry. On the other hand, the near horizon geometry reduces to an AdS to confirm an AdS_4 black hole. However, the deformations in (tt)- and (rr)-components of the metric possibly account for the missing gauge degree in disguise.

It reassures the presence of a nontrivial electric field $\mathbf{E}$ apparently ignored by the A_μ gauge field configurations (22) naively obtained by Kaluza-Klein compactification of the $B_{\mu\nu}$ field in an emergent AdS_5 description. Then, the AdS_4 black hole geometry may be re-expressed as:

$$ds^2 = -\left(1 + \frac{r^2}{b^2} - r_0|\mathbf{E}|\right)dt^2 + \left(1 + \frac{r^2}{b^2} - r_0|\mathbf{E}|\right)^{-1}dr^2 + \left(1 + |\mathbf{B}|^2\cot^2\theta\right)r^2 d\Omega^2 ,$$
(27)

where $|\mathbf{E}| = (r_0/r^2)$ and the constant r_0 is identified with an electric charge. Then the nontrivial braneworld gauge field components take the standard form:

$$A_t = \frac{r_0}{r} \quad \text{and} \quad A_\phi = -q\cos\theta .$$
(28)

Interestingly the AdS_4 braneworld black hole (27) resembles to the one, already, been obtained in the past,[17,19,20] using a noncommutative scaling on a D_3-braneworld. There, the nontrivial metric components in (tt)- and (rr)- sectors have been identified with $|\mathbf{E}| = (r_0/r)$ for a different set of gauge fields and normalizations. Finally we would like to resolve the apparent problem of missing one local degree, leading to trivial electric field on the braneworld. It cropped up due to the naive treatment of radial coordinate r on the braneworld. Since the D_3-brane gauge theory is at the boundary of AdS_5, the radial coordinate r in AdS_5 becomes very large there. As a result, the deformation in the (rt)-sector becomes trivial at the AdS_5 boundary. However, the electric field on D_3-brane remains unaffected in the asymptotic limit of AdS_5. In other words, the AdS_5 and braneworld are governed by two independent radial coordinates. It is important to note that the S^2-deformation on the braneworld is essentially caused due to the magnetic field, not the potential A_θ. The S^2-deformation geometry being independent of the radial coordinate, it remains unaffected by the naive gauge field configurations (22).

4. Concluding Remarks

The subtle notion of torsion due to a background two form sourced by a fundamental string was investigated. Interestingly, the deformation geometries leading to black holes on D_3-braneworld, and its dual world, in an emergent gravity scenario were explored. It would be interesting to analyze the deformed AdS_5 and braneworld black hole geometries in an emergent gravity scenario to emphasize a plausible notion of temperature in the formalism.

Acknowledgments

A.K.S. gratefully acknowledges the warm hospitalities at Nanyang Technological University, Singapore extended during the conference. The research of A.K.S. is supported by a CSIR fellowship, New Delhi, India.

References

1. J.M. Maldacena, Adv.Theor.Math.Phys.**2**, 231 (1998).
2. E. Witten, Adv.Theor.Math.Phys.**2**, 253 (1998).
3. M. Li, Y-S. Wu, Phys.Rev.Lett.**84**, 2084 (2000).
4. E. Verlinde, arXiv:1001.0785 [hep-th] (2010).
5. W.G. Unruh, Phys.Rev. **D14**, 870 (1976).
6. D. Freedman, P.K. Townsend, Nucl.Phys. **B177**, 282 (1981).
7. E. Braaten, T. Curtright and C. Zachos, Nucl.Phys. **B260**, 630 (1985).
8. A. Saa, Class. & Quan.Grav., L85 (1995).
9. A. Saa, Gen.Rel.Grav.**29**, 205 (1997).
10. S. Kar, K.P. Pandey, S. Singh and A.K. Singh, arXiv:1002.1906 [hep-th] (2010).
11. S. Kar, K.P. Pandey, S. Singh and A.K. Singh, arXiv:1002.3976 [hep-th] (2010).
12. N. Seiberg and E. Witten, Jour.High Ener.Phys.**09**, 032 (1999).
13. G.W. Gibbons and K. Hashimoto, Jour.High Ener.Phys.**09**, 013 (2000).
14. M. Mars, J.M.M. Senovilla and R. Vera, Phys.Rev.Lett.**86**, 4219 (2001).
15. G.W. Gibbons and A. Ishibashi, Class. & Quan.Grav.21, 2919 (2004).
16. S. Kar and S. Majumdar, Phys.Rev. **D74**, 066003 (2006).
17. S. Kar and S. Majumdar, Int.J.Mod.Phys. **A21**, 6087 (2006).
18. S. Kar and S. Majumdar, Int.J.Mod.Phys. **A21**, 2391 (2006).
19. S. Kar, Phys.Rev. **D74**, 126002 (2006).
20. S. Kar, Jour.High Ener.Phys.**0610**, 052 (2006).
21. L-H. Liu, B. Wang and G-H Yang, Phys.Rev. **D76**, 064014 (2007).
22. S. Chen, B. Wang and R-K Su, Phys.Rev. **D77**, 024039 (2008).
23. J. Zhang, Phys.Lett. **B668**, 353 (2008).
24. R. Casadio, P. Nicolini, Jour.High Ener.Phys.**0811**, 072 (2008).
25. S. Kar Int.J.Mod.Phys. **A24**, 3571 (2009).
26. E. Spallucci, A. Smailagic and P. Nicolini, Phys.Lett. **B670**, 449 (2009).
27. Y. Nambu, Phys.Reports **23**, 250 (1976).

CLIFFORD GRAVITY — EFFECTIVE QUANTUM GRAVITY FOR FERMIONIC MATTER WITH ARROW OF TIME

V. V. ASADOV[*,‡] and O. V. KECHKIN[*,†,§]

*Neur OK–III, Neur OK LLC,
Podolskih Kursantov 10, 117545 Moscow, Russia
www.neurok.com

†Skobeltsyn Institute of Nuclear Physics,
Lomonosov Moscow State University,
Vorob'jovy Gory, 119899 Moscow, Russia
www.sinp.msu.ru
‡asadov@neurok.com
§kechkin@depni.sinp.msu.ru

Quantum gravity is considered on the base of five-dimensional Dirac equation with complex coordinates. We have used a multi-world Everett-like approach, which allowed us to introduce an arrow of time into dynamics of the theory. Effective metrics arises as coordinate representation of the quantum metric operator, which belongs to the Clifford algebra. We present the results obtained in the study of isotropic and homogenous cosmology in the theory. They include explanation of the matter - antimatter asymmetry in the physical world.

Keywords: Quantum gravity; fermionic matter; arrow of time.

1. Introduction

Fundamental fermions, i.e. quarks (introduced by M. Gell-Mann and G. Zweig in 1964) and leptons, are considered as the most elementary 'building blocks' of physical matter in the modern physics.[1–3] We understand the Universe as a consistent quantum[4] and thermodynamical[5] system, which lives according some nontrivial cosmological scenario.[6,7]

We develop a unified theory of quantum dynamics and thermodynamics, and apply it to relativistic quantum mechanics for the fermions.[8–11] Our approach preserves the main principles of the quantum theory and guarantees the consistency with the Second Law of thermodynamics. Specifically for the five-dimensional Dirac's equation it was established that the modified spinor dynamics with the arrow of time also includes an effective gravity, which can be proposed to the role of the quantum gravity coupled to the fermionic matter. We also demonstrate the results obtained in the study of isotropic and homogenous cosmological models in the theory.

2. Universe as Ensemble of Fermionic Worlds

We introduce a concept of the Universe that consists of the set of fermionic worlds in this section. The fermionic world is understand as a dynamical quantum system constructed of the massless and massive Dirac's particles; it also contains the tachyons. Then, each fermionic world is characterized by its proper time of life. The evolution of the Universe possesses an arrow of time, which is 'directed' to the most long-living fermionic world.[8,9,11]

2.1. *Quantum mechanics of fermionic world*

First of all, let us note that the conventional Dirac's equation can be represented in the form of the stationary Schrodinger's problem:

$$\hat{m}\,\Psi_m = m\Psi_m, \tag{1}$$

where the wave-function and the mass operator read:

$$\Psi_m = \Psi_m\left(x^\mu\right), \qquad \hat{m} = -\frac{1}{c}\left(\gamma^4\right)^{-1}\gamma^\mu\hat{p}_\mu. \tag{2}$$

In fact, the theory possesses two realizations with $\gamma^4 = -I$ and γ^5; for the both of them mass the relation $\hat{m}^2c^2 = \eta^{\mu\nu}\hat{p}_\mu\hat{p}_\nu$ holds. For the corresponding non-stationary Schrodinger's problem one has:

$$i\hbar\Psi_{,\tau} = \hat{m}c^2\Psi, \tag{3}$$

where τ is the additional time parameter which is not related kinematically to the Minkowskian time x^0/c. Then, a spectral analysis shows that the general solution $\Psi = \Psi\left(\tau, x^\mu\right)$ of Eq. (3) is a superposition of the all massless, massive and tachyon-type modes. We name this quantum system the 'fermionic world'.

2.2. *Invariant time and temperature. Expectation value of mass operator*

There is a possibility to consider the evolutionary parameter τ in Eq. (3), which is the Poincare invariant, as the complex one:

$$\tau = t - \frac{i\hbar c}{2}\beta. \tag{4}$$

In this case, its real and imaginary parts (the invariant parameters t and β, respectively), become the time and inverse absolute temperature in the system. We consider the analytical theory in respect to τ, so that the identity $\Psi_{,\tau^*} = 0$ holds for the any quantum state.

Then, using Eqs. (3) and (4), it is easy to prove that the gauge current $J^M = \bar{\Psi}\gamma^M\Psi$ satisfies the null-divergence relation $J^M_{,x^M} = 0$. The norm of the state in the theory is $Z = <\Psi|\Psi> = \int d^4x J^4$; for the probability density one has the relation $\rho = J^4/Z$. The matrix element of the mass operator is defined as $<\Psi_1|\hat{m}|\Psi_2> =$

$\int d^4x \, \bar{\Psi}_1 \gamma^4 \hat{m} \Psi_2$. It is the Hermitian one, i.e. $\hat{m}^+ = \hat{m}$. The expectation value of $\hat{m}$ (i.e. the averaged mass $\bar{m}$, calculated over the quantum state Ψ) is

$$\bar{m} = \frac{< \Psi|\hat{m}|\Psi >}{< \Psi|\Psi >} = c^2 \left(\log |Z|\right)_{,\beta}. \tag{5}$$

From Eq. (5) it follows that β is actually the inverse absolute temperature of the system. It is important to note that for a tachyonic mode $\bar{m}_{\text{tach}} = \bar{m}_{\text{tach}}(\beta) = |p^2|^{\frac{1}{2}}/c \tan\left(|p^2|^{\frac{1}{2}} c\beta + \text{const}\right)$. Thus, in this special case, the averaged mass depends on the absolute temperature (whereas for the massive particle modes $\bar{m}_{,\beta} = 0$).

2.3. *Universe as ensemble of interacting fermionic worlds*

Finally, the modified quantum theory of this type satisfies the dynamical equations

$$i\hbar\Psi_{,\tau} = \hat{H}\Psi, \qquad \Psi_{,\tau^*} = 0, \tag{6}$$

and the additional (stationary) restrictions $\hat{H}_{,\tau} = \hat{H}_{,\tau^*} = [\hat{H}, \hat{H}^+] = 0$. Let us put

$$\hat{H} = \hat{E} - \frac{i\hbar}{2}\hat{\Gamma}, \tag{7}$$

where $\hat{E}$ and $\hat{\Gamma}$ are the energy and decay operators, respectively. For these theories the inequality

$$\frac{d\bar{\Gamma}}{dt} \leq 0 \tag{8}$$

holds for the arbitrary initial state. This means an appearing of the arrow of time in the system dynamics (also the irreversible dynamics was studied in Refs. 12–15).

We identify the Universe with the multi-fermionic world system for the modified Dirac's theory under consideration, i.e. we put

$$\Psi = \begin{pmatrix} \Psi_1 \\ \Psi_2 \\ \Psi_3 \\ \cdot \\ \cdot \\ \cdot \end{pmatrix}, \quad \hat{\mathcal{E}} = \begin{pmatrix} \hat{E} & 0 & 0 & \dots \\ 0 & \hat{E} & 0 & \dots \\ 0 & 0 & \hat{E} & \dots \\ \cdot & \cdot & \cdot & \cdot \\ \cdot & \cdot & \cdot & \cdot \\ \cdot & \cdot & \cdot & \cdot \end{pmatrix}, \quad \hat{\Gamma} = \begin{pmatrix} \gamma_1 & 0 & 0 & \dots \\ 0 & \gamma_2 & 0 & \dots \\ 0 & 0 & \gamma_3 & \dots \\ \cdot & \cdot & \cdot & \cdot \\ \cdot & \cdot & \cdot & \cdot \\ \cdot & \cdot & \cdot & \cdot \end{pmatrix}, \tag{9}$$

where $\hat{E} = \hat{m}c^2$, and $\gamma_1 < \gamma_2 < \gamma_3 < \dots = \text{const}$ are the inverse times of life of the fermionic worlds described by the wave functions Ψ_1, Ψ_2, Ψ_3, ... This relativistic system also possesses the arrow of time in the its dynamics.

3. Observer in the Universe

We give a fully analytical extension of the theory and discuss some elements of its canonical formalism in this section. After that, we formulate a rule for reduction of the all additional dynamical parameters and relate the choice of the physical observer with this procedure.[10,11]

3.1. Completely analytical model of fermionic world. Canonical formalism. Relation between Noether charges

For the fermionic Universe with the analytical wave function $\Psi = \Psi(X^M)$, where

$$X^M = x^M - \frac{i\hbar c}{2}\beta^M, \tag{10}$$

one has the theory with remarkable properties for its canonical formalism. Namely, let us consider the fermionic world number k (where $k = 1, 2, 3, \dots$). It is easy to prove that after the scale transformation $\Psi_k \to \exp\left(\gamma_k X^4/2\right)\Psi_k$ the equations of motion take the following simplest form:

$$\gamma^M \Psi_{,x^M} = 0, \qquad \Psi_{,\beta^M} = -\frac{i\hbar c}{2}\Psi_{,x^M}, \tag{11}$$

where the index k was omitted. The first equation of the system (11) corresponds to the massless five-dimensional Dirac's action with the energy-momentum tensor $T_N^M = i/2\left(\bar{\Psi}\gamma^M\Psi_{,x^N} - \bar{\Psi}_{,x^N}\gamma^M\Psi\right)$, which satisfies the null-divergence relation $T_{N,x^M}^M = 0$. The theory possesses the additional relation $T_N^M = -\hbar c J_{,\beta^N}^M$ between the energy-momentum tensor T_N^M and the gauge current J^M. As a consequence, the relation between the energy-momentum vector and the statistical sum holds: $P_N = \int d^4 x\, T_N^4 = -\hbar c Z_{,\beta^N}$ (parametric generalization of the Noether theorem was also studied in Refs. 16, 17). We choose this special variant of the theory based on the massless five-dimensional Dirac's equation for the following analysis.

3.2. Principle of maximal probability and reduction of extra dynamical parameters

Probability density $\rho = \rho\left(x^N, \beta^M\right)$ depends on the space-time coordinates x^μ and, also, on the extra dynamical parameters x^4 and β^M. We propose the following reductional principle: for the physical observer, the extra parameters take the values $x_{\max}^4(x^\mu)$, $\beta_{\max}^M(x^\mu)$, which maximize the probability density at the given point x^μ:

$$\max_{x^4,\,\beta^M} \rho\left(x^\mu,\, x^4,\, \beta^M\right) = \rho\left(x^\mu,\, x_{\max}^4(x^\mu),\, \beta_{\max}^M(x^\mu)\right). \tag{12}$$

Then, the relations that detect the functions $x_{\max}^4(x^\mu)$ and $\beta_{\max}^M(x^\mu)$ are the equations $\nabla_{x^\mu}\rho_{,x^4} = \nabla_{x^\mu}\rho_{,\beta^M} = 0$ (where ∇_{x^μ} is the total derivative in respect to the coordinate x^μ). This procedure actually fixes the extra dynamical parameters as some functions of the physical coordinates x^μ, i.e. it defines the reductional relations $x^4 = x_{\max}^4\left(x^\mu\right)$, $\beta^M = \beta_{\max}^M\left(x^\mu\right)$.

3.3. Nonlinearity of reduced theory

Finally, one obtains the reduced spinor field $\psi(x^\mu) = \Psi\left[X^N\left(x^\mu\right)\right]$, which is calculated on shell of the fixed reductional mode (12). For the spinor derivatives the following relation reads:

$$\Psi_{,x^\mu} = \left(D^{-1}\right)_\mu{}^\lambda \psi_{,x^\lambda}, \quad \text{where} \quad D_\mu{}^\lambda = X^\lambda_{,x^\mu} - X^4_{,x^\mu}\left(\gamma^4\right)\gamma^\lambda. \tag{13}$$

Note, that the resulting system is highly nonlinear, because the original spinor Ψ defines the reduced one ψ over 'Ψ-dependent' procedure based on Eqs. (12)–(13).

4. Effective Gravity — Clifford Wrapped Gravity

We show that the reduction of extra dynamical parameters generates curvature effects in this section. These effects can be described by a metric operator with some defect, which leads to arising of distribution of a dark mass in the system. Then, a physical metrics coincides with the coordinate representation of the metric operator (we name the corresponding procedure as 'wrapping' here). For simplicity, we present a formalism for the single fermionic world. This situation takes place at the final stage of the evolution – namely, for the longest-living 'part' of the Universe.[11]

4.1. *Local principle and mass distribution*

Let us introduce a local principle, which states that value of any reduced quantity is equal to the corresponding original one calculated on shell of the reductional mode. For our purposes the following reduced distributions

$$m^2\left(x^\mu\right) = m^2\left[x^\mu, x_{\max}^4\left(x^\mu\right), \beta_{\max}^M\left(x^\mu\right)\right], \quad J^4\left(x^\mu\right) = J^4\left[x^\mu, x_{\max}^4\left(x^\mu\right), \beta_{\max}^M\left(x^\mu\right)\right], \quad (14)$$

are important, and also the reduced statistical sum $Z = \int d^4x J^4(x^\mu) = \text{const}$. Note that $m^2\left(x^\mu, x^4, \beta^M\right) = p^2\left(x^\mu, x^4, \beta^M\right)/c^2$, where

$$p^2\left(x^\mu, x^4, \beta^M\right) = \frac{\bar{\Psi}\gamma^4\hat{p}^2\Psi}{\bar{\Psi}\gamma^4\Psi}. \quad (15)$$

Using the quantity $m^2\left(x^\mu\right)$, one can calculate expectation value of the squared mass according to the relation $\overline{m^2} = \int d^4x \rho\left(x^\mu\right) m^2\left(x^\mu\right)$ with $\rho\left(x^\mu\right) = J^4\left(x^\mu\right)/Z$.

4.2. *Geometrization. Metric operator and candidate to dark mass*

We use the General Relativity motivated form

$$p_{\text{geom}}^2 = \frac{\bar{\psi}\gamma^4\hat{p}_\mu\hat{g}^{\mu\nu}\hat{p}_\nu\psi}{\bar{\psi}\gamma^4\psi} \quad (16)$$

to describe 'geometrically' the reduced value $p^2\left(x^\mu\right)$ of distribution of the squared momentum given by Eq. (15). The best choice of the metric operator $\hat{g}^{\mu\nu}$ reads:

$$\hat{g}^{\mu\nu} = \frac{\gamma}{2}\eta^{\alpha\beta}\left\{\left[\left(D^{-1}\right)_\alpha^\mu\right]^+ \gamma\left(D^{-1}\right)_\beta^\nu + \left[\left(D^{-1}\right)_\alpha^\nu\right]^+ \gamma\left(D^{-1}\right)_\beta^\mu\right\}, \quad (17)$$

where $\gamma = \gamma^0\gamma^4$. The corresponding defect of the squared mass distribution, i.e. the quantity

$$m_{\text{dark}}^2\left(x^\mu\right) = \frac{1}{c^2}\left[p_{\text{geom}}^2\left(x^\mu\right) - p^2\left(x^\mu\right)\right], \quad (18)$$

580

can be proposed as candidate to the role of distribution of a (squared) dark mass. Namely, if one calculates the total mass by help of the gravity effects, then the quantity (18) will give difference between 'geometrical' and kinematical inputs to the squared mass distribution.

4.3. *Clifford wrapped gravity and metric calculation*

It is clear, that the metric operator (17) is defined completely by the quantity $\left(D^{-1}\right)^\mu_\lambda$. This can be calculated explicitly using the relation $\left(D^{-1}\right)^\mu_\lambda = \left(F^{-1}\right)^\nu_\lambda \left(E^{-1}\right)^\mu_\nu$, where $E^\mu_\nu = X^\mu_{,x^\nu}$, whereas

$$\left(F^{-1}\right)^\nu_\lambda = \delta^\nu_\lambda + \frac{1+a}{1-\phi^2} A^\nu_\lambda, \tag{19}$$

where $a = A^\alpha_\alpha$, $A^\nu_\lambda = \phi_\lambda \left(\gamma^4\right)^{-1} \gamma^\nu$, $\phi_\lambda = \left(E^{-1}\right)^\nu_\lambda X^4_{,x^\nu}$ and $\phi^2 = \eta^{\lambda\sigma}\phi_\lambda\phi_\sigma$. Thus, the metric operator $\hat{g}^{\mu\nu}$ is an element of the Clifford algebra, i.e. it can be written in as

$$\hat{g}^{\mu\nu} = a^{\mu\nu}I + b^{\mu\nu}\gamma^5 + c^{\mu\nu}_\lambda\gamma^\lambda + d^{\mu\nu}_\lambda\gamma^5\gamma^\lambda + e^{\mu\nu\lambda\sigma}\Sigma_{\lambda\sigma}, \tag{20}$$

where $\Sigma_{\lambda\sigma} = i/2\left[\gamma^\lambda, \gamma^\sigma\right]$. The inverse tensor $\hat{g}_{\mu\nu}$ is defined by the Clifford algebra valued relation $\hat{g}^{\mu\nu}\hat{g}_{\nu\lambda} = \delta^\mu_\lambda \hat{I}$.

For example, in the special quasi-classical (non-quantum) case with weak gravity $\hbar c\beta^M_{,x^\mu} << x^4_{,x^\mu} \equiv \phi^4(x^\mu) << 1$, and the metric distribution $g_{\mu\nu} = g_{\mu\nu}(x^\lambda)$ reads:

$$g_{\mu\nu} \approx \eta_{\mu\nu} - \phi_{,x^\mu}j_\nu - \phi_{,x^\nu}j_\mu, \tag{21}$$

where $j_\mu = J_\mu/J^4$. It is seen, that the flat Minkowskian metrics $g_{\mu\nu} = \eta_{\mu\nu}$ is supported by equations of this theory.

5. Isotropic and Homogenous Cosmology

We study an isotropic and homogenous solution of the theory in this section. Its physical consequences include mass generation mechanism and dynamical explanation of the matter - antimatter asymmetry. Also, we show that the asymptotical metrics is flat, so that the corresponding cosmological model is the realistic one.[9,11]

5.1. *General solution*

The isotropic and homogenous ansatz is defined by the relation $\Psi_{,x^k} = 0$, where $k = 1, 2, 3$. The corresponding general solution of the motion equations reads:

$$\Psi = \Psi_+ + \Psi_-, \tag{22}$$

where (we consider, for definiteness, the $\gamma^4 = \gamma^5$ variant of the theory)

$$\Psi_\pm = \frac{1}{\sqrt{2}}\begin{pmatrix} f_\pm \\ \mp if_\pm \end{pmatrix}, \tag{23}$$

and $f_\pm = f_\pm \left(X^0 \mp X^4\right)$. Note, that the spinors Ψ_+ and Ψ_- are waves which propagates forward and backward in the Minkowskian time x^0 when the invariant time x^4 increases. They are natural candidates to the role of wave functions of the matter and antimatter 'sub-worlds', respectively.

5.2. *Coherent states and mass formula*

The single special solution $\Psi = \Psi_\pm$ with

$$f_\pm = d_\pm \exp\left[-\frac{\left(x^0 \mp X^4\right)^2}{4D_{x^0}^2}\right], \tag{24}$$

where $d_\pm$ and D_{x^0} are the arbitrary constants, describes a class of the coherent states in this set of solutions. Actually, for dispersions of the rest energy D_{mc^2} and the Minkowskian time $D_{x^0/c}$, it holds the exact equality $D_{mc^2} D_{x^0/c} = \hbar/2$, which minimizes the uncertainty relation. The mass formula for this solution reads:

$$\bar{m} = \left(\frac{\hbar}{2D_{x^0}}\right)^2 \beta = D_{mc^2}\beta, \tag{25}$$

i.e. the mass expectation value is inverse proportional to the absolute temperature of the system. This result seems interesting in the cosmological framework.

5.3. *Matter-antimatter asymmetry and asymptotical flatness*

Now let us consider the solution (22)–(23) in the classical case of $\hbar c \beta_{,x^\mu}^M << x_{,x^\mu}^4$ at the limit $x^0 \to +\infty$. Also, let us suppose that each of the probability densities $\rho_\pm$, calculated over the wave functions $\Psi_\pm$, possess exactly one maximum. These maximums define the observer and anti-observer (for ρ_+ and ρ_-, respectively) in the Universe under consideration. Note that at the big values of the time x^0 the observer 'lives' at

$$x^4 \approx x^0. \tag{26}$$

Of course, this relation reduces the extra coordinate x^4 here. For the anti-observer one obtains $x^4 \approx -x^0$ at the same limit $x^0 \to +\infty$. Thus, the observer and anti-observer are 'located' in the different points at the time axis and, moreover, they 'scatter' monotonously at the limit under consideration. It can be said that at the big times x^0 the observer does not see anti-matter, whereas the anti-observer could not detect matter. This situation takes place for the general case of solutions (22)–(23), icluding the 'symmetrical' situation with $\rho_+ = \rho_-$ at the initial time $x^0 = 0$ (at the Big Bang).

Performing the calculations based on Eq. (17), it is possible to show that the asymptotical of the metrics $g_{\mu\nu}$ coincides with the Minkowskian metrics $\eta_{\mu\nu}$ after some rescaling of the spatial coordinates x^k. Thus, at the big times the isotropic and homogenous cosmological model describes matter living in the flat four-dimensional space-time.

Conclusions

Let us summarize the results of this paper. It was shown that the theory based on the totally analytical and massless five-dimensional Dirac's field detects:

(1) the arrow of time in context of the multi-world approach;
(2) the mass generation mechanism;
(3) the effective metrics for the four-dimensional physical space-time;
(4) the nontrivial dark mass distribution;
(5) the cosmological scenario for the isotropic and homogenous Universe, which explains asymmetry between particles and antiparticles.

Acknowledgments

Authors would like to thank Profs. B. Ishkhanov, Yu. Rybakov, G. Costa and M. Matone for fruitful discussions while this investigation was carried out. We are really grateful to the organizers of the Conference in Honour of Murray Gell-Mann's 80th Birthday for the provided possibility to present and discuss these results of our activity.

References

1. F. Halzen and A.D. Martin, *Quarks and Leptons: An Introductory Course in Modern Particle Physics* (John Wiley and Sons, New York, 1984).
2. K. Huang, *Quarks, Leptons and Gauge Fields* (World Scientific, Singapore, 1982).
3. E. Sudbery, *Quantum Mechanics and Particles of Nature* (Cambridge University Press, Cambridge, 1986).
4. P.A.M. Dirac, *Principles of Quantum Mechanics* (Oxford University Press, Oxford, 1958).
5. K. Huang, *Statistical Mechanics* (John Wiley and Sons, New York, 1987).
6. A.D. Linde, *Inflation and Quantum Cosmology* (Academic Press, Boston, 1990).
7. A.D. Linde, *Particle Physics and Iinflationary Cosmology* (Harwood Academic Publishers, Chur, Switzerland, 1990).
8. V. V. Asadov and O. V. Kechkin, *Moscow University Physics Bulletin* **2**, 105 (2008).
9. V. V. Asadov and O. V. Kechkin, *Gravit. Cosmol.* **15**, 295 (2009).
10. V. V. Asadov and O. V. Kechkin, *Bulletin of the Peoples' Friendship University of Russia (Mathematics. Informatics. Physics)* **3**, 99 (2009).
11. V. V. Asadov and O. V. Kechkin, Modified Noether theorem and arrow of time in quantum mechanics, in *Proc. of Symmetries in Science XIV, Journal of Physics, Conference Series* **3** (2010).
12. H. Dekker, *Phys. Rep.* **80**, 1 (1981).
13. V. E. Tarasov, *Phys. Lett.* A **288**, 173 (2001).
14. E. Celeghini, M. Rasetti and G. Vitiello, *Annals Phys.* **215**, 156 (1992).
15. H. Feshbach and Y. Tikochinsky, *Trans. New York Acad. Sci.* **38**, 44 (1977).
16. O. Castanos, R. Lopez-Pena and V. I. Manko, *Mod. Phys. Lett.* A **9**, 1785 (1994).
17. O. Castanos, R. Lopez-Pena and V. I. Manko, *J. Phys.* A **27**, 1751 (1994).

STRINGY STABILITY OF DILATON BLACK HOLES IN 5-DIMENSIONAL ANTI–DE SITTER SPACE

YEN CHIN ONG

Department of Mathematics, National University of Singapore,
Block S17 (SOC1), 10, Lower Kent Ridge Road, Singapore 119076
yenchin@nus.edu.sg

Flat electrical charged black holes in 5-dimensional Anti–de Sitter space have been applied to the study of the phase diagram of quark matter via AdS/CFT correspondence. In such application it is argued that since the temperature of the quark gluon plasma is bounded away from zero, the dual black hole cannot be arbitrarily cold, but becomes unstable due to stringy instability once it reaches sufficiently low temperature. We study the stringy stability of flat dilaton black holes with dilaton coupling $\alpha = 1$ in asymptotically Anti–de Sitter space and show that unlike the purely electrically charged black hole, dilaton black holes do not suffer from stringy instability.

Keywords: Dilaton black hole stability; AdS/CFT correspondence.

1. Holography of Flat Black Holes

AdS/CFT correspondence has been employed to study the phase transitions of quark matter, where the temperature of the quark gluon plasma corresponds to electrical charges on a flat black hole in the 5-dimensional AdS bulk. Adding electrical charges to cool down the black hole eventually causes the black hole to become unstable.[1,2] This instability, first discovered by Seiberg and Witten,[3] is due to the presence of brane in the bulk that modifies the geometry of the black hole (Euclidean) spacetime at large distances. For a friendly introductory account of the Seiberg–Witten instability, see Ref. 4. For simplicity, all electrical charges will be taken as positive in the subsequent discussion.

The *Seiberg–Witten action* is given by

$$S_{\text{SW}} = \Theta \, (\text{Brane Area}) - \nu \, (\text{Volume Enclosed by Brane})$$

where Θ is the tension of the brane and ν relates to the charge enclosed by the brane. Instability arises if S becomes negative since brane nucleation leads to large branes having energy which is unbounded from below as they approach the boundary. Equivalently, stability requires that the scalar curvature at conformal infinity should remain non-negative. This is clearly the case for AdS space itself, with positively curved conformal infinity. The most dangerous case is when the charge on the brane is saturated, called the BPS case. In 5 dimension, this means that $\nu_{\text{BPS}} = 4\Theta/L$

where L is the curvature scale of the AdS space. For the case of flat black hole, adding charges up to about 96% of the extremal value effectively makes the Seiberg–Witten action becomes negative at large r, and destroys the stability of the black hole.[1] Consequently on the field theory side of the story, it means that quark gluon plasma cannot be arbitrarily cold, which is what we should expect.

In low energy limit of string theory, scalar field called *dilaton*, denoted by ϕ, can couple to the Maxwell field, giving the action

$$S = \int d^4x \sqrt{-g} \left(R - 2(\nabla\phi)^2 - e^{-2\phi}F^2 \right).$$

We will assume that the dilaton decays and vanishes at infinity. Note that because of the coupling between dilaton field and Maxwell field, the dilaton is not an independent "hair" of the black hole.

Recently, dilaton black holes have also been explored for its holography and applications in AdS/CFT correspondence.[5-7] Notably Gubser and Rocha[7] argued that dilaton black hole in AdS$_5$ or a relative of it with similar behavior might be dual to Fermi liquid. Unlike quark gluon plasma, Fermi liquid can attain zero temperature. Therefore the dual black hole should be allowed to reach extremal charge without subjected to the Seiberg–Witten instability. In the following sections, we will first recall the properties of dilaton black holes, followed by computation of its Seiberg–Witten action to study its stability.

2. Spherically Symmetric Dilaton Black Holes in Asymptotically Flat Spacetime

We first recall 4-dimensional dilaton black holes in asymptotically flat case. For the sake of comparison we shall recall the Reissner–Nordström solution:

$$g(\text{RN}) = -\left(1 - \frac{2M}{r} + \frac{Q^2}{r^2}\right) dt^2 + \left(1 - \frac{2M}{r} + \frac{Q^2}{r^2}\right)^{-1} dr^2 + r^2 d\Omega^2.$$

We assume $Q > 0$ for simplicity.

For $0 < Q < M$, we have two horizons

$$r_\pm = M \pm \sqrt{M^2 - Q^2}$$

where $r = r_+$ is the event horizon and $r = r_-$ is the inner horizon. The corresponding black hole solution in the presence of dilaton field in low energy limit of string theory is remarkably simple. Known as the *Garfinkle–Horowitz–Strominger* or GHS black hole,[8,9] its metric is:

$$g(\text{GHS}) = -\left(1 - \frac{2M}{r}\right) dt^2 + \left(1 - \frac{2M}{r}\right)^{-1} dr^2 + r\left(r - \frac{Q^2}{M}\right) d\Omega^2.$$

As mentioned before, the dilaton is coupled to the electric field and hence is not an independent parameter of the black hole. The precise relation between the dilaton and the electric charge Q is given by

$$e^{-2\phi} = 1 - \frac{Q^2}{Mr}.$$

Note the absence of dependence on electrical charge in the g_{tt} and g_{rr} terms. The r-t plane is thus similar to the Schwarzschild black hole, but the spherical horizon is smaller in area for any nonzero electrical charge.

Interestingly the GHS black hole behaves differently compared to the Reissner–Nordström black hole when electrical charge is increased. In the latter case, the event horizon moves *inward* while the Cauchy horizon moves outward, finally the two horizon coincide when extremality is reached. For the GHS black hole however, the event horizon stays *fixed* at $r_+ = 2M$ and it has *no inner horizon*. The effect of decreasing electrical charge on the GHS black hole is to decrease its area, which goes to zero at $Q^2 = rM$. This gives the extremal limit: the *event horizon becomes singular* at $Q^2 = 2M^2$, i.e. $Q = \sqrt{2}M$, unlike the extremal limit of Reissner–Nordström black hole that satisfies $Q = M$. One can understand that the dilaton black hole has larger charge over mass ratio in the extremal limit because the scalar field contributes an extra attractive force, and so for any fixed M, we need a larger Q to balance it.

In general, we can introduce a free parameter $\alpha \geq 0$, that governs the strength of coupling between the dilaton field and the Maxwell field. This yields the *Garfinkle–Maeda* or GM black hole solution:[10]

$$g(\text{GM}) = -\left(1 - \frac{r_+}{r}\right)\left(1 - \frac{r_-}{r}\right)^{\frac{1-\alpha^2}{1+\alpha^2}} dt^2 + \left(1 - \frac{r_+}{r}\right)^{-1}\left(1 - \frac{r_-}{r}\right)^{\frac{\alpha^2-1}{\alpha^2+1}} dr^2$$
$$+ r^2\left(1 - \frac{r_-}{r}\right)^{\frac{2\alpha^2}{1+\alpha^2}} d\Omega^2$$

where

$$e^{-2\phi} = e^{-2\phi_0}\left(1 - \frac{r_-}{r}\right)^{\frac{2\alpha}{1+\alpha^2}}, \quad F = \frac{Q}{r^2} dt \wedge dr$$

with the asymptotic value of the dilaton field ϕ_0 taken to be zero in the following discussion. The horizons are at

$$r_\pm = \frac{1+\alpha^2}{1\pm\alpha^2}\left[M \pm \sqrt{M^2 - (1-\alpha^2)Q^2}\right], \alpha \neq 1 \text{ for } r_-.$$

When $\alpha = 1$ (the coupling strength that appears in the low energy string action), the GM solution reduces to the GHS solution, and there ceases to be an inner horizon, while $\alpha = 0$ case reduces to the Reissner–Nordström solution.

3. Topological Dilaton Black Holes

Gao and Zhang[11] generalized the GM solution to include dilatonic topological black hole in asymptotically AdS spacetime in n-dimension,

$$ds^2 = -U(r)dt^2 + W(r)dr^2 + [f(r)]^2 d\Omega^2_{k,n-2}$$

where $k = -1, 0, +1$ and

$$U(r) = \left[k - \left(\frac{r_+}{r}\right)^{n-3}\right]\left[1 - \left(\frac{r_-}{r}\right)^{n-3}\right]^{1-\gamma(n-3)} - \frac{1}{3}\Lambda r^2 \left[1 - \left(\frac{r_-}{r}\right)^{n-3}\right]^r$$

$$W(r) = \left\{\left[k - \left(\frac{r_+}{r}\right)^{n-3}\right]\left[1 - \left(\frac{r_-}{r}\right)^{n-3}\right]^{1-\gamma(n-3)} - \frac{1}{3}\Lambda r^2 \left[1 - \left(\frac{r_-}{r}\right)^{n-3}\right]^r\right\}^{-1}$$

$$\times \left[1 - \left(\frac{r_-}{r}\right)^{n-3}\right]^{-\gamma(n-4)}$$

with

$$[f(r)]^2 = r^2 \left[1 - \left(\frac{r_-}{r}\right)^{n-3}\right]^{\gamma}; \qquad \gamma = \frac{2\alpha^2}{(n-3)(n-3+\alpha^2)}.$$

Note that in this notation Λ is the effective cosmological constant $|\Lambda| = 3/L^2$ where L is the curvature scale of de Sitter or Anti–de Sitter space, independent of dimensionality. We also note that for $n \geq 5$, $\alpha \neq 0$, we have $U(r)W(r) \neq 1$ in general. This is not surprising since the presence of scalar field contributes to the stress energy tensor and thus affects the geometry of spacetime leading to $g_{tt}g_{rr} \neq -1$ in general.[12]

The mass of the black hole, and the *charge parameter* q, are, with $L = 1$ for simplicity,

$$M = \frac{\Gamma_{n-2}}{16\pi}(n-2)\left[r_+^{n-3} + k\left(\frac{n-3-\alpha^2}{n-3+\alpha^2}\right)r_-^{n-2}\right]$$

and

$$q^2 = \frac{(n-2)(n-3)^2}{2(n-3+\alpha^2)}r_+^{n-3}r_-^{n-3}.$$

respectively. The charge parameter q is directly proportional to the black hole electrical charge Q.[13]

Let us consider $k = 0$, $L = 1$, $n = 5$, $\alpha = 0 = \gamma$ which should reduce to the case of flat charged black hole in Ref. 1, 2. Using the above formula, we compute that

$$M = \frac{8\pi^3 K^3}{16\pi}3r_+^2 = \frac{3}{2}\pi^2 K^3 r_+^2$$

and

$$U(r) = \left(-\left(\frac{r_+}{r}\right)^2\right)\left(1 - \left(\frac{r_-}{r}\right)^2\right) - \frac{1}{3}\Lambda r^2$$

$$= -\frac{r_+^2}{r^2} + \frac{r_+^2 r_-^2}{r^4} + \frac{r^2}{L^2} = W(r)^{-1}.$$

If we compare this with the explicit form of metric for flat electrically charged black hole,

$$U(r) = \frac{r^2}{L^2} - \frac{2M}{3\pi^2 K^3 r^2} + \frac{Q^2}{48\pi^5 K^6 r^4}$$

we see that

$$r_+^2 = \frac{2M}{3\pi^2 K^3}, \quad r_+^2 r_-^2 = \frac{Q^2}{48\pi^5 K^6}.$$

The event horizon is the solution of $U(r) = 0$ which is *not* $\frac{2M}{3\pi^2 K^3}$. Thus in the notation of Gao and Zhang, r_+ and r_- are merely parameters that relate to the horizons instead of the horizons themselves. The authors in Ref. 13 for example, use the symbols c and b in place of r_+ and r_- and refer to them as "integration constants".

4. Seiberg–Witten Action for Flat AdS Dilaton Black Holes

Consider the 5-dimensional flat dilaton black hole in AdS with $L = 1$. As above,

$$r_+^2 = \frac{2M}{3\pi^2 K^3}.$$

Since Q^2 is proportional to q^2, we have

$$Q^2 \equiv Q^2(\alpha) = Q^2(\alpha = 0)\frac{n-3}{n-3+\alpha^2} = Q^2(\alpha = 0)\frac{2}{2+\alpha^2}.$$

That is,

$$Q^2 = \frac{2}{2+\alpha^2}(48\pi^5 K^6 r_+^2 r_-^2).$$

I.e.

$$r_-^2 = \frac{Q^2(2+\alpha^2)}{96\pi^5 K^6 r_+^2} = \frac{Q^2(2+\alpha^2)}{96\pi^5 K^6 \left(\frac{2M}{3\pi^2 K^3}\right)}.$$

We study the comparatively easy case of $\alpha = 1$ in which we have

$$Q^2 = 32\pi^5 K^6 r_+^2 r_-^2, \quad r_-^2 = \frac{3Q^2}{64M\pi^3 K^3},$$

$$\gamma = \frac{2(1)}{2(2+1)} = \frac{1}{3}, \quad f(r)^3 = r^3 \left[1 - \left(\frac{r_-}{r}\right)^2\right]^{\frac{1}{2}}.$$

Thus the Euclidean metric satisfies

$$g_{\tau\tau} = \left[-\left(\frac{r_+}{r}\right)^2\right]\left[1 - \left(\frac{r_-}{r}\right)^2\right]^{1-\frac{2}{3}} + r^2\left[1 - \left(\frac{r_-}{r}\right)^2\right]^{\frac{1}{3}}$$

$$= \left[-\left(\frac{r_+}{r}\right)^2 + r^2\right]\left[1 - \left(\frac{r_-}{r}\right)^2\right]^{\frac{1}{3}}$$

and

$$g_{rr} = \left\{ \left[-\left(\frac{r_+}{r}\right)^2 + r^2 \right] \left[1 - \left(\frac{r_-}{r}\right)^2 \right]^{\frac{1}{3}} \right\}^{-1} \left(1 - \left(\frac{r_-}{r}\right)^2 \right)^{-\frac{1}{3}}.$$

with the horizon at

$$r_{\text{eh}} = (r_+^2)^{\frac{1}{4}} = \left(\frac{2M}{3\pi^2 K^3} \right)^{\frac{1}{4}}$$

which is fixed independent of the electrical charge, just like its asymptotically flat GHS cousin.

For any fixed dilaton coupling α, varying the electrical charge means equivalently, varying the parameter r_-, via the relationship

$$r_-^2 = \frac{3Q^2}{64M\pi^3 K^3}.$$

At extremal limit, the horizon becomes singular with $r_{\text{eh}} = r_+^{\frac{1}{2}} = r_-$, i.e.

$$\left(\frac{2M}{3\pi^2 K^3} \right)^{\frac{1}{4}} = \left[\frac{3Q^2}{64M\pi^3 K^3} \right]^{\frac{1}{2}}.$$

Therefore the extremal charge is

$$Q_E = \frac{8 \times 2^{\frac{1}{4}}}{3^{\frac{3}{4}}} \pi M^{\frac{3}{4}} K^{\frac{3}{4}} \approx 13.11 \ M^{\frac{3}{4}} K^{\frac{3}{4}}.$$

Again this is greater than $Q_E \approx 9.96(KM)^{\frac{3}{4}}$ for flat AdS Reissner–Nordström black hole,[1] a similar behavior as its asymptotically flat counterpart.

The Seiberg–Witten action S_{SW} takes the form:

$$\alpha \left\{ r^3 \left[1 - \frac{3Q^2}{64\pi^3 K^3 M r^2} \right]^{\frac{2}{3}} \left[r^2 - \frac{2M}{3\pi^2 K^3 r^2} \right]^{\frac{1}{2}} - 4 \int_{r_{\text{eh}}}^{r} dr'(r')^3 \left[1 - \frac{3Q^2}{64\pi^3 K^3 M(r')^2} \right]^{\frac{1}{3}} \right\}$$

where $\alpha = 2\pi\Theta PLA_k$. Here $2\pi P$ is the period imposed on the imaginary time after Wick-rotation and A_k the area of the event horizon. The action vanishes at the horizon. See details of the technique in Ref. 1.

Taking typical values of the parameters we can obtain a plot of the action as function of r as in Fig. 1.

We note that for $Q = 0$ the action reduces to that of uncharged flat AdS black hole (the dilaton, being a secondary hair coupled to the Maxwell field, also vanishes when electrical charge is zero), which asymptotes to a positive value.[1] Unlike flat AdS Reissner–Nordström black hole with action increases to a maximum before plunging to negative, the action of flat AdS dilaton black hole is always positive. In particular, $\lim_{r\to\infty} S(r, Q_E) = +\infty$ where Q_E denotes the extremal charge. For any fixed charge Q, increasing the charge makes the action starts out with smaller value than the one with charge Q, but subsequently takes over at some finite value of r.

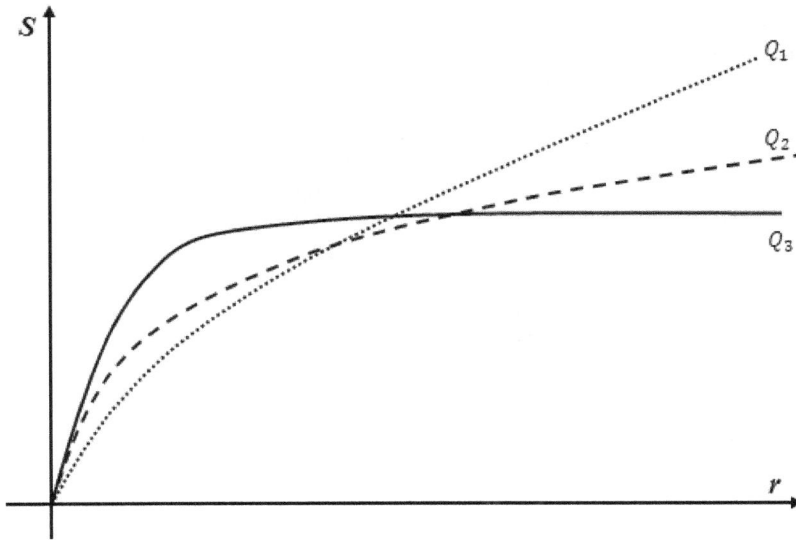

Fig. 1. The Seiberg–Witten action with typical values of parameter, where $Q_1 > Q_2 > Q_3 \approx 0$. The curve for $Q = 0$ would correspond to flat uncharged black hole case where the action asymptotes to a constant positive value. $S = 0$ axis corresponds to the horizon.

The value of r in which this take over occurs decreases with increasing charge. Thus the presence of dilaton stabilizes the black hole (at least in this special case with $\alpha = 1$) against non-perturbative instability in the Seiberg–Witten sense.

5. Conclusion

If AdS/CFT correspondence were to make sense between flat dilaton black hole and condensed matter system which can attain zero temperature, then the black hole should not become unstable in the Seiberg–Witten sense when electrical charges are increased. We have shown that this is indeed so at least for the case of dilaton black hole with coupling $\alpha = 1$. More works are needed to study the stringy stability of dilaton black holes with general coupling strength, especially for those as yet unknown range of values that are of application interest in AdS/CMT. The work of Hendi and Sheykhi[13] suggests that dilaton black holes possess unstable phase for large value of α, at least thermodynamically speaking. The conjecture of Gubser and Mitra[14] would suggest that such black holes are also dynamically unstable, although more works are still required to settle this stability issue definitively.

Acknowledgments

The author wishes to express his deep gratitude to his thesis supervisor Professor Brett McInnes for useful discussions and advises without which this work would not have been possible.

References

1. Brett McInnes, *Bounding the Temperatures of Black Holes Dual to Strongly Coupled Field Theories on Flat Spacetime*, JHEP09 (2009) 048. http://arxiv.org/abs/0905.1180.
2. Brett McInnes, *Holography of the Quark Matter Triple Point*, Nuclear Physics B832 (2010) pp. 323-341. http://arxiv.org/abs/0910.4456.
3. Nathan Seiberg and Edward Witten, *The D1/D5 System and Singular CFT*, JHEP9904 (1999) 017. http://arxiv.org/abs/hep-th/9903224.
4. M. Kleban, M. Porrati and R. Rabadan, *Stability in Asymptotically AdS Spaces*, JHEP0508 (2005) 016. http://arxiv.org/abs/hep-th/0409242.
5. Kevin Goldstein, Shamit Kachru, Shiroman Prakash, and Sandip P. Trivedi, *Holography of Charged Dilaton Black Holes*. http://arxiv.org/abs/0911.3586.
6. Chiang-Mei Chen and Da-Wei Pang, *Holography of Charged Dilaton Black Holes in General Dimensions*, http://arxiv.org/abs/1003.5064.
7. Steven S. Gubser and Fabio D. Rocha, *Peculiar Properties of a Charged Dilatonic Black Hole in AdS$_5$*, Phys. Rev. D 81:046001, (2010). http://arxiv.org/abs/0911.2898.
8. D. Garfinkle, G.T. Horowitz and A. Strominger, *Charged Black Holes in String Theory*, Phys Rev D, 43(10): 31403143, (1991).
9. Gary T. Horowitz, *The Dark Side of String Theory: Black Holes and Black Strings*, http://arxiv.org/abs/hepth/9210119.
10. Cheng Zhou Liu, *The Information Entropy of a Static Dilaton Black Hole*, Science in China Series G: Physics, Mechanics & Astronomy, Volume 51, Number 2, February (2008), pp.113-125.
11. Chang Jun Gao and Huang Nan Zhang, *Topological Black Holes in Dilaton Gravity Theory*, Physics Letters B. Volume 612, Issues 3-4, (21 April 2006) pp. 127 - 136.
12. Ted Jacobson, *When is $g_{tt}g_{rr} = -1$?*, Class. Quant .Grav. 24:5717-5719, (2007). http://arxiv.org/abs/0707.3222.
13. S. H. Hendi and A. Sheykhi, *Thermodynamics of Higher Dimensional Topological Charged AdS Black Holes in Dilaton Gravity*. http://arxiv.org/abs/1002.0202.
14. Steven S. Gubser and Indrajit Mitra, *Instability of Charged Black Holes in Anti-de Sitter Space*. http://arxiv.org/abs/hep-th/0009126.

AN INTERNAL SCHWARZSCHILD-DE SITTER SOLUTION?

BORZOO NAZARI

Faculty of Physics, University of Tehran,
End of North Kargar Street Tehran 1439955961, Iran
borzoo.nazari@gmail.com

We reinterpret a class of exact solutions of Einstein's field equations for spherically symmetric perfect fluids (SSPFs) to be an internal Schwarzschild-de Sitter solution. Via this way we will see also that some of well known exact SSPF solutions inspect each other.

1. Introduction

In the passage of spherically symmetric perfect fluid solutions of the Einstein field equations (EFEs) there are some open problems which are very interesting in point of view of mathematical methods. But major problem in the field of exact solutions certainly is that how many SSPF solutions to EFEs exist at all and how many of them are physically acceptable. If there exist even one more physical solution that give us some new physical data, like the ones which related to neutron stars, then it is worthwhile for researchers to spent more time to find a good adjusted physical solution.

There are some recent attempts to generate all of the SSPF solutions via generating function methods and some general methods are developed reaching the above stated goal but none of them can answer the question:

Is there any method to generate all of SSPF solutions via a deterministic and apparently controllable method? By controllable, we mean that the method must give us a criterion that enables us to classify all the solutions. However because of lack of such criterion there are plenty of re-derivations of some solutions in the literature. We denote here such duplication via a simple example.

2. Einstein's Field Equations for a Spherically Symmetric Perfect Fluid

Suppose we have a spherical symmetric perfect fluid with stress-energy-momentum tensor

$$T_{\mu\nu} = \text{diag}(\rho, P, P, P)$$

which is relative to rest frame of a proper time observer and $\eta^{\mu\nu} = \text{diag}(-1,1,1,1)$. The Einstein field equations are

$$G^{\mu\nu} + \Lambda g^{\mu\nu} = \kappa_0 T^{\mu\nu}.$$

By assuming the most general spherically symmetric form

$$ds^2 = -e^{2v}dt^2 + e^{2\lambda}dr^2 + r^2 d\Omega^2,$$

where ϑ and λ are functions of r, we have

$$\begin{aligned}
G_0^0 &: [r(1 - e^{-2\lambda})]'/r^2 = \kappa_0(\rho + \Lambda/\kappa_0) \\
G_3^3 &: \kappa_0(P - \Lambda/\kappa_0) = e^{-2\lambda}[v'' + v'^2 - v'\lambda' - (v' - \lambda')/r] \\
G_1^1, G_2^2 &: [-1 + e^{-2\lambda}(1 + 2rv')]/r^2 = \kappa_0(P - \Lambda/\kappa_0)
\end{aligned} \tag{1}$$

with $'$ denote differentiation with respect to r. Now we investigate a solution to these equations putting density of luminous matter $\rho = \alpha_0/r^2$. This density maybe thought of as distribution of luminous mass in a typical spiral galaxy when we are going distant points from the centre. In fact a more real distribution for this purpose is

$$\rho = \frac{\rho_0}{(1 + (r/r_0)^2)}.$$

So the constant α_0 is recognized as $\rho_0 r_0^2$ in the limit $r \to \infty$. By the way we put the above mass distribution into the right hand side of the first equation of Eq. (1) to be

$$[r(1 - e^{-2\lambda})]'/r^2 = \kappa_0 \rho' \tag{2}$$

where the following definitions has been developed

$$\begin{aligned}
a &\equiv 1 - \kappa_0 \alpha_0 \\
\Lambda &\equiv -3b \\
\rho' &= \frac{1}{\kappa_0}((1 - a)r^{-2} - 3b).
\end{aligned} \tag{3}$$

We have done so far some definitions to recast the form of our equations into the general internal Schwarzschild solution. In fact any internal Schwarzschild solution may be interpreted as an internal Schwarzschild-de Sitter solution provided matching of the internal solution with the external one is possible.

We will see in the appendix that for a wide range of galaxies $1/2 \leq a < 1$. This is a Kuchowicz solution.[5,6]

3. Kuchowicz Solution

If we calculate $e^{-2\lambda}$ from Eq. (2) we have

$$e^{-2\lambda} = 1 - \frac{2m(r)}{r}, \quad m(r) = \frac{\kappa_0}{2}\int_0^r \rho'(r')r'^2 dr'.$$

In fact $m(r)$ can be interpreted as gravitational mass inside the galaxy up to the radial distance r, taking into account the cosmological constant, as same as in the case of the famous external Schwarzschild solution.

Thus we can write down $e^{-2\lambda}$ from Eq. (3) as Ref. 10,

$$e^{-2\lambda} = a + br^2 . \tag{4}$$

This form of $e^{-2\lambda}$ is well-known ansatz of R.C. Tolman, the first one who used this as an ansatz, but it was Kuchowicz that studied it extensively. The form in Eq. (4) has some interesting characteristics. It just recently proved that all known explicit solutions with a linear equation of state satisfy also Eq. (4), except for the Whittaker solution,[7] and this property, individually stress the importance of those forms of $e^{-2\lambda}$ which satisfy Eq. (4).

However if we want to find e^{v} we must know the parameters a and b leading to the Kuchowicz's solution

$$e^{-2\lambda} = a + br^2 , \quad e^{v} = r^{1-\gamma}[c_1(1+V)^\gamma + c_2(1-V)^\gamma] \tag{5}$$

where

$$V(r) = (1 + b/ar^2)^{\frac{1}{2}} , \quad \gamma = \sqrt{2 - a^{-1}}, \quad \frac{1}{2} \le a < 1 .$$

And c_1, c_2 are some constants to be determined.

4. Calculating the Pressure

In order to calculate the pressure we begin with G_1^1 component of Eq. (1). First we must determine the term $1 + 2rv'$. If we simply differentiate both side of Eq. (5) and hold in mind these definitions

$$B(r) = c_1[1 + V(r)]^{\gamma-1} - c_2[1 - V(r)]^{\gamma-1}$$

$$A(r) = c_1[1 + V(r)]^\gamma + c_2[1 - V(r)]^\gamma .$$

We have

$$v'e^{v} = (1 - \gamma)r^{-\alpha}A(r) + \gamma V'(r)r^{1-\alpha}B(r) .$$

And finally arrive at

$$1 + 2rv' = 3 - 2\gamma + \frac{2b\gamma r^2}{aV(r)} \frac{B(r)}{A(r)} .$$

Putting this in G_1^1 component of Eq. (1) the explicit form of the pressure would be as follows

$$\kappa_0 P(r)r^2 = -1 + (a + br^2)\left[3 - 2\gamma + \frac{2b\gamma r^2}{aV(r)} \frac{B(r)}{A(r)}\right] + \Lambda r^2 . \tag{6}$$

This equation can be simplified in a more concise form

$$\kappa_0 P(r) = 2b\gamma V(r)\left[-K(r) + \frac{B(r)}{A(r)}\right]$$

where we defined $K(r)$ as

$$K(r) = \frac{aV(r)}{2b\gamma r^2} \left[-3 + 2\gamma + \frac{1 - \Lambda r^2}{a + br^2} \right].$$

At this stage the two constant c_1 and c_2 are not determined yet and must be determined.

5. Matching the Internal and External Solutions

As we discussed above any full Schwarzschild solution, consist of a set of internal and external solution, which has a correct matching of the two in the surface can be interpreted as a full Schwarzschild-de Sitter solution. However until now we interpret general form in Eq. (4) as an internal Schwarzschild-de Sitter solution and the explicit matching of the internal and external metrics must be done for consistency of the equations. Also some educational points are of interest.

The two metrics are

$$ds_{\text{in}}^2 = -e^{2v}dt^2 + (a + br^2)^{-1}dr^2 + r^2 d\Omega^2$$

$$ds_{\text{out}}^2 = -\left(1 - \frac{2M}{r} - \frac{\Lambda}{3}r^2\right)dt^2 + \left(1 - 2M/r - \frac{\Lambda}{3}r^2\right)^{-1}dr^2 + r^2 d\Omega^2.$$

First we check out the radial component somewhere on the surface which defined by $r = R$:

$$(a + bR^2)^{-1} = \left(1 - \frac{2M}{R} - \frac{\Lambda}{3}R^2\right)^{-1}.$$

If definitions of Eq. (3) are reminded we have

$$M(R) = \frac{\kappa_0 \alpha_0 c^2}{2G} R$$

$$bR^2 = -\frac{\Lambda}{3}R^2$$

$$(7)$$

where the second one is trivial. If we know $\kappa_0 \alpha_0$ we are capable of determining mass distribution of the galaxy.

Secondly we check out the time component. This give us a relation consist of constants c_1 and c_2

$$e^{2v}|_{r=R} = R^{2-2\gamma}A(R)^2 = 1 - 2M/R - \frac{\Lambda}{3}R^2 = a + bR^2,$$

or in a more explicit form we have

$$c_1[1 + V(R)]^\gamma + c_2[1 - V(R)]^\gamma = A(R) = R^{\gamma-1}(a + bR^2)^{\frac{1}{2}}. \qquad (8)$$

To determining c_1 and c_2 completely we must have another relation which is apparently a condition on pressure $P(r)|_{r=R} = 0$. This and the Eq. (6) gives us

$$B(R) = \frac{aV(r)}{2b\gamma r^2} \left[-3 + 2\gamma + \frac{1 - \Lambda r^2}{a + br^2} \right] A(R) = K(R)A(R),$$

or in a more explicit form we have

$$c_1[1 + V(r)]^{\gamma-1} - c_2[1 - V(r)]^{\gamma-1} = K(R)A(R).$$ (9)

6. A Special Case of the Above Solution and Tolman's Solution

If we set $\Lambda = 0$ $(b = 0)$ in the above processes we have

$$e^{-2\lambda} = a$$

$$e^{\nu} = c_1 2^{\gamma} r^{1-\gamma}$$

$$P(r) = \frac{(1-\gamma)^2}{2-\gamma^2} \frac{1}{\kappa_0 r^2},$$

while the equation of state behaves like

$$\frac{P}{\rho} \approx \frac{1}{4}\varepsilon.$$

And according to Appendix, $1 - a = \varepsilon$ is a positive infinitesimal number. The above special case is is exactly the Tolman VI solution,[1] which is given as follows

$$e^{-2\lambda} = (2 - n^2)^{-1}, \quad e^{\nu} = c_3 r^{1-n} - c_4 r^{1+n}$$

$$\kappa_0 \rho(r) = \frac{1-n^2}{2-n^2} \frac{1}{r^2}, \quad \kappa_0 P(r) = \frac{(1-n)^2 A - (1+n)^2 B}{(2-n^2)(A - Br^{2n})} \frac{1}{r^2}.$$

If we set $n = \sqrt{2 - 1/a}$ and $B = 0$ the above special case will recovered.

Acknowledgments

This work is supported by University of Tehran.

Appendix

If we assume the profile

$$\rho = \frac{\rho_0}{1 + (r/r_0)^2} = \frac{\rho_0 r_0^2}{r^2 + r_0^2} \xrightarrow{r \to \infty} \frac{\rho_0 r_0^2}{r^2}.$$ (A.1)

For a typical spiral galaxy then we have a rotation curve depicted by Ref. 11,

$$\nu^2 = \frac{4\pi G \rho_0 r_0^2}{c^2} \left(1 - \frac{r_0}{r} \tan^{-1} \frac{r_0}{r}\right), \quad \nu_\infty^2 = \frac{4\pi G \rho_0 r_0^2}{c^2}.$$

And if we choose the limiting value of velocity for distant points from centre of our galaxy to be typically 1.4×10^5 m/s we see simply that the parameter $a = 1 - 0.46 \times 10^{-6}$ satisfy the relation

$$\frac{1}{2} \leq a < 1.$$

596

References

1. R. C. Tolman, *Phys. Rev.* **55**, 364 (1939).
2. M. Wyman, *Phys. Rev.* **70**, 74 (1946).
3. M. Wyman, *Phys. Rev.* **75**, 1930 (1949).
4. H. A. Buchdahl, *Phys. Rev.* **116**, 1027 (1959).
5. B. Kuchowicz, *Acta Phys. Pol.* **33**, 541 (1968).
6. B. Kuchowicz, *Acta Phys. Pol.* **34**, 131 (1968).
7. M. S. R. Delgaty and K. Lake, arXiv:gr-qc/9809013.
8. B. V. Ivanov, "Relativistic static fluid spheres with a linear equation of state," arXiv:gr-qc/0107032
9. M. R. Finch and J. E. F. Skea, "A review of the relativistic static fluid sphere" (unpublished); http://www.dft.if.uerj.br/usuarios/JimSkea/papers/pfrev.ps
10. H. Stephani, D. Kramer, M. MacCallum, C. Hoenselaers and E. Herlt, *Exact Solutions of Einstein's Field Equations* (Cambridge University Press, Cambridge, England, 2003).
11. D.-E. Liebscher, *Cosmology* (Springer-Verlag, 2005), Chapter 5.

THE ROBUSTNESS OF THE GALAXY DISTRIBUTION FUNCTION TO EFFECTS OF MERGING AND EVOLUTION

ABEL YANG

Department of Astronomy, University of Virginia, Charlottesville, VA 22904, USA

WILLIAM C. SASLAW

Institute of Astronomy, Madingley Road, Cambridge CB3 0HA, UK
and
Department of Astronomy, University of Virginia, Charlottesville, VA 22904, USA

AIK HUI CHAN

Department of Physics, National University of Singapore,
2 Science Drive 3, Singapore 117542
and
Institute of Advances Studies, Nanyang Technological University,
#02-18 60 Nanyang View, Singapore 639673

BERNARD LEONG

Senatus Pte. Ltd, 19 Amoy Street, 02-01, Singapore 069584

We examine the evolution of the spatial counts-in-cells distribution of galaxies and show that the form of the galaxy distribution function does not change significantly as galaxies merge and evolve. In particular, bound merging pairs follow a similar distribution to that of individual galaxies. From the adiabatic expansion of the universe we show how clustering, expansion and galaxy mergers affect the clustering parameter b. We also predict the evolution of b with respect to redshift.

Keywords: Galaxies; statistics; cosmology; theory; large-scale structure of universe.

1. Introduction

The galaxy spatial distribution function $f(N, V)$ is a simple but powerful statistic which characterizes the locations of galaxies in space. It includes statistical information on voids and other underdense regions, on clusters of all shapes and sizes, on filaments, on the probability of finding an arbitrary number of neighbors around randomly located positions, on counts of galaxies in cells of arbitrary shapes and sizes randomly located, and on galaxy correlation functions of all orders. These are just some of its representations.[1,2] Moreover it is also closely related to the distribution function of the peculiar velocities of galaxies around the Hubble flow.[3,4]

A physically motivated form of the distribution function for galaxies in quasi-equilibrium was derived and generalized using a statistical mechanical approach.[5,6] It has also been generalized to systems containing particles of two different masses.[7] In its simplest form, the probability of finding N galaxies in a cell of volume V is given by the gravitational quasi-equilibrium distribution(GQED):

$$f_V(N) = \frac{\overline{N}(1-b)}{N!}\left(\overline{N}(1-b) + Nb\right)^{N-1} \exp\left(-\overline{N}(1-b) - Nb\right) \qquad (1)$$

where $\overline{N}$ is the average number of galaxies in a cell and the clustering parameter $b = -W/2K$ is the ratio of the gravitational correlation energy W to twice the kinetic energy K of peculiar velocities relative to the Hubble flow. Although other distributions have been proposed (e.g. negative binomial; for an early review see Ref. 8), they are generally not physically motivated. Extensive computer simulations (summarized in Ref. 1) designed to test the GQED agree closely with the analytical results. Observations at both low[9] and high[10] redshifts suggest that the form of the distribution function $f_V(N)$ is essentially unchanged over a wide range of redshifts. However, mergers among galaxies could have modified the form of $f_V(N)$ significantly compared to the simple model which involves no mergers at all, and yet they did not. In this paper we examine the robustness of the GQED to merging galaxies.

2. The Positions of Merging Galaxy Pairs

Merging galaxy pairs are often extended structures that would have a different interaction potential than simple spheres or point masses. With a different interaction potential, they may also be distributed differently. Extended structures resulting from mergers will not only have a different interaction potential, they also have a different mass and should be treated differently. To consider how these extended structures influence the GQED, we first consider how they modify the interaction potential. Using the modified potential we can then derive the distribution function for an ensemble of species with a range of masses.

The generalized interaction potential between particles (galaxies) each of which has an isothermal halo is given by[5]

$$\phi(r) = -\frac{Gm^2}{(r^2 + \epsilon^2)^{1/2}} \qquad (2)$$

where G is the gravitational constant, r is the separation between a pair of particles, m is the mass of each particle, and ϵ is a parameter related to the finite size of a galaxy in proper coordinates. We can generalize this interaction potential by factoring out the ϵ terms to get

$$\phi(r) = -\frac{Gm^2}{r}\kappa(\epsilon/r) \qquad (3)$$

where $\kappa(\epsilon/r)$ represents a modification to the Newtonian potential. This modification only affects the potential energy part of the configuration integral given by equation (3) of Ref. 5

$$Q_N(T, V) = \int \cdots \int \exp\left[-\phi(\mathbf{r}_1, \mathbf{r}_2, \ldots, \mathbf{r}_n)T^{-1}\right] d^{3N}\mathbf{r} \qquad (4)$$

where T is the kinetic temperature of the ensemble in units where Boltzmann's constant is unity, and ϕ is the interparticle potential energy of the ensemble. By following the procedure in section 2 of Ref. 5, the potential energy part of the Hamiltonian for a 2-galaxy system becomes

$$Q_2(T, V) = V^2 \left[1 + \frac{3Gm^2}{2T(\overline{n})^{-1/3}}\zeta\left(\frac{\epsilon}{R_1}\right)\right] \qquad (5)$$

where R_1 is the scale where the two-galaxy correlation function is negligible, $\overline{n}$ is the number of particles per unit volume, and

$$\zeta\left(\frac{\epsilon}{R_1}\right) = \int_0^{R_1} \frac{2r}{R_1^2}\kappa\left(\frac{\epsilon}{r}\right) dr \qquad (6)$$

describes how a modification to the potential changes the partition function.

The modification given by equation (6) is analogous to $\alpha(\epsilon/R_1)$ given by equation (16) of Ref. 5, but can describe a generalized modification of the potential rather than the particular modification that arises from an isothermal halo. The effects of a modified potential enter into the distribution function only through the parameter b, which now becomes

$$b_\epsilon = \frac{(3/2)G^3m^6\overline{n}T^{-3}\zeta(\epsilon/R_1)}{1 + (3/2)G^3m^6\overline{n}T^{-3}\zeta(\epsilon/R_1)}. \qquad (7)$$

For an attractive potential, $\kappa(\epsilon/r)$ is is always positive and hence $\zeta(\epsilon/R_1) > 0$ for all values of R_1 so from equation (7), $0 \leq b_\epsilon \leq 1$.

From the thermodynamic variables of the system given by equations (26)-(30) of Ref. 5, we see that the forms of the distribution function $f_V(N)$ and the thermodynamic functions of the system are essentially unchanged by a modified potential although their values are different. This shows that the form of the galaxy distribution function is robust to modified potentials. This also allows us to treat merging galaxy pairs as a separate species and extend the GQED to describe the distribution of merging galaxy pairs.

A merging galaxy pair which is bound will have its dynamical center at its center-of-mass, and will relax to form a single galaxy at its center-of-mass. Hence the position of the centers-of-mass of these merging pairs will be related to the positions of the merged galaxies. While they merge, these merging pairs form an extended system with an external potential given by the vector sum of the external potentials of both galaxies in the merging pair:

$$\phi_2(\mathbf{r}) = -\frac{Gm_1M}{\|\mathbf{r} + \mathbf{x}_1\|} - \frac{Gm_2M}{\|\mathbf{r} + \mathbf{x}_2\|} = -\frac{Gm^2}{\|\mathbf{r}\|}\left(\frac{\|\mathbf{r}\|}{\|\mathbf{r} + \mathbf{x}_1\|} + \frac{\|\mathbf{r}\|}{\|\mathbf{r} + \mathbf{x}_2\|}\right) \qquad (8)$$

where $\mathbf{r}$ is the distance from the center of mass of the merging pair to a more distant galaxy of mass M, and $\mathbf{x}_1$ and $\mathbf{x}_2$ are the distances from each component to the center-of-mass of the system. To approximate the case when mergers between similar-sized galaxies have the dominant effect on the distribution function, we assume that all galaxies have the same mass m so that $m = m_1 = m_2 = M$. From equation (8), the first order approximation to the external potential contains a factor of $2m$. In the case where $\|\mathbf{r}\| \gg \|\mathbf{x}\|$, the first order term dominates and we obtain the modification factor $\kappa_2 \approx 2$ to the potential that arises from treating these merging pairs as single extended particles.

The universe however does not contain only merging pairs. To model the presence of individual galaxies that are not currently merging, we consider a two-species distribution[7] where one species consists of merging pairs, and the other species consists of galaxies that are not merging. For simplicity we consider a system containing two species of different extended particles where species 1 represents individual galaxies with an average mass of m, each with a halo, and species 2 represents bound pairs of merging galaxies with a total mass of $2m$. Using the modification to the GQED described above, the potential between a pair of particles of species 1 (each of which is a galaxy) is

$$\phi_1(r) = \frac{Gm_1^2}{r} \kappa_1(\epsilon_1/r) \tag{9}$$

where $\kappa_1(\epsilon_1/r)$ is a softening factor that arises from an extended halo with a physical extent described by ϵ_1 (e.g. isothermal halo[5]). Likewise, the external potential between a particle of species 1 (a galaxy) and a particle of species 2 (a bound merging pair) is given by

$$\phi_2(r) = \frac{Gm_1 m_2}{r} \kappa_2(\epsilon_2/r) \tag{10}$$

where $m_2 = 2m_1$ because there are two galaxies of mass m_1 in each merging pair, and the modification to the potential is given by $\kappa_2(\epsilon_2/r)$. Since a modified potential only changes b in the single-species distribution function, a modified potential only changes the two-species clustering parameter b_m in the two-species distribution function

$$b_m = \frac{N_1}{N} \frac{\beta \bar{n} T^{-3}}{1 + \beta \bar{n} T^{-3}} + \frac{N_2}{N} \frac{\beta_{12} \bar{n} T^{-3}}{1 + \beta_{12} \bar{n} T^{-3}} = \frac{b}{1 + N_2/N_1} \left(1 + \frac{(N_2/N_1)(\beta_{12}/\beta)}{1 - b + (\beta_{12}/\beta)b} \right) \tag{11}$$

where β and β_{12} are given by

$$\beta = \frac{3}{2}(Gm_1^2)^3 \zeta_2 \left(\frac{\epsilon_1}{R_1} \right) = \frac{3}{2}(Gm_1^2)^3 \int_0^{R_1} \frac{2r}{R_1^2} \kappa_1 \left(\frac{\epsilon_1}{r} \right) dr \tag{12}$$

$$\beta_{12} = \frac{3}{2}(Gm_1 m_2)^3 \zeta_2 \left(\frac{\epsilon_2}{R_1} \right) = \frac{3}{2}(Gm_1 m_2)^3 \int_0^{R_1} \frac{2r}{R_1^2} \kappa_2 \left(\frac{\epsilon_2}{r} \right) dr \tag{13}$$

and b is the single-species clustering parameter given by[5]

$$b = \frac{\beta \bar{n} T^{-3}}{1 + \beta \bar{n} T^{-3}}. \tag{14}$$

The two-species distribution function is thus[7]

$$
f_V(N) = \frac{\overline{N}(1-b)}{N!} \left[\overline{N}(1-b) + Nb\right]^{N_1-1} \left[\frac{\overline{N}(1-b) + (\beta_{12}/\beta)Nb}{1-b + (\beta_{12}/\beta)b}\right]^{N_2}
$$
$$
\times \exp[-\overline{N}(1-b_m) - Nb_m]. \tag{15}
$$

The two-species distribution reduces to the single-species distribution function in the limit $N_2/N_1 \to 0$ and $\beta_{12}/\beta \to 1$. When $N_1 \gg N_2$ and $\overline{N}$ is large, the deviation from the GQED is small because the $\left[\overline{N}(1-b) + Nb\right]^{N_1-1}$ term dominates. Measurements of the fraction of merging pairs in the VVDS catalog[11] suggest that for redshifts of $z \lesssim 1$, $N_2/N_1 \approx 10\%$. Assuming these mergers are between galaxies that are similar in mass, $m_2 \approx 2m_1$ so $\beta_{12}/\beta \approx 8$. For large cells which are a representative sample of the universe, $\overline{N} \gtrsim 100$ and the difference between the two-species distribution and the single-species distribution is small on the level of about 5%. Hence under these conditions, the only significant effect of galaxy mergers in this context is a change in the average mass of a galaxy.

3. Redshift Evolution of b

To determine the change in the clustering parameter b we consider a merging pair of galaxies each of mass m which approach each other with velocities of $\mathbf{v}_1$ and $\mathbf{v}_2$. Since momentum is conserved, the merged galaxy follows the trajectory of the center of mass of the progenitors, and has a final velocity after the merger of $\mathbf{v}_f = (\mathbf{v}_1 + \mathbf{v}_2)/2$. The final velocity of the merged galaxy depends on the detailed dynamics of the system, but by averaging over all orientations, we find that mergers will not change the average kinetic energy of an ensemble. Hence the more important contribution to the evolution of b comes from the change in the positions of galaxies and the expansion of the universe.

We extend the analysis of Ref. 12 to describe an ensemble where galaxies merge by considering the effect of the adiabatic expansion of the universe. The equations of state for internal energy U and pressure P are[1,5]

$$
U = \frac{3}{2}\overline{N}T(1-2b) \tag{16}
$$

and

$$
P = \frac{\overline{N}T}{V}(1-b) \tag{17}
$$

where V is the volume and T is the kinetic temperature of peculiar velocities in units where the Boltzmann constant is 1. Equations (14) and (17) can be combined to get[12]

$$
b = \frac{b_0\overline{n}T^{-3}}{1 + b_0\overline{n}T^{-3}} = b_0 P T^{-4}. \tag{18}
$$

The analysis of section 2 generalizes equation (18) to extended objects with the generalized form of b given in equation (7). By comparing equations (18) and (7) we see that b_0 is given by

$$b_0 = \frac{3}{2}G^3\overline{m}^6\zeta\left(\frac{\epsilon}{R_1}\right) \tag{19}$$

where $\overline{m}$ is the average mass of a galaxy, and ζ is a function of order unity that depends weakly on ϵ/R_1.

Because R_1 is defined as the scale where the two-galaxy correlation function becomes negligible, the universe is approximately uniform averaged over scales $\gtrsim R_1$. Here we take R_1 to be the scale at which the two-galaxy correlation function begins to decrease faster than a power law. Measurements from the 2DFGRS[13] have indicated that R_1 is about $12h^{-1}$ Mpc at which the two-galaxy correlation function is of the order 10^{-2}. Here $h = H_0/100$ is the reduced Hubble constant. We note that such cells are large enough to contain individual field galaxies and clusters of galaxies, and hence would be an approximately representative sample of the universe. Assuming that on such scales, galaxies have isotropic average velocities, then for cells with a radius larger than R_1, galaxies are as likely to enter a cell as they are to leave a cell. With this assumption, the total mass in each comoving cell M_c would be approximately constant, and each cell would have on average $\overline{N} = M_c/\overline{m}$ galaxies and $d(\overline{m}\overline{N}) = 0$. Therefore $\overline{N} \propto \overline{m}^{-1}$ and b_0 can be written as a function of $\overline{N}$ instead of $\overline{m}$. This transforms equation (18) into the form

$$b = b_0(\overline{N})PT^{-4} = \frac{3}{2}G^3\left(\frac{M_c}{\overline{N}}\right)^6\zeta\left(\frac{\epsilon}{R_1}\right)PT^{-4}. \tag{20}$$

Differentiating equation (20) with respect to $\overline{N}$ gives

$$\frac{db}{b} = -6\frac{d\overline{N}}{\overline{N}}\left(1 + \frac{\overline{N}}{6R_1}\left(\frac{\partial\ln\zeta(\epsilon/R_1)}{\partial\overline{N}}\right)_{T,P}\right) = -6\frac{d\overline{N}}{\overline{N}}(1 + \zeta_\star) \tag{21}$$

from which we define the term

$$\zeta_\star = \frac{\overline{N}}{6R_1}\left(\frac{\partial\ln\zeta(\epsilon/R_1)}{\partial\overline{N}}\right)_{T,P}. \tag{22}$$

In general, we do not rule out the possibility that $\zeta(\epsilon/R_1)$ may indirectly depend on $\overline{N}$, hence the $\partial\ln\zeta(\epsilon/R_1)/\partial\overline{N}$ factor in $\zeta_\star$ may be nonzero.

In the case of adiabatic expansion, equations (16) and (17), give

$$0 = dU + PdV = \frac{3}{2}(1 - 2b)\left[Td\overline{N} + \overline{N}dT|_{\overline{N},P}\right] - 3\overline{N}Tdb + \overline{N}T(1-b)\frac{dV}{V}. \tag{23}$$

Equation (20) implies

$$dT|_{\overline{N},P} = -\frac{Tdb}{4b} \tag{24}$$

and hence using $dV/V = 3dR/R$ where R is the scale length of the universe, we have

$$0 = \frac{3}{2}(1 - 2b)Td\overline{N} - \frac{3}{2}(1-2b)\overline{N}T\frac{db}{4b} - 3\overline{N}Tdb + 3\overline{N}T(1-b)\frac{dR}{R}. \tag{25}$$

Rearranging the terms and using equation (21), we find in terms of redshift $z \propto 1/R - 1$

$$\frac{db}{dz} = -\frac{1-b}{1+z} \left(\frac{1+6b}{8b} + \frac{1-2b}{12b\zeta_\star} \right)^{-1}. \tag{26}$$

To illustrate how mergers contribute to the time evolution of b, we consider a simple model of a galaxy. In our model, galaxies have isothermal halos with a characteristic radius ϵ, and all galaxies have the same density so that in a cell of total mass M_c

$$\epsilon = a \, (\overline{m})^{1/3} = a \left(\frac{M_c}{\overline{N}} \right)^{1/3} \tag{27}$$

for some constant of proportionality a such that ϵ depends on the average mass of a galaxy. We use the GQED and form of ζ for such a case from Ref. 5 to obtain the constraint $-1/18 \leq \zeta_\star \leq 0$. The extremes of this constraint gives

$$\frac{db}{dz} = -\frac{1-b}{1+z} \left(\frac{24b}{5+14b} \right) \tag{28}$$

for the case where $\zeta_\star = 0$ and

$$\frac{db}{dz} = -\frac{1-b}{1+z} \left(\frac{136b}{29+78b} \right) \tag{29}$$

for the case with $\zeta_\star = -1/18$ with all other cases occurring in between equations (28) and (29).

Since $0 \leq b \leq 1$, we compare the two cases numerically and find that the difference between the two cases is small at less than 2%. This result tells us that although galaxy mergers can influence the time evolution of b, their influence mostly results from changes of the number of galaxies in a cell.

4. Conclusion and Future Work

We have established that the effects of galaxy mergers leave the form of the galaxy distribution function essentially unchanged and just alter the parameters of the counts in cells distribution. In particular, by describing bound merging pairs as objects with a modified gravitational potential, we obtain a modified form of the two-species counts in cells distribution[7] and show that it only changes the counts in cells distribution slightly from the single-species result given by equation (1).

As a result of mergers, the clustering parameter b increases with time, and we have shown that it depends very weakly on the physical extent of a galaxy and the scale R_1 at which the two point correlation function is negligible. The effect of the physical extent of a galaxy changes db/dz by less than 2%, which shows that the evolution of b depends mainly on the adiabatic expansion of the universe and the change in the number of galaxies from mergers.

These results show that even when we take galaxy mergers into account, we can not only reproduce the GQED but also trace the evolution of the clustering parameters. However, an analysis of the GOODS catalog[10] indicates a large variation between the North and South fields and suggests that the sample is probably too small to draw any meaningful conclusions about the evolution of b at high redshift. Future surveys however may provide sufficiently large samples at high redshifts to test our predicted evolution of b.

Acknowledgments

A. Yang is grateful for the support from the National University of Singapore and the Institute of Astronomy of the University of Cambridge where part of this work was done. A. H. Chan would like to thank the department of History and Philosophy of Science, Cambridge University and Nanyang Polytechnic for kind hospitality where part of the initial work was done.

References

1. W. C. Saslaw, *The distribution of the galaxies : gravitational clustering in cosmology* (Cambridge, UK: Cambridge University Press, 2000).
2. W. C. Saslaw and A. Yang, Statistical Mechanics of the Cosmological Many-body Problem and its Relation to Galaxy Clustering, in *Lecture Notes of the Les Houches Summer School: Long-Range Interacting Systems*, eds. T. Dauxois, S. Ruffo and L. F. Cugliandolo (Oxford, UK: Oxford University Press, February 2010).
3. W. C. Saslaw, S. M. Chitre, M. Itoh and S. Inagaki, *Astrophys. J.* **365**, 419(December 1990).
4. B. Leong and W. C. Saslaw, *Astrophys. J.* **608**, 636(June 2004).
5. F. Ahmad, W. C. Saslaw and N. I. Bhat, *Astrophys. J.* **571**, 576(June 2002).
6. F. Ahmad, W. C. Saslaw and M. A. Malik, *Astrophys. J.* **645**, 940(July 2006).
7. F. Ahmad, M. A. Malik and S. Masood, *Intl. J. Modern Physics D* **15**, 1267 (2006).
8. J. N. Fry, *Astrophys. J.* **306**, 358(July 1986).
9. G. R. Sivakoff and W. C. Saslaw, *Astrophys. J.* **626**, 795(June 2005).
10. H. Rahmani, W. C. Saslaw and S. Tavasoli, *Astrophys. J.* **695**, 1121(April 2009).
11. L. de Ravel, O. Le Fèvre, L. Tresse, D. Bottini, B. Garilli, V. Le Brun, D. Maccagni, R. Scaramella, M. Scodeggio, G. Vettolani, A. Zanichelli, C. Adami, S. Arnouts, S. Bardelli, M. Bolzonella, A. Cappi, S. Charlot, P. Ciliegi, T. Contini, S. Foucaud, P. Franzetti, I. Gavignaud, L. Guzzo, O. Ilbert, A. Iovino, F. Lamareille, H. J. Mc-Cracken, B. Marano, C. Marinoni, A. Mazure, B. Meneux, R. Merighi, S. Paltani, R. Pellò, A. Pollo, L. Pozzetti, M. Radovich, D. Vergani, G. Zamorani, E. Zucca, M. Bondi, A. Bongiorno, J. Brinchmann, O. Cucciati, S. de La Torre, L. Gregorini, P. Memeo, E. Perez-Montero, Y. Mellier, P. Merluzzi and S. Temporin, *Astronomy & Astrophysics* **498**, 379(May 2009).
12. W. C. Saslaw, *Astrophys. J.* **391**, 423(June 1992).
13. E. Hawkins, S. Maddox, S. Cole, O. Lahav, D. S. Madgwick, P. Norberg, J. A. Peacock, I. K. Baldry, C. M. Baugh, J. Bland-Hawthorn, T. Bridges, R. Cannon, M. Colless, C. Collins, W. Couch, G. Dalton, R. De Propris, S. P. Driver, G. Efstathiou, R. S. Ellis, C. S. Frenk, K. Glazebrook, C. Jackson, B. Jones, I. Lewis, S. Lumsden, W. Percival, B. A. Peterson, W. Sutherland and K. Taylor, *Monthly Notices of the Royal Astronomical Society* **346**, 78(November 2003).

SEARCH FOR COSMIC STRINGS IN THE COSMOS SURVEY — STUDIES PERTAINING GALAXY-GALAXY CORRELATION AND CROSS-CORRELATION

TENG PO-WEN IVAN

Department of Physics, National University of Singapore,
2 Science Drive 3, Singapore 117542

The COSMOS survey is investigated for the presence of cosmic strings, indicated by observations of galaxy pairs that are consistent with the gravitational lensing signature of a cosmic string. Employing a new technique combining the simulation of cosmic strings at the catalog level and relying on the correlation and cross-correlation of galaxy pairs to deduce their morphological similarity, no evidence indicating the existence of straight and static cosmic strings is observed at tile position 55 of the COSMOS survey for a string with energy-density/relative-tilt of 5" and redshift of 1.

1. Introduction

1.1. *Formation of cosmic strings*

Cosmic strings are hypothetical one-dimensional topological defects thought to have been formed in the early universe during the first fraction of a second after the Big Bang, as a result of phase transitions in different regions of spacetime. The existence of cosmic strings will undoubtedly provide solid evidence towards the understanding of the early universe, especially during the Planck epoch and earlier.

Present theories about the early universe propose an intensely hot, dense and violent environment, whereby such a dense primordial soup of matter (the first matter to exist in the universe) underwent a series of phase transitions at high temperatures in different regions of spacetime present after the expansion from the singularity.[1] These phase transitions involve the breaking and restoration of symmetry, as described by quantum field theory.[2] The effects of high temperature should be accounted for, as a result of the early universe being very hot and dense immediately after the Big Bang. In this section, we shall consider the Abelian Higgs model[2] and assume that the cosmic strings formed are non-superconducting and local (i.e. the Lagrangian describing the energy density of the string is invariant under a symmetry transformation which may be different at every point),[1] and in addition take into account that the phase transitions are occurring in a radiation-dominated FRW universe.[1]

The early universe may be thought of as being in the symmetric phase when it was very hot and dense, with no strings being formed initially. However, when the temperature dropped below what was known as the critical temperature, T_c, as it cooled, the Higgs field (a field hypothetically responsible for the genesis of inertial mass in matter)[2] in most of the expanding spacetime regions of the universe would have acquired a non-zero average value, and the symmetric state of the universe became unstable. Such instability in the symmetric state then led to universe going into a broken-symmetric state, giving rise to phase transitions. It may hence be said that T_c is the temperature for triggering cosmological phase transitions.

After phase transitions, cosmic strings possess string tension μ and are moving in a very dense environment, which causes their motion to be heavily damped. Such heavy damping on the motion of cosmic strings may be attributed to friction, which may be accounted for by the relatively large energy density of the surrounding hot dense matter in the radiation-dominated era of the early universe, as well as momentum transfer due to scattering between the strings and the matter.[1] Then with the temperature decreasing further with the expansion of the universe, the heavy damping experienced by the cosmic strings soon becomes negligible. The string motion becomes independent of anything else in the universe, and then the strings acquire relativistic speeds.

The cosmic strings are expected to be stretched or shrunken but they cannot break, and additionally they may exchange partners or intercommute when they meet.[1] While the stretching of the cosmic strings result in straightening of any kinks that may be present on the strings, intercommuting between strings gives rise to the formation of new kinks. It is proposed that cosmic strings are expected to dominate the energy density of the universe,[1] but however such do not take place to the formation and decay of small loops of the strings themselves, giving rise to the loss of energy. When a string intersects itself, it cuts off a closed loop, and once formed the loop is doomed until it reconnects with another longer piece of string. After that, the loop oscillates, gradually losing energy by gravitational radiation until it disappears.

Cosmic strings are also expected to be involved in generating primordial density perturbations, and such perturbations are important for the formation of clusters and galaxies.

1.2. Gravitational effects of cosmic strings

Any plausible observational evidence for the existence of cosmic strings is based heavily on the interactions of cosmic strings with gravity. While the detection of cosmic strings may also be done through searching for discontinuities in the temperature of the CMBR,[3] however, possible observations of cosmic strings at present depend on their distinctive gravitational lensing effects, which this paper will focus on.

The space-time around a straight and static cosmic string is very unusual, such that the equality between energy per unit length μ and tension of the string mean both quantities are cancelled out, and nearby masses experience no gravitational acceleration towards the string. The space-time near a cosmic string is locally flat

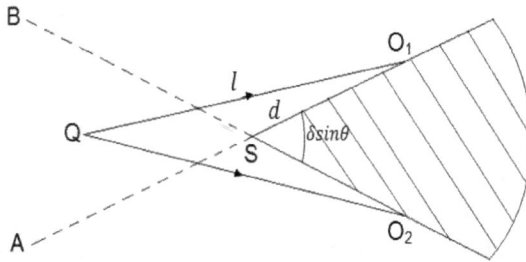

Fig. 1. Gravitational lensing by a cosmic string (Wedge is shaded with black lines).

and globally curved, and the space around the string itself is conical in shape with the curvature localised entirely within the string. Assuming that $\frac{G\mu}{c^2} \ll 1$, the metric for a string along the z-axis in space is given by Eq. (1) as shown:

$$ds^2 = dt^2 - dz^2 - d\rho^2 - \left(1 - 8\frac{G\mu}{c^2}\right)\rho^2 d\varphi^2 \tag{1}$$

where the local flatness of space-time near the string can be made explicit by introducing $\bar{\varphi} = (1 - 4G\mu)\varphi$ and re-expressing the metric according to Eq. (2):

$$ds^2 = dt^2 - dz^2 - d\rho^2 - \rho^2 d\bar{\varphi}^2 \ . \tag{2}$$

Since $0 < \bar{\varphi} < 2\pi - \delta$, where δ = conical deficit angle given by Eq. (3):

$$\delta = \frac{8\pi G\mu}{c^2} \ . \tag{3}$$

For gravitational lensing by cosmic strings, the above relations suggest that a straight static cosmic string acts like a cylindrical gravitational lens, and the conical nature of space around the string results in the formation of lensed double images of objects behind the string, one on either side of it. The angular separation between the images is also comparable to the deficit angle δ. A figure of gravitational lensing by a cosmic string is shown in Fig. 1 as follows:

As observed in Fig. 1, the observer is represented by points O_1 and O_2 on opposite sides of the shaded wedge, which represents the conical space around a string. The half-space above the line ASO_1 or below the line BSO_2 is seen when the observer looks above or below the string S respectively. Intuitively, it can be deduced that all objects falling within the area BSA have duplicated images. For example, taking Q to be an object within area BSA, light rays from the object intersect after passing on opposite sides of the string and the observer will see two images of the object. The angular separation between the images is then given by the sum of the angles QO_1S and QO_2S, and letting α = angular separation as shown in Eq. (4):

$$\alpha = \frac{8\pi G\mu}{c^2} \frac{l}{d+1} \sin\theta \ , \tag{4}$$

where l and d are the distances from the string to the object and to the observer respectively, and $\theta = $ the angle between the string and the line of sight of the observer. This relationship is also applicable to a string at rest with respect to the Hubble flow in an expanding universe, with both l and d then taken to be co-moving distances. For $\theta \neq \frac{\pi}{2}$, the plane of Fig. 1 does not cross the string at right angles and $\angle O_1 S O_2 = \delta \sin \theta$.

Supposing an extended object, for instance a galaxy, crosses the boundary of area ASB, the observer should observe one full image of the object, while the other image should include only the part lying inside ASB, and thus possess a sharp edge.[1] It should also be noted that in comparison with the lensing patterns of ordinary gravitational lenses, gravitational lensing by cosmic strings induces no magnification or demagnification, and the two images are of equal magnitude unless one of the two is only a partial image itself. Commonly, Eq. (4) is written in the form of Eq. (5) as shown:

$$\theta = \frac{D_{ls}}{D_s} \delta \sin \beta, \qquad (5)$$

where $\frac{D_{ls}}{D_s} = \frac{l}{d+l}$ and $\delta = \frac{8\pi G \mu}{c^2}$. More accurately, D_s is the angular diameter distance of the object from the observer, and D_{ls} is the angular diameter distance of the source from the string.[1]

A pictorial diagram showing gravitational lensing by a straight static cosmic string is shown in Fig. 2 as follows.

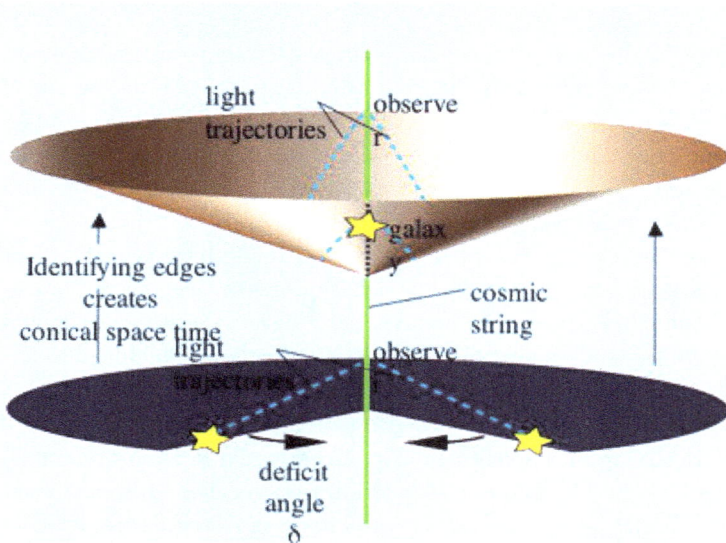

Fig. 2. Gravitational lensing of a galaxy by a straight static cosmic string;[4] if a string lies between an observer and a galaxy, light from the galaxy travels in two paths around the string, hence the observer will see an identical pair of galaxies, which are two distinct images of the same object.

There are several features of gravitational lensing by cosmic strings. One important feature is the linear nature of the lens itself which, owing to a long cosmic string, is likely to induce many lensed pairs arranged in a roughly linear array, while loops of cosmic strings would induce several lensed pairs within a small region of the sky.[1]

This paper will specifically focus on cosmic strings of a static and straight nature. Therefore, to detect straight and static cosmic strings, the methodology employed in this paper would be to search for all lensed galaxy pairs that are morphologically similar with opening angles smaller than 15". This will be further discussed in Sec. 2.

2. Methodology

2.1. *Data sample*

The COSMOS survey comprises 81 fits images taken using Advanced Camera for Surveys (ACS) onboard the Hubble Space Telescope (HST), which are observed using the F814W (I-band) filter for the Version 1.3 release.[5] In the analysis, only the mosaic image at tile position 55 and its accompanying weight map would be used. This position is chosen as it is located at the centre of the entire survey field and assumed to possess a homogeneous number of objects present.

2.2. *Identification of sources*

For the identification of sources SExtractor (Source Extractor) is used, namely for processing the fits image file to identify required sources based on parameters set, and also to calculate photometric data associated with the identified sources. It has been the software of choice in the astronomical and cosmological community for locating objects such as stars and galaxies based on variables such as their shape and brightness.

Essentially, the catalog generated by SExtractor v2.8.6[6] is used for the analysis after running the fits image with it, based on the following input parameters.

Table 1. Input parameters for SExtractor's default.sex configuration file.

Parameter	Value
PIXEL_SCALE	0.05 (according to COSMOS 0.05"/pixel)
DETECT_MINAREA	9
DETECT_THRESH	1.0
ANALYSIS_THRESH	1.0
DEBLEND_NTHRESH	32 (default value)
DEBLEND_MINCONT	0.1
MAG_ZEROPOINT	25.937
WEIGHT_TYPE	MAP_WEIGHT

2.3. *Selection of galaxies and galaxy–galaxy correlation*

The catalog generated by SExtractor for the fits image based the input parameters as laid out in Sec. 2.3. contains 35,467 objects in total. To remove stars and spurious detections from the catalog, only galaxies with MAG_AUTO less than 26.5 was accepted and objects with CLASS_STAR greater than 0.9 are removed. Setting the criterion follows the assumption that stars are not gravitationally lensed by cosmic strings.

To determine the morphological similarity between each potential pair of galaxies that has been determined in the previous section, the correlation and cross-correlation of the two galaxy images in the pair would be calculated. The galaxies would be assessed for similarity, based on their brightness and shape as determined by SExtractor's photometric calculations. The correlation(CORR) and cross-correlation(XCORR) are calculated based on the following Eqs. (6) and (7):

$$\mathrm{CORR} = \frac{\sum I_1(x_i, y_i)^2 - \sum I_2(x_i, y_i)^2}{\sum I_1(x_i, y_i)^2 + \sum I_2(x_i, y_i)^2} \,, \tag{6}$$

and

$$\mathrm{XCORR} = \frac{2\sum I_1(x_i, y_i)^2 * I_2(x_i, y_i)^2}{\sum I_1(x_i, y_i)^2 + \sum I_2(x_i, y_i)^2} \,, \tag{7}$$

where $I(x_i, y_i)$ is the intensity of each pixel in the galaxy and subscripts 1 and 2 refer to the galaxies being correlated or cross-correlated. It may be noted that identical galaxies will give a perfect cross-correlation of 1, while those which are totally dissimilar will have a cross-correlation of 0. The distribution for CORR vs XCORR is as shown in Fig. 3.

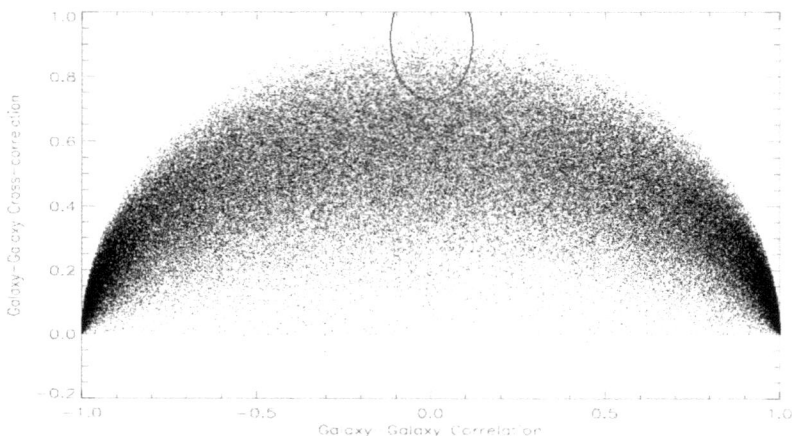

Fig. 3.   CORR-XCORR distribution for random galaxy pairs in the mosaic image of tile position 55 of the COSMOS survey. Galaxy pairs with optimal CORR and XCORR values within the half-elliptical line are defined as matched galaxy pairs.

2.4. *Selection of matched galaxy pairs*

The matched galaxy pairs are defined as the galaxy pairs with optimal CORR and XCORR values within the half-elliptical line, as shown in Fig. 3. Such a cut on Fig. 3 serves to maximize the cosmic string signal pairs relative to the background pairs, and also fits into the definition of morphologically similar galaxies by selecting galaxy pairs with optimal and reasonable corresponding CORR and XCORR values. Additionally, only galaxy pairs with opening angles smaller than 15" would be considered. After applying both cuts, 9,341 matched galaxy pairs are obtained, as compared to the total 117,996 pairs that have been randomly generated with objects from the catalog.

2.5. *Simulation of cosmic string signals*

The simulation of lensed galaxy pairs in the presence of a cosmic string is done by overlaying cosmic strings of a specific redshift and energy-density/relative-tilt across a region of interest on the fits image. This is simply achieved by using the density of galaxies in the catalog to monte carlo the number of pairs predicted to exist from the cosmic strings crossing the corresponding image statistically. For this paper, only a simulated cosmic string with energy-density/relative-tilt of 5" and redshift of 1would be generated for the analysis.

2.6. *Distribution of matched galaxy pairs*

The binned distribution of the matched galaxy pairs is as shown in Fig. 4. The shape of the background in Fig. 4 is characterized by the distribution of all possible pairs of galaxies generated from the catalog, and does not take into account the size and shape of the galaxies. Additionally, it should be noted that cosmic strings with masses large enough to create opening angles greater than 7" have been ruled out,[7,8]

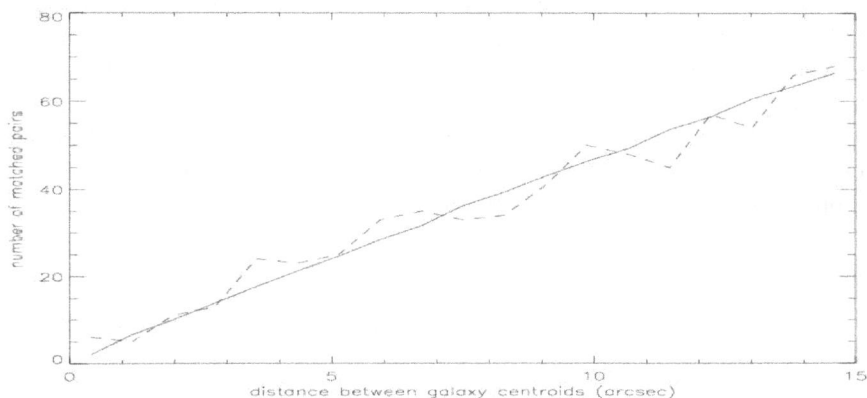

Fig. 4. Binned distribution of matched galaxy pairs (dashed line) for tile position 55 of the COSMOS survey, as compared to the background (solid line).

612

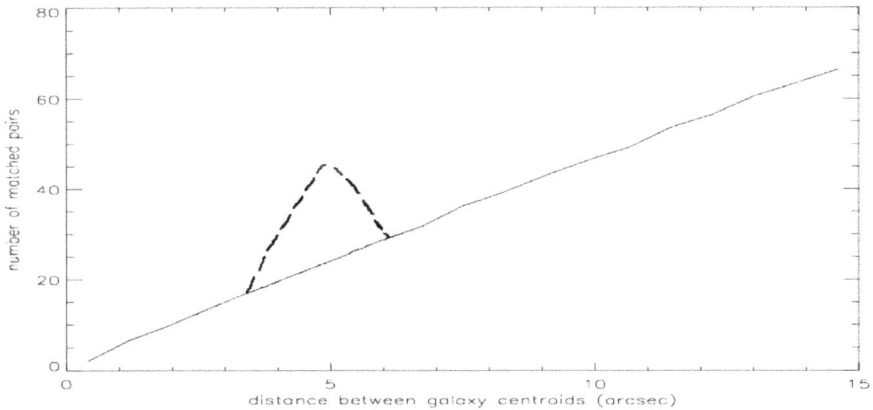

Fig. 5. Binned distribution of simulated matched galaxy pairs (dashed line) for tile position 55 of the COSMOS survey, as compared to the background (solid line).

hence the background distribution in Fig. 4 has been normalized to the number of measured matched pairs between 7" and 15". This serves to give a more accurate estimation of the background at opening angles smaller than 7". Comparing this with the binned distribution of galaxy pairs based on the simulation described in Sec. 2.5 and as shown in Fig. 5 where it may be observed that there is an excess of galaxy pairs at approximately 5", it may be concluded that there is no significant evidence for an excess of pairs at opening angles in the region of 5" for the catalog.

3. Conclusion

Based on the methodology employed, there is no evidence for the existence of the gravitational lensing signature of a straight and static cosmic string with energy-density/relative-tilt of 5" and redshift of 1 in tile position 55 of the COSMOS survey. However, it may be noted that further investigations on other regions of the COSMOS survey are still being carried out, and accompanying results would be reported in future. Additionally, this analysis has highlighted the advantage of simulating cosmic strings at the catalog level, which is the quick embedding of many cosmic strings as required in order to get an accurate estimation of the average number of lensed galaxies observed from a specific set of cosmic string parameters, based on their redshift and energy-density/relative-tilt.

References

1. A. Vilenkin and E. P. S. Shellard, *Cosmic Strings and Other Topological Defects* (Cambridge University Press, 1980).
2. C. Itzykson and J. Zuber, *Quantum Field Theory* (McGraw-Hill, 1995).
3. J. C. Mather *et al.*, *Astrophys. J.* **397**(2): 420, Preprint No. 92-02 (1992).
4. Smoot Cosmology Group, Gravitational lensing by a cosmic string.
   http://aether.lbl.gov/CSbackgrnd.html

5. COSMOS Public Archives, Caltech. http://irsa.ipac.caltech.edu/data/COSMOS
6. E. Bertin, SExtractor v2.8.6 User Manual http://terapix.iap.fr/rubrique.php?id_rubrique=91 (2006)
7. M. Wyman, L. Pogosian and I. Wasserman, *Phys. Rev. D* **68**, 023506 (2003).
8. E. Jeong and G. Smoot, *Astrophys. J.* **624**, 21 (2005).

SEARCH FOR RADIO BURSTS FROM UHE NEUTRINOS

R. G. STROM[*]

Astronomy Group, ASTRON, Postbus 2,
7990 AA Dwingeloo, Netherlands
strom@astron.nl

O. SCHOLTEN[†], K. SINGH[‡] and R. AL YAHYAOUI[§]

Kernfysisch Versneller Instituut, University of Groningen,
Zernikelaan 25, 9747 AA Groningen, Netherlands
[†]*scholten@kvi.nl*
[‡]*k.singh@astro.ru.nl*
[§]*R.el.Yahyaoui@student.rug.nl*

S. BUITINK[¶] and H. FALCKE[‖]

Deptartment of Astrophysics, IMAPP, Radboud University,
P O Box 9010, 6500 GL Nijmegen, Netherlands
[¶]*sbuitink@lbl.gov*
[‖]*falcke@astro.ru.nl*

J. BACELAR

ASML Netherlands BV, De Run 6501,
5504 DR Veldhoven, Netherlands
bacelar@kvi.nl

R. BRAUN[**], A. G. DE BRUYN[††] and B. STAPPERS[‡‡]

Astronomy Group, ASTRON, Postbus 2,
7990 AA Dwingeloo, Netherlands
[**]*Robert.Braun@csiro.au*
[††]*ger@astron.nl*
[‡‡]*ben.stappers@manchester.ac.uk*

Particle cascades initiated by ultra-high energy (UHE) neutrinos in the lunar regolith will emit an electromagnetic pulse with a time duration of the order of nanoseconds through a process known as the Askaryan effect. It has been shown that in an observing window around 150 MHz there is a maximum chance to detect this radiation with radio

[*]Also affiliated with the University of Amsterdam, Netherlands, & James Cook University, Australia.
[**]Present address: ATNF, Astrophysics Group, P O Box 76, Epping NSW 1710, Australia.
[††]Also affiliated with the Kapteyn Institute, University of Groningen, Netherlands.
[‡‡]Present address: School of Physics & Astronomy, The University of Manchester, Alan Turing Building, Manchester, M13 9PL, United Kingdom.

telescopes commonly used in astronomy. In 50 hours of observing time with the Westerbork Synthesis Radio Telescope array we have set a new limit on the flux of neutrinos, summed over all flavors, with energies in excess of 4×10^{22} eV.

Keywords: Ultra-high-energy neutrinos; cosmic rays; radio astronomy.

1. Introduction

Energetic cosmic rays (CRs) have been detected up to energies of some 10^{21} eV.[1] The CR particle spectrum is approximately a power law over most of the relevant energy range, $N(E)\mathrm{d}E \propto E^{-2.7}\mathrm{d}E$, although there are deviations near 10^{16} eV (the "knee"), and in the range 10^{18}–10^{19} eV (the "ankle"). Near the upper end of the energy range, for $E > 6 \times 10^{19}$ eV, ultrahigh energy (UHE) protons interacting with photons from the cosmic microwave background (CMB) will produce pions which carry away a large fraction of the proton energy, a process called the Greisen-Zatsepin-Kuzmin (GZK) effect.[2,3] The charged pions decay to produce UHE neutrinos, the detection of which is our goal. One consequence of the GZK effect is that any UHE protons must have a "local" origin.

Attempts to detect UHE CRs have looked for the Čerenkov radiation such energetic particles produce in various transparent media. The low expected particle fluxes require the use of physically large detectors, and a number of attempts have turned to natural environments: the ice caps of Antarctica and Greenland,[4,5] and our thick atmosphere at low elevations.[6] The emission of coherent microwave radiation by the Askaryan effect[7] has been the observational goal of Refs. 4 and 5. It has been pointed out[8] that the Moon offers an even larger detector, and the Askaryan coherent radiation should peak at a wavelength equal to the CR shower width, which is about 10 cm. Most searches have been done near this wavelength, thus far only providing upper limits.[9–11]

2. Advantages of Searching at Lower Frequencies

On the face of it, there would appear to be little reason to search for radio pulses from UHE CRs at lower frequencies. Near a wavelength of 10 cm (3 GHz) very sensitive radio astronomy receivers exist, a number of large (diameter > 50 m), highly efficient parabolic antennas are available, and the background radiation level (from our Galaxy and the atmosphere) is relatively low. For pencil beam instruments, however, the brightness temperature of the Moon itself largely determines the system sensitivity.

At low frequencies, on the other hand, the lunar regolith is more transparent to radio waves. As the wavelength approaches the shower length (of several meters) in the regolith, the Čerenkov radiation becomes practically isotropic.[12] The result of this is that more of the emission from particles striking the Moon will be observable. We have consequently made a series of observations at frequencies near 150 MHz, using the Westerbork Synthesis Radio Telescope (WSRT) with its standard low-frequency receiver system (LFFE) and pulsar backend (PuMa II).

The WSRT is an east-west array of fourteen 25 m parabolic dishes, mainly used for mapping radio sources by employing the technique of aperture synthesis.[13] For our observations, eleven of the elements equally spaced over a distance of 1.4 km have been utilized, giving the equivalent of a 70 m dish. The LFFE system was set up to observe at four frequencies, 123, 137, 151 and 165 MHz, with full polarization. In this "tied-array" observing mode, signals from the eleven elements are coherently combined for each polarization channel, and then sent to the pulsar backend. PuMa II[14] provides up to eight 20 MHz bands, each of which can be further subdivided into narrow-band channels in a very flexible way. The signals have been combined so as to provide us with two independent fan beams and full polarization information (all four Stokes parameters). The beams were pointed at opposite halves of the Moon, and each covers about one-third of the lunar disk. In addition to thereby monitoring more of the Moon's surface, this provides some discrimination against interference. The effective bandwidth for each beam was 65 MHz.

3. Results and Discussion

A major concern at all radio frequencies is the presence of man-made interference (RFI). Its nature is that it occupies narrow bandwidths and is quite variable. The first step in the data reduction process is therefore to split the recorded signals into blocks of 0.1 s duration, Fourier transform the time series to produce a spectrum and apply a polynomial fit to the resulting amplitudes. The raw spectral data are then compared with the fit, and any peaks which differ by 50% or more are considered to be RFI. Since the impulsive signal from UHE particle radio bursts should be broadband, this filtering should selectively remove only unwanted man-made signals. The resulting loss of data amounts to less than 2%. In four observation runs, the number of RFI triggers was very high; they have been excluded from further consideration.

The Earth's ionosphere disperses any signals which pass through it, imposing an additional signature upon extraterrestrial radio bursts. Using measurements of total electron content (TEC), an estimate is made of the amount of dispersion expected at the time and location of every observation, and this is applied to each data set. Allowance is made for uncertainties in the TEC values used, which will have the effect of broadening any pulses found. The RFI-filtered and dispersion-corrected time series are then searched for broadband pulse candidates. Any triggers, if they are genuine UHE-induced lunar bursts, have to have counterparts in all four frequency bands, within a time interval set by the uncertainty in the TEC value, and without corresponding pulses in the other beam. In the first instance, a handful of events seemed to pass all these requirements, and they were investigated further. Closer examination showed, however, that none of them actually satisfied all of the requirements: a weak counterpart in the other beam had been missed; a probable noise peak had been picked up as a signal; etc. For a more detailed discussion, see Refs. 15 and 16.

In the final analysis, we are only able to set an upper limit to the strength of bursts of lunar origin. Because we have no direct method for estimating the intensity of impulsive signals in our data, we have to compare their strength to the average noise level in our observations, and estimate the expected noise on the basis of receiver sensitivity, lunar thermal emission, etc. The result is that for UHE neutrinos with energies ranging from about 2×10^{23}–2×10^{24} eV, we obtain a flux density limit of 10^{-6} GeV cm^{-2} sr^{-1} s^{-1}. This is nearly an order of magnitude improvement upon the limit set by the FORTE experiment,[5] and is similar to the upper limit found by ANITA'08 at an energy of 10^{21} eV.[4] Both of these upper limits remain over an order of magnitude above the Waxman-Bahcall limit for UHE neutrinos.[17]

4. Conclusions

In slightly less than fifty hours of usable observing time with the WSRT, we have been able to improve upon earlier limits to the UHE neutrino flux density. The value obtained, 10^{-6} GeV cm^{-2} sr^{-1} s^{-1}, has a considerable uncertainty which we estimate at $\pm50\%$. This includes measurement errors, but also uncertainties which arise from the neutrino cross section; the density, attenuation length and stopping power of the lunar regolith; and geometrical factors. Although RFI constitutes a significant contaminant of the signals detected, our observations show that it usually damages only a small fraction of an observation. Its typical characteristics (narrow band, time variability) enable us to identify and eliminate it. In addition, dispersion of cosmic radio emission in the ionosphere provides another way of discriminating between lunar bursts and manmade signals.

To significantly improve upon our limit with additional observations using the WSRT would require a substantial increase in observing time, probably by at least a factor of ten. Even then, success would be far from guaranteed, as low-level RFI and systematic instrumental effects could also begin to play a role. We look, rather, to the next generation of radio telescopes. The recently opened Low Frequency Array (LOFAR)[18] will cover the same frequency range as the WSRT, but with a sensitivity improvement of a factor of 25.[16] Further down the road, the Square Kilometer Array (SKA) will push the sensitivity by another order of magnitude or so (for general information on SKA, see Ref. 19).

Acknowledgments

RGS is grateful to the organizers of the Gell-Mann 80 Conference for assistance in covering local costs. The WSRT is operated by ASTRON with financial support from the Nederlandse Organisatie voor Wetenschappelijk Onderzoek (NWO). This work was performed in the context of the research programs of the Stichting voor Fundamenteel Onderzoek der Materie (FOM) and of ASTRON, both with financial support from NWO.

References

1. J. Abraham *et al.*, *Phys. Rev. Lett.* **101**, 061101 (2008).
2. K. Greisen, *Phys. Rev. Lett.* **16**, 748 (1966).
3. G. Zatsepin and V. Kuzmin, *Pis'ma Zh. Eksp. Teor. Fiz.* **4**, 114 (1966).
4. P. Gorham *et al.* (ANITA Collaboration), *Phys. Rev. Lett.* **103**, 051103 (2009).
5. N. G. Lehtinen *et al.*, *Phys. Rev. D* **69**, 013008 (2004).
6. J. Abraham *et al.* (Pierre Auger Collaboration), *Phys. Rev. Lett.* **100**, 211101 (2008).
7. G. Askaryan, *Sov. Phys. JETP* **14**, 441 (1962).
8. R. Dagesamanskii and I. Zheleznyk, *Sov. Phys. JETP* **50**, 233 (1989).
9. T. Hankins, R. Ekers and J. O'Sullivan, *Mon. Not. R. Astron. Soc.* **283**, 1027 (1996).
10. P. Gorham *et al.*, *Phys. Rev. Lett.* **93**, 041101 (2004).
11. A. Beresnyak *et al.*, *Astronomy Reports* **49**, 127 (2005).
12. O. Scholten *et al.*, *Astropart. Phys.* **26**, 219 (2006).
13. J. A. Högbom and W. N. Brouw, *Astron. Astrophys.* **33**, 289 (1974).
14. R. Karuppusamy, B. Stappers and W. van Straten, *Pub. Astron. Soc. Pacific* **120**, 191 (2008).
15. O. Scholten *et al.*, *Phys. Rev. Lett.* **103**, 191301 (2009).
16. S. Buitink *et al.*, *Astron. Astrophys.* (in press); arXiv: astr0-ph.HE/1004.0274v1.
17. E. Waxman and J. N. Bahcall, *Phys. Rev. D* **59**, 023002 (1998).
18. H. D. E. Falcke, LOFAR — The Low Frequency Array, in *Long Wavelength Astrophysics, 26th meeting of the IAU, Joint Discussion 12*, 21 August 2006, Prague, Czech Republic, #16.
19. http://www.skatelescope.org

PROTEIN FOLDING STAGES AND UNIVERSAL EXPONENTS

KERSON HUANG

Physics Department, Massachusetts Institute of Technology,
Cambridge, MA 02139, USA
and
Institute of Advanced Studies,
Nanyang Technological University, Singapore 639673

We propose three stages in protein folding, based on physical arguements involving the interplay between the hydrophobic effect and hydrogen bonding, and computer simulations using the CSAW (conditioned self-avoiding walk) model. These stages are characterized by universal exponents $\nu = 3/5, 3/7, 2/5$ in the power law $R \sim N^\nu$, where R is the radius of gyration and N is the number of residues. They correspond to the experimentally observed stages: unfolded, preglobule, molten globule.

Keywords: Protein folding; molten globule; CSAW model; critical exponent.

Protein folding is a stochastic process by which a polypeptide folds into its characteristic and functional 3D structure from random coil. We describe it from the point of view of statistical physics by means of the CSAW (conditioned self-avoiding walk) model,[1,2] which combines the features of self-avoiding walk (SAW) and the Monte Carlo method. In this model, the unfolded protein chain is treated as a random coil described by a statistical ensemble of SAW. Folding can be accounted for through a Monte Carlo procedure that picks out a sub-ensemble, according to a Boltzmann weight involving interaction energies. Implementing this model on a computer is equivalent to solving a generalized Langevin equation with given forces. The simplest choice for interactions is to including only the hydrophobic force and hydrogen bonding. Despite its simplicity, such a model can describe universal aspects of protein folding, among which are the formation of the alpha helix[3] and the beta sheet,[4] elastic energy and various stages in the folding process, which we shall briefly describe in this report.

Using physical arguments and CSAW simulation data, we can construct an elastic energy $E(R, N)$ for proteins,[5] where R is the radius of gyration, and N is the residue number. The analysis is a generalization of that used in polymer physics by Flory.[6] The idea is that the hydrophobic force is responsible for the compactness of the protein, and hydrogen-bonding supplies rigidity. The energy is postulated to have the form

$$E(R, N) = aN^{4/5} + b(NR)^{1/2} + c(\rho)N^2R^{-3},$$

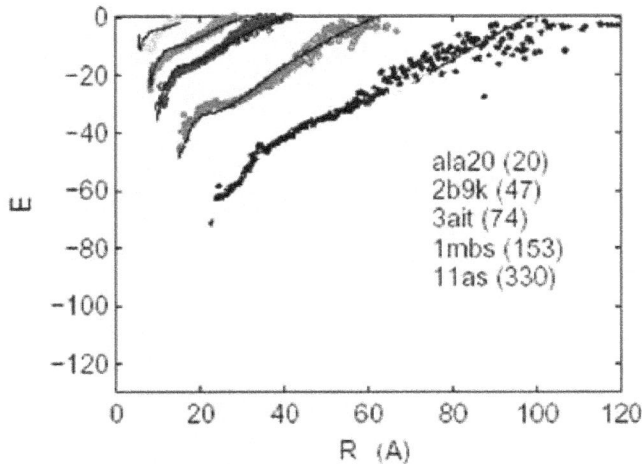

Fig. 1.   Simulation Data from CSAW simulations, for five proteins. The five curves correspond to the proteins listed, from top down. Parenthesis after the name of each protein gives the residue number N. The energy $E(R, N)$ is the ensemble average of the potential energy over protein conformations that have the same radius of gyration R.

where a, b are adjustable parameters, and $c(\rho)$ represents an excluded volume, which depends only on the combination $\rho = RN^{-2/5}$. The coefficients a, b should be universal for all proteins, but $c(\rho)$ is not. We try to find the best average fit by parametrizing $c(\rho)$ in terms of inverse powers of ρ. The parameters are determined by fitting the formula to CSAW data for five proteins, with N ranging from 20 to 330, as displayed in Fig. 1.

The elastic energy contains terms with different powers of N, and the relative importance of these terms changes with R. In different ranges of R, however, the energy to a good approximation obeys a power law $R \sim N^{\nu}$, with different exponents ν. We find three stages[5] corresponding to $\nu = 3/5, 3/7, 2/5$. The results are summarized in Fig. 2, which sketches R as a function of real time in a typical case. The various stages correspond to the experimentally observed unfolded stage, the pre-globule, and the molten globule. Physcially, they correspond to different degrees of hydrogen bonding, as indicated in the notes in Fig. 2. The pre-globule may be too short-lived to be observed in specific proteins, since the lifetimes of the stages are not universal.

It is known experimentally that the molten globule is as compact as the native state, which differs from the former only in the locking of side chains. This locking process cannot be described in our simulation, because the side chains were approximated by hard spheres. As far as our model is concerned, therefore, the native state is not distinct from the molten globule. There exist computer code for an all-atom CSAW model that includes side chains,[7] but the detailed description increases the computer time considerably.

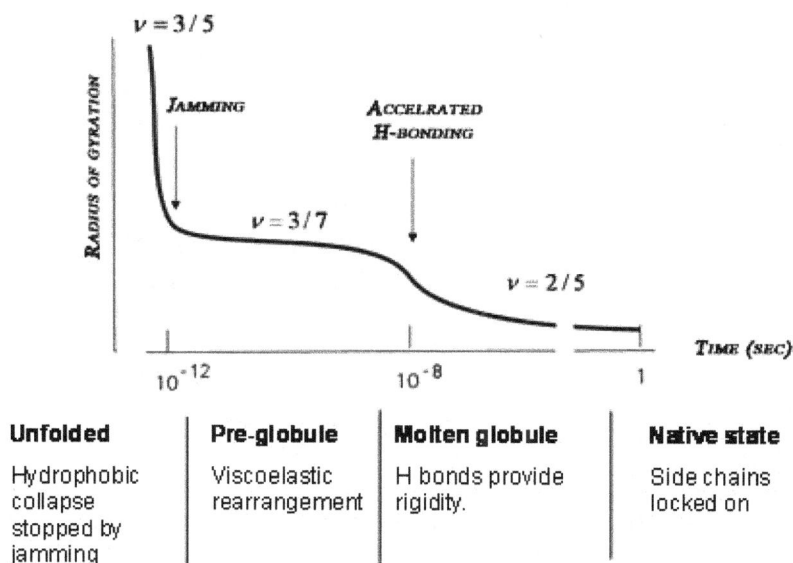

Fig. 2. Time development of the radius of gyration for a typical protein. The mechanisms governing the various stages are indicated. The precise lifetimes of these stages depend on the specific prote.n.

References

1. K. Huang, *Biophys. Rev. Lett.* **2** (2007) 139.
2. J. Lei and K. Huang, Protein folding: A perspective from statistical physics, in *Protein Biochemistry, Synthesis, Structure and Cellular Functions*, ed. E. C. Walters (Nova Publishers Hauppauge, New York, 2010).
3. J. Lei and K. Huang, *Eur. Phys. J. E* **27** (2008) 197.
4. H. W. Leong and L. Y. Chew, Unpublished report, Nanyang Technological University, Singapore.
5. J. Lei and K. Huang, *Europhys. Lett.* **88** (2009) 68004.
6. P. Flory, *Principles of Polymer Chemistry* (Cornell University Press, New York, 1953).
7. W. Sun, unpublished report, Tsinghua University, Beijing.

ANY ANSWERS?
SELF-ORGANISED CRITICALITY IN
THE THIRD DECADE AFTER BTW

GUNNAR PRUESSNER

Department of Mathematics, Imperial College London,
180 Queen's Gate, London SW7 2AZ, UK
g.pruessner@imperial.ac.uk
www.ma.imperial.ac.uk/~pruess/

Self-Organised Criticality, for more than a decade the most popular subject within Complexity, or even within Statistical Mechanics, was expected to be found everywhere, explaining not only earthquakes and forest fires, but hospital waiting times, wars and consciousness.

Taking stock more than twenty years later, what are the core findings and what the open questions? I will attempt to give a fair assessment of the state of affairs. The understanding of Self Organised Criticality has advanced very far, not only on the basis of numerics, but also using technically more demanding methods developed in Statistical Mechanics. A few robust models have been identified and studied in detail, some even solved analytically. A number of puzzling phenomena and tempting problems remain, which I will try to highlight.

Keywords: Self-organised criticality; generic scaling; sandpiles.

1. Introduction

Self-Organised Criticality (SOC) has been around for almost 25 years. During its first decade it has attracted a remarkable amount of attention. The article by Bak Tang and Wiesenfeld [1] introducing the subject as the physics of fractals [2] has been cited well over 3 000 times. It is fair to say that SOC was one of the dominating themes in statistical mechanics during the 90's. What has happened? Is SOC understood? Are any of the original questions answered?

In the following, some of SOC's history is retraced, by briefly introducing some of the key models and experiments, which have guided the development of some very powerful theories mostly in the form of Langevin equations of different types. Taking a more sceptical stance, it looks as if there were only very few systems that display SOC as originally envisaged. The fact that experimentalists have struggled for a long time to find any evidence of SOC at all, undermines the notion of the "ubiquity of SOC". Yet, there remains a hard core of evidence; SOC does exist, but it might not be as common and universal as initially thought.

2. Models and Experiments

The famous "sandpile model", introduced by Bak, Tang and Wiesenfeld [1] should not be regarded as a model of a sandpile (very much like Thomson's plum pudding model is not a model of a plum pudding). It contains some very basic features of a sandpile, which led some authors to suspect that it actually contains all relevant interactions of a real sandpile, producing the same universal behaviour as the latter. The basic setup of the "BTW model", as it is refered to in the following, is shown in Figure 1. Sites on a d-dimensional hypercubic lattice have a height associated with them, which indicates the number of sand grains resting at them.

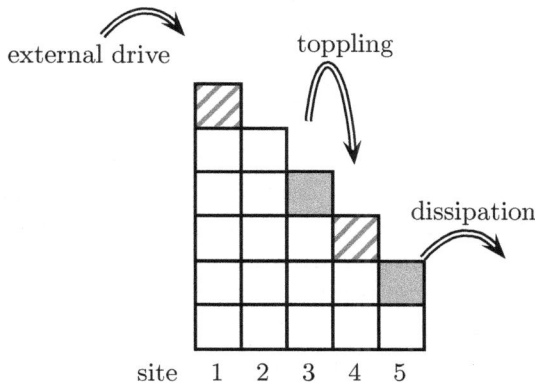

Fig. 1.   The one dimensional BTW model consisting of 5 sites. Each site carries a stack of "grains", which topple downhill if their slope exceeds the threshold value of 1. The slope is measured as the height difference to the right column. The grain added by the external drive is shown hatched. The two filled grains are about to topple, one in the bulk, giving rise to a hatched grain to the right, one at the right boundary, where grains are dissipated. The BTW model is normally considered in dimensions higher than 1, as it is essentially trivial otherwise [3].

The height difference between a column and its right neighbour defines its slope. Whenever the slope exceeds a **threshold**, normally 1, one particle is moved to the column on the right. That column might in turn exceed the threshold, triggering another grain movement *etc.* The process is repeated until none of the slopes exceeds the threshold any longer. Grains are lost when the rightmost site topples.

The system is **driven** by adding grains either at randomly chosen sites or only at, say, the leftmost one. Driving takes place only when the system is quiescent, which gives rise to a **separation of time scales** between driving and toppling. The totality of the topplings between driving and quiescent defines an **avalanche**.

Bak, Tang and Wiesenfeld introduced their model as an explanation of $1/f$ **noise**, which had been observed in natural and experimental systems for many years and whose long temporal correlations inspires researchers to this day. In the spirit of van der Ziel [4], Bak, Tang and Wiesenfeld regarded them as the product of superimposed events (avalanches), whose size distribution followed a power law distribution. Their spatial features being fractal, Bak, Tang and Wiesenfeld also addressed the need for a **physics of fractals** which had been addressed by Kadanoff [2] only a year earlier.

An enormous number of experiments have been performed in the past. Starting with Jaeger, Liu and Nagel's famous **sandpile experiments** [5], initially they focused on the dynamics of granular matter, in particular on sand in various setups. The findings were somewhat disparate, with most authors agreeing that avalanching in real sandpile suffered from a certain degree of periodicity [6]. The experiment by Frette *et al.* [7] was the first to display the expected behaviour more clearly, which has been confirmed in higher dimensions [8].

A different route was taken by Ling *et al.* [9] who observed vortex avalanches in **superconductors**. The experimental setup has been improved several times [10–12] and to this day avalanching in type II superconductors remains one of the most solid example of SOC in experimental systems.

SOC has also been proposed to play a major role in **evolution** [13, 14], which remains disputed, not least because most biologists agree that big extinction events (avalanches) are caused by external factors, such as climate change and meteors, rather than the internal dynamics. Although experimentally much better accessible, the role of SOC in **earthquakes** [15] remains somewhat contentious as well. Olami, Feder and Christensen [16] proposed a model of the dissipative dynamics of an earthquake fault, very much in the spirit of the BTW model. Since then, various scaling results [17] found in empirical data suggest a degree of **universality** in earthquakes which supports a link to SOC.

The effort and the time it took for solid evidence of SOC in Nature to surface contradicts its notion of its being ubiquitous feature of natural systems. Frequently, the analysis of data from experiments and observation failed to produce collapsable distributions governed by a clean power law. Instead, a large characteristic time or length scale was revealed, and thus a break down of the supposed scaling at larger system sizes. This ties in with the critique raised about the observation of fractal features in natural systems, where claims generally seem to be based on a fairly narrow range of observation, normally less than two orders of magnitude [18, 19]. One might thus wonder, whether Nature is fractal and whether SOC is an appropriate explanation.

Nevertheless, as mentioned above, some experimental systems display scaling and SOC very convincingly. What is more, some *numerical* models show all the SOC features as originally envisaged: **thresholds, interaction, spatial structure, separation of time scales, transport, finite size scaling and algebraic spatio-temporal correlations.** Probably the two most prominent models in this respect are the **Manna** [20] and the **Oslo Model** [21],[a] which were later found to belong to the same, large universality class [23], which also comprises models that display ordinary critical phenomena requiring tuning.

---

[a]But maybe also the Bak-Sneppen Model [22].

3. Theory

Very early on, it was suggested that the BTW Model is related to **logarithmic conformal field theory** [24] and subsequently, a large number of beautiful exact results were determined for this model [25, 26]. In this respect, the BTW Model is by far the best understood SOC model. With the introduction of the **wave decomposition** [27], significant progress has been made to understand the scaling of avalanches. However, it remains controversial how to extend these results to the scaling of *entire* avalanches and whether the BTW model displays scaling in this respect at all.

A lot of effort has gone into the formulation of a viable **renormalisation group** scheme, which is the most significant paradigm in the understanding of ordinary critical phenomena. Renormalisation approaches to SOC are either based on some (effective) equation of motion of the degrees of freedom [28–30] or a derived quantity as discussed below, or on the description of a model in terms of some effective dynamics, such as parameterising the probability that a toppling site triggers further topplings on n other sites [31].

Some models are governed by an equation of motion that has been subject to detailed analysis in the past, such as the Oslo Model, which is a discrete realisation of the quenched Edwards-Wilkinson equation [32, 33]. While notoriously difficult to analyse [34, 35], the Manna model (in the same universality class) has provided a link to **absorbing state phase transitions** [36] and thus to **Reggeon field theory** [37].

Some of the schemes mentioned above have been validated on the basis of exactly known solutions [38, 39]. Unfortunately, **exact solutions** are available only for directed models which generally develop into a spatially uncorrelated product state, due to the lack of back avalanches [40]. Depending on whether they are deterministic or stochastic, different sets of exponents characterise them, which can all be traced back to features of random walkers in various dimensions [41]. Despite of that, some models have been linked to the enormous universality class of directed percolation [42–44], by introducing sticky sand grains [45, 46].

4. Explanations of SOC

Hwa and Kardar coined the term **dissipative transport in open systems** [28] as an explanation of SOC, which in the early 1990s started to become synonymous to SOC. The key idea was to describe the surface evolution of a sandpile by means of a Langevin equation of motion, such as

$$\partial_t \phi(\mathbf{r}, t) = (\nu_\parallel \nabla_\parallel^2 + \nu_\perp \nabla_\perp^2)\phi(\mathbf{r}, t) + \eta(\mathbf{r}, t) , \qquad (1)$$

where $\phi(\mathbf{r}, t)$ is the height of the surface at time t and at position $\mathbf{r}$ over the substrate, $\nu_\parallel$ and $\nu_\perp$ are two diffusion constants and η is white, Gaussian noise. The two operators $\nabla_\parallel^2$ and $\nabla_\perp^2$ operate on two distinct sets of components in the substrate space, reflecting a preferred transport direction.

626

In this simple, bilinear form, the propagator of Eq. (1) is easy to find and has a pole at vanishing wave vector, because there is no dissipative or mass-like term of the form $-\epsilon\phi$ on the right. Adding non-linearities, Hwa and Kardar were able to show that the surface develops nontrivial scaling behaviour. The clash between the non-conserved noise and the conservative bulk dynamics led Grinstein *et al.* [47] to consider conserved noise, where the noise correlator is

$$\langle\eta(\mathbf{r},t)\eta(\mathbf{r}',t')\rangle = 2(D_\parallel\nabla_\parallel^2 + D_\perp\nabla_\perp^2)\delta(\mathbf{r}-\mathbf{r}')\delta(t-t') \ . \tag{2}$$

Provided that $D_\parallel/\nu_\parallel \neq D_\perp/\nu_\perp$, this system develops, again, into a scale invariant state, this time, however, with fully conserved noise.

This result suggests that the governing features of SOC are, again, symmetries, either in the form of time invariance (**conservation**) or in the form of spatial **anisotropy**. In any case, bulk dissipation would generate an effective cutoff in the scaling behaviour, and thus conservation was widely regarded as a necessary condition for self-organised criticality. The advent of non-conservative models, such as the Bak-Sneppen Model [22], the OFC Model [16] and the Forest Fire Models [48, 49], challenged this view. Given that the scaling behaviour of some of these models has been questioned in the recent past [50], it might be worthwhile to revisit dissipative transport in open systems. One characteristic it does not seem to account for, however, is the separation of time scales and the very existence of avalanches.

Another approach that has gained a lot of pace over recent years is the interpretation of SOC in terms of **absorbing state phase transitions** by Vespignani *et al.* [51, 52], which goes back to earlier work by Tang and Bak [53, 54]. In this interpretation, self-organised criticality is precisely that: An ordinary non-equilibrium phase transition (namely one into an absorbing state, such as the contact process or directed percolation mentioned above), whose control parameter is subject to a **feedback loop** which tunes it such that it reaches the critical point of the transition. More concretely, in systems like the Manna model, the control parameter is the number of particles present. Activity leads to dissipation at the open boundaries, reducing the particle number in turn, which, in the stationary state, is balanced by the external drive, keeping the system close to the critical point where activity sets in. The scaling relations of critical exponents found in the two regimes, that of the ordinary absorbing state phase transition and SOC, confirm this link and there is a remarkable amount of numerical work corroborating it [55, 56].

What surprises is the fact a *linear* feedback loop is all that is apparently needed to drive a system to its critical point. Moreover, a careful analysis of the finite size scaling in systems subject to such a linear feedback loop shows that it might not be enough to explain *universal* finite size scaling [57]. More recently, Fey *et al.* have found that the feedback loop would *not* work in the original BTW model on various different lattices [58, 59].

5. Any Answers?

Taking a very sceptical stance, one might question the very existence of SOC. Numerical models like the Manna model or the Oslo model, which display robust, nontrivial, universal scaling behaviour, show that SOC is possible. But whether it is as widespread and easy to find as originally thought [60] is questionable. If SOC is seen as the "physics of fractals", one might even ask whether fractals are as commonly found as initially suspected [2].

Although a number of mechanisms of SOC have been proposed, even the most successful ones are still incomplete. Eventually they might provide a recipe to turn any phase transition into a self-organised one or explain why this is not possible. And even if neither materialises, an explanation of SOC can still help to explain some of the long range scaling observed in Nature. It could well be, however, that a completely novel approach is needed: Scaling in Nature is not as well behaved as phase transitions in classical statistical mechanics suggest, so it might be underpinned by something very different.

Acknowledgments

The author would like to wish Murray Gell-Mann a very happy 80th birthday and many happy returns. He thanks the organisers of the conference for their invitation and Lock Yue Chew and Hoai Nguyen Huynh of NTU for their hospitality and generosity. He gratefully acknowledges the support of a Royal Society Travel Grant.

References

1. P. Bak, C. Tang and K. Wiesenfeld, *Phys. Rev. Lett.* **59**, 381 (1987).
2. L. P. Kadanoff, *Phys. Today* **39**, 6 (1986).
3. P. Ruelle and S. Sen, *J. Phys. A: Math. Gen.* **25**, L1257 (1992).
4. A. van der Ziel, *Physica* **16**, 359 (1950).
5. H. M. Jaeger, C.-h. Liu and S. R. Nagel, *Phys. Rev. Lett.* **62**, 40(Jan 1989).
6. M. Bretz, J. B. Cunningham, P. L. Kurczynski and F. Nori, *Phys. Rev. Lett.* **69**, 2431(Oct 1992).
7. V. Frette, K. Christensen, A. Malthe-Sørenssen, J. Feder, T. Jøssang and P. Meakin, *Nature* **379**, 49(Jan 1996).
8. C. M. Aegerter, R. Günther and R. J. Wijngaarden, *Phys. Rev. E* **67**, p. 051306(May 2003).
9. X. S. Ling, D. Shi and J. L. Budnick, *Physica C* **185–189**, 2181 (1991).
10. S. Field, J. Witt, F. Nori and X. Ling, *Phys. Rev. Lett.* **74**, 1206(Feb 1995).
11. E. R. Nowak, O. W. Taylor, L. Liu, H. M. Jaeger and T. I. Selinder, *Phys. Rev. B* **55**, 11702(May 1997).
12. C. M. Aegerter, *Phys. Rev. E* **58**, 1438(Aug 1998).
13. K. Sneppen, P. Bak, H. Flyvbjerg and M. H. Jensen, *Proc. Natl. Acad. Sci. USA* **92**, 5209(May 1995).
14. T. Gisiger, *Biol. Rev.* **76**, 161 (2001).
15. S. Hergarten, *Self-Organized Criticality in Earth Systems* (Springer-Verlag, Berlin, Germany, 2002).

628

16. Z. Olami, H. J. S. Feder and K. Christensen, *Phys. Rev. Lett.* **68**, 1244 (1992).
17. P. Bak, K. Christensen, L. Danon and T. Scanlon, *Phys. Rev. Lett.* **88**, p. 178501(Apr 2002).
18. O. Malcai, D. A. Lidar, O. Biham and D. Avnir, *Phys. Rev. E* **56**, 2817 (1997).
19. D. Avnir, O. Biham, D. Lidar and O. Malcai, *Science* **279**, 39(Jan 1998).
20. S. S. Manna, *J. Phys. A: Math. Gen.* **24**, L363 (1991).
21. K. Christensen, Á. Corral, V. Frette, J. Feder and T. Jøssang, *Phys. Rev. Lett.* **77**, 107 (1996).
22. P. Bak and K. Sneppen, *Phys. Rev. Lett.* **71**, 4083 (1993).
23. H. Nakanishi and K. Sneppen, *Phys. Rev. E* **55**, 4012(Apr 1997).
24. S. N. Majumdar and D. Dhar, *Physica A* **185**, 129 (1992).
25. E. V. Ivashkevich, *J. Phys. A: Math. Gen.* **27**, 3643 (1994).
26. M. Jeng, G. Piroux and P. Ruelle, *J. Stat. Mech.* **2006**, P10015 (2006).
27. E. V. Ivashkevich, D. V. Ktitarev and V. B. Priezzhev, *Physica A* **209**, 347 (1994).
28. T. Hwa and M. Kardar, *Phys. Rev. Lett.* **62**, 1813 (1989).
29. A. Díaz-Guilera, *Phys. Rev. A* **45**, 8551 (1992).
30. A. Díaz-Guilera, *Europhys. Lett.* **26**, 177 (1994).
31. L. Pietronero, A. Vespignani and S. Zapperi, *Phys. Rev. Lett.* **72**, 1690 (1994).
32. M. Paczuski and S. Boettcher, *Phys. Rev. Lett.* **77**, 111 (1996).
33. G. Pruessner, *Phys. Rev. E* **67**, p. 030301(Mar 2003).
34. T. Nattermann, S. Stepanow, L.-H. Tang and H. Leschhorn, *J. Phys. II (France)* **2**, 1483 (1992).
35. H. Leschhorn, T. Nattermann, S. Stepanow and L.-H. Tang, *Ann. Physik* **6**, 1 (1997).
36. R. Dickman, M. A. Muñoz, A. Vespignani and S. Zapperi, *Braz. J. Phys.* **30**, 27 (2000).
37. M. Alava and M. A. Muñoz, *Phys. Rev. E* **65**, p. 026145(Jan 2002).
38. J. Hasty and K. Wiesenfeld, *J. Stat. Phys.* **86**, 1179 (1997).
39. J. Hasty and K. Wiesenfeld, *Phys. Rev. Lett.* **81**, 1722(Aug 1998).
40. M. Paczuski and K. E. Bassler, Theoretical results for sandpile models of soc with multiple topplings, eprint arXiv:cond-mat/0005340v2, (2000).
41. N. Z. Bunzarova, On the statistical properties of directed avalanches, personal communication.
42. H. K. Janssen, *Z. Phys. B* **42**, 151 (1981).
43. P. Grassberger, *Z. Phys. B* **47**, 365 (1982).
44. H. Hinrichsen, *Adv. Phys.* **49**, 815 (2000).
45. B. Tadić and D. Dhar, *Phys. Rev. Lett.* **79**, 1519 (1997).
46. P. K. Mohanty and D. Dhar, *Phys. Rev. Lett.* **89**, 104303 (2002).
47. G. Grinstein, D.-H. Lee and S. Sachdev, *Phys. Rev. Lett.* **64**, 1927 (1990).
48. P. Bak, K. Chen and C. Tang, *Phys. Lett. A* **147**, 297 (1990).
49. B. Drossel and F. Schwabl, *Phys. Rev. Lett.* **69**, 1629(Sep 1992).
50. G. Pruessner and H. J. Jensen, *Phys. Rev. E* **65**, 056707 (2002).
51. A. Vespignani, R. Dickman, M. A. Muñoz and S. Zapperi, *Phys. Rev. Lett.* **81**, 5676 (1998).
52. R. Dickman, A. Vespignani and S. Zapperi, *Phys. Rev. E* **57**, 5095 (1998).
53. C. Tang and P. Bak, *J. Stat. Phys.* **51**, 797(Jan 1988).
54. C. Tang and P. Bak, *Phys. Rev. Lett.* **60**, 2347 (1988).
55. S. Lübeck, *Int. J. Mod. Phys. B* **18**, 3977 (2004).
56. K. Christensen, N. R. Moloney, O. Peters and G. Pruessner, *Phys. Rev. E* **70**, 067101 (2004).

57. G. Pruessner and O. Peters, *Phys. Rev. E* **73**, 025106(R) (2006).

58. A. Fey, L. Levine and D. B. Wilson, *Phys. Rev. Lett.* **104**, p. 145703(Apr 2010).

59. A. Fey, L. Levine and D. B. Wilson, The approach to criticality in sandpiles, eprint arXiv:1001.3401v1, (2010).

60. P. Bak, *how nature works* (Copernicus, New York, NY, USA, 1996).

DISORDER IN COMPLEX HUMAN SYSTEM

K. GEDIZ AKDENIZ

Department of Physics, Istanbul University, Vezneciler, Istanbul, 34118, Turkey
gakdeniz@istanbul.edu.tr

Since the world of human and whose life becomes more and more complex every day because of the digital technology and under the storm of knowledge (media, internet, governmental and non-governmental organizations, etc. . .) the simulation is rapidly growing in the social systems and in human behaviors. The formation of the body and mutual interactions are left to digital technological, communication mechanisms and coding the techno genetics of the body. Deconstruction begins everywhere. The linear simulation mechanism with modern realities are replaced by the disorder simulation of human behaviors with awareness realities.

In this paper I would like to introduce simulation theory of "Disorder Sensitive Human Behaviors". I recently proposed this theory to critique the role of disorder human behaviors in social systems. In this theory the principle of realty is the chaotic awareness of the complexity of human systems inside of principle of modern thinking in Baudrillard's simulation theory. Proper examples will be also considered to investigate the theory.

Keywords: Complexity; simulation theories; cyborg.

1. Introduction

Modern thinking ideology (modernity) defines human systems (society) as complex and far from disintegration, but also limits the dynamics of society with paradigms dependent on modernity. As a result of this limitation, disordered human dynamics can not have any contribution to the progression and sustainable development of the society. On the other hand, establishments and unitary organizations which are purified from their disorders by modernity, determine the progression of the society.

With the objective of humanity's normalization (modern life), modernity isolated the disordered and/or unorthodox human dynamics from politics, law, science, philosophy, religion, morality, esthetics and art, while reforming the historical process of the society. Modernity called this process "enlightenment" and the reformation (linearization) and redirection (normalization) of this process was called "the history of humanity".

Meanwhile, modernity considered complexity, disorder, chaos, uncertainty, coincidence and surprise as the enemies of its domination since the gaining of its power. It worked to isolate the dynamics of these concepts from the behaviors of human beings by re-defining the concepts such as prostitution, homosexuality, fortune-telling, witchcraft and addictions, and excluded them by using some illnesses defined by the modern medical science, such as madness, schizophrenia and hysteria.

And it has isolated people who opposed to its authority in various places, used violence against them and even destroyed them in different ways in the name of linear and normalizing constitutions such as law and justice. During the cold war, in order to expand their hegemony, two super-power states of the era declared "searchers for new ideas" as potential criminals of the humanity and science, and suspended those who opposed their constitutions and hegemony from the academic environment.

In the 1960s, emerging events, which would shake the hegemony that the modern domination had been trying to establish for 400 years appeared one after another; while the claims of the western world to be the progression of a predictable and rational human history began vanishing. Their mission in re-writing human beings' behaviors, economical, social and political constitutions, history and psychology with quantum theory –which was thought to be the wonder of modern science– did not work out. On the contrary, they fell in their own quantum traps as quantum metaphors triggered the emergence of radical and weird ideas that modernity had previously altered.

Thanks to the journey to the moon, it was seen with bare eyes that the world was a living organism and that every part of it was significant (environment philosophy). Secondly, the spread of information exchange accelerated through internet and media. Most important of all, in prestigious universities, chaos theory had proved that Newtonian mathematical technology had narrow and limited applications.[1] This revolution opened the way for non-linear thinking. Metaphorical paradigms such as simulation theory, complexity, coincidence, self-organization and surprise became significant.[2] The hegemonies of simulation mechanisms such as nationalism, science, globalism, realism, as well as religious and ideological dynamics rapidly started to lose power.

2. Post-Modern Science to Post-Modern World

We are living in a new age of polemical nature. The differences between those identities which were used to spread the hegemony of modernity are getting meaningless, and information theories which were presented as global thoughts for years have been easily demolished and redefined. Increasing number of social and psychology scientists and philosophers, called post-modernists, criticize modernity while being well equipped with Western knowledge.[3] Polarized classes such as old vs. young, women vs. men, alive vs. dead and bad vs. good will be completely replaced and will eventually disappear in this new age; leaving a world virtualized by simulation and devoid of its reality. Such a simulation world has already started and this progression will end with all of us being trans-sexual cyborgs (human-machine crossbred) as metaphors.

One has to remind these post-modernists who are set out to new journeys without leaving their modernist compasses aside: what will they find for us and others at the end of these unknown journeys, what will they hear for those who they define as easterners and what will they see for those who did not have any role in constructing the modernity.

One has to remind these post-modernists that the simulations of identities structured by modernity's impositions, such as the one between transsexualism and the cyborg body, will never replace reality.[4]

The following facts can not be left unnoticed: we are living in a world where people of our time, whether they are white or not, westerners or not, civilized or not, are surrounded with a complex web of civil societies, national-international organizations and digitalized information. In this age, human beings constructed by machine-like bodies (cyborg) are losing their object status. The behavioral dynamics of bodiless systems are becoming virtualized and the hyper-reality part in the simulation world of these dynamics is rapidly expanding. The external part of the human being in the world is gradually building a complex whole with its surrounding and this complexity is flowing inside its body.

3. The Simulation Theory of Disorder Sensitive Human Behaviors and Critics

Cyborg is not only a machine-like body, it is the crossbred of a machine and a human being. Even the view of a cyborg, in human being classification of Schopenhauer,[5] being completed with the will (meaning spirit in Human being classification of Aristotle, meaning mind in modernity) of the identity a of human being is a submission to the ideas of the modern domination in today's world of more and more complex human beings.

Recently, I proposed a simulation theory for complex human beings and the society.[6] Following this theory, one can state that the machine-like body (its possessions and representations) with will (identity, spirit, quantum consciousness[7]) and external beings (culture, traditions and beliefs) form a complex human system, and this can suggest the paradigm that complex human systems form a chaotic awareness. This "chaotic awareness" will be the reality principle of disordered simulations of a cyborg emergence in a cyborg world.

In this theory, emergences unlike images (simulakr,[8] do not happen with the concern of demolition, and while reacting to a reality do not claim to have a continuous surreal contribution to civilization (the domination of modernity). On the other hand, small differences in the reality principle of chaotic awareness will cause major unpredictable differences in disordered simulations. They are the essential (chaos edge) transformational dynamics not for the continuity of an emergence system, but for the replacement of a previous state by an unidentified structure. From this point on, when cyborg emergence is considered, one can presume that cyborg emergence will play a more important role in the change and transformation of a complex cyborg system; especially in times when the human body becomes more machine-like. Cyborg, in its new perception of existence can gain freedom due to cyborg emergences. Cyborg emergences can discover the other part of the human being belonging to a pre-cyborg age. They can also cause the human being within cyborg, which prevents the cyborg from losing its will of living, to commence new unpredicted fractal journeys.

This statement only is enough to realize that for the future there is only one solution for the software engineers of a globalized technology. In the software they submit to their bosses they will include: to demolish the human being within cyborg by rapidly mechanizing the body of cyborg, and to prevent the cyborg from observing its body by breaking its relationship with external beings. For this reason, the simulation mechanisms of Baudrillard that they had once developed to control human beings and their systems are being replaced by consuming dynamics dependant on globally marketed high technology. They use all of their power to transfer the contact of human being with its own body and others' bodies to digital inter parts (interfaces). They are coding the techno-genetics of the body. In a virtual world, where the project of turning the human being into the other is in action, they do not feel any risk about the disappearance of the awareness of the local values by making the human being more machine-like, while using the same values that they once tried to identify as the other. In order to take these actions mentioned, their dependency on energy is rapidly increasing. Unfortunately, our world is on their side; it has unlimited energy sources such as wind, nuclear and solar. As long as this energy production increases to achieve the phase equilibrium, it will be unavoidable for proportions of machinery to increase so rapidly that it accepts the human being within the cyborg non-existent. For this reason, they have transferred the developmental methods and technological projects of forming a cyborg-like being, as well as the control of a cyborg's communication and behavior to a new science called "cyborg science", which claims computer as a paradigm.[9]

4. Image (Simulakr) and Emergence in Cyborg Bodies "Cyborg-Like Cyborg Scientist" and Post-Physicist

Modernity always saw enlightenment and science as the most imposing and glamourous constitutions and used them openly in its border expansion projects without any hesitation. Today, while all the identities given by the modernity are diminishing, one can not claim that intellectuals and scientists will stay away from the cyborg metaphor and lose the identities that they had gained possessions with. Particularly, one needs to ask: will the reformation of a physicist's body show such divergence as in the example of an invading USA soldier in Iraq vs. the resisting suicide bomber? Their bodies represent the same technology with the digital equipment and the bombs they carry. The American soldier shows predicted physical behaviors under the influence of camouflaged simulations of saving humanity, bringing justice and peace for humanity; the concepts which he believes to be his possessions (ordered simulation). And resisting suicide bomber shows behaviors of chaotic awareness that comes from a culture and tradition that due to historical facts did not play any role in forming modernity (disordered simulation).

The complexity that arises due to the impact of an emerging body on his surroundings acts as a source to the freedom dynamics of modernity's domination that has non-linearity within. Western civilization with its domination, contributes to the rise of personal movements which resist its global hegemony and/or formation of new civilizations. One can observe under the light of "disordered-sensitive human behaviors" simulation theory that the techno-shaped force against human body can not be only evaluated as a clever control of the human body through media with global dominations or through alienation of the human being from himself and nature.

For instance; how far the physicists will resist the reformation and mechanization of their bodies by the thinking methods and scientific ethics that they have claimed to be universal since the time of Galileo? They clearly know that, with or without their will, their bodies have been already mechanized with the most illegitimate divergence, and in those bodies, genes of globalization and militarism have been rapidly increasing.

Many scientists work for a big (powerful) science,[10] and show scientific behaviors with the motivation of progressing in science, under the influence of camouflaged simulations of saving humanity and bringing justice and peace for humanity'; the concepts which they believe to be their possessions. They can be stated as "simulakr" with cyborg science at the virtual world that agrees with Jean Baudrillard's simulation theory.[8] Through Baudrillard's simulations, they will produce realities to cover up the hyper realities that camouflages their identities as scientists, which they will eventually lose. Their behaviors can be described as a hyper-reality; one which camouflages the true will (global technology) that is based on the reality concept of a modernist hegemony. This simulakr will remove the human being away from militarism, fast money exchange in the economy, big money consuming space and high energy experiments (powerful science), and it will camouflage the reality of searching ways to replace the human being by a computer dependent "cyborg scientist". On the other hand, cyborg science will demand from them machines that produce greater energy in order to get stronger and to expand their global domination area further.

According to the theory of Baudrillard,[8] physicists, who are not involved in the powerful science, can not be "simulakrs" as they choose vanishing. They can not be the hyper-reality of a simulation world, can not replace a reality in modern world, and can not contribute to the continuity of the powerful science in the simulation world. "Disorder-Sensitive Human Behaviors" simulation theory accepts this rising as a surprising emergence for the Western world.

According to the "Disorder-Sensitive Human Behaviors" simulation theory, the emergence of a different cyborg scientist (post-physicist) is possible; one who is interested in the complex whole and lives in a region like Middle East, where the reality principle stands excluded from modernity and disorder-sensitive dynamics, fed by the non-Western culture and tradition, are still alive.

Post-physicists will not be the products of the cyborg science process, which is the project of reformation and imposition of modernity's hegemony simulated by the informational systems of international research and exchange centers.

5. The Manifestation of Post-Physicist

The manifestation of the post-physicist[6] is written with the aim that physicists will not be a product of the concern of the exclusion from high energy international research, such as the CERN experiment which is camouflaged by the virtual simulation of finding "god's particle". And they will not participate in schizophrenically uncertain experiments for the global technology and energy domination. It is not manifested emotionally, with the expectation of scientists who will achieve freedom from global threat of cyborg science, be in search of new ways in all areas, and devote themselves to honesty with humanity and perversion with modernity, without any hesitation. The manifest, with the "disordered sensitive human behaviors" simulation theory, is a post-utopic text in salvation of the humanity as well as being a critic of the science world of the future.

It can also be stated from my simulation theory that the region, where the emergence of post physicists is possible among the scientists who are made cyborglike by disordered sensitive simulations with chaotic awareness, is the Middle East where Iranian and Turkey is placed. However, it should not be understated that simulation mechanisms of global science, camouflaged by the images of academic work, are also placed in this region. How can it be possible for post-physicists to stay alive alongside the ones who, as the copies of cyborg scientists, work on improving cyborg science using the own realities of this region? It is definitely not possible to predict the outcome of this condition, today. This region, having the complex whole for the emergence of post-physicists, also has the most versatile environment for them to stay alive. The number of scientists and intellectuals in this region, will even loose their ability to copy the modern world, living with the difficulty of finding models that are built on the basis of simulation theories of enlightenment and knowledge of colonists in order to control the divergence and change, that will eventually end the globalism of western civilization.

6. Conclusion

Simulation mechanisms of global technology and communication (media) will contribute to the process of undermining human cyborg. Besides, global climate change will make the human being within the complex whole of cyborg disappear, and also make the resistance of emergence, against the reformation and adaptation of new conditions of cyborg, diminish. For instance, when sea levels rise, the fish-human emergence, undeprived of human awareness might not happen, which is accepted as the first cyborg as Jean de hire predicted in his book in 20th century. This schizophrenic condition that might halt the possibility of the emergence to stay alive will end with cyborg community, mercilessly throwing human being within themselves away. The very new cyborg scientists of energy vampires who think they will gain even more power from the world's unlimited energy sources, have commenced working on to control the birth and formation of cyborg. Undoubtedly, post-physicists too, will be cyborgs. But, they are our only chance to find chaos that will enable us to continue observing our machinelike bodies and nature as a human beings. People who refuse to be part of international big sciences that is formed to spread the new informational technology of hegemony and the unlimited energy production, who raise attention to all these with their acts can be called as ancestors of tomorrow's post-physicists. Therefore the modern world, in all aspects, has started the process of destruction. Virtual simulations are replacing the mechanical simulations of modern domination rapidly. However, I emphasize, referencing to my "disordered sensitive human behaviors" simulation theory that the destruction in the world of virtual simulation will not happen as the image, which only has the reality principle of western knowledge, opposing the expectations of post-modernists fed by the awareness of opening western civilization. According to my theory, emergences with the reality principles of chaotic awareness will play a major role in the destruction in the world of virtual simulations. The competition between emergences and simulakrs in the destruction in the world of virtual simulation has already started within the human, family, society, science, art and literature world, national and international constitutions. The most hostile of the conflicts between the emergence and simulakr is the war of civilizations which has a global scale.

The simulation theory of "Disorder Sensitive Human Behaviors" states that it is hopeless for post-modernity to lead the destruction in order to keep the western civilization strong. And Baudrillard's simulakr that replaces the reality, will not be sufficient enough for the continual progression of the western civilization. Referring to my theory, one can have hope for the emergence of long living chaotic awareness, particularly of post-physicists, for the chance of creating a new world where all living things and humans within cyborg can live in.

Acknowledgments

I thank to Miss Bige Akdeniz and Mr. Nazmi Yılmaz reading the manuscript. This work was supported by Scientific Research Projects Coordination Unit of Istanbul University. Project number 3534.

References

1. M. M. Waldrop, *Complexity: The Emerging Science at the Edge of Order and Chaos*, (Simon&Schuster, New York, 1992).
2. M. Mason, *Complexity Theory and the Philosophy of Education*, (John Wiley and Sons, United Kingdom, 2008); C. Tsallis, M. Gell-Mann and Y. Sato, *Europhysics News*

36, 186 (2005); H. Jensen, *Self-organized Criticality: Emergent Complex Behavior in Physical and Biological Systems*, (Cambridge University Press, New York, 1998).

3. P. Anderson, *The Origins of Post Modernity*, (Verso Books, London, 1998).
4. D. Haraway, *Socialist Review* **80**, 65 (1985); D. Haraway, *Simians, Cyborgs, and Women: The Reinvention of Nature*, (Routledge, New York, 1991).
5. A. Schopenhauer, *Essays and Aphorisms*, (Penguin Books, London, 1970).
6. K. G. Akdeniz, *Istanbul University Sociology Journal* **3**, 15 (2007).
7. D. Zohar, *The Quantum Self*, (Quill/William Morrow, New York, 1990).
8. J. Baudrillard, *Simulacra and Simulation*, (University of Michigan Press, Michigan, 1995).
9. P. Mirowski, *Machine Dreams: Economics Becomes a Cyborg Science*, (Cambridge University Press, New York, 2002).
10. S. Ziauddin, *Thomas Kuhn and The Science Wars, Postmodernism And Big Science*, eds. R. Appignanesi, (The Icon Books, United Kingdom, 2002); K. G. Akdeniz, Globalization in Physics and it's Role in South, in *Proc. of the International Conference on New Technologies in Physics Education*, eds, J. Huo and S. Xiang, Hefei, China, 1999, p. 221.

OBSERVER LOCALIZATION IN MULTIVERSE THEORIES

MARCUS HUTTER

RSISE @ ANU and SML @ NICTA, Canberra, ACT, 0200, Australia
marcus@hutter1.net
www.hutter1.net

The progression of theories suggested for our world, from ego- to geo- to helio-centric models to universe and multiverse theories and beyond, shows one tendency: The size of the described worlds increases, with humans being expelled from their center to ever more remote and random locations. If pushed too far, a potential theory of everything (TOE) is actually more a theories of nothing (TON). Indeed such theories have already been developed. I show that including observer localization into such theories is necessary and sufficient to avoid this problem. I develop a quantitative recipe to identify TOEs and distinguish them from TONs and theories in-between. This precisely shows what the problem is with some recently suggested universal TOEs.

Keywords: World models; observer localization; predictive power; Ockham's razor; universal theories; computability.

1. Introduction

A number of models have been suggested for our world. They range from generally accepted to increasingly speculative to apparently bogus. For the purpose of this work it doesn't matter where you personally draw the line. Many now generally accepted theories have once been regarded as insane, so using the scientific community or general public as a judge is problematic and can lead to endless discussions: for instance, the historic geo↔heliocentric battle; and the ongoing discussion of whether string theory is a theory of everything or more a theory of nothing. In a sense this paper is about a formal rational criterion to determine whether a model makes sense or not. In order to make the main point of this paper clear, below I will briefly traverse a number of models.[1,5,7] The presented bogus models help to make clear the necessity of observer localization and hence the relevance of this work.

Egocentric to Geocentric model. A young child believes it is the center of the world. Localization is trivial. It is always at "coordinate" (0,0,0). Later it learns that it is just one among a few billion other people and as little or much special as any other person thinks of themself. In a sense we replace our egocentric coordinate system to one with origin (0,0,0) in the center of Earth. The move away from an egocentric world view has many social advantages, but dis-answers one question: Why am I this particular person and not any other?

Geocentric to Heliocentric model. While being expelled from the center of the world as an individual, in the geocentric model, at least the human race as a whole remains in the center of the world, with the remaining (dead?) universe revolving around *us*. The heliocentric model puts Sun at (0,0,0) and degrades Earth to planet number 3 out of 8. The astronomic advantages are clear, but dis-answers one question: Why this planet and not one of the others? Typically we are muzzled by semi-convincing anthropic arguments.[3,10]

Heliocentric to cosmological model. The next coup of astronomers was to degrade our Sun to one star among billions of stars in our milky way, and our milky way to one galaxy out of billions of others, according to current textbooks. Again, it is generally accepted that the question why we are in this particular galaxy in this particular solar system is essentially unanswerable.

Multiverses. Many modern more speculative cosmological models (can be argued to) imply a multitude of essentially disconnected universes (in the conventional sense), often each with its own (quite different) characteristic: Examples are Wheeler's oscillating universe, Smolin's baby universe theory, Everett's many-worlds interpretation of quantum mechanics, and the different compactifications of string theory.[11] They "explain" why a universe with our properties exist, since the multiverse includes universes with all kinds of properties, but they cannot *predict* these properties. A multiverse theory *plus* a theory predicting in which universe we happen to live would determine the value of the inter-universe variables for our universe, and hence have much more predictive power. Again, anthropic arguments are sometimes evoked but are usually vague and unconvincing.

Universal TOE (UTOE). Taking the multiverse theory to the extreme, Schmidhuber[9] postulates a universal multiverse, which consists of *every* computable universe. Clearly, if our universe is computable (and there is no proof of the opposite[9]), the multiverse generated by UTOE contains and hence perfectly describes our own universe, so we have a Theory of Everything (TOE) already in our hands. Unfortunately it is of little use, since we can't use UTOE for prediction. If we knew our "position" in this multiverse, we would know in which (sub)universe we are. This is equivalent to knowing the program that generates *our* universe. This program may be close to any of the conventional cosmological models, which indeed have a lot of predictive power. Since locating ourselves in UTOE is equivalent and hence as hard as finding a conventional TOE of our universe, we have not gained much.

All-a-Carte models. Champernowne's normal number glues the natural numbers, for our purpose written in binary format, 1,10,11,100,101,110,111,1000,1001,... to one long string.

$$110111001011101111000 1001...$$

Obviously it contains every finite substring by construction. The digits of many irrational numbers like $\sqrt{2}$, π, and e are conjectured to also contain every finite substring. If our space-time universe is finite, we can capture a snapshot of it in a truly gargantuan string u. Since Champernowne's number contains every finite string, it also contains u and hence perfectly describes our universe. Probably even $\sqrt{2}$ is a perfect TOE. Unfortunately, if and only if we can localize ourselves, we can actually use it for predictions. (For instance, if we knew we were in the center of universe 001011011 we could predict that we will 'see' 0010 when 'looking' to the left and 1011 when looking to the right.) Locating ourselves means to (at least) locate u in the multiverse. We know that u is the u's number in Champernowne's sequence (interpreting u as a binary number), hence locating u is equivalent to specifying u. So a TOE based on normal numbers is only useful if accompanied by the gargantuan snapshot u of our universe. In light of this, such an "All-a-Carte" TOE (without knowing u) is rather a theory of nothing than a theory of everything.

Localization within our universe. The loss of predictive power when enlarging a universe to a multiverse model has nothing to do with multiverses per se. Indeed, the distinction between a universe and a multiverse is not absolute. For instance, Champernowne's number could also be interpreted as a single universe, rather than a multiverse. It could be regarded as an extreme form of the infinite fantasia land from the NeverEnding Story, where everything happens somewhere. Champernowne's number constitutes a perfect map of the All-a-Carte universe, but the map is useless unless you know where you are. Similarly but less extreme, cosmological inflation models produce a universe that is vastly larger than its visible part, and different regions may have different properties.

Predictive power. The exemplary discussion above has hopefully convinced the reader that we indeed lose something (some predictive power) when progressing to too large universe and multiverse models. Historically, the higher predictive power of the large-universe models (in which we are seemingly randomly placed) overshadowed the few extra questions they raised compared to the smaller ego/geo/heliocentric models. But the discussion of the (physical, universal, and all-a-carte) multiverse theories has shown that pushing this progression too far will at some point harm predictive power. We saw that this has to do with the increasing difficulty to localize the observer.

Contents. Classical models in physics are essentially differential equations describing the time-evolution of some aspects of the world. A Theory of Everything (TOE) models the whole universe or multiverse, which should include initial conditions. As argued above, it can be crucial to also localize the observer. I call a TOE with observer localization, a *Complete TOE* (CTOE). Section 2 gives an informal introduction to the necessary ingredients for CTOEs, and how to evaluate and compare them using a quantified instantiation of Ockham's razor. Section 3 gives a formal definition of what accounts for a CTOE, introduces more realistic observers with limited perception ability, and formalizes the CTOE selection principle. The Universal TOE is a sanity critical point in the development of TOEs, and will be investigated in more detail in Section 4. Important extensions listed in Section 5 are detailed in Ref. 7.

2. Complete TOEs (CTOEs)

A TOE by definition is a perfect model of the universe. It should allow to predict all phenomena. Most TOEs require a specification of some initial conditions, e.g. the state at the big bang, and how the state evolves in time (the equations of motion). In general, a TOE is a program that in principle can "simulate" the whole universe. An All-a-Carte universe perfectly satisfies this condition but apparently is rather a theory of nothing than a theory of everything. So meeting the simulation condition is not sufficient for qualifying as a Complete TOE. We have seen that (objective) TOEs can be completed by specifying the location of the observer. This allows us to make useful predictions from our (subjective) viewpoint. We call a TOE plus observer localization a Complete TOE. If we allow for stochastic (quantum) universes we also need to include the noise. If we consider (human) observers with limited perception ability we need to take that into account too. So

A complete TOE needs specification of

- (i) initial conditions
- (e) state evolution
- (l) localization of observer
- (n) random noise
- (o) perception ability of observer

We deal with limited perception ability (o) in Section 3. Space prevents discussing stochastic theories (n); they are dealt with in Ref. 7. Next we need a way to compare TOEs.

Predictive power and elegance. Clearly we can never be sure whether a given TOE makes correct predictions in the future. After all we cannot rule out that the world suddenly changes tomorrow in a totally unexpected way. We have to compare theories based on their predictive success in the past. It is also clear that this is not enough: For every model we can construct an alternative model that behaves identically in the past but makes different predictions from, say, year 3000 on. Popper's falsifiability dogma is little helpful. Beyond postdictive success, the guiding principle in designing and selecting theories, especially in physics, is elegance (and mathematical consistency). The predictive power of the first heliocentric model was not superior to the geocentric one, but it was much simpler. In more profane terms, it has significantly less parameters that need to be specified.

Ockham's razor suitably interpreted tells us to choose the simpler among two or more otherwise equally good theories. For justifications of Ockham's razor, see Ref. 8. Some even argue that by definition, science is about applying Ockham's razor, see Ref. 6. For a discussion in the context of theories in physics, see Ref. 4. It is beyond the scope of this paper to repeat these considerations. In Ref. 7 I prove that Ockham's razor is suitable for finding TOEs.

Complexity of a TOE. In order to apply Ockham's razor in a non-heuristic way, we need to quantify simplicity or complexity. Roughly, the complexity of a theory can be defined as the number of symbols one needs to write the theory down. More precisely, write down a program for the state evolution together with the initial conditions, and define the complexity of the theory as the size in bits of the file that contains the program. This quantification is consistent with our intuition, since an elegant theory will have a shorter program than an inelegant one, and extra parameters need extra space to code, resulting in longer programs.

Standard model versus string theory. To give an example, let us pretend that the standard model of particle physics + gravity (P) and string theory (S) both qualify as TOEs. P is a mixture of a few relatively elegant theories, but contains about 20 unexplained parameters that need to be specified (although some regularities can be explained[2]). String theory is truly elegant, but ensuring that it reduces to P needs sophisticated extra assumptions (e.g. the right compactification). It would require a major effort to quantify which theory is the simpler one in the sense defined above, but I think it would be worth the effort. It is a quantitative objective way to decide between theories that are (so far) predictively indistinguishable.

CTOE selection principle. It is trivial to write down a program for an All-a-Carte multiverse A. It is also not too hard to write a program for the universal multiverse U, see Section 4. Lengthwise A easily wins over U, and U easily wins over P and S, but as discussed A and U have serious defects. Given all of the above, it now nearly suggests itself that we should include the description length of the observer location in our TOE evaluation measure. That is,

among two CTOEs, select the one that has shorter overall length

$$\text{Length}(i) + \text{Length}(e) + \text{Length}(l) \tag{1}$$

For an All-a-Carte multiverse, the last term contains the gargantuan string u, catapulting it from the shortest TOE to the longest CTOE.

TOE versus UTOE. Consider any (C)TOE and its program q, e.g. P or S. Since U runs all programs including q, specifying q means localizing (C)TOE q in U. So U+q is a CTOE whose length is just some constant bits (the simulation part of U) more than that of (C)TOE q. So whatever (C)TOE physicists come up with, U is nearly as good as this theory. This essentially clarifies the paradoxical status of U. Naked, U is a theory of nothing, but in combination with another TOE it excels to a good CTOE, albeit slightly longer=worse than the latter.

Localization within our universe. So far we have only localized our universe in the multiverse, but not ourselves in the universe. Assume the about $10^{11} \times 10^{11}$ stars in our universe are somehow indexed. In order to localize our Sun we only need its index, which can be coded in about $\log_2(10^{11} \times 10^{11}) \approx 73$ bits. To localize earth among the 8 planets needs 3 bits. To localize yourself among 7 billion humans needs 33 bits. These localization penalties are tiny compared to the difference in predictive power (to be quantified later) of the various theories (ego/geo/helio/cosmo). This explains and justifies theories of large universes in which we occupy a random location.

3. Complete TOE - Formalization

Objective TOE. Since we essentially identify a TOEs with a program generating a universe, we need to fix some general purpose programming language on a general purpose computer. In theoretical computer science, the standard model is a so-called Universal Turing Machine (UTM).[8] It takes a program coded as a finite binary string $q \in \{0,1\}^*$, executes it and outputs a finite or infinite binary string $u \in \{0,1\}^* \cup \{0,1\}^\infty$. The details do not matter to us, since drawn conclusion are typically independent of them. In this section we only consider q with infinite output

$$\mathrm{UTM}(q) = u_1^q u_2^q u_3^q \ldots =: u_{1:\infty}^q$$

In our case, $u_{1:\infty}^q$ will be the universe (or multiverse) generated by TOE candidate q. So q incorporates items (i) and (e) of Section 2. Surely our universe doesn't look like a bit string, but can be coded as one as explained in Ref. 7. We have some simple coding in mind, e.g. $u_{1:N}^q$ being the (fictitious) binary data file of a high-resolution 3D movie of the whole universe from big bang to big crunch, augmented by $u_{N+1:\infty}^c \equiv 0$ if the universe is finite. Again, the details do not matter.

Observational process and subjective complete TOE. As we have demonstrated it is also important to localize the observer. In order to avoid potential qualms with modeling human observers, consider as a surrogate a (normal not extra cosmic) video camera filming=observing parts of the world. The camera may be fixed on Earth or installed on an autonomous robot. It records part of the universe u denoted by $o = o_{1:\infty}$. (If the lifetime of the observer is finite, we append zeros to the finite observation $o_{1:N}$).

In a computable universe, the observational process within it, is obviously also computable, i.e. there exists a program $s \in \{0,1\}^*$ that extracts observations o from universe u. Formally

$$\mathrm{UTM}(s, u_{1:\infty}^q) = o_{1:\infty}^{sq} \tag{2}$$

where we have extended the definition of UTM to allow access to an extra infinite input stream $u_{1:\infty}^q$. So $o_{1:\infty}^{sq}$ is the sequence observed by subject s in universe $u_{1:\infty}^q$ generated by q. Program s contains all information about the location and orientation and perception abilities of the observer/camera, hence specifies not only item (l) but also item (o) of Section 2.

A Complete TOE (CTOE) consists of a specification of a (TOE,subject) pair (q, s). Since it includes s it is a Subjective TOE.

CTOE selection principle. So far, s and q were fictitious subjects and universe programs. Let $o_{1:t}^{true}$ be the past observations of some concrete observer in our universe, e.g. your own personal experience of the world from birth till today. The future observations $o_{t+1:\infty}^{true}$ are of course unknown. By definition, $o_{1:t}$ contains *all* available experience of the observer, including e.g. outcomes of scientific experiments, school education, read books, etc.

The observation sequence $o_{1:\infty}^{sq}$ generated by a correct CTOE must be consistent with the true observations $o_{1:t}^{true}$. If $o_{1:t}^{sq}$ would differ from $o_{1:t}^{true}$ (in a single bit) the subject would have 'experimental' evidence that (q,s) is not a perfect CTOE. We can now formalize the CTOE selection principle as follows

Among a given set of perfect $(o_{1:t}^{sq} = o_{1:t}^{true})$ CTOEs $\{(q,s)\}$

$$\text{select the one of smallest length } \text{Length}(q) + \text{Length}(s) \tag{3}$$

Minimizing length is motivated by Ockham's razor. Inclusion of s is necessary to avoid degenerate TOEs like U and A. The selected CTOE (q^*, s^*) can and should then be used for forecasting future observations via $...o_{t+1:\infty}^{forecast} = \text{UTM}(s^*, u_{1:\infty}^{q^*})$.

4. Universal TOE - Formalization

Definition of Universal TOE. The Universal TOE generates all computable universes. The generated multiverse can be depicted as an infinite matrix in which each row corresponds to one universe.

To fit this into our framework we need to define a single program $\breve{q}$ that generates a single string corresponding to this matrix. The standard way to linearize an infinite matrix is to dovetail in diagonal serpentines though the matrix:

q	$\text{UTM}(q)$				
ϵ	u_1^ϵ	u_2^ϵ	u_3^ϵ	u_4^ϵ	$u_5^\epsilon \cdots$
0	u_1^0	u_2^0	u_3^0	u_4^0	$\cdots\cdots$
1	u_1^1	u_2^1	u_3^1	$\cdots\cdots$	
00	u_1^{00}	u_2^{00}	$\cdots\cdots$		
$\vdots$	$\vdots$	$\vdots$	$\vdots$	$\vdots$	

$$\breve{u}_{1:\infty} := u_1^\epsilon u_1^0 u_2^\epsilon u_3^\epsilon u_2^0 u_1^1 u_1^{00} u_2^1 u_3^0 u_4^\epsilon u_5^\epsilon u_4^0 u_3^1 u_2^{00} ...$$

Formally, define a bijection $i = \langle q, k \rangle$ between a (program, location) pair (q, k) and the natural numbers $I\!N \ni i$, and define $\breve{u}_i := u_k^q$. It is not hard to construct an explicit program $\breve{q}$ for UTM that computes $\breve{u}_{1:\infty} = u_{1:\infty}^{\breve{q}} = \text{UTM}(\breve{q})$.

Remarks. Cutting the universes in bits and interweaving them into one string might appear messy, but is unproblematic for two reasons: First, the bijection $i = \langle q, k \rangle$ is very simple, so any particular universe string u^q can easily be recovered from $\breve{u}$. Second, such an extraction will be included in the localization/ observational process s, i.e. s will contain a specification of the relevant universe q and which bits k are to be observed.

TOE versus UTOE. We can formalize the argument in the last section of simulating TOE by UTOE as follows: If (q, s) is a CTOE, then $(\breve{q}, \tilde{s})$ based on UTOE $\breve{q}$ and observer $\tilde{s} := rqs$, where program r extracts u^q from $\breve{u}$ and then o^{sq} from u^q, is an equivalent but slightly larger CTOE, since $\text{UTM}(\tilde{s}, \breve{u}) = o^{qs} = \text{UTM}(s, u^q)$ by definition of $\tilde{s}$ and $\text{Length}(\breve{q}) + \text{Length}(\tilde{s}) = \text{Length}(q) + \text{Length}(s) + O(1)$.

The best CTOE. Finally, one may define the best CTOE (of an observer with experience $o_{1:t}^{true}$) as

$$UCTOE := \arg\min_{q,s}\{\text{Length}(q) + \text{Length}(s) : o_{1:t}^{sq} = o_{1:t}^{true}\}$$

where $o_{1:\infty}^{sq} = \text{UTM}(s, \text{UTM}(q))$. This may be regarded as a formalization of the holy grail in physics; of finding such a TOE.

5. Conclusions

Respectable researchers, including Nobel Laureates, have dismissed and embraced each single model of the world mentioned in the introduction, at different times in history and concurrently. (Excluding All-a-Carte TOEs which I haven't seen discussed before.) As I have shown, Universal TOE is the sanity critical point. The most popular (pseudo) justifications of which theories are (in)sane have been references to the dogmatic Bible and Popper's limited falsifiability principle. This paper contained a more serious treatment of world model selection. I introduced and discussed the usefulness of a theory in terms of predictive power based on model *and* observer localization complexity. Extensions to more practical and realistic (partial, approximate, probabilistic) theories (rather than TOEs), more motivation and examples, and a proof that Ockham's razor is suitable for finding TOEs can be found in Ref. 7.

References

1. J. D. Barrow, P. C. W. Davies, and C. L. Harper, editors. *Science and Ultimate Reality*. Cambridge University Press, 2004.
2. A. Blumhofer and M. Hutter. Family structure from periodic solutions of an improved gap equation. *Nuclear Physics*, B484:80–96, 1997. Missing figures in B494 (1997) 485.
3. N. Bostrom. *Anthropic Bias*. Routledge, 2002.
4. M. Gell-Mann. *The Quark and the Jaguar: Adventures in the Simple and the Complex*. W.H. Freeman & Company, 1994.
5. E. Harrison. *Cosmology: The Science of the Universe*. Cambridge University Press, 2nd edition, 2000.
6. M. Hutter. *Universal Artificial Intelligence: Sequential Decisions based on Algorithmic Probability*. Springer, Berlin, 2005.
7. M. Hutter. A complete theory of everything (will be subjective). 2009. arXiv:0912.5434.
8. M. Li and P. M. B. Vitányi. *An Introduction to Kolmogorov Complexity and its Applications*. Springer, Berlin, 3rd edition, 2008.
9. J. Schmidhuber. Algorithmic theories of everything. Report IDSIA-20-00, quant-ph/0011122, IDSIA, Manno (Lugano), Switzerland, 2000.
10. L. Smolin. Scientific alternatives to the anthropic principle. Technical Report hep-th/0407213, arXiv, 2004.
11. M. Tegmark. Parallel universes. In *Science and Ultimate Reality*, pages 459–491. Cambridge University Press, 2004.

LINKS AND QUANTUM ENTANGLEMENT

ALLAN I. SOLOMON

Department of Physics and Astronomy, Open University,
Milton Keynes, MK7 6AA, UK
and
LPTMC, University of Paris VI, 75252, Paris, France
a.i.solomon@open.ac.uk

CHOON-LIN HO

Department of Physics, Tamkang University, Tamsui 251, Taiwan, R.O.C.
and
Department of Physics, and Center for Quantum Technologies,
National University of Singapore, 117543 Singapore
hcl@mail.tku.edu.tw

We discuss the analogy between topological entanglement and quantum entanglement, particularly for tripartite quantum systems. We illustrate our approach by first discussing two clearly (topologically) inequivalent systems of three-ring links: The Borromean rings, in which the removal of any one link leaves the remaining two non-linked (or, by analogy, non-entangled); and an inequivalent system (which we call the NUS link) for which the removal of any one link leaves the remaining two linked (or, entangled in our analogy). We introduce unitary representations for the appropriate Braid Group (B_3) which produce the related quantum entangled systems. We finally remark that these two quantum systems, which clearly possess inequivalent entanglement properties, are locally unitarily equivalent.

Keywords: Quantum entanglement; braid groups; topological links.

1. Introduction: The Borromean Rings and the NUS Link

In this note we shall explore the analogy between topological links and the quantum entanglement of tripartite systems. In the figures Fig. 1(a) and Fig. 1(b), we give examples of two different three-ring links.

The first, Fig. 1(a), represents the celebrated Borromean rings. This link has the property that removing any ring leaves the remaining two rings unlinked (non-entangled). The second, Fig. 1(b) , which we call for brevity the NUS link as it is part of the logo of the National University of Singapore, has the converse property; removing any ring still leaves the two remaining linked (entangled).

These two links recall the following tripartite quantum states: The Greenberger-Horne-Zeilinger (GHZ) state,[1] which is simply a tripartite extension of the bipartite

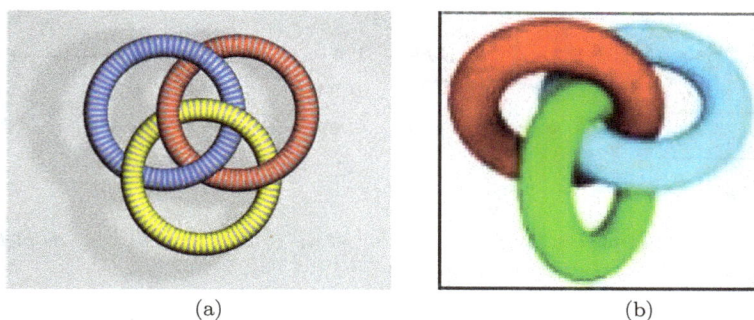

(a)   (b)

Fig. 1.   Two three-ring links. (a) Borromean Rings (b) NUS Link.

Bell state $(1/\sqrt{2})(|0,0,\rangle + |1,1\rangle)$,

$$|GHZ\rangle = (1/\sqrt{2})(|0,0,0\rangle + |1,1,1\rangle),   \qquad (1)$$

and

$$|\phi\rangle = (1/2)(|0,0,0\rangle + |0,1,1\rangle + |1,0,1\rangle + |1,1,0\rangle).   \qquad (2)$$

In the first case, measuring any subspace state as $|0\rangle$ (*resp.* $|1\rangle$) leads to the non-entangled state $|0,0\rangle$(*resp.* $|1,1\rangle$); while in the second case a similar determination always leads to a (maximally) entangled bipartite state (Bell state).

The mathematical representation of links is made via Braid Groups, introduced by Artin.[2] To pursue the quantum entanglement analogy further, we first discuss braid groups, with an introductory reminder of a presentation of the closely-related symmetric group. Then, in order to apply these ideas in quantum theory, we discuss their unitary representations, which we take to act on the qubit spaces.

2. Braid Groups and Links

2.1. *Symmetric group*

The symmetric group S_n (sometimes called the permutation group) is defined as the the set of $n!$ permutations on n distinct objects, combining according to the rule illustrated by

$$\begin{pmatrix} 1 & 2 & 3 & 4 \\ 3 & 1 & 2 & 4 \end{pmatrix} \begin{pmatrix} 1 & 2 & 3 & 4 \\ 1 & 3 & 2 & 4 \end{pmatrix} = \begin{pmatrix} 1 & 2 & 3 & 4 \\ 2 & 1 & 3 & 4 \end{pmatrix}   \qquad (3)$$

for the case of S_4. A diagrammatic representation of the resultant permutation is found in Figure 2(a). The symmetric group S_n has a presentation in terms of $n-1$ adjacent transpositions,[a] $\{s_i \ i = 1 \ldots n-1\}$ where s_i sends the i to $i+1$ and $i+1$

---

[a]The right-hand side of Eq.(3) is an adjacent transposition.

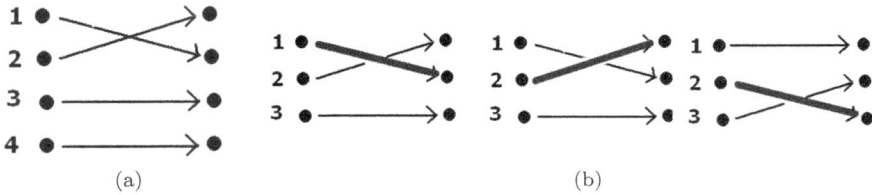

Fig. 2. Elements of S_4 and B_3. (a) A transposition in S_4. (b) Elements $\sigma_1, \sigma_1^{-1}, \sigma_2$ of B_3.

to i. This rather mysterious presentation is:

$$s_i s_j = s_j s_i \quad |i - j| > 1 \tag{4}$$

$$s_i s_i = I \tag{5}$$

$$s_i s_{i+1} s_i = s_{i+1} s_i s_{i+1} \tag{6}$$

where Eq.(6) plays an important role in the generalization to the *Braid group*, in which context it is known as the *braiding relation* or the *Yang-Baxter condition*.

2.2. Braid group

The braid group is like the symmetric group, but in three dimensions, so one must imagine the arrows joining the elements of a permuted set of points to go "over" or "under" each other. Intuitively, each element of the braid group B_n is one way of joining n points to another n points by strings. (For an expanded version of this intuitive definition see Ref. 3.)

The braid group B_n has a presentation in terms of $n - 1$ generators σ_i. This (defining) presentation is:

$$\sigma_i \sigma_j = \sigma_j \sigma_i \quad |i - j| > 1 \tag{7}$$

$$\sigma_i \sigma_{i+1} \sigma_i = \sigma_{i+1} \sigma_i \sigma_{i+1} . \tag{8}$$

Note that the constraint Eq.(5) is absent; this absence leads to all the Braid groups being of infinite order. Eq.(8) is known as the *braiding relation* or the *Yang-Baxter condition*. A diagrammatic representation of the elements σ_1 and σ_1^{-1}, as well as the second generator σ_2, of B_3 is given in Figure 2(b). This group is the main example that we discuss in this note, although for simplicity and illustration we start by discussing the group B_2, which has only one generator, and no braiding condition to satisfy; it is isomorphic to the infinite cyclic group, equivalently $\mathcal{Z}$, the set of integers under addition.

2.3. Knots and links

Of particular interest to us is the fact that, as shown by Alexander,[4] *all* knots and links may be obtained from elements of a braid group by the simple expedient of joining the the "dots"; that is, join 1 to 1, 2 to 2, and so on.

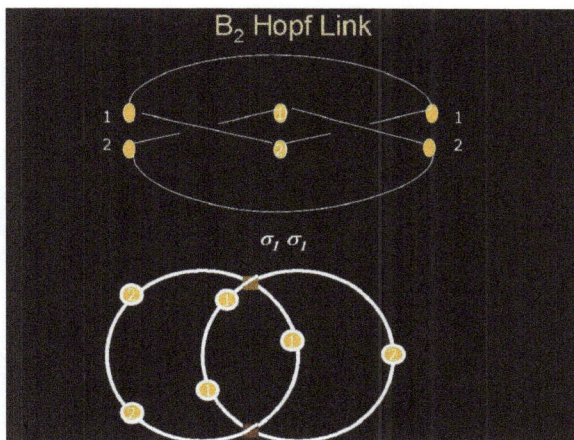

Fig. 3. In B_2, σ_1^2 produces the Hopf Link.

For the braid group B_2 with one generator σ_1, we can see that performing the action using the element σ_1^2 gives the Hopf Link, as in Figure 3.

For the braid group B_3 with two generators σ_1 and σ_2, we can see that performing this action with the braid element

$$\sigma_1\sigma_2^{-1}\sigma_1\sigma_2^{-1}\sigma_1\sigma_2^{-1} \tag{9}$$

produces the Borromean rings, as in Figure 4(a).

On the other hand, the braid element

$$\sigma_1\sigma_2\sigma_1\sigma_2\sigma_1\sigma_2 \tag{10}$$

corresponds to the NUS link, as in Figure 4(b).

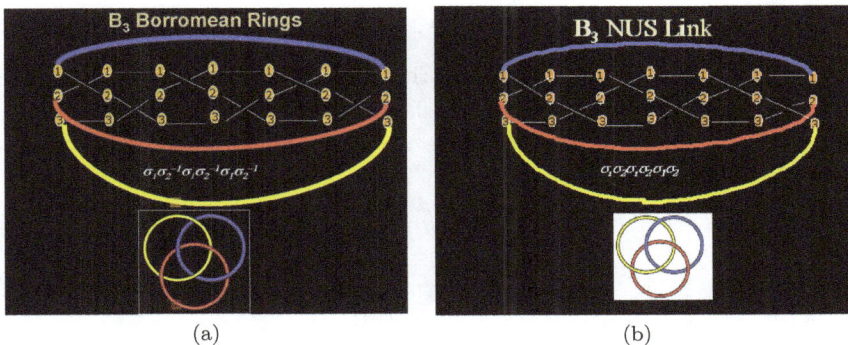

Fig. 4. Two links from B_3. (a) Borromean Rings produced from an element of B_3. (b) NUS Link produced from an element of B_3.

3. Unitary Representations of Braid Groups and Entanglement

In order to relate the action of the braid group to unitary transformations on quantum systems, we adopt the following procedure:

(i) we associate each initial point of the braid group with a qubit (e.g. for B_3 there are 3 initial points and therefore we may represent unitary action on a three-qubit system);

(ii) for a braid word of the form g^n we shall assume that the quantum entanglement is generated by the unitary representative $\hat{g}$;

(iii) to simulate the closure of the action of a braid word, say g^n, to form a *link*, the unitary matrix $\hat{g}^n$ must equal I (up to a phase factor).

A generic unitary representation of the braid group which satisfies the relation Eq.(7) can in principle be obtained from the following:

$$\hat{\sigma}_i = I \times \cdots \times U \times I \cdots \times I \tag{11}$$

where $I = \begin{bmatrix} 1 & 0 \\ 0 & 1 \end{bmatrix}$ and U is a 4×4 unitary matrix occupying the $(i, i+1)$ position in the product. Of course it is more difficult to satisfy Eq.(8), the braiding, or Yang-Baxter, relation. We describe representations for B_2 and B_3 in the following.

3.1. *The Hopf link and entanglement*

In a sense finding a unitary representation for B_2 is a trivial exercise, as in this case there are effectively no relations on the single generator σ_1. Thus any unitary matrix will do. For our purpose we require a 4×4 unitary matrix - since it is acting on the two-qubit space. We define a unitary transformation matrix as follows:[b]

$$\hat{\sigma}_1 \equiv \frac{e^{i\theta}}{\sqrt{2}} \begin{bmatrix} 1 & 0 & 0 & 1 \\ 0 & 1 & 1 & 0 \\ 0 & 1 & -1 & 0 \\ 1 & 0 & 0 & -1 \end{bmatrix} \qquad (\theta/\pi \text{ irrational}). \tag{12}$$

The braid word word corresponding to the Hopf link is $\sigma_1{}^2$ so following the procedure as in 3(ii) outlined above, our choice of unitary representative $\hat{\sigma}_1$ is the generator of entanglement, and produces a maximally entangled (Bell) state from a (generic) non-entangled state,

$$\hat{\sigma}_1 |0, 0\rangle = \frac{\exp(i\theta)}{\sqrt{2}} (|0, 0\rangle + |1, 1\rangle). \tag{13}$$

Note that $\hat{\sigma}_1^2 = e^{2i\theta} I$, satisfying condition 3(iii).

---

[b]The multiplicative phase factor is necessary to ensure a genuine representation of B_2, as in its absence the representation would be *non-faithful*, and finite dimensional (Z_2).

3.2. *Unitary representations for* B_3

3.2.1. *The NUS link and entanglement*

Using the matrix U of Ref. 5 (where it is defined however without the phase factor) we define

$$U \equiv \frac{e^{i\theta}}{\sqrt{2}} \begin{bmatrix} 1 & 0 & 0 & -1 \\ 0 & 1 & -1 & 0 \\ 0 & 1 & 1 & 0 \\ 1 & 0 & 0 & 1 \end{bmatrix} \tag{14}$$

where θ/π is irrational but otherwise arbitrary, as above. The representation for B_3 is

$$\hat{\sigma}_1 = U \times I, \quad \hat{\sigma}_2 = I \times U. \tag{15}$$

One may verify that the braiding relation Eq.(8) is satisfied. As in Eq.(10), the braid word $(\sigma_1\sigma_2)^3$ produces the NUS link. Following the recipe above, we note that $(\hat{\sigma}_1\hat{\sigma}_2)^3$ is indeed the 8×8 unit matrix (up to a non-vanishing phase factor); and the generator of entanglement for this link $\hat{\sigma}_1\hat{\sigma}_2$ produces the state $|\phi\rangle$ of Eq.(2) (up to the phase factor $e^{2i\theta}$)

$$\hat{\sigma}_1\hat{\sigma}_2|0,0,0,\rangle = \exp(2i\theta)|\phi\rangle. \tag{16}$$

3.2.2. *Entanglement and the Borromean rings*

We use a different representation for the Borromean rings in order to to obtain the GHZ state directly. Following the procedure detailed in Ref. 6 we use the Jones representation[c]

$$\hat{\sigma}_i = Ah_i + A^{-1}I,$$
$$\hat{\sigma}_i^{-1} = A^{-1}h_i + AI. \tag{17}$$

We choose $A = \exp(3\pi i/8)$, and the matrices h_1 and h_2 as follows:

$$h_1 = \sqrt{2} \begin{pmatrix} 1&0&0&0&0&0&0&0 \\ 0&1&0&0&0&0&0&0 \\ 0&0&1&0&0&0&0&0 \\ 0&0&0&1&0&0&0&0 \\ 0&0&0&0&0&0&0&0 \\ 0&0&0&0&0&0&0&0 \\ 0&0&0&0&0&0&0&0 \\ 0&0&0&0&0&0&0&0 \end{pmatrix}. \tag{18}$$

---

[c]In what follows we omit the explicit irrational phase factor needed to ensure the faithfulness of the representation.

and

$$
h_2 = \frac{1}{\sqrt{2}}
\begin{pmatrix}
1 & 0 & 0 & 0 & 0 & 0 & 0 & -1 \\
0 & 1 & 0 & 0 & 0 & 0 & -1 & 0 \\
0 & 0 & 1 & 0 & 0 & -1 & 0 & 0 \\
0 & 0 & 0 & 1 & -1 & 0 & 0 & 0 \\
0 & 0 & 0 & -1 & 1 & 0 & 0 & 0 \\
0 & 0 & -1 & 0 & 0 & 1 & 0 & 0 \\
0 & -1 & 0 & 0 & 0 & 0 & 1 & 0 \\
-1 & 0 & 0 & 0 & 0 & 0 & 0 & 1
\end{pmatrix}. \tag{19}
$$

Then it may be verified that $\hat{\sigma}_1$ and $\hat{\sigma}_2$ satisfy Eq.(8). The Borromean link is defined by the braid word given in Eq.(9), and additionally the criterion of 3(iii) is satisfied, since $(\hat{\sigma}_1 \hat{\sigma}_2^{-1})^3$ equals the identity up to a phase factor.

Applying the braid word entanglement generator, in this case $\hat{\sigma}_1 \hat{\sigma}_2^{-1}$, to the fiducial ground state $|0,0,0\rangle$, we obtain the GHZ state

$$
\hat{\sigma}_1 \hat{\sigma}_2^{-1} |0,0,0\rangle = \frac{1+i}{2}(|0,0,0\rangle + |1,1,1\rangle). \tag{20}
$$

4. Conclusions: Local Unitary Equivalence

This note has emphasized the analogy between topological entanglement in the form of links, and quantum entanglement.[d]

We introduced a recipe whereby we could relate a topological link to an appropriate entangled quantum state, via a unitary representation of the braid word producing the link. For the two cases of links produced by B_3, the Borromean rings link and the one we dubbed the NUS link, we used two different unitary representations of B_3. It should come as no surprise that different unitary representations produce different pictures of entanglement, as quantum entanglement is not invariant under unitary transformations. And indeed, from our description of the Borromean rings link and the NUS link in the Introduction, we can see that the topological entanglement properties of these two links are quite different. Similarly, from the discussion following Eqs.(1) and (2) we also see that the quantum entanglement properties of the states $|GHZ\rangle$ and $|\phi\rangle$ are similarly distinct.

Further, it would appear that the entanglement properties in the 3-qubit case are not invariant under *local unitary transformations* either. It has been pointed out[7] that in fact the two states $|GHZ\rangle$ and $|\phi\rangle$ are locally unitarily equivalent, since by use of the *local transformation* $V = v \otimes v \otimes v$ where $v = \frac{1}{\sqrt{2}}\begin{pmatrix} 1 & 1 \\ -1 & 1 \end{pmatrix}$,

$$
V \frac{1}{\sqrt{2}}(|0,0,0\rangle + |1,1,1\rangle) = -\frac{1}{2}(|0,0,0\rangle + |0,1,1\rangle + |1,0,1\rangle + |1,1,0\rangle).
$$

---

[d]This analogy has also been remarked upon by, among others, Kauffman and Lomonaco,[8] and one of the authors.[9]

Thus, in the case of *tripartite* states, at least, local unitary equivalence does not preserve the entanglement properties.

Acknowledgments

Both authors wish to acknowledge discussions with Professor Choo Hiap Oh of the Department of Physics whom we thank for his warm hospitality, as well as that of the Centre for Quantum Technologies at the National University of Singapore. This work is supported in part by Singapore's A*STAR grant WBS (Project Account No.) R-144-000-189-305, and in part by the National Science Council (NSC) of the R.O.C. under Grant No. NSC 96-2112-M-032-007-MY3.

References

1. D.M. Greenberger, M.A. Horne, and A. Zeilinger, *Going Beyond Bell's Theorem*, in *Bell's Theorem, Quantum Theory, and Conceptions of the Universe*, M. Kafatos (Ed.), pg. 69-72 (Kluwer, Dordrecht, 1989).
2. E. Artin, *The theory of braids*, Annals of Mathematics (2) 48 (1947),101 126
3. http://en.wikipedia.org/wiki/Braid_group, http://mathworld.wolfram.com/BraidGroup.html
4. J.W. Alexander, *A lemma on systems of knotted curves*, Proc. Nat. Acad. Sci. USA,9 (1923) pp. 9395
5. J-L. Chen, K. Xue and M-L. Ge, Phys. Rev. A **76**, 042324 (2007).
6. C.-L. Ho, A.I. Solomon and C.-H. Oh, "Quantum entanglement, unitary braid representation and Temperley-Lieb algebra", NUS and Tamkang preprint, May 2010.
7. Y. Zhang, N. Jing, and M.-L. Ge, J. Phys. **A41**, 055310 (2008).
8. L.H. Kauffman and S.J. Lomonaco, New Journal of Physics, 4, 73.1 - 73.18 ,(2002).
9. A. I. Solomon, " Bell, Group and Tangle", Physics of Atomic Nuclei **73**, 524 (2010). (Presents in preliminary form some of the ideas of the current paper. However, due to an algebraic error the generator of entanglement for the Borromean Rings system was taken to be $(\hat{\sigma}_1\hat{\sigma}_2^{-1})^3$ instead of $\hat{\sigma}_1\hat{\sigma}_2^{-1}$.)

NONLINEAR SCHRÖDINGER-PAULI EQUATIONS

WEI KHIM NG* and RAJESH R. PARWANI†

Department of Physics, National University of Singapore,
2 Science Drive 3, Singapore 117542
*phynwk@nus.edu.sg
†parwani@nus.edu.sg*

We obtain novel nonlinear Schrödinger-Pauli equations through a formal non-relativistic limit of appropriately constructed nonlinear Dirac equations. This procedure automatically provides a physical regularisation of potential singularities brought forward by the nonlinear terms and suggests how to regularise previous equations studied in the literature. The enhancement of contributions coming from the regularised singularities suggests that the obtained equations might be useful for future precision tests of quantum nonlinearity.

Keywords: Nonlinear Dirac equations; nonlinear quantum mechanics; non-relativistic limit; regularisation.

1. Introduction

Several nonlinear extensions of Schrödinger's equation have been constructed to probe the accuracy of quantum linearity.[1-3] For example, Weinberg proposed a class of equations which were then used in several experimental tests, see Refs. 4–7 and references therein. The results indicated that any potential non-linearity in those systems had to be smaller than some bounds.

Ignoring external fields, the nonlinear Schrödinger equations may be written in the form

$$i\hbar \frac{\partial}{\partial t}\psi = -\frac{\hbar^2}{2m}\nabla^2\psi + f_{NR}(\psi)\psi \tag{1}$$

where the nonlinearity f_{NR} depends in general on the wavefunction, its conjugate and their derivatives. f_{NR} may be written as a ratio of two terms, $N(\psi)/D(\psi)$, with equal factors of ψ in the numerator and denominator to keep the scale invariance $\psi \to \lambda\psi$, λ a constant, of the linear Schrödinger equation. This scale invariance ensures the wavefunction can be freely normalised. The denominator is typically a monomial in $\psi^\star\psi$ so that the nonlinear term may be made separable for independent systems. As the nonlinearity must be weak on phenomenological grounds, the solutions of the linear equation must be very close to some solutions of the modified equation. However, any solutions of the linear equation that have nodes would make $D(\psi)$ vanish at some points. Thus, the nonlinearity would generally be singular and ill-defined at those points.

Weinberg,[3] discusses classes of nonlinear Schrödinger-Pauli equations where the nonlinearity turns out to be finite at the nodes because the numerator vanishes faster than the denominator there. However this will not happen for general classes of nonlinearities where $N(\psi)$ has derivatives, such as the equations studied in Ref. 8.

It was suggested in Ref. 9 that quantum nonlinearity might be linked to the breaking of space-time symmetry. This idea was supported by a study in Ref. 10 in the relativistic regime: That is, a deviation from quantum linearity is associated with a violation of Lorentz symmetry. Thus, in this paper, we discuss how to construct novel classes of nonlinear Schrödinger-Pauli equations, which have the above-mentioned scale invariance, starting from nonlinear Dirac equations. However, we will keep our option open by considering both Lorentz invariant and Lorentz violating nonlinear Dirac equations.[11]

As we shall see, our procedure of obtaining the Scrodinger-Pauli equations by taking the limit of relativistic equations has the advantage of indicating a natural and physical regularisation of the singularities. We remark that we focus on genuine nonlinear Dirac equations that cannot be linearised by performing a nonlinear gauge transformation.[11]

In the next section, we discuss, in general terms, the formal non-relativistic limit of a subset of nonlinear Dirac equations constructed in Ref. 11. For conciseness, in this paper we only consider the case where $F = fI$ in (2), I being the identity matrix in spinor space. Explicit examples of the lowest order nonlinearities, corresponding to one factor of $\psi^\star \psi$ in $D(\psi)$ are exhibited in Section 3, other cases being similarly handled. The singularity resolution is discussed in Section 4 and we end with a discussion in Section 5.

We note in passing that nonlinear Schrödinger equations of other types have been constructed from Levy-Leblond's "non-relativistic Dirac equation" which is itself the non-relativistic limit of the usual Dirac equation.[12,13]

2. Non-Relativistic Limit

We start from nonlinear Dirac equations of the form

$$(i\hbar\gamma^\mu\partial_\mu - mc + \epsilon F)\psi = 0, \tag{2}$$

where $F = F(\psi, \bar\psi) = fI$ and where we have made the small parameter ϵ explicit. We demand that F has certain properties so that desirable characteristics of the linear Dirac equation, such as locality, conservation of probability, separability and invariance under $\psi \to \lambda\psi$, are retained (we are adopting the standard kinematical structure of quantum mechanics, in particular the standard inner product). The other symbols in (2), such as those for the gamma matrices, have their usual meanings; our conventions are similar to those in the textbook[14] and in Ref. 11.

In Hamiltonian form the equation is

$$i\hbar\frac{\partial}{\partial t}\psi = \left(i\hbar c\boldsymbol{\alpha} \cdot \boldsymbol{\nabla} + \beta mc^2 - \epsilon c\beta f\right)\psi \tag{3}$$

where $\alpha^i = \gamma^0 \gamma^i$ and $\beta = \gamma^0$. It maybe be decomposed into two equations by introducing upper and lower components of the wavefunction,

$$\psi = \begin{pmatrix} \varphi \\ \chi \end{pmatrix} e^{-imc^2 t/\hbar} \tag{4}$$

where the rest energy has been extracted as it is the largest component in the non-relativistic limit. We adopt the standard textbook procedure in obtaining the leading non-relativistic limit, but for clarity we repeat some steps below. In order to make the algebra manageable, we simply take $1/c$ to be the same order of magnitude as the nonlinearity scale ϵ and keep only the leading nonlinear term in the standard non-relativistic expansion. Thus we can isolate the leading order nonlinear contribution. However, in realistic applications, ϵ will be much smaller than $1/c$: This will introduce higher order, $1/c$, terms which *do not affect the leading order nonlinear contribution.*

Substituting (4) into (3) we get

$$i\hbar \frac{\partial}{\partial t} \begin{pmatrix} \varphi \\ \chi \end{pmatrix} = i\hbar c \begin{pmatrix} \boldsymbol{\sigma} \cdot \boldsymbol{\nabla} \chi \\ \boldsymbol{\sigma} \cdot \boldsymbol{\nabla} \varphi \end{pmatrix} + mc^2 \begin{pmatrix} 0 \\ -2\chi \end{pmatrix} - \epsilon c f \begin{pmatrix} \varphi \\ -\chi \end{pmatrix}. \tag{5}$$

(As in the usual textbook procedure, the *ansatz* (4) removes the mass term for the upper component). From the lower component of (5) we have,

$$\chi = \frac{i\hbar \boldsymbol{\sigma} \cdot \boldsymbol{\nabla} \varphi}{2mc} - \frac{i\hbar}{2mc^2} \frac{\partial \chi}{\partial t} + \frac{\epsilon f \chi}{2mc}. \tag{6}$$

Let $\chi_0 = \frac{i\hbar \boldsymbol{\sigma} \cdot \boldsymbol{\nabla} \varphi}{2mc}$. Then expanding (6) about χ_0, we obtain $\chi = \chi_0 + O\left(\frac{\epsilon}{c^2}, \frac{1}{c^3}\right)$. That is, χ is the same as that in the linear theory. Substituting (6) into the upper component of (5), we arrive at

$$i\hbar \frac{\partial}{\partial t} \varphi \simeq -\frac{\hbar^2}{2m} \nabla^2 \varphi - \epsilon c f_{NR} \varphi \tag{7}$$

where f_{NR} means that the state dependence of f has been simplified using (4, 6) and higher order terms dropped. Below we look at some explicit examples.

3. Examples

3.1. *Lorentz invariant f with one derivative*

A Lorentz invariant f with one derivative and which is odd under the parity transformation is

$$f_1 = \epsilon \frac{\partial_\mu j_5^\mu}{\bar{\psi}\psi}, \tag{8}$$

where $j_5^\mu = \bar{\psi} \gamma^\mu \gamma_5 \psi$ is the usual chiral current. The non-relativistic limit is

$$i\hbar \frac{\partial \varphi}{\partial t} \simeq -\frac{\hbar^2 \nabla^2 \varphi}{2m} - \epsilon c \varphi \frac{\boldsymbol{\nabla} \cdot (\varphi^\dagger \boldsymbol{\sigma} \varphi)}{|\varphi|^2}. \tag{9}$$

The factor $\boldsymbol{\nabla} \cdot (\varphi^\dagger \boldsymbol{\sigma} \varphi)$ appears often in parity odd equations;[11] it couples the spin components of the two-component spinor.

3.2. *Lorentz invariant f with two derivatives*

For an example of a Lorentz invariant f with two derivatives consider

$$f_2 = \frac{\epsilon \left(\partial_\mu \partial^\mu \bar{\psi}\psi\right) + \delta \left(\partial_\mu \bar{\psi}\right)\left(\partial^\mu \psi\right)}{\bar{\psi}\psi} , \tag{10}$$

where ϵ and δ are two independent small parameters taken to be of order $1/c$ below. The non-relativistic equation is

$$i\hbar \frac{\partial \varphi}{\partial t} \simeq -\frac{\hbar^2 \nabla^2 \varphi}{2m}\left(1 + \frac{mc\delta}{2\hbar^2}\right) + \frac{\varphi}{|\varphi|^2}\left\{-\frac{\delta imc}{\hbar}\left[\varphi^\dagger \frac{\partial \varphi}{\partial t} - \left(\frac{\partial \varphi^\dagger}{\partial t}\right)\varphi\right]\right.$$
$$\left. + \epsilon c \left(\nabla^2 \varphi^\dagger \varphi\right) + \delta c \left(\nabla \varphi^\dagger\right)\cdot\left(\nabla \varphi\right)\right\} . \tag{11}$$

3.3. *Lorentz violating, parity even f*

Lorentz violating non-linear Dirac equations are of some interest.[9,11,15–17] An example of such an f with no derivatives and even under parity is

$$f_3 = A_\mu \frac{\bar{\psi}\gamma^\mu \psi}{\bar{\psi}\psi} \tag{12}$$

where A_μ is a constant vector background field. The non-relativistic limit is

$$i\hbar \frac{\partial \varphi}{\partial t} \simeq -\frac{\hbar^2 \nabla^2 \varphi}{2m} - cA_0\varphi + \frac{i\hbar\varphi}{2m}\frac{\boldsymbol{A}\cdot\left[\varphi^\dagger \nabla \varphi - \left(\nabla \varphi^\dagger\right)\varphi\right]}{|\varphi|^2} . \tag{13}$$

3.4. *Lorentz violating, parity odd f*

A Lorentz violating f which is odd under parity is

$$f_4 = A_\mu \frac{\bar{\psi}\gamma_5 \gamma^\mu \psi}{\bar{\psi}\psi} . \tag{14}$$

The non-relativistic equation is

$$i\hbar \frac{\partial \varphi}{\partial t} \simeq -\frac{\hbar^2 \nabla^2 \varphi}{2m} - \frac{c\varphi^\dagger \boldsymbol{A}\cdot\boldsymbol{\sigma}\varphi}{|\varphi|^2}\varphi + \frac{A_0 i\hbar\varphi}{2m}\frac{\left[\varphi^\dagger \boldsymbol{\sigma}\cdot\nabla \varphi - \left(\nabla \varphi^\dagger\right)\cdot\boldsymbol{\sigma}\varphi\right]}{|\varphi|^2} . \tag{15}$$

4. Apparent Singularities

From the above examples, we see the appearance of the following structures in the non-linear Schrödinger-Pauli equations,

$$X = \frac{\varphi^\dagger \boldsymbol{\sigma}\cdot\nabla \varphi}{|\varphi|^2} , \quad Y = \frac{\left(\nabla \varphi^\dagger\right)\cdot\left(\nabla \varphi\right)}{|\varphi|^2} , \quad Z = \frac{\varphi^\dagger \nabla^2 \varphi}{|\varphi|^2} . \tag{16}$$

Clearly, at the nodes of φ, these forms are singular. However, we can avoid these singularities in a natural way. For our nonlinear Dirac equations,[11] the nonlinearities have the common structure $\frac{N(\bar{\psi},\psi)}{(\bar{\psi}\psi)^n}$, the $n = 1$ case being discussed here. In terms of the two component spinors this is $\frac{N}{|\varphi|^2 - |\chi|^2}$ where the lower (small) component contribution $|\chi|^2$ is usually dropped in the non-relativistic limit. However, at the nodes of φ, we must keep the small component in the denominator. This regulates the above mentioned singularity for the following reason: From (6), the lower component is proportional to the slope of φ (i.e. $\nabla\varphi$), which is unlikely to vanish simultaneously at the nodes except for special cases. In such extreme cases, one would need to retain the smaller terms (higher order in $1/c$) in the non-relativistic expansion of the denominator.

For the specific examples illustrated above, the replacement $|\varphi|^2 \to |\varphi|^2 - |\chi|^2$ in the denominator makes $X = Z = 0$ at a node of φ while Y becomes finite and actually enhanced because of the small denominator. Note that, at the level of the equation of motion, there is an extra factor of φ which multiplies the nonlinearity f. It is clear that X and Z contributions in the equation of motion are not singular even at the nodes but the Y contribution is, unless regularised as discussed above.

So far we have discussed singularities in f and at the level of equations of motion. As for observables, let us consider shifts in the energy levels given by first-order perturbation theory,

$$\delta E = \int d^3x \ <\varphi|F|\varphi> = \int d^3x |\varphi|^2 f(\varphi) \tag{17}$$

where the unperturbed (linear equation) wavefunctions are used. We see that the X, Y, Z structures give finite shifts. Singularities will appear in $n \geq 2$ classes of nonlinearities discussed in Ref. 11, two examples of which are given by

$$V = Y^2 = \frac{\left[(\boldsymbol{\nabla}\varphi^\dagger)\cdot(\boldsymbol{\nabla}\varphi)\right]\left[(\boldsymbol{\nabla}\varphi^\dagger)\cdot(\boldsymbol{\nabla}\varphi)\right]}{|\varphi|^2|\varphi|^2}, \tag{18}$$

$$W = YZ = \frac{\left[(\boldsymbol{\nabla}\varphi^\dagger)\cdot(\boldsymbol{\nabla}\varphi)\right](\varphi^\dagger\nabla^2\varphi)}{|\varphi|^2|\varphi|^2}. \tag{19}$$

It is clear that the energy shifts will be singular for such terms unless the regularisation is implemented.

The above discussion has ignored external potentials which must be included in realistic experiments. For example, in the presence of an external gauge field and for a particular spin component $\varphi = \begin{pmatrix} 1 \\ 0 \end{pmatrix} \varphi_0$, the lower component is modified from its previous form χ_0 to become

$$\chi_0 = \frac{i\hbar}{2mc}\left(\frac{\partial}{\partial z} - \frac{e}{c}A_z\right)\varphi_0. \tag{20}$$

Setting $\varphi_0 = g(\boldsymbol{x} - \boldsymbol{x_0})$ near a node we have

$$|\chi_0|^2 = \frac{\hbar^2}{4m^2c^2}\left[\left(\frac{\partial g}{\partial z}\right)^2 - 2\frac{e}{c}A_z g\frac{\partial g}{\partial z} + \frac{e^2}{c^2}A_z^2 g^2\right]. \tag{21}$$

In this case, at the node of φ, $|\chi_0|^2$ has exactly the same form as the case when the gauge field is absent.

We can see that the contributions from nonlinear effects are largest (if non-zero) at the nodes. This suggests that future tests for quantum nonlinearity should focus on systems containing nodes in their wavefunctions.

We remark that the nonlinear equations discussed in Ref. 8 have been applied to the hydrogen system[18] but the physical consequences of singularities at the nodes of wavefunctions ($\varphi \to 0$) was not discussed.

5. Discussion

We have illustrated how to obtain novel classes of nonlinear Schrödinger-Pauli equations starting from the nonlinear Dirac equations constructed in Ref. 11, the latter equations themselves being more general than previous constructions.[19–21] For example, we have cases where the time-derivatives appear in the nonlinearity, and cases where the two components of the spinor are coupled through parity violation. We remark that probability is conserved for all of our non-relativistic equations. Also, the equations that are descended from Lorentz covariant equations are Galilean invariant.

An interesting point to note is that certain Lorentz-violating nonlinear Dirac equations have non-relativistic limits that are Galilean invariant. For example, for f_3, if the background field has only a time component, the leading non-relativistic limit actually becomes linear and invariant under Galilean transformations. For f_4, choosing a space-like background field will cause the non-relativistic equation to be still nonlinear but invariant under Galilean transformations.

We had taken the nonlinearity parameter ϵ to be the same order of magnitude as $1/c$ for ease of power counting, as our main aim was to isolate the leading nonlinear structure in the formal non-relativistic limit. We saw that potential singularities in the Schrödinger-Pauli equations are regularised by keeping the subleading lower components of the four component Dirac spinor in the denominators of the nonlinear terms. Thus, physically, it is the relativistic corrections that regulate the singularities. Precisely at a node, if the numerator is is nonzero, the nonlinearity is actually enhanced by the small denominator.

The situation here is qualitatively similar to a previous study[22] of an information-theoretic motivated nonlinear Schrödinger equation,[9] where the contribution to energy shifts from states with nodes was enhanced relative to states which had no nodes. Note also that in replacing the potentially singular denominator $|\varphi|^2$ by $|\varphi|^2 - |\chi|^2$ as in Section 4, one has introduced an infinite number of derivatives, through a formal expansion of the denominator, into the nonlinear terms even though we had started with a finite number of derivatives. This again is qualitatively similar to the situation with the information-theoretic nonlinearity.[9]

In actual applications, such as tests of quantum linearity, one would have to set ϵ much smaller than $1/c$ in the constructed nonlinear Schrödinger-Pauli equations even though they were formally derived from the nonlinear Dirac equations assuming $\epsilon \sim 1/c$.

The main suggestion from this study is that future precision low-energy experiments, probing deviations from quantum linearity, should focus on systems which have nodes in their limiting linear wavefunctions. It is there that the nonlinearity, if nonzero, will be enhanced.

References

1. I. Bialynicki-Birula and J. Mycielski, *Ann. Phys.* **100**, 62 (1976).
2. T. W. B. Kibble, *Comm. Math. Phys.* **64**, 73 (1978).
3. S. Weinberg, *Ann. Phys.* **194**, 336 (1989).
4. J. J. Bolinger, et. al., *Phys. Rev. Lett.* **63**, 1031 (1989).
5. T. Chupp and R. Hoare, *Phys. Rev. Lett.* **64**, 2261 (1990).
6. R. L. Walsworth, et. al., *Phys. Rev. Lett.* **64**, 2599 (1990).
7. P. K. Majumder, et.al., *Phys. Rev. Lett.* **65**, 2931 (1990).
8. H. D. Doebner, G. A. Goldin and P. Nattermann, *J. Math. Phys.* **40**, 49 (1999).
9. R. Parwani, *Ann. Phys.* **315**, 419 (2005).
10. W. K. Ng and Rajesh R. Parwani arXiv:0908.0180.
11. W. K. Ng and R. Parwani, *SIGMA* **5**, 023 (2009).
12. C. Duval, P. A. Horvathy and L. Palla, *Phys. Rev. D***52**, 4700 (1995).
13. C. Duval, P. A. Horvathy and L. Palla, *Ann. Phys.* **249**, 265 (1996).
14. C. Itzykson and J. B. Zuber, *Quantum field theory* (New York: McGraw-Hill International Book Co. 1980).
15. W. K. Ng and R. Parwani, arXiv:0805.3015.
16. L. H. Haddad and L. D. Carr, arXiv:0803.3039.
17. W. K. Ng, *Int. J. Mod. Phys. A* **24**, 3476 (2009).
18. H. D. Doebner and G. A. Goldin, *IOP: Conf. Ser.* **185**, 243 (2004).
19. H.D. Doebner and R. Z. Zhdanov, arXiv:quant-ph/0304167.
20. W. I. Fushchich and R. Z. Zhdanov, *Phys. Repts.* **172**, 123-174 (1989).
21. W. I. Fushchich and R. Z. Zhdanov, *Symmetries and Exact Solutions of Nonlinear Dirac Equations* (Ukraina Publishers 1997).
22. R. Parwani and G. Tabia, *J. Phys. A: Math. Gen.* **40**, 5621-5635 (2007).

STATISTICAL MECHANICS OF DAVYDOV-SCOTT'S PROTEIN MODEL IN THERMAL BATH

A. SULAIMAN,[*,a,b,d] F. P. ZEN,[a,d] H. ALATAS[c,d] and L. T. HANDOKO[e,f]

[a] *Theoretical Physics Laboratory, THEPI Research Division,*
Institut Teknologi Bandung, Jl. Ganesha 10, Bandung 40132, Indonesia

[b] *Badan Pengkajian dan Penerapan Teknologi, BPPT Bld. II (19th floor),*
Jl. M.H. Thamrin 8, Jakarta 10340, Indonesia

[c] *Theoretical Physics Division, Department of Physics, Bogor Agricultural University,*
Kampus Dermaga, Bogor, Indonesia

[d] *Indonesia Center for Theoretical and Mathematical Physics (ICTMP),*
Jl. Ganesha 10, Bandung 40132, Indonesia

[e] *Group for Theoretical and Computational Physics,*
Research Center for Physics, Indonesian Institute of Sciences,
Kompleks Puspiptek Serpong, Tangerang 15310, Indonesia

[f] *Department of Physics, University of Indonesia,*
Kampus UI Depok, Depok 16424, Indonesia
* asulaiman@webmail.bpp.go.id

The Davydov-Scott monomer contacting with thermal bath is investigated using Lindblad open quantum system formalism. The Lindblad equation is discussed through path integral method. It is found that the environmental effects contribute destructively to the specific heat, and large interaction between amide-I and amide-site is not prefered for a stable Davydov-Scott monomer.

Keywords: Davydov-Scott; open quantum system; specific heat.

1. Introduction

Does the Davydov-Scott's soliton exist at biological temperature? The question has attracted more interest in the last decades.[1–4] Earlier studies using finite temperature molecular dynamics showed that the Davydov soliton lifetime is only few picoseconds which is too short at the biological temperature. The reason is the random thermal prevents Davydov self-trapping from occurring as, for example as discussed in Ref. 5 which showed that the two-quantum state might be more stable than the one-quantum state. Furthermore, using the standard Davydov model, some numerical calculations also indicated that soliton is stable at 310K. On the other hand, the analytic calculation based on trial function or perturbation methods obtained that soliton is stable at 300K.[6,7]

The above-mentioned calculations were performed using the equilibrium quantum system at finite temperature. The finite temperature means that the quantum system is in contact with environment such as thermal bath. The interaction of a system with its environment is given by the dissipation effect in quantum system. However, the dissipation effect leads to a serious problem for quantization procedure. The most appropriate theory to resolve this problem is the Quantum State Diffusion (QSD) based on Lindblad formulation.[8] The first application of QSD to the protein model has been done by Cuevas $et.al.$[4] Their calculation on Davydov-Scott monomer showed that at room temperature the semi classical approach might be a good approximation compared to the corresponding full quantum system. However the study was focused on the dynamical aspect of the system, $i.e.$ the solution of Heisenberg's picture and its wave function based on the QSD equation. Recent studies of the anharmonic effect for the monomer has also been done by us using path integral and the thermodynamics function.[9] The advantage of calculating thermodynamics function is also relevant in another approaches such as the models describing the phenomena in term of elementary matter interactions using lagrangian.[10,11] It was argued that the thermodynamical properties should be easier to observe than another ones based on the wave function. Therefore, it is important to study such system from statistical mechanics point of view.

The paper is organized as follows. In Sec. 2 the Davydov Hamiltonian for a one-dimensional molecular monomer is described and the thermal bath effect is studied in Lindblad equation. In Sec. 3 the thermodynamic properties are investigated using path integral method. The paper is ended by a short summary.

2. Lindblad Open Quantum System Formulation

We use Davydov-Scott's model of the alpha-helix protein. The Davydov-Scott monomer is a coupled of the amide-I oscillator that expressed by the coordinate (x) and momentum (p) operators, and the amide-site is expressed by the displacement and momentum operators, Q and P, respectively. Hamiltonian in the model has the form,[9]

$$H = \frac{p^2}{2m} + \frac{1}{2}m\omega^2 x^2 + \frac{P^2}{2M} + \frac{1}{2}\kappa Q^2 + \chi x Q \ , \tag{1}$$

where ω is the intrinsic frequency of amide-I oscilation, χ' is the coupling constant between two oscilators, m (M) is the amide-I (amide-site) mass and δ is the anharmonic coefficient. Throughput the paper we also use the notations $\chi = \chi'\sqrt{\hbar/(2M\Omega)}$ and $\Omega = \sqrt{\kappa/M}$. The Hamiltonian describes the Davydov-Scott's monomer as a coupled harmonic oscillator.

If the environmental or dissipation effect can not be ignored, the physical system is not reversible. In another words, the irreversibility implies the dissipation effect in a quantum system under consideration. One of the basic tools to introduce dissipation in quantum mechanics is the dynamical semi groups, and called as the Lindblad open quantum system formalism. In particular, the quantum system whose the wave function equation can be obtained from the Lindblad equation is called as QSD.[8]

According to Lindblad formalism, the usual von Neumann-Liouville equation is replaced by Lindblad equation or master equation in the form of,

$$\frac{\partial \rho}{\partial t} = -\frac{i}{\hbar}[H, \rho] + \sum_j (L_j \rho L_j^\dagger - \frac{1}{2} L_j^\dagger L_j \rho - \frac{1}{2} \rho L_j^\dagger L_j) . \tag{2}$$

ρ is the density function and L_j denotes the Lindblad operator which may neither Hermitian nor unique. In this formalism, the operator H describes internal dynamics, while L represents the environmental effects in the system.

Throughout the paper, let us assume that the environmental strength of the amide-site is stronger than the amide-I excitation. Since L must be the first order in Q and P, we choose the Lindblad operators as follow,

$$L_1 = \sqrt{\gamma(1+\nu)} \left(\sqrt{\frac{M\Omega}{2\hbar}} Q + i\sqrt{\frac{1}{2M\hbar\Omega}} P + \frac{\chi'}{\hbar\Omega}\sqrt{\frac{m\omega}{2\hbar}} x \right) , \tag{3}$$

$$L_2 = \sqrt{\gamma\nu} \left(\sqrt{\frac{M\Omega}{2\hbar}} Q - i\sqrt{\frac{1}{2M\hbar\Omega}} P + \frac{\chi'}{\hbar\Omega}\sqrt{\frac{m\omega}{2\hbar}} x \right) . \tag{4}$$

γ is a damping parameter and $\nu = \left(e^{\hbar\Omega/k_B T} - 1\right)^{-1}$ is Bose-Einstein distribution function with the Boltzman coefficient k_B.

The master equation in Eq.(2) is calculated using Feynman path integral. Assuming the diffusion term is dominant over the frictional damping rate ($[Q, [Q, \rho]] \gg [Q, [P, \rho]]$), Eq. (2) is rewritten in a differential representation,

$$\frac{\partial \rho}{\partial t} = \frac{i\hbar}{2m} \left(\frac{\partial^2}{\partial x^2} - \frac{\partial^2}{\partial x'^2} \right) \rho - \left(\frac{im\omega^2}{2\hbar} + \frac{\delta_3}{2\hbar^2} \right) \left(x^2 - x'^2 \right) \rho$$

$$+ \left(\frac{i\hbar}{2M} + \frac{\delta_1}{2\hbar} \right) \left(\frac{\partial^2}{\partial Q^2} - \frac{\partial^2}{\partial Q'^2} \right) \rho - \left(\frac{iM\Omega^2}{2\hbar} + \frac{\delta_2}{2\hbar} \right) (Q^2 - Q'^2)\rho$$

$$- i \left(\frac{\chi}{\hbar} + \frac{\delta_4}{2\hbar} \right) (xQ - x'Q') \rho . \tag{5}$$

Here, the Lindblad's coefficients are $\delta_1 = \gamma(1+2\nu)/(2M\Omega)$, $\delta_2 = \gamma(1+2\nu)M\Omega/2$, $\delta_3 = \gamma(1+2\nu)(\chi^2 m\omega)/(\hbar\Omega)^2$ and $\delta_4 = \sqrt{m\omega M\Omega}\chi'/(\hbar\Omega)$. The propagator of Eq. (5) is given by,

$$K(x, x'; Q, Q') = \int \int \mathcal{D}[x]\mathcal{D}[Q]\exp \left[\frac{i}{\hbar} \int dt \left(\frac{1}{2}m\dot{x}^2 - \frac{1}{2}m\tilde{\omega}^2 x^2 \right. \right.$$

$$\left. \left. + \frac{1}{2}\bar{M}\dot{Q}^2 + \frac{1}{2}\bar{M}\tilde{\Omega}^2 Q^2 - \tilde{\chi}'xQ \right) \right] , \tag{6}$$

where $\tilde{\omega}^2 = \omega^2 + i\delta_3/(m\hbar)$, $\tilde{\Omega}^2 = \Omega^2 + i\delta_2/(M\hbar)$, $\bar{M} = M + i\hbar^2/\delta_1$ and $\tilde{\chi} = \chi + \delta_4/2$. Making use of the Gaussian approximation, only the classical path of amide-site ($\bar{Q}$)

contributes to the interaction term.[12,13] It yields,

$$K(x, x'; Q, Q') = K_x K_Q = \int \mathcal{D}[x] \exp\left[-\frac{i}{\hbar} \int dt \left(\frac{1}{2}m\dot{x}^2 - \frac{1}{2}\tilde{\omega}^2 x^2 - \tilde{\chi}x\bar{Q}\right)\right]$$
$$\times \int \mathcal{D}[Q] \exp\left[-\frac{i}{\hbar} \int dt \left(\frac{1}{2}\bar{M}\dot{Q}^2 - \frac{1}{2}\bar{M}\tilde{\Omega}^2 Q^2\right)\right]. \quad (7)$$

The propagator K_Q is just a harmonic oscillator, and the solution is well known,[13,14]

$$K(Q, t; Q', 0) = \exp\left[-i\frac{\pi}{2}\left(\frac{1}{2} + \left\|\frac{\tilde{\Omega}t}{\pi}\right\|\right)\right]\sqrt{\frac{\bar{M}\tilde{\Omega}}{2\pi\hbar|\sin(\tilde{\Omega}t)|}}$$
$$\times \exp\left\{\frac{i\bar{M}\tilde{\Omega}}{2\hbar|\sin(\tilde{\Omega}t)|}\left[(Q'^2 + Q^2)\cos(\tilde{\Omega}t) - 2QQ'\right]\right\}. \quad (8)$$

where $\|x\|$ denotes the largest integer smaller than x.

The propagator of K_x is a driven harmonic oscillator and the solution is also known.[13,14] The driven function $\bar{Q}$ is the classical solution of equation of motion (EOM) of Q, that is a harmonic oscillator. Taking the solution $\bar{Q} = \bar{Q}_0 \sin(\tilde{\Omega}t)$ and substituting it into the path integral solution of a driven harmonic oscillator, then the propagator becomes,[13,14]

$$K(x, t; x', 0) = \exp\left[-i\frac{\pi}{2}\left(\frac{1}{2} + \left\|\frac{\tilde{\omega}t}{\pi}\right\|\right)\right]\sqrt{\frac{m\tilde{\omega}}{2\pi\hbar|\sin(\tilde{\omega}t)|}}\, e^{\frac{i}{\hbar}S_{\text{cl}}}, \quad (9)$$

where S_{cl} is given by,

$$S_{\text{cl}} = \frac{m\tilde{\omega}}{2\sin(\tilde{\omega}t)}\left[(x^2 + x'^2)\cos(\tilde{\omega}t) - 2xx'\right]$$
$$+ \frac{\tilde{\chi}\bar{Q}_0 x'}{(\tilde{\Omega}^2 - \tilde{\omega}^2)\sin(\tilde{\omega}t)}\left[\tilde{\omega}\cos(\tilde{\omega}t)\sin(\tilde{\Omega}t) - \tilde{\Omega}\sin(\tilde{\omega}t)\cos(\tilde{\Omega}t)\right]$$
$$+ \frac{\tilde{\chi}\bar{Q}_0 x}{(\tilde{\Omega}^2 - \tilde{\omega}^2)\sin(\tilde{\omega}t)}\left[\tilde{\Omega}\sin(\tilde{\omega}t) - \tilde{\omega}\sin(\tilde{\Omega}t)\right] \quad (10)$$
$$+ \frac{\tilde{\chi}^2\bar{Q}_0^2}{m\tilde{\omega}(\tilde{\Omega}^2 - \tilde{\omega}^2)\sin(\tilde{\omega}t)}\left[A\cos(\tilde{\omega}t)\sin^2(\tilde{\Omega}t) + B\sin(\tilde{\omega}t)\sin(2\tilde{\Omega}t) - \frac{\tilde{\omega}}{4}t\right].$$

with $A = \tilde{\omega}(\tilde{\omega}\tilde{\Omega} + 1)/(4(\tilde{\Omega}^2 - \tilde{\omega}^2))$ and $B = -\tilde{\omega}(\tilde{\Omega}^2 + \tilde{\omega}^2)/(8\tilde{\Omega}(\tilde{\Omega}^2 - \tilde{\omega}^2))$.

Combining Eqs. (9) and (10), the propagator $K(x, x'; Q, Q'; t, t')$ becomes,

$$K(Q, x, t; Q', x', 0) = \exp\left[-i\pi\left(\frac{1}{2} + \frac{1}{2}\left\|\frac{(\tilde{\Omega} + \tilde{\omega})t}{\pi}\right\|\right)\right]$$
$$\times \sqrt{\frac{\bar{M}\tilde{\Omega}}{2\pi\hbar|\sin(\tilde{\Omega}t)|}}\sqrt{\frac{m\tilde{\omega}}{2\pi\hbar|\sin(\tilde{\omega}t)|}} \quad (11)$$
$$\times \exp\left\{\frac{i\bar{M}\tilde{\Omega}}{2\hbar|\sin(\tilde{\Omega}t)|}\left[(Q'^2 + Q^2)\cos(\tilde{\Omega}t) - 2QQ'\right] + \frac{i}{\hbar}S_{\text{cl}}\right\}.$$

3. Thermodynamics Properties

The discussion of a system interacting with heat bath is characterized by the temperature T. The state of those systems is therefore given by an equilibrium density matrix which can be obtained by performing a transformation $t \to \tau = -i\hbar\beta$ in the propagator with $\beta = 1/(k_B T)$.[13,14] The density matrix is actually the propagator with $\rho(x, x, Q, Q, \tau, 0) = K(Q, x; Q, x, \tau, 0)$. Substituting $t \to \tau = -i\hbar\beta$ into Eq. (11), one obtains,

$$
\rho_\beta(x, Q) = \exp\left[-i\pi\left(\frac{1}{2} - \frac{i}{2}\left\|\frac{(\tilde{\Omega} + \tilde{\omega})\hbar\beta}{\pi}\right\|\right)\right]\sqrt{\frac{\bar{M}\tilde{\Omega}}{2\pi\hbar\sinh(\tilde{\Omega}\hbar\beta)}}\sqrt{\frac{m\tilde{\omega}}{2\pi\hbar\sinh(\tilde{\omega}\hbar\beta)}}
$$

$$
\times\exp\left\{-\frac{\bar{M}\tilde{\Omega}}{\hbar}\tanh(\tfrac{1}{2}\tilde{\Omega}\hbar\beta)Q^2 - \frac{m\tilde{\omega}}{\hbar}\tanh(\tfrac{1}{2}\tilde{\omega}\hbar\beta)x^2\right.
$$

$$
+ \frac{2\tilde{\chi}\bar{Q}_0}{\hbar(\tilde{\Omega}^2 - \tilde{\omega}^2)}\left[\tilde{\Omega}\sinh^2(\tfrac{1}{2}\tilde{\Omega}\hbar\beta) - \tilde{\omega}\sinh(\tilde{\Omega}\hbar\beta)\sinh(\tfrac{1}{2}\tilde{\omega}\hbar\beta)\right]x
$$

$$
+ \frac{\tilde{\chi}^2\bar{Q}_0^2}{m\tilde{\omega}\hbar^2(\tilde{\Omega}^2 - \tilde{\omega}^2)}\left[A\Omega\sinh^2(\tilde{\Omega}\hbar\beta)\coth(\tilde{\omega}\hbar\beta) + B\sinh(2\tilde{\Omega}\hbar\beta)\right.
$$

$$
\left.\left.- \frac{\tilde{\omega}\hbar\beta}{4\sinh(\tilde{\omega}\hbar\beta)}\right]\right\}. \tag{12}
$$

Having the density matrix at hand, in the statistical mechanics one can consider partition function,[12]

$$
Z(\beta) = e^{-\beta F} = \int dx \int dQ\, \rho_\beta(x, Q). \tag{13}
$$

Substituting Eq. (12) into Eq. (13) and using the hyperbolic manipulation yield,

$$
Z(\beta) = \frac{1}{2\sinh(\tfrac{1}{2}\tilde{\Omega}\hbar\beta)}\frac{1}{2\sinh(\tfrac{1}{2}\tilde{\omega}\hbar\beta)}\exp\left[-\frac{i\pi}{2}\left(1 - \left\|\frac{\tilde{\Omega}\hbar\beta}{\pi}\right\| + \left\|\frac{\tilde{\omega}\hbar\beta}{\pi}\right\|\right)\right]
$$

$$
\times\exp\left[\frac{\tilde{\chi}^2\bar{Q}_0^2\left\{\tilde{\Omega}\sinh^2(\tfrac{1}{2}\tilde{\Omega}\hbar\beta) - \tilde{\omega}\sinh(\tilde{\Omega}\hbar\beta)\sinh(\tfrac{1}{2}\tilde{\omega}\hbar\beta)\right\}^2}{\hbar\bar{M}\tilde{\Omega}(\tilde{\Omega}^2 - \tilde{\omega}^2)\tanh(\tfrac{1}{2}\tilde{\omega}\hbar\beta)}\right.
$$

$$
\left.+ \frac{\tilde{\chi}^2\bar{Q}_0^2\left\{A\Omega\sinh^2(\tilde{\Omega}\hbar\beta)\cosh(\tilde{\omega}\hbar\beta) + B\sinh(2\tilde{\Omega}\hbar\beta)\sinh(\tilde{\omega}\hbar\beta) - \tfrac{1}{4}\tilde{\omega}\hbar\beta\right\}}{m\tilde{\omega}\hbar^2(\tilde{\Omega}^2 - \tilde{\omega}^2)\sinh(\tilde{\omega}\hbar\beta)}\right], \tag{14}
$$

by borrowing the Gaussian integral, $\int dx \exp(-ax^2 + bx + c) = \sqrt{\pi/a}\exp[b^2/(4a) + c]$.

For an open quantum system, such as the Davydov-Scott monomer, the changes of the surrounding environment entropy must be taken into account. The effect is conveniently incorporated by the specific heat which is also experimentally measurable. It describes the quantities of heat that must be added to a system in order to increase its temperature, and defined as,[12]

$$
C = k_B\beta^2\frac{\partial^2\ln Z}{\partial\beta^2}. \tag{15}
$$

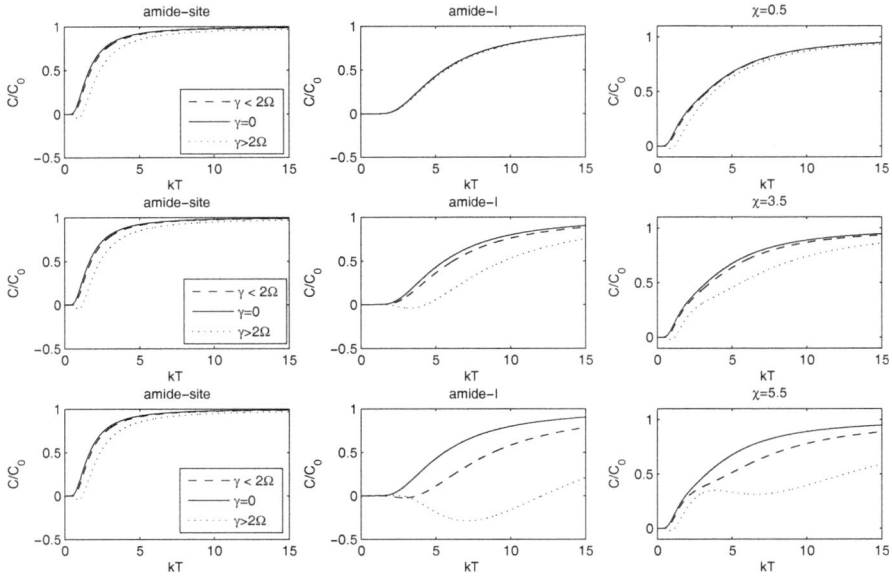

Fig. 1. The temperature dependence of normalized specific heat for the first (left), second (middle) and total (right) terms in Eq. (19) for various values of γ.

Bringing Eq. (14), one gets,

$$\frac{C}{k_B} = \frac{I^2}{\sinh^2 I} + \frac{w^2}{\sinh^2(w)} + \tilde{\chi}^2 \Lambda(\beta) \, , \tag{16}$$

where $I = \tilde{\Omega}\hbar\beta/2$, $w = \tilde{\omega}\hbar\beta/2$ and,

$$\Lambda(\beta) = \frac{\bar{Q}_0^2 \beta^2}{2\hbar} \frac{\partial^2}{\partial\beta^2} \left\{ \frac{\left[\tilde{\Omega}\sinh^2 I - \tilde{\omega}\sinh(2I)\sinh w\right]^2}{\bar{M}\tilde{\Omega}(\tilde{\Omega}^2 - \tilde{\omega}^2)\tanh w} \right.$$
$$\left. + \frac{A\Omega\sinh^2(2I)\cosh(2w) + B\sinh(4I)\sinh(2w) - \frac{1}{2}w}{m\tilde{\omega}\hbar(\tilde{\Omega}^2 - \tilde{\omega}^2)\sinh(2w)} \right\} \, . \tag{17}$$

For small coupling case we have approximately $\sinh(x) \sim x$, $\cosh(x) \sim 1$ and $\tanh(x) \sim x$ to yield,

$$\Lambda(\beta) = \frac{\bar{Q}_0^2}{2\hbar(\tilde{\Omega}^2 - \tilde{\omega}^2)} \left(\frac{3}{4}\tilde{\Omega}^3 - \frac{3}{2}\tilde{\Omega}^2\tilde{\omega}^2 - 3\tilde{\omega}^4\tilde{\Omega}\right)\beta^3 \, . \tag{18}$$

The specific heat in Eq. (16) can be read as,

$$C = C_{\text{amide-site}} + C_{\text{amide-I}} + C_{\text{mixing}} \, . \tag{19}$$

If there is no coupling between amide-site and amide-I, i.e. $\tilde{\chi} = 0$, this is just the total specific heat of two independent harmonic oscillators.

The result is depicted in Fig. 1 for three cases corresponding to the values of γ, that is $\gamma = 0$, $\gamma < \Omega$ and $\gamma > \Omega$. The case of non-zero but small γ has similar behavior as non-damping case. All of them coincide each other at low temperature and tend to be asymptotically constant at high temperature. On the other hand, large γ reduces the specific heat for the whole region of temperature. This means the environtmental effect contributes destructively to the specific heat, and the system requires less energy to increase the temperature to reach the equilibrium.

The application of the present model to the $\alpha-$helix protein requires proper knowledge on the coupling constant between amide-I and amide-site.[3] There are some attempts to determine its allowed range through several methods like the Ab initio calculation and also the extraction from the experimental data as well. The value is found to be within 7 pN and 62 pN.[3] Considering the amide-site mass $5.7 \times 10^{-25} kg$ and the string constant $\kappa = 58.5$ Nm^{-1}, one immediately obtains the coupling constant $\chi' = 1.7, 11.6, 18.2$ corresponding to $\chi = 0.5, 2.5, 5.5$. Note that most of previous works takes $\chi = 62$ pN. However, those works do not take into account the thermal bath effect. On the other hand, including the thermal bath as done in the present paper enhance the contribution of amide-I and amide-site interaction.

The result is depicted in Fig. 1 showing the temperature dependencies of the normalized specific heat for various values of γ. The left, middle and right figures correspond to the contribution of the first term, the second term and the total in Eq. (19) respectively. From the figures one can conclude that the results are sensitive to the size of coupling constant χ at intermediate temperature. Especially, the amide-I is more affected than amide-site. The reason is because amide-I has higher frequency than amide-site. Moreover, amide-I is also suppressed significantly by thermal bath contribution which indicates the dependencies of system frequency on the effect of thermal bath as already pointed out by Ingold $et.al.$through Caldiora-Lenggets formalism.[14]

It should be remarked that at low temperature region large environmental effect induces an anomaly, that is the specific heat is getting negative. This anomaly has also been observed by Ingold $et.al.$[15] for free harmonic oscillator using Caldiora-Lenggets formalism, and by us using full-quantum approach and the Lindblad formulation of master equation.[9]

4. Summary

The interaction of Davydov-Scott monomer with thermal bath is investigated using the Lindblad open quantum system formalism. In contrast with previous work by Cuevas $et.al.$,[4] the statistical partition function is calculated instead of solving the EOM itself.

Using path integral one can calculate the propagator of the Lindblad equation. Under an assumption that the diffusion term is dominant, we have shown that the environment contributions shift the kinetic, potential and interaction terms in the lagrangian. The mixing term is survive only if the frictional damping rate is taken into account.

In the open quantum system the damping coefficient γ represents the relaxation time due to interaction with the environment. Non-zero γ contributes destructively to the specific heat. The higher value of γ corresponds to the shorter relaxation time, and it induces specific heat anomaly as pointed out in previous works.[9,15] It is also found that large interaction between amide-I and amide-site is not prefered for a stable Davydov-Scott monomer.

Acknowledgments

AS thanks the Group for Theoretical and Computational Physics LIPI for warm hospitality during the work. This work is funded by the Indonesia Ministry of Research and Technology and the Riset Kompetitif LIPI in fiscal year 2010 under Contract no. 11.04/SK/KPPI/II/2010. FPZ is supported by Riset KK 2010 Institut Teknologi Bandung.

References

1. L. Cruzeiro, J. Halding, P. L. Christiasen, O. Skovgard and A. C. Scott, *Physical Review A* **37**, 880 (1988).
2. L. Cruzeiro-Hansson and S. Takeno, *Physical Review E* **56**, p. 894 (1997).
3. A. C. Scott, *Physics Report* **217**, p. 167 (1992).
4. J. Cuevas, P. A. S. Silva, F. R. Romero and L. Cruzeiro, *Physical Review E* **76**, p. 011907 (2007).
5. L. Cruzeiro-Hansson, *Physical Review Letters* **73**, 2927 (1994).
6. J. P. Cottingham and J. W. Schweitzer, *Physical Review Letter* **62**, 1792 (1989).
7. D. V. Kapor, M. J. Skrinjar and S. D. Stojanovic, *Physical Review A* **41**, 5694 (1990).
8. I. Percival, *Quantum State Diffusion*, 2 edn. (Cambridge Univ. Press, 1998).
9. A. Sulaiman, F. P. Zen, H. Alatas and L. T. Handoko, *Physical Review E* **81**, p. 061907 (2010).
10. A. Sulaiman and L. T. Handoko, *Journal of Comptational and Theoretical Nanoscience* **in press** (2010).
11. M. Januar, A. Sulaiman and L. T. Handoko, *International Journal of Modern Physics A* **in press** (2010).
12. R. P. Feynmann, *Statistical Mechanics* (W. A. Benjamin Inc., 1965).
13. H. Kleinert, *Path Integrals in Quantum Mechanics, Statistics, and Polymer Physics, and Financial Markets* (World Scientific, 2009).
14. P. Hanggi, G.-L. Ingold and P. Talkner, *New Journal of Physics* **10**, p. 115008 (2008).
15. G.-L. Ingold, P. Hanggi and P. Talkner, *Physical Review E* **79**, p. 061105 (2009).

q-GAUSSIAN ANALYSIS IN COMPLEX POLYMERS

G. CIGDEM YALCIN

Department of Physics, Istanbul University, Vezneciler, Istanbul, 34118, Turkey
gcyalcin@istanbul.edu.tr

YANI SKARLATOS

Department of Physics, Bogazici University, Bebek, Istanbul, 34342, Turkey
sakarlat@boun.edu.tr

K. GEDIZ AKDENIZ

Department of Physics, Istanbul University, Vezneciler, Istanbul, 34118, Turkey
gakdeniz@istanbul.edu.tr

Recently, we analyzed the temperature dependent q-Gaussian characteristics of weak chaotic transient currents through thin Aluminum-Polymethylmethacrylate-Aluminum films under voltages in the range of ± 10V at $22°$C temperature.

In this work we investigate the role of q-Gaussian characteristics in the conductivity mechanism of the Polymetylmethacrylate polymer in order to understand the electron self-organization criticality in complex polymers.

Keywords: Complex polymers; conductivity mechanism; q-Gaussian analysis.

1. Introduction

q-statistics has a multidisciplinary role in science, and it is employed in a multitude of diverse fields ranging from earthquake data to brain electroencephalograph signals and finance movements, while new fields of application are continually found by q-statistics researchers.

q-Gaussian analysis, in the context of q-statistics, is used to interpret weak chaotic systems which have approximately zero Lyapunov exponents.

This work is based on an analysis of the conductivity mechanism of the Polymethylmethacrylate (PMMA) polymer, whose time evolution has weak chaotic properties. The transient current data of Aluminum-Polymethylmethacrylate-Aluminum (referred to as Al-PMMA-Al in this work) thin films subjected to an electric field at $22°$C temperature were considered.

The maximum Lyapunov exponents of the above films were found to be in the range between 0.003 and 0.023, indicating the weak chaotic nature of the transient currents through them.[1]

We consider the weak chaotic behavior of the electrical conductivity of PMMA thin films through q-Gaussian analysis. We show that its probability density function (PDF) for the current magnitude differences at 22°C has long tails with a q-Gaussian shape, and compare the fitted q-Gaussian curve of the probability density function with a normal Gaussian distribution.

2. Calculations

Hacinliyan, A. *et al.*[1] have analyzed the chaoticity in the transient current though Polymethylmethacrylate (PMMA) thin films for times ranging up to 30,000 s, and for applied voltages in the range 10–80V. Fig. 1. is a semi logarithmic graph of the time evolution of a typical transient current through thin Al-PMMA-Al at 22°C and 10V in units of 10 seconds.

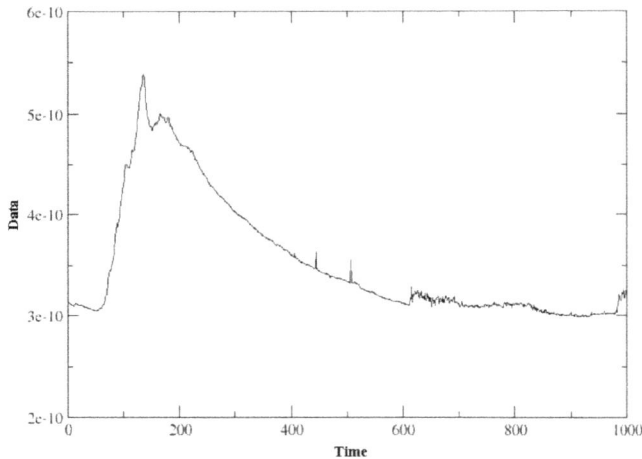

Fig. 1. The time evoluation of the transient current through thin Al-PMMA-Al at 22°C.

The PDF of the current magnitude differences (returns) $I[t] = S(t+1) - S(t)$ for the transient current through the thin film Al-PMMA-Al structure of Fig. 1 is shown in Fig. 2. The curve has been fitted with a q-Gaussian (red dashed line) with an exponent $q = 2.3$. This and all subsequent curves were normalized so as to have unitary area. Returns were normalized to the standard deviation σ.

We investigate the probability density functions (PDFs) for the obtained data sets at 22°C. We see that they have long tails with a q-Gaussian shape. Then, one can well fit these data sets by a q-Gaussian curve typical of Tsallis statistics.[2]

$$f(x) = A[1 - (1 - q)x^2/B]^{1/(1-q)} . \tag{1}$$

This function generalizes the standard Gaussian curve, depending on the parameters A, B and on the exponent q. For $q = 1$ the normal distribution is obtained again; thus, $q \neq 1$ indicates a departure from Gaussian statistics.

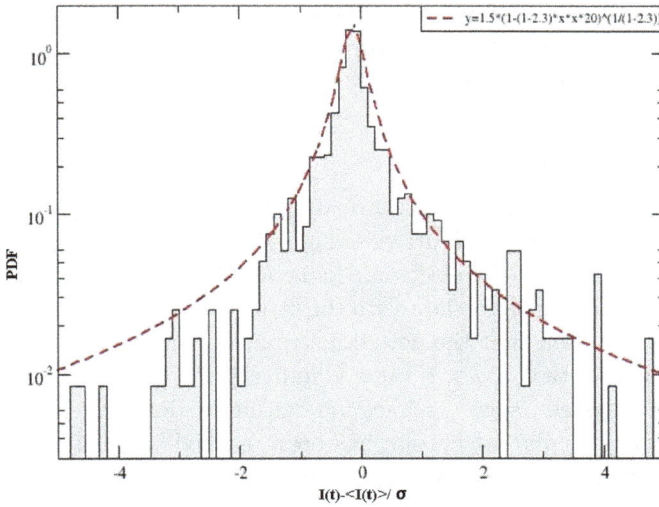

Fig. 2. PDF of the current magnitude differences (returns) $I[t] = S(t+1) - S(t)$ for the transient current through thin Al-PMMA-Al at $22°$C. The curve has been fitted with a q-Gaussian (red dashed line) with an exponent $q = 2.3$.

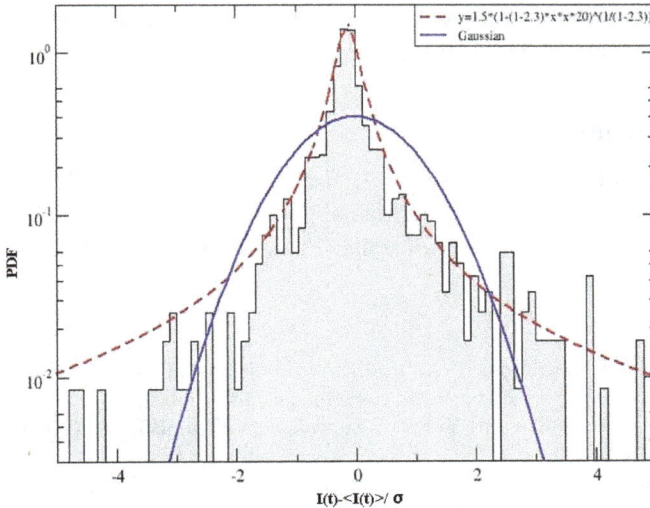

Fig. 3. PDF of the current magnitude differences (returns) $I[t] = S(t+1) - S(t)$ for the transient current through thin Al-PMMA-Al at $22°$C. The curve has been fitted with a q-Gaussian (red dashed line) with an exponent $q = 2.3$. A standart Gaussian (blue full line) is also reported for comparison.

In Fig. 3. We plot the PDF's of the current magnitude differences (returns) $I[t] = S(t+1) - S(t)$ for the transient current through thin Al-PMMA-Al at $22°$C temperature. The curve has been fitted with a q-Gaussian (red dashed line) with an exponent $q = 2.3$. A behavior very different from a Gaussian shape (plotted as a blue full curve) is observed. The data are peaked, with long tails; and non-Gaussian probability density functions are observed.

3. Discussion and Conclusion

In the case of $q \neq 1$ values of the probability distribution function the relation between normal Gaussian distribution and q-Gaussian statistics was investigated in an attempt to understand the electronic behavior of the system.

From the comparison of the data with the fitted q-Gaussian and normal Gaussian curves, it is clearly seen that the actual distribution is q-Gaussian, with a q-value of 2.3. To the extent that a $q > 1$ value is indicative of the presence of long range interactions in a system, such a suggestion can be made for the case of a PMMA polymer. In fact, a similar conclusion has been reached by Hacinliyan et al.[1] from standard chaotic behavior analysis of the same I-V characteristics of PMMA.

On the other hand, if a sand pile is considered to represent a critically self organized system,[3] then, since the conductivity mechanism of the Polymethylmethacrylate polymer has weak chaotic properties, and thus q-statistics analysis is applicable as this work shows, the electronic behavior of the polymer could be modelled after critically self organized systems.

Further calculations from similar data obtained at higher temperatures are in progress and will be reported shortly.

Acknowledgments

We would like to thank A. Robledo, A. Rapisarda and A. Pluchino for useful discussions and comments. This work was supported by Scientific Research Projects Coordination Unit of Istanbul University. Project number 7094 and 3534.

References

1. A. Hacinliyan, Y. Skarlatos, H. A. Yildirim and G. Sahin, *Fractals-Complex Geometry, Patterns and Scaling in Nature and Society*, **14**, 125 (2006).
2. C. Tsallis, M. Gell-Mann and Y. Sato, *Europhysics News*, **36**, 186 (2005) and references therein.
3. H. Jensen, *Self-organized Criticality: Emergent Complex Behavior in Physical and Biological Systems* (Cambridge University Press, New York, 1998).

MORPHOLOGIC ECONOMICS

KEN NAITOH

Faculty of Science and Engineering, Waseda University,
3-4-1, Ookubo, Sinjuku, Tokyo, 169-8555, Japan
k-naito@waseda.jp

It has been difficult to predict various classes of boom-and-bust economic cycles and these cyclic catastrophes systematically, because they are related to several biological phenomena. In this report, we will show that our theory on the morphogenetic process and the brain with a rhythm of about seven beats can explain several economic system cycles, because different types of economic cycles are about seven times the length of the fundamental production cycles or durable periods. We will also outline the spatial structure underlying economic systems on the basis of the fluid dynamic theory that describes subatomic systems, biological systems, human network systems, and stars.

Keywords: Morphology; economic cycle; fluid dynamics.

1. Introduction

Little is known about the physical mechanisms controlling several classes of boom-and-bust economic cycles: the Kondratev cycle of about 50 years, the Kuznets cycle of about 20 years, the Juglar cycle of about 7 years, the Kitchin cycle of about 3.5 years, and one week as the shortest one.[1-4] In 2008, the world slid into an economic crisis again. The longest type of economic catastrophe repeats with an interval of about 70 years: 1870-1880 in England, 1930-1940 in the USA, and 2000-2009 in the USA, Japan, and Europe. Very little is known about such cyclic catastrophes, because the problem is related to biological phenomena, including cognition in the human brain.

Evolutionary economics has tried to include a biological aspect, which is expected to become a new stage of economics.[5] However, cyclic economic phenomena cannot be revealed systematically by the simple concept of evolution based on mutation and natural selection, because there are still unknown aspects underlying living beings.

It is well known that a cycle of about seven cell divisions is the basic morphogenetic cycle determining the bifurcation points of organs in human beings. (see Fig. 1) For example, about seven cell divisions of a fertile egg produce blast cysts and the next seven divisions lead to another new stage with the complex structures of ectodermal and endoderm cells. In the later part of the morphogenesis of human beings, the first heart beat (beating cardiomyocytes) occurs at about seven

Phase	Fertile egg		Diploblastics		Heartbeat	Arms , legs	Unborn baby	
		Blastocysts		Germ layers		and face		Neonate

The number of divisions

0	6	11	19	26	32	37	40-45

6   5   8   7   6   5

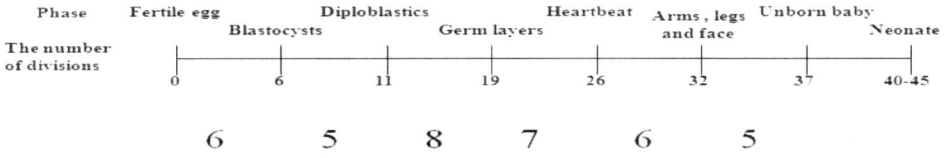

Fig. 1.   Seven-beat cycle in the morphogenetic process of human beings.

divisions after mesoderm formation and hands and legs also emerge at seven divisions after the first heart beat (Fig. 1). Recently, we proposed a fundamental model for clarifying the reason why organs are generated in a process of sevenfold cell divisions.[6-8]

In this report, we will show that the morphogenetic process with a sevenfold rhythm for the emergence of organs is similar to the repetition of economic catastrophes. We will show that new organs in the morphogenetic process correspond mathematically to the rise of new companies or new business categories.

Finally, we will also outline the spatial structure underlying economic systems.

2. Bio-Temporal Cycles

In biological systems, at least two types of enzyme categories x21 and x22, which are for replicating gene groups and enzyme systems respectively, are necessary for achieving a closed reaction cycle. Examples of the former and latter are DNA replicase and ribosome protein, respectively.

This leads to the conclusion that two types of gene groups x11 and x12 for generating the two types of enzyme systems are inevitable and serve to code the two enzyme systems (Fig. 2). The core cycle for self-replication can be modeled by these four molecular categories, i.e., two gene groups and two enzyme systems, which are denoted as categories x11, x12, x21, and x22 shown in Fig. 2.[6] This four-stroke system of the four categories works as a closed loop, if elements such as nTP and amino acids are input from the outside with energy. This is the minimum 4-stroke engine (minimum hypercycle[6]), which produces molecules exponentially in cases where there are no degradations.

It yields the following essential equation system for the densities of the four categories $x_{ij}(N)$, with N denoting time (or generation).

$$x_{ij}{}^{N+1} - x_{ij}{}^{N} = \alpha_{ij} * x_{1j}{}^{N} * x_{2i}{}^{N}, \qquad x_{ij} \geq 0, \quad (i = 1, 2, j = 1, 2) \qquad (1)$$

where the first subscript i in $x_{ij}(N)$ on the left hand side signifies the molecular type: $i = 1$ for the gene group and $i = 2$ for the enzyme system; the second subscript j in $x_{ij}(N)$ on the left hand side indicates function: $j = 1$ for generating the gene group and $j = 2$ for generating the enzyme system; and the operator $*$ denotes multiplication. The notation α_{ij} indicates an arbitrary constant.

It is stressed that variable x21 on the left-hand side of Eq. (1), i.e., $i = 2$ and $j = 1$, never appears on the right-hand side of the equation, although common

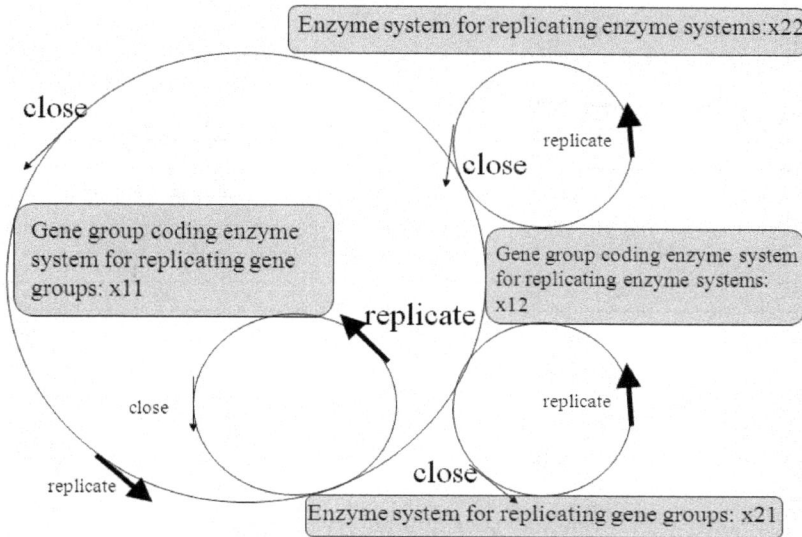

Fig. 2. Minimum hyper-cycle: two gene groups and two enzyme systems.

variables can be observed on both sides of Eq. (1) when x11, x12, and x22 are on the left-hand side. Equation (1) and Fig. 2 essentially show the asymmetry topologically. Results computed with this equation system show nearly exponential increases in $x_{ij}(N)$.

The negative enzyme system and its gene group (x23 and x13) are then incorporated into the foregoing core cycle of x11, x12, x21, and x22, because the morphogenetic process of multi-cellar systems in mammals must include negative controllers such as Oct-4 and SOX2 for producing tissues and organs.[7,8]

Hence, we get

$$x_{ij}^{N+1} - x_{ij}^{N} = \alpha_{ij} * (x_{1j}^{N} - \beta_{ij} * x_{23}^{N}) * x_{2i}^{N} ,$$
$$x_{ij} \geq 0, \quad x_{1j}^{N} - \beta_{ij} * x_{23}^{N} \geq 0, \quad (i = 1-2, j = 1, -3) \tag{2}$$

where β_{ij} is also an arbitrary constant. (Only β_{22} is set to be between 1.0 and 2.0, while the others are set to 0.0 in this report.)

The density ratio of x12 and x23 signifies the amount of DNA uncovered by protein group x23. The condition of x12/x23 > 1.0 means that a part of x12 is not covered by x23. The condition of x12/x23 < 1.0 means that x12 is completely blocked by redundant x23. When x23 is dense in stem or induced pluripotent stem (iPS) cells, it means that these cells can be reprogrammed by the presence of much Oct-4. This oscillation of x12/x23 will lead to changes in the gene combination for expressions, i.e., the emergence of new organs.

Computational results obtained with Eq. (2) clearly show that the antagonism between the negative controller x23 and the positive replication factors x21 and

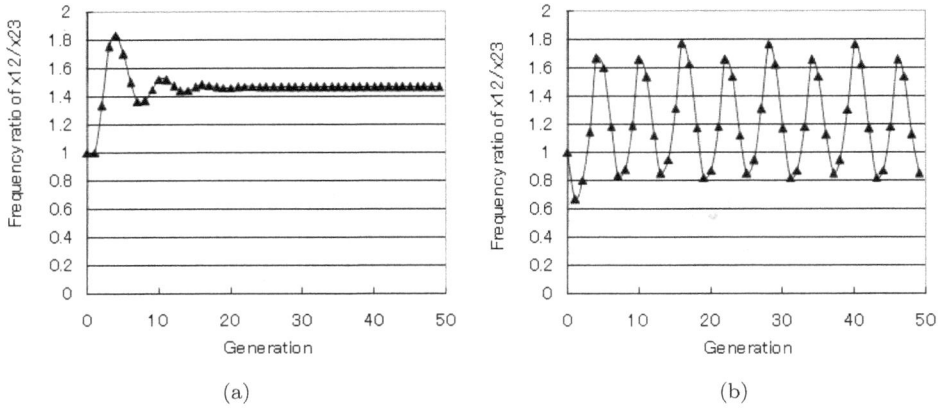

Fig. 3.   Density oscillations: (a) $\alpha_{ij} = 1$ for all i and j. (b) $\alpha_{1j} = 1$, $\alpha_{2j} = 2$ for all j.

x22 induces bifurcation events at rhythmic intervals constituting about seven divisions, although the intervals are slightly chaotic and the vibrational amplitude is attenuated[7,8] (Fig. 3).

It is stressed that this cycle of about seven divisions, i.e., the branching time between periodic bifurcation events, corresponds to the emergence timing of blast cysts, embryos, germ layers, tissues, and organs, which can be observed for about every seven cell divisions. Emphasis is placed on the fact that the parameters α_{ij} have less influence on the time cycle. The cycle of about seven divisions is fairly stable.

Next, let us consider the circadian clock.[9] We know that when suprachiasmatic nuclei (SCN) in the brain are removed, the clock shows a cycle of about 3-4 hours.[10] This means that the sevenfold-beat cycle based on the six-stroke engine of molecules is also present in the brain. Excitation and suppression also occur in several phases and places inside the brain system. Thus, there will also be other oscillation processes inside the brain, which are related to memory, learning, and instinct. Japanese have a custom of honoring the dead on the seventh day and forty-ninth day after death. It should be stressed that this sevenfold beat also comes from the human brain.

The speed of information transfer inside the brain is relatively faster than those of the clocks in the brain. Thus, spatial variations can be negligible in the model.

3. Economic Cycles

There are several types of boom-and-bust economic cycles, which are related to human network systems regulated by social rules generated by the human brain. The shortest economic cycle is one week of "seven" days. This shortest cycle is important for the human brain, because a week always has a rhythm of seven days, although a month may vary between 28 to 31 days. The longest repetitive catastrophes occur about every 70 years, which is seven times the actual durable period of automobiles of about 10 years. Several types of economic cycles will naturally be about seven times the length of the fundamental production cycles (durable periods).

In order to examine the economic cycles, we classify all things in our economic system into two major categories: information and functional objects. Information includes CAD data for designing many different functional products, while computers and industrial robots are examples of functional objects. Functional objects can be further divided into two subgroups: those such as computers for generating information and those such as industrial robots for generating functional products. These four categories, i.e., information necessary for functional objects that generate information, information necessary for functional objects that generate function, functional objects that generate information, and functional objects that generate functional products, correspond to x11, x12, x21, and x22 in Eqs. (1) and (2).

Next, let us think about the functional objects such as popular foods and alcohol as x23 and x13 for coding x23. These often suppress the generation of functional objects because of the excessive supply of products. The fifth category of x23 for depressing x12 and a sixth category of x13 for coding x23 lead to the cycles. These are then incorporated into the foregoing core cycle of x11, x12, x21, and x22. This results in a macroscopic model having six types of categories, or in other words, a six-stroke engine. The corresponding points between biological and economic systems are listed in Table 1.

Table 1. Corresponding points between biological and economic systems.

	Biological system	Economical system
For information production	DNA polymerases, topoisomerase, helicases, ligases, and primases	Computers, CAD systems, Internet systems
For functional production	Ribosomes, RNA polymerase, mRNAs, and tRNAs	Automobiles, metal-cutting machines and industrial robots
For functional depression	Transcription factors such as Oct-4	Popular foods and alcohol

For economic systems, x_{ij} in Eqs. (1) and (2) denote the amounts of information and functional objects, which signify the cost of business investment and profit, including labor cost. One generation corresponds to the fundamental production cycle (durable period), with N indicating the generations after the initial condition. A condition of $x_{1i} = x_{2i}$ means that the benefit to companies dealing in objects x_{2i} is zero, while $x_{1i} < x_{2i}$ denotes a surplus. Figure 3 also shows the sevenfold-beat cycle of economic systems.

Let us consider an appropriate economic growth rate of α_{ij} and β_{ij}. Here, we assume that automobiles are one of the critical factors influencing economic cycles of boom and bust and that their durable period is on the order of 10 years. If the economic growth rate of a country is 10 percent per year, the size of its economy will double in 10 years. Thus, the twofold rate of increase over 10 years implies values of α_{ij} and β_{ij} on the order of 1.0.

Computations made with values of α_{ij} and β_{ij} between 1.0 and 2.0 in Fig. 3 show strong density oscillations, leading to economic crises. Thus, we should opt for a lower rate of economic growth, if people want a monotonic increase rather than strong oscillations.

We can conclude that new organs in the morphogenetic process correspond mathematically to the rise of new companies or new business categories. The transfer speed of information and products in terms of days or weeks is relatively faster than that of economic cycles on the order of years. Thus, spatial variations can be negligible in the model.

Will the other cycles have intervals between 50 years (Kondratev cycle), 20 years (Kuznets cycle), 7 years (Juglar cycle), and 3.5 years (Kitchin cycle)? There are a lot of production cycles or durable periods ranging from 0.5 year to 10 years for various commercial products. Here, we define the production cycles and durable periods as m and assume that m = 0.5, 1, 2, 3, 4, 5, 6, 7, 8, 9, and 10 years. Let us make a set of integral multiples of each value of m. Then, we calculate the frequency distribution of the new integers obtained. When we count the numbers between 1 and 70, relatively low frequencies appear for the integers of 2, 3, 5, 7, 11, 13, 17, 19, 23, 29, 31, 41, 43, 47, 53, 59, 61, and 67 years. These are the prime numbers. As a result, we can see a new mysterious order generated in an orderly manner of integral multiples. This simple thought experiment qualitatively explains the reason why cycle lengths such as 70, 50, 20, 7, and 3.5 continually appear.

4. Discussion on Spatial Structure

In the foregoing sections, time-dependent ordinary differential equations (Eqs. (1) and (2)) unrelated to three-dimensional space revealed the cycle lengths. However, we should also clarify the spatial structure, i.e., the spatial inhomogeneity of money. In this section, we will show that the fluid dynamic model for revealing the spatial aspects of subatomic systems, biological systems, and stars[11-13,16] can also be applied to the spatial structure underlying economic systems such as the size variations of companies.

The theory of fluid dynamics[11-13] reveals the inevitability of asymmetry of about 1:1.5 for the molecular weights of purine and pyrimidine in Watson-Crick base pairs of DNA and 1:3.5 for cysteine as the maximum and glycine as the minimum, while the symmetric ratio is 1:1 in several base pairs in RNAs (Table 2). A ratio of about 1:1.5 close to the golden ratio and the silver ratio is also possible in the human brain, because we feel that the ratio is comfortable.[12,13]

Table 2. Quasi-stable molecules fundamental in living beings.

Quasi-stable ratio of weights	Molecules
1:1 & about 1.1.5	Nitrogenous bases
about 1:3.5	Amino acids

There are some similarities between subatomic particles, biological systems, and stars. There are mesons having "two" quarks and baryons with "three". The frequency ratio of neutrons and protons in the core of atoms is between 1:1 and about 1:1.5, which is very similar to that of pyrimidines and purines in nucleic acids (DNA of 1:1, tRNA of about 1:1-1.3, and rRNA of about 1:1-1.5). It is also stressed that a larger atomic core such as that of thorium and larger RNA such as rRNA have a more asymmetric frequency ratio, closer to 1:1.5. A neutron impacting uranium 235 often leads to an asymmetric weight ratio of about 2:3 in resultant smaller child atoms, while varying the impact speed of neutrons also results in a nearly symmetric division of uranium 235.[14] The size ratio of about 3.5 can also be seen in He 10.[15]

Data on the largest known stars in the cosmos show that their sizes have bimodal frequency peaks in a range of approximately 1,400–2,000 times greater than the sun and of less than 1,000 times.[16,17] (There are no stars between 1,000 and 1,400 times the size of the sun.) This may be similar to the bimodal frequency peaks for the sizes of purines and pyrimidines in nucleic acids.

Here, we examine whether or not this model also explains the variation in size of companies. First, we classify Japanese companies into several categories. The ratio of sales of the top and second companies in each category is mostly between 1:1 and 1:3.5, while about half of companies show ratios of less than 1:1.5. These results verify that the fluid model for explaining the size ratios seen at several levels of living beings can also reveal the size ratios of companies.

The direct reason why the ratio of sales of the top and second companies is mostly less than 1:3.5 may be related to the patent law and antitrust law. However, such laws, including the written standards for optimizing the economic world, come from the human brain, i.e., living beings. Thus, the principle controlling living beings is applicable to human economic systems. [The fluid model[11–13] includes the forces of internal convection, surface tension, and interacting force between two particles. In the economic world, the interaction force appears as competition between companies, while internal convection corresponds to the fact that people and money move between several sections inside a company, and surface tension corresponds to the principle of confidentiality in companies.]

An interesting thing is that the spatial structures of both living beings and companies in the economic world also have symmetry and asymmetry. Living beings have a bipolar order of symmetry in three-dimensional space: asymmetric Watson-Crick base pairs in DNA and symmetric ones in RNA as the origin of information, asymmetric and symmetric divisions of microorganisms and stem cells, and the left-right asymmetric liver and symmetric kidneys.

5. Conclusion and Outlook

As research on complexity has shown the direction of the road ahead,[18] economic policy should be based on an understanding of the overall network system related to living beings. The present approach may reveal the fundamental principle for controlling economic systems.

However, there are still mysterious phenomena underlying the human economic world. First, is the economic trouble in Greece, which may be mathematically related to immunological rejection or absence of an immune system, because the circulating medium called the euro represents external aggression for the body of Greece. Thus, an immune-suppressing drug or an immune system is necessary within the economic system of the euro. A second mystery concerns the reason why economic growth propagates clockwise on the Earth (Europe → USA → Japan → Korea → Singapore → China → India). This may be related to the rotation of the Earth, because more eastward countries can get information for business earlier and also because the timing when economic growth stops in one country is good for the start of growth in the next country.

References

1. J. A. Schumpeter, *Business Cycles: A Theoretical, Historical, and Statistical Analysis of the Capitalist Process* (McGraw-Hill, New York, 1939).
2. N. Kondratieff, *The Long Wave Cycle* (Richardson and Snyder, New York, 1984).
3. M. Kato, *Economic Fluctuation and Time* (Iwanami, Tokyo, 1-218, 2006).
4. K. Asako and S. Fukuda, Editors, *Business Cycle and Business Forecast* (University of Tokyo Press, Tokyo, 2003).
5. R. Nelson and S. G. Winter, *An Evolutionary Theory of Economic Change* (Belknap Press of Harvard University Press, London, 1982).
6. K. Naitoh, *Artificial Life Robotics* **13**, 10 (2008).
7. K. Naitoh, *Proceedings of 13th Int. Conf. on Biomedical Engineering*, Springer-Verlag (2008).
8. K. Naitoh, *Proceedings of the 2009 Conference on Chemical, Biological, and Environmental Engineering (CBEE)*, Singapore (2009). (also to be published in *Japan J. of Industrial and Applied Mathematics* (2010)).
9. J. C. Leloup and A. Goldbeter, *J. Biol. Rhythms* **13**, 70 (1998).
10. M. F. Bear, B. W. Connors and M. A. Paradiso, *Neuroscience* (Lippincott Williams & Wiklins Inc. USE, 2007).
11. K. Naitoh, *Oil & Gas Science and Technology* **54**, 205 (1999).
12. K. Naitoh, *Japan Journal of Industrial and Applied Mathematics* **18-1**, 75 (2001).
13. K. Naitoh, *Gene Engine and Machine Engine* (Springer-Japan, Tokyo, 2006).
14. E. M. Henley and A. Garcia, *Subatomic Physics* (World Scientific, 2007).
15. T. Nakamura *et al.*, Halo Structure of the Island of Inversion, *Nucleus Phys. Rev. Lett.* **103**, 262501 (2009).
16. K. Naitoh, *J. of Cosmology* **5**, 999 (2010).
17. List of the largest known stars: http://en.wikipedia.org/wiki/List_of_largest_known_stars.
18. M. Gell-Mann, *The Quark and the Jaguar: Adventures in the Simple and the Complex* (Owl Books, NewYork, 2002).

THE COMPLEXITY OF RESEARCH MANAGEMENT

JAN W. VASBINDER

Institute Para Limes, Vordenseweg 36, 7231 PC Warnsveld, Netherlands
jan.w.vasbinder@paralimes.org

Four factors and the relationships between them dominate the complexity of research management: selecting people, setting agenda's, being relevant and guaranteeing quality. Filling in each of these factors is determined by expectations that are formed both inside and outside the direct influence of the research manager. The challenge to research management is to make sure that the confluence of these factors moves the boundaries of knowledge in a way that excites and satisfies both the inside and the outside world.

Keywords: Research management; people; agenda; relevance; quality; the table; IUCRC; expectations.

1. Introduction

In 1984 two colleagues and I visited the Los Alamos National Laboratories in New Mexico on a mission to uncover the secrets of the American research management. At Los Alamos our host was George Cowan, senior fellow at the laboratory and, not known to me at that time, in the process of setting up the Santa Fe Institute. We raised the question how Los Alamos managed to be a leader in nuclear research for so long. George's answer was simple:

> "It is all a matter of selecting the right people [···]. What you are looking for as a leader of a lab is a scientist who has the intuition to know where to look for the boundaries of knowledge and how to cross those boundaries [···]. You look for the best and then you give him what he needs, and you give him five years [······]. If he has not changed the field within that time, you take him off the job and tell him what to do."

I have never considered research management to be a complex process since then.

2. People

Successful labs are led by scientists who dare to let their excellent research capacities be directed by their intuition. These are the people that open up new horizons and define new areas of research. There are other scientists, who are equally good researchers, but who lacks the intuition to open up new horizons. They should not be given the responsibility to lead, but be shown the direction. Given the direction, these excellent researchers should explore. So you have leaders and excellent researchers who should follow the lead. To the research manager the challenge is to find the right people, to motivate them to come, and to make sure they are happy and can do the things they know best: to move the boundaries of knowledge.

But there is more to research than just excellent researchers. And there is more to research management than selecting the right people. Selecting the right people is the first *sine qua non*. When you have selected them, you must give them what they need. That is the second *sine qua non*. If scientists drive the science, these are more or less all the conditions you need to meet. Research management becomes a simple matter that organizes itself as long as you know how to find the right people.

That is if you are dealing with science driven by scientists. But if the need for science is driven by commercial ambitions or societal needs, then other aspects become important as well. Selecting the right people and giving them what they need are still the dominant factors in achieving success. But then there are three other aspects that determine to what extent expectations can be met: setting agenda's, being relevant and guaranteeing quality.

It is the research manager's responsibility to make sure that these aspects are addressed in such a way that expectations are realistic and realizable.

3. Setting Agenda's

Basically a research agenda is a list of questions to be addressed. If there are no questions, there is no agenda, unless formulating the question is the agenda. I will discuss **that** type of research agenda, although it may be the most basic agenda in science.

The questions that form the research agenda are not immediately obvious. If the source of the questions is an industrial ambition to create a competitive advantage, or a societal need such as changing the climate, these questions are not formulated in terms of research questions, but in terms of commercial goals or long-term survival strategies. *Research can serve these goals and strategies only if they are translated into questions that have a meaning in the research arena.*

It is the research manager's task to make sure that these translations take place. This task is not a trivial, but it is often overlooked. However, such translations are essential for matching the expectations that companies or society foster towards the outcome of research with the expectations of the researchers to do exciting research.

I will discuss two methods with which this translation can be made. I will refer to them as *The Table* and the *Industry University Cooperative Research Center* (*IUCRC*).

3.1. *The Table*

The Table is a tool developed by Prisma & Partners to map systematically relations between strategic questions, knowledge domains and the content of a research program (see Ref. 1). An example will explain the concept. Let us say that our strategic

knowledge domain	knowledge area	application area / research question	knowledge question	strategic question

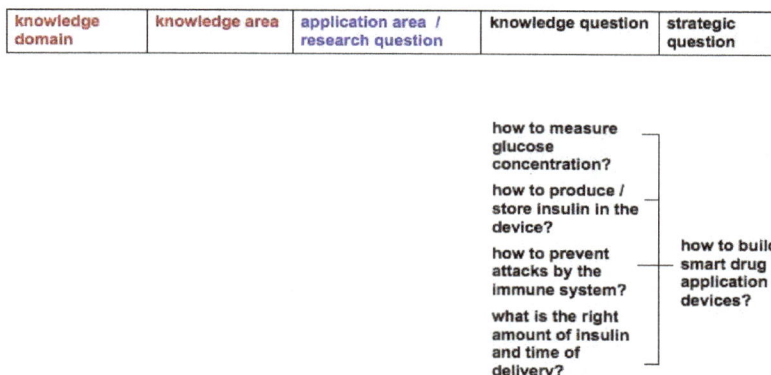

Fig. 1. The Table, step 1: Decomposition of the strategic question into a set of knowledge questions.

goal is *to build small devices that can be implanted in our body and release medicines to keep us healthy.* This could be a device for a diabetic that automatically releases insulin in the right amounts and at the right moment. Such a device would pose a tremendous improvement in the quality of life for diabetics, as it eliminates the need for blood samples, laboratory measurements, pills, and insulin injections.

Stating such a goal really implies the use of knowledge to achieve it. So the underlying questions are *knowledge* questions. To answer these, knowledge from different knowledge domains is necessary, such as endocrinology, biochemistry and immunology. And for designing, building and producing such a device we need knowledge from the domain of sensor technology.

The Table is a tableau for a stepwise decomposition of the strategic questions and the relevant knowledge domains, in order to develop an agenda with research questions that fit within the knowledge domains and that may lead to answers that serve the strategic goals. So the decomposition at the Table is done in two directions.

The strategic question is: how can we build small, implantable devices that respond to metabolic irregularities by controlled and optimized release of drugs? For diabetics, we can decompose this strategic question into knowledge questions like (Fig. 1):

– How to measure glucose concentrations with the device?
– How to produce/store insulin in the device?
– How will the device calculate the right amount of insulin?
– How to prevent attacks by the immune system?
– Can we use the device to function as an alarm (hypoglycemia)?

In order to answer such questions we must find and combine knowledge from different relevant knowledge domains. For that we need to decompose knowledge domains such as biochemistry, immunology and sensor technology into areas of

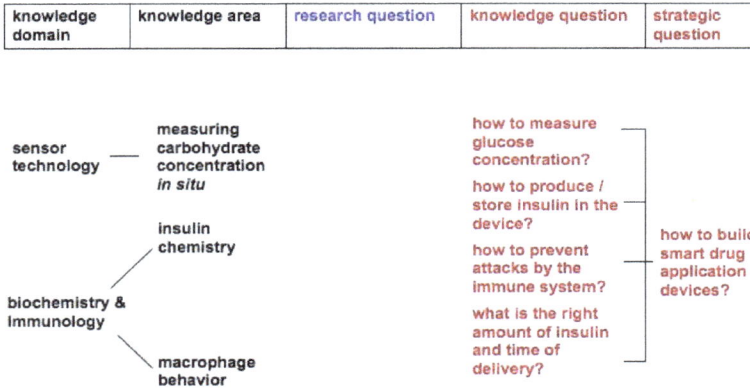

knowledge domain	knowledge area	research question	knowledge question	strategic question

Fig. 2. The Table, step 2: Decomposition of the knowledge domains into sets of specialized knowledge areas.

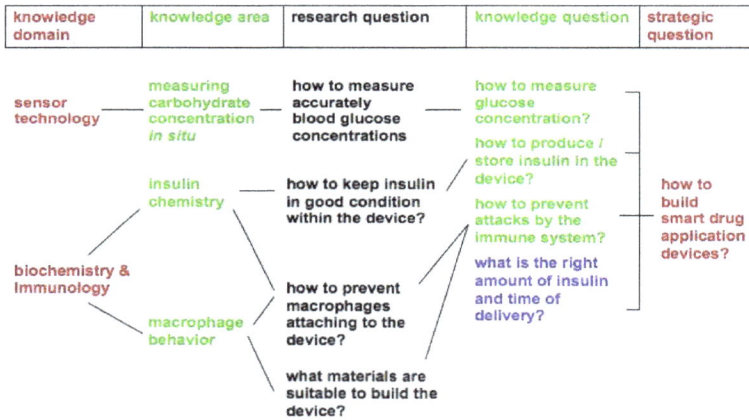

Fig. 3. The Table, step 3: Further decomposition of knowledge areas and knowledge questions, leading to sets of connecting research questions that are sufficiently concrete to develop research projects or programs.

research such as insulin chemistry, macrophage behavior against implants and measuring carbohydrate concentration *in situ* (Fig. 2).

The last step is to make the connection between (matching) the knowledge questions and the more specific research areas. These connections must be made as specifically as possible. In fact they must lead to research questions that may answer the knowledge questions that are derived from the strategic question (Fig. 3).

Example of such research questions might be

– How to keep insulin in good condition within the device?
– How to prevent macrophages attaching to the device?
– What materials are suitable to build the device?

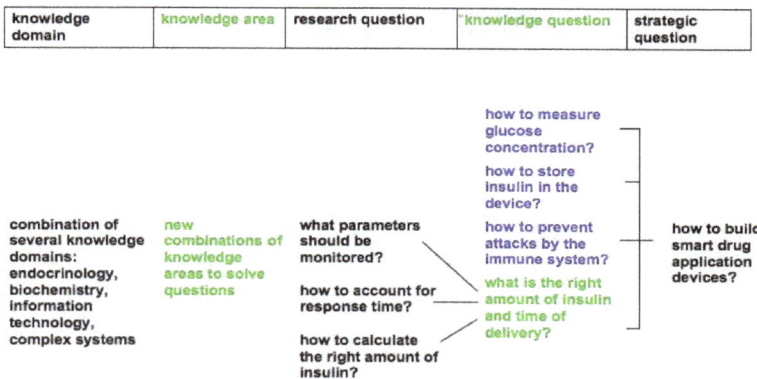

Fig. 4. The Table, step 4: finding missing research areas. Research questions, as generated by decomposing the strategic question, show the need for additional knowledge domains to feed the research agenda.

Fig. 5. The Table, step 5: construction of the research agenda as the total of all relevant research questions, explicitly linking the strategic question with the knowledge domains that provide the knowledge needed.

If we succeed to make these matches and answer these questions we might be able to design and produce a device that in fact will supply diabetics with a new system to regulate their insulin levels.

By the way, in our example we may find that we need additional knowledge domains. The question whether the device can function as an alarm involves the use of information technology and maybe even psychology, because we need to understand the patient's reaction to the alarm (Fig. 4).

The sum of all these specific matches is the research agenda or the core of the innovation or research program (Fig. 5).

Thus the use of the Table directly leads to a research agenda that is relevant to the strategic goals that are pursued. Making sure that the research is done by very good researchers guarantees the quality of the research.

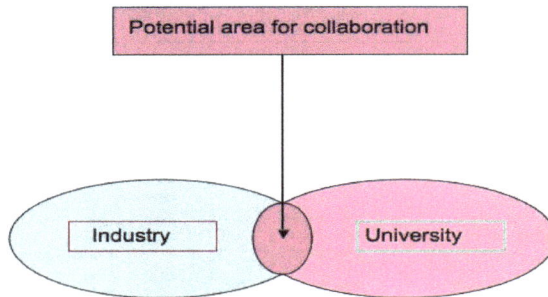

Fig. 6. Industry University Cooperative Research Center: potential area of collaboration lies in the overlap between the interests of industry and university.

When strategic goals are pursued, The Table is a very useful tool for a research manager to match expectations of industry or society about the relevance of research, and the expectations of researchers about doing exiting and high quality research. The Table is particularly useful in a context of industrial research or large laboratories.

Another tool that is more fitting for a university setting achieves the same goal of creating a relevant and exiting research agenda. It is especially designed to develop and maintain long-term collaboration between a university and a group of industries. A major problem in such collaboration is the totally different arenas in which universities and industries work.

3.2. *Industry University Cooperative Research Centers (IUCRC)*

Industry works in the arena of the market. The currency for success in that arena is profit. Universities (that is research at universities) work in the academic arena. The currency for success in that arena is peer recognition. For these arenas to work together, industry must see the relevance of the research in terms of profits to be made, and universities must see that they can do high quality and exiting research.

A number of conditions must be met to develop such collaboration. The first one is the most trivial: there must be an area where academia and industry have overlapping interests (Fig. 6). Within that area industry must think that new knowledge and insights may lead to new products and new profits, while university must think that research in that area will lead to exiting projects and publications.

The second condition is that the university has shown leadership in that area and in fact that the best researcher in that field is in charge. That way industry will never have to feel that they should also look elsewhere.

The third condition is that a number of industries and/or organizations must be interested in the area. That condition is relatively easy to meet if the second condition is met. It is our experience that leaders in academic research often have large networks in industry that they already work with.

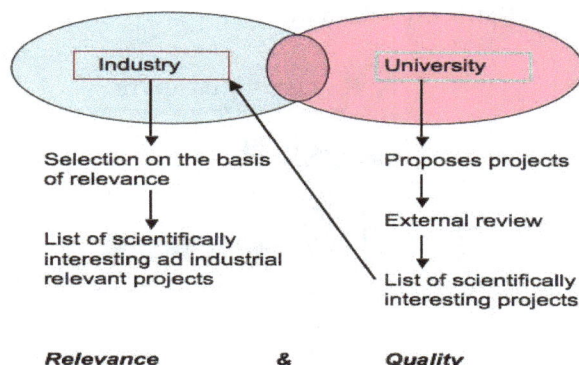

Fig. 7. Industry University Cooperative Research Center: Ensuring quality and relevance. Developing a program that is both interesting to the researchers and relevant for industry.

Having met these conditions, a research agenda must be developed and continuously refreshed. Ignoring the details of the process, the basic trick is to have the researchers propose research projects, have outside scientists review these and pre-select the scientifically most interesting ones. From that pre-selection the industries selects the portfolio of proposals to be funded. This way the two pillars for success are established: the proposals have the interest of the researchers, and are relevant to industry (Fig. 7).

Setting up such an IUCRC is not trivial. Among others, it requires leadership, discipline and persistence. Yet such centers are good examples how research can be managed in a way that the expectations of parties that live in completely different arenas are met.

The National Science Foundation has developed the methodology and regime to do that in the late 1970s. The way to set up and operate IUCRC's has been extensively monitored and evaluated.[2] We have copied that formula to the Netherlands with success, and I am sure that the formula can be copied to Singapore as well, because interests of universities and industries are the same all over the world.

3.3. *The matrix organization*

I have discussed four aspects that I consider essentials of research management: people, agenda's, relevance and quality. I have not discussed the organizational aspects of research management because combining the requirements of research with the requirements of organizations tends to lead to highly complicated structures and equally complicated philosophical arguments that I personally find less inspiring. However, I will shortly discuss one example of an organizational approach to research management, the matrix organization (Fig. 8). Sandia National Laboratories is an example of a large laboratory that adopted the matrix organization as its structure of managing projects.

688

Matrix organization

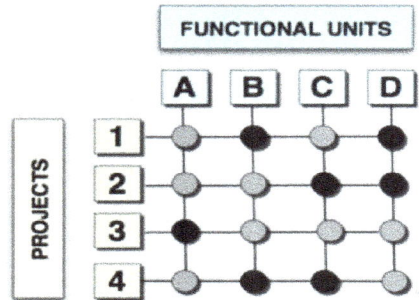

Fig. 8. The matrix organization reconciles the built in differences between scientific departments (A to D) driven **along** disciplinary lines, and projects that aim to solve problems that are typically defined **across** disciplinary lines. At some of the junctions problems (symbolized as black dots) will arise; these are the places where discussions arise and priorities must be set.

The matrix organization was created in the business world to create synergism through shared responsibility between project and functional management.[3] The equivalent in research would be to reconcile the built in differences between scientific departments that typically are driven **along** disciplinary lines, and projects that aim to solve problems that are typically defined **across** disciplinary lines. Projects aimed at solving such interdisciplinary problems need to recruit expertise from different disciplines, thus from different departments. This raises issues about responsibilities, priorities, budgets from which people and activities are paid, about how individuals get the appreciation they deserve, how they should set out career paths, etc. The places for resolving such issues become visible immediately in figure below.

During our 1984 mission to uncover the secrets of the American research management, we also visited Sandia National Laboratories. We asked Dick Andes, at that time member of the top management team of Sandia how he managed to manage both the science and the projects, while solving the problems that inevitably rise at the junctions of the research departments and the projects. His answer was:

Quote
We do not manage the science or the projects. We just watch the junctions. That is where problems in the organization show up. We stimulate discussions at these junctions.

Discussions at junctions are essential for the management of Sandia. They are also essential to involve everybody in setting the priorities of the organization. Thus the problems and the people allow us to run an organization *where the best people want to work and want to stay.*
Unquote

So I am back at were I started my discussion. There is no complexity in research management, if you go for the best people and give them what they need.

References

1. Prisma & Partners, URL:
   http://www.prismaenpartners.nl/gereedschap-e.htm#gotokop3.
2. D. O. Gray and S. G. Walters (1998). *Managing the Industry/University Cooperative Research Centre*, Batelle Press, Columbus, Ohio, USA.
3. K. Knight (1977). *Matrix Management*, PBI, New York, New York, USA.

APPLICATION OF PERCOLATION THEORY TO COMPLEX INTERCONNECTED NETWORKS IN ADVANCED FUNCTIONAL COMPOSITES

P. HING

Physics, Faculty of Science, University of Brunei Darussalam, Brunei
peter.ng@ubd.edu.bn
peter.hing@brunet.bn

Percolation theory deals with the behaviour of connected clusters in a system. Originally developed for studying the flow of liquid in a porous body, the percolation theory has been extended to quantum computation and communication, entanglement percolation in quantum networks, cosmology, chaotic situations, properties of disordered solids, pandemics, petroleum industry, finance, control of traffic and so on. In this paper, the application of various models of the percolation theory to predict and explain the properties of a specially developed family of dense sintered and highly refractory Al_2O_3-W composites for potential application in high intensity discharge light sources such as high pressure sodium lamps and ceramic metal halide lamps are presented and discussed. The low cost, core-shell concept can be extended to develop functional composite materials with unusual dielectric, electrical, magnetic, superconducting, and piezoelectric properties starting from a classical insulator. The core shell concept can also be applied to develop catalysts with high specific surface areas with minimal amount of expensive platinium, palladium or rare earth nano structured materials for light harvesting, replicating natural photosynthesis, in synthetic zeolite composites for the cracking and separation of crude oil. There is also possibility of developing micron and nanosize Faraday cages for quantum devices, nano electronics and spintronics. The possibilities are limitless.

1. Introduction

Percolation theory deals with the behaviour of connected clusters in a system. Originally developed for studying the flow of liquid in a porous body, the percolation theory has been extended to quantum computation and communication, entanglement percolation in quantum networks, cosmology, chaotic situations, properties of disordered solids and pandemics. It is also applied in the petroleum industry, finance, control of traffic, instability of financial systems, chaotic events following minute perturbations, the collapse of Bose Einstein Condensates, quantum computation, cosmology, chaotic situations, disordered solids, on set of avalanche and so on.

The low cost composite core shell concept developed by the author can also be used to design cost effective catalysts with minimal amount of super expensive metals like platinum, palladium or rare earth nano structured materials for light harvesting, replicating natural photosynthesis, in synthetic zeolite composites for the cracking and separation of crude oil. There is also possibility of developing micron and nanosize Faraday cages, and other types of cages for quantum devices, nanoelecronics and spintronics susceptible to low level of electromagnetic interference.

In this paper, the application of various models of the percolation theory to predict and explain the properties of a specially developed family of dense sintered and highly refractory Al_2O_3-W composites for applications in high intensity discharge light sources such as high pressure sodium lamps and ceramic metal halide lamps are presented and discussed.

Ceramic-metal composites, also known as cermets have been used extensively as cutting tools for example in WC–Co system.[1–4] Cermets in the ZrO_2–Ni[5–8] are also used as anode for solid oxide fuel cells.[6] Cermets containing metallic phase in glass matrix is commonly used as thick film resistors. (TFR).[9] Cermets have also been developed for high temperature heating elements and electrodes for MHD. Metallic phases in insulating matrix such as polymer constitute a class of electrically conducting composites operating at ambient and not too high temperatures.

The development of these refractory cermets was granted several patents.[9–11] The distributions of the metallic phases in the Al_2O_3-W cermets on the thermal expansion electrical properties, strengths and fracture behaviour were reported by the author in the science of ceramics.[12,13] In this paper, the author describes the application of various percolation models to account for the 8 orders of magnitude drop in the electrical resistivity of these refractory Al_2O_3-W and Al_2O_3-Mo cermets developed for high intensity discharge light sources. The following percolation models have been applied:

(1) A Core-Shell model developed by Yanagida and Kawarada.[15] These authors considered hexagonal insulating islands coated with a continuous conducting film. The model gave resistivity of at least an order of magnitude lower than experimental dataobtained by the author.

(2) Gilbert's model using alumina spheres coated with a continuous film deposited by Chemical Vapour Deposition.[16] Each coated sphere was considered a star network of resistors. The theory of random network of resistors was then applied to the coated spheres in contact to arrive atan expression for the resitivity in terms of the volume fractions of the composites.

(3) P. Hing extended Gilbert's model using ellipsoidal islands coated with discrete particles in contact.[12] The model analysed the resistivity in term of the texture developed. The model gave considerably lower resistivity than that obtained experimentally.

(4) The Turner and Malliaris's model was also applied to the experimental result. In the TM model, the insulating polymer was considered to have smooth spherical surface, the diameter of the sphere is of the order of tens of microns coated with micron size metallic particles. It was assumed that the conducting particles did not penetrate the polymer surface of the spheres. The critical probability for the

formation of an infinitely long sequences of various types of occupied lattice sites was applied. The model was also found inadequate to explain the experimental results obtained by the author.

(5) Pike's bond percolation model was also used to analyse the experimental results. In this model, the probability of forming infinitely long sequences of occupied lattice sites was takenloped. The author was able to fit the experimental result with a power law model but with considerably lower value of the critical exponent in the power relationship between the resistivity and above the threshold loading.

(6) The Ewen and Robertson's model[19] was reviewed. In this model, a segregated system of insulating sphere and much smaller conducting phase was considered. The model excludes the penetrability of the conducting phase into the non conducting phase, but takes into account conducting phase located at the interstices of the insulating islands. The percolation model, moreover, assumed that the conducting phase are distributed on a sloping insulating surface.

2. Core-Shell Model of Electrical Resistivity based on Continuous Electrical Conducting Film Surrounding Two Dimensional Lattice

According to the model developed by Yangida and Kawarada[15] for a hexagonal insulating island surrounded by a continuous layer of conducting phase as shown in Fig. 1, the composite resistivity ρ_c for thin layer is given by the relation

$$\rho_c = \rho_m \left[\frac{2 - t + t^2}{t(2 + t^2 - 3t^2)} \right], \tag{1}$$

where t is half the thickness of the thin metallic layer separating For a thick layer, the relation is given by

$$\rho_c = \rho_m \left[\frac{2 - t + t^2}{3t} \right] \tag{2}$$

ρ_m the resistivity of the metallic phase.

Since the metal room temperature resistivity of tungsten is 5.65×10^{-6} $\Omega \cdot$ cm, the composite resitivity ρ_c is of the order of 1×10^{-2} $\Omega \cdot$ cm for $2t$ taken as 4 micron as the thickness Fig. 2(a) of the conducting layer separating the insulating islands. This is at least an order of magnitude lower than the experimental resistivity of the Al_2O_3-W cermet shown in Fig. 2(b), which is typically of the order of 1 Ω cm.

For a thicker metallic layer of 20 micron separating the insulating islands, the estimated composite resistivity is of the order about 4×10^{-3} $\Omega \cdot$ cm. Thus the experimental resistivity is at least two orders of magnitude higher than the theoretical estimate for a continuous conducting layer. Examination of the microstructures shows that the single continuous layer is not realized by the fabrication technique we have adopted.

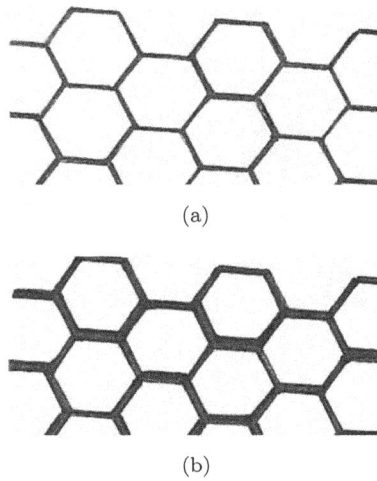

(a)

(b)

Fig. 1. Sketches illustrating (a) thin and (b) thick continuous conducting layer surrounding hexagonal insulating lattice.

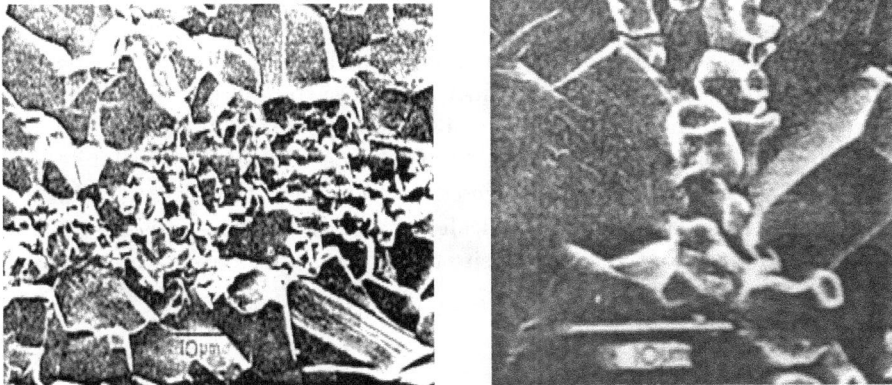

Fig. 2. Microstructure of sintered Al$_2$O$_3$-W cermets with thick (left) and thin layer (right) of conducting phases between the insulating alumina islands.

A continuous layer can be effected by chemical vapour deposition (CVD) and other thin film deposition techniques. These processes are expensive and not amenable to the large scale fabrication processes. The powder technology disclosed in this paper is, however, a very low cost high volume production technique with potential for drastically reducing the fabrication cost, particularly when other metallic and or conducting phase with much lower relative density than tungsten.

The above models with continuous metallic layer was not quite appropriate for the cermets where the metallic phases are linked together to form a chain of metallic particles in contact, which develop a three dimensional insulating islands surrounded by the conducting phase as shown in Figs. 3(a) and 3(b).

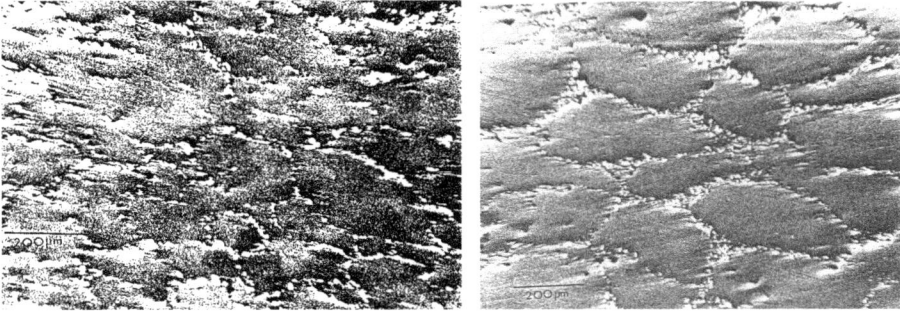

Fig. 3. Microstructure of sintered Al_2O_3-W cermets with three dimensional network of electrically conducting path. The network is formed from individually coated alumina granules which were subsequently processed and sintered. Left: Micrograph: shows thin three dimensional interconnected network. Right: Micograph exhibits thick three dimensional interconnected network for same volume content of metallic phase. Note the complexity of the network due to particle to particle contact.

3. Model of Resistivity of Continuously Coated Film on a Sphere

Examination of the microstructures in Fig. 3, moreover, shows that the cermets exhibit a truly three dimensional network structure, with insulating islands that depart from sphericity. We can apply a core shell model of electrical resistivity based on continuous electrically conducting film surrounding insulating spheres. A more appropriate model would be to use that proposed by Gilbert,[16] and had been applied by Pleass and Schuimmel[16] to CVD coated spheres of molybdenum on alumina. Each coated spheres is considered as a star network consisting of a central point through a resistance. This is basically a theory of random resistors applied to coated spheres in contact. The composite resistivity is given approximately by

$$\rho \approx \rho_m \left(\frac{\ln\left(1 + \frac{6}{V_f}\right)}{V_f} \right). \tag{3}$$

In Eq. (3), V_f is the volume fraction of the metallic coating. V_f is given by $3T/R$, where T is the thickness of the coating and R the radius of the insulating sphere. Substituting the value of resistivity ρ_m for tungsten, and various volume fraction of tungsten in the cermets developed, one can compare the experimental values of the composite resistivity with the core shell model based on continuous conducting film surrounding the insulating sphere. The results are summarized in Figs. 4 and 5.

4. Model based on Texture of the Shell

It can be seen in Fig. 5 that the electrical resistivity of the composite based on the core-shell model with various volume fractions of metal loading are several orders of magnitude lower than experimental resistivity. To obtain a more realistic model to fit the experimental results, it is thought useful to consider whether there is a

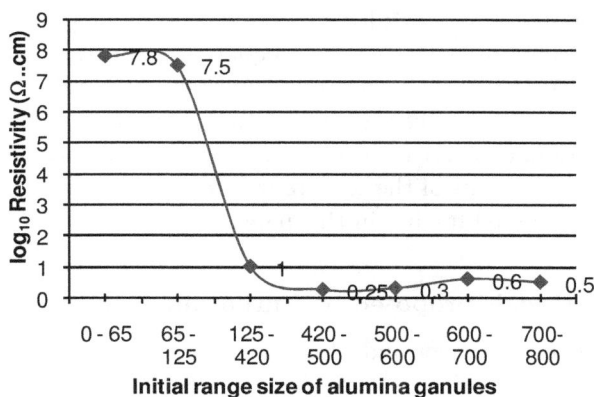

Fig. 4.   Effect of initial range of alumina granules on the resitivity of the cerments.

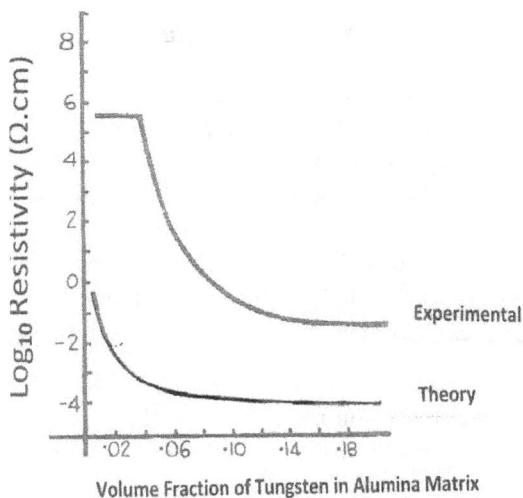

Fig. 5.   The resistivity of Al$_2$O$_3$-W cermets with various loadings of W. The top plot is experimental data from various volume fractions of metal loading used, es, namely: V$_f$: 0.01, 0.038,
0.065, 0.107, 0.132, and 0.167. The bottom plot is calculated resistivity using Eq. (3) in Sec. 3.

correlation between the ratios of the major and minor axes of the oblong ellipsoid
and the thickness of the shells of metallic layer surrounding the oblong ellipsoid.

For an oblong ellipsoid, one needs to consider three axes. If R_1, R_2 and R_3
are the semi major and semi minor axes, and assuming that the metallic layer is
uniform, we can write and expression for the composite resistivity in terms of the
parameter R_i/T where i represents 1, 2, and 3 given by[10–12]

$$\rho_c \approx \rho_m \frac{R_i}{3T} \ln \left(1 + \frac{2R_i}{T} \right). \tag{4}$$

The insulating alumina islands have ellipsoidal shape as a result of the uniaxial compaction of the coated granules, and subsequent densification at elevated temperatures typically around 1900°C in the alumina refractory metal cermets. Because of the anisotropy one would expect the composite resistivity to be different in the direction of compaction, and across the other 2 orthogonal directions. The analysis of electrical resistivity in terms of the texture was reported in previous publication,[12] and will not be considered further in this paper.

5. Percolation Models proposed by Turner and Malliaris

The model proposed by Turner and Malliaris assumes insulating polymer as large smooth spheres (tens of microns) and conducting phases in micron range. It is assumed that the conducting particles do not penetrate the polymer surface of the spheres. The critical volume fraction is shown to be[17]

$$V_{fm} = \frac{P_c}{2} \frac{1}{\left\{1 + \phi/4 \left(\frac{R_I}{R_m}\right)\right\}}, \tag{5}$$

where P_c the critical probability for the formation of infinitely long sequences of occupied lattice sites, Φ a factor dependent on mode of packing: $\Phi = 1.110, 1.250$ and 1.375 for hexagonal, square and triangular lattices respectively; R_I the radius of the insulating alumina island (sphere), R_m radius of metallic phase such as W, Mo, Nb, Ni ....

From the analysis in Fig. 6, it is shown that the critical volume fractions estimated is 2 to 3 orders of magnitude lower than that found in the Al_2O_3-W investigated To account for these discrepancies, more realistic models should include the following factors.

(1) Penetrability of the particles on the surface of the insulating islands.
(2) The surface roughness of the insulating spheres.

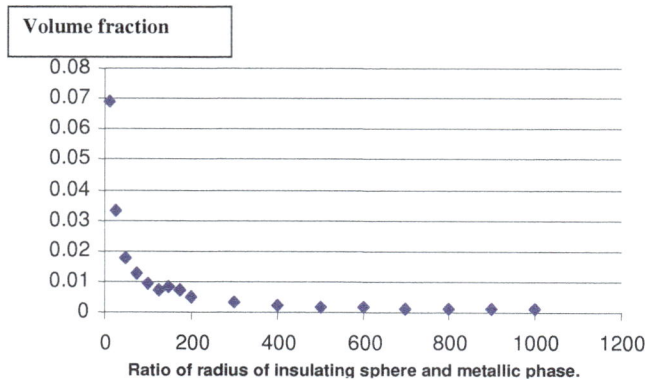

Fig. 6.   Calculated volume fraction of metallic loading to form infinitely long chains.

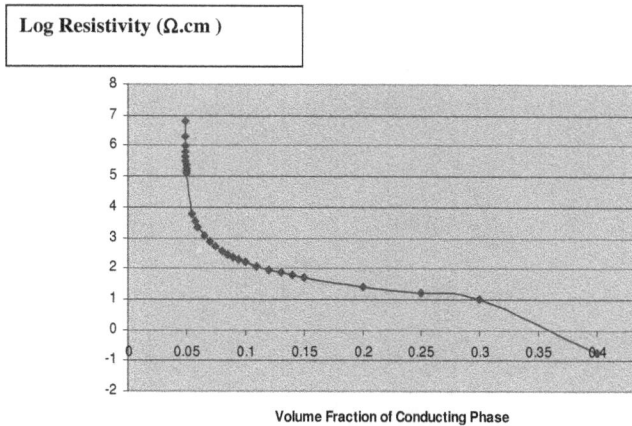

Fig. 7. The computed resistivity using Pike's relation for electrical conductivity of Al₂O₃-W with a critical exponent of 1.7.

(3) The agglomeration of the small metallic phase.
(4) The segregation of metallic phase at the interstices.
(5) The departure of the insulating islands from spheres to oblate spheroids.

Models taking all these factors into account are areas for further studies.

6. Basis of Pike's Bond Percolation Model[18]

In this section, we will apply Pike's bond percolation model based on a segregated system to the Alumina — tungsten system. In the Pike's model, the glass spheres are assumed to sinter into cubes. Along edges of cube, conducting phase are segregated Rectangular channels along cube edges are formed due the constraint. The resistance of such a system is described using bond — percolation approach. The probability of forming infinitely long sequences of occupied lattice sites need to be taken into account. This leads to the Pike's relations for the resistivity of the composite as shown in Fig. 7.

Pike's equation to describe the resistivity of cermets for thick film resistors (TFR) is given by[18]

$$\rho = \rho_0(V - V_c)^{-\mu} \tag{6}$$

ρ the electrical resistivity of the composite, ρ_0 a measure of the effective resistivity taking into account contact resistance between metallic phase. V is the volume fraction of the conducting phase above the threshold or critical volume fraction V_c. The critical exponent μ is assumed to indicate the extent of the blending of the mixture of different phases, and resitivity of the composite is related to $(p - p_c)^{-\mu}$, where p represents the probability of forming an infinitely long sequences of occupied lattice sites; p_c the critical dramatic drop in property or behaviour. Replacing p by V and p_c by V_c leads to Pike's relation.

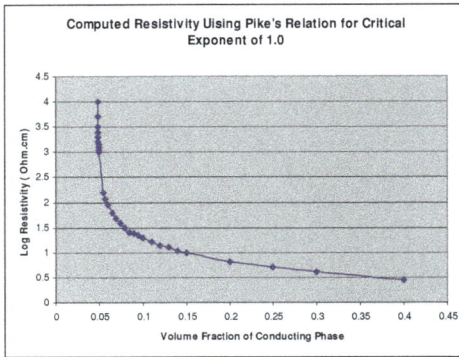

Fig. 8. (a) Computed resistivity of Al$_2$O$_3$-W for exponent of 1.

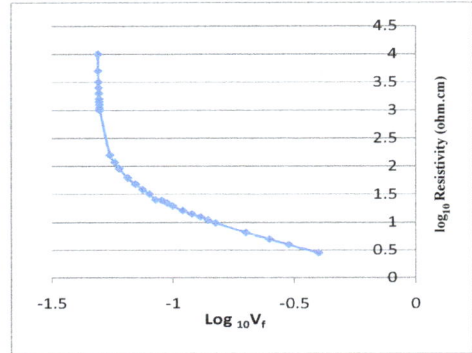

Fig. 8. (b) Calculated Log Resistivity vs log V_f for a critical exponent of 1.

The resisivity computed with μ value of 1.7 is much higher than experimntal result at and near threshold loading in Al$_2$O$_3$-W cermet.

The plot with μ of 2 also follows rather similar feature compared with of 1.7 and is not included in the paper. At critical loading, the resistivity is, however, is much higher than experimental results.

In Fig. 8(a), the plot for μ of 1 shows that at the critical loading of around 0.05 to 0.06, the resistivity has dropped to around $100\ \Omega\cdot$cm, still somewhat higher than the experimental results obtained in the Al$_2$O$_3$-W. however, the rend shows that if the critical exponent is further reduced, to say, 0.05, the computed resistivity would fit the experimental result fairly closely.

Comparing the plots using Pike's relation for electrical resistivity, we can see that the plot for the computed resistivity approaches experimental values at the critical metal loading only when the critical exponent approaches 1.0 or below. Plotting the results on a log − log scale as shown in Fig. 8(b) gives similar trend in the resistivity as in the case of log −lin plot.

For critical exponent of 1.0, the computed resistivity is of the order of 10 to Ω ohm $\cdot$ cm, still somewhat higher than the experimental results. For volume fraction around 0.06, slightly above the critical volume fraction of 0.049, the experimental resistivity is of the order of 1 to 10 Ω ohm $\cdot$ cm. The critical exponent μ in the power law proposed by Pike can thus be refined further to fit the experimental results. This is achieved by compressing the plot towards the axis representing the volume of the metallic phase. It could be argued that the critical exponent indicates the extent of the blending, segregation and possibly some degree of penetrability of the metallic phase around and in the surface of the insulating islands.

For resistivity intermediate between the high and low values, it can be seen that the slope of the plot of experimental resistivity below the critical volume fraction is steep, and hence unstable. This steep slope leads to difficulties in obtaining accurate resistivity values as slight variation in the metal loading could lead to large and unacceptable change in the resistivities.

For more reliable resistive components, one needs to load the metallic phase well above the critical volume fraction of metallic loading threshold. However, higher loading of metallic phase increases cost particularly if platinium, gold, palladium are used.

In Al_2O_3-W system, we are not designing an on and off state. If one can design the slope to be more gentle slope through materials processing, this could be useful. The model proposed by Ewen and Robertson show appreciable differences in the slopes below and above the critical loading for the large number of resistivity curves generated.[19]

It is also of interest to mention that the resistivity R_0 in the Pike's power law is not a constant parameter. The author found that using the experimental values of the resistivity for the Al_2O_3-W for various volume fractions above the threshold increases from about 6×10^{-6} to 70×10^{-6} $\Omega \cdot$ cm for a critical exponent of 3.

However, in both Pike and Ewen and Roberson's models, an arbritary value for R_0 is used. In future studies, the author will analyse the implications of a varying R_0 with volume fractions of W above the threshold. It is possible that the increase in the value of R_0 implies the possibility of lower sinterability, and hence poorer densification of the metallic phase with higher metal loading.

7. A Review of Ewen and Robertson's Percolation Model based on a Sloping Surface for the Non-conducting Phase

In this section, the percolation model proposed by Ewen and Robertson[19] is reviewed. A segregated system of insulating sphere and much smaller conducting phase, which may be metallic or nonmetallic, was considered. The model excludes penetrability of the conducting phase into the non conducting phase, and hence a significant amount is located at the interstices. Percolation on a sloping non conducting lattice was considered. This differs from Pike's model. Ewen and Robertson considered the cubes to be sliding on the surface. ice, the surface is frictionless and the cubes will be able to slide down the slope. Figure 9 illustrates fraction of filled sites in the square lattice. This model is of interest and we note the following features. For $p = 1.0$ or 100% sites filled (say with electrically conducting phase; hence fully connected (see white squares — all connected). 0% black squares in Fig. 9(a). For p_c of 0.62 or 62% sites filled; 62% white squares still connected, hence still conducting, but resistivity higher. Path more tortuous. 38% black squares. Black squares can be the insulating phase. For $p = 0.34$ or 34% connected; see white squares now not connected. Not conducting. 66% black squares.

Ewen and Roberson obtained a relation for the electrical resistivity of the conducting composite[19] given by

$$R = R_0[(\{V/[f_m(1 - f_g)]\}n + p_c^n)^{1/n} - p_c]^{-1.7} \qquad (7)$$

R_0 is a measure of the effective resistivity which takes the contact resistance between particles into account; n is the exponent relating fraction of sites filled in

700

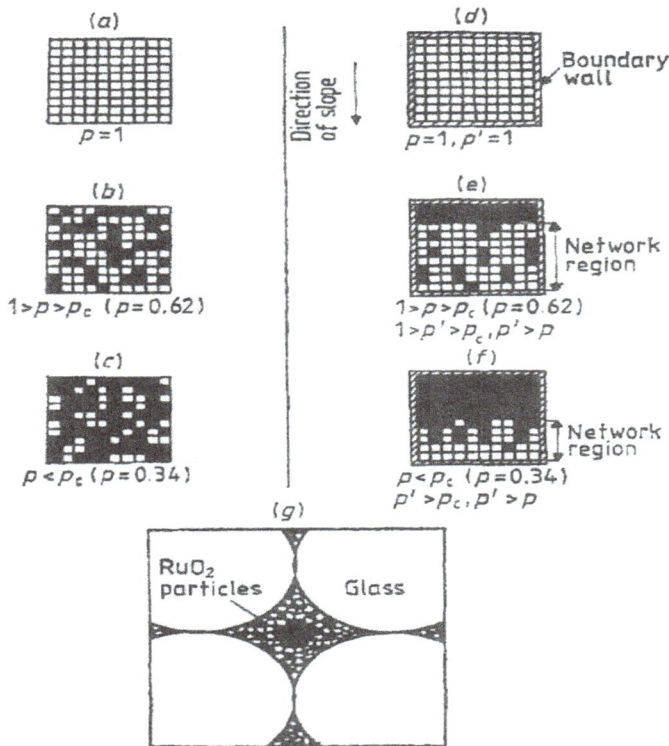

Fig. 9. Ewen and Robersson's percolation model for segregated system of insulating spheres and much smaller conducting phase.[19] Left hand side of Fig. 9(a)–(c) represents standard percolation situation where identical metallic cubes placed on lattice points on a horizontal insulating surface. Right hand side of Fig. 9(d)–(e) describes the modified percolation model where the cubes representing the metallic phase lie on a sloping insulating surface. Figure 9(g) shows segregation of the metallic phase at the interstices, and very little conducting phase between the insulating surfaces.

sloping cube lattice model.

$$p' = (p^n + p_c^n)^{1/n} \tag{8}$$

Using Eq. (7), Ewen and Robertson was able to generate series of curves with very steep to a fairly gentle slopes above the threshold values, and not so steep slope before the threshold value. which can also be used beneficially. The slope m of the function for the resistivity of composite developed by Ewen and Robertson[19]

$$m = \frac{n\mu 1ge}{f_m(1 - f_g)p_c} \tag{9}$$

n is exponent from the relation between p', the fractions of filled sites in a sloping square lattice and, p, fraction of sites filled in a normal square lattice; μ the critical exponent discussed earlier; f_m packing fraction for the conducting phase (pf_m) becomes volume fraction of the conducting phase), f_g, packing fraction of nonconducting phase; then $(1 - f_g)$ is the fraction of sample volume through which conducting phase is distributed; p_c critical fraction of filled sites; f_m p_c is then the volume fraction of conducting phase.

If f_g the packing fraction of non-conducting phase; $(1 - f_g)$ = fraction of sample volume through which conducting phase is distributed in the interstitial region, V the overall volume fraction of conducting phase is $V = V_i (1 - f_g)$ where V_i is the volume fraction of conducting phase in the interstitial region.

The author is studying the ER's percolation model to see whether modifications can be made to the slopes of the percolation behaviour in the Al_2O_3-W, and other ceramic — metal systems.

8. Conclusions

The dramatic drop in the electrical resistivity of the Al_2O_3-W can be modeled by various percolation models. Most of the models can explain the dramatic drop of the electrical resistivity with critical metal loading. For highly segregated system with large ratios of the insulating islands and metallic phase, the critical volume fractions predicted is a factor of 2 to 3 smaller than those observed experimentally in the Al_2O_3-W cermets. For random distribution of metal loading, the critical volume fraction of metallic phase to form continuous network is at least 0.35.

The analysis shows that the bond percolation model proposed by Pike with a critical exponent of 1 or lower can be fit the experimental results fairly well. The author has reviewed briefly the Ewen and Robertson's model based on a sloping lattice for the insulating surface, and plan to analyse critically the ER's model and its applicability Al_2O_3-W and Al_2O_3-Mo cermets.

The causes of discrepancies have been identified and attributed to penetrability of the metallic particles on the surfaces, accumulation of particles between the interstices of the insulating islands, surface roughness, agglomeration of particles etc. The author is also developing models to incorporate these features. In view of its implications.

References

1. Sun, Cheng-Chang Jiao, Chen-Gang, Ruiz-Jun CIO, VC Addition Prepared Ultra fine WC-Co Composites by Spark Plasma Sintering, Journal of Iron and Steel Research, International, Volume 14, Issue 5, Supplement 1, September 2007, Pages 85-89.
2. Xingqing Wang, Yingfang Xi, Hailiang Gu, Van der Biest, J. Vleugels, Sintering of WC-Co powder with nanocrystalline WC by spark plasma sintering, Rare Metals, Volume 25, Issue 3, June 2006, Pages 246-252.
3. B. Huang, L. D. Chen, S. Q. Bai, Bulk ultrafine binderless WC prepared by spark plasma sintering, Scripta Materialia, Volume 54, Issue 3, February 2006, Pages 441-445.
4. J. E. Sundeen, R. C. Buchanan, Thermal sensor properties of cermet resistor films on silicon substrates, Sensors and Actuators A: Physical, Volume 90, Issues 1-2, 1 May 2001, Pages 118-124.
5. BCH Steele, Electrical Ceramics for Fuel Cells and High Energy Batteries, pp. 203–225, Electronic Ceramics Ed: BCH Steele, Elsevier.

6. R. Campana, A. Larrea, J. I. Peña, V. M. Orera, Ni–YSZ cermet micro-tubes with textured surface, Journal of the European Ceramic Society, Volume 29, Issue 1, January 2009, Pages 85-90.

7. Wen Lai Huang, Qingshan Zhu, Zhaohui Xie, Gel-cast anode substrates for solid oxide fuel cells, Journal of Power Sources, Volume 162, Issue 1, 8 November 2006, Pages 464-468.

8. A percolation model of conduction in segregated systems of metallic and insulating materials, J. Phys.D: Appl. Phys., 14 (1981) 2253-68.

9. P. Hing, US Patent 4155757.

10. P. Hing, D. T. Evans and R. Marshall, US Patent 51555758.

11. P. Hing, US Patent 4585972.

12. P. Hing, Spatial Distribution of tungsten in alumina and its effect on the electrical and mechanical properties of the alumina — W Cermets, Science of Ceramics 9, pp. 135-143, 1977.

13. P. Hing, The Strengths and Fracture Properties of Al_2O_3-W Cermets, Science of Ceramics 10, pp. 521-528, 1980

14. P. Hing, Spatial Distribution of Tungsten on the Physical Properties of Al_2O_3 Cermets, Science of Ceramics, Proceedings of the 12th International Conference on Science of Ceramics, Vol. 12, pp. 87-94, 1984.

15. H. Yanagida and H. Kawavada, Estimation of electrical conductivity of composite materials. I. Conductivity along grain boundary, Jap. J. App. Phys. Vol. 13, No. 3, Feb. 1974, pp. 244-248.

16. G. M. Pleass and D. G. Schidmmel, Proc. Int. Conf. Chemical Vapour Deposition. Ed. F. A. Glaski.

17. A. Malliaris and D. T. Turner, "Influence of Particle Size on the Electrical Resistivity of Compacted Mixtures of Polymeric and Metallic Powders," J Appl. Physics 42 (20) 614-68, 1971.

18. G. E. Pike, AIP Conference, 40, 366-71, 1978.

19. P. J. S. Ewen and J. M. Robertson, A percolation model of conduction in segregated systems of metallic and insulating materials, application to thick film resistors, J. Phys. D: Appl. Phys. 14, 2253, 1981.

LIST OF INVITED SPEAKERS

Steven L ADLER	Institute of Advanced Study, USA	adler@ias.edu
Ignatios ANTONIADIS	CERN, Switzerland	ignatios.antoniadis@cern.ch
Myron BANDER	University of California, Irvine, USA	mbander@uci.edu
Thomas BANKS	Rutgers, USA	banks@scipp.ucsc.edu
Itzhak BARS	University of Southern California, USA	bars@usc.edu
Johannes BLUEMLEIN	DESY – Zeuthen, Germany	Johannes.Bluemlein@desy.de
Lars BRINK	Chalmers University of Technology, Sweden	lars.brink@chalmers.se
Pisin CHEN	National Taiwan University, Taiwan	pisinchen@phys.ntu.edu.tw
Cesareo DOMINGUEZ	University of Cape Town & Stellenbosch University, South Africa	cesareo.dominguez@uct.ac.za
Georgi Dvali	New York University, USA	georgi.dvali@cern.ch
Jonathan ELLIS	CERN, Switzerland	John.Ellis@cern.ch
Sergio FERRARA	CERN, Switzerland	sergio.ferrara@cern.ch
Victor FLAMBAUM	University of New South Wales, Australia	v.flambaum@unsw.edu.au
Paul H. FRAMPTON	University of North Carolina, USA	frampton@physics.unc.edu
Harald FRITZSCH	Ludwig-Maximilians-University, Munich, Germany	fritzsch@mppmu.mpg.de
Murray GELL-MANN	Sante Fe Institute, USA	
John Francis GUNION	University of California at Davis, USA	gunion@physics.ucdavis.edu
Gabriel KARL	University of Guelph, Canada	gk@physics.uoguelph.ca
Karliner MAREK	Tel Aviv University, Israel	marek@proton.tau.ac.il,
Hagen KLEINERT	Free University of Berlin, Germany	kleinert@physik.fu-berlin.de
Dieter LUEST	Ludwig-Maximilians-University, Munich, Germany	dieter.luest@lmu.de
Peter MINKOWSKI	University of Bern, Switzerland	mink@itp.unibe.ch
Rabindra MOHAPATRA	University of Maryland, USA	rmohapat@umd.edu
Vitatcheslav MUKHANOV	Ludwig-Maximilians-University, Munich, Germany	mukhanov@physik.lmu.de
Serguey Todorov PETCOV	SISSA, Italy	petcov@sissa.it
Nicholas SAMIOS	Brookhaven National Laboratory, USA	samios@bnl.gov
John SCHWARZ	Caltech, USA	JHS@Theory.Caltech.edu
Goran SENJANOVIC	ICTP, Italy	goran@ictp.it
Mikhail SHIFMAN	University of Minnesota, USA	shifman@umn.edu
Anthony THOMAS	University of Adelaide, Australia	anthony.thomas@adelaide.edu.au

Gerardus 't Hooft	Utrecht University, The Netherlands	g.tHooft@uu.nlx
Spenta WADIA	Tata Institute of Fundamental Research, India	wadia@theory.tifr.res.in
Kenneth WILSON	Ohio State University, USA	kgw@maine.rr.com,
Koichi YAMAWAKI	Nagoya University, Japan	yamawaki@eken.phys.nagoya-u.ac.jp
Chen-Ning YANG	Tsinghua University, China	cnyang@tsinghua.edu.cn
Anthony ZEE	University of California, Santa Barbara, USA	zee@kitp.ucsb.edu
Antonino ZICHICHI	CERN, Switzerland	Esthel.Laperriere@cern.ch
George ZWEIG	MIT, USA	Zweig@mit.edu

LIST OF CONTRIBUTED SPEAKERS

Kamil Gediz AKDENIZ	Istanbul University, Turkey	gakdeniz@istanbul.edu.tr
Fatemeh ARBABIFAR	Semnan University, Iran	Arbabifar_f@yahoo.com
Vadim V. ASADOV	Moscow State University, Russia	asadov@neurok.com
Steven D. BASS	University of Innsbruck, Austria	steven.bass@uibk.ac.at
CHEN Yu	National University of Singapore, Singapore	g0600513@nus.edu.sg
Andreas DEWANTO	National University of Singapore, Singapore	phyda@nus.edu.sg
Tjong Po DJUN	Indonesian Institute of Sciences, Indonesia	tpdjun@mail.lipi.go.id
Manmohan GUPTA	Panjab University, India	mmgupta@pu.ac.in
Yuan K. HA	Temple University, USA	yuanha@temple.edu
L.T. HANDOKO	Indonesian Institute of Sciences, Indonesia	laksana.tri.handoko@lipi.go.id
Andreas HARTANTO	Indonesian Institute of Sciences, Indonesia	hartanto@mail.lipi.go.id
HE Haitao	National University of Singapore, Singapore	U0602691@nus.edu.sg
Peter N K N Y HING	Universiti Brunei Darussalam, Brunei	peterhing1@gmail.com
Kerson HUANG	MIT, USA	kersonhuang@aol.com
Marcus HUTTER	Australian National University, Australia	marcus.hutter@anu.edu.au
Jayesh Kumar JASVANTLAL	National University of Singapore, Singapore	U0603020@nus.edu.sg
Hamzeh KHANPOUR LEHI	Semnan University, Iran	hazmeh_khanpour@nit.ac.ir
Supriya KAR	University of Delhi, India	skkar@physics.du.ac.in
Oleg V. KECHKIN	Moscow State University, Russia	kechkin@gmail.com
Alinaghi KHORRAMIAN	Semnan University, Iran	khorramiana@theory.ipm.ac.ir
LOW Lerh Feng	National University of Singapore, Singapore	U0602574@nus.edu.sg
Parthasarathi MAJUMDAR	Saha Institute of Nuclear Physics, India	parthasarathi.majumdar@saha.ac.in
Zahra G. MOGHADDAM	Islamic Azad University, India	zahra_ghmoghaddam@yahoo.com
Ken NAITOH	Waseda University, Japan	k-naito@waseda.jp
Borzoo NAZARI	University of Tehran, Iran	borzoo.nazari@gmail.com
Moosavi Nejad	Yazd University, Iran	mmoosavi@yazduni.ac.ir
NG Wei Khim	National University of Singapore, Singapore	phynwk@nus.edu.sg
Wei-Tou NI	National Tsing Hua University, Taiwan	weitou@gmail.com
Ulrich NIERSTE	Universitaet Karlsruhe, Germany	ulrich.nierste@kit.edu
ONG Yen Chin	National University of Singapore, Singapore	yenchin@nus.edu.sg
Premana W. PREMADI	Bandung Institute of Technology, Indonesia	premadi@as.itb.ac.id
Gunnar PRUESSNER	Imperial College London, UK	g.pruessner@imperial.ac.uk
Jagmohan Singh RANA	H N B Garhwal University, India	ranajms@gmail.com

Fabio SCARDIGLI	National Taiwan University, Taiwan	fabio@phys.ntu.edu.tw
Keshav N. SHRIVASTAVA	University of Malaya, Malaysia	keshav1001@yahoo.com
Abhishek KUMAR SINGH	University of Delhi, India	Abhishek151983@gmail.com
Ikuo S. SOGAMI	Kyoto Sangyo University, Japan	sogami@cc.kyoto-su.ac.jp
Maryam SOLEYMANINIA	Semnan University, Iran	Maryam.soleimaninia@gmail.com
Allan SOLOMON	Open University, UK	a.i.solomon@open.ac.uk
Richard Gordon STROM	Univ of Amsterdam & Netherlands Inst for Radio Astronomy, The Netherlands	strom@astron.nl
Albertus SULAIMAN	Indonesian Institute of Sciences, Indonesia	sulaiman@mail.lipi.go.id
Sara TAHERI MONFARED	Semnan University, Iran	sara_taherimonfared@yahoo.com
TENG Po-Wen Ivan	National University of Singapore, Singapore	g0900767@nus.edu.sg
Jan Wouter VASBINDER	Insitute Para Limes, The Netherlands	Jan.w.vasbinder@paralimes.org
G. Cigdem YALCIN	Istanbul University, Turkey	cigdem_yalcin@yahoo.com
YANG Jiahui Abel	University of Virginia, USA	ajy6n@virginia.edu
ZHOU Yu-Feng	Institute of Theoretical Physics, CAS, China	yfzhou@itp.ac.cn

www.ingramcontent.com/pod-product-compliance
Lightning Source LLC
Chambersburg PA
CBHW081209220326

41598CB00037B/6718